BMDP Statistical Software Manual

Volume 1

To accompany BMDP Release 7

Much of the development of techniques underlying these programs was sponsored under grant RR-3 and continuation contract number N01-RR-8-2107 of the Biotechnology Resources Branch of the National Institutes of Health. The time series programs, 1T and 2T, were supported in part by National Science Foundation grant number CDP 80-20837. Program 5V is based on a program designed with support from NIMH grant number 37188 in the Department of Biomathematics at UCLA.

BMDP Statistical Software Manual

Volume 1

To accompany BMDP Release 7

W. J. Dixon, chief editor

UNIVERSITY OF CALIFORNIA PRESS

BERKELEY • LOS ANGELES • OXFORD

Orders for this publication may be directed to

California/Princeton Fulfillment Services
1445 Lower Ferry Road
Princeton, N.J. 08618
Orders: 800-822-6657

UNIVERSITY PRESS OF CALIFORNIA
c/o John Wiley & Sons, Ltd.
Distribution Centre
1 Oldlands Way
Bognor Regis, West Sussex
PO22 9SA, England
Telephone: 44-243-842165
Fax: 44-243-842167
Telex: 8611 Wiley G

Correspondence and orders for programs should be addressed to

BMDP Statistical Software, Inc.
1440 Sepulveda Boulevard
Suite 316
Los Angeles, CA 90025
Telephone: (310) 479-7799
FAX: (310) 312-0161
Telex: 4972934 BMDP UI

BMDP Statistical Software
Cork Technology Park
Model Farm Road
Cork, Ireland
Telephone: 353-21-542722
FAX: 353-21-542822
Telex: 852 75659 SSWL EI

Library of Congress Cataloging-in-Publication Data

BMDP statistical software manual : to accompany the 7.0 software release / W.J. Dixon, chief editor. — Completely rev. ed.
v. < >
Volume 3 has a distinctive title: The data manager.
Includes bibliographical references and index.
ISBN 0-520-08138-2 (v. 1). — ISBN 0-520-08139-0 (v. 2). — ISBN 0-520-08140-4 (v. 3)
1. Mathematical statistics—Data processing. 2. BMDP-79.
I. Dixon, Wilfrid Joseph, 1915- . II. Title: Data manager.
Volume 3.
QA276.4.B58 1992
519.5'0285'5369—dc20 92-34074

Contents

Programs

Preface

This new edition of the BMDP Manual covers changes and additions to the BMDP programs for Release 7. Model examples cover program features from the basic to the most sophisticated, and the examples are thoroughly explained and illustrated. Graphical and statistical diagnostic output are emphasized so that the user can decide if the data meet assumptions.

The programs have been written to run with less-than-perfect data; missing data, outliers, and nonnormal data can be detected and handled. Default options have been chosen so that the new user need enter only a few commands to successfully run the programs. New features include additional printing options and new statistical tests, as well as the capability of performing more analyses interactively. All programs now allow the user to enter and execute commands, view the output at a terminal, and then alter the commands without exiting the program. A subset of the programs that enter variables one or more at a time respond immediately to instructional changes, so the user can intervene in the choice of variables and see the results.

Volume 1 introduces BMDP and describes the most commonly used programs. It includes a series of introductory chapters describing the main components of BMDP. Volume 2 covers the more specialized or advanced programs. This new edition is styled in the same manner as the 1990 *BMDP Statistical Software Manual*; therefore those familiar with that edition will find this one very similar.

The manual is written for new users of computer programs as well as for experienced statisticians. Chapter 1 provides instructions on how to use the manual, an overview of BMDP programs, advice on how to choose among them, and the basics of how they work. Chapters 2 through 9 describe BMDP instructions in detail. Chapter 10 discusses how to run the programs in batch and interactive modes, and Chapter 11 contains useful shortcuts and convenience features for writing BMDP instructions. Statistical formulas, computational algorithms, specialized applications, descriptions of the data sets used in the program chapters, FORTRAN subroutines, and a discussion of problem size requirements are provided in appendices.

The professionals at BMDP are continually developing and releasing new programs. Each program is identified by a two-character code: the first is a number or letter, and the second is a letter, e.g., 3D. The number does not indicate increasing complexity, and some numbers do not appear. The letters classify the programs loosely into series:

D – data description
R – regression analysis
V – analysis of variance
M – multivariate analysis
L – life tables and survival analysis
T – time series

Four programs do not belong to any series:

CA – correspondence analysis
LE – maximum likelihood estimation
4F – frequency tables
3S – nonparametric statistics

Other documentation provided by BMDP may be useful for your needs. Major changes in the programs and novel ways to use them are included in the *BMDP Communications* newsletter. The *BMDP User's Digest* contains an abbreviated summary of the BMDP program commands. *BMDP Statistical Software Manual*, Vol. 3 describes the Data Manager, a powerful interactive data manipulation system that is fully compatible with BMDP's data description and analysis procedures. (Platform-specific *User's Guides* provide instructions for using BMDP on PCs, workstations, and VAX installations).

New Program in Release 7.0

5M (Volume 2) – Linear and Quadratic Discriminant Analysis

New features

General enhancements in BMDP Release 7:

- many of the programs on VAX/VMS and VAX/UNIX platforms now produce high resolution plots, which can be saved in either HPGL or PostScript formats.
- many programs have new print features which supply printout of means, standard deviations, extreme cases with case numbers, *z*-scores, skewness, and kurtosis, as well as correlation and covariance matrices.
- five new transformation functions have been added.
- interactive capability has been added to 1D, 2D, 4M, 5M, 7M, and LE.
- additional interactive paragraphs have been added to a number of programs.
- the instructions for grouping variables have been made uniform throughout all BMDP programs.

Specific program enhancements:

1D – Now works interactively.
2D – Now works interactively.
2L – Competing risk analysis was added.
4M – Now prints matrix of eigenvalues, goodness-of-fit chi-square test, Cronbachs' alpha, and log-likelihood at each iteration.
7M – Additional interactive capabilities have been added.
1R – The Durbin-Watson test is now calculated and printed.
2R – The Durbin-Watson test and serial correlations are now calculated and printed.
3R – The Durbin-Watson test is now calculated and printed.
4R – Now prints coefficients for all variables, including those not included the model.
LR – Receiver Operator Characteristic (ROC) plots are now available.
3S – Now calculates and prints the Kolmogorov-Smirnov test.
1T – Supports additional SAVE features.

Acknowledgments

Early and continuing development of the BMDP programs was made possible through grants from the Division of Research Resources of the National Institute of Health. In particular, Bruce Walman and later William Raub were instrumental in guiding these activities from NIH.

Robert Jennrich proposed and designed many of the programs in regression analysis, analysis of variance, and discriminant analysis. He made significant contributions in nonlinear regression, factor analysis, and the analysis of variance. He supervised Mary Ralston's Ph.D. thesis, which provides the algorithm for the derivative-free nonlinear regression.

Laszlo Engelman designed the basic framework of the BMDP programs, including the language used to specify instructions, the methods used for transformations, and the method of saving data and results between analyses (the BMDP File). He also programmed many of the subroutines common to all programs and designed and programmed 14 of the BMDP analyses.

Jim Frane made significant contributions in the area of multivariate analysis. Morton Brown developed the frequency table programs, and John Hartigan contributed greatly in the area of cluster analysis.

Many statisticians have made contributions and suggestions for developing the BMDP program library: R. L. Anderson in regression and analysis of variance, Virginia Clark and Robert Elashoff in survival analysis, Robert Ling in the method of pictorially representing a matrix, and John Tukey in data analysis. Valuable comments have been received from David Andrews, Peter Claringbold, Cuthbert Daniel, Michael Davidson, Charles Dunnett, Janet Elashoff, Ivor Francis, George Furnival, Don Guthrie, Robert Jennrich, Henry Kaiser, Sir Maurice Kendall, Eliot Landau, H. L. Lucas, John Nelder, Shayle Searle, Frank Stitt, Max Woodbury, Karen Yuen, and Coralee Yale. Even the list of references is, of necessity, incomplete.

In recent years, our Advisory Committee has helped consider directions for our research, making suggestions for improvements in numerical methods, program design and standards, statistical techniques, and exchange of information with statisticians and the statistical computing community.

The development of each individual BMDP program was the product of a great deal of time and effort from a wide variety of individuals. These are summarized below for each program:

1D – was programmed by Koji Yamasaki and Jerome Toporek.

2D – was designed by Laszlo Engelman, with contributions made by Peter Mundle. Improvements were made by Charles Lin and Jeffrey Meyer.

3D – the original version was programmed by Sandra Fu and Jerry Douglas, with latter revisions by David Sookne, Peter Mundle, Alan Forsythe, and R. J. Little.

4D was designed by W. J. Dixon and Laszlo Engelman and was programmed by Laszlo Engelman.

5D – was designed by Steve Chasen, with later revisions by Lanaii Kline.

6D – was designed and programmed by Steve Chasen, with later revisions by Laszlo Engelman.

7D – was programmed by Paul Sampson, with later revisions by Peter Mundle. The original design was proposed by W. J. Dixon.

8D – was programmed by Laszlo Engelman. The original version was programmed by Peter Mundle.

9D – was designed and programmed by Laszlo Engelman.

CA – was programmed by Aidan Moran, Laszlo Engelman, and Elaine Stephen, with assistance from Gilbert FitzGerald.

LE – was designed and programmed by Laszlo Engelman and Barbara Leake.

4F – was developed by Morton Brown, with contributions made by Jacqueline Benedetti, and Camil Fuchs.

- **1L** – was designed by Jacqueline Benedetti and Karen Yuen with major contributions from Virginia Clark, Robert Elashoff, and Ray Mickey. It was programmed by Larry Young, with program improvements made by Charles Lin, Alan Hopkins, and Calvin Chun. Robert Wolfe and Morton Brown contributed the formulas for the asymptotic standard errors of the quantiles.
- **2L** – was designed and programmed by Alan Hopkins. Contributions to the design were made by Nancy Flournoy, Jacqueline Benedetti, and Michael Tarter. Assistance in developing the program was obtained from James Frane, Jerry Toporek, Larry Young, and Laszlo Engelman. Improvements to the program were added by Calvin Chun.
- **1M** – was designed by John Hartigan and programmed by Howard Gilbert and Steve Chasen.
- **2M** – was designed and programmed by Laszlo Engelman and Sandra Fu, with later additions by James Frane and Albyn Jones.
- **3M** – was designed by James Frane with assistance from Paul Sampson.
- **4M** – was designed by James Frane with major contributions from Paul Sampson and Robert Jennrich.
- **5M** – was programmed by Mori Jamshidian and Shahriyar Dadkhah with later additions made by David Sookne.
- **6M** – was designed by James Frane with assistance from Paul Sampson.
- **7M** – was designed by Robert Jennrich and Paul Sampson and programmed by Paul Sampson.
- **8M** – was designed by M. R. Mickey and Laszlo Engelman, and was programmed by Peter Mundle and Laszlo Engelman.
- **9M** – was designed by Laszlo Engelman and Ray Mickey and programmed by Laszlo Engelman, with later contributions by Peter Mundle.
- **AM** – was programmed by James Frane, with later additions from Laszlo Engelman.
- **KM** – was designed by John Hartigan and Laszlo Engelman, and programmed by William Eddy and Laszlo Engelman. Supported in part by NSF Contact MCS-7906218, Yale University.
- **1R** – was programmed by Douglas Jackson and Jerry Douglas.
- **2R** – was programmed by Jerry Douglas, based on BMDO2R developed by Robert Jennrich. Charles Lin, Janis Hardwick, and David Sookne added later program improvements.
- **3R** – was programmed by Steve Chasen and substantially enhanced by Paul Sampson.
- **4R** – was programmed by Jerry Douglas, Laszlo Engelman, and Chyan-Ji Wang.
- **5R** – was designed by Robert Jennrich and programmed by Peter Mundle.
- **6R** – was designed and programmed by James Frane.
- **9R** – was designed and programmed by James Frane. FORTRAN coding for the best possible subset algorithm was obtained from George Furnival and Robert Wilson.
- **AR** – was designed and programmed by Mary Ralston. Later improvements were by Paul Sampson, Jerry Toporek, Calvin Chun, Laszlo Engelman, and Barbara Leake.
- **LR** – was designed and programmed by Laszlo Engelman, with contributions from Alan Forsythe, Ray Mickey, Frank Massey, and Virginia Clark.
- **PR** – was designed by Aidan Moran and Laszlo Engelman, with contributions from Gilbert FitzGerald and Brendan Lynch of BMDP Statistical Software, Cork Ireland. PR was programmed by Brendan Lynch.
- **3S** – was programmed by Steve Chasen, with improvements made by Calvin Chun and Jian-Shen Chen.

1T – was programmed by Tony Thrall and Laszlo Engelman. John Tukey, Peter Bloomfield, David Brillinger, and Bert Davis made valuable comments on the program's design. David Brillinger and Gordon Sande contributed subroutines to the program.

2T – was based partially on a time series program by David J. Pack, was designed by Lon-Mu Liu and revised by James Frane, Jeffrey Mayer, Chinh Le, and Chyan-Ji Wang.

1V – was designed by Laszlo Engelman and programmed by Laszlo Engelman and Koji Yamasaki. Improvements were made by Charles Lin.

2V – was designed by Robert Jennrich and Paul Sampson, with contributions from Alan Forsythe, James Frane and Jian-Shen Chen. 2V was programmed by Paul Sampson.

3V – was designed by Robert Jennrich and Paul Sampson, with major contributions from R. L. Anderson. It was programmed by Paul Sampson.

4V – was designed and written by Michael Davidson and Jerome Toporek. Initial work was done as URWAS, the University of Rochester Weighted Means Analysis of Variance System.

5V – was designed by Mark Schluchter. The program was based on a design by Robert Jennrich and Mark Schluchter with support from NIMH grant number 37188 in the Department of Biomathematics at UCLA. Contributions to the design and/or methodology were made by W. J. Dixon, Janet Elashoff, Alan Forsythe, Donald Guthrie, Elliot Landaw, and Rod Little. Additional programming was done by Taskin Atilgan, Kwan Lee, David Sookne, and Jian-Shen Chen.

8V – was designed by Robert Jennrich and Paul Sampson and was programmed by Paul Sampson.

BMDP has benefited from the work of many conversion centers that have customized BMDP for use on a variety of computer systems. For a complete list of systems, call BMDP at (310) 479-7799. Some of the many who have been responsible for conversions include B. Maillot (CDC), Victor Pietrzak (Data General), James Krupp (PDP–11), David Wilson (Harris/VOS), Rob Sciuk (HP 3000, 9000), Dr. Aenea Reid (Honeywell GCOS), Thierry P. Roget (Perkin Elmer), R. Valder (Siemens), Kari Gluski (IBM MTS), and David Muxworthy (ICL). At BMDP, Ben Lo was responsible for all UNIX-related conversions; Christian Vye for IBM versions. Elaine Lucey is responsible for the various conversions in Europe.

BMDP's Documentation staff, led by Virginia Lawrence, revised and edited the new material for this edition of the manual. Jeffrey Leeds and Michael Dula were responsible for significant portions of the manual production. The writers benefited from the technical advice of many programmers, statisticians, and other BMDP staff. In addition to the people mentioned above, particular thanks are due to David Sookne, Jennifer Row, Gilbert FitzGerald, Jian-Shen Chen, and the BMDP Technical Support Staff.

1

Introduction to BMDP

The computer programs that make up the BMDP system provide flexible and convenient procedures for data analysis, ranging from simple data display and description to advanced multivariate statistical techniques. BMDP runs on many different computer systems, from mainframes and minicomputers to PCs. Features include:

Clear, straightforward instructions. BMDP instructions are based on English for easy use. Because BMDP provides commonly used options automatically, a few lines of instructions can produce an extensive analysis, but you can add additional control language to tailor options to meet your own requirements.

Helpful output. In addition to the numerical result of a computational problem, output includes descriptive information that helps you evaluate the results. Instructions are reprinted in the output so you can easily review the options selected and keep a valuable record of the analysis performed.

Interactive or batch execution. When you invoke BMDP interactively, you can run a series of analyses without exiting the program. After each analysis you can insert new instructions or make corrections. Many programs allow paragraph by paragraph interaction, making it easy to scan results before proceeding to subsequent steps of an analysis. Stepwise procedures in some programs allow you to intervene at each step of the variable selection process.

Data management features. Select a subset of cases and/or variables during a run without permanently altering your data. Flexible transformation functions (addition, square roots, logarithms, etc.) allow you to recode, transform, and create new variables. These new variables may be retained in BMDP SAVE files.

Help for data analysis problems. Automatic data screening features and optional diagnostics help you detect coding errors, data outliers, and other anomalies in your data. The wide variety of available plots, histograms, and graphical displays make the screening process easier and more accurate.

Robust statistical methods permit more reliable analysis when classical assumptions are violated (e.g., if data are nonnormal, or variances are not equal). Alternative analyses are offered for nearly every classical test.

Power to handle large data sets. BMDP programs can analyze very large problems with many cases and variables. For detailed information about program capacity, see Appendix A, Problem Size.

In this chapter we offer some advice on how to use this manual, show how to specify BMDP instructions and run a simple example, and provide an overview of the BMDP programs.

Where to Find It

1.1 How to use this manual

The BMDP Manual is a two-volume set. Volume 1 contains introductory chapters on how to use BMDP as well as instructions on the most popular BMDP programs. Volume 2 contains advanced or specialized programs. Introductory manual chapters are numbered 1 to 11; program chapters have labels such as 2R and 4M. Each program chapter contains numerous examples—with input, output, and output commentary—that guide you step-by-step through the program's most important features. The overview in section 1.5 describes each BMDP program and the types of data analysis for which it is designed. The following describes the remaining introductory chapters in this volume:

Chapter 2 discusses data—what it is and how to use it.

Chapters 3–9 describe in detail the paragraphs that are common to all BMDP programs: INPUT, VARIABLE, GROUP, TRANSFORM, etc.

Chapter 10 discusses batch and interactive operation of BMDP in some detail and provides an interactive tutorial.

Chapter 11 describes a number of convenience features to make specifying BMDP instructions quicker and easier.

Appendices contain formulas and computational procedures, information about problem size, descriptions of the data sets used in the manual, and other useful information.

Beginning users should read sections 1.2 to 1.4 in this chapter, all of Chapter 2, sections 3.1 to 3.4, 4.1 to 4.5, all of Chapter 5, and 6.1 to 6.4. Then go directly to a program chapter such as lD or 3D to run some simple examples. Experienced users may want to examine Chapters 3 through 11 in some detail to familiarize themselves with BMDP's advanced capabilities and latest features.

1.2 A simple example

For our introductory example we show a very simple data analysis. Exam scores were recorded for a group of students taking the same course from three different professors. Some exams were administered in the morning, some in the afternoon. We use data from only thirteen respondents (Data Set 1.1). Data for each case, or respondent, contains three variables: time, professor, and score. Note that the time and professor variables are grouping variables, which are not intrinsically numeric. In this example, the three professors are Williams, Chang, and Nelson; and the two times are AM and PM. These values are recorded as numeric nominal codes in the data set. These are often referred to as dummy codes. (Chapter 2 describes how to handle nonnumeric data.)

Data Set 1.1
Exam score data

Williams		Chang		Nelson	
Time	Score	Time	Score	Time	Score
1	69	1	89	1	95
2	70	2	90	2	70
1	79	1	75	1	75
2	55	2	69	2	80
				1	70

Our first goal is to obtain a computer listing of the data for each case with names instead of codes. We also want to obtain simple descriptive statistics for the exam score for each professor and answer such questions as: How many people surveyed took exams in the afternoon and how many in the morning? What is the average exam score for professor Nelson's students? For Williams'? We will use program lD to find the answers to these questions.

Entering instructions and data

You can run BMDP programs either interactively at the terminal or by executing a file of instructions that you have previously created. When running BMDP interactively, you can use your local full screen editor or the BMDP line editor to type in both instructions and data directly at the keyboard. (In the PC version, you can use the BMDP full screen editor or your word processor in the ASCII text mode.)

In the following example we demonstrate the noninteractive batch mode. To begin, type the lines of instructions and data shown in Figure 1.1 into a system file. The blank lines between instructions are optional but may help you when reading your BMDP file. You may use any text editor or word processor that creates an ASCII file (see Chapter 2). The PC program has a built-in full screen editor that may be used. The number of spaces between words does not matter, but be careful to punctuate exactly as shown. Any alteration of syntax may cause the program to be read improperly or cause an error to occur. Once you have entered the information, save your instructions and data in a file called EXAM.INP.

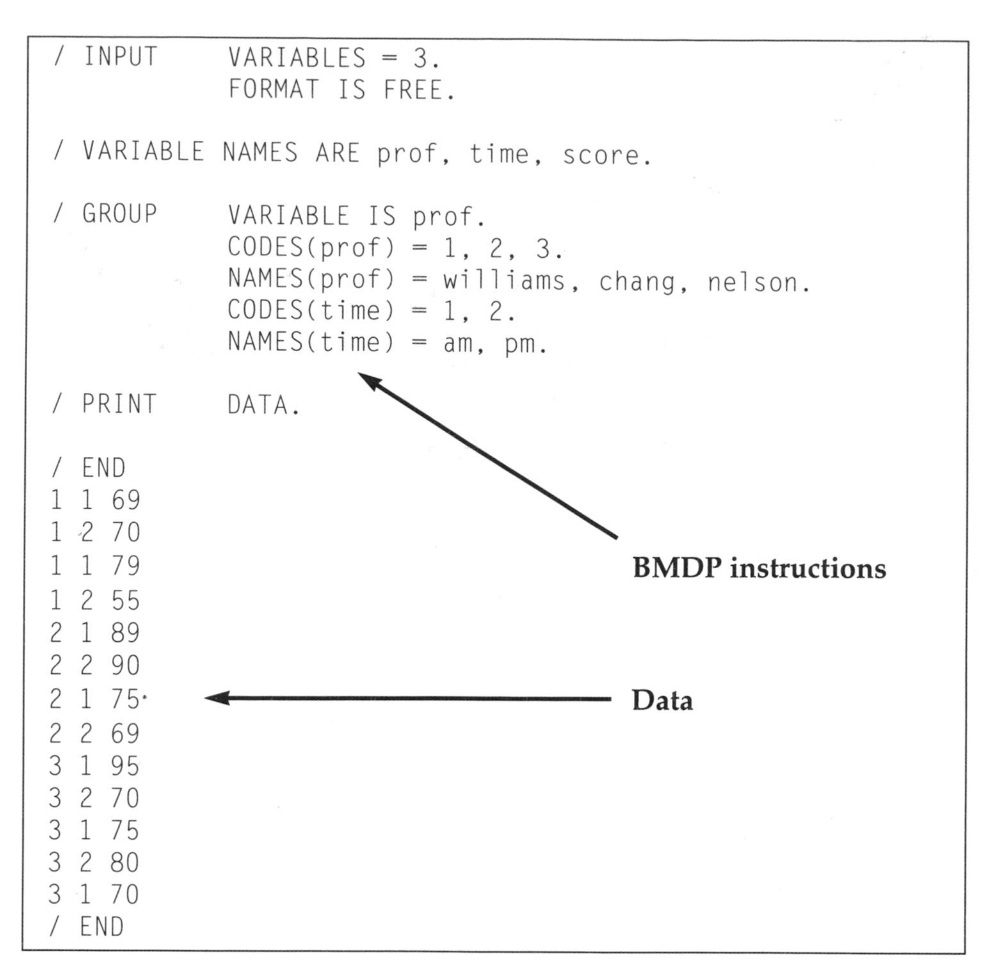

```
/ INPUT      VARIABLES = 3.
             FORMAT IS FREE.

/ VARIABLE NAMES ARE prof, time, score.

/ GROUP      VARIABLE IS prof.
             CODES(prof) = 1, 2, 3.
             NAMES(prof) = williams, chang, nelson.
             CODES(time) = 1, 2.
             NAMES(time) = am, pm.

/ PRINT      DATA.

/ END
1 1 69
1 2 70
1 1 79
1 2 55
2 1 89
2 2 90
2 1 75
2 2 69
3 1 95
3 2 70
3 1 75
3 2 80
3 1 70
/ END
```


Figure 1.1 BMDP instructions for exam data

The BMDP instructions. BMDP instructions are modeled after English and consist of *paragraphs and commands*. All BMDP programs utilize a basic set of paragraphs, each with a different function. We begin each paragraph with a slash (/). Each paragraph (except END) consists of at least one command ending with a period. Each command MUST end in a period. For some programs, only the basic set of paragraphs and commands is needed; other programs require more special paragraphs and commands. You will find it easy to analyze the same data using different programs once you learn the following basic set of instructions:

- The INPUT paragraph tells the program how many variables to read per case (VARIABLES = 3) and how the data are formatted in the file (FORMAT IS FREE). It also tells the program where the data can be found. In this example, since we do not mention a data file, the program will look for the data following the instructions.
- The VARIABLE paragraph allows you to name variables (TIME, PROF, and SCORE). In this example, one variable (PROF) is a grouping variable . Later in this manual we will show you how to select variables for analysis, specify acceptable range limits, and assign codes or characters to flag missing data.
- The GROUP paragraph specifies the grouping variable to use and classifies and names the groups. Specifically, we choose the grouping variable PROF, assign names for PROF (WILLIAMS, CHANG, NELSON) to the numeric codes in the data file so these names will appear in the output. We also name the two TIME codes (AM, PM).
- The PRINT paragraph instructs the program to include the data printout with the DATA command. The first ten cases are printed by default by 1D.
- The END paragraph is required to terminate BMDP instructions.

These are the most universal BMDP instructions. All programs use the INPUT, VARIABLE, and END paragraphs, while many use the GROUP paragraph. The INPUT, VARIABLE, and GROUP paragraphs are discussed in Chapters 3, 4, and 5, respectively.

Data entry: We chose to enter the data following the END paragraph which closes the program instructions. The data can also be stored in a separate system file, in which case you specify the name of the file (or its location) in the INPUT paragraph. See Chapter 3 for details.

Starting the Program

Suppose you have typed the above instructions and data into a system file called EXAM.INP. Your next step is to call the desired BMDP program, in this case program 1D, Simple Data Description. How you do this depends on the system you are using. BMDP runs on many different computer systems. IBM mainframes are discussed briefly in Chapter 3; see your Installation Guide or system manager for more information. Here we show how to run BMDP in batch mode on a VAX. After logging onto the system, simply type

```
BMDPRUN 1D IN=exam.inp OUT=exam.out [Return]
```

Remember to press the RETURN or ENTER key at the end of the line. The program output will be stored in the file named EXAM.OUT. You can examine this file through a system editor and/or print it using a system print command. We show the output stored in EXAM.OUT below.

Program output

```
BMDP1D - SIMPLE DATA DESCRIPTION                                        [1]


Copyright 1977, 1979, 1981, 1982, 1983, 1985, 1987, 1988, 1990
                 by BMDP Statistical Software, Inc.

       BMDP Statistical Software, Inc.| BMDP Statistical Software
       1440 Sepulveda Blvd            | Cork Technology Park, Model Farm Rd
       Los Angeles, CA 90025 USA      | Cork, Ireland
          Phone (310) 479-7799        |    Phone +353 21 542722
          Fax   (310) 312-0161        |    Fax   +353 21 542822
          Telex 4972934 BMDP UI       |    Telex 75659 SSWL EI

Version: 7.0     (VAX/VMS)            DATE:    25-FEB-92       AT 08:45:21
 Manual: BMDP Manual Vol. 1 and Vol. 2.
 Digest: BMDP User's Digest.
Updates: State NEWS. in the PRINT paragraph for summary of new features.

PROGRAM INSTRUCTIONS                                                    [2]

/ INPUT      VARIABLES = 3.
             FORMAT IS FREE.

/ VARIABLE   NAMES ARE prof, time, score.

/ GROUP      VARIABLE = prof.
             CODES(prof) = 1, 2, 3.
             NAMES(prof) = williams, chang, nelson.

             CODES(time) = 1, 2.
             NAMES(time) = am, pm.

/ PRINT      DATA.

/ END

PROBLEM TITLE IS
   25-FEB-92           08:45:21                                         [3]

NUMBER OF VARIABLES TO READ . . . . . . . . . . .       3
NUMBER OF VARIABLES ADDED BY TRANSFORMATIONS. .         0
TOTAL NUMBER OF VARIABLES . . . . . . . . . . .         3
CASE FREQUENCY VARIABLE . . . . . . . . . . . .
CASE WEIGHT VARIABLE. . . . . . . . . . . . . .
CASE LABELING VARIABLES . . . . . . . . . . . .
NUMBER OF CASES TO READ . . . . . . . . . . . . TO END
MISSING VALUES CHECKED BEFORE OR AFTER TRANS. . NEITHER
BLANKS IN THE DATA ARE TREATED AS     . . . . . MISSING
NUMBER OF INTEGER WORDS OF MEMORY FOR STORAGE .   19998

VARIABLES TO BE USED                                                    [4]
     1 prof          2 time          3 score


DATA FORMAT:  FREE                                                      [5]

THE LONGEST RECORD MAY HAVE UP TO   80 CHARACTERS.

NUMBER OF CASES READ. . . . . . . . . . . . . .      13                 [6]
```

```
 VARIABLE        STATED VALUES FOR          GROUP CATEGORY    INTERVALS                       [7]
NO.   NAME    MINIMUM MAXIMUM MISSING   CODE  INDEX   NAME     .GT.   .LE.
---- -------- ------- ------- -------  ------ ----- --------  ------ ------

  1  prof                               1.000    1  williams
                                        2.000    2  chang
                                        3.000    3  nelson

  2  time                               1.000    1  am
                                        2.000    2  pm
------------------------------------------------------------------------------

GROUPING VARIABLE. . . prof
                               CATEGORY    FREQUENCY
                               --------    ---------
                               williams        4
                               chang           4
                               nelson          5

DESCRIPTIVE STATISTICS OF DATA
----------- ---------- -- ----

VARIABLE     TOTAL         STANDARD  ST.ERR   COEFF      SMALLEST        LARGEST               [8]
NO. NAME      FREQ.   MEAN    DEV.   OF MEAN  OF VAR  VALUE  Z-SCR   VALUE  Z-SCR  RANGE

  2 time        13  1.4615  .51887   .14391   .35502  1.0000 -0.89  2.0000  1.04  1.0000
  3 score       13  75.846  10.800   2.9954   .14239  55.000 -1.93  95.000  1.77  40.000

 PRINT SUMMARY STATISTICS OVER ALL CASES AND
 BROKEN DOWN BY INDIVIDUAL CATEGORY ON prof

CASE     1         2        3                                                                  [9]
 NO. prof      time      score
----- -------- -------- --------
   1 williams am          69.00
   2 williams pm          70.00
   3 williams am          79.00
   4 williams pm          55.00
   5 chang    am          89.00
   6 chang    pm          90.00
   7 chang    am          75.00
   8 chang    pm          69.00
   9 nelson   am          95.00
  10 nelson   pm          70.00
  11 nelson   am          75.00
  12 nelson   pm          80.00
  13 nelson   am          70.00

VARIABLE        GROUPING          TOTAL              STANDARD    ST.ERR    COEFF. OF     S M A L L E S T      L A R G E S T              [10]
NO. NAME    VARIABLE LEVEL      FREQUENCY      MEAN  DEVIATION   OF MEAN   VARIATION     VALUE    Z-SCORE     VALUE    Z-SCORE     RANGE

  3 score                           13        75.846   10.800     2.9954    .14239      55.000    -1.93       95.000    1.77       40.000
            prof      williams       4        68.250   9.9121     4.9561    .14523      55.000    -1.34       79.000    1.08       24.000
                      chang          4        80.750   10.404     5.2022    .12885      69.000    -1.13       90.000    0.89       21.000
                      nelson         5        78.000   10.368     4.6368    .13293      70.000    -0.77       95.000    1.64       25.000

NUMBER OF INTEGER WORDS USED IN PRECEDING   PROBLEM     692
CPU TIME USED       4.070 SECONDS                                                             [11]

END OF INSTRUCTIONS

PROGRAM TERMINATED
```

Explaining the output

Each portion of BMDP program output is formatted and labeled to be largely self-explanatory. In later examples we will usually show only the portions of output that are of current interest.

Panels printed by every program include:

[1] A banner that tells which program you called. The banner tells when your version of the program was last revised and reminds you to specify NEWS in the PRINT paragraph for information about updates.

[2] Your program instructions, just as you entered them.

[3] An interpretation of the instructions, including default values assigned by the program. (BMDP preassigns values for many commands so that you will not need to specify them.)

[4] A list of the variable names; each variable is assigned a sequence number which can be used to specify commands in later analyses.

[5] Format of the data and further interpretation of program instructions.

[6] The number of cases read by the program.

[7] Grouping information. Check to see if you assigned the correct names to the correct codes. Just below is the number of cases included in each group. You can check this panel to be sure the correct number of subjects have

been assigned to each group.

[8] Panel of descriptive statistics without grouping breakdown. This includes the mean, standard deviation, standard error of the mean, coefficient of variation, and the highest and lowest values found with corresponding standardized z-scores for those extreme cases.

[9] A printout of the data. 1D supplies case numbers, variable headings, and, when specified, group names derived from the code and name statements.

[10] Descriptive statistics for SCORE, grouped by PROF. We see that the WILLIAMS group received a lower mean rating, 68.250, than the other two groups, 80.75 and 78. This table is specific to program 1D.

The last item **[11]** is printed by all programs; it provides information about the size of the problem analyzed, and lets you know that the program has terminated correctly. If the program does not run correctly, you may see an error message.

1.3 Manipulating BMDP paragraphs

Moving from program to program. Investigators usually analyze data through a series of steps, first examining the data graphically and numerically for unreasonable values, then checking such values and, if possible, correcting them before continuing the analysis. Moving between different programs is easy to do in BMDP and often requires only minor changes in the instructions. For example, you might want to analyze all the exam data, not just 13 cases and, study the average rating within each professor and time combination. To do this, simply insert a HISTOGRAM paragraph in the instructions and call program 7D.

Figure 1.2 Adding a HISTOGRAM paragraph to the exam analysis

```
/ INPUT      VARIABLES = 3.
             FORMAT IS FREE.

/ VARIABLE   NAMES ARE prof, time, score.

/ GROUP      VARIABLE = prof.
             CODES(prof) = 1, 2, 3.
             NAMES(prof) = williams, chang, nelson.
             CODES(time) = 1, 2.
             NAMES(time) = am, pm.

/ HISTOGRAM GROUPING = prof, time.

/ PRINT      DATA.

/ END
[data for the complete sample]
```

Analysis paragraphs. HISTOGRAM is just one of many specialized analysis paragraphs used by BMDP programs. Other analysis paragraphs include:

PLOT – in 6D, the bivariate scatterplot program

REGRESS – in 2R and other regression programs

All BMDP programs ignore paragraphs specific to other programs. For example, you could still run 1D using the input above, even though 1D does not use the HISTOGRAM paragraph. BMDP will indicate that it has ignored this paragraph.

Paragraphs common to all programs

Along with INPUT, VARIABLE, GROUP, and END, several other paragraphs are used by all programs:

TRANSFORM – (Chapter 6) used to create new variables using arithmetic operators (+, –, *, /) and functions (logarithms, square roots, dates, etc.). You can also select cases (e.g., males over twenty years old) and edit data.

PRINT – (Chapter 7) used to control width and length of output. The NEWS feature lists information about BMDP program changes.

SAVE – (Chapter 8) saves data, variable names, range limits, and grouping information in a BMDP File. Some programs allow you to save computed results: predicted values, factor scores, etc.

CONTROL – (Chapter 9) used to set program environment (interactive or noninteractive, workspace size, debugging level, etc.).

FINISH – (Chapter 11) used to terminate a program when running interactively.

1.4 Rules for writing BMDP instructions

The instructions shown in Input 1.1 are written according to the following set of general rules. These rules apply to both basic paragraphs and paragraphs added for specific analyses.

The paragraph name comes first (e.g., INPUT).

- Paragraphs are separated by a slash (/). We prefer to put the slash before the paragraph so additional instructions are easy to insert for later runs. (In some interactive situations, the slash must come after the paragraph—see Chapter 11.)
- A given paragraph can be used only once in each problem unless otherwise stated in the program description.
- Except for END and where otherwise noted, paragraphs may be in any order in BMDP instructions. END terminates BMDP instructions for a problem or subproblem and consists only of the paragraph name (END). No other instructions should be typed after END on the same record. Note that a period should not follow END.
- Additional information is given in commands. One type of command takes the form "keyword" followed by IS, ARE, or =, followed by a list of items. IS, ARE, and = are interchangeable; use whichever seems most natural. Another type is the one word command, such as DATA in the PRINT paragraph.
- Every command must end with a period (.) or semicolon (;). Commands may be in any order within a paragraph.
- Values or names in a list are separated by commas.
- Names may contain up to eight characters. If a name contains characters other than letters and digits, that name must be enclosed in single quotes (apostrophes). More specific rules for names are given in Chapter 4.
- Instructions may be in upper or lower case.
- Although the above examples are nicely indented, instructions can be written in free format. The following is equivalent to Figure 1.2:

```
/INPUT VARIABLES=3. FORM is FREE. /VARIABLE NAMES=prof,
time, score. /GROUP VARIABLE = prof. CODES(prof) = 1,2,3.
NAMES(prof) = williams, chang, nelson. CODES(time)=1,2.
NAMES(time)=am, pm. /HISTOGRAM GROUPING = prof, time. /
PRINT DATA. / END
```

We strongly recommend that you use an organized format to make reading and editing your instructions easier. Note that we abbreviated FORMAT. In the formal command definitions at the end of each chapter, we use upper case letters to show the required part of each instruction; lower case portions are optional (e.g., FORMat).

1.5 An overview of data analysis with BMDP

The BMDP program library encompasses a wide range of data analysis procedures. This section is designed to help you decide which programs are applicable to your study. For a detailed description of what each program provides, read the chapter describing that program.

All programs allow you to:

- Name variables
- Specify missing value codes and minimum and maximum range limits
- Define and name groups
- Re-express data (e.g., take the log of body weight)
- Save data with names, grouping information, flags for missing data and values outside range limits, group information
- Select a subset of cases for analysis

Many programs provide:

- Data printouts
- Descriptive statistics (mean, standard deviation, sample size, etc.)
- Case weights
- Double precision computations

Screening data—listing, histograms, plots, descriptive statistics

An important first step in data analysis is to find and correct recording or data entry errors and to locate outliers or extreme values which could affect further results. To assist with data screening, many programs provide data listings, histograms, cumulative histograms, and normal probability plots.

You may also want to review means, standard deviations, and other descriptive statistics to assess characteristics of the distributional shape of the data. These statistics can help you decide whether to transform the data to improve agreement with assumptions (constant variance, normality, etc.) of the statistical model you plan to use.

It is often a good idea to screen data after the cases have been sorted into categories or groups. Unusual data values that are masked in a total population may stand out when the data are classified into groups by such factors as sex or age. For example, if the data are classified by age, a height of six feet in an over–twenty group does not arouse suspicion, but six feet in an under-ten group is almost certainly an error. The data within each group can be examined for recording or data entry errors, for outliers or extreme values, and to assess the normality of the distributions. Data listings, histograms, cumulative histograms, and normal probability plots can assist you. By examining means, standard deviations, and other descriptive statistics, you can obtain an initial overview of the data structure and determine whether a transformation to stabilize variances is advisable.

Data screening and the search for outliers do not end with univariate screening; several multivariate programs provide distance measures and diagnostic statistics to assist in identifying extreme cases.

Lists:	**Program**
• Sorted data lists with missing and out of range values highlighted	1D, 7D
• Listings with user-specified format	1D, 3D, most programs
• Display showing missing data pattern	AM
• Frequencies for each distinct value	2D

Histograms and Plots:	
• Histogram	2D, 5D, 7D
• Stem and leaf display	2D

- Cumulative frequency distribution or histogram — 2D, 5D
- Normal, half-normal, and detrended normal probability plots — 5D
- Line plot of measures of location — 2D
- Quantile/Quantile plots — 3D
- Miniplots of group means — 9D
- Box-Cox plot to assist in choosing transformation — 7D
- Plots of regression diagnostic statistics — 2R

Descriptive statistics:

Most programs report various descriptive statistics; the following programs are especially useful for obtaining complete descriptive reports.

- Mean, standard deviation, minimum, maximum — 1D, 2D*, 7D*, 9D*
- Robust estimators of location, e.g., trimmed mean — 2D*, 3D*, 7D*
- Other estimators of distribution shape such as minimum and maximum standard scores, skewness, kurtosis — 1D, 2D, others
- Cell frequencies — 9D, 4F
- Distance of case from group mean — 4M, 7M, 9R, 2R
- Diagnostic statistics for regression — 2R

*may be requested for each group

Tests for means or location—*t* test, analysis of variance, and nonparametric tests

You may want to test the significance of a difference in means (or other measures of location) between two or more groups. The *t* test is used for matched or two-group tests, the analysis of variance *F* test for testing means between several groups; both are useful when it can be assumed that the data are normally distributed with equal variances. When the data are not normally distributed, the nonparametric equivalents to the *t* test and analysis of variance may be useful. Note that other assumptions, such as equal distributional shapes, are often required by nonparametric procedures. Sometimes the data can be made more nearly normal by appropriate choice of a transformation (see 7D), or robust methods based on trimming may be helpful (3D, 7D). You may want to use Levene's test to assess whether one group has higher variability than another. When variances are unequal, transforming the data may help; tests robust to inequality of variances (such as the Brown-Forsythe and Welch statistics) are also useful. Multiple comparison or contrast methods show which groups differ from which other groups. An analysis of covariance provides tests of differences between group means after adjustment for related variables.

***t* tests and related statistics:**

- Two-group *t* test — 3D, 7D, 9D, 1V, 2V
- Two-group Mann-Whitney rank sum test — 3D, 3S
- Trimmed *t* test — 3D
- Multivariate T^2 and D^2 — 3D, 4V
- Matched *t* test — 3D
- Trimmed matched — 3D
- Sign test and signed rank test — 3D, 3S
- One-group *t* test — 3D
- Trimmed one-group *t* test — 3D

Analysis of variance and covariance:

- One-way analysis of variance — 7D, 1V, 2V, 9D
- One-way analysis of variance for unequal variances — 7D
- One-way analysis of variance using trimming — 7D
- Kruskal-Wallis nonparametric analysis of variance — 3S
- One-way analysis of covariance — 1V, 2V, 4V
- Test for parallelism of regression between groups — 1V

- Two-way analysis of variance 7D, 2V
- Repeated measures (and split-plot) analysis of variance 2V, 4V
- Unbalanced repeated measures models with structured covariance matrices 5V
- Friedman test 3S
- Mixed model analysis of variance 8V, 4V, 3V
- Nested analysis of variance 8V, 4V
- Multiple comparisons 7D
- Nonparametric multiple comparisons 3S
- Multiple comparisons with covariate adjustments 1V
- User-specified contrasts 1V, 7D, 2V, 7M, 4V
- Contrasts with covariate adjustments 1V, 2V
- Contrasts over within groups 2V
- General analysis of variance 2V, 4V, 8V
- General analysis of covariance 2V, 4V, 3V
- Multivariate analysis of variance 4V, 7M
- Multivariate analysis of covariance 4V

Nonparametric tests:

- Sign and signed rank 3D, 3S
- Two-group Mann-Whitney rank sum test 3D, 3S
- Friedman analysis of variance 3S
- Kruskal-Wallis nonparametric analysis of variance 3S

Test for variability:

- Levene's test 3D, 7D

Relationships between continuous variables—scatterplots, correlations, and linear regression

Researchers often want to know whether one continuous variable predicts or is associated with the values of another. Scatterplots, correlations, regression equations and residuals from fitted regression equations assist this process.

Scatterplots that take advantage of known information can be designed to display unusual cases or outliers—for example, to show whether a subject's pulse before exercise is higher than after exercise. A scatterplot of these two variables will show whether the data coding is mistakenly reversed for some cases.

Regression methods are useful in exploring the relationship between a "dependent" or Y variable and "independent" or X variables. They assess which X variables are most related to the Y variable and estimate a prediction equation. Most of the regression programs described here examine linear relationships between the X and Y variables, but programs are also available for describing nonlinear relationships. Several programs provide diagnostic statistics and plots to help assess the fit of the equations and the degree to which assumptions are satisfied.

1R, 2R, and 9R print and plot residuals and predicted values. The plots are useful in detecting lack of linearity, lack of constant variance, outliers, gross errors, an unusual subpopulation that should be separated from the analysis, etc. The plots may also indicate that transformations of the data are advisable or that an inappropriate model was chosen. The residual analysis in 2R is the most extensive of the three. 9R allows easy cross-validation of the regression model by testing it on a subset of cases excluded from the analysis.

Plots:

- Bivariate plots with grouping information and regression line 6D
- Diagnostic and residual plots 7D, 1R, 2R, 3R, 9R, AR

Correlations:

• Correlations (Pearson product-moment)	6D, 7D, 1R, 2R, 4R, 9R, LR, PR
• Rank correlations (Spearman)	3D, 3S, 4F
• Rank correlations (Kendall)	3S
• Correlations with missing data	8D, AM

Regression:

• Linear regression on one x variable	6D, 1R, 2R, 9R
• Multiple regression	1R, 2R, 9R
• Multiple regression within subgroups and test for equality of regressions	1R
• Stepwise regression—forward and/or backward	2R
• All-subsets regression (double precision)	9R
• Regression on principal components	4R
• Regression diagnostics—residual analysis	2R, 1R, 9R
• Tests of equality of regressions between groups	1R
• Tests of parallelism between groups	1V
• Ridge regression	3R, 4R, AR
• Multivariate regression	6R
• Regression weights	1R, 2R, 3R, 4R, 5R, 6R, 9R, AR
• Polynomial regression	5R
• Nonlinear regression	3R, AR, LR, PR

Frequency tables, correspondence analysis, and log-linear models

Program 4F allows you to cross-tabulate and analyze data in frequency tables. Cross-tabulation is useful as a form of final reporting to give a picture of the number of cases in specified categories. Frequency table methods are appropriate for analysis of the relationships among categorical variables. You can use data or cell frequencies as input, and you can create separate tables for each level of a third variable (such as sex). 4F computes 23 statistics appropriate for the analysis of frequency tables. You can also use 4F to fit a log-linear model to the cell frequencies and test the model's fit. Program CA performs simple and multiple correspondence analysis.

Two-way tables:

• Tests of independence (chi-square, likelihood ratio, Fisher's exact, Cramér's V, Yule's Q, etc.)	4F
• Measures of prediction	4F
• McNemar's test of symmetry, kappa test of reliability, and test of linear trend across ordered proportions	4F

Multiway tables:

• Log-linear models, stepwise models	4F
• Tests of marginal and partial association for screening effects	4F
• Test-of-fit, estimates of parameters and their standard errors, expected frequencies	4F
• Specification of structural zeros (cells to be excluded from the model)	4F
• Stepwise identification of unusual cells and strata	4F

Simple correspondence analysis:

• Printouts and plots of row and column profiles	CA
• Breakdown of total inertia (c^2/n)	CA
• Measures of the CA representation quality	CA
• User specified supplementary points for summarizing information in one plot	CA

Multiple correspondence analysis

- Printout of BURT matrix and plot of each variable in matrix — CA
- Printout of case data for each variable, label variable, weighting variable, and case coordinate — CA
- Supplementary plots of continuous data — CA
- Correlation coefficient for continuous variable and case coordinates of each axis — CA

Nonlinear regression

Four programs allow you to fit models that are nonlinear in the parameters, such as the simple exponential,

$$p_1 e^{p_2 t}$$

Programs 3R and AR provide six general built-in functions and allow you to specify other functions in a special paragraph. Mainframe users can specify functions in a FORTRAN subroutine. These programs offer many special features, including estimating functions of the parameters, maximum likelihood estimation, and ridge estimation.

The logistic function is a special nonlinear model which can be used to predict a binary outcome (response or no-response) using categorical (e.g., sex) and continuous (e.g., age, blood pressure) variables and the interactions between categorical variables. LR provides maximum likelihood estimates of the parameters in a stepwise logistic regression procedure. PR provides maximum likelihood estimation of the parameters in a stepwise multi-valued logistic regression.

- Nonlinear regression — AR, 3R
- Stepwise logistic regression — LR, PR
- Ridge regression — AR, 3R, 4R
- Robust regression — 3R
- Logistic bioassay — AR, 3R
- Case-control computation — LR
- Scatterplots — 3R, AR, LR
- Confidence curve plots — 3R
- Standardized residual plots — PR
- Histograms of predicted probabilities — LR, PR

Maximum likelihood estimation

Five programs perform maximum likelihood estimation. They differ in the types of models they fit and the way functions are specified. The most general of the programs, LE, can perform maximum likelihood estimation for any given density function. LE allows you to specify functions directly. The AR and 3R programs can estimate functions in any exponential family, and perform maximum likelihood through iteratively reweighted least squares. The LR and PR programs are appropriate for estimating logistic functions.

Missing values and cases used

Frequently the data are not complete—some values may be missing and others may lie outside specified range limits (see MIN and MAX in Chapter 4). Usually when the data are being entered and a value is missing, the user inserts a code or symbol (e.g., an asterisk) in the data file to mark its position and identifies this flag to BMDP by using the MISS command (Chapter 4) or the MCHAR command (Chapter 3). All BMDP programs check for values missing and outside specified range limits. A case is considered *complete* if the values of all variables are *acceptable* (there are no values missing or out of range). The treatment of incomplete cases depends on the analysis. For example, ID computes descriptive statistics for each variable using all acceptable values for that variable (regardless of whether other variables may be missing for some cases), and 2R uses only complete cases to estimate the coefficients in the regression equations.

If few cases have missing data, it may be reasonable simply to drop those cases from subsequent analyses. However, if a considerable number of cases have missing data, you will want to see whether deletion of one or more variables substantially reduces the number of missing values. You will also need to assess the patterns of missing data, since methods for estimating missing values or estimating covariance matrices in the presence of missing values require the satisfaction of assumptions about whether "missingness" is related to the value of the variable or to the values of other variables.

Most regression and multivariate analyses require *complete* cases (i.e., no missing values). Many of those analyses can begin from a correlation or covariance matrix rather than from the original data; methods are available to estimate the correlations from all acceptable values. Both AM and 8D can estimate correlations for cases with some data missing; the correlation matrix can then be stored in a BMDP File and used as input to other programs. AM ensures that the resulting correlation matrix is numerically appropriate (positive semidefinite) for a regression or factor analysis. In addition, AM can replace missing values with estimates based on data present for the case; the *completed* data can be saved and used as input to additional analyses. Subsequent analyses are affected by either of these procedures, as are the usual tests of significance reported. However, the results can be useful in exploratory data analysis.

- Displays for patterns of missing data — AM
- Computation of correlations with missing data — 8D, AM
- Computation of covariances with missing data — 8D, AM
- Estimation of missing values — AM
- Repeated measures analysis with missing data — 5V

Multivariate analysis

The term "multivariate analysis" encompasses a variety of statistical procedures, including factor analysis, correspondence analysis, canonical correlation analysis, discriminant analysis, and preference pairs.

Factor analysis is useful in exploratory data analysis. It has three general objectives: to study the correlations of a large number of variables by clustering the variables into factors, such that variables within each factor are highly correlated; to interpret each factor according to the variables belonging to it; and to summarize many variables by a few factors.

Correspondence analysis is an exploratory multivariate technique that converts frequency table data into graphical displays in which rows and columns are depicted as points. The distances between points are approximate chi-square distances between two individual row profiles. Correspondence analysis provides a method for comparing row or column proportions in a two-way or multiway table.

Canonical correlation analysis examines the relationship between two sets of variables, and can be viewed as an extension of multiple regression analysis or of multiple correlation.

Discriminant analysis provides linear or quadratic functions of the variables that best separate cases into predefined groups (7M and 5M). Logistic regression (LR) can also be used when there are two groups. Polychotomous regression (PR) can be used for multiple groups.

9M, Linear Scores from Preference Pairs, makes use of a set of preference judgments between pairs of cases provided by "experts" and a set of objective measures. These are used to create a linear function of the objective measures that reproduces as closely as possible the expert preferences between pairs of cases. This linear rating scale can then be applied to other cases.

Multivariate screening can identify outliers. For each case, you can print the

Mahalanobis distance squared from the case to the center of all cases (4M), or from the case to the center of each group (7M).

- Factor analysis 4M
- Correspondence analysis CA
- Boolean factor analysis 8M
- Canonical correlation analysis 6M
- Linear and quadratic discriminant analysis 5M
- Stepwise discriminant analysis 7M
- Multivariate T^2, ANOVA, and ANCOVA 3D, 4V, 7M
- Mahalanobis distances 7M
- Linear scores from preference pairs 9M

Cluster analysis

Cluster analysis is useful for studies in which many variables are observed for each case and you wish to explore the interrelationships among variables and cases. Clustering can be used in exploratory data analysis to assess whether there are natural groupings of cases or variables. (Factor analysis is the more commonly used technique related to clustering of variables, but different assumptions are required for factor analysis.) Cluster analysis is usually used to cluster cases.

Clustering performs a display function for multivariate data similar to graphs or histograms for univariate data. Clustering provides a multivariate summary—a description of clusters instead of individual cases.

- Cluster analysis of variables 1 M
- Cluster analysis of cases 2M, KM
- K-means clustering KM
- Block clustering of both cases and variables 3M

Survival analysis

The techniques in these programs are appropriate when analyzing the time to occurrence of some event or response; e.g., time to death, time to disease recurrence, time to conception, time to light bulb failure. Behavioral scientists sometimes call this technique event history analysis. Survival analysis techniques have been designed to use data from cases with incomplete or censored observations in which the response had not yet occurred before loss to follow-up.

- Life-tables and survivor functions 1L
- Product limit and life-table plots of survival curves for each group 1L
- Plots of hazard and cumulative hazard 1L
- Mantel-Cox, Breslow, Tarone-Ware, and Peto-Prentice tests for equality of survival curves 1L
- Trend tests and user-specified contrasts between groups 1L
- Regression with incomplete survival data 2L
- Cox proportional hazards model 2L
- Accelerated failure time model 2L
- Survival analysis with covariates 2L
- Stepwise prediction of survival 2L
- Time-dependent covariates 2L
- Wald, likelihood ratio, and score tests 2L
- Plots of Cox-Snell and standardized residuals 2L
- Competing Risks Analysis 2L

Time series analysis

A time series is a sequence of measurements made at regular time intervals, such as yearly gross national product or daily temperature. Other data, such as thread thickness every millimeter along its length, might also be analyzed with time series methods. Most statistical methods require that each case be statistically independent of other cases. However, in time series data we expect that neighboring cases may be correlated with each other. You may simply want useful graphical methods to visualize patterns in the time series data, or you may want to build and verify a model describing the series. Such modeling may be preparatory to making predictions (forecasts) about the future values of the time series or attempting to assess the impact of some intervention. The frequency domain (or spectral) approach to analysis of time series involves representing the data by a superposition of sinusoidal waves of different frequencies (1T). The time domain (finite parameter or Box–Jenkins) approach consists of fitting autoregressive and/or moving average models to data after differencing and/or seasonal adjustment. Time series techniques are most appropriate when there are numerous time intervals.

Univariate and bivariate spectral analysis:

- Graphical and numeric displays of periodograms and spectra of individual or paired time series — 1T
- Filtering and optional re-coloring — 1T
- Estimates of degree of coherence and regression relation between two time series in different frequency bands — 1T
- Smoothing by removing seasonal means or linear trends, or by constructing filters — 1T

Box–Jenkins analysis:

- Time series plots — 2T
- Autocorrelations and partial autocorrelations — 2T
- Parametric time domain models (ARIMA, intervention, transfer function models) — 2T
- Selection, estimation, and testing of models (residual analysis) — 2T
- Forecasting of future observations — 2T

1.6 Common problems and error messages

You may occasionally encounter errors that will cause your program to run improperly or fail to finish. The most common sources of errors are data problems, incorrect BMDP instructions, and system errors.

Data problems.

1. Alphabetic characters in numeric data
2. Omitting space in free format or unequal number of variables
3. Negative values when a log or square root is called for in the TRANSFORM paragraph
4. Blank lines in data matrix

You should also look at the panel in the output that tells you the number of acceptable cases and the number of missing values. An obviously incorrect number of acceptable cases or missing values may indicate a problem with the data.

BMDP instructions. Frequent causes of problems are:

1. Missing periods, commas, slashes, or apostrophes
2. Missing END paragraph
3. Misspelled command
4. Unbalanced parentheses or single quotes (apostrophes)

System problems. See your computer representative.

Program bugs. If you have exhausted all other possibilities, and suspect a program bug, call BMDP Technical Support at (310) 479-7799 during business hours, or, for European and African users, BMDP Statistical Software, Cork, Ireland, at 353-21-542822.

2

Data

If you are not experienced with computer analysis you should read this chapter for information about how to organize your data and communicate it to BMDP. After providing some basic definitions, we discuss how to enter and access data on the computer, what to do when values are missing from your data set, and how to store your data in ways that BMDP can read. Finally, we offer some advice on how to collect data in a way that makes data entry and analysis as easy as possible.

Where to Find It

2.1 What are data?

We generally think of data as information of any sort. In terms of computers and statistical analysis, however, data are usually numbers. For purposes of analysis, data are arranged as *cases* and *variables* (see Figure 2.1).

Some basic definitions

A data file often contains more than the measurements you plan to analyze; you may also need variables to define groups (such as male and female) and identify cases (id numbers, names). A variable is any characteristic that can be recorded and stored for a subject (sex, coded male or female; a pulse measurement; a case identification number). A case contains values for all the variables for one subject or sampling unit. The subject may be a person, an animal, a hospital, a family, or a plot of land—anything on which you can collect information.

Figure 2.1
Data Organization

VARIABLES (columns)
– id
– measurements

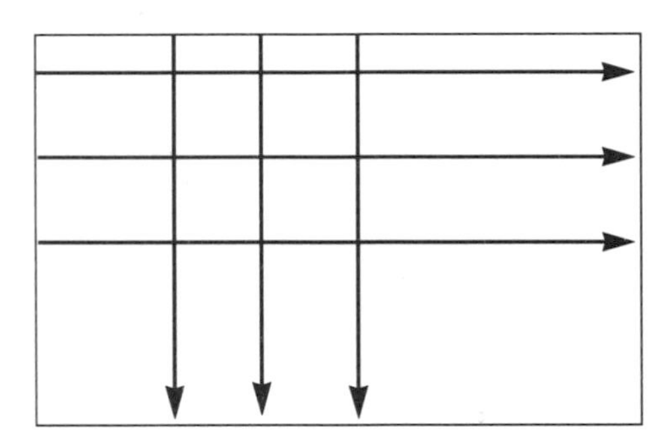

To review some important definitions before going on:

case – a subject or sampling unit
variable – a characteristic, label, or measurement collected for each case
value – the number or characters stored for a particular variable for a particular case

Let's make this clear by collecting some data. Suppose 40 people are asked to take their pulse, run a mile, and then take their pulse again. We enter the results into the computer by typing one line for each person, giving pulse values plus such additional information as age, sex, smoking status (yes/no), and an id number. We enter the information in the same order for each person (see Figure 2.2). Subject number 12 is a male (denoted by the code 1) nonsmoker (denoted by the code 2), who is 22 years old; 74 is his pre-exercise pulse value, and 134 is his post-exercise pulse value:

```
12  1  2  22  74  134
```

Data typed in with blanks between values are called *free format* data.

The Exercise data include examples of the three basic types of data used by BMDP:

Measurements are what we usually have in mind when we think of data. Measurements are numeric quantities such as age and pulse rate.

Case identifiers are codes such as hospital patient numbers, social security numbers, or sequence numbers used to mark cases. Include a case identification number on each record to identify all data pertaining to one subject. The id number allows you to refer to the original data to correct errors. Later we discuss how to use alphanumeric case labels, such as a subject's name.

Categorical values. Some variables, like sex and smoking status, are not intrinsically numeric. These are examples of grouping variables, which can be used to separate the subjects into two or more groups. For example, you might use the following codes: 1 = MALE, 2 = FEMALE; 1 = SMOKER, 2 = NONSMOKER. Many BMDP programs use grouping information to analyze data; for now you just need to remember that these variables are coded as numeric values to simplify data entry and analysis. There are also categorical response variables. For example, three scores could be coded 1, 2, 3 for mild, moderate, and severe.

Note that BMDP can also use continuous variables like age to group cases. You should always record each subject's actual age rather than coding age in categories (such as code 2 for age 30–35). It is easy in BMDP to collapse data into categories, but impossible to go from a few category codes to more refined values for a particular analysis.

Figure 2.2 presents the Exercise data. There are six variables and 40 cases. The record for each subject includes an ID number, sex, smoking status (yes/no), age, and pulse rates before and after exercise. For some simple examples using the Exercise data, see the chapter describing program 1D. The basic instructions for these data are shown in Input 2.1.

Figure 2.2
The Exercise data

id	sex	smoke	age	pulse_1	pulse_2
1	1	1	31	62	126
2	2	1	20	78	154
3	1	2	28	64	128
4	2	2	29	96	155
5	1	1	21	66	128
6	2	1	27	96	265
7	1	2	21	68	120
8	2	2	42	72	138
9	2	1	22	88	160
10	1	1	28	90	144
11	2	2	21	82	140
12	1	2	22	74	134
13	2	1	43	66	148
14	2	2	19	68	142
15	1	1	23	92	134
16	1	2	41	68	112
17	1	2	24	76	158
18	2	2	21	86	146
19	2	1	21	88	156
20	1	1	20	66	132
21	1	1	38	70	122
22	1	2	20	80	136
23	2	1	33	76	148
24	2	2	25	78	148
25	2	2	37	76	136
26	2	2	22	80	158
27	1	2	32	68	116
28	1	2	22	70	120
29	1	1	22	68	126
30	1	1	19	70	144
31	2	2	21	86	144
32	1	2	26	72	126
33	2	2	32	84	136
34	2	2	24	72	142
35	2	2	28	80	138
36	1	1	34	62	132
37	1	2	35	74	116
38	1	1	21	90	138
39	1	2	21	66	142
40	1	2	30	70	132

Input 2.1
The basic BMDP instructions for the Exercise data set

```
/ INPUT      VARIABLES = 6.
             FORMAT IS FREE.
             FILE IS 'exercise.dat'.

/ VARIABLE   NAMES = id, sex, smoke, age, pulse_1, pulse_2.

/ GROUP      CODES(sex) = 1, 2.
             NAMES(sex) = male, female.
/ END
```

Character data

Numeric data are easiest to work with, but there may be times when you need to include character variables in your file. A character variable is any variable that contains nonnumeric values. For example, suppose the Exercise data included the name of each subject and M and F for male and female instead of 1 and 2:

```
1  M  1  31  62  126  ALAN
2  F  1  20  78  154  MARY
```

With variables like this, you must use special instructions to tell BMDP that they are not numbers and should not be analyzed like numbers. In some cases, you may want to recode values so they can be used as numeric variables for grouping or analysis.

Reading character variables. In most cases you should use fixed format (see Chapter 3) to read character variables. However, if you have no more than two such variables of no more than four characters each, you can read them in free format by using the LABEL command in the VARIABLE paragraph (see Chapter 4). The LABEL command tells the program to read the variables as character data.

Uses for character variables. The LABEL command allows you to print names or other character variables as case labels at the beginning of each case in data printouts. Character variables may also make data easier to enter and check (you don't have to remember that 1 means male and 2 means female). You can use the ALPHA command described in Chapter 5 to define groups with such variables (e.g., M and F for male and female), or you can transform their values to numbers using the CHAR function in the TRANSFORM paragraph (see Chapter 6).

2.2 Entering and accessing data

Data may be typed into a computer file just as you type in text, or you may access data in a pre-existing file. Either way, one of the most basic tasks you need to understand before using BMDP is how to enter and save a file on your computer system. These files are system files, where the computer stores information for easy retrieval. The easiest way to store data is to type it into a file in regular columns with spaces between the numbers, like the Exercise data. A record is a line in a computer file. A record often contains one case, but a case may extend across several records if necessary, or a single record may contain more than one case.

Text editors

The text editor is a basic tool for computer use. It lets you create or modify files in the computer. These files may contain BMDP instructions, data, or both. Learn how to access your text editor and save files. You can also use BMDP's built-in line editor, described in Chapter 10.

To store data in the computer, simply enter your text editor and type in columns of numbers, like those of the Exercise data. When entering data from forms, it is important to use a consistent coding scheme, such as 1 for male, 2 for female. Numeric coding simplifies data entry and analysis. You should also line up the columns of numbers for easy inspection. Each space on a line is a column; terminal screens usually display 80 columns, but some printers have more. When you finish entering the data, save the file using your text editor's save command.

Caution: *Do not leave a blank line (record) at the beginning of the file or before the end-of-file marker when you create your data file. This is a common mistake leading to a failed run and an error message.*

Using BMDP with a word processor, spreadsheet, or database

Transferring data from other programs to BMDP. Both the SAS™ and P-STAT™ software systems can place data in BMDP Files. Data may be transferred from many other types of software to BMDP programs. One popular example on the PC is Lotus™ 1-2-3™. It is possible to exchange data between Lotus and BMDP through the use of ASCII or character image files. Check your documentation for these programs to find out how they can create the appropriate file type.

ASCII stands for the American Standard Code for Information Interchange. This code specifies a standard method for representing characters in a computer system. A file composed of these characters is called an ASCII file. Some ASCII codes have special meanings that would create problems when read into BMDP or any program other than that used to create the file. A file without these special ASCII characters is often referred to as a character image file. Character image files contain only the normal character set (numbers, upper and lower case letters, and other symbols that can be seen on a display). Backspaces, form feeds and the like are not present. A character image file can often be created using specific commands within a program, as in the following examples:

WordPerfect™ – use the text in-text out function (Ctrl-F5) to read in and write out ASCII files

WordStar™ – create the file in non-document mode

Lotus™ 1-2-3™ – use the /print (unformatted) command

Symphony™ – use the {Services} Print command; you must "print" the contents of the worksheet to the disk

dBASE III or IV™ – use the COPY command: COPY dbfile TO character SDF (SDF stands for Standard Data Format; in this example it creates a character image file called character.txt from the dBASE III file named dbfile.)

Check the documentation of these programs for details.

Generally (but not always), if you display a file using a system editor and see no unusual characters, the file is a character image file and can be read by BMDP using a FORMAT command. For example, on an IBM PC, use the DOS command TYPE to display a file. If the values in the data file have blanks or commas between them, you can read them in FREE format. If the data are entered in compact form, you will need to use FIXED format (see Chapter 3). If your data are not organized in the desired way, you may need to use the BMDP Data Manager program to restructure your file.

Where to store your data

Data may be stored in three basic ways for analysis:

1. Together with your BMDP instructions, as illustrated in Chapter 1
2. In a separate system file (see Chapter 3)
3. In a BMDP File (see Chapter 8)

The easiest way to access data is the first: simply type in each case following your BMDP instructions and run the whole setup. This procedure is suggested for small data sets. However, you will usually want to store your data in a separate file that can be accessed by any set of instructions and by other users. Two types of files can be used. If data are stored in a separate system file, you must use FILE or UNIT in your BMDP INPUT instructions to identify the location of the file in the computer system. If you intend to run several BMDP programs using the same set of data, it is recommended that you save your data in a BMDP File (see Chapter 8).

2.3 Identifying data to BMDP: format

Before the BMDP programs can analyze your data, the data must be entered into the computer in some format, and you must supply the computer with information about the format. BMDP can read data stored in a variety of different formats, so you should be familiar with the options available. Chapter 3 describes formats in detail.

The simplest alternative is to enter the data using the FREE format option we have already mentioned. The major requirement for using FREE format is that values for individual variables are separated by blanks or commas within each record. If the data set is large or complex or if the data have already been entered, you may need to use FIXED format to describe the data to BMDP. That is, you may need to specify exactly which columns in the record contain what types of data.

Format is important to consider on two occasions:

1. When you are entering data, you should choose a format in advance and type in the data to match.

2. When you are using a pre-existing file of data, you should look at the data records and specify a format to match.

If you are entering a small data set with no more than two character variables, you should use FREE format. This will be simple, and you will not have to learn about FIXED format. You need to read about FIXED format only if presented with a data set not in FREE format or if you have space problems, several character variables, etc.

FREE format

You have already seen FREE format in Chapter 1 and in the Exercise data in this chapter. FREE format is simple and easy to use. Just enter data values with at least one space (or comma) between them. You can enter the data however you like, but the file will be easier to read if you line up the columns. If you see that a data file has blanks between the variables, you can read it in FREE format. Two special types of FREE format (STREAM and SLASH) allow more than one case on a line (see Chapter 3).

Data read in FREE format are assumed to be numeric data. Except for LABEL variables, only numeric data and a specified nonnumeric missing value code are permitted in FREE format. If a nonnumeric code appears in the data, the program will produce an error message and consider the value missing.

FIXED format

FIXED formats are based on FORTRAN conventions and require some getting used to. Unlike FREE formats, where a data value may appear anywhere on a line as long as there are blanks on either side, in FIXED format each piece of information must be in the same column locations for each case. Data for every record must be entered precisely to conform to the set format. Instead of stating FORMAT = FREE in the INPUT paragraph, you write a format specification:

FORMAT = 'F2, 2Fl, F2, 2F3'.

This tells the data reader which columns to group together as one number.

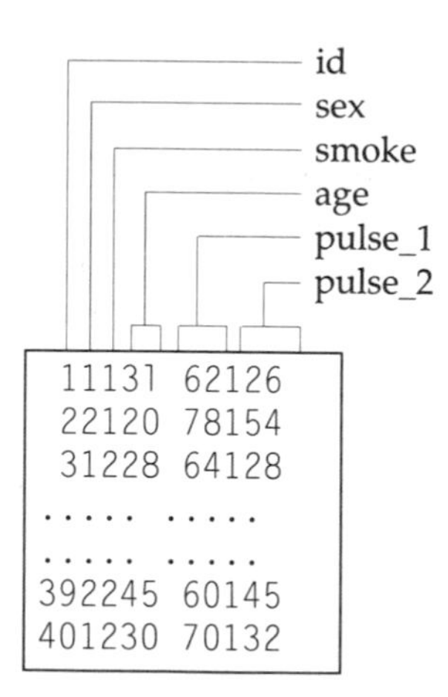

The rules for FIXED format are explained in some detail in Chapter 3. For now you need only know that it offers certain advantages to compensate for its complexity. FIXED format allows you to compactly store large amounts of data because blanks are not necessary to separate variables. It also allows you to use more than the two character variables allowed in free format.

Data characteristics to consider in choosing format type

1. Records do not have to be lined up perfectly in FREE format; in FIXED format every record must be formatted identically.
2. FREE format allows only two four-column character variables. There are no restrictions on character variables in FIXED format.

3. FREE format data in columns with blanks in between are easier to read for error checking.
4. FIXED format is more compact.
5. FIXED format can be used to read variables out of order and skip variables and records. However, this usage is recommended only for experienced users. The same effect can usually be obtained by reading the data in FREE format and then using BMDP commands to rearrange it.

2.4 Missing data

It is often impossible to obtain complete data on every variable for every subject. A data set may contain cases with *missing* (unrecorded) values for some variables. You should insert a special code or symbol in your data file to mark the position of each missing value. The special code or symbol is called a *missing value flag*. Missing value flags serve three purposes: (1) they inform the computer that a value is missing so the computer will not skip ahead to the next value and assign it to the wrong variable; (2) they indicate to you that the value is known to be missing and has not been inadvertently left out of the data file; and (3) they notify the computer to exclude the missing value from all computations.

The easiest way to deal with missing data in BMDP is to use an asterisk (*) as the missing value flag. All BMDP programs assume that asterisks in your data file indicate missing values. You can insert one or more consecutive asterisks for each value that is missing. BMDP also interprets blanks in a FIXED format field as missing. In FREE format, blanks are not recognized as missing values, but consecutive commas can be used instead:

```
25,  ,2 37 75 151 or 25 * 2 37 75 151
```

You may also specify your own missing value flags by using:

- a *symbol* such as a period or dollar sign instead of asterisks (use the MCHAR command in the INPUT paragraph, section 3.4)

or

- *numeric* codes (such as 999) for each variable in your data set (use the MISSING command in the VARIABLE paragraph, Chapter 4).

In the figure below we display a modified version of the Exercise data with two contiguous four-character label variables. We have underlined codes and symbols that represent missing data values. We also flag three out-of-range values (discussed below).

Figure 2.4
The Exercise data with missing value flags and case labels

id	sex	smoke	age	pulse 1	pulse 2	name1name2
1	1	1	11 ←	62	126	ALLAN
2	2	1	20	78	<u>***</u>	MARY
3	1	1	<u>99</u>	64	128	BILL
4	2	2	29	<u>*</u>	155 ↙	LINDA
5	1	1	21	66	28 ↙	MICHAEL
6	2	1	27	96	265	CATHY
7	1	2	21	68	120	HARVEY
8	<u>9</u>	2	42	72	<u>999</u>	JENEE

BMDP automatically recognizes the asterisks as missing value flags and excludes them from all computations. The MISSING command in the VARIABLE paragraph (Chapter 4) allows you to specify one code per variable as an additional missing value flag. For each variable we use a code such as 9, 99, or 999 that is outside the range of possible values.

How BMDP handles missing data and values out of range

Many BMDP programs (e.g., 1D, 2D, 3D, 7D) use all acceptable values for each variable being analyzed. An *acceptable value* is one that is neither flagged as missing nor out of range. (BMDP allows you to specify a range of acceptable range values for any variable; in Figure 2.4, three out-of-range values are flagged—see MIN and MAX commands, Chapter 4.) For example, assuming that the program reads only the eight cases in Figure 2.4, 2D will analyze six AGE values, seven PULSE_1 values, and four PULSE_2 values. 2D analyzes all acceptable values for each variable.

Other BMDP programs use only complete cases. A complete case is a case in which the values of all the variables are acceptable (there are no missing or out-of-range values). The USE list in the VARIABLE paragraph becomes an important part of the instructions in these programs: by selecting the variables that the program scans for missing data, you can maximize the number of cases used for the analysis. BMDP has two methods for determining complete cases:

1. **The USE list.** Some programs include cases in the analysis if they have acceptable values for all variables specified in the USE list. If you want to use all possible values of AGE, PULSE_1, and PULSE_2 in 2R, you should include a USE list specifying only these three variables. If you omit the USE list, 2R will use only those cases with acceptable values for *all* input variables. Omitting the USE list can drastically reduce your number of cases. This method is used in programs 1R, 2R, 3R, 4R, AR, LR, 1V, 3S, 1M, 2M, 4M, 7M, 9M, and 2T.
2. **Analysis lists.** These programs use cases that have acceptable values for the variables listed in their analysis paragraphs; other variables are disregarded, whether or not they appear in a USE list. For example, program 2V checks variables listed in the DESIGN paragraph for missing values; program 6R checks variables listed in the REGRESS paragraph. This method is used in programs 3D, 5R, 6R, 9R, 2V, 3V, 4V, 8V, 6M, 1L, and 2L.

Four programs (3D, 3S, 8D, and AM) allow you to choose explicitly between using all acceptable values and using only complete cases. 8D computes estimates of correlation matrices using all observed values instead of only those values from complete cases. AM does the same and provides estimates to fill in missing observations. AM also has features useful for describing the pattern of missing values.

2.5 Collecting and recording data: helpful hints

The method you choose to record the data will depend on the quantity and sources of the data, as well as on the intended analyses. The values for some measurements may come directly from instruments, but most frequently you will record information on a research form or coding sheet, or directly into a separate system file. For a small study it may be sufficient to lay out a simple data collection sheet using ruled graph paper. The values for each variable are recorded in a fixed set of columns on the sheet.

Design of research forms. For other studies, especially those involving questionnaires or many subjects, it may be desirable to invest considerable effort in creating a compact, easy to fill in form as in Figure 2.5.

In Figure 2.5 the data to be entered are organized in a column to the right of the form. Below or beside each line you enter the numbers into the columns. For example, AGE occupies the 9th and 10th character positions of each record. If you will be using fixed format, you must carefully consider the number of character positions required for each measure. What will the smallest and largest values be? Do you expect to interview subjects over 100 years old (thus requiring three character positions for age instead of two)? Could the value of some measure be as small as .004 (requiring three digits to the right of the decimal point)? In this example, the subject identification number can be one, two, or three digits, so three columns must be allowed for the number. Values less than 100 should be written with a leading zero (e.g., 024) or right-justified (a blank followed by 24).

Figure 2.5
Example of a research form

```
                         EXERCISE STUDY

1. Subject Identification Number               |0|2|4|    1-3
2. Sex (circle one):
     Male . . (1)  Female . . . 2                 |1|      5
3. Smoking Status (circle one):s
     Yes . . .1    No . . .(2)                    |2|      7
4. Age in Years                                 |4|6|    9-10
5. Pre-exercise Pulse                          |0|7|6|   12-14
6. Post-exercise Pulse                         |1|2|4|   16-18
```

Automated Data Collection. For some studies, much of the data collection process is already taking place either on a PC or a mainframe computer. In these instances, the ASCII file format can be used to save data, and, with some minor modifications, these same files can then be submitted to BMDP. The modifications to these files can be accomplished in either a spreadsheet or a word processing program. Once the conditioning of the data has taken place, these files should then be saved again as an ASCII file which can be read directly into BMDP. Note that these files **must** be saved in a ASCII format for BMDP to read the data. The process of creating an ASCII files in a spreadsheet or word processor is discussed in detail in section 2.2.

In addition, automated data entry programs are now available to directly enter data into a personal computer. One such program is BMDP Data Entry, which is an easy to use PC-based program that provides a simple and efficient way to enter data into the computer. This program will automatically set up various data parameters, and then data input is performed according to the prompts given by the program.

Conditioning Data Sets. Many complex studies generate records of more than one type (collected from several different forms). If you use more than one type of form, assign a number to each and place it in the same location for each form (e.g., always in columns 5 and 6). You can then use the form type as a variable to select or sort records for analysis. Keep each subject's ID the same on all forms. See the BMDP Data Manager Manual for helpful commands to handle such data sets. The Data Manager can read such files even when they are nonrectangular; that is, when records from different forms contain different variables.

Sometimes, especially when the data file contains repeated measurements on the same subjects, analyses using different programs like 2V and 6D (for example) require different layouts for the data. See the Data Manager for packing and unpacking options useful in such cases. In general, use the Data Manager for any file handling jobs such as joining records or merging files. The Data Manager can also produce "maps" of missing values for data screening.

Consult with a technical support staff member and/or statistician for suggestions before you begin a large, complex, and costly project. If no experts are available, generate several mock data records from your form or from your computer using values that you might expect to collect. Next, considering the key questions for your research project, select the BMDP programs required for data screening and analysis. Attempt to write the BMDP instructions for these programs and try a prototype computer run using the mock data collected on your form. You may need to revise your form or your data collection techniques.

3

Describing Your Data: The INPUT Paragraph

The INPUT paragraph is one of the basic paragraphs you will use to describe your data. Figure 3.1 presents the INPUT paragraph for the Exercise data described in Chapter 2. The record for each subject includes an ID number, sex, smoking status (yes/no), age, and pulse rates before and after exercise. For this example we assume that the data set is stored on disk in a file named EXERCISE.DAT (see Chapter 2 for an explanation of how to enter and save data).

Figure 3.1
INPUT paragraph for the Exercise data

```
/ INPUT     VARIABLES = 6.
            FORMAT IS FREE.
            FILE IS 'pulse2.dat'.
```

An INPUT paragraph is required by all BMDP programs. Three INPUT commands are essential:

VARIABLES – to specify the number of variables for each case

FORMAT – to specify the layout of data values on each record

FILE or UNIT – to identify the location of a data file on disk (not needed when data are appended to the instructions)

These commands are all you need for most analyses. Other INPUT commands allow you to limit the number of cases read, set flags for missing data, terminate the program if too many data errors are encountered, etc. **Note**: If your data are already stored in a BMDP File, see section 3.3 for special INPUT instructions and information. See Chapter 8 for complete information about using BMDP files as input.

Where to Find It

3.1 The number of variables: VARIABLE

The VARIABLE command is required by all BMDP programs to indicate how many variables to read for each case. In BMDP, a variable is any characteristic that is recorded for a subject, such as a code for male or female, a pulse measurement, or a case identification number. For each subject in the Exercise data, six variables are recorded: (1) an ID number, (2) a code for sex, (3) a code for smoking status (yes/no), (4) age, (5) a pre-exercise pulse rate, and (6) a post-exercise pulse rate. The VARIABLE command for our Exercise data is:

```
VARIABLES = 6.
```

This command refers only to variables present in the input file. Do not include in this count the variables derived in a TRANSFORM paragraph (for example, DIFFRNCE = PULSE_2 – PULSE_1). See Chapter 6 for information about the TRANSFORM paragraph.

If all data are generated by the TRANSFORM paragraph, state VARIABLES = 0. (This feature is rarely used; see Chapter 6 for an example.)

3.2 The layout of the data: FORMAT

Use the FORMAT command to specify the layout or arrangement of the data on each record (see Chapter 2 for an introduction to data and formats). You can choose the format you prefer and enter the data to match. If you are using a pre-existing data file, examine the records to see which format is appropriate for reading the data. Four types of format are available:

1. **FREE field.** Each data value is separated from the next by one or more blanks and/or a comma. Three types of FREE field format are available: FREE, STREAM, and SLASH (explained below).
2. **FIXED field or FORTRAN-type.** The data values are in FIXED locations (fields) on each record (line). If, for the first case, AGE is in the eighth and ninth character positions, then AGE must be entered in the same location for all remaining cases. Fixed format allows you to enter data in a compact fashion.
3. **BMDP Files** are special binary files written by BMDP programs. In addition to the data, BMDP Files store variable names, grouping information, and the results of some analyses. PORTABLE BMDP Files are character files that may be moved between different computer systems. Section 3.3 explains how to use these files as input. (See Chapter 8 for complete instructions).
4. **Binary files** are written by a computer program using binary format. This format is rarely used. See FORMAT = BINARY in the Summary Table.

FREE and FIXED format data files can be entered, viewed, and edited using your system editor. Both FREE and FIXED format data may be stored in a file following your BMDP instructions or in a separate file. In the latter instance you must use FILE or UNIT to identify the location of the data file in the system. BMDP Files and binary files come from earlier computer runs, and cannot be viewed with your system editor.

Free-field formats

BMDP provides three methods for reading data in FREE format: FREE, STREAM, and SLASH. FREE is used frequently; the others are used less often. For each method, you separate each variable by one or more blanks and/or a comma.

Caution: *Before you enter data in FREE format, you must plan for missing data. When a value is missing, you must flag its position in the record with a code or symbol (an asterisk is the default symbol). Use the MCHAR command in the INPUT paragraph to identify an alternative symbol and/or the MISSING command in the VARIABLE paragraph to identify codes. Two consecutive commas also indicate a missing value.*

In FREE format, separate the variables by one or more blanks and/or a single comma. Use as many lines as you need to enter data for a case. All your cases must have the same number of lines as your first case. Each case must start on a new line, and each case must have the same number of variables. The program checks the number of variables you have entered for each case against the number you have specified in the VARIABLE command. If the numbers do not match for a particular case, the program produces an error message and deletes the case from the analysis.

In Figure 3.2 we display the Exercise data entered in FREE format. We have added labels and case numbers.

Figure 3.2 The Exercise data in FREE format

	id	sex	smoke	age	pulse_1	pulse_2
Case 1	1	1	1	31	62	126
Case 2	2	2	1	20	78	154
Case 3	3	1	2	28	64	128
.		.			.	
.		.			.	
.		.			.	
Case 40	40	1	2	30	70	132

The Exercise data can be entered without lining up the variables, and commas may be used to separate the variables. The first three lines of Figure 3.2 might look like:

```
31 62 126                          1, 1, 1, 31 62 126
2  2 1   20 78    154     or      2,2,1.  20. 78.154
3 1 2 28   64   128               3 , 1, 2, 28, 64, 128
```

We recommend aligning the variables as in Figure 3.2 to simplify data scanning and editing. The INPUT paragraph for these data (with or without the commas)

```
/ INPUT    VARIABLES = 6.
           FORMAT IS FREE.
           FILE IS 'pulse2.dat'.
```

We assume that the data are stored in a separate file named PULSE2.DAT (see section 3.3).

The values in the Exercise data set are integers. BMDP programs also read decimal values. Figure 3.3 displays instructions and data for three cases stored with the BMDP instructions. For the third case, the data are 154.91, missing, 6.2, and 3. We inserted an asterisk to mark the position of the missing observation (see section 3.5). You can also use scientific notation (E-notation). That is, .000218 can be written as 21.8E–5, or as .218E–3 (.218 times 10^{-3} = .000218). When FORMAT = FREE, STREAM, OR SLASH, BMDP ignores any blanks between the E (or D) of scientific notation and its following exponent.

Figure 3.3
Data with decimals and a missing value to read in FREE format

```
/ INPUT     VARIABLES = 4.
            FORMAT = FREE.
/ END
 135.67  2     5.4    3
 168.23  1     7.3    2
 154.91  *     6.2    3
```

In STREAM format (FORMAT = STREAM), each case need not start on a new line. The program uses the number of VARIABLES specified in the INPUT paragraph to determine the end of one case and the beginning of the next. For the Exercise data set we might enter cases 1, 2, and half of case 3 on the first line, and continue with the rest of case 3 on the second line:

```
1 1 1 31 62 126   2 2 1 20 78 154   3 1 2
28 64 128 . . .
```

In **SLASH format**, each case need not start on a new line, but you must use slashes to separate cases. Using the previous data, the first line of the data file (and part of the second line) would look like:

```
1 1 1 31 62 126/ 2 2 1 20 78 154/ 3 1 2
28 64 128/ . . .
```

SLASH format is a useful safeguard when you want to enter more than one case per line. Cases are delimited in two ways: by the slashes and by the VARIABLES command. If there are discrepancies between the two for a particular case, the program produces an error message and deletes the case from the analysis.

Data sets with many variables. Lines of data in a file do not have to correspond to cases. Cases with many variables may take up more than one line, as long as the values are separated by blanks and/or commas. If you use FREE format and your lines are longer than 80 characters, see RECLENGTH later in this chapter.

Reading characters in free format. Data read in FREE, STREAM, or SLASH format are assumed to be numeric data. The only exception is that case labeling variables are assumed to be characters. BMDP allows two variables to be declared case labeling variables, each of which may contain up to four characters. You must identify these variables by using the LABEL command in the VARIABLE paragraph (see Chapter 4). (Note that there is no limit to the number of character variables allowed in FIXED format.) You can use LABEL variables as case labels, as grouping variables (see ALPHA in Chapter 5), or you can transform their values to numbers using the CHAR function in the TRANSFORM paragraph (see Chapter 6). Except for LABEL variables, only numeric data and specified missing value codes are permitted in FREE format. If a program finds a nonnumeric code while reading the data, it will produce an error message and consider the value missing.

Fixed field FORTRAN-type format

Fixed field formatting is useful when you have large amounts of data and want to save file space, and when you want to read more than two character variables. In fixed format, you enter variables in fixed fields for each case; that is, a given variable must occupy the same character positions on each case line. You must also write a format specification that tells BMDP which columns of the data record to group together as one number.

For example, the Exercise data could be entered in a compact manner as in Figure 3.4.

Figure 3.4
The Exercise data set in fixed format (we have added labels)

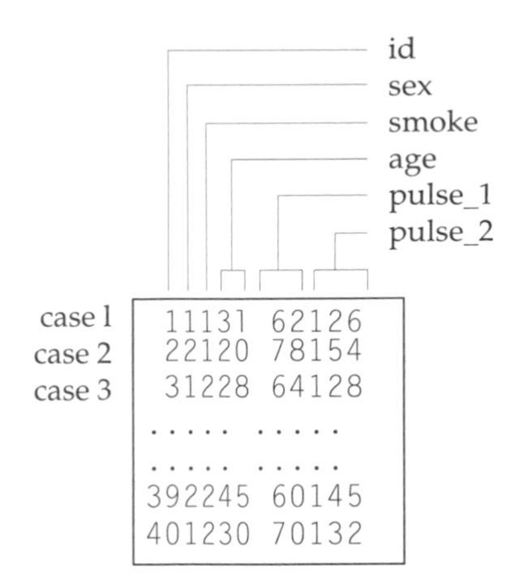

We have drawn vertical lines to show the three-character field for the variable named PULSE_1. Note that the two-digit pulse values are entered in the last two character positions within the field (they are "right-justified"). BMDP interprets the first pulse value as '062' (if we had entered it in the first two character positions within the field, BMDP would have complained, unless BLEVEL is set to 3 as described below). The FORMAT specification for these data is:

```
FORMAT = 'F2, 2F1, F2, 2F3'.
```

This command directs the program to read for each case

- one two-digit number (id) or F2
- two one-digit numbers (sex and smoke) or 2F1
- one two-digit number (age) or F2
- two three-digit numbers (pulse_1 and pulse_2) or 2F3

When the Exercise data are read in FIXED format, the program prints a codebook that identifies where each variable is located on the data record. In Figure 3.5 we see that AGE, for example, is located in columns 5 and 6.

Figure 3.5
Codebook printed when Exercise data are read in FIXED format

```
I N P U T   V A R I A B L E S . . . . .
 VARIABLE  RECORD  COLUMN   INPUT      VARIABLE  RECORD  COLUMN   INPUT
NO.   NAME   NO.   BEG  END FORMAT    NO.   NAME   NO.   BEG  END FORMAT
--- -------- ---   ---  --- ------    --- -------- ---   ---  --- ------
  1 id         1     1    2  F2.0       4 age        1     5    6  F2.0
  2 sex        1     3    3  F1.0       5 pulse_1    1     7    9  F3.0
  3 smoke      1     4    4  F1.0       6 pulse_2    1    10   12  F3.0
```

Format notation. If we specify 2F5.1, we describe two numeric variables following one another. Each is in a field five character positions wide with an assumed decimal point one digit before the end of the five digit field. That is,

- a number before the F specifies how many consecutive variables are described by the format specification
- the first number after the F specifies the maximum number of columns in which the number can appear (be written); i.e., the "field width"
- the last number specifies the number of digits after the decimal point.

Decimals. If you use FIXED format, you need not enter decimal points as long as all cases have the same number of digits to the right of the decimal. The number 256 in F3.1 format is read as 25.6. When a data value being read contains a decimal point, the decimal point in the data overrides the position for the decimal point stated in the F-type format (e.g., the value 25.6 can be read using F4.1, F4.0, or F4.3 format). Plus signs (+) are not entered, but minus signs (–) are, so allow an extra space if negative values occur.

Skipping columns or records. Columns in the record can be skipped (ignored) by specifying the number to be skipped followed by an 'X'. For example, 20X specifies that 20 columns are to be skipped. A slash (/) instructs the program to skip the remainder of the record and begin the next record. Use slashes when the data for a case are recorded on more than one line so that you can read any number of records as a single case.

The rules for fixed-field formatting are the standard rules for FORTRAN formatting, except that BMDP allows more flexibility. For example, BMDP allows you to specify integers in the form F*n* (e.g., F4), where *n* is the number of characters in the field. FORTRAN requires you to specify integers in the form F*n*.0 (e.g., F4.0), where the zero indicates that there are no digits to the right of the decimal. FORTRAN also requires that parentheses surround the format specification: e.g., '(F2.0, 2F1.0, F2.0, 2F3.0)'. In BMDP the format is surrounded by apostrophes; e.g., 'F2, 2F1, F2, 2F3' (parentheses are optional).

Here are additional FIXED format examples:

1. FORMAT = '12F3, F4, 11F2'.
2. FORMAT = '2A4, 2F6, F1, 3X, F5.2 / 5X, F6'.
3. FORMAT = '12F3, F4, 11F2 /'.
4. FORMAT = 'A3, T1, F3.0, 12F4.2'.

1. In the first example, for each case the program reads 12 three-digit numbers, followed by one four-digit number, then 11 two-digit numbers.
2. The program reads two four-character alphanumeric (A) fields, two six digit numbers, one single-digit number, then skips three spaces and reads one five-digit number with two digits to the right of the decimal. Then the program skips to the next record (as indicated by the slash), skips five spaces, and reads one six-digit number.
3. The program reads the first data record as in the first example above and then skips the next record (a slash at the end of a record means skip the following record). The third record is read using the same format as the first, and the program continues skipping every other record.
4. The program reads a three-character alphanumeric field, then tabs back to column 1 and rereads the field in F3.0 format before continuing. The first field is read twice, first as a label variable, and then in an analysis.

Reading letters and characters in fixed format. You can read any number of variables in alphanumeric (A) format, and one or two such variables can be specified as case LABELS in the VARIABLE paragraph. Each variable read in (A) format must be no more than four characters in length. For example, the Exercise data might contain M and F for male and female, and YES and NO for smoking and nonsmoking, as well as subject names:

```
                    1 M YES 31 62 126 ALAN
                    2 F YES 20 78 154 MARY

        FORMAT = 'F1, 1X, A1, 1X, A3, 2F3, F4, 1X, A4'.
or      FORMAT = 'F1, A2, A4, 2F3, F4, 1X, A4'.
```

You can read these records in FIXED format but not in FREE because there are more than two character variables. You can use alphanumeric variables as grouping variables (see ALPHA in Chapter 5), or you can transform their values to numbers using the CHAR function in the TRANSFORM paragraph (see Chapter 6).

Reading data stored in scientific notation or integer format (E-format or I-format). If your data come to you stored in E notation, you can still use F-format. For example, to read the data value 1.5E-8, you can use either F6.0 or E6. D6 and G6 formats are also acceptable. If your data file contains the integer 15836, you can use an I5 format to read the value (internally it will be stored as a floating point (F) number).

This discussion of fixed format specifications applies only to data input. For writing data in FIXED fields (SAVE FORMAT = '...'), BMDP uses your system's FORTRAN conversion routines.

3.3 Reading a BMDP File

Any BMDP program can read the BMDP File created in Figure 3.6; you need only supply the appropriate FILE and CODE statements in the INPUT paragraph. A FORMAT statement can **not** be used with a BMDP File. If the BMDP File contains results other than data, you also need to specify the CONTENT.

Figure 3.6
Reading a BMDP File

```
/ INPUT     FILE = 'exercise.sav'.
            CODE = exercise.

  — program-specific instructions follow —
```

The absence of FORMAT and the presence of CODE in Input 3.6 tell the program to read a BMDP File as input. Because no LABEL is specified, the program uses the first BMDP File it encounters with the specified CODE. You need not state the number of VARIABLES and the number of CASES in the INPUT paragraph.

The VARIABLE, GROUP, and TRANSFORM paragraphs are no longer necessary unless you want to change your instructions. For example, you may:

- restate GROUP with new ways of grouping the data
- perform additional TRANSFORMATIONS
- give VARIABLES new NAMES
- tighten the range between MINIMUM and MAXIMUM
- identify additional values as MISSING

For more information on modifying information stored in a BMDP File, see Special Features.

In Figure 3.7, the program prints information about the BMDP File and restates the variable and grouping information that was saved with the data. Note that the LABEL is the date and time the file was created.

Figure 3.7 lists nine variables, including DIFFRNCE, the new derived variable we created in section 8.1. However, variables 7 and 8 (NAME1 and NAME2) are omitted from the list of VARIABLES TO BE USED because we designated NAME1 and NAME2 as case label variables in Input 8.1.

Figure 3.7
Reading a BMDP File

```
INPUT BMDP FILE
CODE. . . IS     exercise
CONTENT . IS     DATA
LABEL . . IS
   1-JUL-92            12:57:07
VARIABLES
     1 id          2 sex          3 smoke        4 age          5 pulse_1
     6 pulse_2     7 name1        8 name2        9 diffrnce

VARIABLES TO BE USED
     1 id          2 sex          3 smoke        4 age          5 pulse_1
     6 pulse_2     9 diffrnce

NUMBER OF CASES READ. . . . . . . . . . . . . .      15

   VARIABLE        STATED VALUES FOR             GROUP CATEGORY      INTERVALS
  NO.   NAME    MINIMUM MAXIMUM MISSING   CODE  INDEX   NAME       .GT.    .LE.
 ---- -------   ------- ------- -------  ------ ----- --------    ------ -------
   2  sex                                 1.000    1  male
                                          2.000    2  female
   3  smoke                               1.000    1  yes
                                          2.000    2  no
   4  age                                          1  under_25            24.00
                                                   2  _25to34     24.00   34.00
                                                   3  over_34     34.00
-------------------------------------------------------------------------------

— we omit program output —
```

3.4 Where your data are stored: FILE or UNIT

You can enter and store your data after your BMDP instructions in the same file (as we illustrated in Chapter 1), or you can store the data in a separate file. When you store your data in a separate system file, use the FILE or UNIT command to tell the program where to find the data. Use the FILE command if you are running BMDP on any of the following systems: VAX, PDP-11, IBM PC or compatible, or any UNIX system. Use the UNIT command if you are running on an IBM mainframe. If you are using a system other than the ones mentioned above, check with your computing facility or installation instructions for the proper command.

To illustrate the use of separate files for BMDP instructions and data, we enter the Exercise data into a file named PULSE2.DAT. We enter the BMDP instructions into another file named EXERCISE.INP (see Figure 3.8).

Figure 3.8
BMDP instructions are stored in the file EXERCISE.INP; data in file EXERCISE.DAT

EXERCISE.INP

```
/ INPUT     VARIABLE = 6.
            FORMAT IS FREE.
            FILE = 'exercise.dat'.

/ VARIABLE NAMES = id, sex, smoke.
              age, pulse_1, pulse_2.
/ END
```

EXERCISE.DAT

```
 1    1    1    31    62    126
 5    1    1    21    66    128
10    1    1    28    90    144
 .    .    .     .     .      .
 .    .    .     .     .      .
 .    .    .     .     .      .
```

FILE. The FILE command stored with the instructions in EXERCISE.INP tells the program that the data are stored in the file named EXERCISE.DAT. We specify EXERCISE.INP as the BMDP instruction file when we call the program. If you omit the FILE command, the program assumes that your data follow the BMDP instructions. This is equivalent to FILE = CONTROL.

UNIT. When you use UNIT, you must include a system job control language (JCL) statement that links the unit number to the system file where your data are stored. JCL is specific to your computer system and is not part of the BMDP instructions. On an IBM mainframe, for example, you include a line that might look like this:

```
//FT11F001 DD DSN=[account number].EXERCISE.DATA,DISP=SHR
     ↑↑
```

and replace the FILE command with the command UNIT = 11. If you omit the UNIT command or state UNIT = 5, the program assumes that your data follow the BMDP instructions.

There are some restrictions on the choice of a UNIT number: 30–49 should be avoided as they may be used by BMDP for internal storage. At some installations there may be additional conventions with respect to the unit number.

3.5 Commands you may need

In this section we discuss commands that are not required but are occasionally needed. They are:

TITLE	– specifies a title for each problem
CASE	– specifies the number of cases to be read
MCHAR	– identifies a symbol placed in the data file to flag missing values
RECLENGTH	– increases or decreases the number of characters scanned in each data record
ERRMAX	– halts data reading after a predetermined number of cases with data errors have been encountered

Specifying a title: TITLE

The TITLE command allows you to specify a title of up to 160 characters for each problem. All 160 characters will be printed after the program instructions. However, at page breaks, BMDP will print up to 112 characters when LINESIZE = 132, or up to 64 characters when LINESIZE = 80. The title will be printed at the top of each page of program output. You should enclose the title in single quotes. For example, we can specify a title for a 1D run using the Exercise data set:

```
TITLE = 'Summary statistics for age and pulse
         before and after exercise'.
```

If you do not include a TITLE, the program prints the date and time across the top of each output page.

The number of cases to read: CASES

Normally, the program reads your data until it finds an END paragraph or reaches the end of the file. Alternatively, you can tell the program exactly how many cases to read. The CASES command is useful when you have a large data set and want to test your program on a subset of the cases. For example, specify

```
CASES = 50.
```

to use only the first 50 cases.

Caution: *When you use the CASES command and have instructions for another problem following your data (see Multiple Problems, Chapter 10), you should not use an END paragraph to terminate data reading.*

The global missing value flag: MCHAR

Some values in a case may not be recorded. When a value is missing in your data file, you should insert

– a symbol (*, ., $, etc.)
or – a code (9, 999, etc.)

in its place in the data file. The special code is called a missing value code, and the unrecorded value is called a missing value. Missing values are excluded from all computations. Here we describe how to tell BMDP what symbols you have inserted as flags for missing data.

All BMDP programs assume that asterisks (*) in your data file indicate missing values (see Figure 3.3). You can insert one or more contiguous asterisks for each value that is missing. The MCHAR command allows you to specify a symbol other than an asterisk as a missing value indicator. Only one such symbol may be specified. For example, you could state

```
MCHAR = '.'. or MCHAR = '$'.
```

Note that if you specify an alternative MCHAR, asterisks are no longer recognized as missing value symbols. For a given variable, both a symbol and a numeric code can be used to represent missing values. See Chapter 2 for further discussion of missing data and the MCHAR command.

See MISSING in the VARIABLE paragraph (Chapter 4) for how to identify numeric codes. Note that, in free format data, two commas can also denote a missing value; in fixed-field data a blank can indicate a missing value.

The length of the data record: RECLENGTH

BMDP programs normally read 80 characters per line when data are read in free format. However, when your data file contains records that are more than 80 characters long, you can extend the number of characters the program reads. If your records are 120 characters long, for example, you can state

```
RECLENGTH = 120.
```

You can also use RECLENGTH to limit the number of characters the program reads when you want to read only the first part of a record or when you have comments at the end of data records. To specify the number of characters of BMDP instructions to read, see COLUMNS in Chapter 9.

How to stop the data reader after too many errors: ERRMAX

Normally, the program stops reading your data after it encounters ten cases containing errors (such as too many data values on a record). See Chapter 1 for more information about errors and error messages. When you want to avoid unnecessary reading of the data before all errors are corrected, you can make the program stop reading data after encountering more or fewer than ten cases with errors. For example, to direct the program to stop reading after finding one case with errors, state

```
ERRMAX = 1.
```

3.6 Rarely used commands

We now briefly summarize three rarely used commands:

MULTIPLE – allows the program to read more than one case per record when data are read in fixed format

BLEVEL – alters how embedded and trailing blanks are treated when data are read in fixed format

REWIND – NO REWIND tells the program not to rewind to the beginning of the data before each problem

Reading more than one case per record in FIXED format: MULTIPLE

The STREAM and SLASH FREE format commands allow you to have more than one case per record (line). In FIXED format each new case begins on a new record. The MULTIPLE command directs the program to read more than one case per record when the data are read in FIXED format. For example, suppose you have data on 25 pairs of twins. Each FIXED format record contains five variables for both twins (ten variables per record). If you want to analyze the data on all 50 individuals (that is, to treat each member of the twin pairs as a single case), you can state:

```
/ INPUT  VARIABLES = 5.
         FORMAT = '10F4.2'.
         MULTIPLE = 2.
```

These instructions direct the program to read five VARIABLES for the first case, five for the second, and so on. MULTIPLE = 2 directs the program to read two cases from each record. Note that the VARIABLE FORMAT command refers to all ten variables (both members of the pair), rather than to one case.

Embedding and trailing blanks in FIXED format: BLEVEL

You have a choice of five methods for treating embedded and trailing blanks in FIXED format data reading. See INPUT commands for definitions of these options. If you do not specify a BLEVEL, the program assumes BLEVEL = 2, and treats embedded and trailing blanks as illegal, except for trailing blanks in a numeric field with a decimal. These are treated as zeros (e.g., 1.2bbb in a six column field is treated as 1.2000). Leading blanks are ignored.

Telling the program not to rewind the file: NO REWIND

BMDP programs assume that data reading begins with the first record in the file. Normally the program rewinds the file after each problem, so reading begins with the first record. NO REWIND leaves the file positioned exactly where reading stopped at the end of the previous problem. (See Chapter 10 for how to specify multiple problems.) If you have organized your data into sequential subsets, this allows you to use the first subset (say, the first 100 cases) for the first problem, the second subset in the second problem, and so on.

INPUT Commands

Commonly used commands

VARiables = #. `VAR = 6.`

Required. Specify the number of variables to be read for each case. This statement must be omitted if data are from a BMDP File (see section 3.3). Example: The Exercise data contain six variables per case.

FORMat = (*one only*) `FORM = FREE.`
FREE, STREAM, SLASH,
'c', *or* **BINARY.** `FORMAT = 'A2, 2F1, F2, 2F3'.`

Specify the layout of the data for each case. This statement must be omitted if data are read from a BMDP File, otherwise this statement is required.

FREE – Variables are separated by one or more blanks and/or a comma. Each case begins on a new record (more than one record may be used for each case). No excess data are allowed on any record. Each case must use the same number of records.

STREAM – Allows more than one case per record. Variables are separated by blanks or commas.

SLASH – Allows more than one case per record. Variables are separated by blanks or commas. Cases are separated by slashes. (Place a slash after the last variable for each case.)

'c' – The format 'c' specifies the layout of the data (in fixed fields). Use the usual rules for FORTRAN formatting, or the abbreviations allowed by BMDP (see Chapter 1). The format specification cannot exceed 800 characters.

BINARY – Use when your data are on a file in binary (unformatted) form (not a BMDP File).

FILE = 'name'. `FILE = 'EXERCISE.DAT'.`
or **UNIT = #.** `UNIT = 11.`

Identify the location of the data file. Use FILE for VAX, PDP-11, IBM PC and compatibles, and UNIX systems; UNIT for IBM mainframes. (If your system is not mentioned, check with your computing facility or installation instructions about whether to use FILE or UNIT.) You need not specify FILE or UNIT if you are entering data interactively from a terminal (see Chapter 11). If both FILE and UNIT are omitted, the program assumes that the data follow the BMDP instructions in the same file; this is equivalent to FILE = CONTROL or UNIT = 5.

Commands you may need

TITLE = 'text'. `TITLE = 'Run 2'.`

Assign a title of up to 160 characters to be printed after the program instructions for each problem. Text is printed in the output. For LINESIZE = 80 the title at page breaks is restricted to 64 characters, while for LINESIZE = 132, the title can be up to 112 characters. Default: the program prints the date and time across the top of each output page.

CASEs = #. `CASE = 50.`

Specify the number of cases to be read. If the data are in the same file as the BMDP instructions, and more problems follow the data, you must either specify the number of cases or include an END paragraph after the data, but not both. (An END paragraph must also follow the BMDP instructions.) Default: the program reads data until it finds an END paragraph or the end-of-file marker.

MCHAR = 'c'. `MCHAR = '$'.`

Identify a symbol that you have inserted in your data file to flag missing values. BMDP programs automatically assume that asterisks (*) in your data file indicate missing values. The MCHAR command allows you to use any other symbol as a missing value indicator (such as a dollar sign or a letter). Only one such symbol may be specified. You may, however use both a symbol and codes in your data file to flag missing values. For the use of codes for missing data see MISS in the VARIABLE paragraph, Chapter 4.

RECLEN = #. `RECLEN = 60.`

Specify the number of characters per data record to be read. You must state the record length if your FREE format data records exceed 80 characters, or if you want to limit the number of characters the program scans. Default: calculated from FORMAT command, if present. Otherwise, the default will be 80 characters if data are in external data file. If data are in same file as BMDP instructions, the record length has the value of CONTROL COLUMN in Chapter 9.

ERRMAX = #. `ERRMAX = 3.`

Specify the number of cases with errors that you will allow the program to read. If ERRMAX is not stated, the program will stop reading after finding ten cases containing one or more data errors. **Example:** The program will stop reading data after finding three cases containing one or more data errors.

Rarely needed commands

MULTiple = #. `MULT = 2.`

Specify the number of cases described in a FIXED format specification. This allows you to read more than one case per record in FIXED format. Default: the program assumes that the format specification applies to one case. **Example:** The FORMAT describes two cases.

BLEVEL = #. `BLEVEL = 4.`

Use to specify treatment of embedded and trailing blanks in input fields. An embedded blank is a blank in the middle of a number; e.g., weight is recorded as 13 instead of 153. Trailing blanks appear at the end of a field.

= 1 – All embedded and trailing blanks are illegal.
2 – (the default). Trailing blanks in a field with a decimal point are null or zero. In fields without decimal points, trailing blanks are not allowed.
3 – Trailing blanks in a field without a decimal point are ignored (i.e., an integer number need not be right-justified).
4 – All trailing blanks are converted to zero.
5 – All trailing blanks and embedded blanks are converted to zero.

REWIND. `NO REWIND.`

State NO REWIND if you do not want the tape or disk to rewind to the beginning of the data after each problem when running multiple problems. Otherwise, the input unit automatically rewinds after the data are read.

INPUT commands for reading BMDP Files (see Chapter 8)

CODE = name. `CODE = EXERCISE.`

Required to specify a BMDP File as input. Use the same name as was used to create the BMDP File.

CONTent = parameter name. `CONTENT = COVA.`

Specify the CONTENT of the BMDP File. Not needed for DATA or if only one CONTENT exists in the BMDP File. Default: If more than one CONTENT exists, the program reads the first CONTENT stored in the BMDP File.

DIRectory. `DIR.`

Directs the program to print the CODE, CONTENT, LABEL, variable names, and number of cases for each BMDP File stored in the system FILE (or UNIT).

LABEL = 'text'. `LABEL = 'INCLUDES TRANSFORMED VARIABLES'.`

Use to distinguish between two BMDP Files with identical CODE and CONTENT in the same system file. If LABEL was not specified in the SAVE paragraph when the BMDP File was created, the program normally stores the date and time of the file creation as a LABEL. To check the exact LABEL stored, use the DIRECTORY command.

PORTable. `PORT.`

Include when the BMDP File was created with PORT in the SAVE paragraph.

Summary Table for Input Commands

	Commands	Defaults	Multiple Problems	See
■	/ **INPut**			
■	VARiable = #.	required unless BMDP File input	●	3.1
■	FORMat = (*one only*) FREE, STREAM, SLASH, BINARY, *or* 'c'.	required unless BMDP File input	●	3.2
▲	FILE = 'name'.	CONTROL	●	3.4
	or UNIT = #.	5	●	3 4
	TITLE = 'text'.	date and time	–	3.5
	CASEs = #.	read to end-of-file marker or END	●	3.5
	MCHAR = 'c'.	'*' indicates missing data	●	3.5
	RECLENgth = #.	calculated from FORMAT = 'c'; if no FORMAT, 80 for external data file, value of CONTROL COLUMN for data in input file	–	3.5
	ERRMAX = #.	10	●	3.5
	MULTiple = #.	read 1 case per record (fixed format)	●	3.6
	BLEVEL = #.	2 (no embedded blanks; blanks after decimal are zero)	●	3.6
	REWIND. *or* NO REWIND.	REWIND.	●	3.6
	INPUT commands for reading BMDP Files (see Chapter 8)			
▲	CODE = name.	(required for BMDP File input)	–	Chpt 8
	CONTent = parameter.	first CONTENT saved in BMDP File	–	Chpt 8
	DIRectory.	no directory printed	–	Chpt 8
	LABEL = 'text'.	label generated at time of BMDP	–	Chpt 8
	PORTable.	(required to read BMDP File saved with PORT)	–	Chpt 8

Key:
- ■ Required paragraph or command
- ▲ Frequently used command
- ● Value retained for multiple problems
- – Default reassigned

4

Describing Your Data: The VARIABLE Paragraph

The VARIABLE paragraph is part of the basic set of instructions used by all BMDP programs. Figure 4.1 shows an example of the INPUT and VARIABLE paragraphs for the Exercise data described in Chapter 2. The INPUT paragraph is discussed in Chapter 3.

Figure 4.1
INPUT and VARIABLE paragraphs for the Exercise data

```
/ INPUT      VARIABLES = 6.
             FORMAT = FREE.
             FILE IS 'exercise.dat'.

/ VARIABLE   NAMES = id, sex, smoke, age, pulse_1, pulse_2.
```

The VARIABLE paragraph includes several important commands with which you should become familiar:

NAME – to assign names to your variables for use as labels in the output and in other instructions

USE – to select a subset of your variables for analysis

MISSING – to identify missing value codes in your data file

MAX, MIN – to restrict the range of data used in an analysis by setting upper and lower limits for each variable

LABEL – to label the cases in data listings and to identify one or two alphanumeric variables in free format

Other commands in the VARIABLE paragraph allow you to state how to treat blanks in FIXED format (as zeros or as missing), the number of variables added through transformations, whether to check for missing and extreme values before or after transformations, and whether to retain values of TRANSFORMED variables from one case to the next (see Chapter 6).

Where to Find It

4.1 Naming the variables: NAME

Use the NAME command to name the variables in your data set. The names appear in the output and can be used in other instructions (e.g., /GROUP VARIABLE IS SEX). The NAME command in Figure 4.1 is:

```
NAMES = id, sex, smoke, age, pulse_1, pulse_2.
```

The first name is that of the first variable, the second that of the second variable, etc. The names in the list are separated by commas, and the list ends with a period. If you only want to name the third variable SMOKE and the sixth variable PULSE_2, you can use the tabbing feature and specify

```
NAMES = (3)smoke, (6)pulse_2.
```

Naming conventions are as follows:

- Names must contain eight or fewer characters.
- Names must be enclosed in single quotes (') when
  - they do not begin with a letter
  - they contain a character that is not a letter or a number (i.e., a blank, parenthesis, comma, period, etc.) (e.g., '% INCR', '1986 WT'). Single quotes can be used around names even when they are not required.
- Underbars (_) can be used in any part of a name without enclosing the name in single quotes (e.g., WT_IN_87).
- Upper/lower case differences are ignored: AGE = age.

If you do not assign names, the program labels unnamed variables X(1), X(2), X(3), etc. in output. To generate lists of similar names, see the FOR % notation explained in Chapter 10.

Reserved words. IS, ARE, TO, BY, FOR, XMIS, TOOSMALL, and TOOBIG are special words in the BMDP instruction language. These should not be used as names unless enclosed in single quotes.

4.2 Selecting variables for analysis: USE

For a particular analysis, you may want to use only a few variables from a large data file. The USE command allows you to select a subset of variables without rewriting your format and other specifications. Note that variables omitted from the USE list are still available for use in the TRANSFORM paragraph (see Chapter 6). The USE list also performs two other functions:

1. You can limit the number of variables checked by the program for missing data and values out of range. This is important for analyses that require complete cases (cases with no missing data or values out of range). See Chapter 2 for how BMDP handles missing data and values out of range.
2. You can change the order in which variables are analyzed or printed. BMDP follows the order specified in the USE list rather than the order in which variables were read into the program. The original variable sequence number is not changed.

To select the two pulse values from the Exercise data, for example, state the variable names or variable numbers:

```
   USE = pulse_1, pulse_2.
or USE = 5, 6.
```

To omit the first two variables, ID and SEX, specify

```
USE = smoke TO pulse_2.
```

The word TO is a convenient notation for writing lists of consecutive variables (see Chapter 11, section 11.1).

4.3 Identifying flags for missing data: MISSING

Data may contain some cases with missing (unrecorded) values. You should insert a special code or symbol in your data file to mark the position of each missing value. Note that BMDP automatically recognizes that a value is missing when it reads:

- one or more contiguous asterisks (*) in a field.
- blanks in a field read using FIXED format (see BLANKS in the Commands list if you want to treat blanks as zeros). In FREE format, two consecutive commas indicate a missing value. The value between 25 and 6 is missing:

```
25, ,6, 32, 4, 75, 51
```

You may also identify missing value flags that you inserted in the data file by using:

- **numeric codes** (such as 999) for each variable (use the MISSING command discussed below)

or

- a **symbol** (other than *) such as a period or dollar sign (use the MCHAR command in the INPUT paragraph, Chapter 3).

In Figure 4.2 we display a modified version of the Exercise data. We have inserted and underlined codes and symbols to flag missing data values. (Three flagged values are out of range; we explain these in the next section.) We also add subjects' names; these can be read as two contiguous four-character label variables (see Section 4.5).

Figure 4.2 The Exercise data set with missing value flags and case labels

id	sex	smoke	age	pulse_1	pulse_2	name1 name2
1	1	1	11 ←	62	126	ALLAN
2	2	1	20	78	...	MARY
3	1	1	99	64	128	BILL
4	2	2	29	.	155	LINDA
5	1	1	21	66	28 ←	MICHAEL
6	2	1	27	96	265 ←	CATHY
7	1	2	21	68	120	HARVEY
8	9	2	42	72	999	JENEE

BMDP recognizes these codes and symbols as missing value flags and excludes them from all computations when we insert the instructions:

```
MCHAR = '.'.                              in the INPUT paragraph
MISSING = 99, 9, 9, 99, 999, 999.         in the VARIABLE paragraph
```

MCHAR identifies a global symbol that can be used when any value is missing. The MISSING command allows one numeric code per listed variable in the order that the variables appear on the input record. For each variable we select a code that is outside the range of possible values. The first variable, ID, will be considered missing if its value is 99. The second missing value code is 9, used for the second variable (SEX), etc. You can use tabbing to skip over variables:

```
     MISS = (sex)9, (age)99, 999, 999.
or   MISS = (2)9, (4)99, 999, 999.
```

We tab to the 2nd variable (SEX), specify 9, then tab to the 4th variable (AGE) and specify 99, then specify 999 for PULSE_1 and PULSE_2. You may use either the variable name or number when tabbing. You can also use repetition to specify the same code for several variables:

```
MISS = (sex)9, (age)99, 2*999.
```

See Chapter 10 for more information on tabbing and repeating numbers.

The program converts all missing value flags to a very large number (on many machines, the number is 16^{31}), which is known to BMDP by the name XMIS in the TRANSFORM paragraph (see Chapter 6).

4.4 Setting range limits: MINIMUM and MAXIMUM

A frequent goal is to avoid analyzing extreme values that are likely to be data errors. BMDP allows you to specify acceptable ranges for each variable. In any program you can restrict the analysis of a variable to a specified range by assigning an upper limit (maximum) and/or a lower limit (minimum) for values of the variable. A value that is greater than the upper limit or less than the lower limit is *out of range* and is excluded from all computations. You can list one minimum limit and one maximum limit for each input variable, or you can use tabbing to skip over variables.

In the Exercise study we are sampling from healthy adults and want to consider as unreasonable all pulse values below 35 and above 200. We see in Figure 4.2 that the first AGE value is too young, the fifth PULSE_2 value is too low, and the sixth PULSE_2 value is too high. To omit these values from all computations we add the following instructions to the VARIABLE paragraph:

```
MIN = (age)18, (pulse_2)35.
MAX = (pulse_2)200.
```

We have tabbed to the variables for which we want to specify range limits.

The program converts values less than your specified MINIMUMS and greater than your specified MAXIMUMS to internal flags (–2 * XMIS for MINIMUM, 2 * XMIS for MAXIMUM) which can be referenced in TRANSFORM statements as TOOSMALL and TOOBIG (see Chapter 6).

Incorporating MCHAR, MISS, MIN, and MAX into the basic instructions

Figure 4.3 displays the instructions needed to specify missing value flags, range limits, and case labels for the data in Figure 4.2.

After the program interprets your MIN, MAX, and MISSING commands, it prints a table of the values specified. See Figure 4.4.

Figure 4.3
Basic instructions for the Exercise data including range limits, flags for missing data, and case labels

```
/ INPUT      VARIABLES = 8.
             FORMAT = FREE.
             FILE = 'exercise.dat'.
             MCHAR = '.'.

/ VARIABLE   NAMES = id, sex, smoke, age, pulse_1,
                     pulse_2, name1, name2.
             LABELS = name1, name2.
             MISSING = (sex)9, (age)99, 2*999.
             MINIMUM = (age)18, (pulse_2)35.
             MAXIMUM = (pulse_2)200.
```

Figure 4.4
Printout describing range limits and codes for missing values

```
 VARIABLE               STATED VALUES FOR
NO.   NAME           MINIMUM MAXIMUM MISSING
---- --------        ------- ------- -------

  2  sex                               9.000

  4  age              18.00            99.00

  5  pulse_1                           999.0

  6  pulse_2          35.00   200.0    999.0
----------------------------------------------
```

4.5 Case labels: LABEL

The LABEL command refers to one or two variables in your data file and serves two purposes:

1. It provides case labels for printouts. You can use case identification variables such as names or case id numbers to label the output when case-by-case information is printed. For an example of case labels in a printout see program 1D, Example 1D.2. If you do not specify case LABELS, the program prints the case number (i.e., the sequence of the case in the file) to the left of the case-by-case information.
2. It allows you to read one or two alphanumeric variables in FREE format. Note that if you have to read more than two alphanumeric variables you will have to use FIXED format; there is no limit to the number of alphanumeric variables allowed in FIXED format. See Chapter 2 for more about character variables.

BMDP allows two variables to be declared case labeling variables, each of which may contain up to four characters. The two LABEL variables need not be contiguous; however, one eight-character field can be read as two four-character LABEL variables. To read the labels in Figure 4.2 we used two LABEL variables, NAME1 and NAME2, because the field could contain more than four characters. Label variables need not be alphabetic data; you might, for example, want to use id numbers as labels.

When you read data in **FREE format,** there can be *no* blanks or commas within a four- or eight-character field. Trailing blanks are allowed. The following are valid examples of two contiguous four-character case LABELS (b indicates a blank). We show the characters in the data file on the left, the resulting label on the right:

```
JONESbbb        JONES
ALEXANDER       ALEXANDE
SMITH_JOHN      SMITH_JO
```

The following examples are *not* valid, and will cause the program to produce an error message:

```
SMITH,JOHN          MING,YU          COBBbSAMUEL
```

When you read your data in **FIXED format,** you must read the case-labeling variable(s) in A- (alphanumeric) format (a three-character label is formatted A3; two four-character labels are 2A4). For example, if the first 12 characters in your data file contain case labeling information, you can use the first eight characters to label the output by specifying:

```
/ INPUT      FORMAT = '2A4, 4X, ...'.
/ VARIABLE   NAMES = ID1, ID2, ... .
             LABELS = ID1, ID2.
```

('2A4' in the FORMAT command tells the program to read two four-character id variables; '4X' tells it to skip over the last four characters in the twelve-character name.)

Once you specify a variable as a LABEL, it is not available as a number for analysis, but it can be used in transformations. To use LABELS to select cases in the TRANSFORM paragraph you must use the CHAR function (see Chapter 6). If you read data in fixed format, you can read numeric case labelling variable(s) in A-format, then use the FORTRAN T-notation to tab back and re-read the variable(s) in F-format for use in computations (see FORMAT in Chapter 3).

4.6 The RESET command

Specify RESET when running multiple problems to reset every option in the VARIABLE paragraph to its original default value (see Summary Table at end of chapter). All variables are used, and any minima, maxima, and missing value code specifications are erased. Any commands given in the same VARIABLE paragraph will override these reset values.

4.7 Commands used with transformations

The commands in this section are complex and infrequently used. You may first want to read Chapter 6 on the TRANSFORM paragraph. ADD allows you to use temporary variables in transformations and to delete variables from your data file. BEFORETRANSFORM, AFTERTRANSFORM, and NO CHECK control when the program checks for missing and extreme values. RETAIN allows you to create lagged variables.

Specifying the number of variables added through transformation: ADD

When you create new variables in the TRANSFORM paragraph, the program automatically adds them to the end of each case in its internal version of your data file. They can be analyzed, printed, or saved. For example, we can create a difference score between pre- and post-exercise pulse rates for each case in the Exercise data:

```
/ TRANSFORM  DIFFRNCE = PULSE_2 - PULSE_1.
```

The program adds DIFFRNCE to the end of the data for each case. If you create additional variables, they are added to the data in the order in which they are specified in the TRANSFORM paragraph.

If you want to create temporary variables in the TRANSFORM paragraph (e.g., as building blocks for complex equations) and do not want them added to your data file, tell the program how many variables you want added to the file by stating ADD = #. Then select the variables you want to use by adding their names to the VARIABLE NAMES list. The program will treat unnamed variables as temporary and add only the variables named in the VARIABLE paragraph to the end of the data for each case. See Chapter 6 for more about temporary variables.

You can also delete variables from your data file before analyzing, printing, or saving the data by specifying a negative number in the ADD = # command. For example, state

```
ADD = -3.
```

to delete the last three variables from each case before analyzing, printing, or saving the data. (See also KEEP and DELETE in the SAVE paragraph, Chapter 8.)

When to check for missing and extreme values: BEFORETRANSFORM, AFTERTRANSFORM, or NO CHECK

When you transform variables or create new variables in the TRANSFORM paragraph, you can use the MISSING, MINIMUM, and MAXIMUM commands to specify missing value codes and range limits for them. Normally, the program checks for missing and extreme values *before* executing transformations (BEFORETRANSFORM). If you want missing value codes or range limits to apply to a variable *after* it has been transformed or to a newly created variable, specify AFTERTRANSFORM. For example, if you create a variable called SIZE in the TRANSFORM paragraph, you can add the following instructions to the VARIABLE paragraph to check for a minimum of 7.0 in SIZE:

```
MINIMUM = (size)7.0.
AFTERTRANS.
```

To ignore specified range limits and missing value codes for all variables, state NO CHECK (some programs will still use the range limits to determine the scale of graphical displays). This command is useful when you have specified missing value codes and range limits in a previous problem, and want the program to ignore them in the current analysis. Only missing value codes specified using the MISSING command can be recovered (they will be read as values like 9, 99, etc.); other missing value indicators such as asterisks, MCHAR symbols, consecutive commas, etc. are still treated as missing. Note that once the program reads the data and converts values to XMIS, TOOSMALL, or TOOBIG, as in a BMDP File, these values cannot be recovered by stating NO CHECK.

Retaining TRANSFORM values from case to case: RETAIN

BMDP programs perform all computations specified in the TRANSFORM paragraph for one case at a time and do not retain the computed values any longer than it takes to complete computations and store the values for that one case. At the beginning of each new case, the program initializes all new variables to an internal missing value flag (XMIS) before computing their values for that case. If you want to use the values computed for one case in the computations for the next case, state RETAIN. You can use this command to create lagged variables. See the TRANSFORM paragraph, Chapter 6.

Checking order of records

When there are several records per subject or experimental unit, their proper ordering in a file is essential. Each record should include a case identification number and a record number. The order of records can be checked and deviant records flagged by using program 1D. Suppose that you have two records per subject, the case identification number appears as variables 1 and 21, and the record number (1 or 2) is recorded as variables 2 and 22. In that situation, the setup shown in Figure 4.5 can be used:

Figure 4.5
Instructions for checking order of records

```
/ INPUT      VARIABLES ARE 40.
             FORMAT IS FREE.
/ VARIABLE   MINIMUMS ARE (2)1,(21)0,2.
             MAXIMUMS ARE (2)1,(21)0,2.
             AFTERT.
/ TRANSF     X(21) = X(21)-X(1).
/ PRINT      MAXIMUM.
             MINIMUM.
/ END
```

In the VARIABLE paragraph, we state that the minimum and maximum values for the second identification variable (variable 21) are precisely zero; i.e., before subtracting the first case identification variable, the second case identification variable was the same as the first. To be sure that the records for a subject are in order, we specify that the minimum and maximum values of the record number variables (2 and 22) are precisely 1 and 2. We specify AFTERT because the limits must be checked after variable 21 is transformed. The effect of the PRINT paragraph is to print the data for any deviant cases.

4.8 Frequencies and weights

Three options affect many of the estimates available in BMDP. Two of these options, FREQ and CWEIGHT, are available in a wide range of programs, and the third option, WEIGHT, is available in the REGRESSION paragraph of programs 1R, 2R, 3R, 4R, 5R, 6R, 9R and AR.

Case frequencies

The FREQ option allows you to specify a case frequency variable. For example, specifying

```
/ VARIABLE FREQ = count.
```

tells BMDP that the variable COUNT will be used as a case frequency. The case frequency must be a whole number. A case frequency has the effect of repeating cases. For example, if we were measuring AGE, CITY, and STATE and we had five observations with exactly the same values for all three variables, we could enter five records into our data file. Alternately, we could enter only one record and assign a value of 5 to the variable COUNT for that record. Other records that represent only one case would be required to have a COUNT value of 1. While specifying a case frequency equal to some number s does not literally add s cases to the data file, all of the statistics in the output such as means, standard deviations, correlations and other measures derived from these are reported as though s identical cases were present. Any record with a negative or zero case frequency (COUNT ≤ 0 in this example) is excluded from the calculations as though it were missing.

The FREQ option is available in programs 1D, 2D, 3D, 5D, 6D, 8D, 9D, 1M, 2M, AM, KM, 1R, 2R, 4R, 5R, 6R, 9R, LR, PR, and 4F.

Case weights

Case weights can be given in the VARIABLE paragraph. In the example

```
/ VARIABLE  CWEIGHT = sampwt.
```

the variable SAMPWT serves as a case weight, or "design weight." As with case frequencies, records with negative or zero case weight values are excluded from the computations.

Unlike case frequencies, where the object is to repeat cases, case weights can be used to adjust when sampling from populations with unequal variances. If the ith observation comes from a population with mean μ and variance σ^2/k_i, where k_i is known, the case weight can be set to k_i. One usually assumes equal variances ($k_i = 1$ for all observations). In the example above, the values of SAMPWT provide case weights. The variance estimated when case weights are used is σ^2. As k_i increases, the estimate of σ^2 will increase.

The CWEIGHT option is available in programs 1D, 3D, 5D, 8D, 9D, 1M, 2M, 4M, AM, KM, 1R, 2R, 4R, 5R, 6R, 9R, 3V, and 5V.

VARIABLE Commands

Commonly used commands

NAMES = list.

```
NAMES = AGE, WT_IN_78,
        '% INCR'.
```

List names for each variable in your data set in the order they appear on the input record. Enclose a name in single quotes if it includes blanks or symbols or begins with a number. Underbars (_) can be used in any part of the name without enclosing the name in single quotes. Names must be eight characters or less. You need not name all variables; use tabbing to skip variables. If no names are given for the variables, the program uses X(1), X(2), X(3), etc. to label the output.

USE = list.

```
USE = PULSE_2, AGE, 3, 2.
```

List the names or numbers of the variables to be used in the analysis. If omitted, the program uses all input variables. **Example:** The USE list selects and reorders variables from the Exercise data. PULSE_2 is analyzed or printed first, followed by AGE, then the third and second input variables (SMOKE and SEX, respectively). ID and PULSE_1 are not included in the list and so are not used.

MISSing = # list.

```
MISS = 99, 9, (6)999.
```

Specify missing value codes for each input variable. The first number is for the first variable on the input record (*not* in the USE list), the second for the second variable, etc. Use tabbing to skip variables. **Example:** Any observation coded 99 for the first variable, 9 for the second variable, and 999 for the sixth variable is treated as missing, and will be excluded from all computations.

MAXimum = # list.
MINimum = # list.

```
MAX = (pulse_2)200.
MIN = (age)18, (pulse_2)35.
```

Specify upper and/or lower limits for each input variable. In each list, the first number is for the first variable on the input record, the second for the second variable, etc. Use tabbing to skip variables. Values above the upper limit or below the lower limit are excluded from the computations. **Example:** The program excludes values above 200 for PULSE_2, below 18 for AGE, and below 35 for PULSE_2.

LABEL = 1 or 2 variables.

```
LABEL = ID1, ID2.
```

Specify the names or numbers of one or two variables to be used as case labels in programs that print information for each case. LABEL must be used to identify alphanumeric variables read in FREE format. LABEL variables are not used in numeric analyses, but they can be used to group data (see ALPHA, Chapter 5). If omitted, the programs print case-by-case information using case numbers. When you read data in **FIXED format** (see INPUT FORMAT), you must specify LABELS in A-format, with a maximum of four characters each (A4). In **FREE format,** no blanks or commas may be embedded within a four-character field (or within an eight-character field when you specify two contiguous LABEL variables). BMDP **does** allow trailing blanks (**after** the LABEL) in the four- or eight-character field. See section 4.5.

Commands you may need

ADD = NEW.
or **ADD = #.**

```
ADD = NEW.
ADD = 2.
```

Unless you specify ADD = #, any variable you create in the TRANSFORM paragraph will automatically be added to your data (ADD = NEW). The program adds the names of the new variables (in the order stated in the TRANSFORM paragraph) to your list of NAMES in the VARIABLE paragraph. (Do not include these names in the NAMES list.)

When you specify ADD = #, the program adds only the specified number of variables to your data. You need to add the names of the new variables to the end of your VARIABLE NAMES list in order for the variables to be used in the analysis. Otherwise, the program treats the new variables as temporary and can only use them **within** the TRANSFORM paragraph. You can also use ADD = # to eliminate variables from an analysis. For example, if you state ADD = –2, the program omits the last two input variables from the analysis, and no variables created in the TRANSFORM paragraph are used.

BEForetransform. `AFTER.`
or **AFTERtransform.**
or **NO CHECK.**

Specify when the program should check for missing and extreme values as specified in the MIN, MAX, and MISS statements. By default, the program checks for these values **before** executing transformations stated in the TRANSFORM paragraph (BEFORETRANSFORM). If you create new variables in the TRANSFORM paragraph, and want to specify missing value codes and range limits using MISS, MIN, and MAX in the VARIABLE paragraph, you may need to specify AFTERTRANSFORM. This directs the program to check for missing and extreme values **after** the TRANSFORM paragraph computations are done. NO CHECK directs the program to ignore the MIN, MAX, and MISS statements.

BLANK = MISSing. *or* **ZERO.** `BLANKS = ZERO.`

When you read data in FIXED format, you can specify whether you want the program to treat blank fields in the data set as missing values or as zeros. Default: MISSING.

RETAIN. `RETAIN.`

Specify RETAIN to retain the value of a new variable added in the TRANSFORM paragraph from one case to the next. Use to create lagged variables (see Chapter 6). The default is NO RETAIN: variables added in the TRANSFORM paragraph are initialized as missing values for each new case.

RESET. `RESET.`

Specify RESET when running multiple problems to reset every option in the VARIABLE paragraph to its original default value. Any minima, maxima, and missing value code specifications are erased. Any sentences given in the same VARIABLE paragraph will override these reset values.

FREQuency = variable. `FREQ = COUNT.`

State the name or number of a variable containing the case frequency. A case frequency has the effect of repeating cases. Any record with a negative or zero case frequency (COUNT ≤ 0 in this example) is excluded from the calculations as though it was missing. The FREQ option is available in programs 1D, 2D, 3D, 5D, 6D, 8D, 9D, 1M, 2M, AM, KM, 1R, 2R, 4R, 5R, 6R, 9R, LR, PR, and 4F.

CWeight = variable. `CWEIGHT = SAMPWT.`

Specify a case weight, or "design weight." As with case frequencies, records with negative or zero case weight values are excluded from the computations. Case weights can be used to adjust when sampling from populations with unequal variances. The CWEIGHT option is available in programs 1D, 3D, 5D, 8D, 9D, 1M, 2M, 4M, AM, KM, 1R, 2R, 4R, 5R, 6R, 9R, 3V, and 5V.

Summary Table for VARIABLE Commands

	Paragraph Commands	Defaults	Multiple Problems	See
▲	/ **VARiable**			
▲	NAME = list.	X(1), X(2), etc.	●	4.1
▲	USE = list.	all input variables used	–	4.2
	MISSing = # list.	no missing value codes	●	4.3
	MAXimum = # list.	no maximum limits	●	4.4
	MINimum = # list.	no minimum limits	●	4.4
	LABEL = list.	no case labels	●	4.5
	RESET.	other commands not reset	–	4.6
	ADD = NEW *or* #.	NEW	–	4.7
	BEForetransform, AFTertransform, *or* NO CHECK.	BEFORE.	●	4.7
	RETAIN.	NO RETAIN.	●	4.7
	FREQuency = variable.	no case frequency variable	–	4.8
	CWeight = variable.	no case weights used	–	4.8
	BLANK = ZERO *or* MISSING.	MISSING	●	Cmds

Key: ▲ Frequently used paragraph or command
● Value retained for multiple problems
– Default reassigned

5

Describing Your Data: The GROUP or CATEGORY Paragraph

Some BMDP programs use the GROUP or CATEGORY paragraph to

- classify cases into groups for *t* tests, analyses of variance, etc.
- define categories for frequency tables.
- specify symbols identifying group membership in plots.

You can use the words GROUP and CATEGORY interchangeably as the paragraph name; for simplicity, we will use the word GROUP in our discussion.

You can form groups in two ways:

- by specifying codes for the categories of a discrete variable (e.g., SEX)
- by specifying cutpoints to split a continuous variable into intervals (e.g., AGE)

You can name each category or interval; the program labels the output accordingly. BMDP programs do not require that your data be sorted by groups in advance; they use the code and cutpoint information you specify to place cases in groups during analysis. Figure 5.1 displays the BMDP instructions for the GROUP paragraph for the Exercise data set. The INPUT and VARIABLE paragraphs are discussed in Chapters 3 and 4.

Figure 5.1
GROUP paragraph for the Exercise data

```
/ GROUP     CODES(sex) = 1, 2.
            NAMES(sex) = male, female.
            CUTPOINTS(age) = 20, 40.
            NAMES(age) = '20orless', _21to40, over_40.
```

Previously, eight programs (1D, 5D, 6D, 7M, AM, 1R, 3S, and 1V) required that you identify the GROUPING variable in the VARIABLE paragraph. This is no longer required in any BMDP program. Generally the grouping variable is now defined in the GROUP paragraph with the VARIABLE = NAME command. See the specific program chapters for complete details on defining grouping variables.

Where to Find It

5.1 Defining categories for discrete variables: CODES

You can use the values of a categorical variable as codes to identify groups in an analysis. The Exercise data set presented in Chapter 2 contains two categorical variables: SEX and SMOKE. To classify cases into groups according to the values of these two variables, we state:

```
CODES(sex) = 1, 2.        or        CODE(2) = 1, 2.
CODES(smoke) = 1, 2.                CODE(3) = 1, 2.
```

Any number (positive, zero, or negative) may be used as a code. Codes may be stated in any order (the order affects some analyses):

`CODES(SEX) = 2, 1.` is equivalent to `CODES(SEX) = 1, 2.`

When names are assigned, their order must match the codes (see Section 5.3).

Selecting groups for analysis. You can select codes to be used for analysis by omitting from the CODE list the codes you do not want. For example, in the smoking example above, there could have been a code 3 in the data set to indicate "unknown." By excluding code 3 from the CODE list, we direct the program to exclude cases with unknown smoking status from those analyses that use groups defined by smoking.

Shortcuts. You can specify codes for several variables in a single statement. Since SEX and SMOKE both have codes 1 and 2, we can state:

```
CODES(sex, smoke) = 1, 2.
```

This feature is useful when you have several variables with the same codes. For example, if we want to classify cases into groups according to "yes-no" answers on 20 questions, we state:

```
CODES(Q1 TO Q20) = 1, 2.
```

We have used BMDP's "implied list" feature to specify all 20 question variables (see Chapter 10). Variables in an implied list must be contiguous. You can also use the TO notation in writing CODE lists:

```
CODES(city) = 1,2,3,4,5.    or    CODES(city) = 1 TO 5.
```

Using character data to define and name groups: ALPHA. You can define groups using character codes or alphanumeric fields instead of numbers. For example, suppose we enter the words "YES" and "NO" in the Exercise data file for the variable SMOKE and M and F for SEX instead of 1 and 2:

```
1  M  YES  31  62  126
2  F  YES  20  78  154
          . . .
```

We can then state

```
ALPHA(SEX) = M, F.
ALPHA(SMOKE) = YES, NO.
```

in the GROUP paragraph. The program uses the character values to label output as well as to define the groups. Note: You cannot use ALPHA with NAMES.

In FIXED format, any number of alphanumeric variables can be read using A-format (see FORMAT, Chapter 3); the alphanumeric variables should be left-justified in the data file. In STREAM, or SLASH format you must identify alphanumeric variables as case LABELS in the VARIABLE paragraph (see Chapter 4). In FREE format, the variable must **NOT** be a label. Only two LABEL variables are allowed.

5.2 Defining intervals for continuous variables: CUTPOINTS

You can use CUTPOINTS to define intervals for a continuous variable, such as AGE. Each CUTPOINT is the upper limit of an interval. The first interval extends from the lowest value for the given variable up to and including the value of the first cutpoint. The last interval includes all values larger than the last cutpoint. You will always have one more interval than you have cutpoints; that is, n cutpoints define $n + 1$ intervals. For example, if we want to classify cases into three groups according to AGE (variable 4), we specify

```
CUTPOINTS(age) = 20, 40.    or    CUTPOINTS(4) = 20, 40.
```

This statement creates the three intervals displayed on the number line in Figure 5.2. Note that an AGE recorded as 20.01 will be placed in the second group.

Figure 5.2
Two CUTPOINTS define three AGE intervals

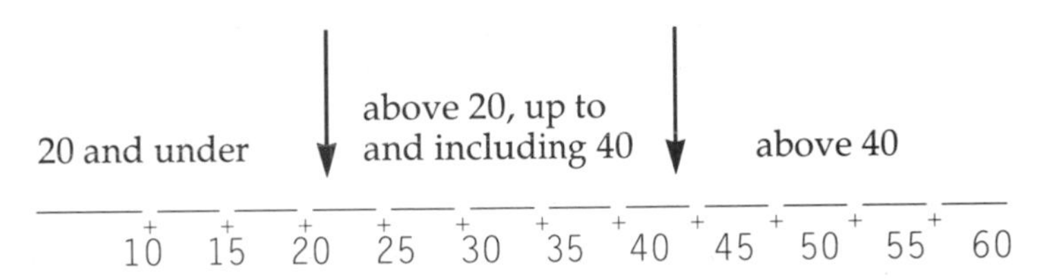

Shortcuts. The same shortcuts available for CODES also work for CUTPOINTS. If the answers to 20 questions are "yes" (code 1), "no" (code 2), and "maybe" (code 3), we can combine the "no" and "maybe" answers for all twenty questions by stating:

```
CUTP(Q1 TO Q20) = 1.
```

This statement creates two intervals: up to and including 1 ("yes" answers), and above 1 ("no" and "maybe" answers). When you have a large number of cutpoints, the "BY" notation (Chapter 10) is convenient:

```
CUTP(age) = 5 TO 80 BY 5.
```

This is equivalent to stating CUTP(age) = 5, 10, 15, 20, etc. If you want to categorize scores from seven tests, you could state

```
CUTP(T1 TO T7) = 60 TO 90 BY 10.
```

5.3 NAMES for categories and intervals

You can assign NAMES to each category defined by a CODE and to each interval defined by a CUTPOINT. Names are used to label the output. (If you do not specify NAMES, the program uses CODE or CUTPOINT values to label the output.) To name the groups identified in:

```
CODES(sex) = 1, 2.
CUTPOINTS(age) = 20, 40.
```

state

```
NAMES(sex) = male, female.
NAMES(age) = '20orless', _21to40, over_40.
```

Use the same naming conventions that you use to name variables in the VARI-

ABLE paragraph (Chapter 4). The NAMES list for a given variable must follow the same order as the CODE or CUTPOINT list for that variable. There must be the same number of names as there are CODES or intervals defined by CUTPOINTS. (A variable with two CUTPOINTS requires three names.) BMDP programs interpret your grouping information in an output panel like the one shown in Figure 5.3.

Figure 5.3. Panel printed when grouping information is specified

```
 VARIABLE         STATED VALUES FOR                 GROUP CATEGORY     INTERVALS
NO.   NAME    MINIMUM MAXIMUM MISSING     CODE  INDEX   NAME        .GT.    .LE.
---- --------  ------- ------- -------    ------ ----- --------    ------  -------

 2  SEX                                   1.000    1  male
                                          2.000    2  female

 4  AGE                                            1  20orless              20.00
                                                   2  _21to40      20.00    40.00
                                                   3  over_40      40.00
--------------------------------------------------------------------------------
```

You can name the categories or intervals for several variables simultaneously. For example:

```
CODES(item1 TO item10) = 1, 2, 3.
NAMES(item1 TO item10) = low, medium, high.
```

Note: You cannot use NAMES with ALPHA.

Using names to combine categories. You can combine different categories or intervals into a single group by giving them the same NAMES. For example, if codes and names for country-of-origin are:

```
CODES(COUNTRY) =      1,        2,      3,          4,      5.
NAMES(COUNTRY) = MEXICO, ENGLAND, CHILE, SCOTLAND, SPAIN.
```

we can change the coding to denote ethnicity by stating:

```
CODES(COUNTRY) =      1,      2,        3,      4,        5.
NAMES(COUNTRY) = HISPAN, ANGLO, HISPAN, ANGLO, HISPAN.
```

Instead of five groups, the program will form two: HISPAN (codes 1, 3, and 5), and ANGLO (codes 2 and 4).

5.4 STACK: combining several grouping variables into one

The STACK feature allows you to combine grouping variables. For example, if SEX and SMOKER are two variables, you can create a STACK variable that contains all possible combinations of sex and smoking status (i.e., male smoker, male nonsmoker, female smoker, female nonsmoker). The variables that comprise each STACK variable must have CODES or CUTPOINTS. Any STACKED variable can be used in an analysis of variance, regression, or any other program using grouping information.

One to five STACK variables are named in the STACK command. Each STACK variable is defined by up to five variables which are listed following the STACK variable name. The sequence of the STACK categories is determined by the order of the variables defining the stack; the first variable in the list moves fastest (see example below). The STACK categories may be named like any group categories, but they may not have duplicate names.

Further stacking of stacked variables is allowed in the 1990 release of BMDP. In Figure 5.4 we first create a stacked variable, SEXSMO, by combining the grouping variables SEX and SMOKE. We then use this stacked variable along with another grouping variable, AGE, to produce the stacked variable SEXSMOLD. In the STACK command, SEXSMO appears before SEXSMOLD since SEXSMOLD builds upon SEXSMO.

Figure 5.4
Creating STACK variables and stacking them

```
/ GROUP    CODES(sex, smoke) = 1, 2.
           NAMES(sex) = male, female.
           NAMES(smoke) = yes, no.
           CUTP(age) = 50.
           NAMES(age) = young, old.
           STACK = sexsmo, sexsmold.
           sexsmo = sex, smoke.
           sexsmold = sexsmo, age.
           NAMES(sexsmold) = ygmalesm, ygfemsm, ygmalens,
                             ygfemns, olmalesm, oldfemsm,
                             olmalens, oldfemns.
```

Special Feature
The RESET command

If you have specified CODES and CUTPOINTS in a previous problem (in the current computer session) or if they are stored in the BMDP File you are using, you may cancel these assignments by stating

```
RESET.
```

If you do not state RESET, then the program retains CODE and CUTPOINT information used in previous problems.

GROUP or CATEGORY Commands

VARiable = list. `VARIABLE = SEX, CIGARET.`

List names or numbers of the grouping variables. You must specify CODES or CUTPOINTS in the GROUP paragraph if the variable takes on more than 10 distinct values. Groups must have unique first letters in order for the names to appear in the analysis of variance table; if not, the groups are labeled G, H,..., etc. in the table. **Example:** The variables SEX and CIGARET contain values for separating the cases into cells (groups). See the specific program chapters for complete information about grouping in that program.

CODEs(*j*) = # list.

```
CODES(2) = 1, 2.
CODE(SEX) = 1, 2.
CODES(Q1 TO Q20) = 1 TO 5.
```

List codes for discrete variable *j*, where *j* is the variable name or number. If several variables have the same values, you can use one statement to specify the codes (as in the third example above). Codes may be stated in any order. Any numbers (positive, zero, or negative) can be used as codes. The program analyzes only those cases with the specified codes. If you do not specify codes, the program will use up to ten distinct values of a variable as codes. **Examples:** Codes for the second input variable are 1 and 2. Codes for the variable named SEX are 1 and 2. Codes for variables Q1 through Q20 are 1 through 5.

CUTPoints(*j*) = # list.

```
CUTP(4) = 20, 40.
CUTP(AGE) = 20, 40.
CUTP(Q1 TO Q20) = 2.
```

List the upper limits of intervals for continuous variable *j*, where *j* is the variable name or number. If several variables have the same CUTPOINTS, you can use one statement as in the third example. If omitted, the program does not assign intervals. **Examples:** The three intervals for the fourth input variable, AGE, are (1) less than or equal to 20, (2) 21 to 40, and (3) over 40. Intervals defined for Questions 1 through 20 are (1) less than or equal to 2, and (2) greater than 2.

ALPHA(*j*) = list. `ALPHA(SMOKE) = YES, NO.`

List the character values stored in character variable *j*. The program uses these characters as codes to define groups and as names to label output. Each element in the list (i.e., each code in your data set) can be up to four characters in length and must be left-justified (i.e., ' NO' is not the same as 'NO '). The variable *j* (either name or number) is used to classify cases; each such variable must be read in A-format. (In FREE format, the /Group Alpha(j) = command will work only if the alpha variable is **NOT** declared a label) You cannot use the ALPHA and NAMES commands simultaneously. **Example:** Based on characters that have been entered in the data set, the program divides the variable SMOKE into two categories (YES and NO), and uses YES and NO to label the output. Note that a variable mentioned in the ALPHA command is assumed to be alphanumeric. If the variable is used in FREE, STREAM, or SLASH format, it is read as characters. Thus, if you recode the variable in the TRANSFORM paragraph, the commands should read `IF (SEX EQ CHAR(1)) THEN SEX = CHAR(M))` rather than `IF(SEX = 1)`...

NAMEs(*j*) = list.

```
NAMES(4) = MALE, FEMALE.

NAMES(SEX) = MALE, FEMALE.
NAMES(Q1 TO Q20) = YES, NO.
```

List names for the categories or intervals of variable *j*, where *j* is the variable name or number. The names are matched in order to the categories or intervals. If names are omitted, the program uses the CODE, CUTPOINT, or ALPHA values to label the output. You cannot use NAMES with ALPHA. **Examples:** Group names for the fourth input variable, SEX, are MALE and FEMALE. Group names for each of the twenty questions (Q1 through Q20) are YES and NO.

RESET.

```
RESET.
```

Cancels previous CODE and CUTPOINT assignments. Use for BMDP File input when you want to change grouping information specified previously or when running several problems during one computer run. Values for CODES and CUTPOINTS remain the same for additional problems unless you specify new values for a variable or state RESET.

STACK = list.
stack variable = variable list.

```
STACK = SEXETHN, SEXETHAG.
SEXETHN = sex, ethnic.
SEXETHAG = sex, ethnic,
           age.
```

Combines grouping variables into up to five new "stack" variables. Variables that comprise each STACK variable must have CODES or CUTPOINTS. Each STACK variable is defined by up to five variables which are listed following the STACK variable name. The sequence of the STACK categories is determined by the order of the variables defining the stack; the first variable in the list moves fastest. STACK categories may be named like any group categories but may not have duplicate names. If no names are specified, default names (*1, *2, etc.) are produced. **Note:** Within the GROUP paragraph, a stacked variable should not begin with the letters "GR," and within the CATEGORY paragraph, it should not begin with "CAT" or "CT."

Summary Table for GROUP Commands

	Paragraphs Commands	Defaults	Multiple Problems	See
▲	/ **GROUP or CATEGory**			
▲	VARiable = variable.	no grouping	●	Cmds
▲	CODEs(*j*) = # list.	up to 10 distinct values (required when program needs groups defined and there are more than 10 values for grouping variable)	●	5.1
▲	CUTPoints(*j*) = # list.	intervals not assigned	●	5.2
▲	NAMEs(*j*) = list.	CODE or CUTPOINT values used to label output	●	5.3
	ALPHA(*j*) = list.	characters in the data set are not recognized as codes or names	●	5.1
	RESET.	CODE and CUTP assignments remain in effect	–	SpecF
	STACK = list.	no stack variables	–	5.4
	stack variable = variable list.	(required with STACK)	–	5.4

Key: ▲ Frequently used command
● Value retained for multiple problems
– Default reassigned

6

The TRANSFORM Paragraph

The TRANSFORM paragraph provides a powerful set of built-in functions, arithmetic expressions, and other options. It allows you to

- **Re-express data.** You can transform variables (e.g., take the log of body weight), create new variables (the difference between pre- and post-exercise pulse), compute summary functions across values on a record (the sum of seven scores), and change dates to a form suitable for sorting records or finding the time between two events.
- **Select cases.** For a particular analysis you can select a subgroup (such as males over 30 years of age).
- **Edit and recode values.** You may want to correct data values or change codes for the current run. The original data file is not changed unless you save the altered data.
- **Generate random numbers.** These may be used to obtain random samples for simulation or cross validation studies.

If you are familiar with TRANSFORM operations, you may need only the brief definitions given in Tables 6.1 through 6.6. Results from the TRANSFORM paragraph may be saved in a BMDP File or a raw data file (Chapter 8). Note that the TRANSFORM commands act on values **within** each case, one case at a time. It is possible to compute differences between data in successive cases (see Lagged Variables, section 6.6). For functions that operate on data across several cases, see the BMDP Data Manager.

Where to Find It

6.1 TRANSFORM capabilities

Before defining the operators, functions, and special instructions that are available, we present an overview of TRANSFORM capabilities. These capabilities are described in detail in section 6.3. Figure 6.1 shows six typical statements in a TRANSFORM paragraph:

Figure 6.1
Typical statements in a TRANSFORM paragraph

```
/ TRANSFORM
    score   = 2 * verbal + math.
    log_wt  = LOG(weight).
    average = MEAN(score_1, score_2, score_3, score_4).
    'sample_CDF' = GCDF(10).
    USE = sex EQ 1 AND age GT 20.
    IF(age GT 65) THEN(group = 2).
```

In order of appearance in Figure 6.1, the six TRANSFORM statements illustrate how you can

- Use **arithmetic operators** (see Table 6.1) to derive new variables:

```
score = 2 * verbal + math.
```

- Use **functions** to transform data. For example, taking the logarithm of WEIGHT could reduce its skewness and makes the data more symmetric for statistical procedures. Functions are defined in Table 6.2.

```
log_wt = LOG(weight).
```

- Use **summary functions** to derive summary variables:

```
average = MEAN(score_1, score_2, score_3, score_4).
```

  Note that the multiple argument function MEAN computes results across variables **within** records, adding a new variable (in this case, AVERAGE) to each record. Multiple argument functions are defined in Table 6.4.

- Use **cumulative distribution functions** (Table 6.5) to obtain probability values:

```
alpha = 1 - GCDF(score).
```

  Inverse cumulative distribution functions are also available:

```
score = GCDI(1 - alpha)
```

- Use **USE** with **logical operators** for case selection (Table 6.6). For example, to restrict an analysis to males over the age of 20, use the logical operators EQ (equal) and GT (greater than):

```
USE = sex EQ 1 AND age GT 20.
```

  USE is an internal variable that controls whether a case is used in an analysis (see section 6.4).

- Use **IF(condition) THEN(action)** statements to edit data for the current run or to deal with special problems (see section 6.3). For example, suppose a set of data has INCOME values incorrectly coded as 400 instead of 4. You can state:

```
IF (income EQ 400) THEN (income = 4).
```

Other special words and commands aid in dealing with **dates, character (non-numeric) data,** and **missing values.** See sections 6.3 and 6.5. **Random numbers** are often used in cross validating, selecting random subsamples, and generating data. See section 6.6.

6.2 Transformation facts and rules

When you use a TRANSFORM statement, the variable on the left side of the equal sign is assigned the value specified or computed on the right side. In the first example in Figure 6.1, BMDP computes a new variable, SCORE, for each case in the file, using two arithmetic operators: multiplication (*) and addition (+):

```
score = 2 * verbal + math.
```

Imagine that a column named "score" is now added to the right of your data file. If you create more variables, they are added as additional columns in the order specified. You can analyze and print new variables while running the program, but your original data file is not changed unless you use a SAVE paragraph to save a new version of it (see Chapter 8).

You can replace existing variables or create new ones. For example, you can replace the values of a variable with the logarithm of that variable (TIME = LOG(TIME)), or you can retain the old variable and store the log values as a new variable (LOG_TIME = LOG(TIME)). A list of other rules follows.

- Variables are referred to by **name** or the form X(*j*), where *j* is the input order. This notation allows the TRANSFORM paragraph to distinguish between a variable and a constant: LOG(2) is not the same as LOG(X(2)).
- Transformations are performed for one record at a time in the order that the statements are listed in the TRANSFORM paragraph. The order is sometimes important; if you need to edit data, you should do the editing **before** performing calculations, or the calculations will be performed on the unedited data.
- When a new variable is created (e.g., LOGVERBL), it is automatically added to the data file unless you specify ADD = # and omit its NAME in the VARIABLE paragraph. See Chapter 4 and section 6.6 on Temporary Variables. New variables that are automatically added must first be referred to by their names before the X(*j*) form can be used.

- Flags set by MIN, MAX, and MISS in the VARIABLE paragraph are recognized in the computations. See Section 6.5.
- You can use any input variable in TRANSFORM operations, even if it is not in the VARIABLE USE list. (However, only variables in the USE list are analyzed.)
- The FOR % notation (see Chapter 11) is available in the TRANSFORM paragraph.
- If in one computer run you want to apply the same transformations to the data in subsequent problems, specify only /TRANSFORM without repeating the entire previous TRANSFORM paragraph. However, if any instructions are specified in the later TRANSFORM paragraph, only those new instructions are performed.

6.3 Functions and operators

In this section we describe the functions and operators used in assembling TRANSFORM paragraph statements. We discuss each component separately, but you often need to combine several components in one statement. The TRANSFORM operators are listed in tables in this section; once you understand how they work, you can use the tables for quick reference.

Arithmetic operators

TRANSFORM statements can contain expressions involving addition, subtraction, multiplication, division, exponentiation, and the modulus, or remainder function (see Table 6.1).

Examples:

```
score = 2 * verbal + math.
POUNDS = OUNCES/16.
E = M*C**2.
```

Table 6.1
Arithmetic operators

Operator[1]		Example
$a + b$	addition	`X(4) = X(5) + X(7).`
$a - b$	subtraction	`WEIGHT = WEIGHT - X(3).`
$a * b$	multiplication	`DISTANCE = RATE * TIME.`
a / b	division $[b \neq 0]$	`X(1) = 1/TIME.`
$a{**}b$	exponentiation a^b $[a \geq 0$ or $b =$ integer$]$	`X(1) = HEIGHT**3.`
a MOD b	The remainder of a/b	`month = age MOD 12.`
$y = a$	Assignment. A constant or the value of another variable can be placed in y.	`X(7) = 10.` `X(7) = X(4).`

[1]If a or b is missing, or is not as indicated in brackets [ ], the result is missing.

Arithmetic operators are executed in standard FORTRAN order: exponentiation first, then multiplication and division, and finally addition and subtraction. Operators of the same type are executed left to right: e.g., A/B*C = (A/B)*C. You may use parentheses to make the formula easier to read or to make sure that operations are carried out in the proper order. If the value of any variable used in an arithmetic function is missing, the result is also missing.

Functions

Functions available in the TRANSFORM paragraph include square root, common logarithm, natural logarithm, exponential, absolute value, and trigonometric functions. The CHAR function for recoding nonnumeric data is discussed at the end of this section. See Table 6.2 for a complete list. Here we reexpress the value of body weight using the LOG function:

```
log_wt = LOG(weight).
```

When you apply the same function to many variables you may use the FOR % notation as a shortcut (see Chapter 11 for more information):

```
FOR k = 5 to 20.  % X(k) = LOG(X(k)). %
```

Complex calculations may be specified in one statement. For example, the following statement is an equation for a response curve over time. The specified transformation uses the value of time (t) on each input record to compute a predicted value "y hat." To obtain

$$\hat{y} = 11.4e^{-.23t} + 7.4e^{-.003t}$$

specify

```
y_hat = 11.4*EXP(-.23*time) + 7.4*EXP(-.003*time).
```

You may prefer to assemble complex equations in stages by using temporary variables (a and b below). For example:

```
a = EXP(-.23*time).
b = EXP(-.003*time).
y_hat = 11.4*a + 7.4*b.
```

For more discussion of temporary variables, see section 6.6.

Table 6.2
Functions

Function[1]		Example
LN(a)	Base e log of a [$a > 0$]	`HEIGHT = LN(HEIGHT).`
LOG(a)	Base 10 log of a [$a > 0$]	`WEIGHT = LOG(WEIGHT).`
SQRT(a)	Square root of a [$a \geq 0$]	`INCOME = SQRT(INCOME).`
EXP(a)	e^a exponential [$\|a\| \leq 60$]	`X(1) = EXP(X(2)).`
ABS(a)	$\|a\|$ absolute value of a	`CHANGE = ABS(X(7)).`
SIN(a)	Trigonometric sine of a in radians	`X(1) = SIN(X(1)).`
COS(a)	Cosine of a in radians	`X(2) = COS(X(2)).`
TAN(a)	Tangent of a in radians	`X(4) = TAN(X(4)).`
ASIN(a)	Arcsine in radians [$\|a\| \leq 1$]	`X(6) = ASIN(X(6)).`
ACOS(a)	Arccosine in radians [$\|a\| \leq 1$]	`X(5) = ACOS(X(5)).`
ATAN(a)	Arctangent in radians	`X(3) = ATAN(X(3)).`
INT(a)	Integer part of a	`AGE = INT(AGE).`
SIGN(a)	-1 if $a < 0$. 0 if $a = 0$ 1 if $a > 0$	`CHANGE = SIGN(X(1)).`
CHAR(c)	One to four characters (c) are available for comparison (EQ or NE) with A-format data.	`USE = SEX EQ CHAR(M).`

[1]If a is missing or is not as indicated in brackets [ ], the result is missing.

The CHAR function: recoding nonnumeric data. Use the CHAR function to recode nonnumeric data (data read in A-format). See Chapter 2 for information about reading character data as input. For example, if the variable SEX has been coded using the letter M to indicate males, you can change the M to the number 1 by creating a new variable "newsex" and then stating

```
IF(sex EQ CHAR(M)) THEN newsex = 1.
```

(See also the ALPHA command in the GROUP paragraph.) If you recode a variable from nonnumeric to numeric, you should create a new variable name making sure to account for all possibilities, including missing values. You may also use the CHAR function for case selection on one or more alphabetic variables (e.g., USE = NEWSEX EQ CHAR(M).).

The CHAR function can use a string of up to four characters. Note that leading **blanks** are counted as characters: CHAR(M) is not the same as CHAR(M) or CHAR(M). The CHAR function takes the characters between the two parentheses and inserts enough blanks on the right side to complete a word on your computer. For computers with four-character words

```
CHAR(M)   becomes Mbbb
CHAR( )   becomes bbbb
CHAR(FEMALE)  becomes FEMA
```

where b means blank. Therefore, in the example above we could not use CHAR(M) to refer to MALE because Mbbb is not the same as MALE.

Date functions

BMDP provides functions to deal with dates for sorting and computing time differences (see Table 6.3). Dates may be stored as a single six-digit variable (mmddyy); as three two-digit variables (mm, dd, and yy); as one variable with five digits (Julian form, where 87046 is the 46th day in 1987); or as the number of days since January 1, 1960. The date functions allow you to convert from one dating system to another. For example, to convert dates stored in the three variables, MONTH, DAY, and YEAR, to NUMDAYS, the number of days since 1/1/60, state

```
numdays = DAYS(month,day,year).
```

Note that they may be in any order in the data file, so the European notation day/month/year can be used. If the date written in month, day, year order in six digits is read as one variable, named DATE_1, then

```
numdays = DAYS(date_1).
```

Suppose each record in your data contains two dates (DATE_1 and DATE_2) entered in month/day/year format:

```
041986     092587
```

To compute the elapsed time between the two dates, convert them to # days form:

```
diffrnce = DAYS(date_1) - DAYS(date_2).
```

The functions MDY, MM, DD, and YY allow conversion from number of days back to month, day, year format. To sort cases by date (e.g., using program 1D), you must convert them to DAYS or JULIAN form.

Table 6.3
Date functions

Function[1]	Result
DAYS (mm,dd,yy)	# of days since Jan. 1, 1960.
DAYS (mmddyy)	# of days since Jan 1, 1960.
MDY (# of days)	mmddyy—6 digit date as one variable.
MM (# of days)	mm—2 digits for the month within the year.
DD (# of days)	dd—2 digits for the day within the month.
YY (# of days)	yy—2 digits for the year.
JULN (# of days)	Julian date form in 5 digits.
DAYJ (Julian)	# of days since Jan. 1, 1960.

If any value is missing, the result is missing. BMDP checks values within the DAYJ and DAYS functions against reasonable limits (e.g., month must be between 1 and 12).

[1]The values of the variables used as date function arguments have specific values. The following notation is used:

- mmddyy – the date is recorded as one variable with six digits (Sept. 17, 1981 is 091781).
- mm,dd,yy – three variables, two digits each, record the date (e.g., the variable mm has the value 09, dd is 17, yy is 81).
- # of days – the number of days since Jan. 1, 1960. Dates before 1/1/60 are converted to negative days.
- Julian – the date is recorded as one variable with five digits (Feb. 15, 1981 is 81046: the 46th day of 1981).

The variables may have any name (e.g., DATE1, X(5), NUMDAYS). We use "mm," etc., as a convenience for these definitions.

Summary functions

Summary (or multiple argument) functions operate upon the values of variables within a case. These functions allow you to compute for each case descriptive statistics such as mean, standard deviation, area under the curve, and so on. BMDP provides more than two dozen functions that operate across variables within a case (see Table 6.4). The elements of the list are separated by commas and can be variable names or numbers (X(j)), explicit values, functions (such as SQRT(income)), or algebraic expressions.

Table 6.4
Summary functions (multiple argument functions)

Function[1]		Result[2]
N	$(a_1,a_2,...,a_n)$	Number of usable values; i.e., the number of values in the list $a_1,a_2,\ldots,a_n$ that are not missing and are within usable range. (11)
NMIS	$(a_1,a_2,...,a_n)$	Number of missing values. (11)
MIN	$(a_1,a_2,...,a_n)$	Smallest usable value. (1)
MAX	$(a_1,a_2,...,a_n)$	Largest usable value. (1)
SUM	$(a_1,a_2,...,a_n)$	Sum of usable values. (1)
SUMC	$(a_1,a_2,...,a_n)$	Sum of values; if any value is missing, the result is missing. (5)
MEAN	$(a_1,a_2,...,a_n)$	Mean of usable values. (1)
MED	$(a_1,a_2,...,a_n)$	Median of usable values: central value or average of two central values. (1)
SD	$(a_1,a_2,...,a_n)$	Standard deviation of usable values. (2)
SEM	$(a_1,a_2,...,a_n)$	Standard error of the mean of usable values. (2)
T	$(a_1,a_2,...,a_n)$	t-value of mean: i.e., mean/(sd/sqrt(n)). (2)
TRIM	$(i,a_1,a_2,...,a_n)$	Trimmed mean; mean after trimming the i largest and i smallest values, where i is an integer. (3)
TT	$(i,a_1,a_2,...,a_n)$	Trimmed t; t-value of the i level trimmed mean. This is a t-value with $(n - 1 - 2*i)$df. (4)
IQR	$(a_1,a_2,...,a_n)$	Interquartile range of usable values:(Q3 – Q1)/2. (2)
RHO	$(y_1,y_2,...,y_n)$	Correlation of $(1,2,\ldots,n)$ and $(y_1,y_2,\ldots,y_n)$. (2)
B	$(x_1,y_1,...,x_n,y_n)$	Coefficient b of the regression line $y = a + bx$. (6)
A	$(x_1,y_1,...,x_n,y_n)$	Constant a of the regression line $y = a + bx$. (6)
R	$(x_1,y_1,...,x_n,y_n)$	Correlation coefficient r (x,y). (6)
TRND	$(y_1,y_2,...,y_n)$	Trend of $(y_1,y_2,\ldots,y_n)$ over $(1,2,\ldots,n)$. (2)
TCON	$(y_1,y_2,...,y_n)$	Intercept of the trend line; i.e., constant of regression line through $(1,y_1),...,(n,y_n)$. (2)
AREA	$(y_1,y_2,...,y_n)$	Area under y: $(y_1 + 2*y_2 +... + 2*y_{n-1} + y_n)/2$. (7)
TRAP	$(x_1,y_1,...,x_n,y_n)$	Area under y via the trapezoidal rule: $\text{sum}((x_i - x_{i-1})*(y_{i-1} + y_i)/2)$. (8)
LIND	$(a_1,a_2,...,a_n)$	Index of last usable value $(1,2,\ldots,\text{or } n)$. (1)
LVAL	$(a_1,a_2,...,a_n)$	Last usable value $(a_1,a_2,\ldots, \text{or } a_n)$. (1)
INDX	$(a,b_1,b_2,...,b_n)$	Index of first b that is equal to a. (11)
REC	$(a,b_1,c_1,...,b_n,c_n)$	A recode function, c_j is stored as the result when a is equal to b_j. (10)
FVAL	$(a_1,a_2, ... a_n)$	First usable value $(a_1,a_2, \ldots a_n)$. (1)
ORD	$(i, a_1, a_2, ... a_n)$	ith smallest value. Any values that are missing or out of range are treated as usable values equal to XMIS, TOOBIG, or TOOSMALL.
ZSCR	$(a_1,a_2, ... a_n)$	Replaces each usable value by its z-score , using the mean and standard deviation of the usable values. (12)
SORT	$(a_1,a_2, ... a_n)$	Permutes the values so that $(a_1 < a_2 < \ldots a_n)$. Missing values replaced by XMIS and is placed at the end.
RANK	$(a_1,a_2, ... a_n)$	Replaces the smallest usable value by 1, the next 2, … etc. Averages are used in ties (e.g. RANK(3, 3, 2) gives (2.5, 2.5, 1). Missing values replaced by XMIS.

[1] a_i,b_i,c_i,x_i,y_i are arguments that may be variable names, explicit values, functions, or algebraic expressions. Note that the notation 3 TO 10 or Q1 TO Q20 cannot be used. Use the FOR% notation (FOR I = 3 TO 10. % X(I), %).

[2] Numbers within parentheses refer to missing values key below. The result is a missing value code if

(1) there are no usable values
(2) there are fewer than two usable values
(3) there are fewer than $(2*i + 1)$ usable values
(4) there are fewer than $(2*i + 2)$ usable values
(5) any value is missing
(6) the variance of x over usable (x,y) pairs is 0.
(7) y_1 or y_n is missing
(8) x_1, y_1, x_n, or y_n is missing
(9) $x_1 = x_2$
(10) no b is equal to a
(11) never yields a missing value code
(12) missing or out of range values not included

For example, here we use the MEAN function to compute the average of four scores stored on each record:

```
average = MEAN(score_1, score_2, score_3, score_4).
```

Note that you may use the implied list feature (e.g., score_1 TO score_4) directly. However, you may use the FOR % notation:

```
     average = MEAN(SCORE_1 TO SCORE_4).
or   average = MEAN(FOR k = 11 TO 14. % X(k), %).
```

where the scores are stored as variables 11 through 14. See Chapter 11 for more about FOR %.

You can use the TRANSFORM paragraph to find the sum of several variables for one case, but **not** to find the sum of one variable across several cases. For that, use the BMDP Data Manager (DM).

Here are some examples of what the multiple argument functions could compute from four original variables named Y1 through Y4. The new values are available for saving, printing, or use in the current analysis. For example, after creating the new variable SLOPE you could perform a two-sample *t* test to see whether the slopes differ between a treatment group and a control group.

Figure 6.2
Multiple argument functions

```
/ TRANSFORM   n_ok     = N(Y1, Y2, Y3, Y4).
              total    = SUM(Y1, Y2, Y3, Y4).
              smallest = MIN(Y1, Y2, Y3, Y4).
              slope    = TRND(Y1, Y2, Y3, Y4).
              area     = AREA(Y1, Y2, Y3, Y4).
```

Example:

id	Y1	Y2	Y3	Y4	n_ok	total	smallest	slope	area
1	2	2	*	3	3	7	2	.357	7
2	2	4	6	2	4	14	2	.200	12
3	1	1	2	1	4	5	1	.100	4

N — For each record, N is the number of **usable** values, that is, how many of the values are neither missing nor outside acceptable range limits (see Chapter 4).

SUM — produces the total of the (N) usable values.

MIN — produces the smallest value of the variables listed.

The TRANSFORM paragraph also offers more complex multiple argument functions, such as slope and area calculations. The slope function for Y, assuming equally spaced x values $1, 2, \ldots, n$, is TRND. For the four-variable example above, TRND computes the slope b of the straight line $y = a + bx$, where the four values are the y's and their order on the record, the respective x's. Similarly, the AREA function calculates the area under equally spaced points. Use B and TRAP instead of TRND and AREA if the x's are not equally spaced.

RECODE. The RECODE function recodes an existing variable or creates a new variable with the recoded values. For example, if x takes on the values 17, 21, and 29 which must be recoded to 1, 2, and 3 respectively, we can specify

```
x = REC(x, 17,1, 21,2, 29,3).
```

Note: *If x takes on additional values besides 17, 21, and 29, and these are not included in the RECODE argument, the result sets x to missing. The following example can be used if only a few values of x need to be recoded:*

```
y = REC(x, 17,1, 21,2, 29,3). IF(y EQ XMIS) THEN y = x.
```

In this example, y equals x unless x = 17, 21, or 29.

Cumulative distribution functions

The TRANSFORM paragraph allows you to specify cumulative distribution functions (CDF). These functions allow you to create new variables. You choose from four functions and their inverses. The distributions available are: Gauss, *t*, chi-square, and *F*. We list and describe the functions in Table 6.5.

As an illustration, say you wanted to obtain the *t* cumulative distribution function value at $t = 2.05$ and degrees of freedom = 26. To do so, you would specify

```
NEWVAR = TCDF(2.05,26).
```

The program reports a value between 0 and 1 as the area to the left of the value of *t*. In this example, it reports the value .98 as a new variable (NEWVAR) for all cases.

To use the inverse CDF, you need the area and the degree or degrees of freedom. Say you wanted to obtain the *t* value where the CDF is .90 and degrees of freedom = 9. To obtain that value, you would specify

```
NEWVAR = TCDI(.90,9).
```

The program would report the value 1.383 as a new variable for each case.

Note that these functions are cumulative (one-sided).

Table 6.5 Cumulative distribution functions

Function[1]	Result[2]
GCDF (s)	Gaussian cumulative distribution function
TCDF (s, d)	*t* cumulative distribution function where *s* is the *t* value and *d* is the degrees of freedom
CCDF (s, d)	Chi-square cumulative distribution function where *s* is the χ^2 value and *d* is the degrees of freedom
FCDF (s, d_1, d_2)	*F* cumulative distribution function where *s* is the *F* value, d_1 is the numerator degrees of freedom, and d_2 is the denominator degrees of freedom
GCDI (a)	Inverse Gaussian cumulative distribution function
TCDI (a, d)	Inverse *t* cumulative distribution function where *a* is the *t* CDF and *d* is degrees of freedom
CCDI (a, d)	Inverse chi-square cumulative distribution function where *a* is the χ^2 CDF and *d* is the degrees of freedom
FCDI (a, d_1, d_2)	Inverse *F* cumulative distribution function where *a* is the *F* CDF, d_1 is the numerator degrees of freedom, and d_2 is the denominator degrees of freedom

[1] *s* and d_1 are arguments that may be variable names, explicit values, functions, or algebraic expressions. *a* is the area under a given distribution to the left of *s*. *a* must fall between 0 and 1. *s* for GCDF and TCDF is the number of standard deviations away from the population mean. This value can be positive, negative or zero. *s* for CCDF and FCDF must be greater than or equal to 0.

[2] Missing value codes are generated if any value is missing, d_1 or d_2 is less than or equal to 0, or *a* is less than or equal to 0 or greater than or equal to 1.

Logical operators

Logical operators (see Table 6.6) are used to compare one or two variables, constants, or functions; the operators yield results that are either true (1), false (0), or missing (XMIS). Logical operators are frequently employed in USE statements for case selection (section 6.4) and in IF–THEN conditional statements. Eight logical operators are available, most with equivalent symbols that you can use interchangeably with the letters:

AND	GT >	LT <	EQ ==
OR	GE >=	LE <=	NE <>
NOT			

- When two (or more) statements are connected by AND, **all** statements must be true for the result to be true.
- With OR, if **either** statement is true, the result is true.

Examples:

`USE = WT87 < WT86.`	USE cases only when WT87 is less than WT86
`USE = SMOKE EQ 1.`	USE cases when SMOKE equals 1
`IF(income EQ 400) THEN(income = 4).`	Change mistaken value of INCOME

Any logical operator (LE, LT, GE, GT, NE, EQ, AND, OR) must be preceded and followed by a blank (e.g., AGE LT 6; not AGELT6). Missing values affect the outcome of logical operators (see Table 6.6).

Table 6.6
Logical operators

Operator	Result[1] 1 if	0 if	Example
a LE b	$a \le b$	$a > b$	`USE = WT85 LE WT86.`
a LT b	$a < b$	$a \ge b$	`'dummy' = X(1) LT X(3).`
a GE b	$a \ge b$	$a < b$	`USE = WEIGHT GE 50.`
a GT b	$a > b$	$a \le b$	`USE = WEIGHT GT 50.`
a NE b	$a \ne b$	$a = b$	`USE = X(2) NE X(3).`
a EQ b	$a = b$	$a \ne b$	`X(1) = X(2) EQ X(3).`
a AND b	$a \ne 0$ and $b \ne 0$	else	`USE = AGE GT 20 AND AGE LE 40.`
a OR b	$a \ne 0$ or $b \ne 0$	else	`USE = CLINIC EQ 1 OR CLINIC EQ 5.`
NOT a	$a = 0$	$a = 1$	`USE = NOT (AGE EQ 99).`

Missing Values[2]

If either of the values before or after a logical operator is XMIS or missing, the result is always missing except in the following cases (A is a nonmissing value):

false AND missing	yields false (0)
true OR missing	yields true (1)
missing EQ XMIS	yields true
missing NE XMIS	yields false
A EQ XMIS	yields false
A NE XMIS	yields true

[1]The result of a logical operation is true (1), false (0), or missing (XMIS).
[2]Out-of-range values are treated as missing.

IF(condition) THEN(action) statements

IF–THEN statements perform an action only when a condition has been met. The command works in the following manner: IF (condition) THEN (action(s)). Here we use an IF–THEN statement with the logical operator EQ to change an incorrect data value:

```
IF(income EQ 400) THEN(income = 4).
```

Whenever income is equal to 400 in the input file, it will be changed to 4 for the current analysis (the changed value can also be saved). We could also use an IF–THEN statement with the logical operator EQ and the function CHAR to change a data value from alphabetic to numeric:

```
IF(sex EQ CHAR(M)) THEN (sex = 1).
```

Logical operators are essential to the IF condition (it may also contain arithmetic operators and functions). The THEN actions can include one or more arithmetic assignments, logical operators, or other IF–THEN conditional statements. When an IF clause contains values that are missing or out of range, the program does not execute the actions.

When you specify two comparisons using the logical operators AND and OR, variables must be stated for both comparisons. Thus, state

```
IF(A EQ B OR A EQ C)...   not   IF(A EQ B OR C)...
```

If the THEN clause contains several statements, the entire list should be enclosed in parentheses.

Examples:

```
IF (AGE LT 21) THEN USE = 0.
IF(A/B+C GT 15 AND AGE GT 50) THEN (GROUP = 2. INDEX = 4.).
IF (Y EQ 0.0) THEN Y = .01.
LOG_Y = LOG(Y).
AGESET = 2.
IF(AGE GE 45) THEN AGESET = 1.
```

In the last example, we create a new variable, AGESET, which we want to set to 1 under certain conditions. We first *initialize* the value of AGESET (AGESET = 2) so that we know its value when the IF conditions are **not** met. Values of new variables are set to XMIS until defined.

6.4 Case selection

In this section we discuss several ways to select cases for analysis. The USE instruction allows you to select cases that fulfill specific conditions, such as SEX EQ MALE or KASE GT 100. The OMIT and DELETE commands allow you to exclude cases by listing their case numbers. Note that you can also restrict the number of cases read for analysis by specifying CASES = # (e.g., CASES = 100.) in the INPUT paragraph.

The USE instruction

The USE instruction in the TRANSFORM paragraph allows you to select **cases** to use in the analysis (the USE command in the VARIABLE paragraph selects **variables**). Imagine that for each case BMDP maintains a USE marker (internally adds a column to your data file named USE). Initially the value of USE is set to 1 for every case. When BMDP processes a file, it checks this value. If it is 1, the case is used. If you change the value of USE to 0, the case is omitted from analysis. If the value of USE is set to –1, the case is omitted from analysis **and** from any file written by the SAVE paragraph.

Value of USE	Case Used in Analysis	Case Stored in SAVE File	
1 (positive)	yes	yes	
0 (zero)	no	yes	
−1 (negative)	no	no	
−100	no	no	(stop reading data)

We employ **logical operators** with USE to select cases. To analyze only those records for subjects over 20 years of age, we include the logical operator GT (greater than) in a USE specification:

```
USE = AGE GT 20.
```

If for a particular case the condition on the right side of the equal sign is **true** (say the subject is 24 years old), a 1 remains stored in USE. If the result is **false** (say the subject is 16), a 0 is stored in USE. If USE is equal to the missing value code after all transformations have been performed, USE is set equal to zero. That is, if USE = AGE GT 20. and AGE is missing, USE is set to zero.

Value of Logical Statement	Value of USE
true	1
false	0
missing	0

Note: USE is reset each time it is specified. If you state USE = SEX EQ 1 followed by USE = AGE LT 25, only the second statement is effective. You must instead state USE = SEX EQ 1 AND AGE LT 25. You can string together as many AND and OR clauses as you like.

Deleting cases with unacceptable values. In a program like 7D, all available cases are used in the analysis of each variable. To restrict analysis of all variables to cases where PULSE_2 is neither missing nor greater than 200, you could state

```
USE = PULSE_2 LE 200.
```

See section 6.6 for selecting a **random subset of cases** from your file.

Other supports for case selection: KASE, OMIT, DELETE

KASE. All BMDP programs store an input sequence number for each case in a variable called KASE. You can use KASE in the TRANSFORM paragraph to refer to specific cases; e.g., for the fifth case in your data file, the value stored in KASE is 5. If you specify ORDER = KASE, the variable ORDER will contain the input case number and will be available for printing. You can select cases by using the USE and KASE variables within logical operations and IF THEN statements:

```
USE = KASE GT 100.
IF(KASE EQ 23) THEN PULSE = 165.
```

OMIT and DELETE. You can use the OMIT and DELETE commands to exclude cases by their KASE numbers.

```
OMIT = 45, 80.
OMIT = 1 TO 19 BY 2.
DELETE = 1 TO 20.
```

OMIT sets USE = 0 for the specified cases. The 45th and 80th cases are excluded from the analysis. DELETE sets USE = – 1. Cases 1 through 20 are excluded from the analysis **and** from inclusion in any data files created in the SAVE paragraph. **Note:** OMIT and DELETE cannot be used in an IF–THEN statement.

6.5 Missing data, range limits, and display commands

If the value of a variable used in an arithmetic operation or function is missing or out of range (see Chapter 4), or if the required arithmetic cannot be performed (e.g., SQRT(–1)), then the program sets the result to an internal missing value indicator (XMIS). Exceptions to this rule are noted in Tables 6.1, 6.2, and 6.3. In this section, we discuss how to use the internal indicators XMIS, TOOSMALL, and TOOBIG to recode values. In addition, we discuss using AFTERTRANSFORM to direct the program to check for missing data **after** executing TRANSFORM operations. The commands REPL and FILL allow you to replace missing values using linear interpolation from surrounding values.

Using XMIS, TOOBIG, and TOOSMALL. XMIS is the name assigned to the code used internally by the program to indicate missing values. TOOBIG and TOOSMALL are the names assigned to values above and below limits (as specified in the VARIABLE MIN and MAX statements). You can use these names to check for missing and out-of-range values, and to recode values to missing.

For each variable, BMDP allows you to specify one missing value code in the VARIABLE paragraph. For example, if the variable STATUS has two additional codes for missing values (say 7 and 8), both can be recoded to missing by using this statement in the TRANSFORM paragraph:

```
IF (STATUS EQ 7 OR STATUS EQ 8) THEN STATUS = XMIS.
```

To set USE = – 1 for all cases with out-of-range values for AGE, state

```
IF (AGE EQ TOOSMALL OR AGE EQ TOOBIG) THEN USE = -1.
```

Flagging unacceptable values: before or after transformations. When MIN, MAX, and/or MISSING are specified in the VARIABLE paragraph, data values that are missing or out of range are internally flagged as XMIS, TOOSMALL, and TOOBIG. If transformations are performed and you state AFTERTRANSFORM in the VARIABLE paragraph (see Chapter 4), the range limits will then apply to the transformed variables. This does not affect the automatic recoding of blanks to missing values. If you do not want any checking either before or after transformations, state NO CHECK.

In summary, as each case is read, its values are checked in the following order:

1. In FIXED format, blanks are replaced by the internal code for missing values (unless BLANKS = ZERO is specified).
2. If BEFORE (the default option for data checking) is in effect, appropriate internal codes replace the user-specified missing value codes (MISS) and values outside the MIN and MAX limits.
3. FORTRAN transformations (see Section 6.7) are executed. Your FORTRAN statements must include any desired checks for extreme or missing values.
4. Transformations in the TRANSFORM paragraph are executed, using information from the MISS, MIN, and MAX statements.
5. If you specify AFTER, variables for which MISS, MAX, or MIN are stated (including new transformed variables) are now checked.

Interpolating missing values: REPL and FILL. The REPL and FILL commands replace missing values using linear interpolation from surrounding usable values. The general form of REPL is REPL(y_1, y_2,...,y_n), where the arguments within parentheses may be variable names or X(i). For example, if your data include the variables SCORE1, SCORE2, SCORE3, and SCORE4, you can replace missing score values of SCORE2 or SCORE 3 by using the command

```
REPL(score1, score2, score3, score4).
```

When y_1 or y_n is missing, leading or trailing missing values remain missing.

FILL works similarly, but uses variable pairs; its general form is FILL(x_1,y_1,...,x_n,y_n). Missing y values are replaced using linear interpolation from surrounding usable value pairs (x,y). The y arguments are variable names or X(i), that are spaced as x arguments indicate; the x arguments may be variable names, explicit values, functions, or algebraic expressions. When x_1, y_1, x_n, or y_n is missing, then leading or trailing missing y values remain missing.

Checking results and displaying unusual values: TEXT, SHOW. The display commands TEXT and SHOW are useful for identifying unusual cases and checking the results of complex transformations or case selection procedures. Display command messages are printed when TRANSFORM statements are executed, not in final data listings.

```
IF (SMOKE EQ XMIS) THEN TEXT('smoking status unknown').
```

If the value of SMOKE is missing, the program prints the message "smoking status unknown" along with the case number and label.

```
IF (PULSE_2 GT 160) THEN SHOW(PULSE_1, PULSE_2).
```

SHOW prints the case number and label, together with any or all requested variables or functions of the variables. For example, SHOW(USE) prints a list of case numbers and labels with each case's USE value.

Using ZSCR, RANK, and SORT functions. Three **new** commands are **ZSCR, RANK, and SORT.** The **ZSCR** command replaces each usable value with z-scores computed using the mean and standard deviation. Values out of range remain unchanged. The **SORT** command permutes values so that $a_1 < a_2 < \ldots < a_n$. The **RANK** command causes data ranking by making the smallest usable value 1, the next 2, etc. If there are ties, averages are taken, e.g. RANK(3,3,2) gives RANK(2.5,2.5,1). For SORT and RANK, values out of range are replaced by missing.

6.6 Special features and examples

Random number generation

The random number generators, RNDU and RNDG, generate one random number for each case. Their operation is summarized below:

- To generate a uniform (0,1) random number   Y = RNDU(*i*).
- To generate a normal (0,1) random number   Y = RNDG(*i*).

In both functions the generator is started by using a large integer (*i*) between 1 and 30000 as a seed number (e.g., RNDU(27327)). RNDU generates uniform random numbers on the interval from zero to one. RNDG generates random numbers from a normal distribution with mean zero and unit variance.

The functions RNDU and RNDG produce pseudo-random numbers. If you specify these functions more than once (i.e., add two or more columns of random numbers to your data file), you should use a different seed each time. The seed must be a number, not a variable. The random number generators are described in Appendix B.3.

The random number generators have several uses:

1. **Cross-validation and random subsamples.** For an example of random numbers in cross-validation studies, see program 7M. When debugging your instructions, you may want to use random numbers to select a subset of data from a big file. Suppose you want to randomly select about 10% of the cases for your analysis. You can use the uniform generator to specify:

   ```
   USE = RNDU(15681) LT .10.
   ```

   The program generates a random number between zero and 1 for each case; approximately 10% of the cases will be assigned values less than .10. However, with this method there is no guarantee that exactly 10% of the cases will be selected—we expect a 10% sample, but any one seed may give more or less than expected. (Randomness strikes again.) This is not usually a problem, but it could be of concern if exact numbers were needed to preserve balance in an ANOVA, for example.

   If we know the file has N = 1000 cases and wish to select exactly S = 100 cases at random, the following can be stated in the TRANSFORM paragraph:

   ```
   IF (KASE EQ 1) THEN (N=1000.  S=100.). # Set initial conditions.
   USE = -1.                              # Reject case unless selected.
   IF (RNDU(1123)*N  LT S)                # If selected, use case and
       THEN (USE = 1.  S = S - 1.).       # decrement selection counter.
   N = N - 1.                             # Decrement case counter.
   ```

   The extra code gives us exactly 100 cases out of the 1000 in the file. Note that you must state RETAIN in the VARIABLE paragraph.

2. **Generating data.** Sometimes it is desirable to get a quick idea of the sensitivity of a statistic to departures from the usual underlying assumptions. You can use the TRANSFORM paragraph to generate data to add to your original data file or to analyze by itself. If no data are read as input, you should specify VARIABLES = 0 in the INPUT paragraph. If the number of input variables is zero (all variables are generated by transformation), you must also specify the number of cases and omit the FORMAT command. The two random number generators are useful if you want to see the effect of violating an assumption of a test. For example, suppose you want to generate a normally distributed variable with mean 50 and standard deviation 3. You can do it this way:

   ```
   NEWVAR = 3 * RNDG(25491) + 50.
   ```

Temporary variables

When you want to perform complex transformations, you can create "temporary" variables that are not analyzed or added to your data file by using the ADD = # command in the VARIABLE paragraph and naming the variables you want to keep. For example, Figure 6.3 presents the instructions to convert measurements to metric units and create a SIZE variable from HEIGHT and WEIGHT.

In the TRANSFORM paragraph, we convert HEIGHT from inches to meters as variable A and convert WEIGHT from pounds to kilograms as variable B. SIZE is computed from A and B. In the VARIABLE paragraph we direct the program to ADD **one** variable to the two read in and name HEIGHT, WEIGHT, and SIZE. SIZE is the only new variable that the program will analyze; A and B are treated as temporary variables. For more information about ADD, see Chapter 4.

Figure 6.3
Temporary and permanent variables

```
/ INPUT       VARIABLES ARE 2.  FORMAT IS FREE.

/ VARIABLE    ADD = 1.   NAMES = height, weight, size.

/ TRANSFORM   A = height * 2.54/100.
              B = weight * .454.
              size = B/(A**3).
```

Lagged variables

Use the RETAIN command in the VARIABLE paragraph to generate lagged variables. Lagged variables are variables added through transformations that are assigned the value set for the previous case. For case *i*, a lagged variable has the value of the variable at case (i – 1), or lag 1.

For example, suppose you have an ordered set of data containing annual prices for a commodity; each case represents a different year. You want to create a new variable containing the difference between successive prices, such as between the PRICE in the current case and the PRICE in the preceding case. The following instructions obtain the difference by first creating the new lagged variable OLDPRICE:

```
/ VARIABLE    RETAIN.
/ TRANSFORM   DIFF = PRICE - OLDPRICE.
              OLDPRICE = PRICE.
```

TRANSFORM operations are performed in the order specified, so DIFF is missing for the first case. Then OLDPRICE is set equal to PRICE. For every case thereafter, DIFF uses the value for OLDPRICE from the preceding case. This is because DIFF is calculated **before** OLDPRICE is reset by the last command. To understand how this works, copy the setup above with a few dummy data cases, and run 1D with a PRINT DATA command.

In a regression analysis, we may want to estimate the weekly incidence of a disease as a linear function of the incidence during the previous two weeks. We could have as our model:

$$\text{Incidence (week } i) = a + b_1 \text{ Incidence (week } i\text{–1)} + b_2 \text{ Incidence (week } i\text{–2)} + e$$

Our data could be one weekly incidence per record. Here are the BMDP instructions for a linear regression program:

Figure 6.4
Lagged variables for linear regression

```
/ INPUT        VARIABLE IS 1.
               FORMAT IS FREE.

/ VARIABLE     NAME IS INCIDENC.
               RETAIN.
               USE = INCIDENC, LAG1, LAG2.

/ TRANSFORM    LAG2 = LAG1.  LAG1 = TEMP.
               TEMP = INCIDENC.

/ REGRESSION   DEPENDENT IS INCIDENC.

/ END
```

Note: The USE statement in VARIABLE paragraph discards the variable named TEMP.

6.7 FORTRAN transformations

Any transformation possible using the TRANSFORM paragraph can be performed using FORTRAN. FORTRAN transformations (previously known as the BIMEDT procedure) can also be used when it is difficult or impossible to achieve the same result with BMDP transformations. Your code can read other files, write to files, or call non-BMDP subroutines. Your statements are processed by the FORTRAN compiler on your system and temporarily become part of the BMDP program. This procedure is not available for PC users. Most users have found little need for FORTRAN subroutines. However, FORTRAN transformations are sometimes more economical than BMDP transformations when you have large quantities of data or transformations. Transformations specified in FORTRAN are processed one case at a time before any in the TRANSFORM paragraph. The BEFORE and AFTER statements in the VARIABLE paragraph apply to both the FORTRAN and the BMDP transformations. Two examples showing FORTRAN transformations follow.

Example 1. You can use FORTRAN to create a new variable from the ratio of two others. For example, if you know that the 3rd and 4th variables are always recorded, specify

```
X(10) = X(4)/X(3).
```

However, if either of the original variables is missing or out of range, you must first check whether the value is present by specifying

```
X(10) = XMIS.
IF(ABS(X(3)).LT.XMIS .AND. ABS(X(4)).LT.XMIS).
     X(10) = X(4)/X(3).
```

The rules of FORTRAN language apply.

Example 2. You can use FORTRAN statements to recode nonnumeric data. For example, let SEX be one of the variables, coded M, F. Then let MONTH be a second variable, coded 1, 2, 3, 4, 5, 6, 7, 8, 9, 0, A, B. If SEX is variable 5 and MONTH variable 7, they can be recoded by the following FORTRAN statements.

```
      DIMENSION AMONTH(12),ASEX(2)
      DATA AMONTH/'1','2','3','4','5','6',        | AMONTH are the valid codes
     *            '7','8','9','0','A','B'/        |   for MONTH (1,2,...,A,B)
      DATA ASEX/'M','F'/                          | ASEX are the valid SEX codes (M, F)
      X5=XMIS                                     |
      IF(X(5).EQ.ASEX(1)) X5=1.                   | Check for 'M' and 'F';
      IF(X(5).EQ.ASEX(2)) X5=2.                   |   replace with 1 and 2
      X(5)=X5                                     |
      X7=XMIS                                     |
      DO 10 I=1,12                                | Check for '1','2',...,'A', 'B';
        IF(X(7).NE.AMONTH(I)) GOTO 10             |   replace the characters '1' through
        X7=I                                      |   '9' with numbers 1 through 9, '0'
        GOTO 15                                   |   with 10, 'A' with 11 and 'B' with 12.
   10 CONTINUE
   15 X(7)=X7
```

The two variables, SEX and MONTH, must each be read in A1 format. You must also specify AFTER in the VARIABLE paragraph so the data in SEX and MONTH will not be checked for missing and out-of-range values until they are transformed to numeric values. To transform nonnumeric values without using FORTRAN, see the CHAR function in the TRANSFORM paragraph. To use nonnumeric values in grouping variables, see ALPHA in the GROUP paragraph.

For information about incorporating FORTRAN transformations and subroutines into your BMDP instructions on various systems, see Appendix E.

TRANSFORM Commands

Functions and Operators

Arithmetic Operators	See Table 6.1
Functions	See Table 6.2
Date Functions	See Table 6.3
Summary Functions	See Table 6.4
Cum. Distribution Functions	See Table 6.5
Logical Operators	See Table 6.6

Commands and key words

USE = # *or* logical expression.

```
IF(pulse_2 EQ 265)
THEN(USE = 0).
```

For each case, the value of USE is initially set to 1. Cases with USE = 0 are omitted from analyses. Cases with USE = –1 are omitted from analyses and from SAVE files. See Section 6.4.

KASE

```
USE = KASE GT 100.
```

All BMDP programs store an input sequence number for each case in a variable called KASE. You can use KASE to refer to specific cases. See Section 6.4.

OMIT = # list.

```
OMIT = 45, 80.
```

Use OMIT to exclude cases by their KASE numbers. OMIT sets USE = 0 for the specified cases.

DELETE = # list.

```
DELETE = 1 TO 20.
```

Use DELETE to exclude cases by their KASE numbers. DELETE sets USE = –1 for the specified cases.

XMIS

```
IF(STATUS EQ 7)
THEN STATUS = XMIS.
```

XMIS is the name assigned to the value used internally by the program to indicate missing values. On many systems, this number is 16^{31}.

TOOSMALL

```
IF(AGE EQ TOOSMALL)
THEN USE = -1.
```

TOOSMALL is the name assigned to the value used internally by the program to indicate values below range limits (as specified in the VARIABLE MIN command). The internal number is –2 * XMIS.

TOOBIG

```
IF(AGE EQ TOOBIG)
THEN USE = -1.
```

TOOBIG is the name assigned to the value used internally by the program to indicate values above range limits (as specified in the VARIABLE MAX command). The internal number is 2 * XMIS.

REPL($y_1,y_2,\ldots,y_n$).

```
REPL(SCORE1, SCORE2,
          SCORE3, SCORE4).
```

Replaces missing values using linear interpolation from surrounding usable values. Arguments within the parentheses may be variable names or X(*i*). **Example:** If one of the score values is missing, it is replaced using linear interpolation from the usable values.

FILL($x_1,y_1,\ldots,x_n,y_n$).

```
FILL(Janvol, Jansales, Febvol,
   Febsales, Marvol, Marsales).
```

Replaces missing *y* values using linear interpolation from surrounding usable value pairs (*x,y*). The *y* arguments are variable names or X(*i*). The *x* arguments may be variable names, explicit values, functions, or algebraic expressions.

TEXT('message')

```
IF(SMOKE EQ XMIS) THEN
TEXT('smoking unknown').
```

The TEXT command prints the case number and label with the specified message.

SHOW($a_1, a_2, \ldots, a_n$)

```
IF(PULSE_2 GT 160) THEN
SHOW (PULSE_1, PULSE_2).
```

Prints the case number and label with the values of $a_1, a_2, \ldots, a_n$ (variables). You can SHOW functions as well, such as SHOW(age, sex, SQRT(weight)). Variable names are not printed.

RNDU(*i*)

```
USE = RNDU(14761) LT .20.
```

Generates uniform (0,1) random numbers. Use a large integer between 1 and 30000 as a seed number.

RNDG(*i*)

```
NEWVAR = 3 * RNDG(25491) + 50.
```

Generates normal (0,1) random numbers. Use a large integer between 1 and 30000 as a seed number.

Summary Table for TRANSFORM Commands

Command or option	Use	See
• Associated commands in the VARIABLE paragraph		
MIN	specify range limits	Ch. 4
MAX	specify range limits	Ch. 4
MISSING	specify missing value codes	Ch. 4
ADD	create temporary variables	Ch. 4 & 6.6
BEFORET, AFTERT, NO CHECK.	check missing and out-of-range values	Ch. 4
RETAIN.	create lagged variables	Ch. 4 & 6.6
• TRANSFORM paragraph components		
arithmetic operators	derive new variables	Table 6.1
functions	derive new variables	Table 6.2
date functions	transform dates	Table 6.3
summary functions	derive summary variables	Table 6.4
cumulative distribution functions	derive cum. distribution values	Table 6.5
logical operators	case selection, conditional statements	Table 6.6
• Commands		
USE = # *or* logical expression.	case selection (set to 0, 1, –1)	6.4
KASE	case selection	6.4
OMIT = # list.	case selection (sets USE = 0)	6.4
DELETE = # list.	case selection (sets USE = –1)	6.4
XMIS	internal missing value flag	6.5
TOOSMALL	internal out-of-range flag	6.5
TOOBIG	internal out-of-range flag	6.5
REPL($y_1, y_2, \ldots, y_n$).	interpolates missing values	6.5
FILL($x_1, y_1, \ldots, x_n, y_n$).	interpolates missing *y* values	6.5
TEXT('message')	print message with case # and label	6.5
SHOW($a_1, \ldots, a_n$)	print variables and functions with case # and label	6.5
• Random number generators		
RNDU(*i*)	generate uniform random numbers	6.6
RNDG(*i*)	generate normal random numbers	6.6

7

The PRINT Paragraph

Commands in the PRINT paragraph allow you to

- control width of the output panel and number of print lines per page
- control the format of printed data
- control verbosity of output; that is, suppress printing of reports and codebooks
- print a list (NEWS) of new program features and recent changes

Other commands allow you to print values instead of group names when data are printed for each case, and to print diagnostics for debugging the program (this is rarely used and invoked only when you think the program has a bug). Some programs provide additional PRINT commands that apply only to the analyses being performed (e.g., predicted and residual values in 2R). Check individual chapters for program-specific PRINT commands. Many programs allow you to print data for all cases; see especially programs 1D and 7D, and the Data Manager (DM).

Figure 7.1
Typical PRINT paragraph for program 1D

```
/ PRINT  LINESIZE = 80.
         NEWS.
         DATA.
```

Where to Find It

7.1 Controlling output width and print lines per page

Output width. The LINESIZE command controls the number of characters printed per line. In noninteractive runs, the program prints output up to 132 characters wide. In interactive runs the default LINESIZE is 80. You can set the LINESIZE to any value up to 132. Some programs respond to any value under 132 by printing output 80 characters wide; other programs can adjust the width of certain portions of printed output to the LINESIZE you specify. If you state LINESIZE = 100, for example, all programs will print output within the 100-character limit; some will print within an 80-character limit. In general, data listings and printouts such as correlation matrices respond to finer control than other portions of output. You can use LINESIZE to control the number of columns printed per panel.

Many programs respond to a reduced LINESIZE by rearranging or deleting portions of the output. For example, 1D omits some statistics (see the 1D chapter for an example of narrow output). In some programs you can use LINESIZE to change the way graphs and plots are presented. Whenever a LINESIZE under 132 results in major changes, we discuss the changes in the individual program chapters.

The LINESIZE command is especially useful when you are running a noninteractive job from a terminal and want to view the output before printing. In this case, you can change the output width from 132 characters (the default) by stating LINESIZE = 80. All output lines will then fit on the terminal screen.

Print lines per page. All BMDP programs print a maximum of 59 lines per page by default. (For clarity, some portions of output are printed with fewer lines per page.) You can control the number of lines printed per page by specifying PAGESIZE. The programs will respond to numbers larger or smaller than 59 (e.g., PAGESIZE = 30). To suppress page skips entirely, state LEVEL = MINIMAL.

7.2 Controlling the format of printed data

Many programs allow you to print data and control the format of the printout. The FIELD command allows you to specify the number of character positions, decimal places, and whether a field is alphanumeric. For example,

```
/ PRINT    FIELD = 3, 2*4, 9.1.
```

specifies the format for the first four variables. The first variable is an integer with a field width of up to three characters; the second and third are integers up to four characters wide; the fourth is a real number with up to nine characters, with one digit after the decimal point. A blank space is added after each variable in the printout so that the variables cannot run into each other.

When nonconsecutive variables are used in the /VAR USE statement, the order of values in the FIELD statement will follow the order of the /VAR NAME statement. In this case, tabbing can be used to skip over unused variables. Here is an example of a variable statement and a corresponding PRINT FIELDS statement for tabbing.

```
/ VAR      NAMES = AGE, SMOKER, DRUGDOSE, WEIGHT, SEX.
           USE = SEX, DRUGDOSE, AGE, WEIGHT.
/ PRINT    FIELD = 2, (3)4, 8.2, -2.
                   :     :      :    :
                  AGE DRUGDOSE WEIGHT SEX
```

Suppose the variables AGE, SMOKER, DRUGDOSE, WEIGHT, and SEX (specified as a label) are printed according to the above format in program 1D. SMOKER (yes, no) and DRUGDOSE (low, high) are grouping variables with values of 0 or 1. The input data look like the following:

```
29   1   0   210.89   m
22   0   1   145.08   f
35   1   1   156.89   m
37   0   0   145.87   f
50   1   0   176.98   m
```

The resulting output is shown in Figure 7.2.

Figure 7.2
Sample output from PRINT FIELD command

```
              a  s    d        w
              g  m    r        e
              e  o    u        i
                 k    g        g
                 e    d        h
                 r    o        t
  C A S E             s
  NO. LABEL           e
----- ----- --- ---- ---- ---------
    1 m      29 yes  low      210.9
    2 f      22 no   high     145.1
    3 m      35 yes  high     156.9
    4 f      37 no   low      145.9
    5 m      50 yes  low      177.0
```

If FIELD is not stated, then the INPUT FORMAT determines each variable's field width. In the FIELD command, alphanumeric fields are indicated by a minus sign. For example, three two-character alphanumeric variables could be indicated by 3*–2. (Case label variables are read as alphanumerics when data are read with FREE format.)

If FIELD is not stated, then FIELD values are derived from FORTRAN-type FORMAT values specified in the INPUT paragraph (see Chapter 3). If FREE format is used, a field of 8.2 is assumed for numeric values (including variables generated in the TRANSFORM paragraph), –4 for case label variables. Note that if the scale of a variable is changed in the TRANSFORM paragraph, then the assumed FIELD value will not be appropriate. FIELD values, whether stated or assumed, are saved in BMDP Files (see Chapter 8).

The FIELD command works in all programs that can list data. All of these programs except 1D and 7D print all variables in a case, possibly using more than one line, before printing the next case. If all variables cannot be printed in one line with the specified FIELDS, then each variable is printed in an 8-character-wide field to permit lining up variables in columns.

7.3 Suppressing reports and codebooks

The following panels are printed by default when running in noninteractive mode:

- interpretation of user specifications and/or default values
- list of variables used
- interpretation of input format (FIXED format only—see Chapter 3, Figure 3.5)
- range limits and missing value flags for variables (see Chapter 4, Figure 4.3)
- grouping information: codes, cutpoints, and names (see Chapter 5, Figure 5.3)
- for some programs, such as 6D, a table of contents for the output

The VNAME, GNAME, and VUSE commands allow you to suppress printing of specific portions of descriptive output. With the LEVEL command, you can control printing of all three types of descriptive output at once.

The interpretation of input format can be deleted by specifying NO VNAME; the list of variables used can be deleted by specifying NO VUSE; and the grouping information can be deleted by specifying NO GNAME. To delete all three at once, plus the report describing the BMDP instructions, state LEVEL = BRIEF. If you ran the example shown in Chapter 1 with LEVEL = BRIEF, the sections of output numbered 3, 4, 5, and 7 would be omitted. To also delete page breaks and the problem TITLE, state LEVEL = MINIMAL (this is the default for interactive runs).

7.4 NEWS

The BMDP programs contain features and updates that have been added since the last release of the software. To find out about these changes, state NEWS in the PRINT paragraph. The text generated by the NEWS command lists new features and problems that have been corrected. The date of the latest version of the BMDP Manual is noted in the banner that appears at the beginning of the output. In all programs other than DM, you can obtain a copy of the News by entering an input file:

```
/ PRINT  NEWS.
/ END
```

Special Features

Printing values instead of group names: VALUES

When you specify NAMES for codes or intervals in the GROUP paragraph, BMDP programs print the NAMES of the codes or intervals rather than actual data values when printing data for cases. If you prefer to have the values printed, state

```
VALUES.
```

Printing diagnostics for debugging: DEBUG

If you suspect you have uncovered a bug in a BMDP program, state

```
DEBUG = ALL.
```

when preparing a run to demonstrate the problem for the BMDP staff. Other values of DEBUG are described in the command definitions.

PRINT Commands

LINEsize = #. `LINE = 80.`

Specify the maximum number of characters printed per line in the output. LINESIZE = 79 is the new lower limit. Anything < 79 will be reset to 79 and you will receive a warning message. Default: 80 for interactive runs, 132 for noninteractive.

PAGEsize = #. `PAGE = 30.`

Specify the maximum number of printed lines per page. Default: 59 for noninteractive runs; interactive runs are not paginated.

FIELDs = list. `FIELD = 3, 2*4, 9.1, -2.`

Controls the format of a data listing by specifying the number of character positions, decimal places, and whether a field is alphanumeric. The order of values in the FIELD command follows the order of the /VARIABLE NAMES. A blank space is added after each variable in the printout so that the variables cannot run into each other. By default, data listings use the FORMAT specified in the INPUT paragraph. If neither FIXED FORMAT nor FIELD are used, a field specification of 8.2 is assumed for numeric values, –4 for case labels. The FIELD command works in programs 1D, 3D, 7D, 8D, 2L, 4M, 6M, 9M, AM, KM, 6R, 9R, LR, 1T, 2T, and 1V. Tabbing may be used to skip variables. **Example:** The first variable is an integer with a field width of up to three characters; the second and third are integers up to four characters wide; the fourth is a number with up to nine characters, with one digit after the decimal point. The fifth variable is alphanumeric with two characters (the minus sign indicates an alphanumeric field).

VNAMe. `NO VNAME.`

State NO VNAME to suppress printing of the codebook interpreting the input format (for FIXED format files) or the list of variable names and file code, content, and label (for BMDP Files). These panels are printed by default for noninteractive runs; NO VNAME is the default for interactive runs.

GNAMe. `NO GNAME.`

State NO GNAME to suppress printing of the codebook of missing value codes (MISSING), range limits (MINIMUM and MAXIMUM), and grouping information (grouping variables, GROUP CODES, CUTPOINTS, and NAMES). These panels are printed by default for noninteractive runs; NO GNAME is the default for interactive runs.

VUSe. `NO VUSE.`

State NO VUSE to suppress printing of variable names you have specified in the VARIABLE USE list. This panel is printed by default for noninteractive runs; NO VUSE is the default for interactive runs.

LEVel = *(one only)* `LEVEL = BRIEF.`
MINimal, BRIEF,
NORMAL, VERBose.

Use the LEVEL command to control the amount of computer output.

NORMAL – The panels of information about variables, grouping information, and variables used are printed. (This is equivalent to stating VNAME. GNAME. VUSE.) This is the default for noninteractive runs.

BRIEF – The panels of information about variables, grouping information, and variables used are not printed. (This is equivalent to stating NO VNAME. NO GNAME. NO VUSE.) The panel describing the BMDP instructions is also deleted.

MINIMAL – The program takes the same action as in LEVEL = BRIEF. In addition, the program suppresses page skips and does not print a TITLE at the top of each output page. This is the default for interactive runs.

VERBOSE – Some programs (e.g., 2R) print additional explanatory output.

NEWs. `NEWS.`

State NEWS to obtain a list of new features and program changes at the beginning of the output. Default: no news.

VALUes. `VALUES.`

When printing data for each case, state VALUES to obtain the values for the grouping variables instead of the category names. By default, the category names are printed.

DEBUg = *(one only)* `DEBUG = ALL.`
NONE, TEST,
INFO, ALL.

If you suspect you have uncovered a bug in a BMDP program, state DEBUG = ALL when preparing a run to demonstrate the problem for the BMDP staff. There are four values for the DEBUG command (TEST and INFO are used by the programming staff at BMDP).

NONE – This is the default. No diagnostic testing is conducted.

TEST – This turns on all internal storage checking. (The program produces results only when it finds a problem.)

INFO – In addition to TEST, the program prints storage management and trace messages.

ALL – In addition to INFO, the program produces program-specific information.

Summary Table for PRINT Commands

Paragraph Commands	Noninteractive Defaults	Interactive Defaults	Multiple Problems	See
/ PRINt				
LINEsize = #.	132	80	●	7.1
PAGEsize = #.	59	no pagination	●	7.1
FIELDs = list.	FORMAT if specified; otherwise 8.2 for numeric, –4 for label variables		–	7.2
VNAMe.	VNAME.	NO VNAME.	●	7.3
GNAMe.	GNAME.	NO GNAME.	●	7.3
VUSe.	VUSE.	NO VUSE.	●	7.3
LEVel = *(one only)* MINimal, BRIEF, NORMal, VERBose.	NORMAL	MIN	●	7.3
NEWs.	no news	no news	–	7.4
VALUes.	NO VALUES.	NO VALUES.	●	SpecF
DEBUg = *(one only)* NONE, TEST, INFO, ALL.	NONE	NONE	●	SpecF

Key: ● Value retained for multiple problems
– Default reassigned

8

Saving Data and Results: The SAVE Paragraph

When you need to use more than one BMDP program to analyze a data set, the SAVE paragraph can come in handy. It allows you to store your data and certain results in a special *BMDP File* for efficient reuse with other programs. In all BMDP programs you can use a SAVE paragraph to create a BMDP File, and then use an INPUT paragraph (or a READ paragraph in the Data Manager) to read the BMDP File for later analyses.

In the SAVE paragraph you can either create a BMDP File or output a raw data file in a format of your own choosing. Unlike a BMDP File, a raw data file can be viewed and edited on-screen, printed in a report, and read by non-BMDP software. However, there are several advantages to a BMDP File:

- BMDP programs read data faster from a BMDP File than from a raw data file. This advantage is particularly noticeable when running a large data set on a PC.
- A BMDP File "remembers" many of the instructions that you specify when you create the File. These instructions need not be restated for additional analyses.
- A BMDP File is an easy way to store results (such as factor scores, sorted data, or covariance matrices) for further analysis by other BMDP programs.

Once you have created a BMDP File, you can use the INPUT and SAVE paragraphs to change the BMDP File into a raw data file whenever you want. In addition, you can use the PORT command to prepare BMDP Files for transfer from one computer system to another.

The SAVE paragraph includes several important commands with which you should become familiar:

CODE – assigns a name that BMDP uses to identify the BMDP File

FILE (or UNIT) – assigns a name (or unit number, on some computers) that the computer system uses to store the file

NEW – indicates that your BMDP File is the first one stored in the system file

FORMAT – specifies a format for a raw data file or PORT file

Where to Find It

8.1 Creating a BMDP File

A BMDP File stores your data and most information from the INPUT, VARIABLE, TRANSFORM, and GROUP paragraphs of your input. If you state any commands from the list below when you create a BMDP File, the resulting file will contain that information. Thus, you will not need to restate those commands when you use the BMDP File in later analyses.

Information stored in a BMDP File

/ INPUT	FORMAT VARIABLE CASES
/ VARIABLE	NAMES MINIMUM and MAXIMUM MISSING LABEL
/ TRANSFORM	USE Transformed variables New derived variables
/ GROUP	CODES CUTPOINTS NAMES
/ PRINT	FIELDS

In Input 8.1 we use program 1D to create a BMDP File from the modified Exercise data presented in Chapter 3 (FILE = 'PULSEMIS.DAT'). The BMDP File will save all information stated in Input 8.1 except the PRINT paragraph instruction (LINESIZE = 80). Once the BMDP File is created, we can use it as input to any BMDP program. None of the saved information need be specified again unless we want to change it. The BMDP File stores all data, as well as the following information from Input 8.1:

- number of variables (VARIABLES = 8.)
- input format (FORMAT = FREE.)
- sorted order of data (SORT = sex, smoke, age.)
- names of all eight variables
- the request to treat the last two variables (NAME1 and NAME2) as labels

- flags for missing and out-of-range data (MCHAR, MISSING, MINIMUM, and MAXIMUM)
- grouping codes, names, and cutpoints
- the derived variable (DIFFRNCE) that shows the change in pulse
- the instruction to omit (USE = –1) cases where PULSE_1 is greater than PULSE_2
- the corrected value (165) for the PULSE_2 error (265)

To create the BMDP File, we include a SAVE paragraph. First we state the name of the system FILE in which the BMDP File is to be stored. (Some systems use UNIT instead of FILE; see Chapter 3.) As required, we also state a CODE (1 to 8 characters) that BMDP programs can use to recognize the file as a BMDP File. We state NEW to indicate that this is the first file written into the system file. (To add a BMDP File to an existing system file, omit NEW.) Output 8.1 shows the results of running Input 8.1 on program 1D.

Input 8.1
Creating a BMDP File

```
/ INPUT       VARIABLES = 8.
              FORMAT = FREE.
              FILE = 'pulsemis.dat'.
              MCHAR = '.'.
              SORT = sex, smoke, age.

/ VARIABLE    NAMES = id, sex, smoke, age, pulse_1,
                      pulse_2, name1, name2.
              LABEL = name1, name2.
              MISSING = (sex)9, (age)2*99, 999.
              MINIMUM = (age)15, (pulse_2)35.
              MAXIMUM = (pulse_2)200.

/ GROUP       CODES(sex,smoke) = 1, 2.
              NAMES(sex) = male, female.
              NAMES(smoke) = yes, no.
              CUTPOINTS(age) = 24, 34.
              NAMES(age) = under_25, _25to34, over_34.

/ TRANSFORM   diffrnce = pulse_2 - pulse_1.
              IF (diffrnce LT 0) THEN USE = -1.
              IF (pulse_2 EQ 265) THEN pulse_2 = 165.

/ SAVE        FILE = 'exercise.sav'.
              CODE = exercise.
              NEW.

/ PRINT       LINESIZE = 80.

/ END
```

Output 8.1
Creating a BMDP File

```
—we omit a portion of the output—
-----------------------------------------
BMDP FILE IS BEING WRITTEN
CODE. . . IS      exercise                                  [1]
FILE NAME IS      exercise.sav
CONTENT . IS      DATA
LABEL . . IS      6-MAY-92           13:49:27               [2]
VARIABLES ARE
      1 id          2 sex        3 smoke      4 age         5 pulse_1
      6 pulse_2     7 name1      8 name2      9 diffrnce

—we omit the default printing of the first ten—

BMDP    FILE HAS BEEN COMPLETED.
FILE NAME IS      exercise.sav
-----------------------------------------                   [3]
NUMBER OF CASES WRITTEN TO FILE        15

—we omit program output—
```

[1] The BMDP File CODE (EXERCISE), the name of the BMDP File (EXERCISE.SAV), and the default CONTENT of the File (DATA). Some programs allow you to save CONTENTS other than DATA; for example, AM can save a correlation matrix (CORR), and 4M can save factor score coefficients (FSCF). For a summary of the types of results that can be saved, see Special Features.

[2] The BMDP File LABEL. You can provide a LABEL of up to 40 characters (surrounded by single quote marks) to distinguish between two BMDP Files with identical CODE and CONTENT in the same system file. A sample LABEL is

```
LABEL = 'RUN 2-OUTLIERS DELETED'.
```

If you do not specify a label, as in this example, the program uses the date and time.

[3] To indicate that the BMDP File has been created, the program reports 'BMDP FILE HAS BEEN COMPLETED' and gives the NUMBER OF CASES WRITTEN TO FILE.

8.2 Reading a BMDP File

Any BMDP program can read the BMDP File created in Output 8.1; you need only supply the appropriate FILE and CODE statements in the INPUT paragraph. If the BMDP File contains results other than data, you also need to specify the CONTENT.

Input 8.2
Reading a BMDP File

```
/ INPUT    FILE = 'exercise.sav'.
           CODE = exercise.

 — program-specific instructions follow —
```

The absence of FORMAT and the presence of CODE in Input 8.2 tell the program to read a BMDP File as input. Because no LABEL is specified, the program uses the first BMDP File it encounters with the specified CODE. You need not state the number of VARIABLES and the number of CASES in the INPUT paragraph.

The VARIABLE, GROUP, and TRANSFORM paragraphs are no longer necessary unless you want to change your instructions. For example, you may:

- restate GROUP with new ways of grouping the data
- perform additional TRANSFORMATIONS
- give VARIABLES new NAMES
- tighten the range between MINIMUM and MAXIMUM
- identify additional values as MISSING

For more information on modifying information stored in a BMDP File, see Special Features.

In Output 8.2, the program prints information about the BMDP File and restates the variable and grouping information that was saved with the data. Note that the LABEL is the date and time the file was created (see Output 8.1).

Output 8.2 lists nine variables, including DIFFRNCE, the new derived variable we created in Section 8.1. However, variables 7 and 8 (NAME1 and NAME2) are omitted from the list of VARIABLES TO BE USED because we designated NAME1 and NAME2 as case label variables in Input 8.1.

Output 8.2
Reading a BMDP File

```
INPUT BMDP FILE
CODE. . . IS       exercise
CONTENT . IS       DATA
LABEL . . IS
   18-MAY-92              14:00:49
VARIABLES
     1 id           2 sex         3 smoke       4 age          5 pulse_1
     6 pulse_2      7 name1       8 name2       9 diffrnce

VARIABLES TO BE USED
     1 id           2 sex         3 smoke       4 age          5 pulse_1
     6 pulse_2      9 diffrnce

— we omit the default printing of the first ten cases —

NUMBER OF CASES READ. . . . . . . . . . . . . .          15

   VARIABLE         STATED VALUES FOR               GROUP CATEGORY     INTERVALS
  NO.   NAME    MINIMUM MAXIMUM MISSING    CODE  INDEX   NAME        .GT.    .LE.
 ---- --------  ------- ------- -------   -----  -----  --------    ------  -------
   2  sex                                 1.000    1  male
                                          2.000    2  female
   3  smoke                               1.000    1  yes
                                          2.000    2  no
   4  age                                          1  under_25              24.00
                                                   2  _25to34      24.00   34.00
                                                   3  over_34      34.00
---------------------------------------------------------------------------------

— we omit program output —
```

8.3 Saving a raw data file

Although you will probably use the SAVE paragraph most frequently to write a BMDP File, at times you may want to save a raw data file. For example, if your data contain many errors, you may prefer to view the data file on screen and correct the errors with your system editor. (You can easily correct a small number of errors with the TRANSFORM paragraph.) Further, you can print a saved raw data file (including transformed variables, new derived variables, and some results) in a report, or input it to other software.

When you want to output a raw data file, you may use either a raw data file (such as PULSEMIS.DAT in section 8.1) or a BMDP File as input. Because BMDP programs save files **after** performing transformation and case selection, the output file will contain data already transformed according to your instructions. However, the raw data output file does not retain your instructions, grouping information, or variable names.

To SAVE data in a raw data file, you must specify the output record FORMAT. Four formats are available: F, G, BINARY, or your own FORTRAN format. (The latter is especially useful for organizing data for reports.) F-format assumes 10F8.3; that is, ten eight-character fields with three digits after the decimal. G-format assumes 5G16.6—that is, five 16-character fields with six digits after the decimal—and writes each variable in either F- or E-type scientific notation as needed.

In this example we show how you can specify your own format to output the exercise data (transformed as specified in Input 8.3) to a raw data file. The FORMAT statement in the SAVE paragraph uses a standard FORTRAN format. Note that you **cannot** use the more flexible rules for BMDP's data reader (described in Chapter 3). For example, the BMDP data reader allows you to state F4 instead of F4.0; the SAVE paragraph does **not** allow you to do this.

We write the LABEL variables using A-format, and we allow one extra column for each of the numeric values in the input file, since the program will write a decimal after each numeric value. The program does **not** allow files to be saved in integer format (without decimals). You can use your system editor to delete unnecessary decimals after the file is written; however, be careful not to change the format or the data.

When you create variables in the TRANSFORM paragraph, the program adds them to the end of each record in the order they were created. Thus DIFFRNCE becomes the ninth variable in the input data set and remains in this position in output unless the KEEP command changes the variable order (see section 8.4).

Output 8.3 provides information about the output file name, format option chosen, and number of cases written to the file.

Input 8.3
Saving a raw data file

```
/ INPUT      VARIABLES = 8.
             FORMAT = FREE.
             FILE = 'exercise.dat'.

/ VARIABLE   NAMES = id, sex, smoke, age, pulse_1,
                     pulse_2, name1, name2.
             LABEL = name1, name2.

/ TRANSFORM  IF(pulse_2 EQ 265) THEN pulse_2 = 165.
             diffrnce = pulse_2 - pulse_1.

/ PRINT      LINE = 80.

/ SAVE       FILE = 'rawexer.dat'.
             FORMAT = '(F4, 2F3.0, 3F5.0, 1X, 2A4, F4.0)'.
/ END
```

Output 8.3
Saving a raw data file

```
OUTPUT FILE IS BEING WRITTEN WITH USER'S FORMAT.
FILE NAME IS      rawexer.dat
CONTENT . IS      DATA
VARIABLES ARE
     1 id          2 sex          3 smoke        4 age          5 pulse_1
     6 pulse_2     7 name1        8 name2        9 diffrnce

USER    FILE HAS BEEN COMPLETED.
FILE NAME IS      rawexer.dat
-------------------------------------------
NUMBER OF CASES WRITTEN TO FILE        40

— we omit remainder of program output —
```

Figure 8.1 displays the raw data file generated by Input 8.3. Compare this version of the exercise data with the original file shown in Chapter 2 (Figure 2.2). This file is more compact, the variable DIFFRNCE has been added, and the erroneous PULSE_2 value of 265 has been changed to 165.

Figure 8.1
Output data file RAWEXER.DAT written with user-specified format

```
 1  1. 1.  31.  62. 126. ALLAN    64.
 2  2. 1.  20.  78. 154. MARY     76.
 3  1. 2.  28.  64. 128. BILL     64.
 4  2. 2.  29.  96. 155. LINDA    59.
 5  1. 1.  21.  66. 128. MICHAEL  62.
 6  2. 1.  27.  96. 165. CATHY    69.
                         .
                         .
                         .
36  1. 1.  34.  62. 132. WILLIAM  70.
37  1. 2.  35.  74. 116. KYLE     42.
38  1. 1.  21.  90. 138. BEN      48.
39  1. 2.  21.  66. 142. GREG     76.
40  1. 2.  30.  70. 132. RICHARD  62.
```

8.4 Saving a raw data file with missing and out-of-range values

Saving a raw data file becomes more complex when data are missing or outside the range limits specified by MINIMUM and MAXIMUM. To flag missing and out-of-range data on many systems, BMDP uses these numbers:

MISSING	16^{31}
over MAX	$2 * 16^{31}$
under MIN	$-2 * 16^{31}$

To replace those numbers, you must use the MISSING instruction in the SAVE paragraph.

In this example we save the data in PULSEMIS.DAT (see Section 8.1) and use the MISSING command to convert BMDP's internal flags for extreme and missing values to more convenient numbers. We also provide a KEEP list to reorder the variables and exclude the ID variable. The FORMAT statement follows the order of the variables as they appear in the KEEP list.

Note that the MISSING list in the SAVE paragraph follows the order of the variables as they were read into the program, **not** the order specified in the KEEP list. Because DIFFRNCE is added to the end of the data, we specify its missing value code last.

If you do not include a MISSING list when writing a raw data file, most systems will write asterisks in place of BMDP's internal flags. When you use the raw data file as input to any BMDP program, the asterisks will automatically be identified as missing values. However, some systems, including the IBM PC, **require** a MISSING list to save data flagged as missing or out of range.

Input 8.4
Saving a raw data file with missing and out-of-range values

```
/ INPUT      FILE = 'pulsemis.dat'.
             VARIABLES = 8.
             FORMAT = FREE.
             MCHAR = '$'.

/ VARIABLE   NAMES = id, sex, smoke, age, pulse_1,
                     pulse_2, name1, name2.
             LABEL = name1, name2.
             MISSING = (sex)9, (age)2*99, 999.
             MIN = (age)15, (pulse_2)35.
             MAX = (pulse_2)200.

/ TRANSFORM  IF(pulse_2 EQ 265) THEN pulse_2 = 165.
             diffrnce = pulse_2 - pulse_1.

/ PRINT      LINE = 80.

/ SAVE       FILE = 'pulseraw.dat'.
             MISSING = (sex)2*9, 2*99, 999, (diffrnce)99.
             KEEP = name1, name2, diffrnce, sex TO pulse_2.
             FORMAT = '(2A4, 1X, F3.0, 2F3.0, 2F4.0, F5.0)'.

/ END
```

Output 8.4
Saving a raw data file with missing and out-of-range value

```
OUTPUT FILE IS BEING WRITTEN WITH USER'S FORMAT.
FILE NAME IS     pulseraw.dat
CONTENT . IS     DATA

THE VARIABLES SAVED
     1 name1         2 name2        3 diffrnce     4 sex            5 smoke
     6 age           7 pulse_1      8 pulse_2

*** N O T E *** THE PROGRAM ASSUMES THAT THE ORDER OF VARIABLES IN THE
                       /SAVE MISSING=LIST
                CORRESPONDS TO THE ORDER OF VARIABLES IN THE INPUT DATA AND NOT
                TO THE LIST OF VARIABLES TO BE SAVED.


USER    FILE HAS BEEN COMPLETED.
FILE NAME IS     pulseraw.dat
_____________________________________
NUMBER OF CASES WRITTEN TO FILE       40
```

—we omit the remainder of program output—

In addition to information about the output file name, format, and number of cases written, the program provides information about minimum, maximum, and missing values. The program also reports means and ranges **before** values have been converted with the MISSING command.

Figure 8.2 displays the new raw data file. Note that the variables have been reordered and that all missing values have been converted to 9, 99, or 999. Compare this file with the original file in Figure 2.2.

Figure 8.2
Raw data file PULSERAW.DAT

```
ALLAN     64. 1. 1. 99. 62. 126.
MARY      76. 2. 1. 20. 78. 154.
BILL      64. 1. 9. 28. 64. 128.
LINDA     99. 2. 2. 29. 99. 155.
MICHAEL   99. 1. 1. 21. 66. 999.
CATHY     99. 2. 1. 27. 96. 999.
HARVEY    52. 1. 2. 21. 68. 120.
JENEE     66. 9. 2. 42. 72. 138.
JEAN      99. 2. 1. 22. 88. 999.
FREDDY    54. 1. 1. 99. 90. 144.
PAT       58. 2. 2. 21. 82. 140.
MARK      60. 1. 2. 22. 74. 134.
SUSAN     99. 2. 1. 43. 66. 999.
DENISE    99. 2. 2. 19. 99. 142.
JOHN      42. 1. 1. 23. 92. 134.
```

Special Features

Saving a subset of the file

BMDP provides several methods for saving a subset of a file:

- To save a subset of your **variables,** you can list either the variables you want to KEEP or the variables you want to DELETE. The subset saved by BMDP will include variable names and labels, grouping information, and flags for missing and out-of-range values.
- To save a subset of your **cases,** you can:
  - specify in the INPUT paragraph the CASES you want to save (e.g., CASES = 100.).
  - specify in the TRANSFORM paragraph the cases you want to omit, by assigning them USE = –1. For example, to eliminate all males from a saved file, you would state IF(sex EQ 1) THEN USE = –1. Note that when you set the value of USE to zero for a case, the case is written to the saved file, and the value of USE is stored with the data. See Chapter 6 for more information.
  - use DELETE and KASE in the TRANSFORM paragraph to list the numbers of the cases you want to delete. If you include a USE statement and a DELETE statement in the same TRANSFORM paragraph, make sure the USE statement comes first. See Chapter 6 for more information.
  - include the COMPLETE command in the SAVE paragraph, to save only those cases with complete data (i.e., cases with no missing or out-of-range values for variables being saved).

Saving results

Many programs allow you to save results in addition to the data. You can save results in a BMDP File or in a raw data file with a format of your own choosing. In some cases you must specify the results to be saved in a CONTENT command (see program chapters). For an example of saving results, see program 4F, Example 4F1.11. The following list summarizes program-specific results you can save.

1D – data in sorted form
CA – data as a table of frequency counts and coordinate scores
8D – correlation or covariance matrix for incomplete data
LE – density functions
4F – data as a table of frequency counts (BMDP File only)
1M – correlation matrix
4M – factor scores and covariance matrix
6M – covariance matrix
7M – predicted group membership, canonical variable scores
AM – estimates of missing values, covariance matrix for incomplete data
KM – cluster membership
2V – predicted values, residuals

Most regression programs can save predicted values, residuals, and the covariance matrix. In addition, 2R and 9R can save regression diagnostics, and LR can save summary cell descriptors and predicted probabilities. For further information, consult individual program writeups.

When the content of your BMDP SAVE file consists of a covariance or correlation matrix, and is saved as an ASCII file, a square matrix is produced. This information is followed by four rows of variable information, which consists of the variance, the mean, the sample size, and the sums of weights (sample size * weights).

Transporting BMDP Files between systems: PORT

Because they are binary, BMDP Files written on one system cannot be read on a different system. For example, a BMDP File created on an IBM PC cannot be read on a VAX. The PORT command translates the data (plus additional information) into a new "portable" text file (character image file) for easy transfer between systems. The portable BMDP File contains the same information as a regular BMDP File, including variable names, grouping information, and flags for missing and out-of-range values. The instructions to save a portable BMDP File are the same as for a standard BMDP File; you simply add the PORT command to the other SAVE instructions.

Once you have saved a portable BMDP File, you can use system commands to copy the file to the new computer system. To use the portable BMDP File as input on the new system, add PORT to the INPUT paragraph.

- Unless you supply a format, the data in a portable BMDP File are written one case at a time with a FORTRAN-type format of '(F13.6)' for numeric variables, '(9X,A4)' for LABEL variables.
- Note that a portable BMDP File may be three times larger than the usual BMDP File.

> ***Caution:*** *On several systems (including VAX/VMS), a file written more than once remains flagged as the original type (formatted raw data file, BMDP File, or portable BMDP File). Hence, you should* ***not*** *write a portable BMDP File over a regular BMDP File. In general, do not overwrite a file with a file of a different type; either delete the old file first or give the new file a different name.*

Modifying information stored in a BMDP File

When you use a BMDP File as input, you can modify several aspects of the file for your current analysis:

- You can reset the value of USE for any case or for all cases. If some cases are stored in the BMDP File with USE = 0, you can specify that all cases be used in the analysis by stating

```
/ TRANSFORM    USE = 1.
```

- You can rename the variables. For example, if variable names in the BMDP

File are ID, SEX, X(3), X(4), X(5), you can rename the last four variables:

```
/ VARIABLE  NAMES = (2)GENDER, SCORE1, SCORE2, SCORE3.
```

(We tab to the second variable by including the 2 in parentheses at the beginning of our list. See Chapter 10.)

- You can respecify case LABEL variables only if both the original and the new LABELS are alphanumeric.
- You can reduce the upper limit (MAXIMUM) or increase the lower limit (MINIMUM) for any variable, but you **cannot** increase the upper limit or reduce the lower limit, since out-of-range values have already been recoded in the BMDP File.
- You can specify additional MISSING value codes, and an additional missing value character (MCHAR). Missing values already recoded in the File will still be treated as missing.
- You can select variables for analysis or for saving in a new BMDP File with the VARIABLE USE list and the KEEP and DELETE commands in the SAVE paragraph.
- You can perform additional transformations and add new variables to the data file.
- You can add or change GROUP information contained in the BMDP File. For example, if the BMDP File stores a cutpoint of 21 for AGE, you can respecify cutpoints:

```
/ GROUP    CUTPOINT(AGE) = 15, 25.
```

Saving more than one BMDP File in a system file

To save the first BMDP File in a system file, include the command NEW in the SAVE paragraph. To add another BMDP File to the same system file, omit the NEW command. It is good practice to use a different CODE name for each BMDP File you store in the same system file. Alternatively, you can use the LABEL command to distinguish between two BMDP Files with the same CODE and CONTENT. The LABEL can include up to 40 characters of text. On most systems, the default label is the date and time the BMDP File was created. You can check the default label generated on your system by printing the DIRECTORY for your first BMDP File. Note that if you have two BMDP Files with the same CODE and CONTENT in a system file, and do not specify a LABEL in the INPUT paragraph, the program will read the first BMDP File with the specified CODE.

> ***Caution:*** *Storing several BMDP Files in a large system file can slow down program execution substantially, since the BMDP Data Reader must scan each individual record in all preceding files as it searches for the current input file. Thus, if you call for the fourth file, all records in the first three files will be scanned.*

When you include the NEW command in the SAVE paragraph, you "initialize" the system file. If you have saved one or more BMDP Files in a system file and then run a job that includes the NEW command, the program overwrites the BMDP Files already saved.

Obtaining a directory of a BMDP File

If you don't remember the contents of a system file that contains one or more BMDP Files, you can obtain a description of the contents by stating DIRECTORY in the INPUT paragraph. To obtain a listing of the contents of the system file PULSE.SAV, for example, state:

```
/ INPUT    FILE = 'pulse.sav'.
           DIRECTORY.
/ END
```

When you run this input on a BMDP program, the program prints the CODE, CONTENT, LABEL, variable names, and number of cases for each BMDP File stored in the system file PULSE.SAV. The program terminates after printing the directory. However, if you also include a CODE name in the INPUT paragraph, the program will analyze the file with this code name after printing the directory.

Saving BMDP Files on the IBM mainframe

Several aspects of saving a BMDP File differ when you are using an IBM mainframe:

- You preallocate a system file before saving a BMDP File.
- You use UNIT instead of FILE in the SAVE paragraph, and you include a system job control language (JCL) statement, which names the system file you want to use.
- You can save data in a temporary BMDP File for use in a single computer run.

Generating multiple output records from a single record

The APPEND command in the SAVE paragraph allows you to restructure data from one input record onto several output records. For example, if each input record contains the variables

ID, SEX, Y1, Y2, Y3

and you want to restructure the data onto three records

ID SEX Y1 1 (the last variable in each record is a
ID SEX Y2 2 sequence number generated by the
ID SEX Y3 3 program)

state:

```
/ SAVE  KEEP = ID, SEX.

        APPEND = Y1.
        APPEND = Y2.
        APPEND = Y3.
```

The BMDP Data Manager can perform this and other functions.

SAVE and INPUT Commands

Creating a BMDP File

/ SAVE

Required to save data, transformations, and analysis results in a raw data file or BMDP File.

FILE = name. `FILE = 'EXERCISE.SAV'.`
or **UNIT = #.** `UNIT = 8.`

Required to identify the system file in which the BMDP File or raw data file will be stored. Use FILE for most computer systems, UNIT for IBM mainframes and some other computers. If you are not sure which to use, check your computer manual.

CODE = name. `CODE = EXERCISE.`

Required to give the BMDP File a name (1 to 8 characters), which will be specified when the File is used as input. Omit CODE when you save a raw data file.

CONTent = parameter names. `CONTENT = CORR.`

Required only in programs that save results other than data. See Special Features for a summary of results that can be saved. See individual program chapters for appropriate parameter names. Default: DATA.

LABEL = 'text'. `LABEL = 'PHASE 2 STUDY'.`

Use up to 40 characters to label a BMDP File when two or more are stored in the same system file with identical CODE and CONTENT. Default: the date and time the saved file was created. (This may vary on some systems. To check your system's default, use the DIRECTORY command in the INPUT paragraph.)

NEW. `NEW.`

Required to initialize a system file for saving a BMDP File. Use the first time you save a BMDP File in a system file, or when you wish to write over data stored in an existing system file. By default, the program stores the current BMDP File after existing BMDP Files. Omit NEW when you save a raw data file.

FORMat = *(one only)* `FORM = '(2A4,1X,F3.0,2F2.0)'.`
'(format)', F,
G, BINARY.

Required to specify a format for a raw data file, including TRANSFORM results and analysis results. (Variable names and other documentation are not written on raw data files.) Omit when data and results are stored in a BMDP File. The default format for a PORTABLE file is F13.6 for numeric variables, 9X,A4 for LABEL variables. Note that I (integer) fields are not allowed. Use FORTRAN conventions, not the BMDP abbreviations allowed in the INPUT paragraph.

'(format)'. – user-specified FORTRAN format
F. – 10F8.3
G. – 5G16.6
BINARY. – unformatted binary file

KEEP = list. `KEEP = DIFFRNCE, 1 TO 3, 5.`

Specify a subset of variables (by name or number) to save. Use also to reorder the variables in the saved file. Default: All variables, including those added in the TRANSFORM paragraph, are saved in their original order. Variable names and labels, grouping information, and flags for missing and out-of-range values are saved with the subset.

DELete = list. `DELETE = 2, 6, 7.`

Specify a subset of variables (by name or number) to omit from the saved file. (See also KEEP.)

MISSing = # list. `MISS = (SEX)9, (AGE)2*99, 999.`

Specify one number for each input variable, for each variable added in the TRANSFORM paragraph, and for each variable generated by the analysis, to replace BMDP's internal flags for missing and out-of-range values when writing a raw data file. Note that the MISSING list follows the order of the variables as they were read into the program, not the order specified in the KEEP list. Default: Most systems write asterisks in place of the internal flags in a formatted file. Some systems, including the IBM PC, do not write asterisks; for these systems a MISSING list is **required.** Not needed when saving data in a BMDP File.

COMPlete. `COMP.`

Directs the program to save cases with complete data only.

APPEND = list.

```
KEEP = ID, SEX.
APPEND = Y1.
APPEND = Y2.
APPEND = Y3.
```

Directs the program to generate multiple output records from a single inputs. Specify one APPEND list for each output record. The program generates one sequence number (named SEQN) per APPEND statement as the last variable on each new record. The output variable takes its name from the first APPEND statement. See Special Features. See also the Data Manager for data file restructuring.

PORTable. `PORT.`

Prepares a BMDP File for transfer from one computer system to another, by writing a character image BMDP File that can be read by BMDP programs in the new system. To save a character image file, you must have a FILE (or UNIT) and a PORT statement. You may also state a FORMAT, which will be used to write the data one case at a time. If FORMAT is not stated, the data are written with a FORTRAN-type format of F13.6 for numeric variables, 9X,A4 for LABEL variables. Do not use an A-format larger than your system word size (usually A4). For variables that contain missing, too large, or too small values, use the MISSING command to recode the flagged values. Each PORT file can contain only one CONTENT (e.g., DATA, CORR, or COVA). Use PORT only when you want to move a file from one system to another. A file created with PORT can be three times as large as a regular BMDP File.

Reading a BMDP File

/ INPUT

The INPUT paragraph is required. See Chapter 3 for commands to input a raw data file. Use the following commands to input a BMDP File.

FILE = name. `FILE = 'EXERCISE.SAV'.`
or **UNIT = #.** `UNIT = 8.`

Required to identify where your input BMDP File is stored.

CODE = name. `CODE = EXERCISE.`

Required to specify a BMDP File as input. Use the same name as was used to create the BMDP File.

CONTent = parameter name. `CONTENT = COVA.`

Specify the CONTENT of the BMDP File. Not needed for DATA or if only one CONTENT exists in the BMDP File. When using CORR or COVA, the variables in the BMDP file will be numbered sequentially 1, 2, 3 ... Thus, if you have used fewer variables than you had available, the BMDP file variable numbers will not correspond to the previous variable numbers. It is important to keep that distinction in mind during future use of the BMDP file, when you should refer to variables by names rather than by number. Default: If more than one CONTENT exists, the program reads the first CONTENT stored in the BMDP File.

TYPE = parameter name. `TYPE = DATA.`

In programs that accept alternate inputs (e.g., DATA or COVA in regression programs), TYPE overrides CONTENT. For example, a BMDP File saved as TABLE may be input as DATA in 1R.

LABEL = 'text'. `LABEL = 'INCLUDES TRANSFORMED VARIABLES'.`

Use to distinguish between two BMDP Files with identical CODE and CONTENT in the same system file. If LABEL was not specified in the SAVE paragraph when the BMDP File was created, the program normally stores the date and time of the file creation as a LABEL. To check the exact LABEL stored, use the DIRECTORY command.

DIRectory. `DIR.`

Directs the program to print the CODE, CONTENT, LABEL, variable names, and number of cases for each BMDP File stored in the system FILE (or UNIT).

PORTable. `PORT.`

Include when the BMDP File was created with PORT in the SAVE paragraph.

Summary Table for SAVE and INPUT Commands

	Paragraphs Commands	Defaults	Multiple Problems	See
	/ SAVE			
■	FILE = name.	(required for first problem)	–	8.1
	or UNIT = #.		•	Cmds
▲	CODE = name.	(required to save a BMDP File)	–	8.1
	CONTENT = parameters.	DATA	–	SpecF
	LABEL = 'text'.	date and time	–	Cmds
▲	NEW.	(required for BMDP Files unless appending saved file to existing file)	–	SpecF
	FORMat = *(one only)* '(format)', F, G, BINARY.	(required to save raw data file)	–	8.3
	KEEP = list.	save all variables in original order	–	8.3
	or DELete = list.		–	SpecF
	MISSing = # list.	write internal flags for missing and out-of-range values to BMDP File (asterisks written to formatted files)	–	8.3
	COMPlete.	all cases saved	–	SpecF
	APPEND = list.	no multiple output records	–	SpecF
	PORTable.	nonportable BMDP File saved	–	SpecF
■	/ INPUT	(required for BMDP File input; see Chapter 3)		
■	FILE = 'name'.	CONTROL	•	8.2
	or UNIT = #.	5	•	8.2
▲	CODE = name.	(required for BMDP File input)	–	8.2
	CONTent = parameter.	first CONTENT saved in BMDP File	–	Cmds
	TYPE = parameter.	(overrides the CONTENT parameter)	–	Cmds
	LABEL = 'text'.	label generated at time of BMDP File creation	–	Cmds
	DIRectory.	no directory printed	–	SpecF
	PORTable.	(required to read BMDP File saved with PORT)	–	SpecF

Key:
- ■ Required command or paragraph
- ▲ Frequently used command or paragraph
- • Value retained for multiple problems
- – Default reassigned

9

The CONTROL Paragraph

The CONTROL paragraph is optional and rarely used. You can use it to reset commands to their default values, specify the number of columns of instructions read by the program, control how carefully the program checks for errors in instructions, access macro files of instructions, limit the amount of storage allocated for a job, and (on a few systems) specify interactive use of BMDP. CONTROL also shares with the PRINT paragraph some commands to control the verbosity of computer output and to print diagnostics for debugging programs.

You must specify the CONTROL paragraph in the first line of your instructions and follow it with an END paragraph, before any other instructions. Your other paragraphs (INPUT, VARIABLE, etc.) begin on the next line and are followed as usual by another END paragraph.

Where to Find It

9.1 Controls for BMDP instruction reader

It is possible to submit multiple sets of instructions (multiple problems) to any program. However, option choices set up by the first run usually carry over to the next run. The RESET command is used to return all options to their default values. Simply state between problems:

Resetting commands to their defaults: RESET

```
/ CONTROL RESET.
/ END
```

Number of columns to read: COLUMN

By default the BMDP instruction reader reads up to 80 columns of information. If you use a text editor (e.g., FSE on IBM) to create your instruction file, the editor may insert sequence numbers in columns 73 through 80. If these sequence numbers are all digits, there is no problem because the instruction reader knows how to deal with them. However, some editors may insert other characters that can interfere with the interpretation of the instructions. You can avoid this problem by specifying the number of columns (characters) in each line that the BMDP program should read and interpret. For example, if the unwanted characters begin in column 73, then state

```
/ CONTROL  COLUMN = 72.          / END
```

You may also want to read only part of a line, reserving the remainder for comments (see also the comment feature in Chapter 10). COLUMN affects both instruction and data reading. See also RECLEN in the INPUT paragraph to limit the number of columns of data read.

Number of errors to allow: ERRLEV

The ERRLEV command controls how the program checks your instructions for errors. Normally, the program scans the instructions for paragraphs and commands specific to the current program, and ignores extraneous information, such as a paragraph not used in the current program. If you specify

```
ERRLEV = STRICT.
```

the program stops when it finds extraneous information and prints an error message. During an interactive run, the program queries you about whether to stop or continue.

Accessing stored instructions: MACRO

MACRO permits you to insert frequently used commands (a macro) into your current BMDP instructions by specifying the system file name or unit number where the macro is stored. Insert the instruction

```
/ CONTROL MACRO = 'mymacro.txt'. / END
```

before the problem in which you want to use the file. Specify a file name for VAX, IBM PC, etc., or specify a unit number for IBM mainframes and other machines. See Chapter 11 for a detailed description of how to use macros.

Location of an instruction file: FILE or UNIT

The FILE and UNIT commands in the CONTROL paragraph are a simpler but less flexible version of the MACRO capability explained above. MACROS can be inserted anywhere in your instructions, but FILE and UNIT can only be used to access the instructions to start a run. Use FILE or UNIT to access instructions stored in another file and then return to your original file for the remainder of the input.

For example, when you analyze the same data repeatedly, it may be convenient to have a BMDP instruction file containing the common paragraphs in a standard configuration. Assume that a file named STANDARD.INP contains instructions for the INPUT, VARIABLE, and TRANSFORM paragraphs. On the current run you state

```
/ CONTROL FILE = 'STANDARD.INP'.      / END
```

followed by the remainder of your instructions. The program reads the instructions from STANDARD.INP, and then returns to the current file or to the terminal for more input. See the BMDP line editor, Chapter 11, for an easier way to store and retrieve files of instructions when running interactively.

9.2 Limiting storage space: LENGTH

If you plan to run the same job or very similar jobs several times, you may want to tailor the number of available words of storage to fit your program. By specifying LENGTH you can limit allocated storage to the amount your job uses so the job will run more efficiently. To determine the proper LENGTH, run the program once, and check the last output page for the number printed after the phrase

```
"number of integer words used in preceding problem"
```

Specify a LENGTH slightly larger than the number printed, N(used), but no larger than the number N(allocated) printed at the beginning of the program, in the first panel after the phrase

```
"number of integer words of memory for storage"
```

N(allocated) is the amount of storage the program allocates automatically. Note that the amount of storage specified in LENGTH must be significantly smaller (say 30% or more) than N(allocated) if you are to affect any economy.

On some systems you can increase the number of integer words allocated for storage when your job is large; see Appendix A.

CONTROL Commands

/ CONTrol

Optional. The CONTROL paragraph is stated before any other BMDP instructions and must be followed by an END paragraph. The other paragraphs (INPUT, VARIABLE, etc.) follow after END.

COLumn = #. `COL = 72.`

Specify the maximum number of characters per line that contain BMDP instructions or data. You may wish to read only part of a line, reserving the remainder for comments. Some system editors (e.g., FSE on IBM mainframes) may use the last seven character positions on a line for sequence numbers; you should specify COL = 72 to avoid reading those numbers. Default: 80 characters, 72 in interactive mode. **Example:** Columns 73 through 80 are not read.

ERRlev = *(one only)* `ERR = STRICT.`
NONE, INTeractive,
NORMal, STRict.

Controls how carefully the program checks your instructions for errors. Default: NORMAL.

NONE – The program will stop only if it is unable to continue because of a system error (not recommended).

INTeractive – The program will stop if calculations cannot be executed or if the instructions make it impossible to continue (e.g., an empty VARIABLE USE list).

NORMal – The program will stop if there is a serious error. Extraneous instructions are ignored.

STRict – The program will stop if there is a warning. For example, the program stops when it finds extraneous information (e.g., when your instructions contain a paragraph from another program).

MACro = name *or* **#.** `MAC = 'MACRO1.TXT'.`
`MAC = 9.`

Specify the system file name or unit number where a macro file is stored. See Chapter 11 for a definition of macro files. Insert the instruction / CONTROL MACRO = name. / END before the problem in which you want to use the file.

Specify a file name for VAX, IBM PC, etc., or specify a unit number for IBM mainframes and other machines.

FILE = name. `FILE = 'MYDATA.DAT'.`
or **UNIT = #.** `UNIT = 9.`

State the file name or unit number with BMDP instructions to start a run stored on disk or tape. FILE is used for VAX, IBM PC, and other machines. UNIT is used for IBM mainframes and other machines. For UNIT, use numbers 7 to 25. At the end-of-file, reading returns to the terminal if the run is interactive.

LENgth = #. `LENGTH = 10000.`

Use to limit the size of storage area (the number of real words of core memory). The default LENGTH is system dependent. To increase storage area on different systems, see Appendix A.

INTeractive. `INT.`

BMDP programs are usually run interactively by invoking them in interactive mode (see Chapter 10). Some systems may not be set up to detect whether you want to run interactively; in that case you can use the command /CONTROL INTERACTIVE. /END. Few systems now use this command, so see your installation guide or system manager for information about your system. Default is NO INT unless running on a system that can detect whether you invoked the program in interactive or noninteractive mode.

DEBUg = *(one only)* `DEBUG = ALL.`
NONE, TEST, INFO, ALL.

See Chapter 7, the PRINT paragraph, for DEBUG definition.

LEVel = *(one only)* `LEVEL = BRIEF.`
MINimal, BRIEF, NORMAL.

See Chapter 7, the PRINT paragraph, for LEVEL definition.

RESET. `RESET.`

Resets all commands to their default when running multiple problems. By default, some commands retain the value you assign to them from problem to problem.

/ END

Required following the CONTROL paragraph.

Summary Table for CONTROL Commands

Paragraphs Commands	Defaults	Multiple Problems	See
/ CONTrol			
COLumn = #.	80; 72 if interactive	•	9.1
ERRlev = *(one only)* NONE, INTeractive, STRict, NORMal.	NORMAL	–	9.1
MACro = name *or* #.	none	•	9.1
UNIT = #.	none	•	9.1
FILE = name.	none	–	9.1
LENgth = #.	storage allocated automatically by the program	–	9.2
INTeractive.	NO INTERACTIVE.	•	Cmds

Summary Table (continued)

Paragraphs Commands	Defaults	Multiple Problems	See
DEBUg = *(one only)* NONE, TEST, INFO, ALL.	NONE	●	Chap. 7
LEVel = *(one only)* NORMAL, BRIEF, MINimal.	NORMAL (batch); MIN (interactive)	●	Chap. 7
RESET.	defaults not reset	–	Cmds
■ / **END**			

Key:
- ■ Required paragraph or command
- ● Value retained for multiple problems
- – Default reassigned

10

Running BMDP

There are two ways to run a BMDP program from a computer terminal. In one, you submit a job to the computer, then look at the results with a system editor or on a printout. This is called *noninteractive* mode. We demonstrated this method in Chapter 1. In the other, you run the programs through an editor that allows you to enter and execute instructions, view the output at the terminal as it is generated, then alter your instructions and resubmit them without exiting the program. This is called *interactive* mode. All BMDP programs can be run interactively in this sense. In addition, a number of programs respond immediately to an instruction or a small subset of instructions, allowing you to intervene in the analysis to add additional instructions or to change the order in which variables enter a model. These special interactive capabilities are discussed in general form in this chapter and in more detail in program chapters.

BMDP programs are easy to use in both interactive and noninteractive modes. In this chapter we provide instructions to get you up and running in either mode. We also provide several sections on running multiple analyses.

Where to Find It

10.1 How to invoke BMDP

The instructions used to invoke BMDP programs vary depending on the computer system and the type of run desired. On multitasking systems like VAX/VMS, there are two kinds of noninteractive runs: *foreground* and *background* (the distinction does not exist on the PC version of BMDP). Foreground operations execute at the terminal while you wait. Background operations (also known as *batch*) allow you to use your terminal for other tasks while the program runs. Background runs may be considerably more convenient than foreground or interactive runs. The table below outlines how to invoke BMDP on VAX/VMS and IBM PCs; see your BMDP Installation Guide or system manager for information about other systems. For more information about running BMDP on a PC, see the *BMDP PC User's Guide.*

Table 10.1
How to invoke BMDP in noninteractive and interactive modes: sample instructions

PC	Interactive	`BMDP` *xx*		`[OUT=outfile]`
	Noninteractive	`BMDPRUN` *xx*	`IN=infile`	`[OUT=outfile]`
VAX/VMS	Interactive	`BMDP` *xx*		
	Nonint. (Backg.)	`BMDPRUN` *xx*	`IN=infile`	`OUT=outfile`
	Nonint. (Foreg.)	`BMDP` *xx*	`IN=infile`	`[OUT=outfile]`

xx is the BMDP program name—e.g., 2D. Instructions in square brackets are optional.

10.2 Running BMDP in noninteractive mode

A noninteractive run uses instructions and data prepared in advance and stored in files. On VAX/VMS, noninteractive runs may be foreground or background; differences between the two modes are summarized below.

Foreground `>BMDP 1D IN=infile [OUT=outfile]`

Output displayed at terminal (scrolls by without pausing) **or** written to outfile if specified (no output on screen)

Background `>BMDPRUN 1D IN=infile OUT=outfile`

Output written to outfile. Terminal freed for other tasks while program runs (also known as batch mode)

The "infile" specification contains the name of a file containing your BMDP instructions. Program output is placed in the file named in the "outfile" specification.

For example, assume that you typed your instructions into a file named JOG.INP using a standard system editor such as EDT.

Figure 10.1
Contents of system file JOG.INP

```
/ INPUT      FILE = 'exercise.dat'.
             VARIABLES = 6.
             FORMAT IS FREE.

/ VARIABLE   NAMES = id, sex, smoke, age, pulse_1, pulse_2.

/ END
```

To run program 1D with these instructions and store the resulting output in a file named JOG.OUT, type

```
BMDPRUN 1D IN=jog.inp OUT=jog.out [Return]
```

Note that two preexisting files are accessed here: the instruction file JOG.INP and the data file EXERCISE.DAT. (Data could also be stored following the instructions in JOG.INP.) After the program runs, the output in JOG.OUT can be examined through a system editor and printed using a system print command. For an example of complete BMDP program output, see Chapter 1.

10.3 Running BMDP interactively

Interactive mode allows you to use a BMDP program and make changes to the instructions and data without exiting. Interactive options are available for certain programs: 2R, 7M, LR, and 2L permit intervention in stepwise procedures, and CA, 1D, 2D, 3D, 5D, 6D, 7D, 9D, 4M, 5M, 1R, 2R, 2L, 7M, LR, LE, 1T, 2T, 1V, 4V, 5V, and DM allow paragraph by paragraph interaction (see the individual program instructions or *User's Digest* for details). On some systems, interactive mode also allows you to execute system commands such as DIR. The program output is displayed on the screen. BMDP output width and extent of output are automatically reduced in interactive mode (see the PRINT paragraph for output controls). In all interactive programs, the CONTROL RESET / command tells the program to stop reading interactive paragraphs and to read a new problem in the batch mode. This switch from the interactive to the batch mode can also be done without resetting all user supplied options, by using an interactive FINISH paragraph, which may be used to terminate any interactive program.

To run interactively, begin by logging onto your system. You will receive a prompt, e.g., $. The format for initiating an interactive run varies from system to system; on an IBM PC or VAX/VMS system it is

```
$ BMDP 1D <Return>
```

Do not specify input or output files. (If you are running on VAX/VMS the system will query you for them—just press <Return>.) Output files cannot be saved when running interactively on a VAX. On a PC, you can specify an output file: all program output is displayed on the screen **and** written to the output file. For economy, you may wish to debug your instructions interactively on a small subset of the records, then run them in noninteractive mode for the whole file.

When you call a program in interactive mode, you enter the BMDP line editor. In the PC and X Windows versions of BMDP, the line editor has been replaced by a full screen editor. See the *User's Guide* for your system details. You will see a banner showing the program you have called, followed by a line number: 1 Com>. You may then use the line editor to type in instructions directly (see the INSERT command) or to **get** an existing file of instructions, e.g.:

```
1 Com> GET JOG.INP <Return>
```

The instructions stored in JOG.INP appear on the screen with line numbers. While in the line editor, you can make changes to existing data and instruction files, or create a completely new set of instructions or data for the run. Use the PUT command to save modified instruction files. (Details of line editor use are given in Section 10.7.) Then, when you are ready to run the BMDP program, type the line editor command E to EXECUTE the contents of JOG.INP:

```
E <Return>
```

The line editor will respond by repeating your instruction file and asking whether you want to enter more data. It then shows the following prompt:

```
Press <RETURN> to continue.
```

At this prompt, you actually have three options. We cover these options next.

If you press <RETURN>: You will usually choose this option. The program executes your current instructions, pausing after each page of output. After all the output is displayed, you are returned to the line editor and can continue editing instructions and data. This ability to loop through a program several times allows you to make successive modifications of your instructions and data. Use the QUIT command to return to the system from the line editor.

If you specify N (for no interaction) at any of the steps where the computer asks for interaction or no interaction, then BMDP will complete the run without pauses between pages. After your output is displayed you will be returned to the line editor.

Invoking operating system commands during interactive execution. On some systems, you can use system commands while in the line editor or during a pause in interactive operation. To use a system command while running BMDP interactively, type !(command). For example !DIR on a PC invokes the system directory command; after displaying the file names in your directory, control returns to where you left off (e.g., in the line editor). The special command !e allows you to re-enter the line editor without executing the rest of the task you have requested.

You can also use this feature to access other editors. Instead of the BMDP line editor, you can use any editor or word processor that produces ASCII or character image files (see Chapter 2) to create and edit input files (e.g., !EDIT). When control returns to the line editor, simply get (G) the new file.

Program control level (silent running). You can enter program control level from the BMDP line editor by using the B command (see Section 10.7). In BMDP program control level, you enter instructions and data directly from the keyboard. There are usually no prompts, and you cannot edit or save instructions. (The BMD> prompt may sometimes appear in program control level.) Program control level is useful for programs 3D, 1T, 2T, 4V, and DM, which allow paragraph by paragraph interaction. You can perform an analysis and add additional paragraphs one at a time without going through the line editor. When you add paragraphs interactively, each paragraph should begin on a new line **and** end with a slash. If your interactive paragraphs are long and complex, you should probably use the line editor to avoid lengthy retyping of errors.

If you decide to return to the **system**, type

```
FINISH/ <Return>
```

FINISH is a special paragraph used only for terminating interactive runs and exiting the program. Before or after running a program in this mode, you can return to the line editor by typing

```
!e <Return>
```

or entering the instruction CONTROL INTERACTIVE. / END (see Chapter 9). The line editor returns the last edited instructions and displays them on the screen.

> ***Caution:*** *Program control level tends to be unforgiving. Unless you are an experienced user, we recommend running programs from the editor.*

The BMD> prompt. After a set of instructions has been submitted from the line editor with the "e" command, sometimes the prompt BMD> appears. This occurs when more instructions are expected or the submitted instructions are incomplete (for example, closing punctuation or an END paragraph are missing). Typing /END will usually resume normal execution and return you to the line editor.

Accidental termination of BMDP and the BMDP.LOG file. In case you neglect to save your current file of instructions, the system saves all instructions and data entered during the last program run in a file called BMDP.LOG, which is erased and rewritten every time you enter BMDP. This information can be retrieved by exiting the program and renaming and editing the BMDP.LOG file for permanent use. When running a BMDP program interactively, it should terminate only upon your request (i.e., FINISH or by the line editor command Q). If a program terminates unexpectedly, copy your BMDP.LOG file into another file and use the name of this file in the line editor "Get" command when you restart the program.

An interactive example

In this section we demonstrate a simple interactive run on a VAX using program 1D and the instruction file JOG.INP described in Section 10.2. The computer output which you would see at the terminal is screened. Arrows mark interventions by the user.

```
bmdp 1d
Name of BMDP Instruction Language file:
Name of file to write output to:
Now running 1D ...

BMDP1D - SIMPLE DATA DESCRIPTION

Copyright 1977, 1979, 1981, 1982, 1983, 1985, 1987, 1988, 1990, 1992
                 by BMDP Statistical Software, Inc.

       BMDP Statistical Software, Inc.| BMDP Statistical Software
       1440 Sepulveda Blvd            | Cork Technology Park, Model Farm Rd
       Los Angeles, CA 90025 USA      | Cork, Ireland
          Phone (310) 479-7799        |    Phone +353 21 542722
          Fax   (310) 312-0161        |    Fax   +353 21 542822
          Telex 4972934 BMDP UI       |    Telex 75659 SSWL EI

Version: 7.0     (VAX/VMS)          DATE:    19-FEB-92       AT 09:22:18
 Manual: BMDP Manual Vol. 1 and Vol. 2.
 Digest: BMDP User's Digest.
Updates: State NEWS. in the PRINT paragraph for summary of new features.

The BMDP output is displayed one screen at a time.  At each screen you can:
  - Press <RETURN> to view the next screen of output
  - Type N <RETURN> to scroll through the output nonstop (not recommended)
  - Type !E <RETURN> to stop executing and return to the editor
  - Type !system-command <RETURN> to execute a system command
After the next <RETURN>, editor is called to prepare your BMDP commands.
When finished editing, type EXIT to execute commands or QUIT to stop BMDP.
Press <RETURN> to continue.

BMDP line Editor.  For further information, Type HELP
Buffer size is  492 lines.

   1 Com>g jog.inp <Return>  <----

Reading:
jog.inp

Buffer now has    7 lines.
   1 >/ input      file = 'pulse2.dat'.
   2 >             var = 6.
   3 >             form = free.
   4->
   5 >/ variable   names are id, sex, smoke, age, pulse_1, pulse_2.
   6 >
   7 >/end
   8 Com>e  <Return> <----
       Lines    1 through    7 are executing ...
/ input      file = 'pulse2.dat'.
             var = 6.
             form = free.

/ variable   names are id, sex, smoke, age, pulse_1, pulse_2.

/end
Press <RETURN> to continue.

 CASE    1         2         3         4         5         6
  NO. id        sex       smoke     age       pulse_1   pulse_2
----- -------- -------- -------- -------- -------- --------
    1     1.00      1.00      1.00     31.00     62.00    126.00
    2     5.00      1.00      1.00     21.00     66.00    128.00
    3    10.00      1.00      1.00     28.00     90.00    144.00
    4    15.00      1.00      1.00     23.00     92.00    134.00
    5    20.00      1.00      1.00     20.00     66.00    132.00
    6    21.00      1.00      1.00     38.00     70.00    122.00
    7    29.00      1.00      1.00     22.00     68.00    126.00
    8    30.00      1.00      1.00     19.00     70.00    144.00
    9    36.00      1.00      1.00     34.00     62.00    132.00
   10    38.00      1.00      1.00     21.00     90.00    138.00

NUMBER OF CASES READ. . . . . . . . . . . . . .          40
Press <RETURN> to continue.  <Return>  <----


DESCRIPTIVE STATISTICS OF DATA
----------- ---------- -- ----

 VARIABLE    TOTAL          STANDARD  ST.ERR    COEFF     SMALLEST       LARGEST
 NO. NAME     FREQ.  MEAN     DEV.    OF MEAN  OF VAR   VALUE  Z-SCR  VALUE  Z-SCR

   1 id          40  20.500  11.690  1.8484  .57027  1.0000 -1.67  40.000  1.67
   2 sex         40  1.4500  .50383  .07966  .34747  1.0000 -0.89  2.0000  1.09
   3 smoke       40  1.6000  .49614  .07845  .31009  1.0000 -1.21  2.0000  0.81
   4 age         40  26.650  6.8894  1.0893  .25851  19.000 -1.11  43.000  2.37
   5 pulse_1     40  75.950  9.5379  1.5081  .12558  62.000 -1.46  96.000  2.10
   6 pulse_2     40  140.50  23.690  3.7458  .16861  112.00 -1.20  265.00  5.26

NUMBER OF INTEGER WORDS USED IN PRECEDING    PROBLEM      648
CPU TIME USED       7.220 SECONDS

        Press <Return> to continue <Return>   <----

BMDP line Editor.  For further information, Type HELP
Buffer size is  487 lines.

Buffer now has    7 lines.
   1 >/ input      file = 'pulse2.dat'.
   2 >             var = 6.
   3 >             form = free.
   4->
   5 >/ variable   names are id, sex, smoke, age, pulse_1, pulse_2.
   6 >
   7 >/end
   8 Com>q  <Return>  <----

PROGRAM TERMINATED
Name of Program to run:  <Return>  <----

All done.
$
```

Switching local editors

You can use your favorite editor in place of the BMDP line editor on most versions of BMDP. We provide instructions for switching between editors on VAX/VMS and UNIX systems below. Consult your system manager for directions on switching editors for other systems.

VAX/VMS version. By default on a VAX/VMS system, the BMDP line editor is called for writing instructions. You can change to your local editor using the ASSIGN command. To switch from the line editor to your local editor, type

```
$ ASSIGN  editor name  BEDIT.
```

For example, to switch to EDIT on the VAX, type

```
$ ASSIGN EDIT BEDIT
```

To switch back to the line editor, type

```
$ ASSIGN BEDIT BEDIT
```

The EXIT command in EDIT runs your current set of instructions. When the BMDP program completes the run, you are returned to EDIT. You may use other EDIT commands as described in your EDIT documentation.

UNIX version. Use of an editor in the UNIX version depends on how you call up BMDP. If you specify only the program name (1D), your local UNIX editor is called. If you specify BMDP and then the name of the program (BMDP 1D), the line editor is called. You can can call up your local editor regardless of invoking commands by using the SETENV command. To switch from the line editor to your local editor, type

```
setenv BMDP_EDIT editor name
```

For example, to switch to VI, type

```
setenv BMDP_EDIT VI
```

Note that such a specification changes the editor for your *current run only*. To change the default editor completely, type

```
BMDP setup
```

and follow the system prompts, making appropriate changes.

10.4 Multiple subproblems

Several BMDP programs permit multiple analyses of the same data **within** a single problem. These multiple analyses are called subproblems; they differ from multiple problems in that the data are not reread. Many commands do not have to be respecified in subproblems; see Section 10.6. You specify subproblems in one of three ways, depending on the program you are using:

- by repeating one or more paragraphs *within* initial instructions
- by repeating one or more paragraphs *after* initial instructions
- by repeating interactive paragraphs that are executed immediately

You can specify all three types of subproblems in interactive or noninteractive mode. To determine how to specify subproblems for a given program, refer to the "Order of Instructions" section at the end of each program chapter.

Repeating paragraphs

Some programs allow you to repeat paragraphs within the instructions to specify repeated analyses. For example, to obtain separate analyses for different dependent variables in the stepwise regression program, 2R, you repeat the REGRESSION paragraph:

```
/INPUT ...
 ...
/REGR  DEPEND = ...
/REGR  DEPEND = ...
/END
```

The following BMDP programs allow you to specify subproblems by repeating a paragraph: 3D, 5D, 6D, 7D, 4F, 1L, 1R, 2R, 4R, 1V, and 3V.

> ***Caution:*** *Programs that use complete cases will eliminate from all subproblems cases having missing data for variables used in any subproblem. To maximize the cases used, run separate* ***problems,*** *not subproblems.*

Paragraphs that follow the problem

Other BMDP programs allow you to repeat a subset of paragraphs after the first problem. These programs require you to include the paragraphs for additional analyses after the data, when your data are stored in the same system file. For example, the setup for the nonlinear regression program, 3R, would be

```
/INPUT ...
 ...
/REGRESS ...
/PARAMETER ...
/END
 (data if not in a tape or disk file)
/REGRESS ...
/PARAMETER ...
/END
```

In 3R the paragraphs REGRESS, PARAMETER, and END may be repeated as many times as you like. Programs that allow subproblems in this manner include 9D, 1M, 3R, 5R, AR, 3S, and 2L.

Interactive subproblems

A third set of programs allows you to insert one or more paragraphs at a time and repeatedly view and modify output without exiting the program. The following programs have these interactive paragraphs: 1D, 2D, 4M, 7M, 3D, 5D, 6D, 7D, 9D, 1R, 2R, 1T, 2T, 1V, 4V, 5V, and LE. In 3D, for example, you can specify any number of MATCHED, ONEGROUP, TWOGROUP, or TRANSFORM paragraphs after you specify INPUT, VARIABLE, and END paragraphs. In each case, the interactive paragraph should end with a slash, and the rest of the line after the slash should be blank. See the descriptions of each program for details.

Specifying an interactive paragraph. In program 3D, you may specify one or several interactive paragraphs between /END and the next problem (or the end of the run). We illustrate how to use an interactive TWOGROUP paragraph in the following example.

After invoking 3D interactively, you are placed in the line editor. In the line editor, you may specify initial instructions as in the 1D example. We add a GROUP, a TWOGROUP, and a PRINT paragraph to the 1D instructions as follows. After entering the instructions, we request 3D to run the instructions using the E command. (See section 10.7 for details on using the line editor.)

```
1 Com>INS
Return to command mode by typing '//'
1 Ins>/INPUT              FILE = 'exercise.dat'.
2 Ins>                    VARIABLE = 6.
3 Ins>                    FORMAT = FREE.
4 Ins>/VARIABLE NAMES = id, sex, smoke, age,
                                   pulse_1,pulse_2.
5 Ins>/GROUP              CODES(smoke) = 1,2.
6 Ins>                    NAMES(smoke) = smoke, nosmoke.
7 Ins>/TWOGROUP GROUPING = smoke.
8 Ins>                    VARIABLE = pulse_1.
9 Ins>/PRINT              LINESIZE = 80.
10 Ins>/END
11 Ins>//
11 Com>e
```

The program prints histogram and table output as follows and returns you to the line editor.

```
NUMBER OF CASES READ. . . . . . . . . . . . . .         40
Press <RETURN> to continue.

pulse_1   VARIABLE NUMBER   5
*****************************

GROUP   1 smoke                     2 nosmoke



                             X
   H                         X X    X
 H H H          HH           XXX XX X   X
 H HHH   HH     HHH H      XXXXX XXXXXX X    X
M---------------------M  M--------------------M
I  AN H=    1 CASES  A  I  AN X=    1 CASES  A
N   (N=   16)        X  N   (N=   24)        X
Press <RETURN> to continue.

GROUP       smoke        nosmoke        TEST STATISTICS    P-VALUE  DF
-------------------------------         ------------------------------
MEAN        76.7500      75.4167        LEVENE F FOR
                                         VARIABILITY     8.67 0.0055 1,   38
STD DEV     11.9972       7.7230
S.E.M.       2.9993       1.5765        POOLED     T     0.43 0.6707    38
SAMPLE SIZE      16           24        SEPARATE   T     0.39 0.6975    23.3
MAXIMUM     96.0000      96.0000
MINIMUM     62.0000      64.0000
Z MAX        1.60         2.67
Z MIN       -1.23        -1.48
CASE (MAX)        6            4
CASE (MIN)        1            3
```

After viewing the output, you decide you want further analysis of the data. You want to test the PULSE_2 variable and obtain additional statistics. So, you use the editor to specify an additional TWOGROUP paragraph. 3D automatically uses the data in the previous problem.You specify the following instructions and request the editor to execute only those instructions.

```
11 Com>INS
Return to command mode by typing '//'
11 Ins>TWOGROUP GROUPING = smoke.
12 Ins>          VARIABLE = pulse_2.
13 Ins>          ROBUST.
14 Ins>          NONPAR.    /
15 Ins>//
16 Com>e 11 14
```

The program next prints the following histogram and table output using the new TWOGROUP instructions.

```
pulse_2   VARIABLE NUMBER   6
*****************************

GROUP   1 smoke                     2 nosmoke


                             X
                             X
                             X
  HH H                    XXXXX
  HHHHHH               H  XXXXXXX
M---------------------M  M--------------------M
I  AN H=    2 CASES  A  I  AN X=    2 CASES  A
N   (N=   16)        X  N   (N=   24)        X

Press <RETURN> to continue.

GROUP       smoke        nosmoke        TEST STATISTICS    P-VALUE  DF
-------------------------------         ------------------------------
MEAN        147.3125     135.9583       LEVENE F FOR
TRIM MEAN   140.7143     136.0455        VARIABILITY     2.01 0.1641 1,   38
STD DEV      33.4758      12.9228
S.E.M.        8.3690       2.6379       POOLED     T     1.51 0.1395    38
SAMPLE SIZE      16           24        SEPARATE   T     1.29 0.2120    18.0
MAXIMUM     265.0000     158.0000       TRIM POOL.T      1.03 0.3096    34
MINIMUM     122.0000     112.0000       TRIM SEP. T      1.03 0.3119    27.7
Z MAX         3.52         1.71
Z MIN        -0.76        -1.85         MANN-WHIT.     228.0 0.3195
CASE (MAX)        6           17         (RANK SUMS      364.0       456.0)
CASE (MIN)       21           16
2ND MAX     160.0000     158.0000
2ND MIN     126.0000     116.0000
```

You are returned to the editor and can continue specifying interactive paragraphs on the same data or start a new instruction file. Be sure, however, to specify the analysis paragraph (i.e., TWOGROUP) with other interactive paragraphs. (We recommend that you include an analysis paragraph in interactive instructions even if you do not specify an alternate design.) You only need to begin a new problem if you want to read in a different data set or change the defaults in the VARIABLE, GROUP, SAVE, or PRINT paragraphs. After the last interactive paragraph, type END/.

You can use interactive paragraphs in either interactive or noninteractive mode. In noninteractive mode, these paragraphs operate like other paragraphs specified after the initial instructions; the difference is that you must specify a slash after the paragraph.

Interactive transformations are available in 3D, 5D, 6D, 7D, 9D, 1R, 2R, 1V, 5V, and LE. Whenever these programs expect an interactive paragraph, you may place a TRANSFORM paragraph.

10.5 Multiple problems

Every BMDP program can analyze multiple problems; that is, you can execute a program several times in one computer run. A problem consists of the complete set of paragraphs needed to do one analysis (e.g., INPUT, VARIABLE, END). When your data are in a separate system file from the BMDP instructions, you can generate multiple problems by repeating consecutive sets of paragraphs, rereading the data each time:

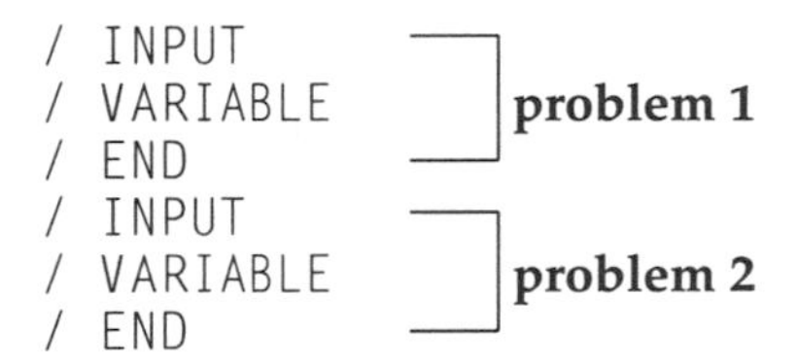

When your data follow the BMDP instructions in the same system file, you must repeat the data after each set of paragraphs, and you must either include an END paragraph after the data **or** specify the number of cases in your data (INPUT CASES = #). If you do both, only the first problem will run.

We recommend that you store your data in a separate system file when you want to perform multiple problems, since repeating the data after each problem is cumbersome. Alternatively, you can save the data from the first problem in a BMDP File and use the BMDP File as input to later problems (see Chapter 8). If you read data from a system file in the first problem and then want to read data following the instructions for a subsequent problem, you must state FILE = CONTROL in the INPUT paragraph (on unit-oriented systems, state UNIT = 5).

10.6 Commands that default from problem to problem

Many of the commands specified in the first problem do not have to be repeated for additional problems or subproblems. The program remembers the values you have assigned in the first problem and applies those values in the next problem. For example, to run two analyses of variance using 2V, one comparing males and females and one comparing smokers and nonsmokers, use the instructions in Figure 10.2:

Figure 10.2 Example of commands that default

```
/ INPUT      FILE = 'exercise.dat'.                    ┐
             VARIABLES = 6.                            │
             FORMAT = FREE.                            │
/ VARIABLE   NAMES = id, sex, smoke, age,              │
                     pulse_1, pulse_2.                 │
/ GROUP      CODES(sex, smoke) = 1, 2.                 │  Problem 1
             NAMES(sex) = male, female.                │
             NAMES(smoke) = yes, no.                   │
/ DESIGN     GROUP = sex.                              │
             DEPEND = pulse_1, pulse_2.                │
             LEVEL = 2.                                ┘
/ END
/ INPUT                                                ┐
/ DESIGN     GROUP = smoke.                            │  Problem 2
             DEPEND = pulse_1, pulse_2.                │
             LEVEL = 2.                                ┘
/ END
```

Most of the assigned values for the commands in Problem 1 remain the same for Problem 2. We only need to restate INPUT. The FILE, FORMAT, and VARIABLE commands remain the same. We respecify DESIGN to request a new GROUP for the second analysis, and we must also restate the DEPEND and LEVEL commands. To determine which commands are required in additional problems and which retain their values, see the Summary Table at the end of each individual program description and at the end of Chapters 3 through 9. Each command acts in one of three ways during additional problems:

- Required commands must be specified for each analysis.
- Commands listed as having value retained for additional problems do not have to be respecified. Respecify these commands only when you want to change the command values.
- Commands listed as having default reassigned must be respecified unless you want to use the default value for the command. (For example, the VARIABLE USE list must be specified for each problem unless you want to use all input variables, the default for the USE command.)

Single word commands with value retained can be countermanded for later problems by using the word NO along with the command word. For example, you could request a printout of your data in a first run of 1D and then state NO DATA in the next problem. The VARIABLE and GROUP paragraphs include numerous commands that retain their value from problem to problem and would be tedious to respecify. State RESET in these paragraphs to reset all options to their original default value. RESET is also available in the CONTROL paragraph (see Chapter 9).

10.7 The BMDP line editor

The line editor is available when you invoke a BMDP program in interactive mode. Its principal function is entering and editing instructions for input to BMDP programs, but it can also be used to enter and edit data files. The line editor prompts you for required information and responds to your answers. You can edit as many files as you choose while in the line editor; use the PUT command to save modified files. When you are ready to run the program, use the GET command to bring in the appropriate file. See Table 10.2 for a quick summary of the commands available; examples and detailed definitions follow the table.

> ***Caution:*** *The line editor can only create and use lines of 80 characters or less. If your data have more than 80 characters on a line, the line editor will truncate the data.*

Basically, the editor functions by inserting, deleting, or replacing entire lines of text. You can also correct text within a line by substituting one text string for another (see SUBSTITUTE command). The line editor makes no distinction between upper and lower case, e.g., EXERCISE and exercise. The tutorial that follows describes the most common line editor operations.

The following prompts appear on the screen and define the current line editor mode (# represents a line number on the screen):

`COM>` – means COMMAND mode. Respond with one of the 17 line editor commands. The command will be executed at the current line.

`INS>` – means INSERT mode. The next line typed will be inserted at the line indicated. Insert mode remains in effect until you type a double slash: 10=>//

`#->` – indicates the current line. The next command will be executed in reference to that line unless you specify another line number.

Table 10.2
Quick summary of line editor commands (see Table 10.3 for detailed definitions)

Getting in and out of the Line Editor	!e	invoking the line editor
	q	QUIT (exit)
	b	BEGIN
Line Editor Help	h	HELP
File Commands	g	GET file
	p	PUT file
	l	LIST lines
Run BMDP Program	e	EXECUTE instructions
Line Editing Commands	i	INSERT mode
	d	DELETE line
	r	REPLACE lines
	m	MOVE line
	c	COPY line
	u	UNDO last delete
Search Commands	f	FIND text
	s	SUBSTITUTE lines
Set up Commands (rarely used)	a	ALL
	n	NUMBER

You can save time by using command abbreviations. You may want to type the entire command, but the first letter is sufficient. The <Return> key must be pressed after each line editor command. With the exception of DELETE, REPLACE, GET, PUT, and NUMBER, line editor commands are repeated when the <Return> key is pressed again.

A line editor tutorial

This tutorial example is designed to demonstrate some basic line editor commands. If you are already familiar with the line editor, skip the tutorial and use the command definitions in Table 10.3.

To run program 1D in interactive mode with a sample data set on a VAX or IBM PC, type

```
$ BMDP 1D <Return>
```

The program loads and transfers control to the command level of the BMDP line editor. Your screen will look like this:

```
BMDP1D SIMPLE DATA DESCRIPTION
[copyright and documentation information]

PLEASE ENTER INSTRUCTIONS
BMDP line Editor. For further information, type HELP if you need it.
Buffer size is  492 lines.

     1 Com>
```

Entering instructions. The last line, 1 Com>, is the prompt line. <Com> indicates that you are at the command level. **Type HELP if you want more information about individual commands.** To start entering your instructions for 1D, use the INSERT command. Begin by typing

```
1 Com>I <Return>
```

The prompt on the screen, 1=> , indicates that you are in INSERT mode at line 1. This prompt is repeated each time you complete a line by pressing the Return key. When you are finished with your entries, type

```
# Ins>// <Return>
```

(see line 11 below); this will return you to command mode.

To practice using the insert and command modes, type the following lines exactly as shown:

```
 1 Ins>INPUT      TITLE = 'LINE EDITOR TUTORIAL'.
 2 Ins>           VAR = 2.
 3 Ins>           FORMAT = FREE.
 4 Ins>/VARIABLE  NAME = A, B.
 5 Ins>/END
 6 Ins>1  2
 7 Ins>3  4
 8 Ins>5  6
 9 Ins>7  8
10 Ins>9  10
11 Ins>//
11 Com>
```

The Com> prompt on the last line indicates that the double slash has taken you back to the command mode. The number, 11, shows where you are in the file.

Reviewing instructions. Now use the LIST command to review your file: type L and press Return. Since the file is only 10 lines long, it fits in a single screen. Your screen will look like this:

Making corrections. Looking over the file, we find two errors in the BMDP

```
11 Com>L
 1 >INPUT       TITLE = 'LINE EDITOR TUTORIAL'.
 2 >            VAR = 2.
 3 >            FERMAT = FREE.
 4 >/VARIABLE   NAME = A, B.
 5 >/END
 6 >1  2
 7 >3  4
 8 >5  6
 9 >7  8
10->9  10
10 Com>
```

instructions:

1. the slash (/) is missing before the INPUT paragraph
2. the word FORMAT is misspelled

Typing errors detected before the Return key is pressed can be corrected by backspacing to the mistake and retyping the remainder of the line. But once the line has been ENTERED, you will have to use line editor commands. You have several options. Try the following (remember to press Return after each command):

1. Use the SUBSTITUTE command to exchange the text string INPUT with the correct text string /INPUT as follows:

   ```
   10 Com>S 1 'INPUT'/INPUT'
   ```

   Note that the single quote mark (') has been used as a separator (see SUBSTITUTE in Line Editor Command Definitions). The line on the screen now reads correctly:

   ```
   1->/INPUT    TITLE = 'LINE EDITOR TUTORIAL'.
   ```

2. Use the DELETE and INSERT commands to correct the spelling of the word FORMAT in line 3. First, delete the line:

   ```
   1 Com>d 3
   ```

The screen display records this deletion as follows:

```
1 Com>d 3
1 >/INPUT      TITLE = 'LINE EDITOR TUTORIAL'.
2 >            VAR = 2.
3->/VARIABLE   NAME = A, B.
4 >/END
5 >1  2
6 >3  4
7 >5  6
8 >7  8
9 >9  10
3 Com>
```

The prompt points to line 3 with a dash (indicating current line). To complete the correction type:

```
3 Com>I          FORMAT = FREE.
```

Now use the LIST command to check your corrections in the file.

```
4 Com>L
```

The corrected file looks like this:

```
 1 >/INPUT      TITLE = 'LINE EDITOR TUTORIAL'.
 2 >            VAR = 2.
 3->            FORMAT = FREE.
 4 >/VARIABLE   NAME = A, B.
 5 >/END
 6 >1  2
 7 >3  4
 8 >5  6
 9 >7  8
10 >9 10
 3 Com>
```

The position of the line marker has not changed.

Saving the instruction file. To save your corrected instruction file, use the PUT command. For example,

```
3 Com>P TUTOR.INP
```

saves the instructions in a file called TUTOR.INP.

BMDP execution. You can now submit these instructions to the BMDP program for analysis, using the EXECUTE command:

```
10 Com>E
```

You will receive the response:

```
Lines 1-10 are executing

        /INPUT      TITLE = 'LINE EDITOR TUTORIAL'.
                    VAR = 2.
                    FORMAT = FREE.
        /VARIABLE   NAME = A, B.
        /END

Press <RETURN> to continue.
```

After you press <Return>, the following message appears:

```
5 CASES HAVE BEEN READ.  TYPE 'END' OR ENTER MORE DATA:
```

Note that this example did not contain an END statement after the data. If it had, this query would not have been printed.

If you are ready to proceed, type END and press Return. You will see:

```
NUMBER OF CASES READ . . . . . . 5

Press <RETURN> to continue.
```

You will normally press <Return>, electing to continue in the interactive mode. The program will run as described in section 10.3. Use the QUIT command to exit from the line editor.

Table 10.3
Line editor commands

Command	Example	Definition
HELP	h	Displays command descriptions.
HELP command	h INSERT	Describes indicated command.
GET file	g MONKEY.DAT	Brings in file for editing, beginning at the current line number. For unit oriented systems (IBM mainframes, etc.), you must use the form G FT17 to access, for example, Unit 17.
GET # file	g 10 MONKEY.DAT	Loads file starting at the specified line number. The example loads MONKEY.DAT from line 10. The first 9 lines are unchanged; all lines after the 9th are deleted.
PUT file	p 1D.INP	Names and stores your current line editor text. In the example, your text is stored in file 1D.INP. For unit oriented systems like IBM mainframes you must use the form P FT17 to access, for example, Unit 17.
LIST	L	Displays 22 lines with the current line in the middle of the screen. (The NUMBER command can be used to change the 22-line default.)

Command	Example	Description
LIST #1 [#2]	`L 5`	Displays the specified line or lines. Example: Line 5 is displayed. L 5 10 would display lines 5 to 10.
EXECUTE	`e`	Concludes editing session and submits instructions to the BMDP program. After execution you are returned to the line editor.
EXECUTE #	`e 20`	Submits instructions, beginning with line 20.
EXECUTE #1 #2	`e 20 25`	Submits lines 20 through 25.
INSERT	`I`	Sets INSERT mode with prompt Ins> at current line. If you then type in a line of data or text and press Return, the data/text will be entered and the insert prompt will move to the next line, ready for more text.
//	`//`	Cancels INSERT mode and sets COMMAND mode with the prompt Com> at current line.
INSERT #	`I 22`	Sets INSERT mode beginning at specified line. In the example, insertion begins at line 22.
INSERT [#] text	`i 5 FORM = FREE.`	Enters data/text at the current or specified line position without leaving command mode. In the example, FORM = FREE. is inserted at line 5.
DELETE #1 [#2]	`d 23 27`	Deletes the designated line(s) and moves subsequent lines up. In the example, lines 23 through 27 will be deleted.
REPLACE #1 [#2]	`R 5`	Deletes the specified line or lines and puts you in INSERT mode. In the example, line 5 is deleted and the insert prompt => points to line 5, ready for text.
REPLACE #1 [#2] text	`R 5 VAR=6.`	Deletes the specified line or lines and places text in the first line deleted without leaving command mode. In the example, line 5 now reads VAR=6.
MOVE #1 #2	`m 5 10`	Moves data/text from line #1 to line #2. In the example, the text is moved from line 5 to line 10. To repeat the command, moving line 6 to line 11, etc., type M or press Return.
COPY #1 #2	`C 3 8`	Copies line #1 to #2. In the example, line 3 is copied to line 8. To repeat the command, copying line 4 to line 9, type C or press Return.
UNDO [#]	`u`	Restores the line(s) most recently deleted (they were kept in a buffer), placing them in the lines starting at the current line number or a specified number. Multiple copies of these lines may be stored in the text by pressing the Return key once for each copy.
FIND [#] text	`F orm`	Searches the file for the specified text string, starting at the current line number or a specified number. Displays the first line containing the string. Upper/lower case is ignored. Note that ORM will match in words such as NORMAL or FORMAT. To continue the search, press Return.

SUBSTITUTE [#] /t1/t2/
't1't2'
"t1"t2"
	`s /fermat/format/`
	`s 'fermat'format'`
	`S "FERMAT"FORMAT"`
		Replaces the original text string, t1, with t2. The strings to be exchanged are delineated by a slash (/), single quote ('), or double quote ("). Searching for t1 begins at the current line or a specified line number. In the example, all occurrences of FERMAT are replaced with FORMAT.

SUBSTITUTE #1 #2 /t1/t2/	`s 5 15 /fermat/format/`	Identical to the SUBSTITUTE command described above except that the search takes place between line #1 and line #2 only and replaces all occurances of t1 with t2 between lines #1 and #2 inclusively.

Getting In and Out of the Line Editor

!E	!e	Returns you to the line editor.
QUIT	q	Stops the line editor and program and returns you to the system.
BEGIN	b	When you enter the editor (or reenter it after running a problem), you are normally in COMMAND mode. The BEGIN command lets you begin in INSERT mode or in BMDP program control level with the editor turned off. In program control level, your instructions enter the program directly, without opportunity for modification and saving. Each time you type B you change the initial mode and receive a message indicating current mode.

Special Set Up Commands

Most users will not need to change the standard editor environment. However, commands are available to modify the selection of lines available for editing and number of lines displayed on the screen.

ALL	a	Each time the editor is entered it contains either the last input to the program, all previous input to the program, or neither (no initial text). The A command allows you to switch among these three modes. Each time you type A you alter the option and receive a message indicating which option is in effect.
NUMBER	n	Displays the current maximum number of lines displayed on the screen (default=22).
NUMBER #	n 10	Sets the maximum number of lines to #.

11

Shortcuts

In this chapter we describe features that help you save time when writing instructions. For example, BMDP allows you to

- abbreviate commands
- specify only a few items in a long list
- generate lists of similar names and commands

We also show how to insert comments into the instruction file to document your variables and commands, and generate multiple versions of an instruction, name, paragraph, or problem.

Where to Find It

11.1 Basic shortcuts

To reduce typing time and avoid errors, you can abbreviate many of the BMDP commands. For example, you can state

```
VAR  instead of VARIABLE
MISS instead of MISSING
```

Abbreviating commands

Refer to the command definition sections and summary tables at the end of each chapter. There, we use upper case to indicate the part of the command you must type. (You can use either upper or lower case when you type the word.) You may complete the command word any way you want to make the sentence easier to read. For example, you can make an abbreviated word plural:

```
VARS = 3 to 6.
```

As an added convenience you can omit vowels, as long as the vowel does not begin the command. For example, you can state

```
VR    instead of    VARIABLE
```

Tabbing: specifying selected entries in a list

Tabbing allows you to skip to one or more items in a list without having to respecify the entire list. For example, the Exercise data set in Chapter 1 has six input variables: ID, SEX, AGE, SMOKE, PULSE_1, and PULSE_2. If you want to name only the pulse values, tab to the fifth location:

```
NAMES = (5)pulse_1, pulse_2.
```

Similarly, to specify missing value codes for AGE and the two pulse values, you can skip to the third and then to the fifth variable, using the variable names or numbers:

```
   MISSING = (3)99, (5)999, 999.
or MISSING = (age)99, (pulse_1)999, 999.
```

Caution: *When using numbers to tab to variables, be careful to count the input variables correctly. The numbers for tabbing are based on all input variables (including LABEL variables, if any), not the VARIABLE USE list.*

Repeating numbers in a list

You can use a multiplier to specify repeated numerical values in a list. For example, when missing value codes are the same for several contiguous input variables:

```
MISSING = 9, 99, 99, 99, 99, 9, 9, 9, 9, 99, 99.
```

you can shorten this list by stating:

```
MISSING = 9, 4*99, 4*9, 2*99.
```

If most of your variables have the same missing value code, you can use the number repeating feature and then reassign values for the exceptions. The program always uses the last value assigned. You can write the above list:

```
MISSING = 11*99, (1)9, (6)4*9.
```

After assigning a missing value code of 99 to all 11 variables, you tab back to the first variable and reassign a code of 9, and then tab to the sixth variable and reassign code 9 to the sixth, seventh, eighth, and ninth variables.

Implied lists: TO and BY

You can use the implied list feature to shorten a list of names or numbers. For example, to use the second through the sixth variables in the Exercise data, SEX, SMOKE, AGE, PULSE_1, and PULSE_2, state

```
USE = sex TO pulse_2.    or    USE = 2 TO 6.
```

Use variable names or numbers; names and numbers cannot be mixed in the same implied list.

Implied lists are useful for specifying CODES, CUTPOINTS, and NAMES in the GROUP paragraph when several variables have the same values. For example, if your variables are yes-no answers to 20 questions, you can state CODES and NAMES for all 20 variables simultaneously:

```
CODES(q1 TO q20) = 1, 2.
NAMES(q1 TO q20) = yes, no.
```

Implied lists may be used on both sides of the equal sign:

```
CODES(item 1 TO item 10) = 1 TO 5.
```

When you want to use every other variable in a list, you can use the BY feature to indicate a "skip factor." This takes the form

#1 TO #2 BY #3 or name *i* TO name *j* BY #.

For example, you can easily specify lists of equally spaced cutpoints in the GROUP paragraph:

```
CUTP(age) = 20 TO 100 BY 10.
```

This feature is also useful if you have four sets of math and reading scores named MATH1, READ1, MATH2, READ2, MATH3, READ3, MATH4, READ4. You can select only the MATH scores by stating

```
USE = math1 TO math4 BY 2.
```

Note that you may not use the implied list feature for summary functions in the TRANSFORM paragraph, unless you use it within the FOR % notation (see Advanced Shortcuts).

Adding comments to document instructions

You can use a pound sign (#) to preface comments anywhere within your BMDP instructions. BMDP programs ignore anything on the record following a pound sign. For example, you can document the location of the instruction file, explain transformations, and describe variables (Figure 11.1):

Figure 11.1
Commenting instructions

```
          #--------------------------------------------------------
          # These instructions are stored in file TEST.INP
          # test1 is a math test of June 12, 1987
          # test2 is a verbal test of June 15, 1987
          # test_sum is the sum of the 2 tests
          #--------------------------------------------------------
/ INPUT . . .
/ VARIABLE  NAMES = id, test1, test2.
/ TRANSFORM test sum = test1 + test2.  # Added July 27, 1987
```

You can also use the pound sign to make BMDP ignore instructions left over from previous runs:

```
/ VARIABLE  NAMES = id, sex, smoke, age, pulse_1, pulse_2.
            GROUPING = sex.
        #   GROUPING = smoke.
```

Here the program will ignore the GROUPING = smoke instruction. Note that BMDP programs also ignore commands that do not apply to them. For example, 2D ignores GROUP and HISTOGRAM paragraphs used in 7D.

Commands with multiple options: ALL and NONE

When a command requires a list of options, you can specify ALL to obtain every option. By using ALL, you avoid listing each option name. Consider the MATRICES command in the PRINT paragraph of 9R. The command allows you to request one or more of the following matrices: CORR, COVA, RREG, CREG, and RESI. To obtain all five matrices, you could specify

```
/ PRINT  MATRICES = ALL.
```

Some commands allow ALL as a shortcut method for specifying a set of plots or statistics. For example, the PLOT command in the ESTIMATE paragraph of 1L contains the following five plots: SURV, LOG, CUM, HAZ, and DEN. To obtain all five plots, you could specify

```
/ ESTIMATE  PLOT = ALL.
```

You may note that some commands have several options, but allow use of *one only*. These commands often specify the method for an analysis or how input data is to be analyzed. For example, the METHOD command in the ROTATE paragraph of 4M indicates the method used in factor rotation. ALL is *not* available for this type of command. ALL is available only for those commands where you can specify *one or more* of the options. Note that ALL works with the CONTENT command in the SAVE paragraph.

> ***Caution:*** In a few programs, you can use ALL for commands requiring variable lists. If you do, note that the program will use *all* variables, including grouping, dependent, and independent variables.

A variation of the ALL option occurs in program 4F. In the STATISTICS paragraph, you could state ALL as a single command. Such a specification would produce all 13 statistics available in the paragraph.

The NONE command serves the opposite function of ALL. A few programs print tables and statistics by default (e.g., PRINT in 4M and 2R). NONE is useful when you do not want the default output. For example, by specifying

```
/ PRINT  NONE.
```

in 4M you could avoid obtaining the default correlation matrices.

11.2 Advanced shortcuts

Generating lists and multiple instructions: the FOR % notation

The FOR % notation allows you to generate multiple versions of an instruction, name, paragraph, or entire problem. The notation takes this general form:

FOR *a* = *b*. % *c* %

- *c* is an instruction, paragraph, name, or problem to be repeated (the % signs define its beginning and end)
- *a* is the text that is to be changed in *c*
- *b* is a list of names or numbers that will replace *a* in the repetitions of *c* (see Chapter 4 rules for naming)

Like parentheses, FOR % loops can be nested to any depth, as shown in examples below. The innermost FOR % loop is resolved first.

Repeating transformations. To replace 16 variables by their logarithms, the required instructions in the TRANSFORM paragraph are

```
X(5) = LOG(X(5)).
X(6) = LOG(X(6)).
X(7) = LOG(X(7)).
    .
    .
    .
X(20) = LOG(X(20)).
```

or

```
FOR I = 5 TO 20.
%   X(I) = LOG(X(I)).  %
```

You can replace more than one symbol in the text to be repeated. For example, to store the logarithms above as new variables (variables 21 to 36), state

```
X(21) = LOG(X(5)).        or        FOR  J = 5 TO 20.
X(22) = LOG(X(6)).                       K = 21 TO 36.
X(23) = LOG(X(7)).                  %  X(K) = LOG(X(J)).  %
      .
      .
      .
X(36) = LOG(X(20)).
```

When repeating transformations, you may want to use the name generation feature as well (described below).

> ***Caution:*** *(1) When more than one symbol is to be replaced (as in the example above), each list must have the same number of elements (one element for each repetition). (2) The index of a default name must be a number (e.g., X(1)) or a symbol to be replaced by a number in a FOR% list (e.g., X(J) in the example above). Thus, expressions such as X(J + 16) are not permitted. (3) The FOR% list cannot contain data dependent values (e.g., J = 5 TO AGE). In other words, the number of times the text within the % signs is repeated cannot depend on the value of a variable.*

Repeating names. In this example, we generate names for 30 scores (within a longer list of names):

```
NAMES ARE age, sex, SCORE1, SCORE2, ..., SCORE30, total.
```

Using the FOR % notation, we specify

```
NAMES ARE age, sex,
FOR  A = 1 TO 30.
%  SCORE|A,  %  total.
```

The vertical bar (|) links the value of A to the text (e.g., SCORE|2 becomes SCORE2). Because we include a comma within the %'s, the program generates commas after each of the 30 names.

When a generated name is last in a list of names, you need to include a period after the last % sign:

```
NAMES ARE  FOR  A = 1 TO 30.
%  SCORE|A,  %.
```

The program will then replace your last generated comma with the period.

To generate names for 30 scores on a MATH test and 30 scores on a VERBAL test, state

```
NAMES ARE  FOR A = math_, verb_.
%    FOR B = 1 TO 30.
%    A||B,  %  %.
```

In this example, two vertical bars between A and B link the value of B (rather than just the letter B) to the value of A. In other words, use two vertical bars when values on both sides of the bars can change. These instructions produce the following list of names:

```
NAMES ARE math_1, math_2, math_3, ..., math_30,
          verb_1, verb_2, verb_3, ..., verb_30.
```

Nesting FOR % statements allows you to form complex sets of names. The following example produces 15 instructions:

```
FOR a = uric, chol, alb.                    loguric1 = LOG(uric1).
% FOR b = 1 TO 5.   %      ----------->     loguric2 = LOG(uric2).
log|a||b = LOG(a||b). % %                     . . .
                                            logalb4 = LOG(alb4).
                                            logalb5 = LOG(alb5).
```

Caution: When using a FOR% loop with variable names as indices, if the names begin with a number instead of a letter, they must be enclosed in three single quotes. For example, FOR z = '''2y1d''', '''2y2d''',....

Abbreviating a list. You can also use the FOR % notation to generate multiple values within a summary function in the TRANSFORM paragraph. To obtain the sum of 30 scores,

sumscore = SUM(score1, score2, score3, ..., score30).

state

```
sumscore = SUM(FOR A = 1 TO 30. % score|A,%).
```

Note that the following form is now **acceptable**:

sumscore = SUM(score1 TO score30).

Repeating a paragraph. Some BMDP programs have paragraphs that may be repeated within one problem. To repeat the REGRESSION paragraph in 2R for different dependent variables, state

```
FOR  A = bp, pulse, endorphn.
%   / REGRESS  DEPEND = A.   INDEPEND = age, sex, activity.  %
```

The complete text within the % signs is repeated three times: first for BP, then for PULSE, and finally for ENDORPHN:

```
/ REGRESS  DEPEND = bp.        INDEPEND = age, sex, activity.
/ REGRESS  DEPEND = pulse.     INDEPEND = age, sex, activity.
/ REGRESS  DEPEND = endorphn.  INDEPEND = age, sex, activity.
```

To use different dependent and independent variables in the example above, you can use nested FOR % statements, e.g.:

```
FOR  A = bp, pulse, endorphn.
%     FOR  B = age, sex.
%  / REGRESS  DEPEND = A.  INDEPEND = B, activity.  % %
```

Repeating a problem. To generate two problems in 2D, one for males (code 1) and one for females (code 2), state

```
FOR  A = 1, 2.
%  / INPUT    FILE = 'exercise.dat'.
              FORMAT = FREE.
              VAR = 6.
   / VARIABLE   NAMES = id, sex, smoke, age, pulse_1, pulse_2.
   / TRANSFORM  USE = sex EQ A.
   / END
```

This generates

```
/ INPUT ...
/ VARIABLE ...
/ TRANSFORM  USE = sex EQ 1.
/ END
/ INPUT ...
/ VARIABLE ...
/ TRANSFORM  USE = sex EQ 2.
/ END
```

***Caution:** The input data must be in a separate system file from the BMDP instructions when you use the FOR % statement to repeat an entire problem. The END paragraph acts as the final % sign when an entire problem is to be repeated. (This is true even when you use nested FOR % statements; the END paragraph will replace any number of ending % signs when you repeat entire problems.)*

Using macros: the INCLUDE file

Lengthy instructions that you use frequently or that you will use several times in an analysis can be written once in a file and included, where needed, as you run BMDP programs. For example, suppose we want to perform the same transformations in several analyses. The first step is to create a file containing the transformations. We enter the instructions in a file named TRANS.DAT:

```
T1      / TRANSFORM  USE = age GT 20.
T1                   IF (pulse_1 GT 100) THEN ...
.                    .
.                    .
.                    .
T1                   IF (smoke EQ XMIS AND ...
```

We start the records with the characters T1, the name given to that set of transformations. You should insert a name (such as T1) using the first eight character positions of each record in the file. If you use fewer than eight characters, be sure to start in the first character position. Begin the instructions themselves starting no sooner than the ninth character position. Use up to 80 character positions to specify the instructions. We use the name T1 to refer to the transformations later. To include the transformations in the current run, we state

```
/ CONTROL   MACRO = 'trans.dat'.
/ END
/ INPUT     FILE = 'exercise.dat'.
            VARS = 6.
            FORMAT = FREE.
/ VARIABLE  NAMES = id, sex, smoke, age, pulse_1, pulse_2.

FOR  %  INCLUDE T1  %
/ REGRESS ...
/ END
```

The CONTROL MACRO command tells where to find the set of transformations we have named T1. The FOR % INCLUDE command inserts the T1 transformations into the current instructions.

You can include other sets of instructions in the same INCLUDE file, give them different names, and refer to them separately when you run the programs. The program scans all records in the INCLUDE file and uses only those prefixed with the name you specify in the FOR % INCLUDE statement.

Note: This technique is useful only for noninteractive runs. When running BMDP interactively, it is much easier to use the line editor GET command to access stored files of instructions.

Note: Nested FOR% statements used for repetition of text are resolved innermost first, while nested FOR% INCLUDE statements are resolved outermost first.

1D

Simple Data Description

1D provides descriptive statistics for all cases or for groups of cases (subpopulations). You can sort and print data in a variety of ways. Statistics include the mean, standard deviation, standard error of the mean, coefficient of variation, standardized and unstandardized extreme values, data range, and frequencies. You can request separate statistics for each level of all specified grouping variables. 1D reports frequency counts for categorical data.

You can request printouts of all data or for selected variables in any order. You can list data for all cases or only for cases with invalid entries. Users often include a SAVE paragraph with this program in order to create a BMDP File (see Chapter 8). When range limits are exceeded or data are missing, 1D prints the words LT MIN, GT MAX, and MISSING in the output. Although program 2D provides more detailed data description, 1D gives the most commonly used statistics and is very useful for listing either all cases or subsets of the data.

Where to Find It

Example 1D.1 Requesting descriptive statistics

We use 1D to examine the Exercise data discussed in Chapter 2. The record for each subject includes an ID number, sex, smoking status (yes/no), age, and pulse rates before and after exercise. The data are stored on disk in a file named EXERCISE.DAT.

For this example, we need no instructions specific to 1D; Input 1D.1 contains only those instructions common to all BMDP programs for describing the data and naming the variables. (The FILE command tells the program where to find the data and is used for systems like VAX and IBM PC. For IBM mainframes, see UNIT, Chapter 3.) We specify a USE list in the VARIABLE paragraph to select the variables for which we want descriptive statistics (we do not want statistics for ID). See Chapters 3 and 4 for further information on the INPUT and VARIABLE paragraphs.

Input 1D.1

```
/ INPUT       FILE = 'exercise.dat'.
              VARIABLES = 6.
              FORMAT IS FREE.

/ VARIABLE    NAMES = id, sex, smoke, age, pulse_1, pulse_2.
              USE = sex TO pulse_2.
/ END
```

Output 1D.1

—we omit the default printing of the first ten cases—

VARIABLE NO. NAME	TOTAL FREQUENCY	MEAN	STANDARD DEVIATION	ST.ERR OF MEAN	COEFF. OF VARIATION	SMALLEST VALUE	SMALLEST Z-SCORE	LARGEST VALUE	LARGEST Z-SCORE	RANGE
2 sex	40	1.4500	.50383	.07966	.34747	1.0000	-0.89	2.0000	1.09	1.0000
3 smoke	40	1.6000	.49614	.07845	.31009	1.0000	-1.21	2.0000	0.81	1.0000
4 age	40	26.650	6.8894	1.0893	.25851	19.000	-1.11	43.000	2.37	24.000
5 pulse_1	40	75.950	9.5379	1.5081	.12558	62.000	-1.46	96.000	2.10	34.000
6 pulse_2	40	140.50	23.690	3.7458	.16861	112.00	-1.20	265.00	5.26	153.00

In this example, 1D reads 40 cases. The computations use all acceptable values for each variable—i.e., values that are neither missing nor out of range. 1D does not delete whole cases just because some variables have missing or out-of range values.

For each variable 1D prints the

- total frequency of acceptable values
- mean
- standard deviation
- standard error of the mean
- coefficient of variation
- smallest acceptable value observed in the data and its standard score (*z*-score)
- Largest acceptable value observed in the data and its standard score (*z*-score)
- range

See Appendix B.1 for definitions of statistics. Note that Output 1D.1 contains a large standard score of 5.26 for a pulse_2 value of 265.

Example 1D.2 Printing data and derived variables

You can use 1D to print input variables and those derived in the TRANSFORM paragraph. We use the TRANSFORM paragraph to compute the difference between the two PULSE values (see Chapter 6). In Input 1D.2 we request a printout of the data with the new variable by specifying DATA in a PRINT paragraph. We use LINESIZE = 80 to obtain narrow output (see Chapter 7). Note that this instruction changes the output format and content, deleting the range. For this example we use a version of the Exercise data which includes each subject's name. The name follows the data on each record and is read as two four-character LABEL variables (see Chapter 2).

Input 1D.2

```
/ INPUT       FILE = 'exercise2.dat'.
              VARIABLES = 8.
              FORMAT IS FREE.

/ VARIABLE    NAMES = id, sex, smoke, age, pulse_1,
                      pulse_2, name1, name2.
              LABELS ARE name1, name2.
              USE = smoke TO diffrnce.

/ TRANSFORM   diffrnce = pulse_2 - pulse_1.

/ PRINT       DATA.
              LINESIZE = 80.

/ END
```

Output 1D.2

```
VARIABLE     TOTAL          STANDARD  ST.ERR   COEFF      SMALLEST        LARGEST
NO. NAME     FREQ.  MEAN     DEV.    OF MEAN  OF VAR   VALUE  Z-SCR   VALUE  Z-SCR

  3 smoke      40  1.6000  .49614   .07845   .31009  1.0000 -1.21  2.0000  0.81
  4 age        40  26.650  6.8894   1.0893   .25851  19.000 -1.11  43.000  2.37
  5 pulse_1    40  75.950  9.5379   1.5081   .12558  62.000 -1.46  96.000  2.10
  6 pulse_2    40  140.50  23.690   3.7458   .16861  112.00 -1.20  265.00  5.26
  9 diffrnce   40  64.550  20.001   3.1625   .30986  42.000 -1.13  169.00  5.22

 C A S E           3        4          5        6        9
 NO. LABEL     smoke     age       pulse_1  pulse_2  diffrnce
----- -------- -------- -------- -------- -------- --------
   1 ALLAN        1.00    31.00    62.00   126.00    64.00
   2 MARY         1.00    20.00    78.00   154.00    76.00
   3 BILL         2.00    28.00    64.00   128.00    64.00
   4 LINDA        2.00    29.00    96.00   155.00    59.00
   5 MICHAEL      1.00    21.00    66.00   128.00    62.00
   6 CATHY        1.00    27.00    96.00   265.00   169.00
   7 HARVEY       2.00    21.00    68.00   120.00    52.00
   8 JENEE        2.00    42.00    72.00   138.00    66.00
   9 JEAN         1.00    22.00    88.00   160.00    72.00
  10 FREDDY       1.00    28.00    90.00   144.00    54.00
  11 PAT          2.00    21.00    82.00   140.00    58.00
  12 MARK         2.00    22.00    74.00   134.00    60.00
  13 SUSAN        1.00    43.00    66.00   148.00    82.00
  14 DENISE       2.00    19.00    68.00   142.00    74.00
  15 JOHN         1.00    23.00    92.00   134.00    42.00
  16 DAVID        2.00    41.00    68.00   112.00    44.00
  17 ROBERT       2.00    24.00    76.00   158.00    82.00
  18 ALISON       2.00    21.00    86.00   146.00    60.00
  19 JILL         1.00    21.00    88.00   156.00    68.00
  20 JACKSON      1.00    20.00    66.00   132.00    66.00
  21 ARTHUR       1.00    38.00    70.00   122.00    52.00
  22 SAMUEL       2.00    20.00    80.00   136.00    56.00
  23 AMY          1.00    33.00    76.00   148.00    72.00
  24 ANNIE        2.00    25.00    78.00   148.00    70.00
  25 JANE         2.00    37.00    76.00   136.00    60.00
  26 BETH         2.00    22.00    80.00   158.00    78.00
  27 CHRIS        2.00    32.00    68.00   116.00    48.00
  28 FRANCIS      2.00    22.00    70.00   120.00    50.00
  29 ERNIE        1.00    22.00    68.00   126.00    58.00
  30 BERTRAM      1.00    19.00    70.00   144.00    74.00
  31 NANCY        2.00    21.00    86.00   144.00    58.00
  32 BRUCE        2.00    26.00    72.00   126.00    54.00
  33 MARGE        2.00    32.00    84.00   136.00    52.00
  34 BARBARA      2.00    24.00    72.00   142.00    70.00
  35 JENNY        2.00    28.00    80.00   138.00    58.00
  36 WILLIAM      1.00    34.00    62.00   132.00    70.00
  37 KYLE         2.00    35.00    74.00   116.00    42.00
  38 BEN          1.00    21.00    90.00   138.00    48.00
  39 GREG         2.00    21.00    66.00   142.00    76.00
  40 RICHARD      2.00    30.00    70.00   132.00    62.00
```

Output 1D.2 shows a labeled printout of the data file, including our new variable, DIFFRNCE. The LABEL variables are printed first, regardless of where they appear in the input file. The case number is not a variable, rather it is assigned by 1D to indicate the order of the cases in the file. Note the extreme value for case number 6 for the variable pulse_2, suggesting that the data value may have been entered incorrectly. If you later discover that the value 265 was a transcription error and should be 165, you could either edit the original data file or use the TRANSFORM paragraph to change the value: /TRANSFORM IF (pulse_2 EQ 265) THEN pulse_2 = 165.

Example 1D.3 Printing cases with unusual values

Instead of printing the complete data file, you can direct 1D to print any one or all of the following:

- only those cases that have missing data values
- only those cases in which the value for any variable is less than its specified lower limit (MINIMUM in the VARIABLE paragraph)
- only those cases in which the value for any variable is greater than its specified upper limit (MAXIMUM in the VARIABLE paragraph)

In any listing of the data, 1D replaces missing values with the word MISSING, values less than the lower limit with LT MIN, and values greater than the upper limit with GT MAX. 1D prints data only if you specify a PRINT paragraph with the command DATA or some combination of MISSING, MINIMUM, and MAXIMUM.

In Input 1D.3, we use a file constructed to illustrate how to use 1D to find extreme and missing values. When a data value is missing, an asterisk (*) has been inserted in its place in the data file. This is the default character for missing data (see Chapter 2 for a discussion of missing data).

Input 1D.3

```
/ INPUT      VARIABLES = 8.
             FORMAT IS FREE.

/ VARIABLE   NAMES = id, sex, smoke, age, pulse_1,pulse_2,
                     name1, name2.
             LABELS ARE name1, name2.
             USE = sex TO pulse_2.
             MINIMUM = (age)18.
             MAXIMUM = (pulse_2)200.

/ PRINT      MAXIMUM.
             MINIMUM.
             MISSING.

 / END
1      1     1      31     62     126     ALLAN
2      1     1      21     66     128     MICHAEL
3      1     1      *      90     144     FREDDY
4      1     1      3      92     134     JOHN
5      1     1      20     66     232     JACKSON
6      2     1      27     96     265     CATHY
```

Output 1D.3

```
 C A S E         2         3         4         5         6
 NO. LABEL     sex      smoke      age     pulse_1   pulse_2
----- -------- -------- -------- -------- -------- --------
   1 ALLAN       1.00      1.00     31.00     62.00    126.00
   2 MICHAEL     1.00      1.00     21.00     66.00    128.00
   3 FREDDY      1.00      1.00   MISSING     90.00    144.00
   4 JOHN        1.00      1.00    LT MIN     92.00    134.00
   5 JACKSON     1.00      1.00     20.00     66.00    GT MAX
   6 CATHY       2.00      1.00     27.00     96.00    GT MAX

NUMBER OF CASES READ. . . . . . . . . . . . . .           6

        —we omit panel of statistics for each variable—
```

In Output 1D.3, four of the six cases listed contain one or more values replaced by MISSING, LT MIN, or GT MAX.

Example 1D.4 Groups and sorting

In this example we demonstrate two important capabilities of 1D: grouping cases into subpopulations, and sorting cases. You can obtain the statistics reported in Example 1D.1 for groups or subsets of the cases. Group names appear in the output. In Input 1D.4 we request separate summary statistics for men and women and for smokers and nonsmokers by specifying the grouping variables SEX and SMOKE. To obtain statistics for groups defined by the combination of levels of two variables, see 7D; for groups defined by two or more variables, see 9D. You can also use STACK in 1D to combine grouping variables (see Chapter 5).

In Input 1D.4 we use a GROUP paragraph to specify a grouping VARIABLE, as well as CODES and NAMES for the grouping variables (see Chapter 5). 1D does not compute statistics (mean, standard deviation, etc.) for the grouping variable itself unless it is a continuous variable that has been divided by CUTPOINTS (e.g., AGE). In this example we obtain statistics for the other variables broken down by sex and smoke, but we do not get statistics for SEX and SMOKE.

1D allows you to sort the data file on specified variables. We specify SORT = SEX, SMOKE, AGE in the INPUT paragraph so that our printout of the data file will be sorted in ascending order on the values of each of these variables. The order in which you list the variables in the SORT command determines how the variables are sorted. Here the file is first sorted by sex. Within each sex group, the records are split into smokers (yes) and nonsmokers (no). Then within each sex and smoke group, the records are ordered by age. Use the ORDER option (1D Commands) if you want to sort one or more variables in descending order. See the BMDP Data Manager (DM) if you need to sort alphabetically.

Input 1D.4

```
/ INPUT          FILE = 'exercise2.dat'.
                 VARIABLES = 8.
                 FORMAT = FREE.
                 SORT = sex, smoke, age.

/ VARIABLE       NAMES = id, sex, smoke, age, pulse_1,
                         pulse_2, name1, name2.
                 LABELS = name1, name2.
                 USE = sex TO age, pulse_2.

/ GROUP          VARIABLE = sex, smoke.
                 CODES(sex, smoke) = 1, 2.
                 NAMES(sex) = MALE, FEMALE.
                 NAMES(smoke) = YES, NO.

/ PRINT          DATA.

/ END
```

Output 1D.4

```
 VARIABLE        STATED VALUES FOR            GROUP CATEGORY     INTERVALS
NO.  NAME    MINIMUM MAXIMUM MISSING   CODE  INDEX   NAME      .GT.    .LE.
---- -------- ------- ------- -------  ----- -----  --------  ------- -------       [1]

   2 sex                               1.000   1   MALE
                                       2.000   2   FEMALE

   3 smoke                             1.000   1   YES
                                       2.000   2   NO
--------------------------------------------------------------------------------

GROUPING VARIABLE. . . sex                                                          [2]
                              CATEGORY   FREQUENCY
                              --------   ---------
                              MALE           22
                              FEMALE         18
GROUPING VARIABLE. . . smoke
                              CATEGORY   FREQUENCY
                              --------   ---------
                              YES            16
                              NO             24

DESCRIPTIVE STATISTICS OF DATA
---------- ---------- -- ----

 VARIABLE    TOTAL          STANDARD  ST.ERR    COEFF     SMALLEST       LARGEST              [3]
 NO. NAME    FREQ.   MEAN     DEV.    OF MEAN  OF VAR  VALUE  Z-SCR  VALUE  Z-SCR  RANGE

   4 age        40  26.650  6.8894  1.0893  .25851  19.000 -1.11  43.000  2.37  24.000
   6 pulse_2    40  140.50  23.690  3.7458  .16861  112.00 -1.20  265.00  5.26  153.00

  PRINT SUMMARY STATISTICS OVER ALL CASES AND
  BROKEN DOWN BY INDIVIDUAL CATEGORY ON sex
                                  AND smoke

********************************************

  SORTING ON                   [4]

    sex          ASCENDING
    smoke        ASCENDING
    age          ASCENDING

  C A S E      2         3        4         6                        [5]
  NO. LABEL   sex      smoke     age       pulse_2
  ---- ------- -------- -------- -------- --------

   30 BERTRAM  MALE     YES        19.00   144.00
   20 JACKSON  MALE     YES        20.00   132.00
    5 MICHAEL  MALE     YES        21.00   128.00
   38 BEN      MALE     YES        21.00   138.00
   29 ERNIE    MALE     YES        22.00   126.00
   15 JOHN     MALE     YES        23.00   134.00
   10 FREDDY   MALE     YES        28.00   144.00
    1 ALLAN    MALE     YES        31.00   126.00
   36 WILLIAM  MALE     YES        34.00   132.00
   21 ARTHUR   MALE     YES        38.00   122.00
   22 SAMUEL   MALE     NO         20.00   136.00
    7 HARVEY   MALE     NO         21.00   120.00
   39 GREG     MALE     NO         21.00   142.00
   12 MARK     MALE     NO         22.00   134.00
   28 FRANCIS  MALE     NO         22.00   120.00
   17 ROBERT   MALE     NO         24.00   158.00
   32 BRUCE    MALE     NO         26.00   126.00
    3 BILL     MALE     NO         28.00   128.00
   40 RICHARD  MALE     NO         30.00   132.00
   27 CHRIS    MALE     NO         32.00   116.00
   37 KYLE     MALE     NO         35.00   116.00
   16 DAVID    MALE     NO         41.00   112.00
    2 MARY     FEMALE   YES        20.00   154.00
   19 JILL     FEMALE   YES        21.00   156.00
    9 JEAN     FEMALE   YES        22.00   160.00
    6 CATHY    FEMALE   YES        27.00   265.00
   23 AMY      FEMALE   YES        33.00   148.00
   13 SUSAN    FEMALE   YES        43.00   148.00
   14 DENISE   FEMALE   NO         19.00   142.00
   11 PAT      FEMALE   NO         21.00   140.00
   18 ALISON   FEMALE   NO         21.00   146.00
   31 NANCY    FEMALE   NO         21.00   144.00
   26 BETH     FEMALE   NO         22.00   158.00
   34 BARBARA  FEMALE   NO         24.00   142.00
   24 ANNIE    FEMALE   NO         25.00   148.00
   35 JENNY    FEMALE   NO         28.00   138.00
    4 LINDA    FEMALE   NO         29.00   155.00
   33 MARGE    FEMALE   NO         32.00   136.00
   25 JANE     FEMALE   NO         37.00   136.00
    8 JENEE    FEMALE   NO         42.00   138.00

 VARIABLE      GROUPING          TOTAL                STANDARD    ST.ERR    COEFF. OF     S M A L L E S T      L A R G E S T                [6]
 NO. NAME    VARIABLE LEVEL      FREQUENCY      MEAN  DEVIATION   OF MEAN   VARIATION     VALUE   Z-SCORE      VALUE   Z-SCORE      RANGE

   4 age                           40         26.650   6.8894     1.0893     .25851      19.000    -1.11      43.000    2.37       24.000
             sex      MALE         22         26.318   6.4689     1.3792     .24579      19.000    -1.13      41.000    2.27       22.000
                      FEMALE       18         27.056   7.5418     1.7776     .27875      19.000    -1.07      43.000    2.11       24.000

             smoke    YES          16         26.438   7.3482     1.8370     .27795      19.000    -1.01      43.000    2.25       24.000
                      NO           24         26.792   6.7243     1.3726     .25098      19.000    -1.16      42.000    2.26       23.000

   6 pulse_2                       40         140.50   23.690     3.7458     .16861      112.00    -1.20      265.00    5.26       153.00
             sex      MALE         22         130.27   10.907     2.3254     .08373      112.00    -1.68      158.00    2.54       46.000
                      FEMALE       18         153.00   28.979     6.8303     .18940      136.00    -0.59      265.00    3.86       129.00

             smoke    YES          16         147.31   33.476     8.3689     .22724      122.00    -0.76      265.00    3.52       143.00
                      NO           24         135.96   12.923     2.6379     .09505      112.00    -1.85      158.00    1.71       46.000
```

Bold numbers below correspond to those in Output 1D.4:

[1] The program prints its interpretation of the grouping information. This panel is produced by the GROUP paragraph. You can assign codes or cutpoints to any number of variables in the GROUP paragraph to generate reports of group frequencies.

[2] 1D prints a table of frequency counts for individual categories, including values missing or out of range.

[3] 1D by default prints a table of descriptive statistics.

[4] The sorting criteria are identified before the data file is sorted and printed.

[5] 1D prints the sorted data file. The program sorts each variable in ascending order, as identified in **[4]**, unless you state otherwise (see ORDER option). Data values are replaced by group names unless you specify VALUES in the PRINT paragraph.

[6] 1D prints descriptive statistics for each group and counts the frequencies of codes or intervals specified in the GROUP paragraph. Statistics are broken down by group only if you include a GROUP paragraph. (The other panels in this output are produced by the NAMES instruction of the GROUP paragraph.)

Special Features

Control over data printouts

1D offers many options that you can use to control the output of your data file. Table 1D.1 summarizes these options.

Controlling the number of columns per line. In noninteractive mode, 1D assumes that the length of a line is 132 characters unless you specify a smaller number using LINESIZE. If you specify a number less than 132, the statistics panel is printed 80 columns wide, and some statistics are omitted. In the printout, 14 columns are reserved for case numbers plus label variables; other variables take up 11 columns each. You can use LINESIZE to adjust the number of variables you want printed on a line, up to ten.

Organizing printouts by case or by variable. You can use the PRINT METHOD command to direct 1D to print data by case or by variable. Using the default (by variable) method, for each case 1D prints only as many variables as fill the print line (up to five plus two label variables in 80 column format, ten in 132 column format) and begins the next line with the same variables for the next case. Thus the data for each variable are in an easy-to-scan column. When the program finishes printing the first subset of variables for all cases, it continues with another subset. Using the by-case method, 1D prints all the data for one case, possibly filling several print lines, before printing any data for the next case.

For greater control over data printouts, see the PRINT FIELDS command (Chapter 7). This command allows you to specify field widths for each variable and prints reports with variable names listed vertically above each column.

Table 1D.1 Controlling data printouts and statistics

Function Desired	Instructions Paragraph	Instructions Command
Print all data	PRINT	DATA.
Print only those cases with missing or out-of-range values	PRINT	MISSING. MAX. MIN.
Select and order variables	VARIABLE	USE = list.
Control width of printout	PRINT	LINESIZE = #.
Add derived variables	TRANSFORM	(see Chapter 6)
Sort records using values of one or more variables	INPUT	SORT = list. ORDER = D or A.
Print group names*	GROUP	VARIABLE=list. NAMES(*i*) = list. CODES(*i*) = list. CUTP(*i*) = list.
Compute statistics for groups	GROUP	VARIABLE = list. CODES(*i*) = list. CUTP(*i*) = list.
Limit number of lines per page	PRINT	PAGESIZE = #.
Organize data by case or variable	PRINT	METHOD = CASE *or* VAR.

*To print code values instead of names in data printout panel, state VALUES in the PRINT paragraph.

1D Commands

/ INPUT

The INPUT paragraph is required. (See detailed description in Chapter 3.) Additional commands for 1D are SORT and ORDER.

SORT = list. `SORT = SEX, SMOKE.`

List names or numbers of variables used to sort the data. If omitted, 1D does not sort the data. The data will be sorted using the values of the variables in the SORT list. Levels of the first listed variable change most slowly; levels of the last change most quickly. (See Example 1D.4.) You may also indicate if the cases should be sorted in ascending or descending order (see ORDER below). You can print the sorted data (see PRINT commands). Sorted data may be saved in a BMDP File or written to a raw data file (see Chapter 8).

ORDER = D or A.
(one for each variable in the SORT list)

```
ORDER = D, A.
SORT = SEX, SMOKE.
```

Use a D or an A for each variable in the SORT list to specify ascending (A) or descending (D) sort order. If omitted, A is assumed. **Example:** The data for the female group (code 2) are listed before those of the male (code 1). Within each sex group the data are listed for smokers first (SMOKE code = 1), and then for nonsmokers (SMOKE code = 2).

/ VARiable

The VARIABLE paragraph is optional. (See detailed description in Chapter 4.) An additional command for 1D is FREQUENCY.

FREQuency = variable. `FREQ = COUNT.`

State the name or number of a variable containing the case frequency. A case frequency has the effect of repeating cases. Any record with a negative or zero case frequency (COUNT ≤ 0 in this example) is excluded from the calculations as though it were missing. See Chapter 4.

CWeight = variable. `CWEIGHT = SAMPWT.`

Specify a case weight variable. As with case frequencies, records with negative or zero case weight values are excluded from the computations. Case weights can be used to adjust for sampling from populations with unequal variances. Means and correlations are unaffected, but variances, standard deviations, and standard errors change. See Chapter 4.

New Feature

MMV = #. `MMV = 10.`

Specify the maximum number of missing values or values out of range that will be allowed before a case is excluded from the analysis. MMV is set to zero for STAND = COV or WCOV (only cases with complete data are used in the analysis). Default: number of variables used.

/ GROUP

Used to specify the name or number of the variable used to classify cases into groups. If you prefer, you can still specify a grouping variable with the GROUPING command in the VARIABLE paragraph as described in the 1990 *BMDP Statistical Software Manual*. To make the printed results easier to read, we suggest that you specify group NAMES. Define NAMES (and CUTPOINTS or CODES) in the GROUP paragraph (see Chapter 5).

New Syntax

VARiable = list. `VARIABLE = IRISTYPE.`

Specify the name or number of one or more grouping variables used to classify cases into groups. Statistics for other variables are reported for each level of each grouping variable. Default: cases are not grouped. If the GROUPING variable takes on more than ten distinct values or codes, CODES or CUTPOINTS for the variable must be specified in the GROUP paragraph (see Chapter 5 and below). Example: Cases are grouped by the two values of the variable BRTHPILL (histograms are labeled A = NOPILL, B = PILL).

CODES (*j*) = #list `CODES(BRTHPILL) = 1, 2.`
NAMES (*j*) = list `NAME(BRTHPILL)= NOPILL,PILL.`

See CODE, CUTPOINT, and NAME commands in Chapter 5. If you specify NAMES in the GROUP paragraph, 1D uses them to label plots. See Example 1D.4.

New Paragraph

SORT ... /

The optional SORT paragraph directs 1D to sort the data by a set protocol.

New Feature

KEY = list. `KEY = SEX, SMOKE.`

List names or numbers of variables used to sort the data. If omitted, lD does not sort the data. The data will be sorted using the values of the variables in the SORT list. Levels of the first variable listed change most slowly; levels of the last change most quickly. (See Example 1D.4.) You may also indicate if the cases should be sorted in ascending or descending order (see ORDER below). You can print the sorted data (see PRINT commands). Sorted data may be saved in a BMDP File or written to a raw data file (see Chapter 8).

New Feature

ORDER = D or A. `ORDER = D. A.`

Use a D or an A for each variable in the ORDER list to specify ascending (A) or descending (D) sort order. If omitted, A is assumed. Example: The data for the female group (code 2) are listed before that of the male (code 1). Within each sex group the data are listed for smokers first (SMOKE code = 1), and then for nonsmokers (SMOKE code = 2).

/ PRINT

The optional PRINT paragraph directs 1D to print cases from the data file. Use DATA or any combination of MIN, MAX, and MISS. No cases are printed unless one of the four is specified. The data printout reflects any grouping instructions in the GROUP paragraph.

New Feature

CASE = #. `CASE = 50.`

Specify the number of cases for which data are printed. Values flagged by MISS, MAX, and MIN in the VARIABLE paragraph specifications are replaced by MISSING, GT MAX, and LT MIN. Default: 10.

DATA. `DATA.`

Prints data for all cases including minimum and maximum cases. Default: the cases are not printed.

MINimum. `MIN.`

Prints data for the first ten cases in which at least one variable has a value less than its lower limit as specified in the VARIABLE paragraph. Default: such cases are not printed unless you specify MIN.

MAXimum. `MAX.`

Prints data for the first ten cases for which at least one variable has a value greater than its upper limit as specified in the VARIABLE paragraph. Default: such cases are not printed unless you specify MAX.

MISSing. `MISS.`

Prints data for cases that have at least one variable with a missing value. Missing values in the input may be identified in several ways: (1) an asterisk (the default missing value character), (2) a missing value code identified using MISS in the VARIABLE paragraph, (3) a character identified as MCHAR in the INPUT paragraph, (4) blanks (applies only to input read with FORTRAN fixed format). Default: such cases are not printed unless you specify MISS.

VALUes. `VALU.`

Use to request a printout of the codes or values for the grouping variables instead of the group names. Default: the group names are printed (e.g., 1D would print "OVER45" instead of the age values 46 or 53).

METHod = VARiable *or* CASE. `METH = CASE.`

Use to determine how the printout is organized. VARIABLE is the default value.

VARIABLE – Prints for each case only as many variables as fill the print line (up to ten variables) and begins the next line with the same variables for the next case. Thus the data for each variable are in an easy-to-scan column. When the program finishes printing the first subset of variables for all cases, it continues with another subset.

CASE – Prints the values of all variables for one case (possibly filling several print lines) before printing data for the next case.

FIELDs = list. `FIELD = 3, 2*4, 9.1, -2.`

Controls the format of a data listing by specifying the number of character positions, decimal places, and whether a field is alphanumeric. A minus sign indicates an alphanumeric field. See Chapter 7 for a full explanation.

LINEsize = #. `LINE = 80.`

Specify the number of columns to use in printing output. The default value is 132 columns in noninteractive mode, 80 in interactive. If you specify any number less than 132, the statistics are printed in 80 columns. When formatting narrow output for the computer terminal, 1D leaves out some of the statistics (standard error of mean, coefficient of variation, and range).

PAGEsize = #. `PAGE = 40.`

Specify the number of printed lines per page. The default is 59 lines per page.

The following commands work only in the batch mode

New Feature

MEAN. `NO MEAN.`

Unless you specify NO MEAN, 1D prints the mean, standard deviation, and frequency of each used variable.

New Feature

EXTReme. `NO EXTR.`

Unless you state NO EXTR, 1D prints the minimum and maximum values for each variable used.

New Feature

EZSCores. `NO EZSC.`

Unless you state NO EZSC, 1D prints *z*-scores for the minimum and maximum values of each used variable.

New Feature

ECASe. `ECAS.`

Prints the case numbers containing the minimum and maximum values of each variable. Default: No extreme case numbers printed.

New Feature

SKewness. `SK.`

Prints skewness and kurtosis for each used variable. Default: No skewness printed.

Order of Instructions

■ indicates required paragraph

```
■ / INPUT
  / VARIABLE
  / TRANSFORM
  / GROUP
  / SAVE                                  Repeat for additional problems.
  / PRINT                                 See Multiple Problems, Chapter 10.
■ / END
  data
  SORT ... /
  GROUP ... /
  PRINT ... /          Repeat for
  SAVE ... /           subproblems.
  TRANSFORM ... /
  CONTROL ... /
  END /
  FINISH /
```

Summary Table for Commands Specific to 1D

Paragraphs / Commands	Defaults	Multiple Problems	See
■ / **INPut**			
SORT = list.	records not sorted	–	1D.4
ORDER = list. *(D or A for each variable)*	A (ascending)	–	Cmds
/ **VARiable**			
FREQuency = variable.	no case frequency variable	–	Cmds
CWeight = variable.	no case weights used	–	Cmds
MMV = #.	# of variables	–	Cmds
▲ / **GROUP**			
▲ VARiable = list.	no grouping (see Chapter 5)	–	Cmds
/ **PRINT**			
CASE = #.	10	●	Cmds
DATA.	NO DATA.	●	1D.2
MISSing.	NO MISS.	●	1D.3
MINimum.	NO MIN.	●	1D.3
MAXimum.	NO MAX.	●	1D.3
VALUes.	NO VALUES.	●	Cmds
MEAN. *or* NO MEAN.	MEAN.	●	Cmds
EXTReme. *or* NO EXTR.	EXTReme. values printed	●	Cmds
EZSC. *or* NO EZSC.	EZSC. (z-scores)	●	Cmds
ECASe. *or* NO ECASe.	ECAS. (Case #'s of Min and Max printed)	●	Cmds
SKewness. *or* NO SKewness.	NO SK. (skewness and kurtosis)	●	Cmds
METHod = VARiables *or* CASE.	VARIABLES	–	SpecF
FIELDs = list.	FORMAT if specified; otherwise 8.2 for numeric, – 4 for label variables	–	Cmds
LINEsize = #.	132 noninteractive; 80 interactive	●	1D.2
PAGEsize = #.	59 lines	–	Cmds
SORT ... /			
KEY = list.	records not sorted	–	Cmds
ORDER = list. *(D or A for each variable)*	A (ascending)	–	Cmds

Key:
- ■ Required paragraph or command
- ▲ Frequently used command
- ● Value retained for multiple problems
- – Default reassigned

2D

Detailed Data Description Including Frequencies

2D computes a wide variety of descriptive statistics and plots a simple histogram for each variable. It also computes the frequency, percent, and cumulative percent of each distinct value. You can use 2D in screening your data to identify outliers, study distributional shape, and provide an initial description of your sample. 2D can also group your data according to a grouping criteria.

For each variable, 2D calculates the mean, median, mode, standard deviation, standard errors for the mean and median, skewness and kurtosis, extreme values, half the interquartile range, and upper and lower 95% confidence intervals for the mean. 2D displays a histogram of cases and a line plot of descriptive statistics. You can request Shapiro and Wilk's *W* statistic for testing normality. Three robust alternatives to the mean are available: the trimmed mean, Hampel, and biweight estimates. You can obtain stem and leaf displays. You can round or truncate data before doing frequency counts and computing statistics. For further reading, see Cramér (1946), Dunn and Clark (1987), and Oja (1983).

Where to Find It

Example 2D.1 Descriptive statistics, frequencies, and a histogram

We analyze survey data from Afifi and Clark (Appendix D). The data include mental and physical health variables as well as demographic variables such as sex, age, education, income, etc. The data are stored on disk in a file named SURVEY.DAT. In Input 2D.1 no instructions specific to 2D are used; the instructions are common to all BMDP programs. (The FILE command tells the program where to find the data and is used for systems like VAX and IBM PC. For IBM mainframes see UNIT, Chapter 3.) We name only the first seven of the 37 variables recorded for each subject. See Chapters 3 and 4 for further information on the INPUT and VARIABLE paragraphs.

2D assumes that you want descriptive statistics for each variable. Here we specify USE = INCOME in the VARIABLE paragraph so that only INCOME will be analyzed. INCOME is recorded in thousands of dollars (20 = $20,000). See Output 2D.1 for results.

Input 2D.1

```
/ INPUT       VARIABLES = 37.
              FORMAT IS FREE.
              FILE IS 'survey.dat'.

/ VARIABLE    NAMES = id, sex, age, marital,
                      educatn, employ, income.
              USE = income.
/ END
```

Output 2D.1

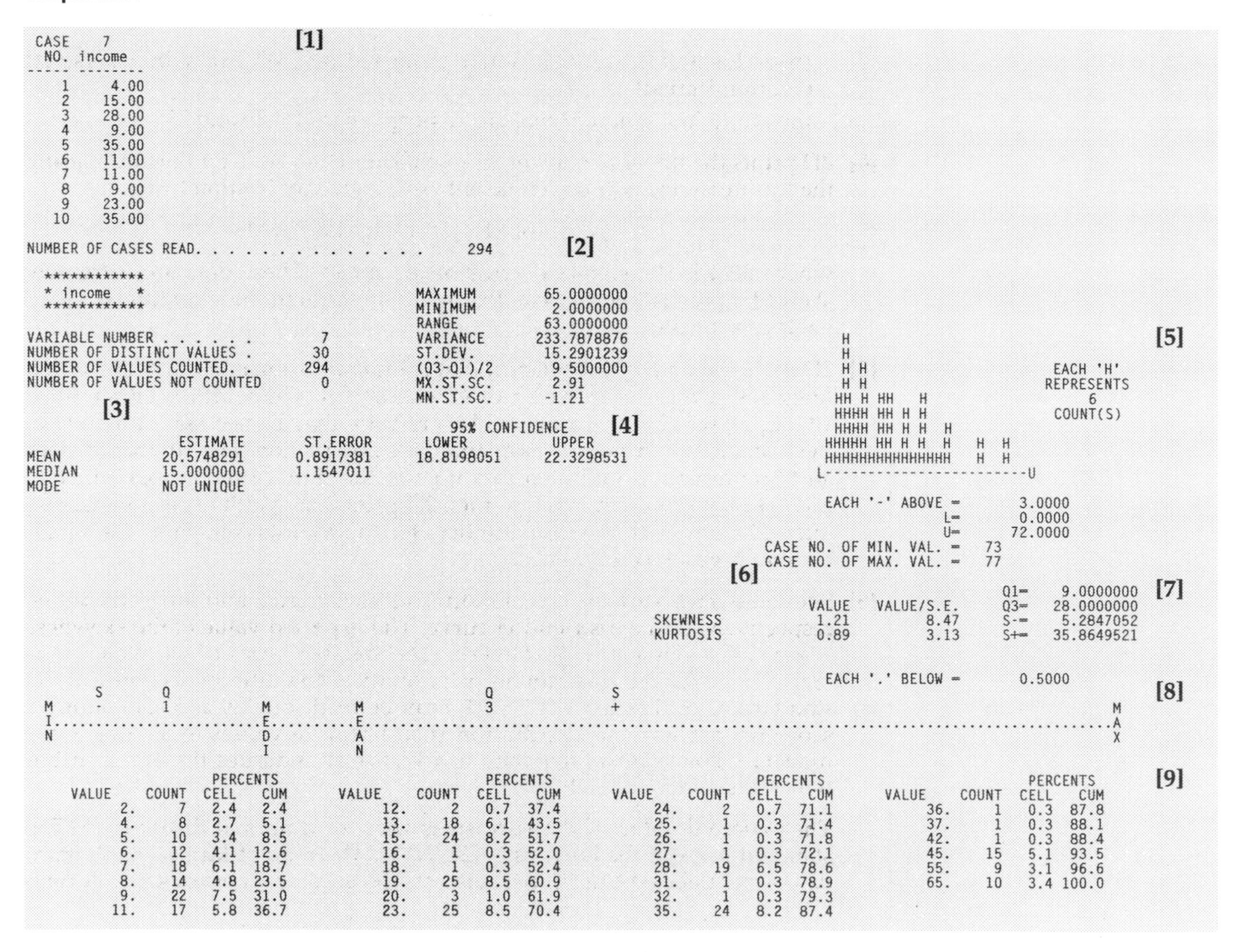

```
 CASE    7                                  [1]
  NO.  income
 ----- --------
     1     4.00
     2    15.00
     3    28.00
     4     9.00
     5    35.00
     6    11.00
     7    11.00
     8     9.00
     9    23.00
    10    35.00

NUMBER OF CASES READ. . . . . . . . . . . . . .      294       [2]

  ************
  * income   *                              MAXIMUM       65.0000000
  ************                              MINIMUM        2.0000000
                                            RANGE         63.0000000
VARIABLE NUMBER . . . . . .        7        VARIANCE     233.7878876          H                          [5]
NUMBER OF DISTINCT VALUES .       30        ST.DEV.       15.2901239          H
NUMBER OF VALUES COUNTED. .      294        (Q3-Q1)/2      9.5000000          H H              EACH 'H'
NUMBER OF VALUES NOT COUNTED       0        MX.ST.SC.      2.91               H H             REPRESENTS
                                            MN.ST.SC.     -1.21              HH H HH   H           6
        [3]                                                                  HHHH HH H H          COUNT(S)
                                             95% CONFIDENCE      [4]         HHHH HH H H  H
               ESTIMATE       ST.ERROR      LOWER          UPPER            HHHHH HH H H  H   H  H
MEAN           20.5748291     0.8917381     18.8198051     22.3298531       HHHHHHHHHHHHHHH   H  H
MEDIAN         15.0000000     1.1547011                                    L-----------------------U
MODE           NOT UNIQUE
                                                                            EACH '-' ABOVE =      3.0000
                                                                                        L=      0.0000
                                                                                        U=     72.0000
                                                                   CASE NO. OF MIN. VAL. =   73
                                                                   CASE NO. OF MAX. VAL. =   77
                                                               [6]
                                                                                            Q1=    9.0000000   [7]
                                                                    VALUE   VALUE/S.E.      Q3=   28.0000000
                                                       SKEWNESS     1.21          8.47      S-=    5.2847052
                                                       KURTOSIS     0.89          3.13      S+=   35.8649521

                                                                            EACH '.' BELOW =      0.5000
       S       Q                                       Q                 S                                       [8]
 M     -       1              M          M             3                 +                                   M
 I............................E..........E...................................................................A
 N                            D          A                                                                   X
                              I          N

                   PERCENTS                    PERCENTS                    PERCENTS                    PERCENTS   [9]
  VALUE     COUNT  CELL   CUM    VALUE  COUNT  CELL   CUM    VALUE  COUNT  CELL   CUM    VALUE  COUNT  CELL   CUM
      2.        7   2.4   2.4      12.      2   0.7  37.4      24.      2   0.7  71.1      36.      1   0.3  87.8
      4.        8   2.7   5.1      13.     18   6.1  43.5      25.      1   0.3  71.4      37.      1   0.3  88.1
      5.       10   3.4   8.5      15.     24   8.2  51.7      26.      1   0.3  71.8      42.      1   0.3  88.4
      6.       12   4.1  12.6      16.      1   0.3  52.0      27.      1   0.3  72.1      45.     15   5.1  93.5
      7.       18   6.1  18.7      18.      1   0.3  52.4      28.     19   6.5  78.6      55.      9   3.1  96.6
      8.       14   4.8  23.5      19.     25   8.5  60.9      31.      1   0.3  78.9      65.     10   3.4 100.0
      9.       22   7.5  31.0      20.      3   1.0  61.9      32.      1   0.3  79.3
     11.       17   5.8  36.7      23.     25   8.5  70.4      35.     24   8.2  87.4
```

Bold numbers correspond to those in Output 2D.1.

[1] The first 10 cases are printed with the values for all used variables.

[2] There are 294 cases in the survey data. All 294 cases are counted (N) and none are missing or out of range. There are 30 distinct values of INCOME.

Univariate statistics contain:
- the maximum and minimum observed values (not out of range)
- the range
- the variance and standard deviation
- half the interquartile range: $(Q_3 - Q_1)/2$, where Q_1 and Q_3 correspond to the 25th and 75th percentiles respectively (first and third quartiles)
- the maximum and minimum standardized scores for cases within range

[3] Location estimates and their standard errors include:
- the mean and standard error of the mean.
- the median ($15,000 per year). The formula for the standard error of the median is given below.

$$\text{S.E.}=(x_{(i)} - x_{(j)})/(2\sqrt{3})$$

where $x_{(i)}$ and $x_{(j)}$ are the ith and jth order statistics (values of rank i and j). i is defined as the integer part of

$$\frac{N+\sqrt{3N}}{2}+1$$

j is the integer part of

$$\frac{N-\sqrt{3N}}{2}+1$$

(pseudostandard error formula proposed by J.W. Tukey in a personal communication).
- the mode (not unique; there are 25 people each for 19 and 23).

[4] 2D prints the upper and lower 95% confidence intervals for the mean, using the t values at the .975 percentile with df (degrees of freedom) = $n - 1$:

$$\bar{x} - t_{.975}(\text{df})^*\text{SEM} \le \mu \le \bar{x} + t_{.975}(\text{df})^*\text{SEM}$$

where SEM is the standard error of the mean. These data are extremely skewed; note that the median does not fall within the 95% confidence interval for the mean.

[5] 2D plots the data in a histogram that is limited to a maximum height of ten lines and a maximum width of 40 characters. The width is either 40 or $10 \log_{10} N$, whichever is smaller. As a result it may be necessary for an H in the histogram to represent more than one observation. The number of cases an H represents is indicated in a note to the right of the histogram. Here each H may represent 1 to 6 counts. 2D prints the width of each histogram interval along with the case numbers for the records containing the smallest and largest income values.

[6] Skewness and kurtosis are measures of asymmetry and long-tailedness, respectively, of the distribution curve. The expected value of the skewness is zero for a symmetric distribution. The standard error of the skewness is $(6/N)^{1/2}$ under the assumption of normality. The ratio of skewness to its standard error (labeled VALUE/S.E.) can be read roughly as a standardized score from a normal distribution (i.e., absolute values exceeding 2 are unusual). For INCOME this ratio is 8.47, which indicates that the distribution is skewed to the right.

The expected value of the kurtosis is zero for a normal distribution. The standard error of the kurtosis is $(24/N)^{1/2}$. The ratio of kurtosis to its standard error (labeled VALUE/S.E.) also can be used as a normal score. A ratio less than –2 indicates shorter tails than a normal distribution; a ratio greater than 2 indicates longer tails than a normal distribution. The formulas for the standard errors of skewness and kurtosis are from Cramér (1946, p. 375).

[7] Quartiles and the mean plus or minus one standard deviation are printed:

Q_1: the 25th percentile

Q_3: the 75th percentile

S–: the mean minus one standard deviation

S+: the mean plus one standard deviation

All four are plotted in a line plot (**[8]** above). Q_1 is defined as $(a + b)/2$ where *a* is the largest value such that no more than 25 percent of the values for the cases are less than or equal to *a*, and *b* is the smallest value such that at least 25 percent of the values of the cases are less than or equal to *b*. Q_3 is defined similarly. The central 50% of subjects in this sample earn between $9,000 and $28,000 per year.

[8] 2D plots the location estimates (such as the mean, median, mode, quartiles, and *S* + and *S* –) on a line. If two measures coincide in their plotting positions, 2D plots only one. In Output 2D.1 the measures are plotted from left to right and occur in the following order: minimum, *S* –, Q_1, mode, median, mean, Q_3, *S* +, and maximum.

[9] 2D prints a table in which each distinct value is listed with its frequency, the percent of observations with this value, and the cumulative percent of observations less than or equal to this value. For example, 31.0 percent of the subjects have an INCOME less than or equal to $9000. This table may be costly to compute and of limited value if there are too many distinct values. You may want to limit the number of distinct values by rounding or truncating the values before the analysis (see Example 2D.3). This table can be omitted by specifying NO COUNT in the PRINT paragraph (Example 2D.2).

Note that this output is considerably wider than your terminal screen. In example 2D.3 we show how to obtain an alternative format more suitable for viewing at the terminal.

Example 2D.2 Basic Descriptive Statistics with one grouping factor

We again look at the Afifi and Clark (Appendix D) survey data as in Example 2D.1. However, in this example we will ask for descriptive statistics grouped by the variable SEX. The standard BMDP GROUP paragraph (see Chapter 5) is used to derive this output. Again, we only ask for the descriptive statistics for the INCOME variable. Note that in Input 2D.2 we must add the variable SEX to the USE statement to allow it to be included in the analysis as a grouping factor.

Input 2D.2

```
/ INPUT      VARIABLES = 37.
             FORMAT IS FREE.
             FILE IS 'survey.dat'.

/ VARIABLE   NAMES = id, sex, age, marital,
                     educatn. employ, income.
             USE = sex, income.

/ GROUP      VARIABLE = sex.
             CODES(sex) = 1, 2.
             NAMES(sex) = Male, Female.

/ END
```

Output 2D.2

—we omit portions of the output—

```
  VARIABLE        STATED VALUES FOR           GROUP CATEGORY    INTERVALS      [1]
 NO.   NAME    MINIMUM MAXIMUM MISSING   CODE  INDEX   NAME     .GT.   .LE.
---- --------  ------- ------- -------  ------ -----  --------  ------ -------

  2  sex                                 1.000    1  MALE
                                         2.000    2  FEMALE
-------------------------------------------------------------------------------

GROUPING VARIABLE. . . sex                                                      [2]
                               CATEGORY    FREQUENCY
                               --------    ---------
                               MALE           111
                               FEMALE         183

  ********************************************                                  [3]
  * ANALYSIS OF income   FOR GROUP MALE      *
  ********************************************
                                                VALUE      ZSCORE  CASE #
                                        MAX   65.0000000    2.511     77
VARIABLE NUMBER . . . . . .       7    MIN    2.0000000   -1.357     73
NUMBER OF DISTINCT VALUES .      21
NUMBER OF VALUES COUNTED. .     111             VALUE     VALUE/S.E.
NUMBER OF VALUES NOT COUNTED      0    SKEWNESS    0.91     3.935
                                       KURTOSIS    0.20     0.423
   H
   H
   H                  EACH 'H' REPRESENTS     2 COUNTS
  HH H H              EACH '-' REPRESENTS  5.00000 UNITS
  HH HHH
 HHHHHHH              L =  0.00000
 HHHHHHH H            U =  100.000
 HHHHHHH H  H
 HHHHHHH H H H
 HHHHHHH H H H
L-------------------U

                  ESTIMATE       ST.ERROR                          ESTIMATE
MEAN            24.1081085      1.5458777          ST.DEV.        16.2868328
MEDIAN          23.0000000      1.1547011          VARIANCE      265.2609253
MODE            35.0000000                         RANGE          63.0000000
                                                   (Q3-Q1)/2      12.0000000
TRIM(.15)       21.6415710
HAMPEL          22.2760868             LOWER 95% C.L. OF MEAN     21.0445404
BWEIGHT         22.0685349             UPPER 95% C.L. OF MEAN     27.1716766
                                                          Q1      11.0000000
TEST OF NORMALITY                                         Q3      35.0000000
W STATISTIC          0.8933                               S-       7.8212757
SIGNIFICANCE LEVEL   0.0000                               S+      40.3949432

              EACH '.' BELOW =       1.0000
             S Q                      Q   S
       M     - 1          HMM         M   +                         M
       I..................AEE.........O.............................A
       N                  MDA         D                             X
                          PIN         E
--------------------------------------------------------------------------------
```

—we omit the table of distinct values for the male group—

```
  ********************************************                                  [4]
  * ANALYSIS OF income   FOR GROUP FEMALE    *
  ********************************************
                                                VALUE      ZSCORE  CASE #
                                        MAX   65.0000000    3.262     77
VARIABLE NUMBER . . . . . .       7    MIN    2.0000000   -1.151     73
NUMBER OF DISTINCT VALUES .      27
NUMBER OF VALUES COUNTED. .     183             VALUE     VALUE/S.E.
NUMBER OF VALUES NOT COUNTED      0    SKEWNESS    1.43     7.878
                                       KURTOSIS    1.56     4.310

   H
   H                    EACH 'H' REPRESENTS     5 COUNTS
   H                    EACH '-' REPRESENTS  3.00000 UNITS
   H H
  HH H                  L =  0.00000
  HH H H                U =  66.0000
  HHHH HH   H
  HHHH HH H H  H
 HHHHHHHHHHHHHHH   H  H
L---------------------U

                  ESTIMATE       ST.ERROR                          ESTIMATE
MEAN            18.4316940      1.0553484          ST.DEV.        14.2764883
MEDIAN          13.0000000      0.8660259          VARIANCE      203.8181152
MODE            NOT UNIQUE                         RANGE          63.0000000
                                                   (Q3-Q1)/2       7.5000000
TRIM(.15)       15.3637781
HAMPEL          14.1989498             LOWER 95% C.L. OF MEAN     16.3494034
BWEIGHT         15.0452795             UPPER 95% C.L. OF MEAN     20.5139847
                                                          Q1       8.0000000
TEST OF NORMALITY                                         Q3      23.0000000
W STATISTIC          0.8271                               S-       4.1552057
SIGNIFICANCE LEVEL   0.0000                               S+      32.7081833

              EACH '.' BELOW =       1.0000
         S  Q               Q        S
       M -  1    MHT  M     3        +                              M
       I.........EAR..E.............................................A
       N         DMI  A                                             X
                 IPM  N
--------------------------------------------------------------------------------
```

—we omit the table of distinct values for the female group—

[1] Interpretations of the GROUP paragraph. Our codes and names instructions separate the data for males and females and label the output accordingly. No data is shown under the intervals GT and LE, since this data is was separated by CODES and not CUTPOINTS.

[2] The raw frequency counts of cases classified into either the male or female groups is shown. In this case, 111 male and 183 female cases are included.

[3] The standard 2D analysis is performed for all of the male data. This output is the same as that shown in Output 2D.1, but with only male data permitted into the analysis.

[4] The standard 2D analysis is performed for all of the female data. This output contains the identical analysis shown in Output 2D.1, with only female data permitted into the analysis.

Example 2D.3 Alternatives to the sample mean and Shapiro and Wilk's test for normality

In Input 2D.3 we study three robust alternatives to the usual sample mean and test normality with Shapiro and Wilk's *W* statistic. When a variable has one or more extreme values or markedly longer tails than the normal distribution, the usual sample mean may be unduly influenced by only a few values and may not be indicative of the bulk of the cases. Robust measures of location are alternatives to the mean which are less influenced by extreme values. We use a TRANSFORM paragraph to derive a new variable, SQRT_INC, by taking the square root of INCOME (square root and log transformations are standard ways to make right-skewed distributions symmetric). We state ESTIMATES in the PRINT paragraph to obtain three robust measures of location for SQRT_INC.

We also request the *W* statistic (WSTAT) to test the normality of the distribution (see Shapiro and Wilk, 1965, and Royston, 1982). We specify NO COUNT to suppress the table of distinct values, and we also specify LINESIZE = 80 to obtain narrow output for the computer terminal; note that this changes the output format.

Input 2D.3

```
/ INPUT      FILE IS 'survey.dat'.
             FORMAT IS FREE.
             VARIABLES = 37.

/ VARIABLE   NAMES = id, sex, age, marital,
                     educatn, employ, income.
             USE = sqrt_inc.

/ TRANSFORM sqrt_inc = SQRT(income).

/ PRINT      ESTIMATES.
             WSTAT.
             NO COUNT.
             LINESIZE = 80.
/ END
```

Output 2D.3

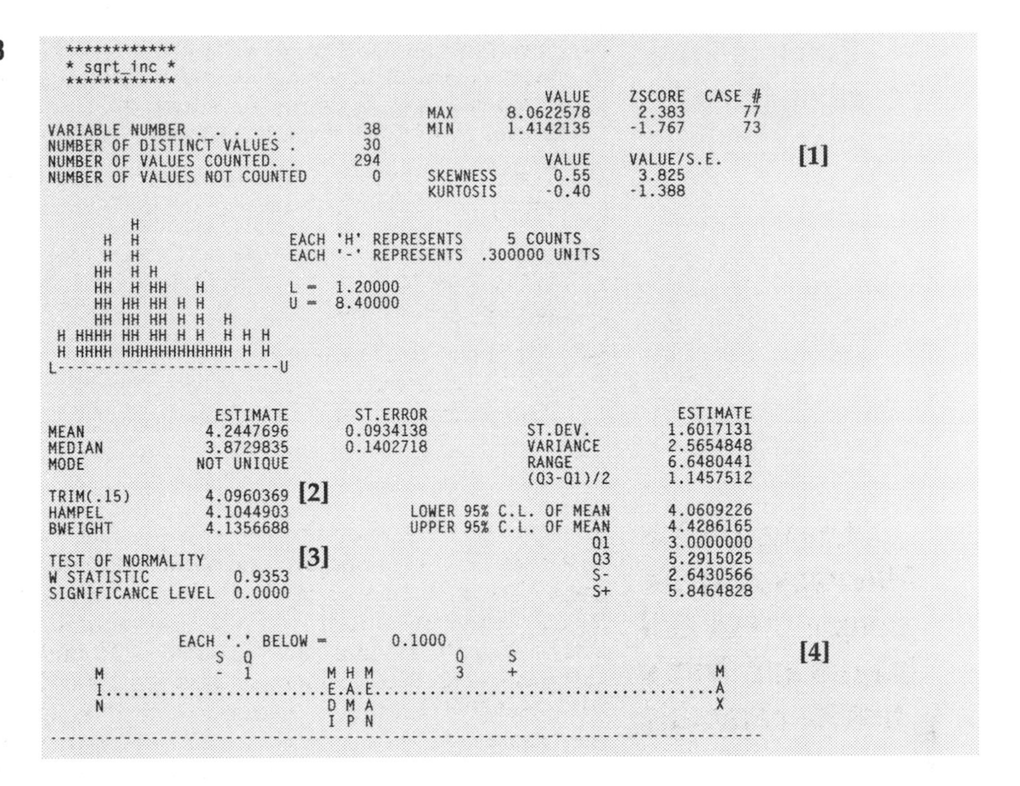

```
************
* sqrt_inc *
************
                                                VALUE      ZSCORE  CASE #
                                   MAX       8.0622578      2.383      77
VARIABLE NUMBER . . . . . .   38   MIN       1.4142135     -1.767      73
NUMBER OF DISTINCT VALUES .   30
NUMBER OF VALUES COUNTED. .  294                VALUE      VALUE/S.E.
NUMBER OF VALUES NOT COUNTED   0   SKEWNESS      0.55       3.825
                                   KURTOSIS     -0.40      -1.388

         H
      H  H                   EACH 'H' REPRESENTS     5 COUNTS
      H  H                   EACH '-' REPRESENTS  .300000 UNITS
     HH  H H
     HH  H HH   H            L =  1.20000
     HH HH HH H H            U =  8.40000
     HH HH HH H H  H
 H HHHH HH HH H H  H H H
 H HHHH HHHHHHHHHHHH H H
L-------------------------U

                    ESTIMATE       ST.ERROR                          ESTIMATE
MEAN               4.2447696      0.0934138          ST.DEV.        1.6017131
MEDIAN             3.8729835      0.1402718          VARIANCE       2.5654848
MODE             NOT UNIQUE                          RANGE          6.6480441
                                                     (Q3-Q1)/2      1.1457512
TRIM(.15)          4.0960369
HAMPEL             4.1044903            LOWER 95% C.L. OF MEAN      4.0609226
BWEIGHT            4.1356688            UPPER 95% C.L. OF MEAN      4.4286165
                                                            Q1      3.0000000
TEST OF NORMALITY                                           Q3      5.2915025
W STATISTIC            0.9353                               S-      2.6430566
SIGNIFICANCE LEVEL     0.0000                               S+      5.8464828

            EACH '.' BELOW =        0.1000
                  S Q                      Q    S
   M              - 1       M H M          3    +                        M
   I......................E.A.E.....................................A
   N                        D M A                                        X
                            I P N
---------------------------------------------------------------------------
```

[1] In Output 2D.2, the amount of positive skewness has been reduced by the square root transformation, but the distribution is still not symmetric. We also specified a log transformation (not shown here) which reduced skewness/S.E. to –1.815; however, the *W* statistic remained significant.

[2] 2D prints three robust measures of location: the 15% trimmed mean, the Hampel, and the biweight (Andrews et al., 1972). These measures estimate the location (mean) of a symmetric distribution with less variability than the mean or median when the distribution has long tails or when the data contain outliers or extreme values.

The 15% trimmed mean is computed by the robust procedures used in program 7D. See Appendix B.9 for details. (The usual sample mean is an estimate in which each observation has the same weight.) The Hampel and biweight estimates are similar in principle in that central values receive higher weights than more extreme cases, but fractional weights are assigned to intermediate cases. See schematic diagram in Special Features.

[3] 2D prints both the *W* statistic and its significance level. The *W* statistic is positive, with a maximum value of one. A small value of *W* indicates departure from normality. That is because the expected value of *W* under normality is always greater than .9 and increases with sample size. Here the results indicate that normality is rejected at the .001 level of significance (the p-value is $< .00005$).

[4] 2D also displays the three robust estimates along the line plot. Only the Hampel estimate is plotted here, since in this example it coincides with the other two estimates in their plotting position.

Example 2D.4 Rounding the data values before analysis

When the variable is continuous and the number of cases is large, the table of distinct values and frequencies may have too many entries. Producing such a table may be time-consuming and costly. Rounding or truncating values to fewer digits produces fewer distinct values and is thus faster to construct. If you request either rounding or truncating, all computations use the rounded or truncated values. Rounding and truncating are specified using the ROUND and TRUNCATE options in the COUNT paragraph.

You can round as many variables as you like, each by a specified amount. In Input 2D.4 we round the INCOME variable its values to the nearest five units ($5000).

Input 2D.4
(add to Input 2D.1 before END)

```
/ COUNT        ROUND = (income)5.
```

Output 2D.4

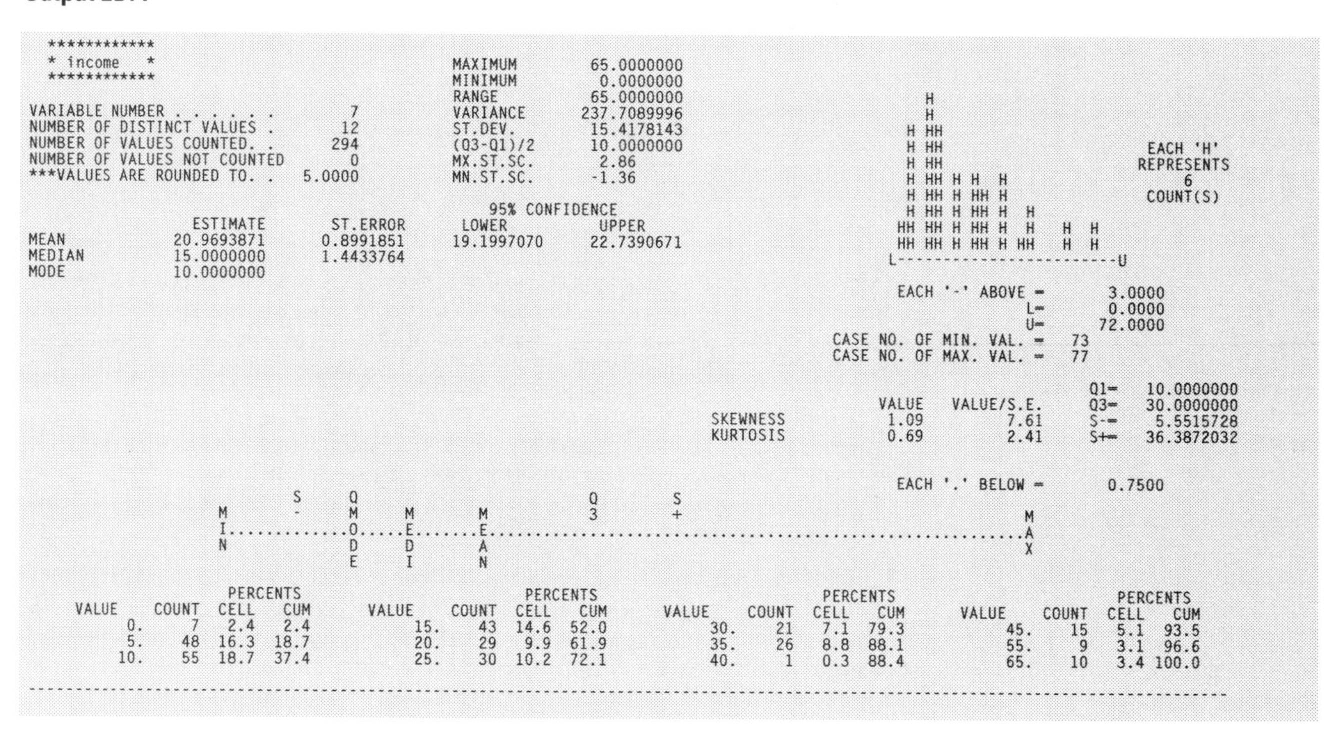

```
 ************
 * income   *                                 MAXIMUM        65.0000000
 ************                                 MINIMUM         0.0000000
                                              RANGE          65.0000000             H
VARIABLE NUMBER . . . . . .         7         VARIANCE      237.7089996             H
NUMBER OF DISTINCT VALUES .        12         ST.DEV.        15.4178143           H HH
NUMBER OF VALUES COUNTED. .       294         (Q3-Q1)/2      10.0000000           H HH                    EACH 'H'
NUMBER OF VALUES NOT COUNTED        0         MX.ST.SC.       2.86                H HH                   REPRESENTS
***VALUES ARE ROUNDED TO. .    5.0000         MN.ST.SC.      -1.36                H HH H H  H                 6
                                                                                  H HH H HH H              COUNT(S)
                                                95% CONFIDENCE                    H HH H HH H  H
                   ESTIMATE        ST.ERROR    LOWER          UPPER              HH HH H HH H  H   H  H
MEAN              20.9693871      0.8991851   19.1997070     22.7390671          HH HH H HH H HH   H  H
MEDIAN            15.0000000      1.4433764                                     L---------------------U
MODE              10.0000000
                                                                                 EACH '-' ABOVE =      3.0000
                                                                                             L=      0.0000
                                                                                             U=     72.0000
                                                                       CASE NO. OF MIN. VAL. =   73
                                                                       CASE NO. OF MAX. VAL. =   77

                                                                                                   Q1=   10.0000000
                                                                             VALUE   VALUE/S.E.    Q3=   30.0000000
                                                                 SKEWNESS    1.09          7.61    S-=    5.5515728
                                                                 KURTOSIS    0.69          2.41    S+=   36.3872032

                                                                                 EACH '.' BELOW =      0.7500
                   S    Q                 Q        S
             M     -    M    M    M       3        +                                    M
             I..........O....E....E.............................................A
             N          D    D    A                                                     X
                        E    I    N

           PERCENTS                       PERCENTS                        PERCENTS                        PERCENTS
VALUE   COUNT  CELL  CUM     VALUE   COUNT  CELL  CUM     VALUE   COUNT  CELL  CUM     VALUE   COUNT  CELL  CUM
   0.       7   2.4   2.4      15.      43  14.6  52.0      30.      21   7.1  79.3      45.      15   5.1  93.5
   5.      48  16.3  18.7      20.      29   9.9  61.9      35.      26   8.8  88.1      55.       9   3.1  96.6
  10.      55  18.7  37.4      25.      30  10.2  72.1      40.       1   0.3  88.4      65.      10   3.4 100.0
----------------------------------------------------------------------------------------------------------------
```

Compare the rounded results for INCOME in Output 2D.4 with the results in Output 2D.1. There are 12 distinct values instead of 30, and each distinct value is divisible by five. All computations in Output 2D.4 are performed using the rounded data. The mean is now 21.0 instead of 20.6.

Example 2D.5 Stem and leaf display

A stem and leaf display is a histogram in which actual data values make up the bars of the histogram (see Tukey, 1977). 2D displays the data values in up to 20 rows (bars). For each value, the digits are separated into two pieces: the stem and the leaf. To demonstrate, we specify STEM in the PRINT paragraph. Here we analyze AGE and INCOME.

Input 2D.5

```
/ INPUT      FILE IS 'survey.dat'.
             VARIABLES = 37.
             FORMAT = FREE.

/ VARIABLE   NAMES = id, sex, age, marital, educatn,
                     employ, income.
             USE = age, income.

/ PRINT      STEM.

/ END
```

Output 2D.5

```
  ************                                          [1]
  * age      *
  ************
             *
DEPTH   STEM * LEAVES
             *
   0       1 *
  10         * 8888899999
  51       2 E 00000011111222222222333333333333344444444
  79         Q 5555556666666666777788889999
 114       3 * 00000011111222222222233333444444444
 132         * 555566666677777889
+ 24       4 M 000001222222222233333344
 138         * 555566677777788889999
 117       5 * 00000111111222233444
  97         Q 55556667777778888888899999999
  68       6 * 000000011111222233444
  47         E 555556667788889
  32       7 * 000001112233444
  17         * 5778899
  10       8 * 011233333
   1         * 9
             *
DEPTH   STEM * LEAVES
             *

MINIMUM =     18.00000
MAXIMUM =     89.00000
 COUNT  =    294
________________________________________________________
  ************                                          [2]
  * income   *
  ************

S T E M  *    0   1   2   3   4   5   6   7   8   9

      0  *            7       8  10  12  18  14  22
      1  *       17   2  18      24   1       1  25
      2  *    3          25   2   1   1   1  19
      3  *        1   1          24   1   1
      4  *            1          15
      5  *                        9
      6  *                       10

S T E M  *    0   1   2   3   4   5   6   7   8   9

MINIMUM =      2.00000
MAXIMUM =     65.00000
 COUNT  =    294
```

[1] The display for age is a standard stem and leaf. The stem contains the digits from the tens position and the leaves contain the digits from the units position. The first bar shows five 18 year olds and five 19 year olds. The single oldest subject is 89 years old. In the display, M marks the bar with the median, Q marks the first and third quartiles, and E marks the lower and upper eighths. The column labeled DEPTH accumulates the number of cases beginning at the top and bottom of the display, moving in towards the median.

[2] This is an alternative stem and leaf display printed when there are too many leaves on a stem to fit on one printer line. For income, the stem contains the first digits and the leaves contain the second digits. In the last row, we see that ten subjects have an income of $65,000: the stem is 6, and the leaf is 5.

Special Features

LINESIZE

To obtain narrow output for the computer terminal you must specify LINESIZE = 80 (or any number less than 132) in the PRINT paragraph, unless you are running in interactive mode. Note that the narrow output has a different format from the default wide output (compare Output 2D.1 with Output 2D.3).

Definitions of robust measures

Both the Hampel and biweight estimates assign higher weights to the observations near the estimate than to those far from the estimate. The weights are a function of

$$u = \frac{\text{observed value} - \text{estimate of location}}{\text{estimate of dispersion}}$$

where the estimate of dispersion (MAD) is the median of the absolute deviations from the sample median:

$$\text{MAD} = \text{median} \mid \text{observation} - \text{median} \mid .$$

The diagrams below show the weights as a function of u.

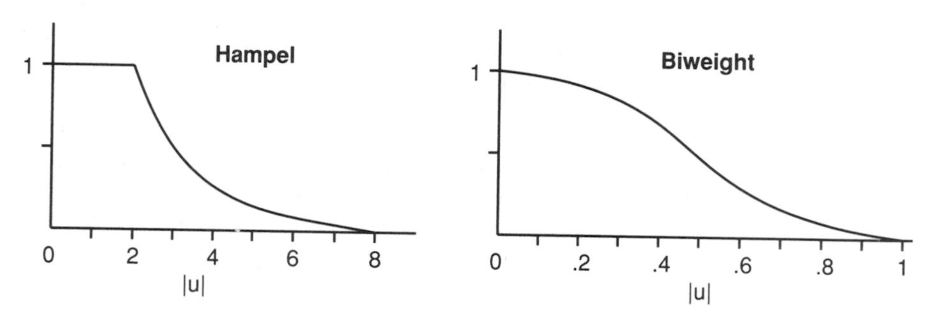

Since the estimate of location is used in the formula for u, the estimate of location must be found by iteration. The 15% trimmed mean and this parameterization of the Hampel estimator are satisfactory for most samples when the tails of the distribution are longer than those from a normal distribution. The biweight estimator is parameterized here to perform well for distributions with extremely long tails. See Appendix B.5.

2D Commands

To get an output panel like that shown in Output 2D.1, you need no instructions unique to 2D.

/ VARiable

The VARIABLE paragraph is optional. (See detailed description in Chapter 4.) Additional commands for 2D are FREQUENCY and MMV.

FREQuency = variable. `FREQ = COUNT.`

State the name or number of a variable containing the case frequency. A case frequency has the effect of repeating cases. Any record with a negative or zero case frequency (COUNT ≤ 0 in this example) is excluded from the calculations as though it were missing. See Chapter 4.

New Feature

MMV = #. `MMV = 10.`

Specify the maximum number of missing values or values out of range that will be allowed before a case is excluded from the analysis. MMV is set to zero for STAND = COV or WCOV (only cases with complete data are used in the analysis). Default: number of variables used.

/ GROUP

New Feature

VARiables = list. `VAR = SEX.`

List names or numbers of the GROUPING variables. You must specify CODES or CUTPOINTS in the GROUP paragraph if the variable takes on more than 10 distinct values. **Example:** The variable SEX contains values for separating the cases into cells (groups). See Chapter 5 for details on the GROUP paragraph.

/ COUNT

The COUNT paragraph is optional.

ROUND = # list. `ROUND = (income)10.`

Specify units for rounding variables. Use one number per variable. You can round data to any positive value, including values less than one, such as 0.1, 0.05, etc. Use tabbing to skip variables. Default: the variables are not rounded. **Example:** The variable INCOME will be rounded to the nearest number divisible by 10.

TRUNCate = # list. `TRUNC = (4)10, 100.`

Specify units for truncating variables. Use one number per variable; tabbing may be used to skip variables. You can truncate data to any positive value, including values less than one. If both ROUND and TRUNCATE are specified for the same variable, the ROUND specification takes precedence. Default: the variables are not truncated. **Example:** The fourth variable will be truncated to the nearest lower number divisible by ten; the fifth variable will be truncated to the nearest lower number divisible by 100 (e.g., 589 will be truncated to 500).

/ PRINT

The PRINT paragraph is optional.

LINEsize = #. `LINE = 80.`

Specify the number of columns of output width. 2D has two output formats. Using a number less than 132 results in a narrower, 80 column page. The default is 132 columns for noninteractive runs, 80 for interactive.

ESTIMate. `ESTIM.`

Specify ESTIMATE to compute and plot on a line plot the 15% trimmed mean, Hampel, and biweight estimates of location.

WSTATistic. `WSTAT.`

Prints Shapiro and Wilk's W statistic for testing normality. The W statistic is available for sample sizes 3 to 2000.

COUNT. `NO COUNT.`

Unless you specify NO COUNT, 2D prints the table of frequencies, percents, and cumulative percents.

STEM. `STEM.`

Requests stem and leaf histograms for each variable. If there are too many leaves on a stem to print on one printer line, 2D prints a stem and leaf frequency count table.

The following commands work only in the batch mode.

New Feature

MINimum. `MIN.`

Prints data for cases in which at least one variable has a value less than its lower limit as specified in the VARIABLE paragraph. Default: such cases are not printed unless you specify MIN.

New Feature

MAXimum. `MAX.`

Prints data for cases in which at least one variable has a value greater than its upper limit as specified in the VARIABLE paragraph. Default: such cases are not printed unless you specify MAX.

New Feature

MISSing. `MISS.`

Prints data for cases that have at least one variable with a missing value. Missing values in the input may be identified in several ways: (1) an asterisk (the default missing value character), (2) a missing value code identified using MISS in the VARIABLE paragraph, (3) a character identified as MCHAR in the INPUT paragraph, (4) blanks (applies only to input read with FORTRAN fixed format). Default: such cases are not printed unless you specify MISS.

New Feature

METHod = VARiable *or* **CASE.** `METH = CASE.`

Use to determine how the printout is organized. VARIABLE is the default value.

VARIABLE – Prints for each case only as many variables as fill the print line (up to ten variables) and begins the next line with the same variables for the next case. Thus the data for each variable are in an easy-to-scan column. When the program finishes printing the first subset of variables for all cases, it continues with another subset.

CASE – Prints the values of all variables for one case (possibly filling several print lines) before printing data for the next case.

New Feature

CASE = #. `CASE = 50.`

Specify the number of cases for which data are printed. Values flagged by MISS, MAX, and MIN in the VARIABLE paragraph specifications are replaced by MISSING, GT MAX, and LT MIN. Default: 10.

New Feature

MEAN. `NO MEAN.`

Unless you specify NO MEAN, 2D prints the mean, standard deviation, and frequency of each used variable.

New Feature

EZSCores. `NO EZSC.`

Unless you state NO EZSC, 2D prints z-scores for the minimum and maximum values of each used variable.

New Feature

ECASe. `NO ECAS.`

Unless you state NO ECAS, 2D prints the case numbers containing the minimum and maximum values of each variable.

New Feature

EXTReme. `NO EXTR.`

Unless you state NO EXTR, 2D prints the minimum and maximum values for each variable used.

New Feature

SKewness. `NO SK.`

Unless you state NO SK, 2D prints skewness and kurtosis for each used variable.

Order of Instructions

■ indicates required paragraph

■ / INPUT
/ VARIABLE
/ GROUP
/ TRANSFORM
/ SAVE
/ COUNT
/ PRINT
■ / END
data
PRINT … /
COUNT … /
GROUP … /
SAVE … /
TRANSFORM … /
CONTROL …/
END /
FINISH /

(PRINT … / through END /) May be repeated interactively.

(/ INPUT through FINISH /) Repeat for additional problems. See Multiple Problems, Chapter 10.

Summary Table for Commands Specific to 2D

Paragraphs Commands	Defaults	Multiple Problems	See
/ **VARiable**			
FREQuency = variable.	no case frequency variable	–	Cmds
MMV = #.	# of variables	–	Cmds
/ **GROUP**			
▲ VARiable = variables.	no grouping	●	2D.2
/ **COUNT**			
ROUND = # list.	no rounding	–	2D.3
TRUNcate = # list.	no truncating	–	Cmds
/ **PRINT**			
ESTIMate.	no robust estimates	–	2D.2
COUNT or NO COUNT.	COUNT.	–	2D.2
STEM.	no stem and leaf histogram	–	2D.4
LINEsize = #.	132 noninteractive, 80 interactive	●	2D.2
WSTATistic.	no *W* statistic	–	2D.2
MISSing.	NO MISS.	●	Cmds
MINimum.	NO MIN.	●	Cmds
MAXimum.	NO MAX.	●	Cmds
METHod = VARiable *or* CASE.	VARIABLE	–	Cmds
CASE = #.	ten cases printed	●	Cmds
MEAN. *or* NO MEAN.	MEAN.	●	Cmds
EXTReme. *or* NO EXTR.	EXTReme. values printed	●	Cmds
EZSCores. *or* NO EZSC.	EZSC. (z-scores)	●	Cmds
ECASe. *or* NO ECASe.	ECAS. (Case #'s of Min and Max printed)	●	Cmds
SKewness. *or* NO SK.	SKewness. (skewness and kurtosis)	●	Cmds

Key:
▲ Frequently used command
● Value retained for multiple problems
– Default reassigned

3D

t Tests

Program 3D performs two-group *t* tests, matched (paired) *t* tests, and one group *t* tests.

For two-group *t* tests, 3D prints descriptive statistics and a histogram for each group, and compares the group means with and without the assumption of equality of variances. 3D performs Levene's test of the equality of variances. You can request a trimmed *t* test to reduce the influence of the largest value and the smallest value in each group (single trimming), and you can request the nonparametric Mann-Whitney (Wilcoxon) rank-sum test. You can use the multivariate Hotelling's T^2 and Mahalanobis' D^2 tests to test several variables simultaneously for equality of means in the two groups. You can also request a matrix of correlations among the variables within each group.

For the **matched (paired) *t* test**, 3D prints descriptive statistics and a histogram for each variable, followed by information on the paired differences. 3D prints the matched *t* statistic to test equality of means and the Pearson correlation coefficient (*r*). You can request a trimmed *t* test or the nonparametric sign test, nonparametric Wilcoxon signed-rank test, and nonparametric Spearman *r*. You can use Hotelling's T^2 and Mahalanobis' D^2 to test the means of several variables simultaneously. Similar output is available for the one-group *t* test.

The program now provides quantile-quantile plots and allows interactive modification of grouping information. For an overview of *t* tests and related statistics see, for example, Dixon and Massey, 1983.

Where to Find It

Examples

Two-group t Tests

Example 3D.1 The two-group t test

The two-group *t* test (Student's two-sample *t* test) compares the means of two groups. We use 3D to analyze the Exercise data discussed in Chapter 2. Here we compare mean pulse rates before exercise for two groups of subjects—smokers and nonsmokers. The record for each subject includes an ID number, sex, smoking status (yes/no), age, and pulse rate before and after exercise. The data are stored on disk in a file named PULSE2.DAT. This data file is identical to the EXERCISE.DAT file shown in Appendix D, without the case label variables (NAME1 and NAME2).

The instructions in the INPUT and VARIABLE paragraphs displayed in Input 3D.1 are part of the basic BMDP instructions common to all BMDP programs. (The FILE command tells the program where to find the data and is used for systems like VAX and IBM PC. For IBM mainframes, see UNIT, Chapter 3.) See Chapters 3 and 4 for an explanation of the INPUT and VARIABLE paragraphs.

To analyze data by groups we specify a grouping variable in the GROUP paragraph with the VARIABLE command:

```
/ GROUP      VARIABLE = SMOKE.
/ TWOGROUP   VARIABLE = PULSE_1.
```

We use the VARIABLE command in the TWOGROUP paragraph to identify the measure under study. If we omitted the command, 3D would analyze all input variables except the grouping variable.

In the GROUP paragraph (see Chapter 5) we specify the name of the grouping variable, as well as the codes and names for the different groups. We specify LINESIZE = 80 in the PRINT paragraph to obtain shorter lines for viewing at the computer terminal.

Input 3D.1

```
/ INPUT      FILE = 'pulse2.dat'.
             VARIABLES = 6.
             FORMAT IS FREE.

/ VARIABLE   NAMES = id, sex, smoke, age, pulse_1,
                     pulse_2.

/ GROUP      VARIABLE = smoke.
             CODES(smoke) = 1, 2.
             NAMES(smoke) = smoke, nosmoke.

/ TWOGROUP   VARIABLE = pulse_1.

/ PRINT      LINESIZE = 80.

/ END
```

Output 3D.1

```
CASE    1         2        3         4         5        6                       [1]
 NO. id        sex      smoke     age      pulse_1  pulse_2
---- --------  -------- --------  -------- -------- --------
   1     1.00     1.00 smoke        31.00    62.00   126.00
   2     5.00     1.00 smoke        21.00    66.00   128.00
   3    10.00     1.00 smoke        28.00    90.00   144.00
   4    15.00     1.00 smoke        23.00    92.00   134.00
   5    20.00     1.00 smoke        20.00    66.00   132.00
   6    21.00     1.00 smoke        38.00    70.00   122.00
   7    29.00     1.00 smoke        22.00    68.00   126.00
   8    30.00     1.00 smoke        19.00    70.00   144.00
   9    36.00     1.00 smoke        34.00    62.00   132.00
  10    38.00     1.00 smoke        21.00    90.00   138.00

NUMBER OF CASES READ. . . . . . . . . . . . .          40                       [2]

  VARIABLE        STATED VALUES FOR           GROUP CATEGORY    INTERVALS       [3]
 NO.   NAME    MINIMUM MAXIMUM MISSING   CODE  INDEX   NAME      .GT.    .LE.
---- --------  ------- ------- -------  ------ -----  --------  ------ -------

  3  smoke                               1.000    1  smoke
                                         2.000    2  nosmoke
-------------------------------------------------------------------------------

GROUPING VARIABLE. . . smoke
                                    CATEGORY    FREQUENCY
                                    --------    ---------
                                    smoke          16
                                    nosmoke        24


DESCRIPTIVE STATISTICS OF DATA                                                  [4]
----------- ---------- -- ----

 VARIABLE     TOTAL          STANDARD  ST.ERR   COEFF
 NO. NAME      FREQ.  MEAN     DEV.    OF MEAN  OF VAR

   1 id          40  20.500  11.690  1.8484  .57027
   2 sex         40  1.4500  .50383  .07966  .34747
   3 smoke       40  1.6000  .49614  .07845  .31009
   4 age         40  26.650  6.8894  1.0893  .25851
   5 pulse_1     40  75.950  9.5379  1.5081  .12558
   6 pulse_2     40  140.50  23.690  3.7458  .16861

 VARIABLE          S M A L L E S T        L A R G E S T
 NO. NAME        VALUE  Z-SCR  CASE    VALUE  Z-SCR  CASE   RANGE

   1 id          1.0000 -1.67     1   40.000  1.67    22   39.000
   2 sex         1.0000 -0.89     1   2.0000  1.09    23   1.0000
   3 smoke       1.0000 -1.21     1   2.0000  0.81    11   1.0000
   4 age         19.000 -1.11     8   43.000  2.37    26   24.000
   5 pulse_1     62.000 -1.46     1   96.000  2.10    24   34.000
   6 pulse_2     112.00 -1.20    14   265.00  5.26    24   153.00
```

—we omit some output—

```
pulse_1    VARIABLE NUMBER    5
*******************************

GROUP    1 smoke                  2 nosmoke                                     [5]


                                        X
   H                                    X X     X
 H H H           HH                    XXX XX X    X
 H HHH    HH     HHH H              XXXXX XXXXXX X      X
M-------------------M   M--------------------M
I  AN H=     1 CASES  A  I   AN X=     1 CASES  A
N    (N=    16)       X  N     (N=    24)       X
```

```
                           [6]
GROUP          smoke             nosmoke         TEST STATISTICS    P-VALUE  DF
-------------------------------------           --------------------------------
MEAN           76.7500           75.4167         LEVENE F FOR
                                                  VARIABILITY      8.67 0.0055 1,    38   [7]
STD DEV        11.9972            7.7230
S.E.M.          2.9993            1.5765         POOLED       T    0.43 0.6707   38
SAMPLE SIZE         16                24         SEPARATE     T    0.39 0.6975   23.3     [8]
MAXIMUM        96.0000           96.0000
MINIMUM        62.0000           64.0000
Z MAX           1.60              2.67
Z MIN          -1.23             -1.48
CASE (MAX)          6                 4
CASE (MIN)          1                 3
```

Output 3D.1 shows a comparison of the variable PULSE_1 for the groups SMOKE and NOSMOKE. Bold numbers below correspond to those in the output.

[1] Prints the first ten cases of data. Grouping names are printed if specified.

[2] Number of cases read. 3D uses only complete cases in all computations. That is, if the value of any variable in a case is missing or out of range, 3D omits the case from all computations. If you specify a USE list in the VARIABLE paragraph, only the variables in the USE list are considered in determining the completeness of each case. When you specify CODES for the GROUPING variables, 3D uses only cases with one of the specified codes.

[3] Interpretation of the grouping variable instructions. The grouping variable in this example is SMOKE, the third variable.

[4] This table gives statistics without grouping information. Note that two tables have been created, since LINESIZE = 80 was specified in the PRINT paragraph. For each variable this table includes:

- mean
- standard deviation
- standard error of the mean
- coefficient of variation
- smallest value with case number and *z*-score for that case
- largest value with case number and *z*-score for that case

[5] 3D plots a histogram of the data for each group. Each histogram is 20 characters wide and a maximum of six lines high. These histograms provide a quick look at the shape of the distribution. Both histograms are plotted on the same scale extending from the minimum value to the maximum value for all the groups analyzed in the paragraph. The label under the histogram specifies the number of observations represented by each H or X. (The number of Hs or Xs to be plotted is rounded to the next higher integer. For example, if each X represents three observations, a single isolated observation is still represented by an X; seven observations are represented by three Xs.) Note that the shape of the histogram for smokers differs from that for nonsmokers (it appears to be in two parts).

[6] For each group (identified by name and number) 3D prints

- mean
- standard deviation
- S.E.M. (standard error of the mean)
- sample size (frequency)
- maximum observed value (not out of range), its *z*-score, and case number
- minimum observed value (not out of range), its *z*-score, and case number

[7] The question of equality of variability between two groups may be of interest. The Levene *F* for variability provides a test of whether the between group difference in variability noted above is statistically significant. Here the *p*-value is less than 0.01, suggesting that smokers' pulse rates vary more than nonsmokers' pulse rates. These statistics are defined in Appendix B.6.

[8] 3D provides two versions of the t test to test the equality of group means. The classical Student's t test is given as pooled t; it assumes that population variances for the two groups are equal, and its denominator contains a pooled sample standard deviation. The separate variance t test (Welch) does not make this additional assumption and tends to be conservative if the population variances are equal. One might think of testing for equality of variances first and choosing the pooled-variance t test when the test of variances is not significant. However, failure to reject the null hypothesis of equal variances is not equivalent to proof of variance equality. Use of the pooled-variance t test should be based on good prior knowledge of variance equality in the populations from which the groups are drawn. Note that here the mean pulse rates of smokers and nonsmokers are not significantly different.

Example 3D.2 The trimmed *t* test and nonparametric statistics for two groups

When there are outliers in the data, the t test can be overly sensitive to the extreme scores. To obtain a trimmed t (reducing the influence of the largest and and smallest values in each group), we specify ROBUST in the TWOGROUP paragraph. 3D reports the trimmed t in both a pooled variance and separate variance version. (See Dixon and Massey, 1983, and Appendix B.6.) If you want to trim more than the single largest and smallest values from your data, see Program 7D.

3D also offers nonparametric statistics. The Mann-Whitney (Wilcoxon) rank-sum test compares two groups by arranging all cases in order of size and assigning rank scores to the individual observations (see Dixon and Massey, 1983, pp. 394-396). The power of the rank-sum test compares favorably with the t test, and many people prefer using nonparametric tests because they do not assume normality. (However, even for the nonparametric tests, distributions for the two groups must be similar; e.g., histograms should have the same shape and variance.)

Here we look to see whether smokers and nonsmokers differ on post-exercise pulse rate. In Input 3D.2 we again use the TWOGROUP paragraph to specify the variable we want to examine (PULSE_2 this time). We use the GROUP paragraph to specify the pair of groups we want to compare (SMOKE and NOSMOKE). Note that 3D can use GROUPING variables with more than two categories. Unless otherwise directed, 3D compares all possible pairs of groups within the specified grouping variable; use the command GLIST (see 3D Commands) if you want to compare only a subset of the groups. If you want to compare more than two groups simultaneously, see program 7D.

Input 3D.2

```
/ INPUT     FILE = 'pulse2.dat'.
            VARIABLES ARE 6.
            FORMAT IS FREE.

/ VARIABLE NAMES = id, sex, smoke, age, pulse_1,
                   pulse_2.

/ GROUP     VARIABLE = smoke.
            CODES(smoke) = 1, 2.
            NAMES(smoke) = smoke, nosmoke.

/ TWOGROUP VARIABLE = pulse_2.
            ROBUST.
            NONPAR.

/ END
```

Output 3D.2

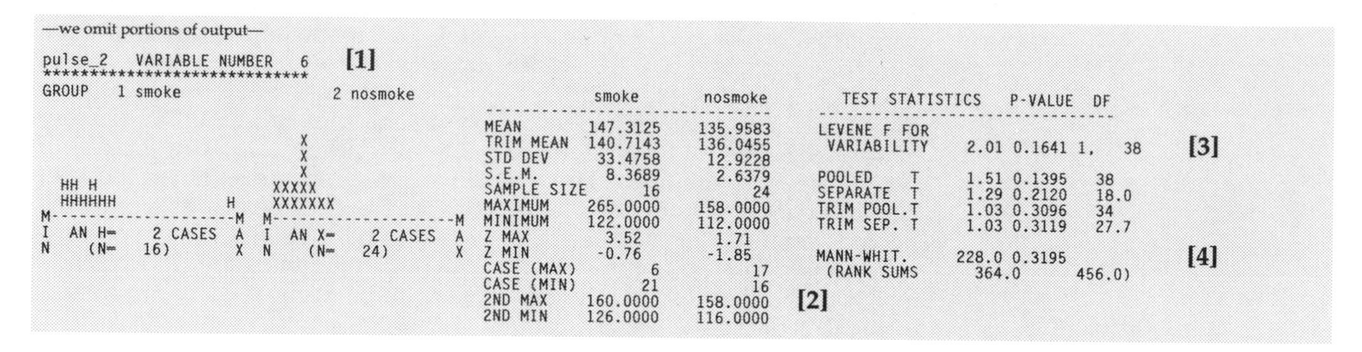

```
—we omit portions of output—
pulse_2   VARIABLE NUMBER   6    [1]
******************************
GROUP   1 smoke                 2 nosmoke              smoke       nosmoke         TEST STATISTICS   P-VALUE  DF
                                                   ----------------------------   --------------------------------
                                                   MEAN        147.3125   135.9583   LEVENE F FOR
                               X                   TRIM MEAN   140.7143   136.0455    VARIABILITY     2.01 0.1641 1,   38    [3]
                               X                   STD DEV      33.4758    12.9228
                               X                   S.E.M.        8.3689     2.6379   POOLED    T      1.51 0.1395   38
  HH H                       XXXXX                 SAMPLE SIZE       16         24   SEPARATE  T      1.29 0.2120   18.0
  HHHHHH               H     XXXXXXX               MAXIMUM     265.0000   158.0000   TRIM POOL.T      1.03 0.3096   34
M--------------------M  M--------------------M     MINIMUM     122.0000   112.0000   TRIM SEP. T      1.03 0.3119   27.7
I  AN H=    2 CASES  A  I  AN X=    2 CASES  A     Z MAX          3.52       1.71
N    (N=   16)       X  N    (N=   24)       X     Z MIN         -0.76      -1.85    MANN-WHIT.     228.0 0.3195          [4]
                                                   CASE (MAX)        6         17     (RANK SUMS     364.0        456.0)
                                                   CASE (MIN)       21         16
                                                   2ND MAX     160.0000   158.0000  [2]
                                                   2ND MIN     126.0000   116.0000
```

Bold numbers below correspond to those in Output 3D.2.

[1] From the histograms we note immediately that one case in the smoking group has a very high pulse rate. The maximum given in the panel to the left is 265, clearly an improbable value. At this point, you should check the data sheets to see whether an error was made when the data were entered into the computer. Both the nonparametric and robust options allow you to reduce the effect of such outliers on the analysis.

[2] When you request the trimmed *t*, 3D reports the second largest and second smallest values in each group (2ND MAX and 2ND MIN) in addition to MAX and MIN.

[3] 3D trims one observation from the top (the largest value) and one from the bottom (the smallest value) of each group. The trimmed *t* test uses a Winsorized standard deviation to adjust for deletion of these cases (see Dixon and Massey, 1983). Note that the trimmed pooled *t* refers to the *t*-table with four fewer degrees of freedom than the pooled *t* (four cases are deleted). The difference between the trimmed means (140.7 minus 136.0) is markedly smaller than between the means of all cases (147.3 minus 135.9), and thus the trimmed pooled *t* value is considerably smaller than the pooled *t*. Other statistics (mean, standard deviation, S.E.M.) are based on all cases without trimming.

[4] The Mann-Whitney (or Wilcoxon) rank-sum test analyzes only the ranks of the values. It thus treats all ordered values as being one unit apart.

After we remove the single extreme value from these data, there is no evidence in either the trimmed *t* test or the rank sum test of a significant difference between smokers' and nonsmokers' post-exercise pulse rates.

Example 3D.3 Correlations and multivariate Hotelling T^2

You can test the equality of group means of several variables simultaneously by using the multivariate Hotelling's T^2 (Morrison, 1967, p. 120) and Mahalanobis' D^2. (If you want multivariate tests for more than two groups or more detailed multivariate output, you may want to use 7M.) The two statistics express the same information about the multivariate distance between the groups, and can be transformed to an *F* statistic to generate *p*-values (see Appendix B.6). In Input 3D.3 we request the multivariate tests on two variables (PULSE_1 and PULSE_2) by specifying HOTEL in the TWOGROUP paragraph. We also request the correlation matrix between the variables within each group by stating CORRELATIONS. Here we are interested in whether males and females differ in their pulse rates before and after exercise.

For this example we assume that the outlier 265 detected in Example 3D.2 was a transcription error and should have been entered as 165. This can be corrected by editing the data set or, as shown here, by using a TRANSFORM statement.

Input 3D.3

```
/ INPUT      FILE = 'pulse2.dat'.
             VARIABLES ARE 6.
             FORMAT IS FREE.

/ VARIABLE   NAMES = id, sex, smoke, age, pulse_1,
                     pulse_2.

/ TRANSFORM IF (pulse_2 EQ 265) THEN pulse_2 = 165.

/ GROUP      VARIABLE = sex.
             CODES(sex) = 1, 2.
             NAMES(sex) = male, female.

/ TWOGROUP   VARIABLES ARE pulse_1, pulse_2.
             CORRELATIONS.
             HOTEL.

/ END
```

Output 3D.3

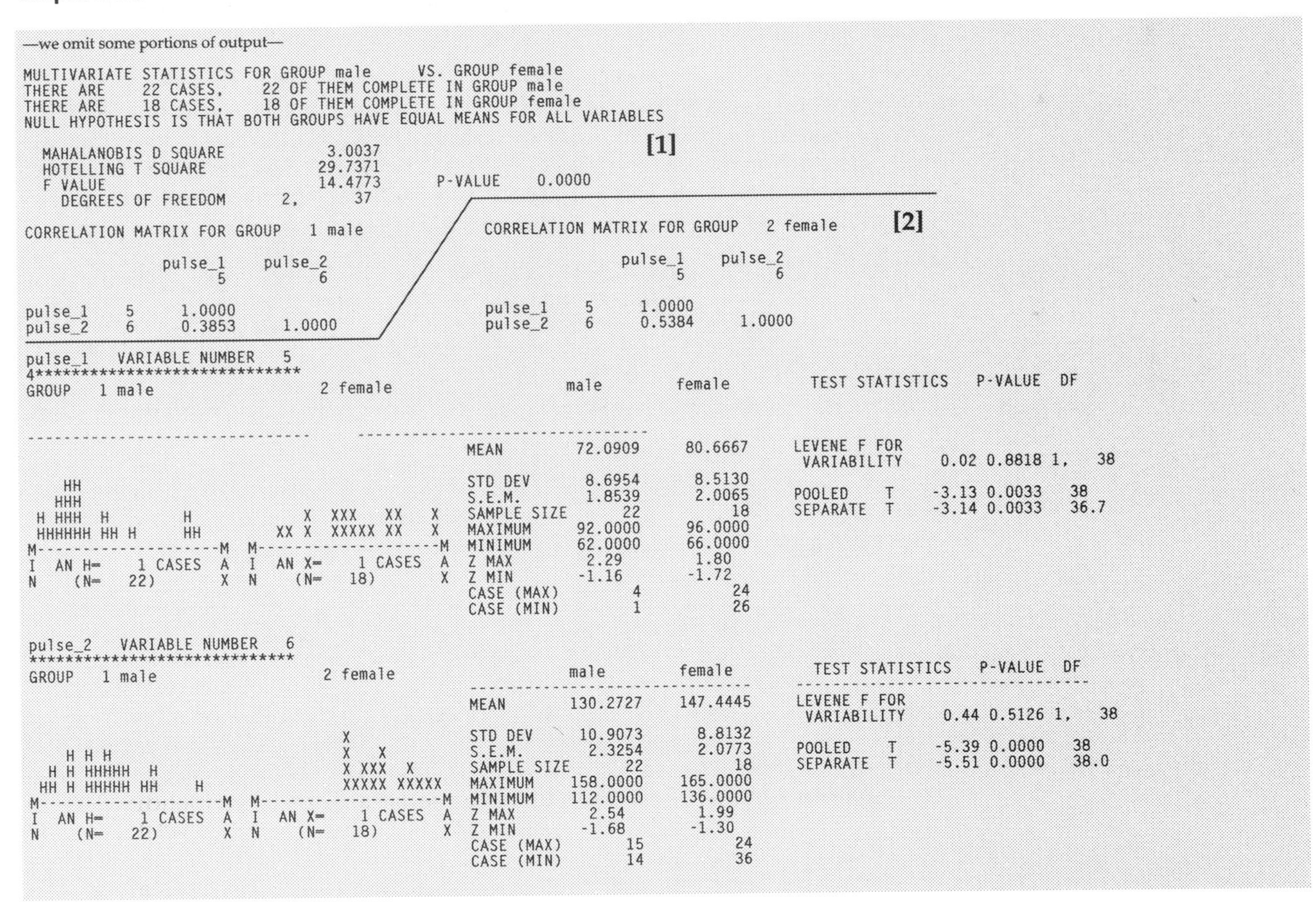

—we omit some portions of output—

```
MULTIVARIATE STATISTICS FOR GROUP male      VS. GROUP female
THERE ARE    22 CASES,    22 OF THEM COMPLETE IN GROUP male
THERE ARE    18 CASES,    18 OF THEM COMPLETE IN GROUP female
NULL HYPOTHESIS IS THAT BOTH GROUPS HAVE EQUAL MEANS FOR ALL VARIABLES

  MAHALANOBIS D SQUARE              3.0037                    [1]
  HOTELLING T SQUARE               29.7371
  F VALUE                          14.4773       P-VALUE     0.0000
    DEGREES OF FREEDOM        2,      37

CORRELATION MATRIX FOR GROUP   1 male        CORRELATION MATRIX FOR GROUP   2 female      [2]

               pulse_1    pulse_2                           pulse_1    pulse_2
                     5          6                                 5          6

pulse_1    5    1.0000                        pulse_1    5    1.0000
pulse_2    6    0.3853     1.0000             pulse_2    6    0.5384     1.0000

pulse_1    VARIABLE NUMBER    5
4******************************
GROUP    1 male                  2 female              male        female         TEST STATISTICS   P-VALUE  DF

-----------------------------    ----------------------------
                                               MEAN        72.0909     80.6667     LEVENE F FOR
                                                                                    VARIABILITY       0.02 0.8818 1,    38
     HH                                        STD DEV      8.6954      8.5130
     HHH                                       S.E.M.       1.8539      2.0065     POOLED     T      -3.13 0.0033   38
  H HHH  H          H                X  XXX   XX   X   SAMPLE SIZE     22          18     SEPARATE   T      -3.14 0.0033   36.7
 HHHHHH HH H        HH          XX X  XXXXX XX   X   MAXIMUM     92.0000     96.0000
M-------------------M  M-----------------------M   MINIMUM     62.0000     66.0000
I  AN H=     1 CASES  A  I   AN X=     1 CASES  A   Z MAX         2.29        1.80
N    (N=   22)        X  N     (N=   18)        X   Z MIN        -1.16       -1.72
                                               CASE (MAX)         4          24
                                               CASE (MIN)         1          26

pulse_2    VARIABLE NUMBER    6
******************************
GROUP    1 male                  2 female              male        female         TEST STATISTICS   P-VALUE  DF
                                               ----------------------------       ------------------------------
                                               MEAN       130.2727    147.4445     LEVENE F FOR
                                                                                    VARIABILITY       0.44 0.5126 1,    38
                                      X        STD DEV     10.9073      8.8132
     H H H                            X    X   S.E.M.       2.3254      2.0773     POOLED     T      -5.39 0.0000   38
   H H HHHHH   H                      X XXX   X   SAMPLE SIZE     22          18     SEPARATE   T      -5.51 0.0000   38.0
  HH H HHHHH HH     H                 XXXXX XXXXX  MAXIMUM    158.0000    165.0000
M-------------------M  M-----------------------M   MINIMUM    112.0000    136.0000
I  AN H=     1 CASES  A  I   AN X=     1 CASES  A   Z MAX         2.54        1.99
N    (N=   22)        X  N     (N=   18)        X   Z MIN        -1.68       -1.30
                                               CASE (MAX)        15          24
                                               CASE (MIN)        14          36
```

The results for the multivariate tests and the correlation matrices are displayed in Output 3D.3. (These statistics are defined in Appendix B.6.) The usual results for the univariate tests follow the correlation matrices.

[1] The multivariate tests show a significant difference between males and females for the variables PULSE_1 and PULSE_2 (the *p*-value is low). Note that the highly significant Hotelling's T^2 in this case reflects a significant difference in both single-variable *t* tests. Pulse rates are higher for females than for males by about 8 points before exercise and 17 points after exercise.

[2] 3D prints the correlations between the variables in each group when you specify CORRELATION. We see here that pulse rates are only modestly correlated and that, given the small sample size, the correlations are similar in the two groups. The assumptions that variances and correlations are similar in the two groups appear satisfied when we compare correlations and standard deviations (the Levene test) for each variable.

Matched Pairs

Example 3D.4 The matched pairs *t* test

The paired *t* test compares two variables rather than comparing one variable for different groups. It is appropriate for two types of studies: (1) those like the exercise data, where the same individuals are measured before and after treatment, and (2) those in which, for example, subjects are matched at the start of the study, one member of each pair is randomly assigned to treatment 1 and the other to treatment 2, and each subject's posttreatment weight is recorded (see Example 3D.6).

To do a matched pairs *t* test, use the MATCHED paragraph. To specify the variables, you may use either the VARIABLE command or the FIRST and SECOND commands. 3D matches the first variable in the VARIABLE list versus the second, the third versus the fourth, etc. When there are two lists (FIRST and SECOND), 3D matches the first variables from each list, then the second, etc. See also CROSS and PAIR (Special Features).

In Input 3D.4 we use the VARIABLE command to compare PULSE_2 and PULSE_1 for the individuals in the exercise data. As in Example 3D.3, we use a TRANSFORM paragraph to correct the outlier PULSE_2 value of 265.

Input 3D.4

```
/ INPUT      FILE = 'pulse2.dat'.
             VARIABLES ARE 6.
             FORMAT IS FREE.

/ VARIABLE   NAMES = id, sex, smoke, age, pulse_1,
                     pulse_2.

/ TRANSFORM IF (pulse_2 EQ 265) THEN pulse_2 = 165.

/ MATCHED    VARIABLES ARE pulse_1, pulse_2.

/ END
```

Output 3D.4

```
pulse_1  VS. pulse_2  (VAR. NO.   5 VS.   6)          [1]
********************************************

      pulse_1                 pulse_2                    pulse_1     pulse_2
                                              ----------------------------
  H                                           MEAN        75.9500    138.0000
  H
 HH                                  X        STD DEV      9.5379     13.1442
 HHHH H                           XXXX        S.E.M.       1.5081      2.0783
 HHHHHH                         XXXXXX X      SAMPLE SIZE      40          40
 HHHHHHH                      XXXXXXXXXXX     MAXIMUM     96.0000    165.0000
M------------------M M-------------------M    MINIMUM     62.0000    112.0000
I  AN H=    2 CASES  A I  AN X=    2 CASES  A    Z MAX        2.10        2.05
N    (N=   40)      X N    (N=   40)        X    Z MIN       -1.46       -1.98
                                              CASE (MAX)       4           6
                                              CASE (MIN)       1          16
```

```
pulse_1 - pulse_2  (VAR. NO.    5 -    6)           [2]
******************************************
                         pulse_1 - pulse_2     TEST STATISTICS   P-VALUE  DF
                         -----------------     --------------------------------
                         MEAN        -62.0500   MATCHED  T    -36.69 0.0000   39

         H     HH        STD DEV      10.6962
         H     HH  H   H S.E.M.        1.6912
  H  HHHH HHHHH HH H H   SAMPLE SIZE      40
  H HHHHHHHHHHHHHHHH H   MAXIMUM     -42.0000   CORRELATION    0.5956 0.0000   39
M-------------------M    MINIMUM     -82.0000
I  AN H=    1 CASES  A   Z MAX         1.87
N    (N=   40)       X   Z MIN        -1.87
                         CASE (MAX)       15
                         CASE (MIN)       13
```

Bold numbers below correspond to those in Output 3D.4.

[1] 3D prints information on each variable, followed by information on the paired differences.

[2] The average difference between the two pulse rates is 62. Since the program subtracted PULSE_1 from PULSE_2, the negative mean difference indicates that PULSE_2 is higher. This could be seen in the individual means in panel 1 above. The matched *t* has a low *p*-value, indicating a significant difference between the matched pairs of values for PULSE_1 and PULSE_2. 3D prints the standard Pearson correlation coefficient between the matched variables. The correlation between pre- and post-exercise pulse is .6, significantly different from zero. (The higher the correlation, the greater the advantage in using a paired rather than a separate-group design.)

Example 3D.5 Matched pairs trimmed *t* test and nonparametric statistics

In Input 3D.5 we again use the uncorrected exercise data set (including a pulse of 265) and specify ROBUST and NONPAR in the MATCHED paragraph to obtain a trimmed *t* test and nonparametric statistics. When you specify NONPAR in the MATCHED paragraph, 3D provides the sign test, the Wilcoxon signed-rank test, parametric statistics, and the Spearman correlation coefficient.

Input 3D.5

```
/ INPUT       FILE = 'pulse2.dat'.
              VARIABLES ARE 6.
              FORMAT IS FREE.

/ VARIABLE    NAMES = id, sex, smoke, age, pulse_1,
                      pulse_2.

/ MATCHED     VARIABLES ARE pulse_1, pulse_2.
              ROBUST.
              NONPAR.

/ END
```

Output 3D.5

```
pulse_1  VS. pulse_2  (VAR. NO.   5 VS.   6)        [1]
*********************************************
       pulse_1                  pulse_2                     pulse_1      pulse_2
                                                -------------------------------
                                                MEAN         75.9500     140.5000
H                                               TRIM MEAN    75.7895     137.9737
H                                               STD DEV       9.5379      23.6903
HH                              X               S.E.M.        1.5081       3.7458
HHH                           XXXXX             SAMPLE SIZE      40           40
HHHH                          XXXXXX        X   MAXIMUM      96.0000     265.0000
M------------------M  M-------------------M     MINIMUM      62.0000     112.0000
I  AN H=    4 CASES A  I  AN X=    4 CASES A    Z MAX         2.10         5.26
N    (N=   40)      X  N    (N=   40)      X    Z MIN        -1.46        -1.20
                                                CASE (MAX)       4            6
                                                CASE (MIN)       1           16
                                                2ND MAX      96.0000     160.0000
                                                2ND MIN      62.0000     116.0000
```

```
pulse_1 - pulse_2  (VAR. NO.   5 -   6)
****************************************

                        pulse_1 - pulse_2      TEST STATISTICS   P-VALUE  DF
                        -----------------     -----------------------------
                        MEAN         -64.5500  MATCHED  T     -20.41 0.0000   39
                  H     TRIM MEAN    -62.3947  TRIMMED  T     -33.74 0.0000   37      [2]
                  H     STD DEV       20.0012  SIGN TEST*            0.0000
              HHHHHH    S.E.M.         3.1625  WILCOXON**        0.0 0.0000
              HHHHHH    SAMPLE SIZE        40
 H           HHHHHHH    MAXIMUM      -42.0000  CORRELATION    0.5580 0.0002   39
M-------------------M   MINIMUM     -169.0000  SPEARMAN R     0.5825 0.0001   38
I   AN H=    2 CASES A  Z MAX         1.13
N     (N=   40)      X  Z MIN        -5.22     * pulse_1  > pulse_2  IN    0
                        CASE (MAX)         15   CASES OF   40 WITH NONZERO DIFS.
                        CASE (MIN)          6  ** TOTAL OF RANKS WITH LESS
                        2ND MAX      -42.0000    FREQUENT SIGN =         0.0
                        2ND MIN      -82.0000
```

Bold numbers below correspond to those in Output 3D.5.

[1] 3D prints trimmed means for each variable and, in **[2]**, for the difference between the paired variables, along with the ordinary matched *t*, the trimmed *t* test, and the Pearson correlation coefficient before trimming.

[2] The nonparametric sign test and Wilcoxon test both indicate a significant difference between the paired variables (the *p*-values are low). The Wilcoxon signed-rank test tests the hypothesis that the differences in the paired variables come from a distribution that is symmetric about zero (Hettmansperger, 1984). The Spearman correlation is a correlation based on ranks (see Dixon and Massey, 1983, p. 402); in this case the Spearman correlation is close to the Pearson correlation.

Example 3D.6 Matched pairs on paired records

Some matched pairs data are not on the same record. In this example, we demonstrate how to perform a matched pairs *t* test on variables in paired records by reading each pair of records as a single record.

The Werner data (see Appendix D) consist of blood chemistry data from 94 pairs of age-matched women. In each pair of records, the first woman does not use contraceptive pills and the second woman does (see Figure 3D.1). We want to perform a paired *t* test on WEIGHT. Each record contains nine variables. In Input 3D.6 we specify 18 variables rather than nine in the INPUT paragraph so that 3D will read two data records as a single case. In the VARIABLE paragraph we use tabbing to supply names for the variables of interest (e.g., ID_P, the id number for the patient on the pill, is the name of the tenth variable).

You must arrange data records for matched pairs carefully. That is, the row of data for a NOPILL woman must come before the corresponding row of data for her age-matched PILL counterpart (see diagram below). Then, when you label the variables WT_NP and WT_P (weight of the patient not on the pill and of the patient on the pill), you will be pointing to the correct data. (The Data Manager is valuable for arranging data. Program 1D can also sort data files.) We do not show output for this example, since it is similar to the output from Example 3D.4.

Figure 3D.1
Reading paired records

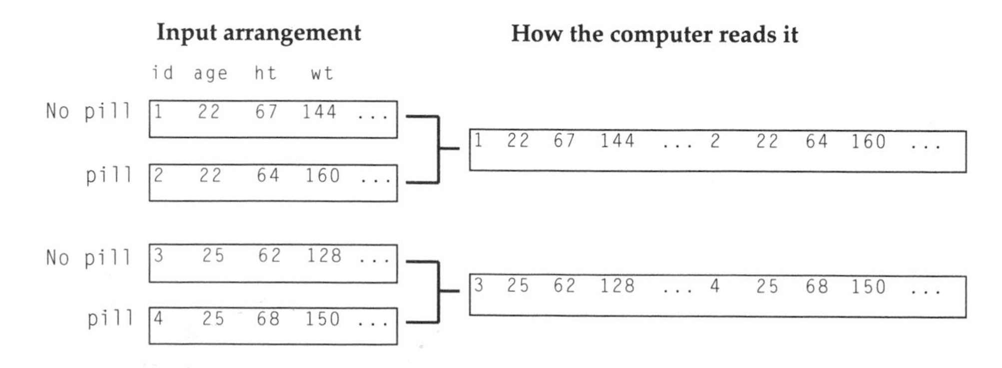

Input 3D.6

```
/ INPUT      FILE = 'werner.dat'.
             VARIABLES ARE 18.
             FORMAT IS FREE.

/ VARIABLE   NAMES = id_np, age_np, ht_np, wt_np,
                     (10)id_p, age_p, ht_p, wt_p.

/ MATCHED    VARIABLES = wt_np, wt_p.

/ END
```

One Group t Test

Example 3D.7 The one-group t test

To perform a one-sample *t* test, testing whether a variable's mean has a specified value, use the ONEGROUP paragraph to specify the variable and its expected mean. You can specify more than one variable for testing; if you do, the list of expected mean values must be indexed by the variable name or number as in the following example. If you do not specify a mean, the default value is zero.

Suppose you want to test whether the diameter of a machine part is equal to the required specification of 5.01 centimeters. A sample of size 10 was taken with the following results (Bowker and Lieberman, 1972):

5.036 5.031 5.085 5.064 4.991 4.942 4.935 5.051 4.999 5.011

The instructions in Input 3D.7 to specify the variable and its expected mean in the ONEGROUP paragraph are quite simple:

```
/ ONEGROUP    VARIABLE = diameter.
              MEAN = (diameter)5.01.
```

Note that we input the data in STREAM format (see Chapter 3).

Input 3D.7

```
/ INPUT         VARIABLE IS 1.
                FORMAT IS STREAM.

/ VARIABLE      NAME IS diameter.

/ ONEGROUP      VARIABLE IS diameter.
                MEAN = (diameter)5.01.
/ END
5.036 5.031 5.085 5.064 4.991 4.942 4.935 5.051 4.999 5.01 1
```

Output 3D.7

```
NUMBER OF CASES READ. . . . . . . . . . . . . .      10

diameter  VAR.   1   VS. MEAN=       5.0100
*******************************************
                                                 TEST STATISTICS   P-VALUE  DF
                         ------------------  ---------------------------------
                         MEAN           5.0145   1-SAMPLE T      0.29 0.7794    9

                         STD DEV        0.0493
                         S.E.M.         0.0156
 H                       SAMPLE SIZE       10
 H     HH H HH H H H     MAXIMUM        5.0850
M------------------M     MINIMUM        4.9350
I  AN H=    1 CASES  A   Z MAX          1.43
N    (N=   10)       X   Z MIN         -1.61
                         CASE (MAX)         3
                         CASE (MIN)         7
```

In Output 3D.7 we see that the mean diameter is 5.0145. The *t* statistic and its *p*-value suggest that this deviation from the specified mean 5.01 is not significant.

The one-group *t* test on a difference variable. A one-group *t* test equivalent to the matched *t* test can be obtained by creating a variable for the difference between two variables and testing whether the mean difference equals zero. For example, you could use the exercise data and create a variable containing the difference between the pulse_1 and pulse_2 values for each case.

```
/ TRANSFORM   diffrnce = pulse_2 - pulse_1.
/ ONEGROUP    VAR = diffrnce.
```

This is equivalent to the matched pairs test shown in Example 3D.4.

Example 3D.8 Running interactively

3D has special features for interactive use. You can specify any number of TRANSFORM, GROUP, MATCHED, ONEGROUP, and TWOGROUP paragraphs after you specify the INPUT, VARIABLE, GROUP, and END paragraphs. Each paragraph should end with a slash, and the rest of the line after the slash should be blank. You can continue performing new transformations and analyses as long as you like. You need to begin a new problem only if you want to read in a different data set or change the defaults in the VARIABLE, SAVE, or PRINT paragraphs. After the last interactive paragraph, type END/.

Interactive mode allows you to perform the tests specified in one paragraph, view the results, and then decide whether to continue to the next paragraph or modify the last. You must put one MATCHED, ONEGROUP, or TWOGROUP paragraph before the first END paragraph. Otherwise the default is a one-group *t* test on all variables.

Imagine that the instructions for Example 3D.1 are stored in a file called 3D.INP. Enter 3D interactively (see Chapter 11). 3D will then print a banner and place you in the BMDP line editor at line 1. You may then type

```
GET 3D.INP
```

to GET your input file (see Chapter 11 for a list of line editor commands). After the ten lines of the file appear on the screen, you type E for Execute to run the program. The program output shown in Output 3D.1 will appear on your screen one portion at a time.

After examining the output, you may want to test a new hypothesis: whether changing the grouping variable from SMOKE to SEX has a significant effect. If you press <Return>, the instructions will be repeated on the screen in the line editor. To test the new hypothesis, use the insert mode (I) to enter the commands shown at the end of Input 3D.8 to the end of the input file. In Input 3D.8 we use interactive paragraphs after the initial analysis to specify new grouping information, correct an outlier, add a transformed variable, and request a new two-group *t* test using the new groups and new variable. We omit the output for this example.

To exit insert mode, type //. When you execute this new set of instructions (E #, where # is the line number of the first new command), 3D will perform another two-group t test, this time with SEX as the grouping variable. You can also run 3D from program control level, one paragraph at a time (see Chapter 11). **Note:** in the BMDP release for the PC, a full-screen editor replaces the line editor. See your *PC User's Guide* for details.

Input 3D.8

```
/ INPUT      FILE = 'pulse2.dat'.
             VARIABLES ARE 6.
             FORMAT IS FREE.

/ VARIABLE   NAME = id, sex, smoke, age, pulse_1,
                    pulse_2.
```

Input 3D.8
(continued)

```
/  GROUP        VARIABLE = smoke.
                CODES(smoke) = 1, 2.
                NAMES(smoke) = smoke, nosmoke.

/ TWOGROUP      VARIABLE = pulse_1.

/ END
GROUP           VARIABLE = sex.
                CODES(sex) = 1, 2.
                NAMES(sex) = male, female.       /
TRANSFORM       IF (pulse_2 EQ 265) THEN pulse_2 = 165.
                diff = pulse_2 - pulse_1.        /
TWOGROUP        VARIABLE = diff.      /

END /
```

Example 3D.9 Quantile-quantile plots

You can use quantile-quantile plots to compare the distribution of two data sets. 3D constructs these plots by plotting the q-quantiles of one data set against the corresponding q-quantiles of the other data set. To get the q-quantile ($0 < q < 1$) of a dataset, we put the data in ascending order, then find the number that divides the data into two groups such that a fraction of the observations fall below the number and another fraction fall above. If the distributions of the two datasets are identical, then all the points should lie on a 45-degree line on the quantile-quantile plot. Alternatively, if the points on a quantile-quantile plot lie on a line parallel to and above or below the 45-degree line, you have an indication that the two distributions differ by an additive constant, a difference in mean. Any angle between the points of the quantile-quantile plot and the 45-degree line is an indication of a multiplicative difference between the two variables, a difference in standard deviation. It is also possible to have a combination of the two situations. See Chambers, Cleveland, Kleiner, and Tukey, 1983.

3D provides quantile-quantile plots when you specify QQPLOT in the TWOGROUP, MATCHED, or TEST paragraph. In Input 3D.9 we add a request for a quantile-quantile plot to our instructions in Input 3D.8.

Input 3D.9
(replaces the last TWOGROUP paragraph in Input 3D.8)

```
TWOGROUP    VARIABLE = diff.
            QQPLOT.          /
```

Output 3D.9

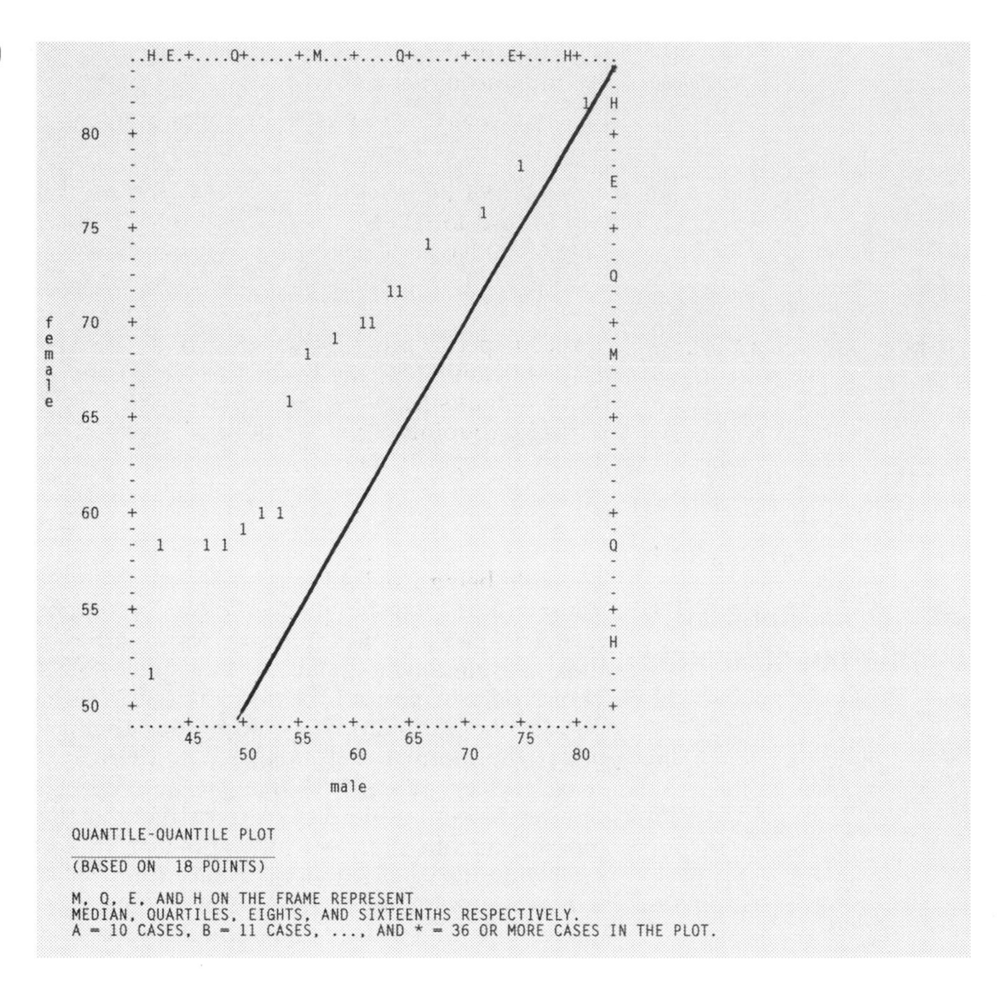

In Output 3D.9, we compare the distribution of the difference of the pulse before and after exercise for the two datasets. Each dataset is a group, male or female. In this example, the plot appears to be a line above the 45 degree line and with somewhat smaller slope. This plot indicates that the two groups, male and female, have a similar distribution of the difference in pulse before and after exercise. However, this difference in pulse has a greater mean for female than male, and a smaller standard deviation.

Special Features

Linesize

When you request narrow output (LINESIZE = 80) or when running interactively, 3D reformats the output. See Example 3D.2 for an example of wide output.

CROSS and PAIR

The MATCHED paragraph allows you to specify variables in one or two lists using the CROSS or PAIR commands. The default method is PAIR. For example:

```
MATCHED  VAR = WEIGHT1, WEIGHT2, ALBUMIN1, ALBUMIN2.    PAIR.
MATCHED  VAR = WEIGHT1, WEIGHT2, WEIGHT3.               CROSS.
```

When there is one list without CROSS, 3D matches the first variable in the list versus the second, the third versus the fourth, etc. With CROSS, all comparisons are made. In the second example, 3D will match WEIGHT1 versus WEIGHT2, WEIGHT1 versus WEIGHT3, and WEIGHT2 versus WEIGHT3.

Instead of the VARIABLE list, you can use two lists: FIRST and SECOND. When there are two lists without crossing, 3D matches the first variable of each list, then the second, etc. If the lists are not of the same length, the shorter is repeated until the longer is exhausted.

```
MATCHED FIRST = A1, B1, C1. SECOND = A2. B2. C2.      PAIR.
MATCHED FIRST = A1 TO A10.  SECOND = B1 TO B6.        CROSS.
```

With CROSS, all possible pairs from the two lists are matched. The second line would lead to 60 tests. Note that PAIR is the default and need not be specified unless you have specified CROSS in a previous problem and want to "turn it off."

Multiple subproblems

You can specify any number of MATCHED, ONEGROUP, and TWOGROUP paragraphs before END, and they will be executed in the order specified. Any command defaults from one paragraph to the next until you respecify it. For example, the grouping variable will be as specified in the last GROUPING paragraph unless you specify a new grouping variable.

Missing data

3D checks for missing values all variables specified in the analysis paragraph currently being used. Missing data affect the computation of correlation matrices as well as Hotelling's T^2 and Mahalanobis' D^2. The COMPLETE command ensures that 3D uses the same cases across all variables. When a data set contains missing or out-of-range values, this procedure may be preferable for the formal presentation of multivariate results, since it complies with the standard definition of Hotelling's T^2 and Mahalanobis' D^2. A large difference between analyses performed with and without the COMPLETE command indicates that your results may be biased because of the pattern of missing values or values out-of-range. You may then want to examine the data by using program AM to study the pattern of values excluded from the analysis. In any event, your results should be viewed with great caution. For more detailed multivariate analysis of group differences, see program 7M.

3D Commands

Which paragraph to use. For a

- two-group *t* test use the TWOGROUP paragraph.
- paired comparisons or matched pairs *t* test use the MATCHED paragraph.
- one-group *t* test use the ONEGROUP paragraph.

/ VARiable

The VARIABLE paragraph is optional. (See detailed description in Chapter 4.) Additional commands for 3D are FREQUENCY, CWEIGHT, and MMV.

FREQuency= variable. `FREQ = COUNT.`

State the name or number of a variable containing the case frequency. A case frequency has the effect of repeating cases. Any record with a negative or zero case frequency (COUNT ≤ 0 in this example) is excluded from the calculations as though it were missing. See Chapter 4.

CWeight = variable. `CWEIGHT = SAMPWT.`

Specify a case weight variable. As with case frequencies, records with negative or zero case weight values are excluded from the computations. Case weights can be used to adjust for sampling from populations with unequal variances. Means and correlations are unaffected, but variances, standard deviations, and standard errors change. See Chapter 4.

New Feature

MMV = #. `MMV = 10.`

Specify the maximum number of missing values or values out of range that will be allowed before a case is excluded from the analysis. Default: number of variables.

/ GROUP

New Syntax

VARiable = variable. `VARIABLE = SMOKE.`

Specify name or number of variable used to divide cases into groups for two-group t test. By default, a one-group test is assumed. If you prefer, you can still specify a grouping variable with the GROUPING command in the TWOGROUP paragraph as described in the 1990 *BMDP Statistical Software Manual*. If the GROUPING variable takes on more than ten distinct values or codes, you must specify CODES or CUTPOINTS in a GROUP paragraph (see Chapter 5). Note that 3D compares all possible pairings of groups unless you specify those you want in the GLIST command.

/ TWOGroup

The TWOGROUP paragraph is optional and may be repeated for additional sub-problems in the same computer run. Use for two-group t tests.

VARiable = list. `VAR = PULSE_1, PULSE_2.`

List variables to be tested, by name or number. By default, 3D tests all variables except the GROUPING variable and any label variables. The list of variables to be tested defaults from one paragraph to the next only if the paragraphs have the same name; i.e., if a MATCHED paragraph follows a TWOGROUP paragraph, you must respecify the VARIABLES list.

ROBust. `ROBUST.`

Requests trimmed t and related statistics. Unless you request them, these statistics are not printed.

NONPar. `NONPAR.`

Requests the Mann-Whitney rank-sum test. Unless you request it, the test is not printed.

CORRelation. `CORR.`

Requests the correlation matrix for each group (if groups are specified). Unless you request it, the matrix is not printed.

HOTelling. `HOTEL.`

Requests Hotelling's T^2 and Mahalanobis' D^2. Unless you request them, these statistics are not printed.

COMPlete. `COMP.`

Specify COMPLETE to use only complete cases for computations. A complete case is one with acceptable data (i.e., not missing or out of range) for all variables in the USE list in the VARIABLE paragraph (not just for the variables in the VARIABLE list). If there is no USE list, a case must have acceptable values for all input variables to be considered complete. COMP must be specified in the first analysis paragraph in a problem and cannot be altered until a new problem begins. By default, a case is excluded only when unacceptable values occur for variables included in the specific test.

Note that the usual definition of Hotelling's T^2 requires that the data values be acceptable for all the variables for any case that is included in the computations. Therefore you may want to use COMPLETE for the formal presentation of multivariate results. If you specify COMPLETE, 3D computes all statistics using complete cases only.

QQplot. `QQPLOT.`

Requests quantile-quantile plots.

TITLe = 'text'. `TITLE = 'CHECK FOR SEX DIFFERENCES'.`

Use to specify a label of up to 160 characters for the output from each paragraph. Default: no title.

GLISt = list. `GLIST = UNDER_20, OVER_40.`

Use to specify a subset of the groups defined by the GROUPING variable. Only the groups listed in the GLIST are tested against each other. By default, 3D compares all possible pairings of groups.

/ MATChed

The MATCHED paragraph is optional and may be repeated. Use for matched pairs or paired comparisons t tests.

VARiable = list. `VAR = PULSE_2, PULSE_1.`

List variables to be tested, by name or number. In the MATCHED paragraph, 3D matches the first variable in the list versus the second, the third versus the fourth, etc. See also FIRST and SECOND below. If both VARIABLE and FIRST and SECOND are omitted, 3D matches 1 versus 2, 3 versus 4, etc., after throwing out the label variables. If you specify CROSS (see below), all possible comparisons are performed. The list of variables to be tested defaults from one paragraph to another only if the paragraphs have the same name.

FIRSt = list. `FIRST = PULSE_1, BP_1.`
SECond = list. `SECOND = PULSE_2, BP_2.`

Use FIRST and SECOND in place of VARIABLES to specify the variables to be tested. The first variable in the first list is matched with the first variable in the second, etc. If you specify CROSS (see below), every variable in the FIRST list is matched with every variable in the SECOND list (see Special Features).

PAIR. *or* **CROSs.** `CROSS.`

Specify CROSS if you want 3D to perform all possible comparisons between the variables in the VARIABLE list or in the FIRST and SECOND lists. Default: PAIR. See Special Features.

NONPar. `NONPAR.`

Specify NONPAR to obtain the sign test, the Wilcoxon signed-rank test, and the Spearman correlation. By default, these statistics are not printed.

ROBust.
HOTelling.
COMPlete.
QQplot.
TITLe = 'text'.

See definitions in TWOGROUP paragraph.

/ ONEGroup

Use to obtain one-group *t* tests. The ONEGROUP paragraph is optional and may be repeated.

VARiable = list. `VAR = DIAMETER.`

List variables to be tested, by name or number. By default, all variables except case LABEL variables are used.

MEANs = list. `MEAN = (DIAMETER)5.01.`

Specify a mean for each variable for which you are requesting a one-group *t* test. 3D tests whether each variable mean differs from the specified value. Default: zeros.

NONPar. `NONPAR.`

Specify NONPAR to obtain the sign test and the Wilcoxon signed-rank test. By default, these statistics are not printed.

ROBust.
HOTelling.
CORRelation. *See definitions in*
COMPlete. *TWOGROUP paragraph.*
TITLe = 'text'.

/ TEST

The TEST paragraph is available for users familiar with the old version of 3D, but we recommend using the new TWOGROUP, MATCHED, and ONEGROUP paragraphs if you have the new version. TEST may be repeated for additional subproblems. Commands available in TEST are VARIABLE=list, TITLE, ROBUST, HOTELLING, COMPLETE, CORRELATION, QQPLOT, and, for the two-group test, GROUP=list. For a two-group *t* test you must include a GROUPING command in the VARIABLE paragraph. See Special Features for details about performing one-group and matched pairs tests with the TEST paragraph.

/ PRINT

Optional.

FIELDs = list. `FIELD = 3, 2*4 , 9.1, -2.`

Controls the format of a data listing by specifying the number of character positions, decimal places, and whether a field is alphanumeric. A minus sign indicates an alphanumeric field. See Chapter 7 for a full explanation.

LINEsize = #. `LINE = 80.`

Specify the number of characters in a line for printing. 3D has two choices: wide (132) or narrow (80). Use the LINESIZE option to obtain narrow output for viewing at the computer terminal. When LINE = 80, 3D prints the statistics of the matched or two-group analyses below the histograms instead of to the side. Default: 132 for batch runs, 80 for interactive.

New Feature

MINimum. `MIN.`

Prints data for cases in which at least one variable has a value less than its lower limit as specified in the VARIABLE paragraph. Default: such cases are not printed unless you specify DATA.

New Feature

MAXimum. `MAX.`

Prints data for cases in which at least one variable has a value greater than its upper limit as specified in the VARIABLE paragraph. Default: such cases are not printed unless you specify DATA.

New Feature

MISSing. `MISS.`

Prints data for cases that have at least one variable with a missing value. Missing values in the input may be identified in several ways: (1) an asterisk (the default missing value character), (2) a missing value code identified using MISS in the VARIABLE paragraph, (3) a character identified as MCHAR in the INPUT paragraph, (4) blanks (applies only to input read with FORTRAN fixed format). Default: such cases are not printed unless you specify DATA.

New Feature

METHod = VARiable *or* **CASE.** `METH = CASE.`

Use to determine how the printout is organized. VARIABLE is the default value.

VARIABLE – Prints for each case only as many variables as fill the print line (up to ten variables) and begins the next line with the same variables for the next case. Thus the data for each variable are in an easy-to-scan column. When the program finishes printing the first subset of variables for all cases, it continues with another subset.

CASE – Prints the values of all variables for one case (possibly filling several print lines) before printing data for the next case.

The following commands work only in batch mode

New Feature

MEAN. `NO MEAN.`

Unless you specify NO MEAN, 3D prints the mean, standard deviation, and frequency of each used variable.

New Feature

EZSCore. `NO EZSC.`

Unless you state NO EZSC, 3D prints *z*-scores for the minimum and maximum values of each used variable.

New Feature

ECASe. `NO ECAS.`

Unless you state NO ECAS, 3D prints the case numbers containing the minimum and maximum values of each variable.

New Feature

EXTReme. `NO EXTR.`

Unless you state NO EXTR, 3D prints the minimum and maximum values for each variable used.

New Feature

SKewness. `SK.`

Prints skewness and kurtosis for each used variable. Default: no skewness printed.

New Feature

CASE = #. `CASES = 25.`

Specify the number of cases to be printed. 3D can print all or a portion of the data file before the analysis. Default: 10.

Order of Instructions

■ indicates required paragraph

```
■ / INPUT
  / VARIABLE
  / GROUP
  / TRANSFORM
  / SAVE
  / TWOGROUP
  / MATCHED      Repeat for
  / ONEGROUP     subproblems.
  / TEST
  / PRINT
■ / END
data
  TRANSFORM ... /
  GROUP ... /
  TWOGROUP ... /    May be repeated
  MATCHED ... /     interactively.
  ONEGROUP ... /
  PRINT ... /
  SAVE ... /
  CONTROL ... /
  END /
  FINISH /
```

Repeat for additional problems. See Multiple Problems, Chapter 10.

Summary Table for Commands Specific to 3D

	Paragraphs Commands	Defaults	Multiple Problems	See
	/ **VARiable**			
	FREQuency = variable.	no case frequency variable	–	Cmds
	CWeight = variable.	no case weights used	–	Cmds
	MMV = #.	# of variables	–	Cmds
	/ **GROUP**			
▲	**VARiable =** variable.	no grouping	●	3D.1, 3D.2
▲	/ **TWOgroup**			
▲	GLISt = list.	all groups compared	●	Cmds
▲	/ **MATChed**			
	FIRSt = list.	all variables compared	●	Cmds
	SECond = list.	all variables compared	●	Cmds
	PAIR. *or* CROSS.	PAIR. (NO CROSS.)	●	SpecF
▲	/ **ONEGroup**			
▲	MEANs = list.	zeros	●	3D.7
	The following commands are available in the TWOGROUP, MATCHED, ONEGROUP, and TEST paragraphs, with the following exceptions: CORR does not apply to MATCHED, QQ does not apply to ONEGROUP.			
▲	VARiable = list.	all variables except grouping and label variables are tested	●	3D.1, 3D.4
	HOTelling.	NO HOTEL.	●	3D.3, SpecF
	COMPlete.	NO COMP.	●	Cmds
	CORRelation.	NO CORR.	●	3D.3
	ROBust.	NO ROBUST.	●	3D.2, 3D.5
	NONPar.	NO NONPAR.	●	3D.2, 3D.5
	QQplot.	NO QQ.	●	3D.8
	TITLe = 'text'.	no title printed	–	Cmds
	/ **TEST**			
	GRoup = list.	all groups compared	●	Cmds
	/ **PRINT**			
	CASE = #.	10	●	Cmds
	MEAN. *or* NO MEAN.	MEAN.	●	Cmds
	EXTReme. *or* NO EXTR.	EXTReme values printed	●	Cmds
	EZSCores. *or* NO EZSC.	EZSC. (z-scores)	●	Cmds
	ECASe. *or* NO ECASe.	ECAS. (Case #'s of Min and Max printed)	●	Cmds
	SKewness. *or* NO SK.	NO SK. (skewness and kurtosis)	●	Cmds
	FIELDs = list.	FORMAT if specified; otherwise 8.2 for numeric, – 4 for label variables	–	Cmds
	LINEsize = #.	132 noninteractive; 80 interactive	●	Cmds
	MISSing.	NO MISS.	●	Cmds
	MINimum.	NO MIN.	●	Cmds
	MAXimum.	NO MAX.	●	Cmds
	METHod = VARiables *or* CASE.	VARIABLES	–	Cmds

Key:
▲ Frequently used command or paragraph
● Value retained for multiple problems
– Default reassigned

5D

Histograms and Univariate Plots

5D prints histograms, cumulative histograms, and other plots, along with simple statistics, for ungrouped or grouped data. The histograms produced by this program provide more detail and are easier to read than those in other programs such as 2D. Beside each histogram, 5D prints frequencies and percentages, and cumulative frequencies and percentages. 5D can plot cumulative frequencies against the data values and provide normal or half-normal probability plots of the data. Normal probability plots can be used to screen data or residuals for nonnormality or for the presence of outliers. 5D can remove the linear trend from a normal probability plot to produce a detrended normal probability plot.

When cases are classified into groups, you can use symbols to identify each group, and the groups can be plotted separately or combined in one plot. You can control the size and scale of the plots, as well as the cutpoints for the histogram intervals, the names of the intervals, and how many observations are represented by each plot symbol. For information about normal probability plots see Bliss, Vol. 1,1967, pp. 101–120, and Dixon and Massey, 1983, pp. 64–66.

Where to Find It

Example 5D.1 Histograms

We use 5D to analyze the Werner blood chemistry data (Appendix D). The record for each subject includes an ID number, age, height, weight, birth control pill user (yes/no), and four blood chemistry measurements: cholesterol, albumin, calcium, and uric acid. The data are stored on disk in a file named WERNER.DAT. In Input 5D.1, the instructions in the INPUT and VARIABLE paragraphs are common to all BMDP programs and are explained in Chapters 3 and 4. (The FILE command tells the program where to find the data and is used for systems like VAX and IBM PC. For IBM mainframes, see UNIT, Chapter 3.)

We use the default options in the PLOT paragraph to produce a histogram for cholesterol. The PLOT paragraph is required in 5D, but you need specify only PLOT if you want only histograms. The height of the histogram can be fixed by specifying CODES or CUTPOINTS in the GROUP paragraph. Each bar then represents a code or an interval (see Example 5D.5).

We also specify CASES = 200 in the INPUT paragraph. The program will operate without this instruction, but histograms will have better scales if you state the number of cases explicitly. We suggest that you provide the number of cases or guess a reasonable upper limit if the exact number is not known due to missing data (in this example, there are actually 188 cases). ***Caution:*** *Be sure your estimate is greater than or equal to the actual number of cases or 5D will stop reading data when it reaches the specified number.* See Example 5D.5 if you want to specify the bins on the vertical axis.

Input 5D.1

```
/ INPUT      FILE = 'werner.dat'.
             VARIABLES = 9.
             FORMAT IS FREE.
             CASES = 200.

/ VARIABLE   NAMES = id, age, height, weight, brthpill,
                     cholstrl, albumin, calcium, uricacid.
             USE = cholstrl.
             MAXIMUM = (cholstrl)400.
             MINIMUM = (cholstrl)150.

/ PLOT

/ END
```

Output 5D.1

```
HISTOGRAM OF VARIABLE    6 cholstrl
                                    SYMBOL  COUNT      MEAN      ST.DEV.
                                       X     186     236.151      42.556
                          EACH SYMBOL REPRESENTS          1 OBSERVATIONS
INTERVAL                                                   FREQUENCY PERCENTAGE
NAME           5    10    15    20    25    30    35    40    45  INT. CUM. INT.  CUM.
            +----+----+----+----+----+---
*162.5      +XXXX                                                    4    4   2.2   2.2
*175        +XXXXXXXXXXX                                            11   15   5.9   8.1
*187.5      +XXXXXXX                                                 7   22   3.8  11.8
*200        +XXXXXXXXXXXXXXXXXXXXXXXXXX                             26   48  14.0  25.8
*212.5      +XXXXXXXXXXXXX                                          13   61   7.0  32.8
*225        +XXXXXXXXXXXXXXXXX                                      17   78   9.1  41.9
*237.5      +XXXXXXXXXXXXXXXXXXX                                    19   97  10.2  52.2
*250        +XXXXXXXXXXXXXXXXXXXXXXXXXXX                            27  124  14.5  66.7
*262.5      +XXXXXXXXXXXXXXXXXXX                                    19  143  10.2  76.9
*275        +XXXXXXXXXXXX                                           12  155   6.5  83.3
*287.5      +XXXXXXX                                                 7  162   3.8  87.1
*300        +XXXXXXXXX                                               9  171   4.8  91.9
*312.5      +XXXXX                                                   5  176   2.7  94.6
*325        +XXXXXX                                                  6  182   3.2  97.8
*337.5      +XX                                                      2  184   1.1  98.9
*350        +X                                                       1  185   0.5  99.5
*362.5      +                                                        0  185   0.0  99.5
*375        +                                                        0  185   0.0  99.5
*387.5      +                                                        0  185   0.0  99.5
*400        +X                                                       1  186   0.5 100.0
            +----+----+----+----+----+---
               5    10    15    20    25    30    35    40    45
```

Output 5D.1 presents the histogram for CHOLSTRL. The number of Xs printed in a line (horizontally) is the frequency of cases in that interval. Each value on the vertical axis represents the upper limit of the interval. For example, there are four cases less than or equal to 162.5.

To the right of the histogram, 5D prints the frequency of observations in each interval and the cumulative frequency of observations up to and including the interval, as well as the percent of observations represented by the frequency and by the cumulative frequency. When there are more observations than can be printed in a single line, the entire line is filled with Xs terminated by an asterisk; the exact frequency is printed to the right of the line. To prevent this from happening, see Controlling plot size and scale, in Special Features.

Example 5D.2 Cumulative frequency plots

In Input 5D.2, we request a cumulative frequency plot for CHOLSTRL by specifying TYPE = CUM in the PLOT paragraph. The default plot size is 45 characters wide and 30 lines high (see SIZE to change the default). Note: These plots are available in high resolution on VAX, PC and X Windows platforms.

Input 5D.2
(substitute for the PLOT paragraph in Input 5D.1)

```
/ PLOT                TYPE = CUM.
```

Output 5D.2

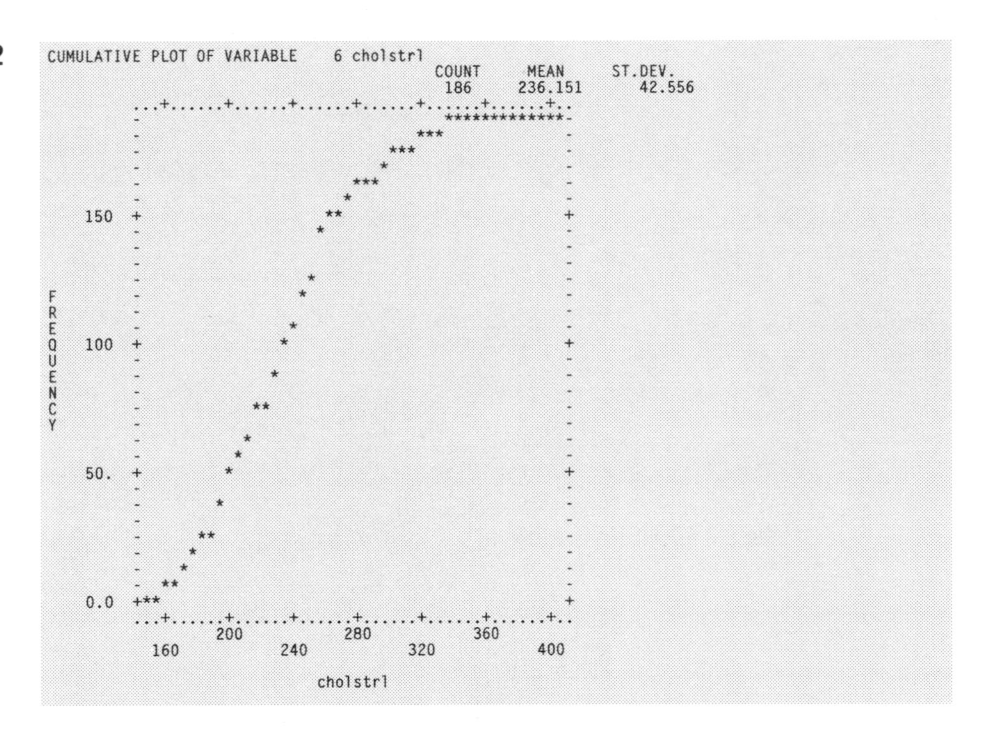

Output 5D.2 presents the cumulative frequency plot of CHOLSTRL. The cumulative frequency distribution is plotted on the vertical axis and the data values on the horizontal axis.

Example 5D.3 Normal probability plots

5D produces three types of normal probability plots: a normal probability plot, a detrended normal plot, and a half-normal plot. We request all three in Input 5D.3.

Input 5D.3
(substitute for the PLOT paragraph in Input 5D.1)

```
/ PLOT              TYPES = NORM, DNORM, HALFNORM.
```

Output 5D.3

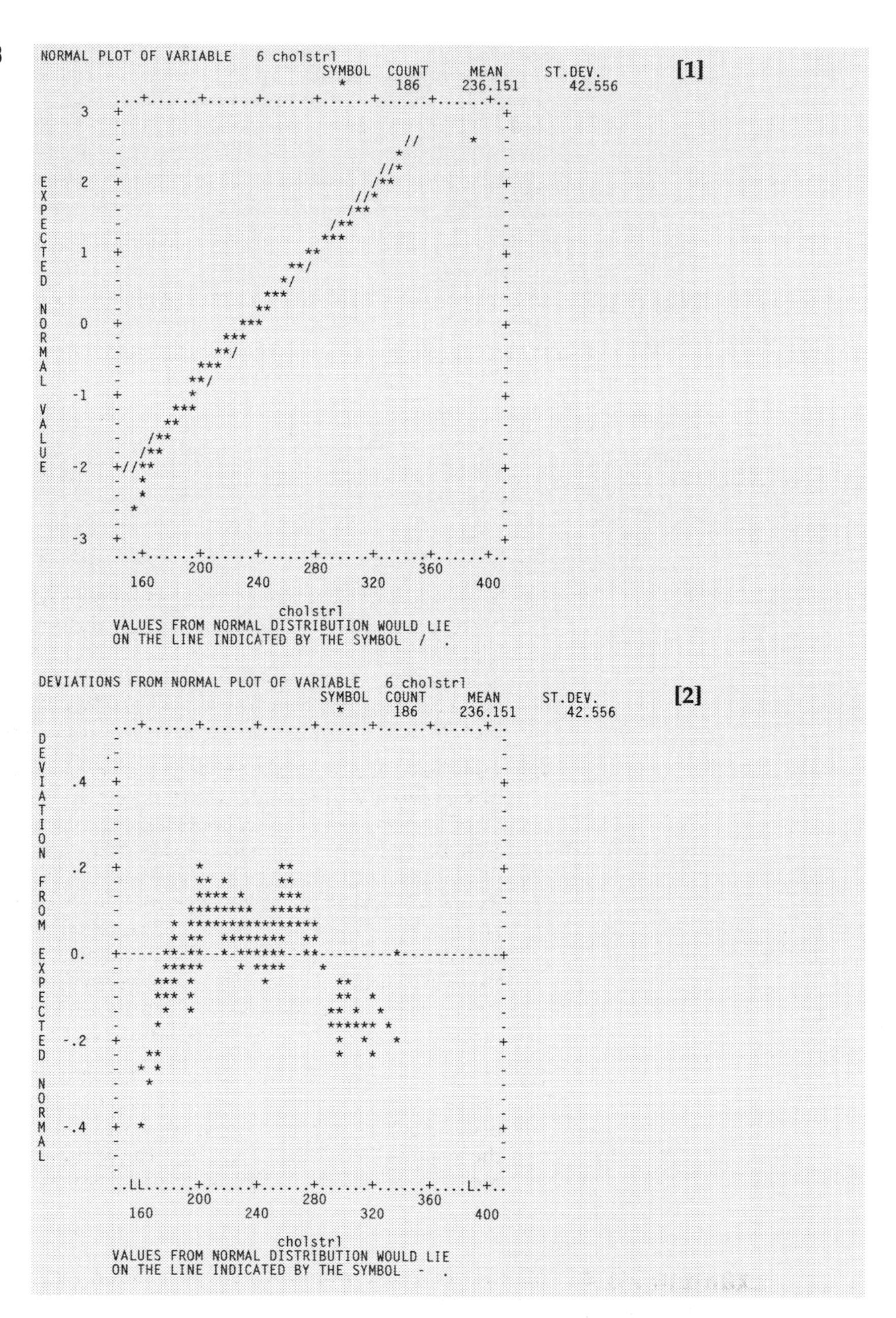

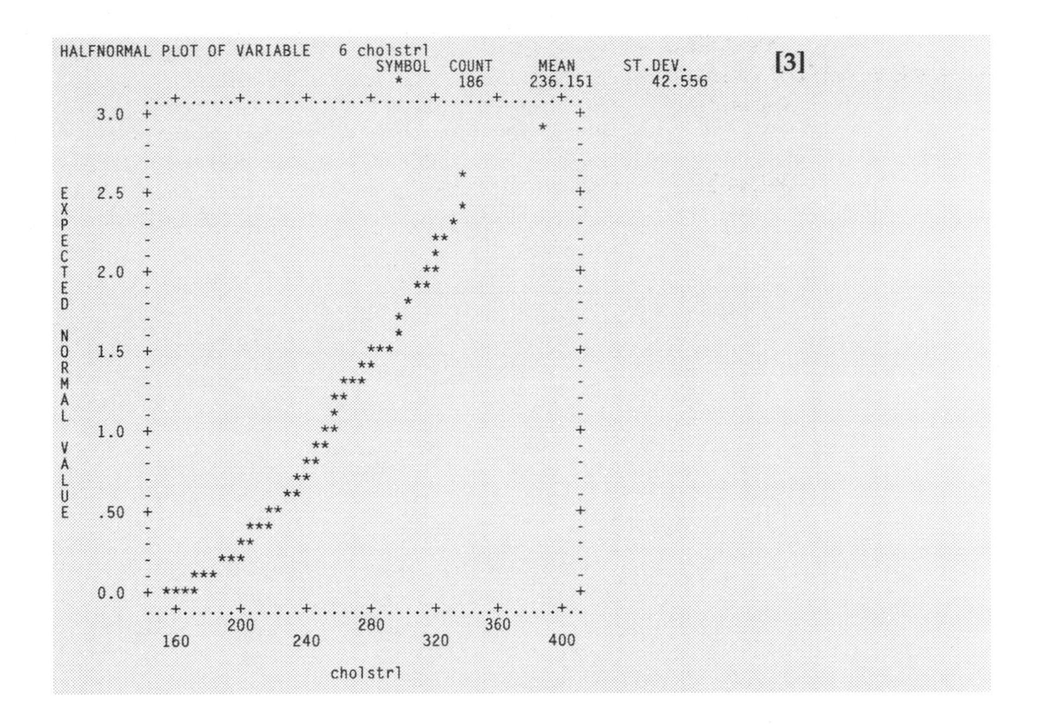

Bold numbers below correspond to those in Output 5D.3.

[1] In the normal probability plot, the observed values are plotted along the horizontal axis. The data values are ordered before plotting: the vertical axis corresponds to the expected normal value based on the rank of the observation. If the data are normally distributed, the normal probability plot should approximate the straight line indicated by the slashes on the plot. Note the extreme points at the high and low ends in this example.

[2] The detrended normal probability plot is similar to the normal probability plot except that the linear trend is removed before the plot is printed. The trend is removed to assist in assessing departures from normality. Imagine rotating the line indicated by the slashes in **[1]** until it is horizontal and then expanding the scale to show deviations from the line more clearly. If the variable cholesterol had a normal distribution, the points should cluster around zero with little apparent pattern. The pattern seen here indicates lack of normality. The vertical scale ranges from –0.5 to 0.5 and represents the differences between the expected normal values and the standardized values of the observations. Deviations outside this range are plotted in the border, as shown by the letter L on the bottom of the plot.

[3] The half-normal plot is similar to the normal plot; however, the expected values are computed using only the positive half of the normal distribution. This plot is primarily used to examine residuals, ignoring the sign of the residual; for example, when the residual is proportional to the square root of a chi-square variate with one degree of freedom (Daniel, 1959).

Example 5D.4 Group or subpopulation identification in plots

5D can print data from each subpopulation (or group) in an individual plot, or combine specified groups into a single plot where members of each group are identified by a different symbol. 5D assigns the symbol A to the first group, B to the second, etc. In Input 5D.4, the grouping variable selected is BRTHPILL, so A represents women who do not use contraceptive pills and B represents women who do. The GROUPING variable is identified in the VARIABLE statement of the GROUP paragraph. We name the groups in the GROUP paragraph so that the output will be labeled appropriately. You must specify codes or cutpoints when the grouping variable is continuous or takes on more than ten distinct values.

In the PLOT paragraph we request HISTOGRAMS and NORMAL probability plots for ALBUMIN. Each GROUP statement in the PLOT paragraph specifies a different plot: first the group NOPILL is plotted, then the group PILL, and finally both (ALL) groups together. You may specify groups by name, as we do here, or state GROUP = EACH. If no GROUPS are specified in the PLOT paragraph, all groups defined by the GROUPING variable are plotted together, each with a different symbol. We also request a plot smaller than the default size.

Input 5D.4

```
/ INPUT       FILE = 'werner.dat'.
              VARIABLES ARE 9.
              FORMAT IS FREE.
              CASES = 200.

/ VARIABLE    NAMES = id, age, height, weight, brthpill,
                      cholstrl, albumin, calcium, uricacid.

/ GROUP       VARIABLE = brthpill.
              CODES(brthpill) = 1, 2.
              NAMES(brthpill) = nopill, pill.

/ PLOT        VAR = albumin.
              TYPES = HIST, NORM.
              GROUP = nopill.
              GROUP = pill.
              GROUPS = ALL.
              SIZE = 40, 25.

/ END
```

Output 5D.4

```
 s HISTOGRAM OF VARIABLE     7 albumin
                                    SYMBOL  COUNT      MEAN      ST.DEV.        [1]
                         nopill        A       92      41.978        3.451
                         EACH SYMBOL REPRESENTS          1 OBSERVATIONS
INTERVAL                                               FREQUENCY PERCENTAGE
NAME          5    10   15   20   25   30   35   40    INT. CUM. INT.  CUM.
         +----+----+----+----+----+----+----+----+
*32      +A                                               1    1   1.1   1.1
*33      +                                                0    1   0.0   1.1
*34      +A                                               1    2   1.1   2.2
*35      +A                                               1    3   1.1   3.3
*36      +A                                               1    4   1.1   4.3
*37      +AAAA                                            4    8   4.3   8.7
*38      +AAA                                             3   11   3.3  12.0
*39      +AAAAAAAAAA                                     10   21  10.9  22.8
*40      +AAAAAAAAAAA                                    11   32  12.0  34.8
*41      +AAAAAAAAAAA                                    11   43  12.0  46.7
*42      +AAAAAAAAA                                       9   52   9.8  56.5
*43      +AAAAAAAAAAAA                                   12   64  13.0  69.6
*44      +AAAAAAA                                         7   71   7.6  77.2
*45      +AAAAAA                                          6   77   6.5  83.7
*46      +AAAAA                                           5   82   5.4  89.1
*47      +AAA                                             3   85   3.3  92.4
*48      +AAAAA                                           5   90   5.4  97.8
*49      +A                                               1   91   1.1  98.9
*50      +A                                               1   92   1.1 100.0
*51      +                                                0   92   0.0 100.0
         +----+----+----+----+----+----+----+----+
              5    10   15   20   25   30   35   40
```

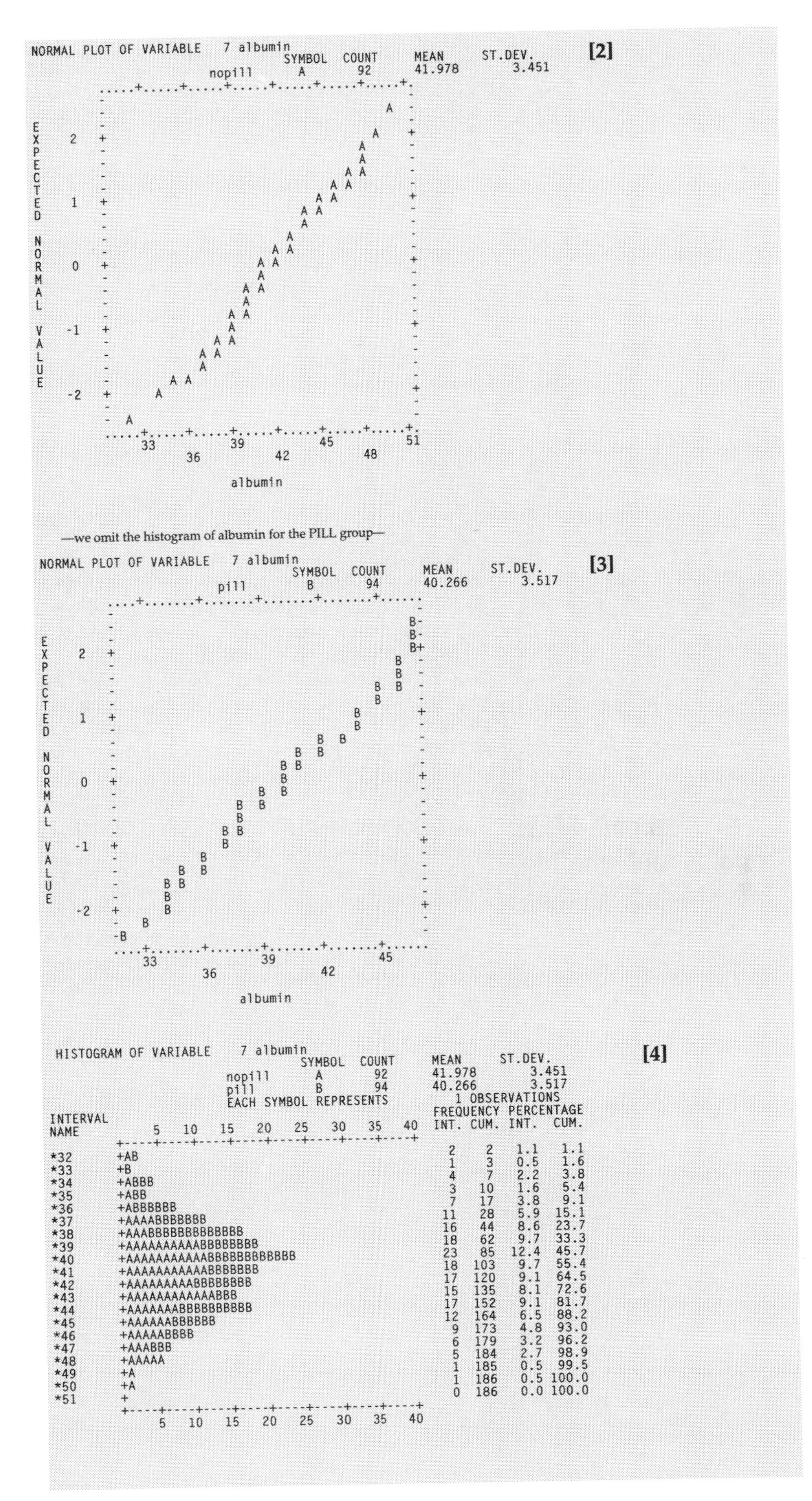

```
NORMAL PLOT OF VARIABLE    7 albumin                                            [2]
                                  SYMBOL  COUNT      MEAN      ST.DEV.
                     nopill         A       92      41.978      3.451

EXPECTED NORMAL VALUE (axis: -2, -1, 0, 1, 2)

          33          39          45          51
                36          42          48
                       albumin
```

—we omit the histogram of albumin for the PILL group—

```
NORMAL PLOT OF VARIABLE    7 albumin                                            [3]
                                  SYMBOL  COUNT      MEAN      ST.DEV.
                     pill           B       94      40.266      3.517

EXPECTED NORMAL VALUE (axis: -2, -1, 0, 1, 2)

          33          39          45
                36          42
                       albumin
```

```
 HISTOGRAM OF VARIABLE     7 albumin                                            [4]
                                 SYMBOL  COUNT      MEAN      ST.DEV.
                      nopill       A       92      41.978       3.451
                      pill         B       94      40.266       3.517
                      EACH SYMBOL REPRESENTS          1 OBSERVATIONS
INTERVAL                                           FREQUENCY PERCENTAGE
NAME           5    10   15   20   25   30   35   40   INT. CUM. INT.  CUM.
           +----+----+----+----+----+----+----+----+
*32        +AB                                           2    2   1.1   1.1
*33        +B                                            1    3   0.5   1.6
*34        +ABBB                                         4    7   2.2   3.8
*35        +ABB                                          3   10   1.6   5.4
*36        +ABBBBBB                                      7   17   3.8   9.1
*37        +AAAABBBBBBB                                 11   28   5.9  15.1
*38        +AAABBBBBBBBBBBBB                            16   44   8.6  23.7
*39        +AAAAAAAAAABBBBBBBB                          18   62   9.7  33.3
*40        +AAAAAAAAAAABBBBBBBBBBBB                     23   85  12.4  45.7
*41        +AAAAAAAAAAAABBBBBB                          18  103   9.7  55.4
*42        +AAAAAAAAABBBBBBBB                           17  120   9.1  64.5
*43        +AAAAAAAAAAAABBB                             15  135   8.1  72.6
*44        +AAAAAAABBBBBBBBBB                           17  152   9.1  81.7
*45        +AAAAAABBBBBB                                12  164   6.5  88.2
*46        +AAAAABBBB                                    9  173   4.8  93.0
*47        +AAABBB                                       6  179   3.2  96.2
*48        +AAAAA                                        5  184   2.7  98.9
*49        +A                                            1  185   0.5  99.5
*50        +A                                            1  186   0.5 100.0
*51        +                                             0  186   0.0 100.0
           +----+----+----+----+----+----+----+----+
               5    10   15   20   25   30   35   40
```

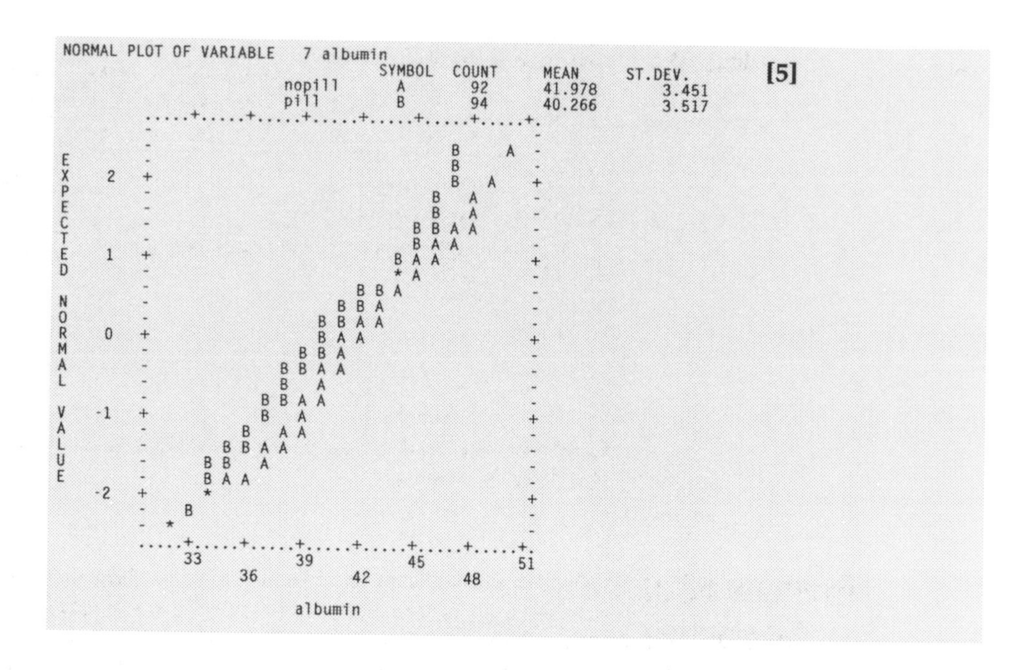

In Output 5D.4, the first histogram **[1]** and normal plot **[2]** contain only A's, the symbol assigned to the first group (NOPILL). The second normal plot **[3]** contains Bs (PILL). Above each plot the names of the groups are printed, along with their symbols, frequencies, means, and standard deviations. A cumulative histogram **[4]** is also printed with symbols representing group membership. In the normal plot with groups combined **[5]**, each group is plotted separately. 5D prints an asterisk when points from different groups fall in the same place. (Slashes are not printed when a GROUP paragraph is used.) Note the slight shift visible between the two groups; group B has a smaller mean, but the spread is similar.

Example 5D.5 Defining and labeling the histogram bins

The histogram bins in Example 5D.1 were determined from the data values. In this example, we show how to use the GROUP paragraph to define and label the histogram bins. Note that we do not specify GROUPING in the HISTOGRAM paragraph.

CUTPOINTS (or CODES) define the bins of the histogram, and NAMES label the bins in the output. 5D is the only BMDP program that allows you to name histogram bins. Remember that a CUTPOINT defines the maximum value of an interval. To define the histogram bins for cholesterol we specify:

```
CUTPOINTS(CHOLSTRL) = 160 TO 340 BY 20.
NAMES(CHOLSTRL) = UPTO160, ‘161-180’, ‘181-200’,...,OVER340.
```

Note that these instructions work for both the regular and the cumulative histogram.

Input 5D.5
(add to Input 5D.1 before end)

```
/ GROUP  CUTPOINTS(cholstrl) = 160 TO 340 BY 20.
         NAMES(cholstrl) = UPTO160, '161-180', '181-200',
                           '201-220', '221-240', '241-260',
                           '261-280', '281-300', '301-320',
                           '321-340', OVER340.
```

Output 5D.5

```
HISTOGRAM OF VARIABLE     6 cholstrl
                                 SYMBOL  COUNT      MEAN      ST.DEV.
                                   X      186     236.151      42.556
                        EACH SYMBOL REPRESENTS        1 OBSERVATIONS
                                                        FREQUENCY PERCENTAGE
INTERVAL                                                INT. CUM. INT.  CUM.
NAME         5    10   15   20   25   30   35   40   45
          +----+----+----+----+----+----+----+----+----+
UPTO160   +XXXX                                            4    4   2.2   2.2
161-180   +XXXXXXXXXXXXXXXX                               16   20   8.6  10.8
181-200   +XXXXXXXXXXXXXXXXXXXXXXXXXXXX                   28   48  15.1  25.8
201-220   +XXXXXXXXXXXXXXXXXXXXXXXXX                      25   73  13.4  39.2
221-240   +XXXXXXXXXXXXXXXXXXXXXXXXXXXXXXX                31  104  16.7  55.9
241-260   +XXXXXXXXXXXXXXXXXXXXXXXXXXXXXXXXXXXXX          37  141  19.9  75.8
261-280   +XXXXXXXXXXXXXXXXXXXX                           20  161  10.8  86.6
281-300   +XXXXXXXXXX                                     10  171   5.4  91.9
301-320   +XXXXXXXXXX                                     10  181   5.4  97.3
321-340   +XXXX                                            4  185   2.2  99.5
OVER340   +X                                               1  186   0.5 100.0
          +----+----+----+----+----+----+----+----+----+
               5    10   15   20   25   30   35   40   45
```

Compare the histogram in Output 5D.5 with that in Output 5D.1, which uses the default intervals.

Example 5D.6 Interactive paragraphs

5D now allows you to transform your data, modify the grouping instructions, and request histograms and plots interactively. You can specify any number of GROUP (with VARIABLE = var.), TRANSFORM, and PLOT paragraphs after you specify INPUT, VARIABLE, and END paragraphs. In each case, the interactive paragraph should end with a slash, and the rest of the line after the slash should be blank.

In Input 5D.6 we use interactive paragraphs after the initial analysis to specify grouping information and request a normal probability plot using the new groups and the variable ALBUMIN. See Chapter 3D for information about interactive transformations. We do not show output for this example.

For information about running BMDP interactively, see Chapter 10.

Input 5D.6

```
/ INPUT      FILE = 'werner.dat'.
             VARIABLES = 9.
             FORMAT = FREE.
             CASES = 200.

/ VARIABLE   NAMES = id, age, height, weight, brthpill,
                     cholstrl, albumin, calcium, uricacid.
             MAXIMUM = (cholstrl)400.
             MINIMUM = (cholstrl)150.

/ PLOT

/ END

GROUP        VARIABLE = brthpill.
             CODES(brthpill) = 1, 2.
             NAMES(brthpill) = nopill, pill.    /

PLOT         VAR = albumin.
             TYPE = NORMAL.                     /

END /
```

Special Features

Controlling plot size and scale

You can use the PLOT SIZE command to control the size of plots and histograms. The SCALE command allows you to adjust the scale of histograms.

Histograms and cumulative histograms. For the histogram and cumulative histogram, you may have more observations than can fit on a line. (This depends on the width of the histogram and is more likely to happen with the cumulative version.) To obtain a histogram with a better scale, use the SCALE command. SCALE allows you to begin the tabulation at the base of the histogram at, for example, 50 instead of zero, and to make each histogram symbol represent more than one observation.

- The number of bars and bin width is controlled by the GROUP paragraph (see Example 5D.5).
- The number of characters printed per bin is controlled by SIZE = #.
- The scale of the histogram is controlled by SCALE = #1, #2, where #1 is the frequency at the base of the histogram and #2 is the number of units represented by each symbol.

HISTOGRAM, CUM, NORM, DNORM, and HALFNORM plots. The default plot size is 45 characters wide and 30 lines high. Use SIZE = #1, #2 in the PLOT paragraph to specify the number of characters for the horizontal axis and the number of lines for the vertical axis.

5D Commands

/ VARiable

The VARIABLE paragraph is optional and is described in detail in Chapter 4. Additional commands for 5D are GROUPING, FREQUENCY, and CWEIGHT.

FREQuency = variable. `FREQ = COUNT.`

State the name or number of a variable containing the case frequency. A case frequency has the effect of repeating cases. Any record with a negative or zero case frequency (COUNT < O in this example) is excluded from the calculations as though it were missing. See Chapter 4.

CWeight = variable. `CWEIGHT = SAMPWT.`

Specify a case weight variable. As with case frequencies, records with negative or zero case weight values are excluded from the computations. 5D reports the number of cases with zero weight (frequency * case weight = 0). Case weights can be used to adjust for sampling from populations with unequal variances. Means and correlations are unaffected, but variances, standard deviations, and standard errors change. See Chapter 4.

/ GROUP

Required when a GROUPING variable is used. If you prefer, you can still specify a grouping variable with the GROUPING command in the VARIABLE paragraph as described in the 1990 *BMDP Manual.*

New Syntax

VARiable = variable. `VAR = BRTHPLL.`

Specify the name or number of a grouping variable used to classify cases into groups. Default: cases are not grouped. If the GROUPING variable takes on more than 10 distinct values or codes, CODES or CUTPOINTS for the variable must be specified in the GROUP paragraph (see Chapter 5 and below). **Example:** Cases are grouped by the two values of the variable BRTHPILL (histograms are labeled A = NOPILL, B = PILL).

CODES (*j*) = #list `CODES(BRTHPILL) = 1, 2.`
NAMES (*j*)= list `NAME(BRTHPILL)= NOPILL,PILL.`

See CODE, CUTPOINT, and NAME commands in Chapter 5. If you specify NAMES in the GROUP paragraph, 5D uses them to label plots (see Example 5D.4). You can also use CUTPOINTS (or CODES) and NAMES to determine and name intervals at the base of histograms and cumulative histograms (see Example 5D.5).

/ PLOT

The PLOT paragraph is required and may be repeated.

TYPE = HIST, CHIST, CUM, NORM, DNORM, HALFNORM. `TYPE = HIST, NORM, CUM.`
(one or more)

Specify the plots to be printed. Default: HISTOGRAM.

HIST – histogram

CHIST – cumulative histogram

CUM – cumulative frequency distribution plot

NORM – normal probability plot

DNORM – detrended normal probability plot

HALFNORM – half-normal plot

VARiable = list. `VAR = CHOLSTRL, ALBUMIN.`

List names or numbers of variables to be plotted. By default, all variables in the VARIABLE USE list are plotted. If there is no USE list, all variables are plotted.

GROUP = list. `GROUP = NOPILL.`
GROW = EACH. `GROUPS = ALL.`
GROUP = ALL.
(one or more)

Specify the groups used in each plot. State EACH to obtain a separate plot for each group. State ALL to plot all groups in a single plot. To list groups, use NAMES or sequence numbers (order) of categories from the GROUP paragraph. The GROUP statement may be repeated. EACH and ALL cannot be used as parts of a list. Default: ALL. **Example:** Group NOPILL is plotted by itself in the first plot, and all groups are plotted together in the second plot.

SIZE = #1, #2. `SIZE = 40, 25.`

Specify the number of characters (#1) for the horizontal axis, and number of lines (#2) for the vertical axis of NORM, DNORM, HALFNORM, and CUM plots. The default plot is 45 characters wide and 30 lines high.

For HIST and CHIST, #1 is the number of characters in the histogram bars (width). See GROUP paragraph for control of histogram intervals. If no codes or intervals are specified in the GROUP paragraph, the number of bins equals 8 $\log_{10}(N) + 2$, where N is the number of cases read. Specification of INPUT CASES improves the selection of histogram intervals. Maximum width is 100 characters.

SCALE = #1, #2. `SCALE = 100, 5.`

Specify the value of the frequency at the base of the histogram and cumulative histogram (#1), and the number of observations represented by each character plotted (#2). By default, 5D begins the tabulation at zero, and each character represents one observation. **Example:** The tabulation begins at 100, and each character represents five observations.

Order of Instructions

■ indicates required paragraph

```
■ / INPUT
  / VARIABLE
  / GROUP
  / TRANSFORM
  / PRINT
  / SAVE
■ / PLOT   ] Repeat for subproblems
■ / END
  data
  GROUP... /
  TRANSFORM... /     May be repeated
  PLOT... /          interactively.
  END /
```

Repeat for additional problems. See Multiple Problems, Chapter 10.

Summary Table for Commands Specific to 5D

Paragraphs Commands	Defaults	Multiple Problems	See
▲ / **VARiable**			
FREQuency = variable.	no case frequency variable	–	Cmds
CWeight = variable.	no case weights used	–	Cmds
▲ / **GROUP**			
▲ VARiable = variable.	no grouping (see Chapter 5)	–	5D.4, 5D.5
■ / **PLOT**			
TYPE = (*one or more*) HIST, NORM, DNORM, HALFNORM, CUM, CHIST.	HISTOGRAM	●	5D.2, 5D.3, 5D.4
VARiable = list.	all variables plotted	●	5D.4
GROUP = list, EACH, *or* ALL.	ALL	–	5D.4
SIZE = #1, #2.	SIZE = 45, 30.	●	5D.1
SCALE = #1, #2.	SCALE = 0,1.	●	SpecF

Key:
- ■ Required paragraph
- ▲ Frequently used command or paragraph
- ● Value retained for multiple problems
- – Default reassigned

6D

Bivariate (Scatter) Plots

6D produces scatterplots of one variable against another, computes descriptive statistics, and calculates the line of best fit, marking its intersections with the plot frame. With this program, you can easily obtain a visual picture of the regression line position. You can classify cases into groups and plot each group in a separate plot or together on the same plot, using characters or symbols to identify group membership.

PLOT paragraph commands allow you to plot all possible pairs of XVAR and YVAR variables or to specify pairs of variables to be plotted. 6D automatically prints the number of points plotted, the correlation between the two variables and its *p*-value, the mean and standard deviation for each variable, the regression line of *Y* on *X* and the residual mean square. More than one pair of variables can appear in the same plot. You can control the symbols used in the plots, the plot SIZE, and the range of values to plot, allowing you to zoom in on a particular area of interest.

Where to Find It

Example 6D.1 Bivariate plots with statistics and regression lines

We use 6D to plot the Fisher iris data (Appendix D). This data set records the length and width of the sepals and petals of 150 flowers belonging to three species of iris. The data are stored on disk in a file called FISHER.DAT. (The FILE command tells the program where to find the data and is used for systems like VAX and IBM PC. For IBM mainframes, see UNIT, Chapter 3.) The PLOT paragraph is specific to 6D; the other instructions in Input 6D.1 are those used by all BMDP programs to read the data and name the variables. See Chapters 3 and 4 for further information on the INPUT and VARIABLE paragraphs.

From the five variables we plot sepal length versus sepal width and petal length versus petal width. In the PLOT paragraph we specify the variables to plot along the vertical, or y-axis (YVAR), and the horizontal, or x-axis (XVAR). 6D plots pairs of variables in the order in which they are listed; thus SEPAL_L is paired with SEPAL_W and PETAL_L is paired with PETAL_W. (If we add the CROSS instruction to the PLOT paragraph, four plots will be printed, one for every possible combination of x and y variables.)

The default plot size in batch mode is 60 characters wide, 42 lines high. We reduce the plots to 44 characters wide and 18 lines high by specifying SIZE = 44, 18. The PLOT paragraph can be repeated several times before END to specify different combinations of variables or to add other instructions.

Input 6D.1

```
/ INPUT      FILE IS 'fisher.dat'.
             VARIABLES = 5.
             FORMAT IS FREE.

/ VARIABLE   NAMES ARE sepal_l, sepal_w, petal_l,
                       petal_w, iris.

/ PLOT       YVAR = sepal_l, petal_l.
             XVAR = sepal_w, petal_w.
             SIZE = 44, 18.
/ END
```

Output 6D.1 contains a table of contents and two plots: SEPAL_L against SEPAL_W and PETAL_L against PETAL_W. Numbers in the plots represent frequencies of points plotted at the same positions. Points are plotted for all cases that contain acceptable values for both variables; i.e., values not missing or out of range. For frequencies greater than 9, A represents 10, B represents 11, etc., and an asterisk indicates a frequency of 36 or more. Missing values are represented by an M on the plot frame. Values out of range (greater than a specified MAXIMUM or less than a specified MINIMUM) are represented on the plot frame by H (too high) and L (too low). On the frame of each plot are two y's which show where the regression line $y = a + bx$ intersects the frame.

Output 6D.1

```
NUMBER OF CASES READ. . . . . . . . . . . . .      150

 TABLE OF CONTENTS

 HORIZONTAL      VERTICAL
 VARIABLE        VARIABLE        GROUP      PLOT                       PAGE
 NO. NAME        NO. NAME        NAME       SYMBOL                      NO.

  1  sepal_l      2  sepal_w                           . . . . . . .     3
  3  petal_l      4  petal_w                           . . . . . . .     4
```

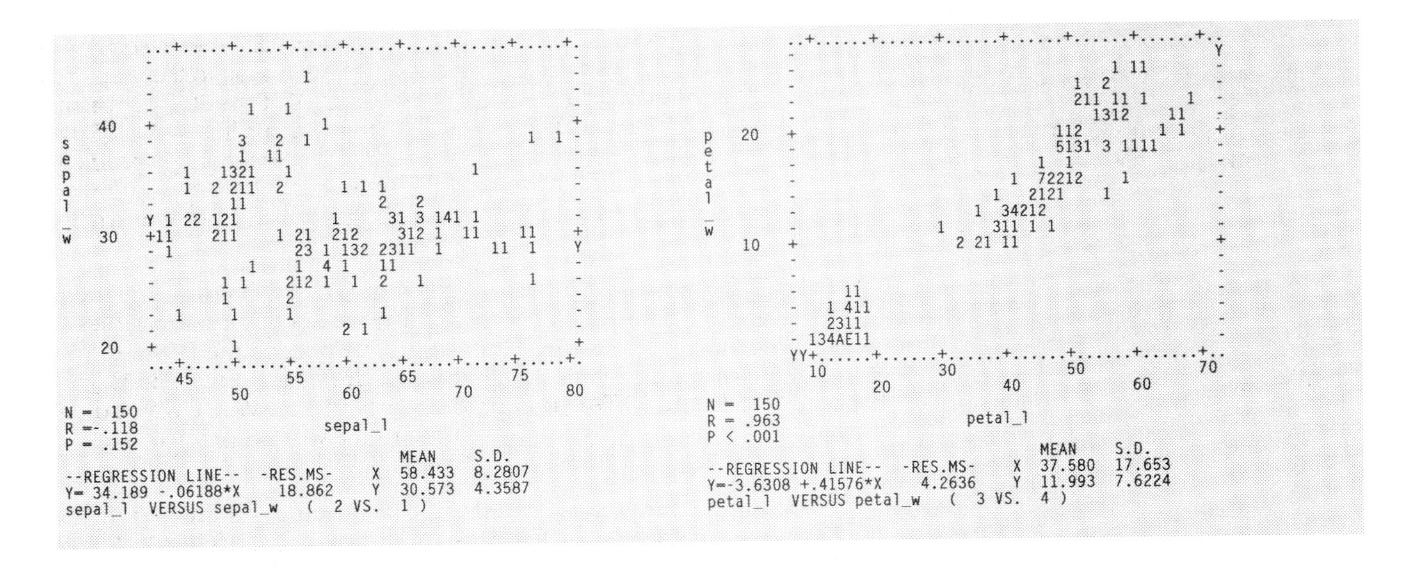

Under each plot 6D produces the following statistics:

- the number (N) of points plotted
- the Pearson product-moment correlation r between the two variables, computed from all pairs of acceptable values
- the p-value for the test of the null hypothesis that the population correlation is zero
- the mean for each variable: $\bar{x}$ and $\bar{y}$
- the standard deviation for each variable: s_x and s_y
- the least squares regression line of Y on X; that is, $y = a + bx$, where $b = rs_y/s_x$ and $a = \bar{y} - b\bar{x}$.
- the residual mean square (labeled RES.MS. in the results):

$$s^2_{y|x} = \frac{\sum (y_j - \hat{y}_j)^2}{N-2} \qquad \text{where } \hat{y}_j = a + bx_j$$

If you do not want to see the statistics, specify NO STATISTICS in the PLOT paragraph.

Output 6D.1 shows no significant correlation between SEPAL_L and SEPAL_W for the iris data, but a strong correlation between PETAL_L and PETAL_W (r = .963, $p < .001$). Note the apparent clusters of points. Example 6D.2 shows how identifying group membership affects these results.

Example 6D.2 Scatterplots with symbols to identify group membership

You can request separate plots for different subpopulations (groups), or you can display values for all groups in one plot, using different plot symbols to identify the groups. Group membership information is specified in three places:

- VARIABLE statement in the GROUP paragraph
- NAMES along with CODES or CUTPOINTS in the GROUP paragraph
- GROUP in the PLOT paragraph

In Input 6D.2 we request symbols identifying the iris species in plots of SEPAL_L versus SEPAL_W and PETAL_L versus PETAL_W.

The GROUP paragraph. We classify the cases into groups by specifying a grouping variable in the GROUP paragraph. We state VARIABLE = IRIS in the GROUP paragraph to separate the cases into three groups according to the values of the variable IRIS (1 to 3). These codes correspond to three species of iris:

setosa, versicolor, and virginica.

Also in the GROUP paragraph, we define the CODES and NAME the groups so that the output will be labeled appropriately. When VARIABLE is specified, the default plot symbol changes from FREQUENCY to GROUP (see CHARACTER command). We use a NAME statement so that 6D will plot the first letters of the group names (S, V, and G); otherwise 6D would assign the plot symbols A, B, and C to the three groups. We change the name of the third group to make it begin with a unique first letter. If two or more group names begin with the same letter, 6D again assigns the symbols A, B, C, etc. (For further discussion of plot symbols see Special Features.) Note that an interval-scaled variable such as AGE may also be used for grouping. For example, if we defined four CUT-POINTS on AGE, the program would sort cases into five age groups, and the symbol plotted for each case would indicate the age group.

The PLOT paragraph. We specify the groups or combinations of groups to be plotted by inserting two GROUP commands in the PLOT paragraph. The command GROUP = ALL produces a plot containing all groups (state GROUP = EACH to obtain a separate plot for each group). In the second GROUP command we use the group NAME specified in the GROUP paragraph.

A less-used alternative. In specifying the groups to plot within a frame you can also use the order in which the GROUP CODES are listed (not the CODES themselves). For example, if the CODES for a grouping variable are 0, 1, 2, you would specify GROUPS = 1, 2, 3.

Input 6D.2

```
/ INPUT        FILE IS 'fisher.dat'.
               VARIABLES = 5.
               FORMAT IS FREE.

/ VARIABLE     NAMES = sepal_l, sepal_w, petal_l, petal_w,
                       iris.

/ GROUP        VARIABLE = iris.
               CODES(iris) = 1 TO 3.
               NAMES(iris) = setosa, versicol, gvirginc.

/ PLOT         YVAR = sepal_l, petal_l.
               XVAR = sepal_w, petal_w.
               SIZE = 44, 18.
               GROUP = ALL.
               GROUP = setosa.
/ END
```

In Output 6D.2 we display two plots of SEPAL_L versus SEPAL_W, one combining all groups and one for only SETOSA irises, and two similarly grouped plots of PETAL_L versus PETAL_W. When grouping is requested, plot scales are determined by the minimum and maximum values across all groups, so the frames are the same for all subpopulations. To set the scale individually for each group, specify SCALE = INDIVIDUAL. Actual 6D output has one plot per page; we group the plots together in Output 6D.2 for easy reference.

Output 6D.2

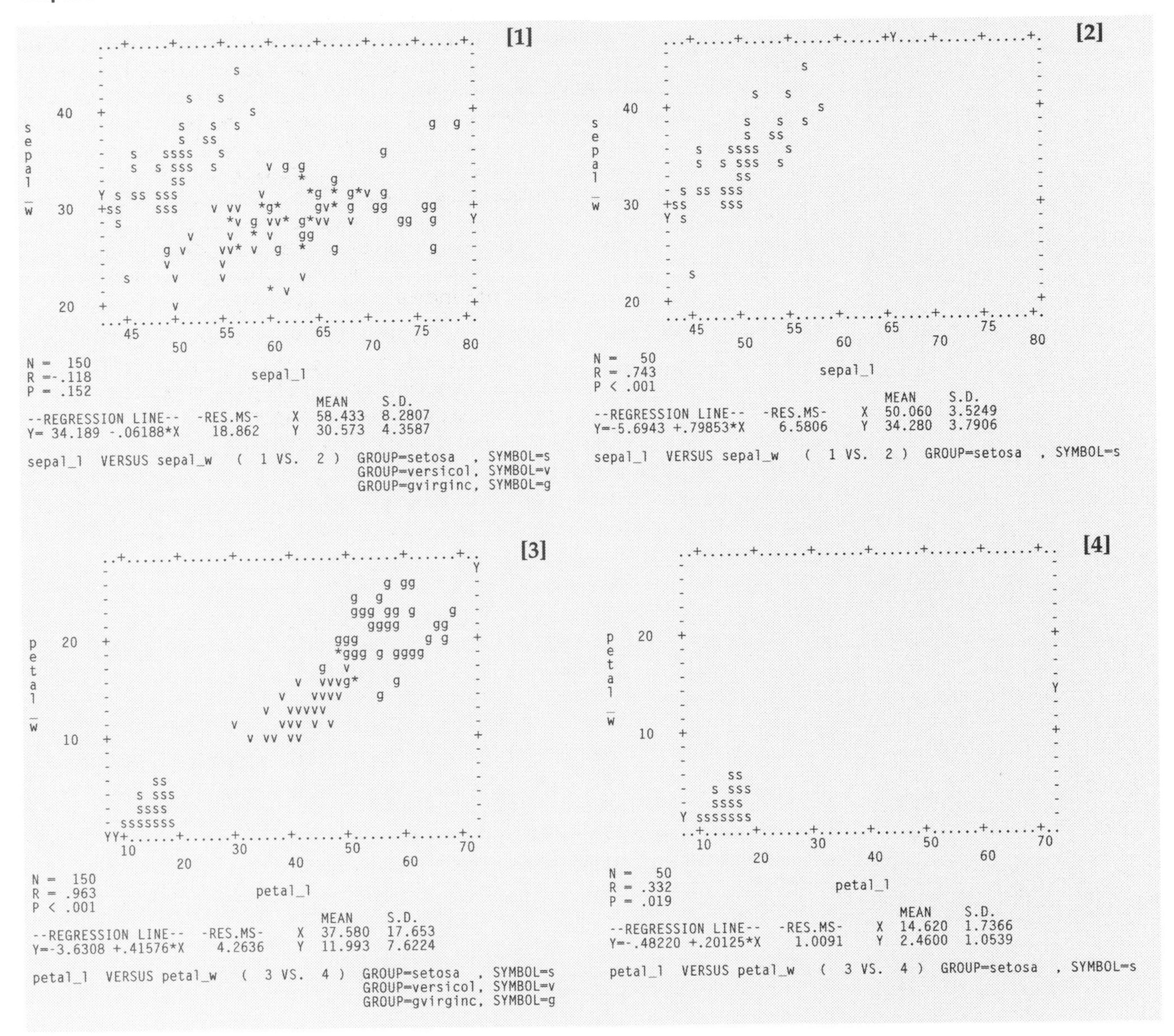

If points from different groups fall in the same place, 6D prints an asterisk, as in the first and third plots. 6D performs computations using all data plotted in a frame without regard to group membership (e.g., STATISTICS for the first and third plots are not broken down by group).

Comparing plots 1 and 3 with the plots in Example 6D.1, we see that grouping supplies us with additional information. In plot 1 SETOSA (S) points fall roughly along a line with a positive slope, although the ungrouped data show no correlation. Plot 2 shows a significant correlation of .743 between SEPAL_L and SEPAL_W within the group SETOSA. By using the grouping information, we have uncovered a correlation that was masked when all three subpopulations were combined in a single sample.

Conversely, plot 3 shows that the strong correlation between PETAL_L and PETAL_W is much less striking within individual groups. Plot 4 confirms this by showing a much smaller correlation within group SETOSA ($r = .332$).

Example 6D.3 Multiple plots in the same frame: YCOMMON

More than one pair of variables can be plotted in the same frame. This option is useful when several variables are measured in the same units or the same variable is measured at different times, and all are to be plotted against a common variable. In Input 6D.3 we plot sales figures for three different salespersons—Smith, Jones, and Brown—against the MONTH in which the sales were made (the common variable).

You can specify whether all XVARIABLES are to be plotted on a common x-axis (XCOMMON), whether all YVARIABLES are to be plotted on a common y-axis (YCOMMON), or whether all plots are to be made in a common frame (COMMON). For any of these instructions, each pair of variables can be plotted using a different symbol. Here we specify YCOMMON, since the y-variables are measured in the same units and we wish to compare them visually. COMMON may be specified with either PAIR or CROSS. When XCOMMON or YCOMMON is specified, the default is CROSS, and the PAIR specification is not allowed.

If more than one variable is plotted on one plot, the default setting for the lower limit of the plot scale is determined by the lowest minimum over all the variables. In the present example, we specify a MINIMUM value of zero for the first variable so that the plot scale on the y-axis starts at zero. In addition, we specify CHAR = VARIABLE so that the first letters of the VARIABLE NAMES, instead of the frequencies, will be used as plot symbols. (See CHARACTER command.)

Input 6D.3

```
/ INPUT        VARIABLES = 4.
               FORMAT IS FREE.

/ VARIABLE     NAMES ARE S_sales, J_sales, B_sales, month.

/ PLOT         YVAR = S_sales, J_sales, B_sales.
               XVAR = month.
               YCOMMON.
               CHAR = VARIABLE.
               MIN = 0.
               SIZE = 44, 18.

/ END
     8    10    20   1
    10    12    25   2
    15    12    26   3
    18    15    27   4
    20    12    22   5
    17     9    21   6
    22    10    22   7
    20    10    23   8
    22    12    24   9
    22    13    25  10
    25    16    28  11
    27    18    30  12
```

Output 6D.3

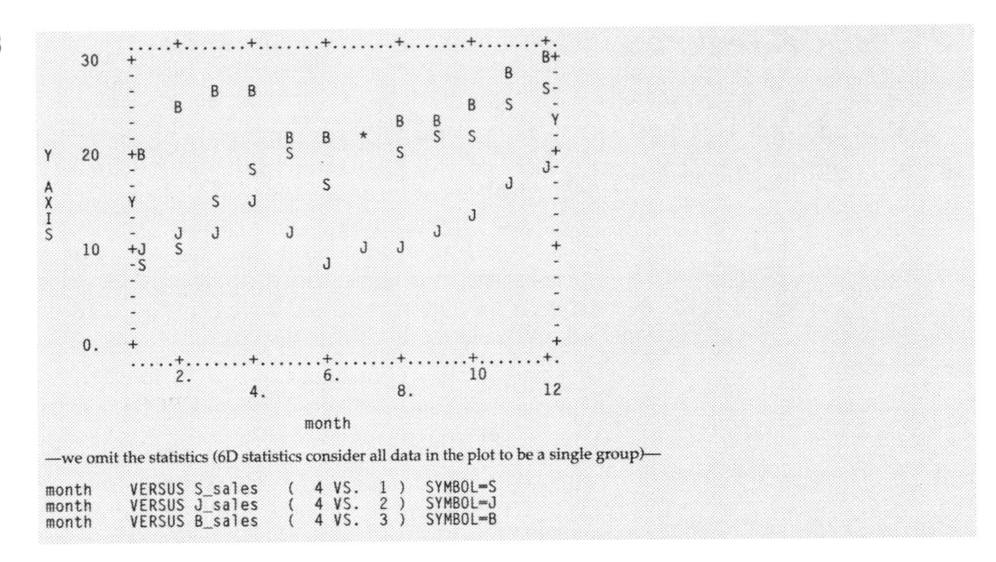

The plot in Output 6D.3 allows us to compare the sales figures for three salespersons over a period of 12 months. By specifying CHARACTER = VARIABLE, we produced a plot using symbols drawn from the VARIABLE NAMES.

Example 6D.4 Interactive paragraphs

6D now allows you to transform your data, modify the grouping instructions, and request plots interactively. You can specify any number of GROUP, TRANSFORM, and PLOT paragraphs after you specify INPUT, VARIABLE, and END paragraphs. In each case, the interactive paragraph should end with a slash, and the rest of the line after the slash should be blank.

In Input 6D.4 we use interactive paragraphs after the initial analysis to specify grouping information and request plots using the new groups. See Chapter 3D for information about interactive transformations.

For information about running BMDP interactively, see Chapter 10.

Input 6D.4

```
/ INPUT      FILE = 'fisher.dat'.
             VARIABLES = 5.
             FORMAT = FREE.

/ VARIABLE   NAMES = sepal_l, sepal_w, petal_l,
                     petal_w, iris.

/ PLOT       YVAR = sepal_l, petal_l.
             XVAR = sepal_w, petal_w.
             SIZE = 44, 18.
/ END

GROUP        VARIABLE = iris.
             CODES(iris) = 1 TO 3.
             NAMES(iris) = setosa, versicol, gvirginc./

PLOT         GROUP = ALL.
             GROUP = setosa.                         /

END /
```

Special Features

Controlling plot size and scale: Zooming in and out

You can use the SIZE command to reduce the size of the plots or to request plots that are several pages in length and width. The printed statistics are also affected; width must be at least 30 to print the means, 38 for the standard deviations, 7 for the regression equation, and 17 for the residual mean squares. The statistics, plot frame, and label for the x-axis occupy 12 lines. You may want to choose the number of lines for the y-axis to allow the statistics and labels to appear on the same page as the plot. By specifying MIN, MAX, and SIZE you can affect the labeling of the plot axes.

6D ordinarily determines plot scale according to the minimum and maximum data values in the plot. You can adjust the scale to zoom in or out on particular areas of the data:

- The MINIMUM and MAXIMUM limits in the VARIABLE paragraph apply to all PLOT paragraphs. The MINIMUM and MAXIMUM limits within a particular PLOT paragraph may override those limits to make the plot scale larger or smaller, as in Example 6D.3.
- Data values excluded by VARIABLE MIN and MAX limits cannot be recovered by stating wider limits in the PLOT paragraph. For example, if you state VARIABLE MIN = (SALES)10, PLOT MIN = (SALES)0, the VARIABLE MIN command will still exclude any values under 10 for the variable SALES from the plot and from all computations.
- By stating narrower MIN and MAX limits in the PLOT paragraph, you can obtain an "enlarged" view of a portion of the data, allowing higher resolution of adjacent points.

When you request plots for several groups, all plots have the same scale by default. Specify PLOT SCALE = INDIVIDUAL if you want each plot to be scaled according to the range of the data values in that plot.

Plot symbol options

Table 6D.1 contains an overview of the options for labeling your plot. See also the definitions of the CHARACTER and SYMBOL commands.

Table 6D.1
Plot symbol options

Symbol	/ PLOT Commands
Ungrouped Data	
Frequency at each point[1]	CHARACTER = FREQ.
Uniform symbol at all points	SYMBOL = list.
First letter of VARIABLE NAME for multiple variables (XCOM or YCOM)[2]	CHARACTER = VAR.
User–specified symbols for multiple variables (XCOM, YCOM, or COM)	CHARACTER = VAR **and** SYMBOL = list.
Grouped Data[3]	
First letter of group NAME[4]	CHARACTER = GROUP.
Specify symbols to identify groups	SYMBOL = list.
A, B, C . . . for multiple variables with groups (XCOM, YCOM, or COM)	CHARACTER = GROUP, VAR.
User-specified symbols for multiple variables with groups (XCOM, YCOM, or COM)	CHARACTER = GROUP, VAR **and** SYMBOL = list.

[1] Default for ungrouped data.
[2] Pairs of variables identified as A, B, C, . . . if NAMES are not specified or if COMMON is specified.
[3] You must identify the GROUPING variable in the VARIABLE paragraph.
[4] Default for grouped data. GROUPS are identified as A, B, C, ... if NAMES are not specified.

Using a case frequency variable

The FREQUENCY command in the VARIABLE paragraph allows one record of the input data matrix to represent more than one case (see Chapter 4). The variable containing the counts is identified in the FREQUENCY command, as shown in Figure 6D.1. The values for the correct sample size are used in the plot frequencies as well as in the computation of statistics.

Figure 6D.1
Using a case frequency variable

```
/  INPUT          VARIABLES = 3.
                  FORMAT IS FREE.

/ VARIABLE        NAMES ARE gross, year, people.
                  FREQ = people.

/ PLOT            YVAR = gross.
                  XVAR = year.
/ END
  10     1    3
  20     1    5
  30     1    6
  40     1    2
  10     2    7
  30     2    3
  40     2    5
  10     3    2
  20     3    2
```

In this example, the variable named PEOPLE contains the number of people having a certain GROSS in a given YEAR. For example, five people grossed $40,000 during the second year. The plot symbol will be 5, the frequency for this point.

6D Commands

/ VARiable

The VARIABLE paragraph is optional and is described in detail in Chapter 4. Additional commands for 6D are COUNT and FREQUENCY.

COUNT = variable. `COUNT = PEOPLE.`

COUNT works the same as FREQ below, but is specific to 6D.

FREQuency = variable. `FREQ = NUMBER.`

State the name or number of a variable containing the case frequency. A case frequency has the effect of repeating cases. Any record with a negative or zero case frequency (NUMBER ≤ 0 in this example) is excluded from the calculations as though it were missing. 6D reports the number of cases with zero weight. FREQUENCY cannot be used interactively. See Chapter 4 for further details.

/ GROUP

The GROUP paragraph is required to group data. If you prefer, you can still specify a grouping variable with the GROUPING command in the VARIABLE paragraph as described in the 1990 *BMDP Statistical Software Manual*. In addition, when a GROUPING variable is continuous or has more than ten distinct values, the values must be named in this paragraph. It may also be used generally to specify NAMES for the GROUPING variable, so that the output will be labeled according to the first letter (or character) of each group NAME. If no names are specified or if the first letters are not unique, 6D uses A, B, C, etc. See Chapter 5 for more information on specifying CODES, CUTPOINTS, and NAMES.

New Syntax

VARiable = variable. `VAR = iris.`

State the name or number of the variable used to classify the cases into groups. By default the cases are not grouped. If the GROUPING variable takes on more than ten distinct values or codes, you must specify CODES or CUTPOINTS in the GROUP paragraph (see Chapter 5).

/ PLOT

The PLOT paragraph is required and may be repeated for additional subproblems if you wish to specify different combinations of variables to be plotted or to modify the size of the plots. The PLOT paragraph specifies the variables to be plotted vertically and horizontally (YVAR, XVAR), the SIZE of the plot, and the GROUPS to be plotted if a VARIABLE statement has been included in the GROUP paragraph. Additional commands are listed below.

XVAR = list. `XVAR = SEPAL_W, PETAL_W.`

List names or numbers of the variables to be plotted horizontally (along the *x-axes*). By default, the first variable in the VARIABLE USE list is plotted (not including GROUPING variables).

YVAR = list. `YVAR = SEPAL_L, PETAL_L.`

List names or numbers of the variables to be plotted vertically (along the *y-axes*). By default, the last variable in the VARIABLE USE list (not including grouping variables) is plotted.

PAIR. *or* **CROSS.** `CROSS.`

By default, 6D will PAIR the YVAR variables with the XVAR variables (the first YVAR against the first XVAR, etc.) and plot each pair in a separate frame. If you specify CROSS, 6D plots all possible pairings of the YVAR and XVAR variables, each in a separate frame. When XCOMMON or YCOMMON is specified, CROSS is the default, and PAIR is not allowed.

STATistics. `NO STAT.`

Unless you specify NO STAT, 6D computes the following: the mean and standard deviation for each variable, the correlation and its *p*-value, the regression line $y = a + bx$, and the residual mean square. The intersections of the line with the frame of the plot are indicated by Y's. Computations are based on data from all the plotted cases. Statistics are not computed separately for individual groups within the same plot. When several variables are plotted on one axis, statistics are computed across all the variables, not for each variable individually. Thus, 6D will consider all x variables as a single variable and all y variables as a single variable.

SIZE = #1, #2. `SIZE = 40, 25.`

Specify the size of the plot as the number of characters for the horizontal axis (#1) and the number of lines for the vertical axis (#2). State the width first. The default plot is 70 characters wide, 42 lines high.

MINimum = #list. `MIN = (petal_w)0, 50.`
(one # per variable)

Specify one number per variable to set a lower limit on the plot scale. Values lower than MIN are omitted from the plot and from all computations. By default, 6D uses the observed minimum to set the scale. **Example:** The lower limit for the variable PETAL_W is 0, and for the next variable, 50.

MAXimum = # list. `MAX = 400.`
(one # per variable)

Specify one number per variable to set an upper limit on the plot scale. Values higher than MAX are omitted from the plot and from all computations. By default, 6D uses the observed maximum to set the scale. **Example:** The upper limit for the first variable is 400.

GROUP = ALL.
GROUP = EACH.
GROUP = list.
(one or more)

```
GROUPS = SETOSA, VERSICOL.
GROUPS = ALL.
```

Each GROUP command defines one plot (except for EACH). GROUP may be repeated within a PLOT paragraph. Use with VARIABLE, CODES, CUTPOINTS, NAMES, and ALPHA in the GROUP paragraph

ALL – Plot all groups in one graph.

EACH – Plot each group in a separate graph.

list – The specified groups are plotted in the same graph. Use group names or numbers (the order of the group in the CODES or CUTPOINTS command) from the GROUP paragraph. Do not include EACH or ALL in the list.

The default value is GROUP = ALL. **Example:** Groups SETOSA and VERSICOL are plotted in one plot, and ALL groups are plotted together in another plot.

CHARacter = FREQuency.
= GROUP, VARiable.
(one or both)

```
CHAR = GROUP, VAR.
```

Specify the plotting symbols used by 6D. The default value is FREQ, where case frequencies are used as plotting symbols (see Example 6D.1). If a GROUPING variable is specified in the GROUP paragraph, the default value becomes GROUP (see below). To define your own plotting symbols, see SYMBOL.

Three or more variables per plot

If you use a GROUP paragraph or if you plot two or more variables on a common axis, 6D can identify each group, each pair of variables, or both. The CHARACTER command allows you to specify the degree of labeling you want.

GROUP – This is the default if you specify a GROUPING variable. The preassigned symbol is the first letter from the NAMES specified in the GROUP paragraph. If no NAMES are specified, or if the first letters are not unique, A, B, C, etc., are used. The SYMBOL command may be used to change these default symbols.

VARiable – Pairs of variables are uniquely identified. The default symbols depend on other instructions specified in the PLOT paragraph: If you specify YCOMMON, or PAIR with COMMON, 6D plots the first letters of the YVARIABLE names (see Example 6D.3). For XCOMMON the first letters of the XVARIABLE names are used. If the first letters of the names are not unique or are not specified, 6D uses the letters A, B, C, etc.

If you specify CROSS with COMMON, 6D uses the characters A, B, C, etc., for each pair of variables. You can use the SYMBOL command to change the default symbols. The number of symbols, if specified, must equal the number of pairs of variables in the plot frame.

GROUP, VAR – Must be used with COMMON. Both groups and pairs of variables are uniquely identified. When x variables and/or y variables and groups are to be identified in one plot frame, the characters A, B, C, ... are assigned in an order where the x variables change most slowly and the group index changes most quickly. For example, in the following list there are two x variables, two y variables, and two possible values of the grouping variable:

```
X1 Y1 G1.........A
X1 Y1 G2.........B
X1 Y2 G1.........C
```

```
X1 Y2 G2.........D
X2 Y1 G1.........E
X2 Y1 G2.........F
X2 Y2 G1.........G
X2 Y2 G2.........H
```

The SYMBOL instruction below may be used to change these symbols. The number of symbols, if specified, must equal the product of the number of groups and the number of pairs of variables (in the list above we have two groups and four pairs of variables, requiring eight symbols).

SYMBol = list. `SYMB = '+', '-', M, F.`

Use to change the default plot symbols. See the CHARACTER command above for an explanation of default symbol selection. Use SYMBOL to define new plot symbols, one for every symbol defined by CHARACTER. You can also specify a uniform symbol, such as an asterisk, for all points by stating SYMB = '*'.

XCOMmon. `COM.`
YCOMmon.
or
COMmon.

Use for multiple plots within the same frame. Use XCOM to plot all the XVARIABLES on a common *x*-axis. Use YCOM to plot all the YVARIABLES on a common *y*-axis. COMMON means the same as XCOM and YCOM together; all variables specified in the PLOT paragraph are plotted in one frame. Unless you state one of these commands, variables are not plotted on a common axis.

SCALE = COMBined. `SCALE = COMB.`
or **INDividual.**

Use to determine the plot scales when GROUPS are specified in the PLOT paragraph. Default: COMB.

COMBINED – Each plot has the same scale, i.e., the scales are determined by the minimum and maximum values across all groups.

INDIVIDUAL – For each plot, the scale is determined by the minimum and maximum values of the data in that plot.

Order of Instructions

■ indicates required paragraph

```
■ / INPUT
  / VARIABLE
  / GROUP                                 Repeat for additional problems.
  / TRANSFORM                             See Multiple Problems, Chapter 10.
  / SAVE
  / PRINT
■ / PLOT     ] Repeat for subproblems
■ / END
  data
  GROUP ... /
  TRANSFORM .../     May be repeated
  PLOT ... /         interactively
  END /
```

Summary Table for Commands Specific to 6D

	Paragraphs Commands	Defaults	Multiple Problems	See
	/ **VARiable**			
	FREQuency = variable.	no case frequency variable	–	Cmds
	COUNT = variable.	no COUNT variable.	●	SpecF
▲	/ **GROUP**			
▲	VARiable = variable.	no grouping (see Chapter 5)	●	5D.4, 5D.5
■	/ **PLOT**			
▲	XVAR = list.	first variable in VAR USE list	●	6D.1
▲	YVAR = list.	last variable in VAR USE list	●	6D.1
	PAIR. *or* CROSS.	PAIR; CROSS if XCOM or YCOM	–	Cmds
	STATistics. *or* NO STAT.	STATISTICS.	●	Cmds
	SIZE = #1, #2.	70, 42	●	6D.1
	MINimum = # list.	observed minimum	●	6D.3
	MAXimum= #list.	observed maximum	●	SpecF
	GROUP = (*one or more*) list, EACH, ALL.	ALL	–	6D.4
	CHARacter = FREQuency. *or (one or both)* GROUP, VARiable.	FREQ; GROUP if a GROUPING variable is specified	–	6D.3
	SYMBol = list.	see CHAR	–	SpecF
	XCOMmon, YCOMmon, *or* COMmon.	no common axes	–	6D.3
	SCALE = COMBined. *or* INDividual.	COMBINED	–	SpecF

Key:
- ■ Required command or paragraph
- ▲ Frequently used command or paragraph
- ● Value retained for multiple problems
- – Default reassigned

7D

One- and Two-Way Analysis of Variance with Data Screening

Program 7D has two main purposes. By displaying and analyzing data in subgroups it provides (1) effective screening of the data for distributional form, outliers, and trends across groups; and (2) an analysis of variance (one- or two-way) to assess significance of group differences. Data screening capabilities include histograms to scan for outliers and skewness, and a test for equality of group variability. You can obtain Box-Cox diagnostic plots to help determine an appropriate variance-stabilizing transformation. Analysis of variance procedures include completely randomized one- and two-way factorial designs based on a fixed-effects model (group sizes need not be equal), and robust tests to assess group differences when the group variances are unequal. Trimmed mean formulas for analysis of variance are also available. You can specify *a priori* contrasts between any combination of means, and select several pairwise multiple comparison tests (Bonferroni, Duncan, Dunnett, Tukey, Scheffé, and Student-Newman-Keuls procedures). The correlations between the variables for each group and for all groups combined can be printed. You can also print out data after the cases are ordered or sorted according to a grouping variable.

Note that 7D does not handle analysis of covariance, repeated measures, nested designs, random effects models, or multivariate analysis of variance. For those models, please refer to programs 1V, 2V, 3V, 4V, 5V, or 8V. For a general introduction to analysis of variance, see Dixon and Massey (1983) and Dunn and Clark (1987). For an introduction to multiple comparisons, see Miller (1981). 7D requires that all data fit into computer memory. If you are using a data set with a large number of variables and cases, you may want to calculate the amount of memory needed. See Appendix A for details.

Where to Find It

Examples

Special Features

Example 7D.1 Data screening within groups

To illustrate the use of 7D, we use data from a community survey to assess how average income may differ for persons with varying levels of education. For the first example, we begin by screening our data set for outliers and distributional within groups consistency; the analysis of variance, which is a part of the output, is covered in the next example. We analyze data from Afifi and Clark (1990; see Appendix D). The data include mental and physical health variables and demographic variables (sex, age, education, income, etc.). For consistency, the same data set is used throughout this chapter. It consists of questionnaire responses from 294 subjects, and is stored on disk in a file named SURVEY.DAT.

In Input 7D.1, only the HISTOGRAM paragraph is specific to 7D; it provides analysis of variance results and side-by-side histograms for each level of the grouping variable (in this case, education). The INPUT and VARIABLE paragraphs for describing the data and variables are common to all BMDP programs. (The FILE command tells the program where to find the data and is used on systems like VAX and the IBM PC. For IBM mainframes see UNIT, Chapter 3.) The GROUP paragraph is required in 7D if group names are desired or if the grouping variable has more than ten distinct values (see Chapter 5). The data include seven codes for education. In Input 7D.1, we collapse the seven levels into the following four by using CUTPOINTS and NAMES in the GROUP paragraph:

1. Those without a high school degree (NO GRAD),
2. High school graduates (HS GRAD),
3. Those completing some college coursework (SOME COL), and
4. College graduates and above (COL GRAD).

Input 7D.1

```
/ INPUT          VARIABLES = 3.
                 FORMAT IS FREE.
                 FILE IS 'survey.dat'.

/ VARIABLE       NAMES ARE
id,       sex,       age,       marital,  educatn,  employ,
income,   religion,  blue,      depress,  lonely,   cry,
sad,      fearful,   failure,   am_good,  hopeful,  happy,
enjoy,    bothered,  no_eat,    effort,   badsleep, getgoing,
mind,     talkless,  unfriend,  dislike,  total,    case,
```

Input 7D.1
(continued)

```
drink,   healthy,  doctor,   meds,     bed_days, illness
chronic.
             USE = income, educatn.

/ GROUP      CUTPOINTS(educatn) = 2, 3, 4.
             NAMES(educatn) = no_grad, hs_grad, some_col,
                              col_grad.

/ HISTOGRAM     GROUPING IS educatn.
                VARIABLE IS income.
/ END
```

Output 7D.1

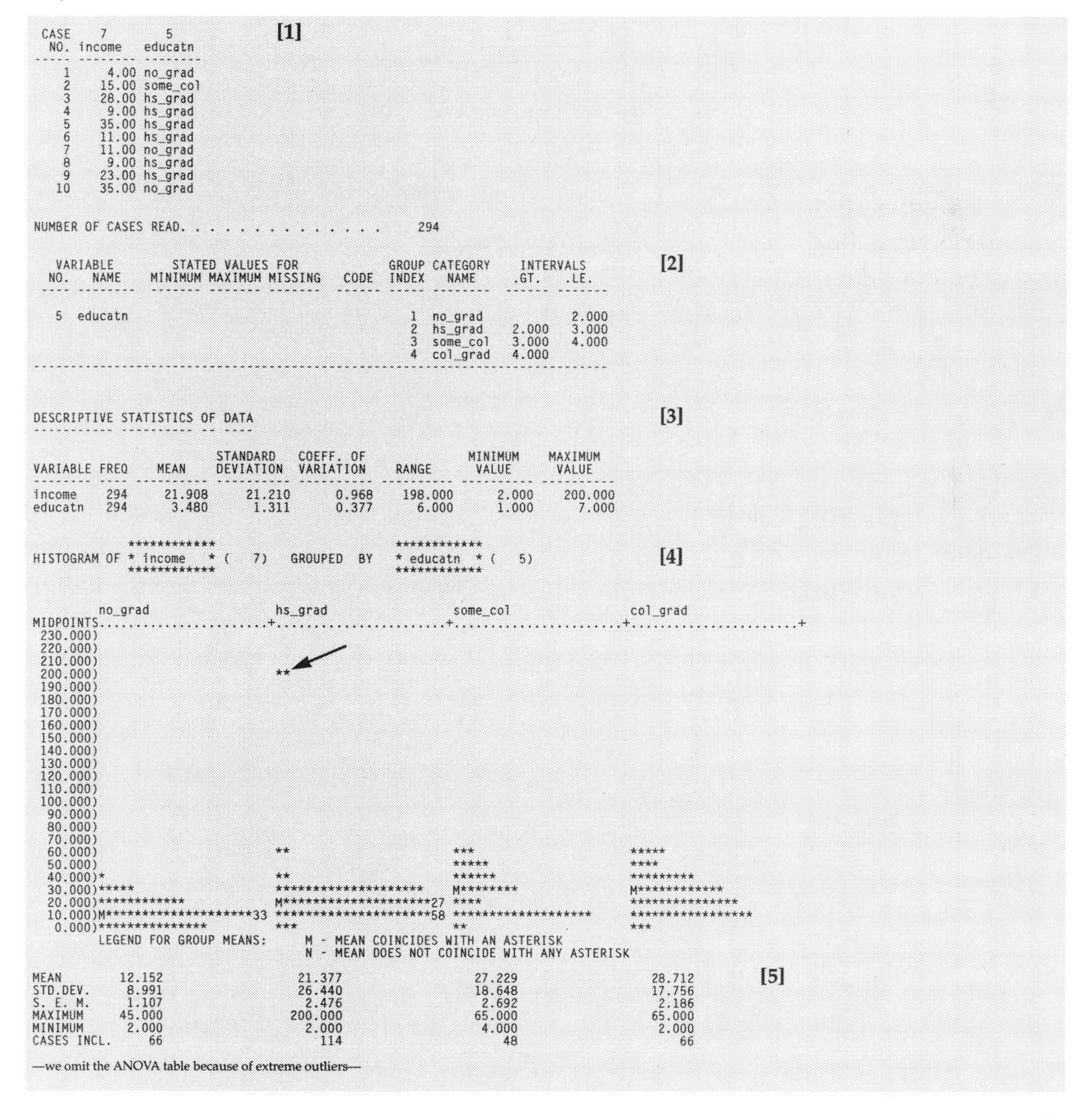

```
 CASE    7        5                [1]
  NO. income    educatn
----- -------- --------
    1     4.00 no_grad
    2    15.00 some_col
    3    28.00 hs_grad
    4     9.00 hs_grad
    5    35.00 hs_grad
    6    11.00 hs_grad
    7    11.00 no_grad
    8     9.00 hs_grad
    9    23.00 hs_grad
   10    35.00 no_grad

NUMBER OF CASES READ. . . . . . . . . . . . . .    294

   VARIABLE       STATED VALUES FOR          GROUP CATEGORY    INTERVALS          [2]
  NO.   NAME   MINIMUM MAXIMUM MISSING   CODE  INDEX   NAME     .GT.    .LE.
  ---- -------- ------- ------- -------  ------ ----- --------  ------ -------

   5  educatn                                    1  no_grad            2.000
                                                 2  hs_grad    2.000   3.000
                                                 3  some_col   3.000   4.000
                                                 4  col_grad   4.000
--------------------------------------------------------------------------------


DESCRIPTIVE STATISTICS OF DATA                                                    [3]
----------- ---------- -- ----

                        STANDARD   COEFF. OF             MINIMUM    MAXIMUM
VARIABLE FREQ    MEAN   DEVIATION  VARIATION    RANGE     VALUE      VALUE
-------- ----  -------- ---------  ---------  --------- ---------  ---------
income    294    21.908    21.210      0.968   198.000     2.000    200.000
educatn   294     3.480     1.311      0.377     6.000     1.000      7.000

             ************                  ************
HISTOGRAM OF * income     * (   7)  GROUPED  BY   * educatn   * (   5)           [4]
             ************                  ************


         no_grad                 hs_grad                  some_col                 col_grad
MIDPOINTS.......................+.......................+.......................+.......................+
 230.000)
 220.000)
 210.000)
 200.000)                        **
 190.000)
 180.000)
 170.000)
 160.000)
 150.000)
 140.000)
 130.000)
 120.000)
 110.000)
 100.000)
  90.000)
  80.000)
  70.000)
  60.000)                        **                      ***                      *****
  50.000)                                                *****                    ****
  40.000)*                       **                      ******                   *********
  30.000)*****                   ********************    M********                M************
  20.000)************            M*******************27  ****                     ***************
  10.000)M*******************33  ********************58  *******************      *****************
   0.000)***************         ***                     **                       ***
        LEGEND FOR GROUP MEANS:      M - MEAN COINCIDES WITH AN ASTERISK
                                     N - MEAN DOES NOT COINCIDE WITH ANY ASTERISK

MEAN        12.152               21.377                  27.229                   28.712         [5]
STD.DEV.     8.991               26.440                  18.648                   17.756
S. E. M.     1.107                2.476                   2.692                    2.186
MAXIMUM     45.000              200.000                  65.000                   65.000
MINIMUM      2.000                2.000                   4.000                    2.000
CASES INCL.     66                  114                      48                       66
```

—we omit the ANOVA table because of extreme outliers—

Bold numbers below correspond to those in Output 7D.1.

[1] Prints the first ten cases of data. Grouping names are printed if specified.

[2] Interpretation of the GROUP paragraph. Our CUTPOINT and NAME instructions collapse the seven original educational levels into four groups. Notice that the column headings under "INTERVALS" display the logical operators GT and LE. The NO_GRAD group, for example, is composed of individuals with an education code Less than or Equal to 2 (i.e., codes 1 and 2).

[3] Descriptive statistics for each used variable without grouping.

[4] These side-by-side histograms display income for each of the four educational groups. Each asterisk represents an observation. The base of each histogram is the vertical axis. The midpoints of the histogram bars range from 0 to 230 thousand dollars. If more cases exist than can fit on a line, the total number is printed to the right of the asterisk row. The mean of each group is indicated by an *M* if it coincides with an asterisk, or an *N* if it does not. There are two subjects (see arrow) in the HS_GRAD group whose incomes fall in a bin with a midpoint of $200,000. Note that if narrow output is requested with a PRINT LINESIZE = 80 command, or obtained in an interactive run, the histograms are spaced closer together, and there may be fewer asterisks in a row.

[5] For each group, 7D prints the mean, the standard deviation, the standard error of the mean, the maximum and minimum income values, and the number of cases. (Note that a CASES EXCLUDED row appears if data are missing or excluded from analysis with MISS, MIN, or MAX statements in the VARIABLE paragraph.) The fact that two cases with INCOME values of $200,000 in the HS_GRAD group are so far from the rest of the data suggests that it would be prudent to check the original data. In fact, a review of the original coding forms reveals that both high values were incorrectly entered as $200,000 rather than $4000. Those incorrect data values can either be changed in the data file or corrected in a TRANSFORM paragraph with the following statement: IF(income EQ 200) THEN income = 4. Note that this statement works because there are no correct data where income = 200.

Example 7D.2 One-way analysis of variance

In this example, we rerun Input 7D.1 after correcting the two erroneous income values. This time we illustrate a standard one-way analysis of variance (ANOVA), with education level as the independent variable and income as the variance dependent variable. In Input 7D.2, we correct the extreme data values by adding a TRANSFORM paragraph (see Chapter 6 for details) to Input 7D.1. Note that although the analysis of variance was not shown in Example 7D.1, it is part of 7D's automatic output, and does not need a special BMDP instruction. For this example, the null hypothesis tested is that the mean incomes of the four educational groups are the same:

$$H_0: \ \mu_{no\_grad} = \mu_{hs\_grad} = \mu_{some\_col} = \mu_{col\_grad}$$

where μ_{no_grad} is the population mean income for persons not graduating from high school, and so on.

A basic assumption underlying the analysis of variance is that of equality of variances. 7D's Levene test provides a test of equal variability among the cells. If that test is significant, the hypothesis of equality of group variances is rejected, and the standard ANOVA does not provide a valid test of the equality of group means.

The Welch and the Brown-Forsythe procedures are two alternate tests of the null hypothesis of equality of group means in which the group variances are not assumed to be equal. Although those two tests allow for unequal group variances, they do so at the expense of a loss of degrees of freedom.

Input 7D.2
(we add the following TRANSFORM paragraph to Input 7D.1)

```
/ TRANSFORM    IF(income EQ 200) THEN income = 4.
```

Output 7D.2

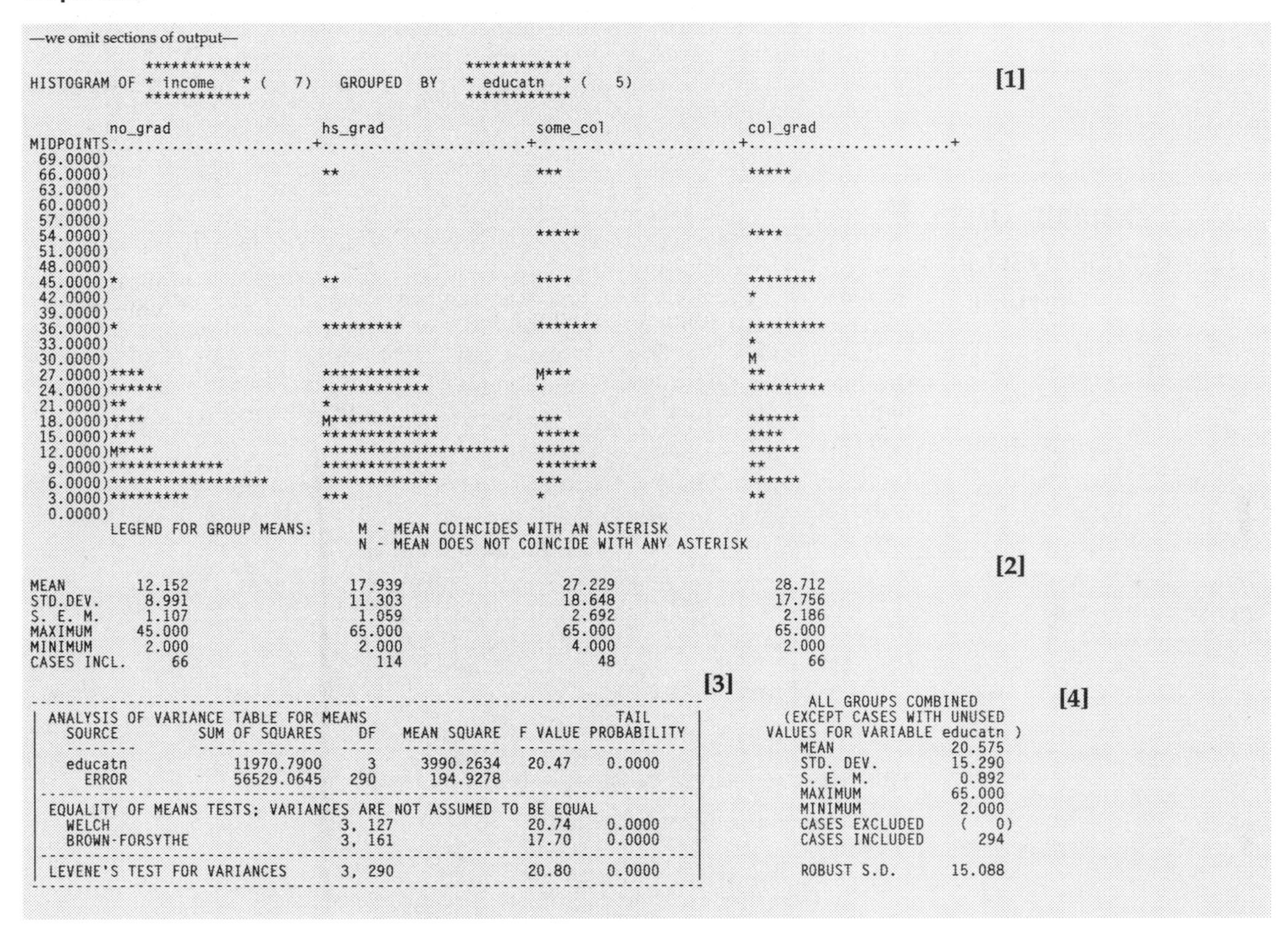

—we omit sections of output—

```
                ************                  ************
HISTOGRAM OF * income    * (   7)  GROUPED  BY   * educatn  * (   5)        [1]
                ************                  ************

         no_grad                  hs_grad                  some_col                 col_grad
MIDPOINTS.......................+.......................+.......................+.......................+
  69.0000)
  66.0000)                        **                       ***                      *****
  63.0000)
  60.0000)
  57.0000)
  54.0000)                                                 *****                    ****
  51.0000)
  48.0000)
  45.0000)*                       **                       ****                     ********
  42.0000)                                                                          *
  39.0000)
  36.0000)*                       *********                *******                  *********
  33.0000)                                                                          *
  30.0000)                                                                          M
  27.0000)****                    ***********              M***                     **
  24.0000)******                  ************             *                        *********
  21.0000)**                      *
  18.0000)****                    M************            ***                      ******
  15.0000)***                     *************            *****                    ****
  12.0000)M****                   *********************    *****                    ******
   9.0000)*************           **************           *******                  **
   6.0000)******************      *************            ***                      ******
   3.0000)*********               ***                      *                        **
   0.0000)
         LEGEND FOR GROUP MEANS:      M - MEAN COINCIDES WITH AN ASTERISK
                                      N - MEAN DOES NOT COINCIDE WITH ANY ASTERISK
                                                                                                 [2]
MEAN          12.152                  17.939                  27.229                  28.712
STD.DEV.       8.991                  11.303                  18.648                  17.756
S. E. M.       1.107                   1.059                   2.692                   2.186
MAXIMUM       45.000                  65.000                  65.000                  65.000
MINIMUM        2.000                   2.000                   4.000                   2.000
CASES INCL.      66                     114                      48                      66
                                                                           [3]
--------------------------------------------------------------------------       ALL GROUPS COMBINED         [4]
| ANALYSIS OF VARIANCE TABLE FOR MEANS                          TAIL     |     (EXCEPT CASES WITH UNUSED
|   SOURCE         SUM OF SQUARES   DF    MEAN SQUARE  F VALUE PROBABILITY |  VALUES FOR VARIABLE educatn )
|   --------       --------------  ----   -----------  ------- ----------- |      MEAN             20.575
|   educatn           11970.7900     3     3990.2634   20.47    0.0000     |      STD. DEV.        15.290
|     ERROR           56529.0645   290      194.9278                       |      S. E. M.          0.892
|------------------------------------------------------------------------ |      MAXIMUM          65.000
| EQUALITY OF MEANS TESTS; VARIANCES ARE NOT ASSUMED TO BE EQUAL           |      MINIMUM           2.000
|   WELCH                          3, 127                  20.74    0.0000 |      CASES EXCLUDED    (    0)
|   BROWN-FORSYTHE                 3, 161                  17.70    0.0000 |      CASES INCLUDED      294
|------------------------------------------------------------------------ |
| LEVENE'S TEST FOR VARIANCES      3, 290                  20.80    0.0000 |      ROBUST S.D.      15.088
--------------------------------------------------------------------------
```

[1] As in Output 7D.1, this panel shows side-by-side vertical histograms of INCOME for each of the four EDUCATN groups. However, with the data entry errors corrected (the two \$200,000 values changed to \$4,000), the histograms are now scaled so that more detail is evident. Notice that the income distributions are positively skewed toward the higher income levels, and that there is more within-group variation for the two higher educational levels.

[2] Descriptive statistics (e.g., the mean, standard deviation, standard error) for the four groups. The standard deviations of groups 1 and 2 are noticeably lower than those of groups 3 and 4 and thus correlated with mean income. Those observations, together with the skewed appearance of the histograms, suggest that a transformation of the data may be warranted (see Examples 7D.3 and 7D.6).

[3] The top panel shows the results of a standard one-way analysis of variance, and indicates a highly significant F value. However, LEVENE's test (bottom panel), which tests for equality of group variability, is also significant. We therefore need to use the WELCH and BROWN-FORSYTHE tests. (When there are only two groups, both of those tests reduce to the separate variance t test discussed in program 3D.) Both of these alternate tests are significant, so the null hypothesis of equal group means is rejected: income levels do vary among educational groups. (See Appendix B.9 for more information on these tests.)

[4] For all groups combined, 7D prints the mean, the standard deviation, the standard error of the mean, the maximum and minimum income values, and the number of cases excluded from and included in the analysis. 7D also prints a robust estimate of the standard deviation (ROBUST S.D.) based on the mean absolute deviation from the mean. (Both this estimate and Winsorized estimates of the standard deviation within each group are included in the trimmed mean panel; see Example 7D.7.)

Example 7D.3 Two-way analysis of variance

The following analysis introduces an additional level of complexity. Using a two way analysis of variance, we reanalyze the survey data to investigate how a second factor, sex, in combination with educational level, affects income. Suppose we believe that men receive significantly higher salaries than women. We can examine these questions with a two-way analysis of variance; specifically, we can test whether: (1) there are sex differences in income across all educational levels; (2) for both sexes combined, the higher education groups have higher incomes; (3) income differences for men and women vary across education levels.

We thus have a 2(sex) x 4(education level) independent groups design with eight cells:

		Educational Level			
		no_grad	hs_ grad	some_col	col_grad
Sex	male	μ_{mn}	μ_{mh}	μ_{ms}	μ_{mc}
	female	μ_{fn}	μ_{fh}	μ_{fs}	μ_{fc}

where μ_{ij} is the mean of the cell in row i and column j.

This two-way analysis of variance tests three hypotheses:

H_1: effects due to **sex** are equal

$$\mu_{mn} + \mu_{mh} + \mu_{ms} + \mu_{mc} = \mu_{fn} + \mu_{fh} + \mu_{fs} + \mu_{fc}$$

there are no differences in average income between males and females

H_2: effects due to **education** are equal

$$\mu_{mn} + \mu_{fn} = \mu_{mh} + \mu_{fh} = \mu_{ms} + \mu_{fs} = \mu_{mc} + \mu_{fc}$$

there are no differences in average income between the different educational groups

H_3: there are no **interaction** effects

$$\mu_{mn} - \mu_{fn} = \mu_{mh} - \mu_{fh} = \mu_{ms} - \mu_{fs} = \mu_{mc} - \mu_{fc}$$

the effect of amount of education on income is the same for males as for females

These hypotheses are not dependent on the group sizes; the method used is the standard parametric model for the analysis of unbalanced factorial designs (Herr, 1986). Computationally, the sums of squares used in testing these hypotheses are obtained by taking the difference in residual sums of squares between regression models fit with and without specific terms. (See Appendix B.9 for more details, and program 4V, Volume 2, for tests of other hypotheses about differences between means.)

In Input 7D.3, we use a TRANSFORM paragraph to create a new variable, LOG_INC, by computing the logarithm of income for every case. A log transformation should reduce positive skewness and help equalize the variances. (For help in choosing a useful transformation, see Example 7D.6.) Since we want to use LOG_INC as the dependent variable, we add it to the VARIABLE command of the HISTOGRAM paragraph. We also restrict the analysis to cases for individuals over age 35. In this example, we want to examine the relationship between education and income after excluding individuals who might still be completing their education (see Chapter 6 for discussion of case selection). In the GROUP paragraph we specify CODES (sex) = 1, 2, and assign the names MALE and FEMALE for easier readability.

In the HISTOGRAM paragraph, we request a two-way analysis of variance by specifying both EDUCATN and SEX as grouping factors. Note that the order of the grouping factors affects the order of the histograms. The first grouping factor entered is the "slowest moving" index; the second grouping factor is the "fastest moving" index. Thus, the level of the first grouping factor changes only after it has been combined with every level of the second grouping factor.

Input 7D.3

(we add a TRANSFORM paragraph and replace the GROUP and HISTOGRAM paragraphs in Input 7D.1)

```
/ TRANSFORM     USE = age GT 35.
                IF(income EQ 200) THEN income = 4.
                log_inc = LOG (income).

/ GROUP         CODES(sex) = 1, 2.
                NAMES(sex) = male, female.
                CUTPOINTS(educatn) = 2, 3, 4.
                NAMES(educatn) = no_grad, hs_grad,
                                 some_col, col_grad.

/ HISTOGRAM     GROUPING = educatn, sex.
                VARIABLE = log_inc.

/ END
```

Output 7D.3

```
               ************                         ************
HISTOGRAM OF * log_inc  * (  38)   GROUPED  BY    * educatn  * (    5)                  [1]
               ************                  AND    * sex      * (    2)
                                                    ************

         no_grad      no_grad      hs_grad      hs_grad      some_col     some_col     col_grad     col_grad
         male         female       male         female       male         female       male         female
MIDPOINTS............+............+............+............+............+............+............+............+
 1.92000)
 1.84000)                          *                                      **           ****
 1.76000)                                                    *            **                        ***
 1.68000)                                       **                        *            *****        *
 1.60000)                                                                                           *
 1.52000)                          ***          ***          ***          *            M***         ***
 1.44000)*                         *****        **           **                        *            M
 1.36000)**           *            **           *******      N                         ****         **
 1.28000)             ****         N            ***          *            N            **           **
 1.20000)             *            *****        ***          *            ***          *            *
 1.12000)             *            *            M***         *                         *            *
 1.04000)**           **                        *******
 0.96000)M*           M****                     ******       *
 0.88000)****         ********                  *****                     ***                       *
 0.80000)*            *                         ******
 0.72000)****         **                        *                         *
 0.64000)*            **           **                                     *
 0.56000)
 0.48000)
 0.40000)
 0.32000)             *                         *
 0.24000)
          LEGEND FOR GROUP MEANS:       M - MEAN COINCIDES WITH AN ASTERISK
                                        N - MEAN DOES NOT COINCIDE WITH ANY ASTERISK

MEAN           0.928        0.938        1.295        1.105        1.379        1.270        1.520        1.437      [2]
STD.DEV.       0.254        0.243        0.301        0.283        0.241        0.439        0.211        0.245
S. E. M.       0.062        0.046        0.069        0.040        0.076        0.117        0.045        0.061
MAXIMUM        1.447        1.362        1.813        1.653        1.740        1.813        1.813        1.740
MINIMUM        0.602        0.301        0.602        0.301        0.954        0.602        1.114        0.903
CASES INCL.       17           28           19           50           10           14           22           16
```

—we omit the descriptive statistics for all groups combined—

[3]

```
ANALYSIS OF VARIANCE                                              TAIL
  SOURCE          SUM OF SQUARES    DF    MEAN SQUARE  F VALUE PROBABILITY
  ----------      --------------   ----   -----------  ------- -----------
  educatn                6.2332      3       2.0777     26.79     0.0000
  sex                    0.3082      1       0.3082      3.97     0.0478
  INTERACTION            0.2417      3       0.0806      1.04     0.3770
    ERROR               13.0284    168       0.0776
---------------------------------------------------------------------------
ANALYSIS OF VARIANCE; VARIANCES ARE NOT ASSUMED TO BE EQUAL
  WELCH                           7,  54                17.39     0.0000
  BROWN-FORSYTHE
    educatn                       3,  63                22.81     0.0000
    sex                           1,  67                 3.95     0.0508
    INTERACTION                   3,  65                 0.68     0.5655
---------------------------------------------------------------------------
LEVENE'S TEST FOR EQUALITY OF VARIANCES
    educatn                       3, 168                 2.69     0.0479
    sex                           1, 168                 3.38     0.0676
    INTERACTION                   3, 168                 2.32     0.0774
```

[1] Side-by-side histograms for each combination of grouping levels. The order of subgroup histograms is determined by the order of the grouping factors in the GROUPING statement. Levels of the second listed factor, SEX, change first. Because of the log transformation and the restriction of cases, the distributions are less skewed than was the case in the previous example. If you run this example interactively, or specify LINESIZE = 80, the histogram panel is split, with four histograms above and four below.

[2] This panel shows separate means, standard deviations, etc., for each sex and education group. The standard deviations of LOG_INC are now fairly homogeneous. The ratio of the largest to smallest (0.439/0.211) is approximately 2.1. For the three highest educational levels, males have higher mean LOG_INCs than females; the difference is particularly large in the case of HS_GRADs.

[3] The top panel shows the classical independent groups pooled-variance, analysis of variance results; the middle panel shows the WELCH and BROWN-FORSYTHE separate-variance analysis of variance results; and the bottom panel shows LEVENE's test for equality of variability. LEVENE's test provides evidence against equality of variance for EDUCATN, so it is not appropriate to use the classical, pooled-variance, analysis of variance results. (The results of LEVENE's test also suggest that you may want to rerun the analysis, using a different transformation.) The Welch statistic provides only a one-way analysis of variance comparing the eight groups. The Brown-Forsythe statistic, however, computes a two-way analysis of variance. It reveals a highly significant difference among the means of the EDUCATN groups, but no significant differences due to SEX or the SEX X EDUCATN interaction. The lack of significant interaction indicates that the difference between male and female incomes is about the same at each education level.

Example 7D.4 Multiple comparison tests

This example illustrates the use of *post hoc* pairwise comparisons in a one-way analysis of variance to identify which group differences account for the significant overall *F* value. You can also request multiple comparisons in a two-way analysis of variance; however, every pair of subgroups will be compared (no comparisons of marginal means are made).

In Input 7D.4, we add three paragraphs to Input 7D.1 and modify the HISTOGRAM paragraph. First, we add a TRANSFORM paragraph to restrict the analysis to cases for males over age 35. As in Example 7D.3, we use the logarithm of income (LOG_INC) to reduce skewness and help equalize the variances. Therefore, we define LOG_INC in the TRANSFORM paragraph and identify it as the dependent variable in the HISTOGRAM paragraph. The COMPARISON and PRINT paragraphs are discussed below.

The COMPARISON paragraph. This paragraph is used to request a variety of *post hoc* pairwise multiple comparison tests. (For user-specified contrasts between means, see Example 7D.5.) The Bonferroni, Tukey, and Scheffé methods produce tests of significance between every pair of means; the three methods differ somewhat in the way they obtain p-values (see Appendix B.9). The Tukey method is the default; when a COMPARISON paragraph is included, the Tukey test is always printed unless you specify NO TUK. The Dunnett procedure is intended for situations in which one group is a control group and you want to compare all group means to the control group mean; however, if the groups have unequal sample sizes, the control group must have the largest sample size. Since the output from these four procedures looks similar, we use only the default Tukey test. We also request confidence intervals for the pairwise differences in means by stating CONFIDENCE. Two multiple range tests are also available. We illustrate the Student-Newman-Keuls method by stating NK. (For more information on the above comparison tests, see Milliken and Johnson, 1984.)

The PRINT paragraph. In addition to the Bonferroni comparisons obtainable by specifying Bonferroni in the COMPARISON paragraph, you can obtain more detailed results for both pooled- and separate-variance pairwise t tests by specifying TTEST in the PRINT paragraph.

The Bonferroni significance levels (indicated by asterisks) will be adjusted for the number of comparisons you specify using COMPARISON in the PRINT paragraph. If you do not specify the number of comparisons, 7D adjusts the significance levels as if all possible pairs of means are tested. (Note that 7D adjusts the Bonferroni significance levels but not the number of comparisons actually printed; you receive all comparisons in the output regardless of the number you specify for significance level adjustment. However, it is important that you look only at those comparisons that you selected **before** the analysis, despite the fact that some of the others may be significant.) Here, we specify TTEST and COMPARISON = 3 because of a decision to make only three specific pairwise comparisons chosen in advance (the NO_GRAD group versus each of the others). Finally, we include LINESIZE = 80 to produce a narrower output than the default width of 132 columns.

Input 7D.4

(we add TRANSFORM, COMPARISON, and PRINT paragraphs to Input 7D.1 and replace the HISTOGRAM PARAGRAPH)

```
/ TRANSFORM     USE = sex EQ 1 AND age GT 35.
                IF(income EQ 200) THEN income = 4.
                log_inc = LOG(income).

/ HISTOGRAM     GROUPING IS educatn.
                VARIABLES ARE log_inc.

/ COMPARISON    NK.
                CONFIDENCE.

/ PRINT         TTEST.
                COMPARISON = 3.
                LINESIZE = 80.

/ END
```

Output 7D.4

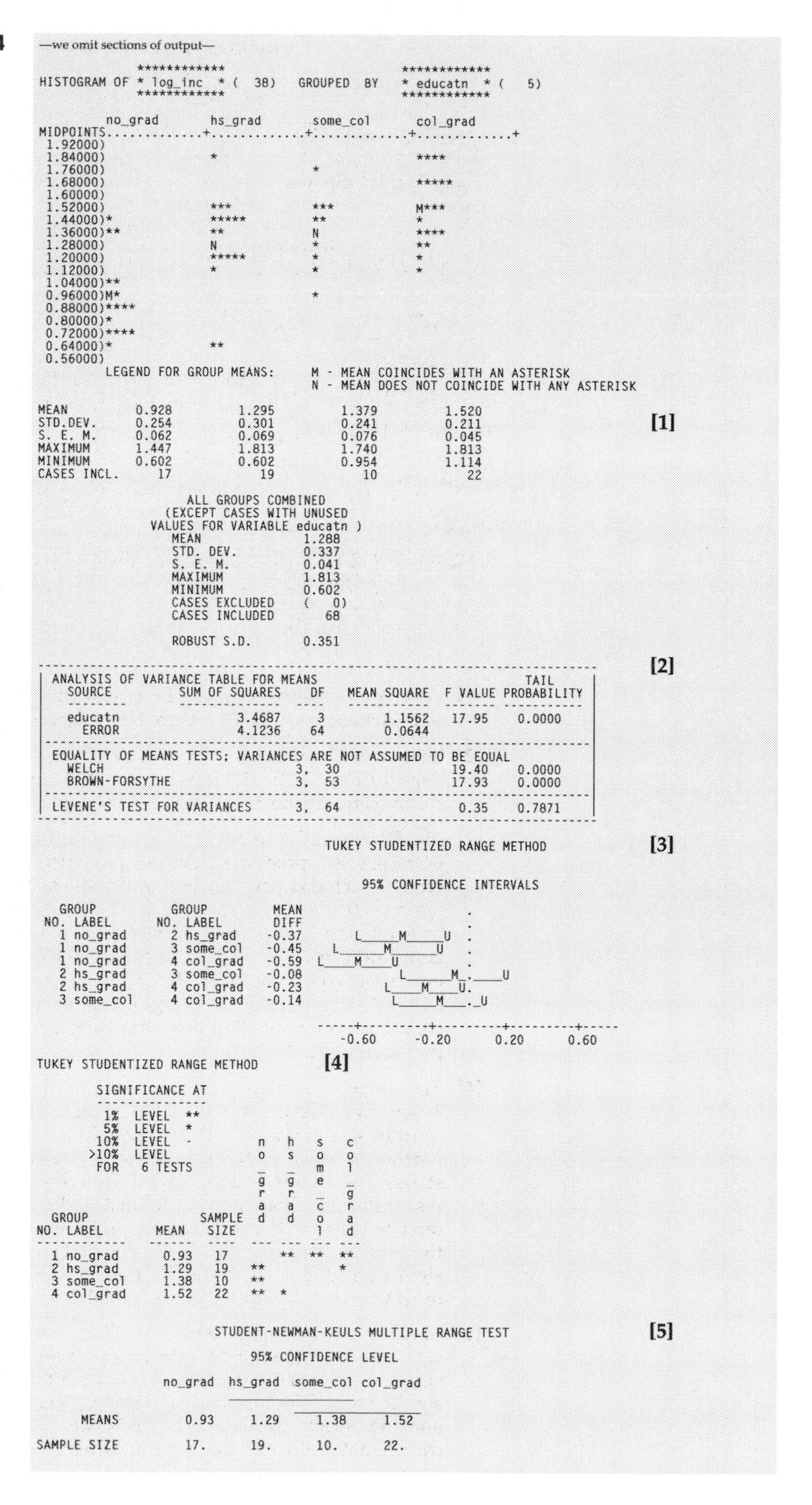

—we omit sections of output—

```
                  ************                           ************
HISTOGRAM OF * log_inc   * (  38)   GROUPED  BY   * educatn  * (    5)
                  ************                           ************

            no_grad          hs_grad          some_col        col_grad
MIDPOINTS.............+.............+.............+.............+
 1.92000)
 1.84000)             *                           ****
 1.76000)                           *
 1.68000)                                         *****
 1.60000)
 1.52000)             ***           ***           M***
 1.44000)*            *****         **            *
 1.36000)**           **            N             ****
 1.28000)             N             *             **
 1.20000)             *****         *             *
 1.12000)             *             *             *
 1.04000)**
 0.96000)M*                         *
 0.88000)****
 0.80000)*
 0.72000)****
 0.64000)*            **
 0.56000)
        LEGEND FOR GROUP MEANS:     M - MEAN COINCIDES WITH AN ASTERISK
                                    N - MEAN DOES NOT COINCIDE WITH ANY ASTERISK

MEAN          0.928         1.295         1.379         1.520
STD.DEV.      0.254         0.301         0.241         0.211                [1]
S. E. M.      0.062         0.069         0.076         0.045
MAXIMUM       1.447         1.813         1.740         1.813
MINIMUM       0.602         0.602         0.954         1.114
CASES INCL.      17            19            10            22

                  ALL GROUPS COMBINED
                (EXCEPT CASES WITH UNUSED
              VALUES FOR VARIABLE educatn )
                  MEAN                1.288
                  STD. DEV.           0.337
                  S. E. M.            0.041
                  MAXIMUM             1.813
                  MINIMUM             0.602
                  CASES EXCLUDED      (   0)
                  CASES INCLUDED        68

                  ROBUST S.D.         0.351

-----------------------------------------------------------------------------  [2]
| ANALYSIS OF VARIANCE TABLE FOR MEANS                              TAIL      |
|   SOURCE         SUM OF SQUARES    DF   MEAN SQUARE  F VALUE PROBABILITY    |
|   --------       --------------   ----  -----------  ------- -----------    |
|   educatn            3.4687        3       1.1562     17.95    0.0000       |
|     ERROR            4.1236       64       0.0644                           |
-----------------------------------------------------------------------------
| EQUALITY OF MEANS TESTS; VARIANCES ARE NOT ASSUMED TO BE EQUAL              |
|   WELCH                            3,  30             19.40    0.0000       |
|   BROWN-FORSYTHE                   3,  53             17.93    0.0000       |
-----------------------------------------------------------------------------
| LEVENE'S TEST FOR VARIANCES       3,  64              0.35    0.7871       |
-----------------------------------------------------------------------------

                                     TUKEY STUDENTIZED RANGE METHOD            [3]

                                          95% CONFIDENCE INTERVALS

   GROUP           GROUP         MEAN                           .
 NO. LABEL       NO. LABEL       DIFF                           .
   1 no_grad       2 hs_grad     -0.37          L_____M_____U   .
   1 no_grad       3 some_col    -0.45       L______M______U    .
   1 no_grad       4 col_grad    -0.59     L____M____U          .
   2 hs_grad       3 some_col    -0.08                 L______M_.____U
   2 hs_grad       4 col_grad    -0.23              L____M____U.
   3 some_col      4 col_grad    -0.14               L_____M___._U

                                        ----+---------+---------+---------+-----
                                          -0.60     -0.20      0.20      0.60

TUKEY STUDENTIZED RANGE METHOD           [4]

        SIGNIFICANCE AT
        ---------------
          1%  LEVEL  **
          5%  LEVEL  *
         10%  LEVEL  -             n   h   s   c
        >10%  LEVEL                o   s   o   o
         FOR   6 TESTS             _   _   m   l
                                   g   g   e   _
                                   r   r   _   g
                                   a   a   c   r
  GROUP                  SAMPLE    d   d   o   a
NO. LABEL         MEAN    SIZE             l   d
-----------      ------   ----    --- --- --- ---
  1 no_grad       0.93    17          **  **  **
  2 hs_grad       1.29    19      **          *
  3 some_col      1.38    10      **
  4 col_grad      1.52    22      **  *

                           STUDENT-NEWMAN-KEULS MULTIPLE RANGE TEST            [5]

                                95% CONFIDENCE LEVEL

                  no_grad  hs_grad  some_col  col_grad
                           __________________
      MEANS         0.93     1.29      1.38      1.52
                                    __________________
SAMPLE SIZE          17.      19.       10.       22.
```

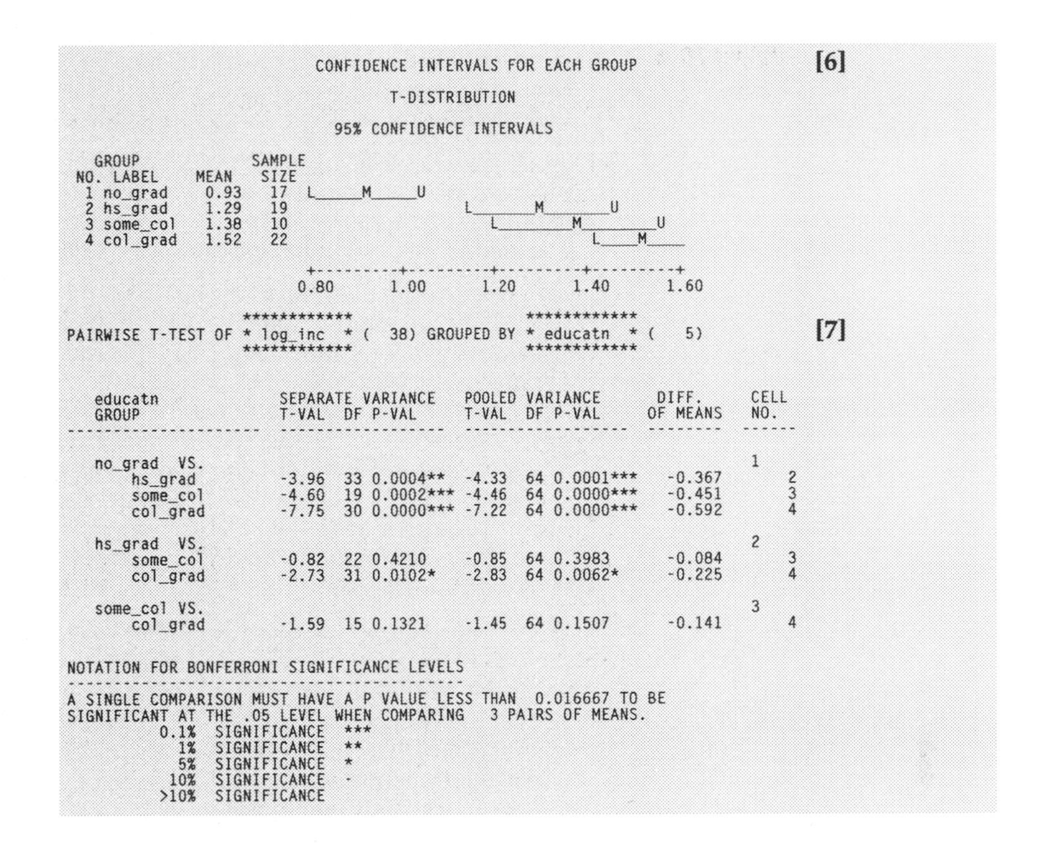

```
                    CONFIDENCE INTERVALS FOR EACH GROUP                          [6]
                              T-DISTRIBUTION
                         95% CONFIDENCE INTERVALS

  GROUP            SAMPLE
NO. LABEL    MEAN   SIZE
 1 no_grad   0.93    17  L_____M_____U
 2 hs_grad   1.29    19                L_______M_______U
 3 some_col  1.38    10                   L________M________U
 4 col_grad  1.52    22                             L____M____

                     +---------+---------+---------+---------+
                   0.80      1.00      1.20      1.40      1.60

                  ************               ************
PAIRWISE T-TEST OF * log_inc   * (  38) GROUPED BY * educatn   * (   5)          [7]
                  ************               ************

  educatn            SEPARATE VARIANCE    POOLED VARIANCE       DIFF.    CELL
  GROUP              T-VAL  DF P-VAL      T-VAL  DF P-VAL      OF MEANS   NO.
--------------------  -----------------   -----------------   --------  ------

  no_grad  VS.                                                              1
      hs_grad        -3.96  33 0.0004**  -4.33  64 0.0001***   -0.367        2
      some_col       -4.60  19 0.0002*** -4.46  64 0.0000***   -0.451        3
      col_grad       -7.75  30 0.0000*** -7.22  64 0.0000***   -0.592        4

  hs_grad  VS.                                                              2
      some_col       -0.82  22 0.4210    -0.85  64 0.3983      -0.084        3
      col_grad       -2.73  31 0.0102*   -2.83  64 0.0062*     -0.225        4

  some_col VS.                                                              3
      col_grad       -1.59  15 0.1321    -1.45  64 0.1507      -0.141        4

NOTATION FOR BONFERRONI SIGNIFICANCE LEVELS
---------------------------------------------------------------------------
A SINGLE COMPARISON MUST HAVE A P VALUE LESS THAN  0.016667 TO BE
SIGNIFICANT AT THE .05 LEVEL WHEN COMPARING   3 PAIRS OF MEANS.
           0.1%  SIGNIFICANCE  ***
             1%  SIGNIFICANCE  **
             5%  SIGNIFICANCE  *
            10%  SIGNIFICANCE  -
           >10%  SIGNIFICANCE
```

[1] Descriptive statistics for the four educational groups. Notice that the standard deviations of LOG_INC are fairly homogeneous.

[2] Since the LEVENE test of equality of variability indicates no significant departure ($p = 0.7871$), it is reasonable to proceed to the *post hoc* pairwise mean comparisons that assume homogeneity of variance.

[3] The 95% confidence intervals are shown for the Tukey method. Specifying CONFIDENCE produces a similar panel for each pairwise method requested. For each comparison, an L indicates the lower limit, an M the mean difference, and a U the upper limit of the interval. Two means differ significantly when the confidence interval for their difference does not contain zero (indicated by the vertical line of dots). Note that although the differences between the NO_GRAD vs. SOME_COL and the NO_GRAD vs. COL_GRAD groups are both significant at the .01 level, we can see from the confidence intervals that the latter difference is more clearly differentiated from zero than the former.

[4] This panel describes the output of the default TUKEY test. The standard Tukey procedure assumes equal sample sizes; however, for the unequal sample size situation, 7D uses the Tukey-Cramér adjustment, taking the harmonic mean of the groups when calculating the studentized range statistic, q. Asterisks at the intersection of a particular column and row represent the significance level (defined in the key on the upper left) of the difference between the respective group means. If no asterisk is present, then the p-value is greater than 0.05. The table shows that the NO_GRADS differ significantly ($p < .01$) from all other groups, and the HS_GRADS differ significantly ($p < .05$) from the COL_GRADS.

[5] The Student-Newman-Keuls test. This method first orders the means from smallest to largest. It then tests for differences between the largest and smallest pairs and connects, with horizontal lines, groups that do not differ significantly. The top line connects the HS_GRAD and SOME_COL groups, and the bottom line connects the SOME_COL and COL_GRAD groups indicating that neither of these mean differences are significant. Since no lines connect the NO_GRAD group with any other group, it differs significantly from the other three. And since neither line connects the HS_GRAD with the COL_GRAD groups, those two groups differ significantly from each other.

[6] Also shown, when CONFIDENCE has been stated, are the 95% confidence intervals for each group mean. The shorter the line, the more accurate the estimate of the group mean. The mean for the NO_GRAD group is more accurately estimated than for the SOME_COL group, for example.

[7] 7D gives pairwise *t* tests, both separate and pooled variance, when you use the TTEST command in the PRINT paragraph. Pooled-variance *t* tests are the same as those obtained using BON in the COMPARISON paragraph. The variance estimate in the denominator of the test statistic uses information from all four groups. Each separate-variance *t* test uses in its denominator only the variances of the two groups being compared; hence, each has fewer degrees of freedom than its corresponding pooled-variance *t* test. However, an advantage of this comparison is that it is more robust when group variances are unequal. The asterisks alongside each test represent the overall Bonferroni-adjusted significance levels: the *p*-values are for *t* tests **without** the Bonferroni correction.

We specified COMP = 3 because only three comparisons (the NO_GRAD vs. the other three groups) were of interest to us (therefore, we should not be scanning the table for the largest differences). Note that if you do not restrict the number of pairwise comparisons, 7D uses α/k to adjust the α levels, where k is the number of all possible pairwise comparisons. The COMP instruction does not influence the calculation of the *t* or *p* values in the display; it influences the adjustments on the significance levels (the asterisks that label the *p*-values). In this example, the conclusions reached using separate-variance and pooled-variance *t* tests are identical. This will not always be true.

Example 7D.5 User-specified contrasts and trend analysis

Often, you will have specific hypotheses about differences between particular groups. If your grouping factor has some structure, you may want to test particular hypotheses with *a priori* contrasts rather than multiple comparison tests. Consider our ongoing example. Suppose you had specific questions such as the following:

- Does completion of a college degree result in a noticeable increase in income over attendance at college without completion?
- Is the relationship between education and income positive and linear?
- Do individuals with at least some college experience have higher incomes than individuals with no college experience?

Rather than using multiple comparisons, you could test those hypotheses directly. Using contrasts, you test whether any linear combination of the means equals 0. You can test for specific contrasts among group means or for trends where

$$H_0: c_1\mu_1 + c_2\mu_2 + ... + c_k\mu_k = 0$$

You supply the contrast coefficients $c_1,....,c_k$. The coefficients $c_1,....,c_k$ must sum to zero. Contrast *F*'s are computed using a pooled variance estimate from all cells in the design.

In Input 7D.5, the contrast names are specified in the HISTOGRAM paragraph as CONTRAST = FIN_COL (finished college), LINEAR, ANY_COL (any college) with the contrast coefficients for each group entered as a list. Note that the order of the coefficients must match the group order. For our first contrast, the null hypothesis is:

$$\mu_{some\_col} = \mu_{col\_grad}$$

We rewrite it as

$$0 * \mu_{\text{no\_grad}} + 0 * \mu_{\text{hs\_grad}} - 1 * \mu_{\text{some\_col}} + 1 * \mu_{\text{col\_grad}} = 0$$

and specify the coefficients using the statement

```
FIN_COL = 0, 0, -1, 1.
```

To test for a linear trend across the ordered levels of education on income, we specify

```
LINEAR = -3, -1, 1, 3.
```

(Tables of polynomial contrast coefficients, often called orthogonal polynomials, can be found in Dixon and Massey, 1983, p. 344.) To test whether individuals with any college experience have higher incomes than individuals with no college experience, we add the name ANY_COL to the CONTRAST list, and specify:

```
ANY_COL = -1, -1, 1, 1.
```

The results for FIN_COL, LINEAR, and ANY_COL appear in Output 7D.5. Note that 7D allows you to request *conditional contrasts*, e.g., a quadratic contrast conditioned on a linear contrast. See Special Features for more information.

Input 7D.5
(replacement for the HISTOGRAM paragraph of Input 7D.4)

```
/ HISTOGRAM      GROUPING IS educatn.
                 VARIABLE IS log_inc.
                 CONTRASTS ARE fin_col, linear, any_col.
                 fin_col   =  0,  0, -1, 1.
                 linear    = -3, -1,  1, 3.
                 any_col   = -1, -1,  1, 1.
```

Output 7D.5

—we omit sections of output—

```
                ************                             ************
CONTRASTS FOR * log_inc   * (  38)   GROUPED  BY   * educatn   * (    5)
                ************                             ************

                    SAMPLE             C O E F F I C I E N T
    G R O U P        SIZE    MEAN  fin_col   linear    any_col
 ---------------   -------  -----  -------  --------  --------
   no_grad           17     0.928    0.0      -3.0      -1.0
   hs_grad           19     1.295    0.0      -1.0      -1.0
   some_col          10     1.379   -1.0       1.0       1.0
   col_grad          22     1.520    1.0       3.0       1.0

        CONTRAST VALUE               0.14      1.86      0.68

                                        F        TAIL
 CONTRAST                    D.F.    VALUE       PROB.
 ----------------------------------------------------
   fin_col                     1       2.12     0.1507
   linear                      1      49.19     0.0000
   any_col                     1      27.62     0.0000
```

The panel in Output 7D.5 shows that the difference between the incomes (LOG_INC—the transformed income) of male college graduates over 35 and those with some college is not significant ($p = .1507$), whereas the difference between the incomes of individuals with some college experience and those with no college experience (the third contrast) is highly significant. In addition, there is a highly significant linear increase in average income across the four ordered levels of education.

Example 7D.6 Box-Cox plots for selecting transformations

The classical analysis of variance assumes equality of group variances. As we have seen in previous examples, the data may not conform to that assumption. On some of those occasions, you may wish to use the Welch and Brown-Forsythe procedures. However, if the distribution within each group is skewed and the group variances appear to be correlated with group means, then a transformation, such as square root or logarithm, may help normalize the distributions and stabilize the variances. In this example, we illustrate the Box-Cox diagnostic plots, a data-screening tool that is designed to help you decide whether transforming the data would result in greater variance homogeneity.

When you request PLOT in the PRINT paragraph, two diagnostic plots are printed. The first plots the group mean versus the group standard deviation. The second plots the trimmed mean versus the winsorized standard deviation. However, if all means are positive, the logs of these values are plotted instead. The slope of the regression equation

$$\log(s_i) = B_0 + B_1 \log(\bar{X}_i)$$

is used to suggest an appropriate transformation. In the above expression, s_i is the estimate of the cell standard deviation, $\bar{X}_i$ is the cell mean, and B_1 is the slope. If a line with slope B_1 describes the relationship between log(s.d.) and log(mean) well, transforming the variable by raising it to the power $(1- B_1)$

$$y^{1-B_1}$$

should stabilize the variances (see Box and Cox, 1964, pp. 211–252).

To compute the Box-Cox diagnostic, 7D obtains the value of B_1 (given beneath the logarithmic plot, under the heading "REGRESSION LINE") and then uses the following table to obtain an approximate transformation:

Table 7D. 1
Box-Cox transformations

B_1	$1 - B_1$	Transformation of y to Stabilize Cell Variances
2	–1	reciprocal
0	0	logarithm
0.5	0.5	square root
0	1	(none suggested)

Since B_1 is only an estimate (perhaps a poor one for small sample sizes), you should generally not regard the suggested transformation, say $y^{.33}$, as an exact prescription, but take into consideration other information about the variable (such as skewness or the presence of outliers), or try the two neighboring transformations from the table. (For $1 - B_1 = .33$, you could try both log and square root transformations.)

In Input 7D.6, we state PLOT in the PRINT paragraph to request the Box-Cox procedure for income. Note that we specify SIZE = 45,18 to decrease the size of the Box-Cox plots.

Input 7D.6
(we add TRANSFORM and PRINT paragraphs to Input 7D.1)

```
/ TRANSFORM        IF(income EQ 200) THEN income = 4.

/ PRINT            PLOT.
                   SIZE = 45, 18.
```

Output 7D.6

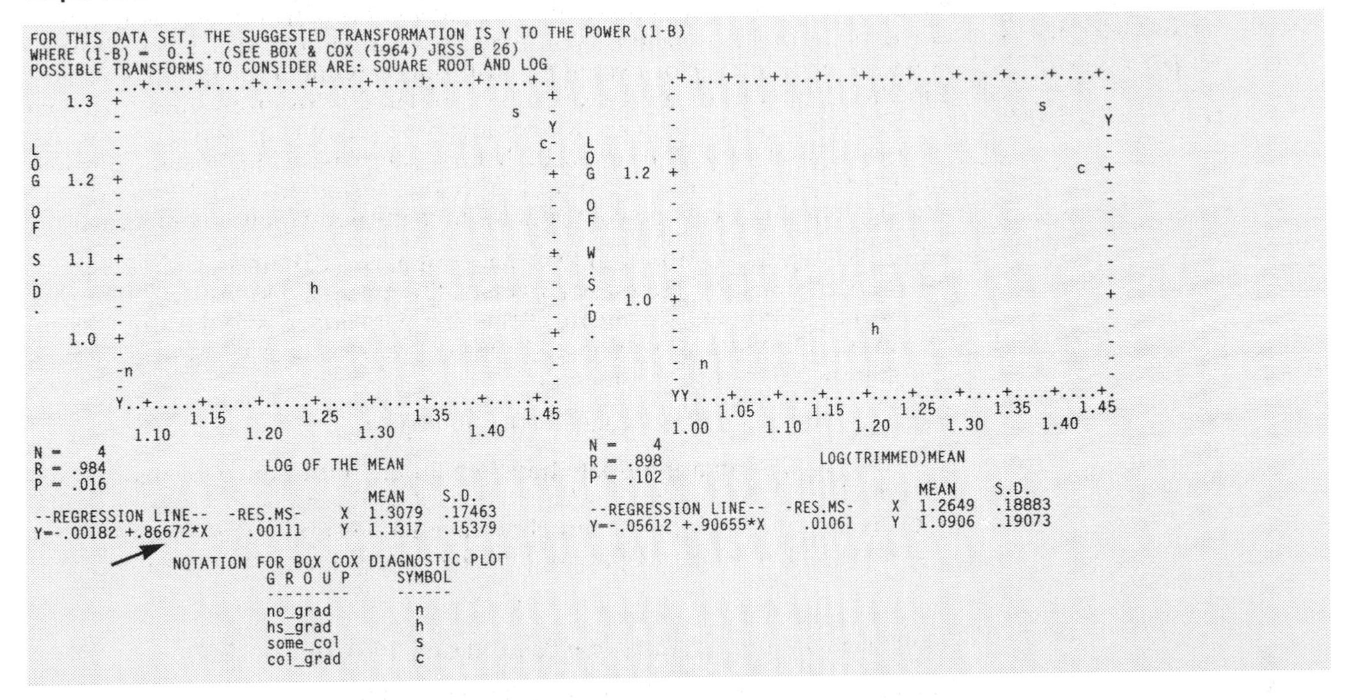

Output 7D.6 shows a plot of the log of estimates of the cell standard deviations against the log of the cell means. (Plot points are represented by the first letter of each group name, if unique; otherwise they are replaced by letters in alphabetical order.) This is the primary Box-Cox diagnostic. The transformation suggested for stabilizing the cell variances is determined by the slope of the regression line. The slope of .86672 (arrow) is close to 1.0, suggesting that a logarithmic transformation is in order. We chose the logarithmic transformation in our examples.

The plot for log trimmed mean versus log winsorized standard deviations may differ from the first plot if the largest and smallest observations differ markedly from the other observations.

Example 7D.7 Trimmed mean procedure

Outliers in a data set may have a marked effect on the value of the *F* ratio. One extremely high value can inflate both the sample mean and standard deviation. The result may be an incorrect rejection of the null hypothesis or a failure to find significance when there is an effect. You may want to see if your conclusions change due to the presence or absence of a few outliers by removing them from the analysis. To obtain a trimmed analysis of variance (reducing the influence of the largest and smallest values in each group), specify TRIM in the HISTOGRAM paragraph. This trims 15% (the default) of the cases from the top and bottom of each group. To trim percentages other than 15% (to a maximum of 25%), specify PERCENT in the HISTOGRAM paragraph (see COMMAND section). 7D then computes a trimmed version of the standard analysis of variance and of the Brown-Forsythe, and Welch analyses of variance, as well as the untrimmed analysis of variance statistics (see Appendix B.9).

To illustrate, we return to the data of Example 7D.1 which contained two outlying data points. We first restrict cases to males over 35 with USE = SEX EQ 1 AND AGE GT 35 (as in Input 7D.4). In Input 7D.7, we specify TRIM in the HISTOGRAM paragraph to do the trimming, request 20% trimming, and specify TRIM in the PRINT paragraph to obtain a trimmed statistic panel. Only the revised section of the input setup is shown below.

Input 7D.7
(we replace the HISTOGRAM paragraph in Input 7D.1 and add TRANSFORM and PRINT paragraphs)

```
/ TRANSFORM     USE = sex EQ 1 AND age GT 35.

/ HISTOGRAM     GROUPING IS educatn.
                VARIABLE IS income.
                PERCENT = 20.

/ PRINT         TRIM.

/ END
```

Output 7D.7

—we omit sections of output—

```
 ------------------------------------------------------------------------------   [1]
| ANALYSIS OF VARIANCE TABLE FOR MEANS                                    TAIL    |
|   SOURCE           SUM OF SQUARES    DF    MEAN SQUARE   F VALUE  PROBABILITY |
|   --------         --------------   ----   -----------   -------  ----------- |
|   educatn            11666.7275       3      3888.9092     3.81      0.0141    |
|     ERROR            65291.0825      64      1020.1732                         |
|---------------------------------------------------------------------------------|
| EQUALITY OF MEANS TESTS; VARIANCES ARE NOT ASSUMED TO BE EQUAL                  |
|   WELCH                              3,  28               17.52     0.0000     |
|   BROWN-FORSYTHE                     3,  24                4.38     0.0136     |
|---------------------------------------------------------------------------------|
| LEVENE'S TEST FOR VARIANCES          3,  64                5.33     0.0024     |
 ------------------------------------------------------------------------------

 ------------------------------------------------------------------------------
| ANALYSIS OF VARIANCE TABLE FOR MEANS WITH 20 PERCENT TRIMMING          TAIL     |
|   SOURCE           SUM OF SQUARES    DF    MEAN SQUARE   F VALUE  PROBABILITY |
|   --------         --------------   ----   -----------   -------  ----------- |
|   educatn             4293.4087       3      1431.1362    12.67      0.0000    |
|     ERROR             4156.5254      37       112.9491                         |
|---------------------------------------------------------------------------------|
| EQUALITY OF MEANS TESTS; VARIANCES ARE NOT ASSUMED TO BE EQUAL                  |
|   WELCH                              3,  15               25.77     0.0000     |
|   BROWN-FORSYTHE                     3,  22               12.79     0.0000     |
 ------------------------------------------------------------------------------

                    ************                          ************
TRIMMED MEANS OF * income     * (   7) GROUPED BY * educatn   * (   5)
                    ************                          ************

               TRIM    no_grad      hs_grad     some_col     col_grad
 ---------     ----   ---------    ---------    ---------    ---------
TRIMMED          0      10.176       44.368       27.200 *     36.864**   [2]
 MEAN            1       9.400       37.059       26.000**     36.650 *
                 2       8.692       27.667       26.667       36.278
                 3       7.727**     25.769**     27.500       35.563
               PCT       7.706 *     25.877 *     26.667       34.682
 ---------     ----   ---------    ---------    ---------    ---------
ROBUST SD                6.857       44.986       13.951       17.869     [3]
 ---------     ----   ---------    ---------    ---------    ---------
WINSORIZED       0       7.282       56.121       13.766 *     16.966**   [4]
  S. D.          1       6.984       59.462       10.944**     17.701 *
                 2       7.544       19.272       12.198       18.200
                 3       3.138**     10.086**     13.084       19.375
               PCT       3.508 *     11.049 *     12.198       13.795
 ---------     ----   ---------    ---------    ---------    ---------
1/2 LENGTH       0       3.744       27.049        9.847 *      7.522**   [5]
  OF THE         1       3.868       30.572        9.148**      8.284 *
95 PERCENT       2       4.559       10.672       12.802        9.050
CONFIDENCE       3       2.108**      6.095**     20.820       10.324
 INTERVAL      PCT       2.697 *      7.422 *     12.802        8.765
 ---------     ----   ---------    ---------    ---------    ---------
LARGEST          0      28.000      200.000       55.000 *     65.000**
 VALUE           1      23.000      200.000       35.000**     65.000 *
                 2      23.000       65.000       35.000       65.000
                 3      11.000**     35.000**     35.000       65.000
        PERCENTILE      11.000 *     35.000 *     35.000       45.000
 ---------     ----   ---------    ---------    ---------    ---------   [6]
SMALLEST         0       4.000       13.000        9.000 *     13.000**
 VALUE           1       5.000       15.000       13.000**     15.000 *
                 2       5.000       15.000       15.000       19.000
                 3       5.000**     15.000**     19.000       19.000
        PERCENTILE       5.000 *     15.000 *     15.000       23.000
 ---------     ----   ---------    ---------    ---------    ---------
 LARGEST         0      13           98          292          103
  CASE           1      39          176           62          162
NO./LABEL        2     211          121          146          166
                 3      79           33          287          270
 ---------     ----   ---------    ---------    ---------    ---------   [7]
SMALLEST         0     195          226          128          262
  CASE           1      28           36          285          129
NO./LABEL        2      51          130            2           31
                 3     173          181           69          123

PERCENT TRIM            20.0         20.0         20.0         20.0
CASES TRIMMED            3.4          3.8          2.0          4.4

NOTATION:  ** INDICATES VALUE CORRESPONDING TO THE SHORTEST CONFIDENCE INTERVAL
            * INDICATES VALUE CORRESPONDING TO THE NEXT SHORTEST

*** N O T E *** THE VALUE LISTED FOR PCT IS THAT WHICH IS OBTAINED AFTER TRIMING
                A PERCENTAGE (IN PERCENT TRIM LINE) OF CASES FROM EACH END.
```

[1] 7D performs a standard analysis of variance, as well as the Welch and Brown-Forsythe separate-variance tests on both untrimmed (first panel) and trimmed (second panel) data. Note that 20% of the cases have been trimmed from the ends of each group (approximately 27 cases have been removed) to yield the second analysis of variance table. The actual number of cases trimmed are shown at the bottom of the trimmed mean table (following **[7]**, below). The Winsorized standard deviations, which are pooled for the error term, are shown in a separate panel in that table (see discussion in **[4]**). Comparing the untrimmed with the trimmed results for the Welch and Brown-Forsythe tests (since the variances are not equal) shows that the *F*-values have been increased from 17.52 and 4.38 to 25.77 and 12.79, respectively.

[2] The TRIMMED MEAN panel gives means of each group after 0, 1, 2, and 3 cases, and 20% (PERCENT is the specified percentage) has been trimmed from the ends of each distribution. Note the very marked drop in mean for the HS_GRAD group when two cases are trimmed from each end. If the percentage requested exceeds the cases available, a warning is printed. The trimmed means with the shortest (**) and next-to-shortest (*) confidence intervals are indicated.

[3] The ROBUST S.D. is a robust estimate of the standard deviation based on the mean absolute deviation from the untrimmed mean.

[4] Winsorized standard deviations for each level of trimming (see Appendix B.9 and Dixon and Massey, 1983). The row labeled PCT shows the Winsorized standard deviations that are pooled for the error term used in the trimmed analysis of variance. (The Winsorized standard deviations are also used in the denominator of the *F* test with trimmed means.) In first level Winsorization, the smallest and largest observations are given the value of their nearest neighbors. Note that for HS_GRADS, the group with the two outliers, the Winsorized standard deviation drops from 56 with no cases trimmed to 19 with two cases trimmed from each end.

[5] Half-lengths of the 95% confidence intervals are given for each group at each level of trimming. The distributions with the shortest (**) and next-to-shortest (*) confidence intervals around their means are indicated. Note that the shortest confidence interval for the HS_GRAD group occurs when three cases are trimmed from each end.

[6] The extreme values remaining in the distributions after the indicated level of trimming are printed in the panels labeled LARGEST VALUE and SMALLEST VALUE. The PERCENTILE row corresponds to the PCT row of the previous panels. It gives the extreme value remaining after the outlying cases have been removed.

[7] The case numbers and labels for the largest and smallest values still present in the distributions after trimming the indicated number of cases. In the HS_GRAD distribution, for example, case number 98 has the highest income before trimming, but after two cases are trimmed, case number 121 has the highest income. Note that cases are identified by their sequence number in the data file. If we had specified a case labeling variable in the VARIABLE paragraph (e.g., LABEL = ID), 7D would print each case number followed by its label.

Example 7D.8 Running interactively

7D allows you to transform your data, modify the grouping instructions, and request HISTOGRAM, COMPARISON, and PRINT paragraphs interactively. You can specify any number of GROUP, TRANSFORM, HISTOGRAM, COMPARISON, and PRINT paragraphs after you specify INPUT, VARIABLE, and END paragraphs. In each case, the interactive paragraph should end with a slash, and the rest of the line after the slash should be blank. You can continue performing new transformations and analyses as long as you like; you only need to begin a new problem if you want to read in a different data set or change the defaults in the VARIABLE or SAVE, paragraphs. After the last interactive paragraph, type END/. Interactive mode allows you to perform the tests specified in one paragraph, view the results, and then decide whether to continue to the next paragraph or modify the last.

Imagine that the instructions for Example 7D.1 are stored in a file called 7D.INP. Enter 7D interactively (see Chapter 10). 7D will then print a banner and place you in the BMDP line editor at line 1. You may then type

```
GET 7D.INP
```

to GET your input file (see Chapter 10 for a list of line editor commands). After your file appears on the screen, type **E** for Execute to run the program. The output shown in Output 7D.1 will appear on your screen one portion at a time.

Use Insert mode in the line editor to add the interactive instructions shown following the first END paragraph in Input 7D.8. To exit insert mode, type //. When you execute this new set of instructions (**E #**, where # is the line number of the first new command), 7D will perform the analysis specified in each interactive paragraph. You can also run 7D from program control level, one paragraph at a time (see Chapter 10). **Note:** in the BMDP release for the PC, a full screen editor replaces the line editor. See your *PC User's Guide* for details.

In Input 7D.8 we use interactive paragraphs after the initial analysis to specify a new transformed variable and request an analysis using it in the HISTOGRAM paragraph. We also request additional output in the COMPARISON and PRINT paragraphs. (This is an interactive version of Example 7D.4.) COMPARISON and PRINT must come before HISTOGRAM. See Chapter 2 for information about interactive transformations. We omit the output for this example.

Input 7D.8

```
/ INPUT          VARIABLES = 3
                 FORMAT = FREE.
                 FILE = 'survey.dat'.

/ VARIABLE   NAMES =
 id,       sex,       age,      marital, educatn,  employ,
 income,   religion,  blue,     depress, lonely,   cry,
 sad,      fearful,   failure,  am_good, hopeful,  happy,
 enjoy,    bothered,  no eat,   effort,  badsleep, getgoing,
 mind,     talkless,  unfriend,dislike,  total,    case,
 drink,    healthy,   doctor,   meds,    bed days, illness,
 chronic.
                 USE = income, educatn.

/ GROUP          CUTPOINTS(educatn) = 2, 3, 4.
                 NAMES(educatn) = no_grad, hs_grad,
                                  some_col,col_grad.

/ HISTOGRAM      GROUPING = educatn.
                 VARIABLE = income.
/ END
 TRANSFORM       USE = sex EQ 1 AND age GT 35.
                 IF(income EQ 200) THEN income = 4.
                 log_inc = LOG(income).              /
 COMPARISON      NK.
                 CONFIDENCE.                /
 PRINT           TTEST.
                 COMPARISON = 3.     /
 HISTOGRAM       VARIABLE = log_inc. /
 END  /
```

Special Features

User-specified contrasts

You can use planned contrasts in both the one- and two-way analyses of variance. In the HISTOGRAM paragraph, include a CONTRAST statement containing a list of the contrast names, and specify the coefficients for each contrast name (see Example 7D.5). For example, in a one-way analysis of variance with four increasing levels of a treatment variable, you can specify a contrast between the means of groups 1 and 2, and the means of groups 3 and 4, by first naming the contrast

```
CONTRAST = LOW_HI.
```

and then specifying the coefficients:

```
LOW_HI = -1, -1, 1, 1.
```

For a two-way analysis of variance, you must pay particular attention to the order of the contrast coefficients. Use one coefficient for each cell in the analysis of variance design (each treatment level combination) in the same order as the cells appear in the side-by-side histogram plots. The order of the grouping variables in the HISTOGRAM paragraph determines the order of those cells (the first grouping variable is the slow moving index, and the second grouping variable is the fast moving index).

Suppose your two-way analysis of variance had two levels of treatment A and four levels of treatment B, and you had stated GROUP = A, B. Any contrast you perform would require 2 x 4 = 8 coefficients. If you wanted to test for the linear component of treatment B across both levels of A, you would specify

```
CONTRAST = LINEAR.
LINEAR = -3, -1, 1, 3, -3, -1, 1, 3.
```

If there was a significant interaction, and you wanted to test for the linear component of treatment B for only the second level of A, you would specify

```
CONTRAST = LINEAR2.
LINEAR2 = 0, 0, 0, 0, -3, -1, 1, 3.
```

You may request conditional contrasts by adding the TEST command in the HISTOGRAM paragraph. For example, to request a cubic contrast conditioned on a linear contrast when there are four ordered groups you would state:

```
/ HISTOGRAM   CONTRASTS = LINEAR, QUAD, CUBIC.
              LINEAR = -3, -1, 1, 3.
              QUAD = -1, 1, 1, -1.
              CUBIC = -1, 1, -1, 1.
              TEST = LINEAR, CUBIC.
```

To request quadratic conditioned on linear, and cubic conditioned on both quadratic and linear, add the statement

```
TEST = LINEAR, QUAD, CUBIC.
```

In addition, a combined test for all contrasts is given. Any number of TEST statements can be included in a HISTOGRAM paragraph. In general, the conditional contrasts do not give the same answer as the unconditional contrasts unless all contrasts are orthogonal and the sample sizes are equal.

Correlations within each group

7D computes correlations between the variables specified in the VARIABLE list group in the HISTOGRAM paragraph and all variables in the USE list of the VARIABLE paragraph. For each variable, you get a panel displaying its correlation with the other available variables. The correlations are computed for all the data combined and for the cases in each group separately.

Listing and sorting the data

When you specify DATA in the PRINT paragraph, the data for variables specified in the USE statement of the VARIABLE paragraph can be listed in the same order as read. When you specify ORDER in the PRINT paragraph, the data are also listed after the cases are ordered or sorted according to the GROUPING variable. If there are two grouping variables, the primary sort is on the first grouping variable. See program 1D and the BMDP Data Manager for more information about sorting.

7D Commands

/ VARiable

The VARIABLE paragraph is optional and is described in detail in Chapter 4.

New Feature

MMV = #. `MMV = 10.`

Specify the maximum number of missing values or values out of range that will be allowed before a case is excluded from the analysis. Default: use all acceptable values.

/ GROUP

Required if you want to name your groups to make the output easier to read, or if a grouping variable has more than ten distinct values. Define NAMES (and CUTPOINTS or CODES) in the GROUP paragraph (see Chapter 5).

/ HISTOgram

The HISTOGRAM paragraph is required and may be repeated before the END paragraph. May also be repeated interactively.

GROUPing = variable(s). `GROUP = SEX.`

Required. Specify one variable name or number for a one-way ANOVA; two for a two-way ANOVA. All analyses specified in subsequent HISTOGRAM paragraphs use these same groupings unless new GROUPINGS are specified.

VARiable = list. `VAR = INCOME, SQRT_INC.`

List the names or numbers of the variables to be analyzed. 7D uses the maximum number of cases available to analyze each VARIABLE listed. Default: 7D analyzes all variables specified in the USE list of the VARIABLE paragraph.

CONTrast = list. `CONT = LINEAR, QUAD.`

Specify names for group contrasts. Specify the coefficient for each contrast in a coefficient list (see below). Default: no group contrasts.

contrast name = #list.

TEST = list.

```
LINEAR = -3, -1, 1, 3.
QUAD = -1, 1, 1, -1.
TEST = LINEAR. QUAD.
```

Required to specify coefficients for each contrast named in the CONTRAST list. Use one coefficient for each cell in the ANOVA design in the same order as the cells appear in the side-by-side histogram plots. The example above shows coefficients for linear and quadratic contrasts across four ordered cell means of a one-way ANOVA. Conditional contrasts are requested by adding a TEST statement; in the above example, a quadratic contrast is conditioned upon a linear contrast.

TRIM. `TRIM.`

Requests trimmed versions of analysis of variance, Brown-Forsythe, and Welch statistics. Unless you state PERCENT, 15% trimming is used. Default: no trimmed tests.

PERCent = #. `PERC = 20.`

Specify the percent of **cases** to trim from the largest and the smallest values within each analysis of variance. (If TRIM is specified, the default is 15%.)

INCRement = # list. `INCR = 10, 1, (6)100.`

Specify the size of each histogram interval, one value per variable. Default: (maximum value – minimum value) number of intervals. **Example:** The width of the histogram intervals for the first variable is set to 10; the width of the histogram intervals for the second variable is set to 1 (good for single digit data); the width of the histogram intervals for the sixth input variable is set to 100.

INTerval = # list. `INTER = (3)25, (5)15.`

Specify the maximum number of intervals, one value for each input variable. The maximum number of intervals allowed is 30. If both INCR and INTER are specified for a variable, INCR is used. Example: The third input variable has 25 intervals, the fifth, 15. Default: ($10 \log_{10} n$), where n is the number of cases.

/ COMParison

The COMPARISON paragraph is optional, and may be repeated following each HISTOGRAM paragraph you specify (before the END paragraph). When repeating paragraphs interactively, COMPARISON must come before HISTOGRAM. It provides multiple comparison tests between group means and/or confidence intervals and multiple range tests. Two-way grouping structures are interpreted as a one-way structure. The TUKEY test is the default.

Pairwise tests and/or confidence intervals

For pairwise tests, specify one or more of the following test names to obtain a table of significance levels (α = .01, .05 and .10 for all tests except Dunnett's, which does not give .10). If you want only one or two levels printed, use ALPHA (see below). See SIGNIFICANCE to omit the table of significance levels. If, in addition, you want adjusted confidence intervals for the differences in group means, see CONFIDENCE below.

TUKey. `TUK.`

Unless you specify NO TUK, 7D prints significance levels for pairwise mean comparisons using the Tukey Studentized Range method. When the number of pairwise comparisons is very large, the Tukey procedure will be more powerful than the Bonferonni method. The opposite may be true when the total number of comparisons is small.

BONferroni. `BON.`

Prints Bonferroni significance levels for pairwise mean comparisons between group means. The Bonferroni tests may also be requested by stating TTEST in the PRINT paragraph. TTEST provides more detailed information than BONFERRONI.

SCHeffe. `SCH.`

Prints significance levels for pairwise mean comparisons using the Scheffé method of adjusting for the number of comparisons being made. (The Scheffé method is usually more conservative than the Tukey test.)

DUNNett. `DUNN.`

Prints significance levels (only α = .01 and α = .05 are available) for Dunnett's mean comparison between a single control group and any number of treatment groups. Groups may have unequal sample sizes if the control group has the largest sample size. Use with the CONTROL option below.

CONTrol = group. `CONT = STANDARD.`

Specify a group name or number to identify the control group with which the DUNNETT comparisons are made. Default: 7D will use the first group (as specified in the GROUP paragraph) as the CONTROL group. If there are two grouping variables, the first cell is used as the control.

Multiple range methods

7D generates displays for the following multiple range tests at the 95 percent confidence level. To specify additional displays at the 99 and/or 90 percent levels, see ALPHA below.

DUNCan. `DUNC.`

Display the Duncan multiple range tests for group means.

NK. `NK.`

Display the Student-Newman-Keuls multiple range test for group means.

VERT. `VERT.`

Print multiple range tests in a vertical format. Default: range tests are printed in horizontal format. If a panel is too long to be printed, the tests are printed in vertical format.

Significance levels and confidence limits

ALPHA = *(one to three values)* `ALPHA = 1, 5, 10.`
1, 5, 10.

Specify one to three alpha values ($\alpha = .01, .05, .10$) to generate displays of confidence intervals (see below) at $1 - \alpha = .99, .95, .90$. (For Dunnett's comparison, $\alpha = .10$ is not available.) Default: 5 ($\alpha = .05, 1 - \alpha = .95$).

CONFidence. `CONF.`

Specifying CONFIDENCE in the COMPARISON paragraph produces a display of confidence intervals (at the $1 - \alpha$ value(s) specified) for each pairwise comparison test requested as well as a display for each group based on the *t*-distribution (again at the $1 - \alpha$ value(s) specified). If ALPHA is not specified, the confidence level defaults to .95. Note that confidence intervals with Bonferroni adjustments are available only when you specify BONFERRONI with CONFIDENCE in the COMPARISON paragraph, not when you simply use the TTEST option in the PRINT paragraph.

SIGnificance. `NO SIG.`

Unless you state NO SIG, 7D displays panels and significance values for each pairwise comparison and ALPHA value specified in the COMPARISON paragraph.

/ PRINT

The PRINT paragraph is optional. All instructions in the PRINT paragraph are executed for each HISTOGRAM and COMPARISON paragraph specified. When repeating paragraphs interactively, PRINT must come before HISTOGRAM.

PLOT. `PLOT.`

Print two diagnostic plots. The first plots the group mean versus the group standard deviation. The second plots the trimmed mean versus the winsorized standard deviation. If all means are positive, the logs of these values are plotted. (see Example 7D.6).

SIZE = #1, #2. `SIZE = 45, 35.`

Specify number of characters for the plot width (#1) and number of lines for the plot height (#2). The default plot size is 50 characters wide and 35 lines high.

TTEST. `TTEST.`

Request pairwise mean differences with Bonferroni-adjusted significance levels. Prints separate and pooled t test values with significance levels adjusted by α/k, where k is the number of all possible pairwise comparisons unless restricted to a subset of pairwise comparisons with COMP, below.

COMP = #. `COMP = 4.`

Use to restrict the number of pairwise comparisons in TTEST, to be a subset of all possible comparisons. The Bonferroni-adjusted significance level then becomes $\alpha/\#$. If you omit COMP and specify TTEST, the Bonferroni adjustment is made for all possible pairwise comparisons.

TRIM. `TRIM.`

Prints up to three levels of trimmed means, and means trimmed according to the PERCENT you have specified in the HISTOGRAM paragraph. (If you do not specify PERCENT, 7D will trim 15 per cent.) For each level of trimming, 7D prints Winsorized standard deviations, half of the 95% confidence interval, and the smallest and largest values with case numbers and case labels.

DATA. `DATA.`

Print data in order of input for all variables specified in the USE statement of the VARIABLE paragraph, plus any new or transformed variables generated in the TRANSFORM paragraph.

ORDER. `ORDER.`

Prints data as ordered by the grouping variable(s). If you have two grouping variables, 7D will sort on the second grouping variable within the first. (For example, if HISTOGRAM GROUP = SEX, EDUCATN., then the data will be listed for the first category of SEX, and, within that, ordered by EDUCATION, and so on.) If you want a case label for the printout, specify a LABEL in the VARIABLE paragraph (e.g., LABEL = ID). Cases are then listed by case number and label. Data can also be sorted and listed in program 1D.

FIELDs = list. `FIELD = 3, 2*4, 9.1, -2.`

Controls the format of a data listing by specifying the number of character positions, decimal places, and whether a field is alphanumeric. A minus sign indicates an alphanumeric field. See Chapter 7 for a full explanation.

LINEsize = #. `LINE = 80.`

Specify the number of characters for the width of output. Default: LINESIZE is set to 132 for batch runs, and 80 for interactive runs. Note that 7D's output is arranged differently for narrow output.

CORRelation. `CORR.`

Print correlations between each VARIABLE in the HISTOGRAM paragraph and all variables in the USE statement of the VARIABLE paragraph for each GROUP and for all groups combined.

New Feature **CASE = #.** `CASE = 50.`

Specify the number of cases for which data are printed. Values flagged by MISS, MAX, and MIN in the VARIABLE paragraph specifications are replaced by MISSING, GT MAX, and LT MIN. Default: 10.

New Feature **MINimum.** `MIN.`

Prints data for cases in which at least one variable has a value less than its lower limit as specified in the VARIABLE paragraph. Default: such cases are not printed unless you specify DATA.

New Feature **MAXimum.** `MAX.`

Prints data for cases in which at least one variable has a value greater than its upper limit as specified in the VARIABLE paragraph. Default: such cases are not printed unless you specify DATA.

New Feature **MISSing.** `MISS.`

Prints data for cases that have at least one variable with a missing value. Missing values in the input may be identified in several ways: (1) an asterisk (the default missing value character), (2) a missing value code identified using MISS in the VARIABLE paragraph, (3) a character identified as MCHAR in the INPUT paragraph, (4) blanks (applies only to input read with FORTRAN fixed format). Default: such cases are not printed unless you specify DATA.

New Feature **METHod = VARiable** *or* **CASE.** `METH = CASE.`

Use to determine how the printout is organized. VARIABLE is the default value.

VARIABLE – Prints for each case only as many variables as fill the print line (up to ten variables) and begins the next line with the same variables for the next case. Thus the data for each variable are in an easy-to-scan column. When the program finishes printing the first subset of variables for all cases, it continues with another subset.

CASE – Prints the values of all variables for one case (possibly filling several print lines) before printing data for the next case.

The following commands work only in the batch mode

New Feature **MEAN.** `NO MEAN.`

Unless you specify NO MEAN, 7D prints the mean, standard deviation, and frequency of each used variable.

New Feature **EXTReme.** `NO EXTReme.`

Unless you state NO EXTR, 7D prints the minimum and maximum values for each variable used.

New Feature **EZSCores.** `EZSCores.`

Prints *z*-scores for the minimum and maximum values of each used variable.

New Feature **ECASe.** `ECAS.`

Prints the case numbers containing the minimum and maximum values of each variable.

New Feature **SKewness.** `SK.`

Prints skewness and kurtosis for each used variable.

Order of Instructions

■ indicates required paragraph

■ / INPUT
/ VARIABLE
/ GROUP
/ TRANSFORM
/ SAVE
■ / HISTOGRAM
/ COMPARISON
/ PRINT
■ / END
data
GROUP ... /
TRANSFORM ... /
COMPARISON ... /
PRINT ... /
HISTOGRAM ... /
END /

(HISTOGRAM and COMPARISON: Repeat for subproblems)

(GROUP ... / through HISTOGRAM ... /: May be repeated interactively)

(INPUT through END /: Repeat for additional problems. See Multiple Problems, Chapter 10.)

Summary Table for Commands Specific to 7D

	Paragraphs Commands	Defaults	Multiple Problems	See
	/ VARiable			
	MMV = #.	# of variables	–	Cmds
	/ HISTogram			
■	GROUPing = list.	none	●	7D.1
▲	VARiable = list.	all in VAR USE statement	●	7D.1
	CONTrast = list.	no contrasts	–	7D.5
	contrast name = # list.	(required with CONTRAST named)	–	7D.5
	TEST = list.	NO TEST. (use only with CONTRAST)	–	SpecF
	TRIM.	NO TRIM.	●	7D.7
	PERCent = #.	15%; use with TRIM	●	7D.7
	INCRement = # list.	(max value – min value)/ # of histogram intervals	●	Cmds
	INTERval = # list.	10 $\log_{10} n$, where n is number of cases; maximum histogram intervals allowed is 30; INCR overrides INTER	●	Cmds
	/ COMParison			
	TUKey *or* NO TUK.	TUK.	●	7D.3
	BONferroni.	NO BON.	●	Cmds
	SCHeffe.	NO SCH.	●	Cmds
	DUNNett.	NO DUNN.	●	Cmds
	CONTrol = group. (*use with DUNNETT*)	first group as specified in GROUP paragraph	●	Cmds
	DUNCan.	NO DUNC.	●	Cmds
	NK.	NO NK.	●	7D.3
	VERT.	NO VERT.	●	Cmds
	ALPHA = (*1 to 3 values*) 1, 5, 10.	5 (α = .05)	●	Cmds
	CONFidence.	NO CONF.	●	7D.3
	SIGnificance *or* NO SIG.	SIG.	●	Cmds

Summary Table (continued)

Paragraphs Commands	Defaults	Multiple Problems	See
/ **PRINT**			
CORRelations.	NO CORR.	•	7D.6
TTEST.	NO TTEST.	•	7D.3
COMParison = #.	all possible pairwise mean comparisons; use with TTEST	•	7D.3
TRIM.	NO TRIM.	•	7D.7
PLOT.	NO PLOT.	•	7D.6
SIZE = #, #.	50 characters wide and 35 lines tall; use with PLOT	•	7D.6
DATA.	NO DATA.	•	Cmds
ORDER.	NO ORDER.	•	Cmds
FIELDs = list.	FORMAT if specified; otherwise 8.2 for numeric, –4 for label variables	–	Cmds
LINEsize = #.	132 noninteractive; 80 interactive	•	7D.3
CASE = #.	10	•	Cmds
MISSing.	NO MISS.	•	Cmds
MINimum.	NO MIN.	•	Cmds
MAXimum.	NO MAX.	•	Cmds
METHod = VARiables *or* CASE.	VARIABLES	–	Cmds
MEAN. *or* NO MEAN.	MEAN.	•	Cmds
EXTReme. *or* NO EXTR.	EXTReme values printed	•	Cmds
EZSCores.	NO EZSC. (*z*-scores)	•	Cmds
ECASe.	NO ECAS. (Case #'s of Min and Max printed)	•	Cmds
SKewness.	NO SK. (skewness and kurtosis)	•	Cmds

Key:
- ■ Required command or paragraph
- ▲ Frequently used command or paragraph
- ● Value retained for multiple problems
- – Default reassigned

9D

Multiway Description of Groups

Program 9D produces small "miniplots" of cell means and computes descriptive statistics for data that are classified into cells by one or more grouping variables. Miniplot options allow you to plot the cell means from a factorial design with two or more factors, means from repeated measures designs, or means of two or more variables simultaneously. Cell means, standard deviations, and frequencies are printed in vertical columns to help detect extreme values and compare cell values across variables. Tests printed for each variable include a one-way ANOVA for testing equality of cell means and a χ^2 test for equality of cell frequencies. You can request plots that show the shift in means from group to group, and you can plot marginal subsets of the data.

9D is a screening device useful for assessing the homogeneity of subgroups defined by the cross classification of several variables. It is also a useful support for an analysis of variance, as the miniplots are helpful for examining trends and interactions among cell means.

Where to Find It

Examples

Special Features

Example 9D.1 Displaying means for cells defined by three grouping variables: the GPLOT miniplot

We use 9D to examine average income for groups of subjects cross-classified by education, sex, and religion. We analyze survey data from Afifi and Clark (see Appendix D). The record for each subject includes an ID number, a set of mental and physical health variables, and a set of demographic variables (sex, age, education, income, etc.). We name only the first eight of the 37 variables recorded for each subject. The data are stored on disk in a file named SURVEY.DAT.

The INPUT and VARIABLE paragraphs in Input 9D.1 contain instructions common to all BMDP programs for describing the data and naming the variables. (The FILE command tells the program where to find the data and is used for systems like VAX/VMS and the IBM PC. For IBM mainframes, see UNIT, Chapter 3.) See Chapters 3 and 4 for more information about the INPUT and VARIABLE paragraphs. We use a GROUP paragraph to specify CODES, CUTPOINTS, and NAMES for the grouping variables (see Chapter 5). By using CUTPOINTS, we collapse the EDUCATN variable from seven original codes to four broader categories. We also include a TRANSFORM paragraph to eliminate retired persons from the analysis.

The TABULATE paragraph is specific to 9D. We use it to specify the type of miniplot desired, the variable to be plotted on the y-axis, and the grouping variables used to form cells. The order of the variables in the GROUPING list determines how the GPLOTS are formed. The first grouping variable (education) defines the levels on the x-axis. The second grouping variable (sex) defines the symbols used in the plots: 9D uses the first characters of the group names. The levels of the remaining grouping variable(s) define the plot frames. If you specify more than one variable in the VARIABLE list, 9D prints one set of plots for each variable.

Input 9D.1

```
/ INPUT      VARIABLES = 37.
             FORMAT = FREE.
             FILE = 'survey.dat'.

/ VARIABLE   NAMES = id, sex, age, marital, educatn,
                     employ, income, religion.

/ GROUP      CODES(sex) = 1, 2.
             NAMES(sex) = male, female.
             CODES(religion) = 1 TO 4.
             NAMES(religion) = protest, catholic,
                               jewish, none.
             CUTP(educatn) = 2, 3, 4.
             NAMES(educatn) = no_grad, hs_grad,
                              some_col, col_grad.

/ TRANSFORM USE = age LE 65.

/ TABULATE   VARIABLE = income.
             GROUPING = educatn, sex, religion.
             GPLOT.
/ END
```

Output 9D.1

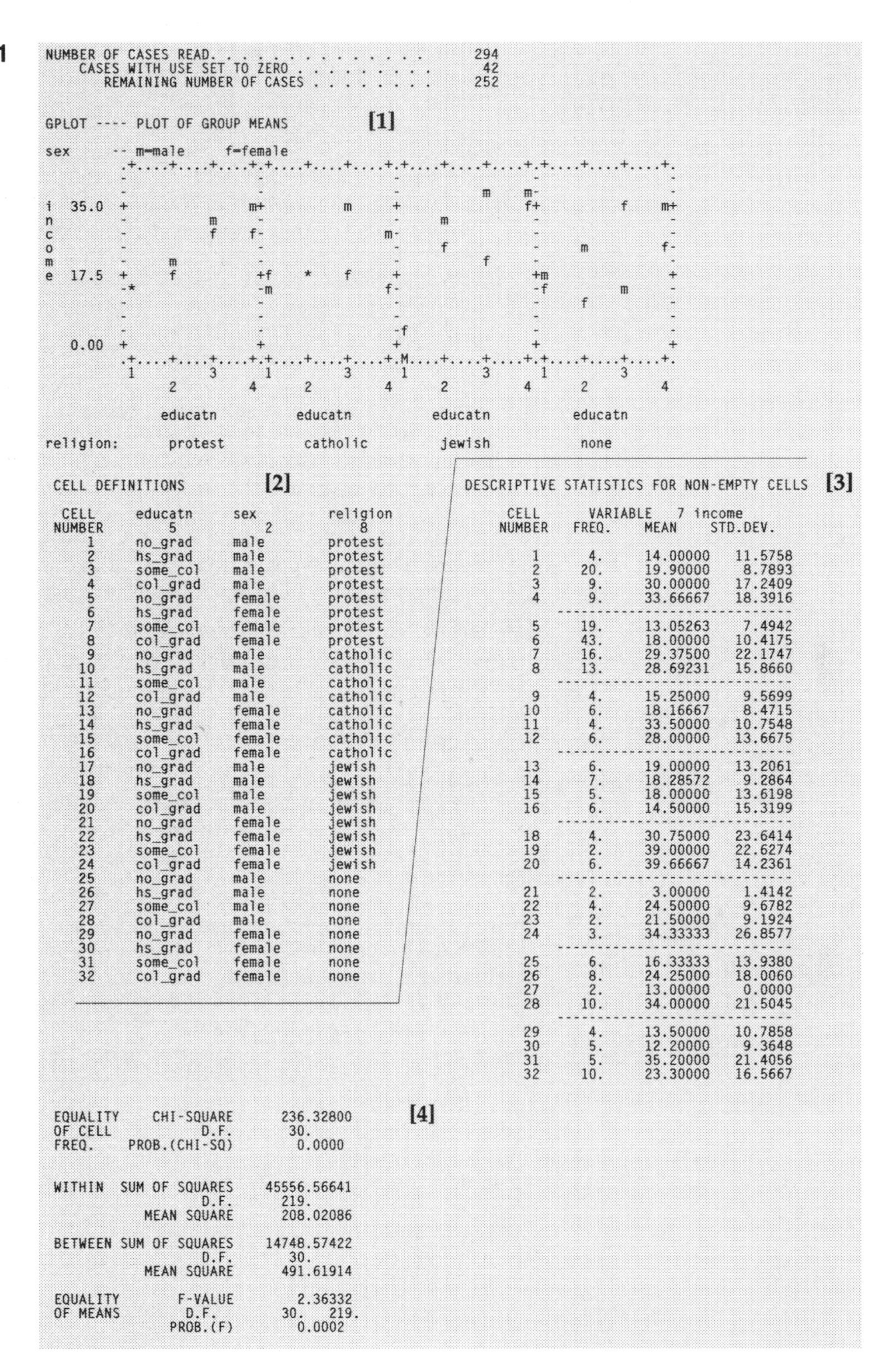

```
NUMBER OF CASES READ. . . . . . . . . . . . . . .        294
    CASES WITH USE SET TO ZERO . . . . . . . . .          42
       REMAINING NUMBER OF CASES . . . . . . . .         252

GPLOT ---- PLOT OF GROUP MEANS            [1]

sex     -- m=male     f=female
         .+....+....+....+.+....+....+....+.+....+....+....+.+....+....+....+.
         -                -                -                -                -
         -                -                -          m    m-                -
i  35.0 +               m+          m     +               f+          f    m+
n        -          m     -                -     m          -                -
c        -          f    f-               m-                -                -
o        -                -                -     f          -     m         f-
m        -     m          -                -          f     -                -
e  17.5 +     f          +f    *    f     +                +m               +
         -*               -m              f-                -f         m     -
         -                -                -                -     f          -
         -                -                -                -                -
         -                -                -f               -                -
   0.00 +                +                +                +                +
         .+....+....+....+.+....+....+....+.M....+....+....+.+....+....+....+.
          1         3      1         3      1         3      1         3
               2         4      2         4      2         4      2         4

              educatn          educatn         educatn           educatn

religion:      protest          catholic        jewish            none
```

CELL DEFINITIONS **[2]**

CELL NUMBER	educatn 5	sex 2	religion 8
1	no_grad	male	protest
2	hs_grad	male	protest
3	some_col	male	protest
4	col_grad	male	protest
5	no_grad	female	protest
6	hs_grad	female	protest
7	some_col	female	protest
8	col_grad	female	protest
9	no_grad	male	catholic
10	hs_grad	male	catholic
11	some_col	male	catholic
12	col_grad	male	catholic
13	no_grad	female	catholic
14	hs_grad	female	catholic
15	some_col	female	catholic
16	col_grad	female	catholic
17	no_grad	male	jewish
18	hs_grad	male	jewish
19	some_col	male	jewish
20	col_grad	male	jewish
21	no_grad	female	jewish
22	hs_grad	female	jewish
23	some_col	female	jewish
24	col_grad	female	jewish
25	no_grad	male	none
26	hs_grad	male	none
27	some_col	male	none
28	col_grad	male	none
29	no_grad	female	none
30	hs_grad	female	none
31	some_col	female	none
32	col_grad	female	none

DESCRIPTIVE STATISTICS FOR NON-EMPTY CELLS **[3]**

CELL NUMBER	VARIABLE 7 income FREQ.	MEAN	STD.DEV.
1	4.	14.00000	11.5758
2	20.	19.90000	8.7893
3	9.	30.00000	17.2409
4	9.	33.66667	18.3916
5	19.	13.05263	7.4942
6	43.	18.00000	10.4175
7	16.	29.37500	22.1747
8	13.	28.69231	15.8660
9	4.	15.25000	9.5699
10	6.	18.16667	8.4715
11	4.	33.50000	10.7548
12	6.	28.00000	13.6675
13	6.	19.00000	13.2061
14	7.	18.28572	9.2864
15	5.	18.00000	13.6198
16	6.	14.50000	15.3199
18	4.	30.75000	23.6414
19	2.	39.00000	22.6274
20	6.	39.66667	14.2361
21	2.	3.00000	1.4142
22	4.	24.50000	9.6782
23	2.	21.50000	9.1924
24	3.	34.33333	26.8577
25	6.	16.33333	13.9380
26	8.	24.25000	18.0060
27	2.	13.00000	0.0000
28	10.	34.00000	21.5045
29	4.	13.50000	10.7858
30	5.	12.20000	9.3648
31	5.	35.20000	21.4056
32	10.	23.30000	16.5667

```
EQUALITY     CHI-SQUARE        236.32800          [4]
OF CELL            D.F.         30.
FREQ.     PROB.(CHI-SQ)          0.0000

WITHIN  SUM OF SQUARES       45556.56641
                  D.F.         219.
           MEAN SQUARE         208.02086

BETWEEN SUM OF SQUARES       14748.57422
                  D.F.          30.
           MEAN SQUARE         491.61914

EQUALITY        F-VALUE          2.36332
OF MEANS           D.F.       30.    219.
              PROB.(F)           0.0002
```

Bold numbers below correspond to those in Output 9D.1.

[1] 9D plots the means for each RELIGION category in separate plots. Within each plot, the cell means for INCOME are plotted along the y-axis, and the four education groups along the x-axis. Note that the values printed on the x-axis are the indices of the groups and not the group codes or cutpoints. The first letters of the SEX groups label their respective cell means. Note: These plots are produced in high resolution on VAX, PC and XWindows platforms.

Examine the miniplots for patterns. From the first plot we see that for Protestants, income appears to increase with level of education. Men tend to make more than women, although an asterisk at one point indicates that their means coincide. Outliers are also important: the third plot shows a very low income for Jewish women with low education. An "M" on the x-axis stands for Missing and indicates that there are no Jewish males in the lowest education group. In the last plot, one point shows an income for females much higher than that for the corresponding males. We examine these cells more closely by looking at their frequencies and standard deviations in the Descriptive Statistics panel **[3]**.

[2] 9D assigns a number to identify each cell and prints a compact description of the cell. The 32 cells are formed using 2 levels of SEX, 4 levels of EDUCATN, and 4 levels of RELIGION (2 x 4 x 4 = 32).

[3] Frequency, mean, and standard deviation for each cell defined in the preceding panel. There are no statistics for Cell 17, Jewish males with low education, since the cell is empty. Cell 21, Jewish women with low education, has only two women; the small sample size may account for the very low average income. Cell 31, which has an atypically high income, has a large standard deviation of 21.4. From this panel we can see that cells with large means tend to have large standard deviations; this does not agree with the assumption of equality of variances for an analysis of variance.

[4] The data in the cells are compared by

- a χ^2 test of the equality of cell frequencies
- a test of the equality of cell means; i.e., a one-way analysis of variance. The components are WITHIN SUM OF SQUARES, WITHIN MEAN SQUARE, BETWEEN SUM OF SQUARES, BETWEEN MEAN SQUARES, F-VALUE, and PROB(F) = probability of exceeding this F.

The F value (ANOVA) is a test of the equality of the 31 cell means (one cell is empty). See Appendix B.9 (program 7D) for an explanation of these statistics.

Example 9D.2 Plotting cell means for a Latin square design: GPLOT

In this example we plot means from a 3-way Latin square design where each factor has 4 levels. If responses had been measured at all possible combinations of the levels, we would have results for 64 cells (4 x 4 x 4). A 4-level Latin square design allows estimates of the main effects when data are collected for only 16 of the 64 cells (see Example 2V.3 for more about Latin square designs and a complete analysis of variance using these data). We use data from an industrial problem evaluating the durability of four rubber compounds used in tires (Kirk, 1982). We want to study thickness of tread remaining after 10,000 miles of driving, measured for 4 rubber compounds on 4 different automobiles and 4 wheel positions. The experiment uses a Latin square to control for two variables: type of automobile and wheel position (right-front, right-rear, left-front, left-rear). We have added names for the four types of car. A diagram of the Latin square is shown in Figure 9D.1.

Figure 9D.1
Latin square design for rubber compound experiment

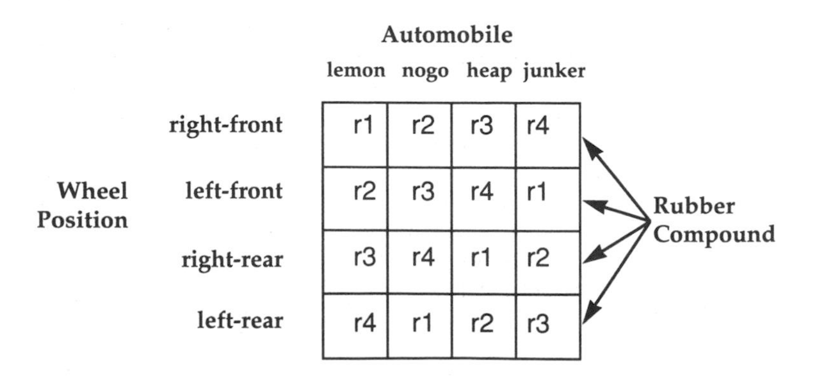

The data file, TIRE.DAT, contains one record for each tread measurement with four variables: codes for the three GROUPING variables RUBBER, WHEEL, and AUTO, and the dependent variable, TREAD. In the GROUP paragraph in Input 9D.2, we specify the codes used to identify the four types of rubber, four wheel positions, and four types of auto.

The plot structure is defined in the TABULATE paragraph as follows:

y-axis	– TREAD (the VARIABLE)
x-axis	– the four levels of RUBBER (RUBBER is first in the GROUP list)
symbols	– the first letters of the AUTO group names (AUTO is second in the GROUP list)
frames	– one for each WHEEL (last in the GROUP list)

Input 9D.2

```
/ INPUT      VARIABLES = 4.
             FORMAT = FREE.
             FILE = 'tire.dat'.

/ VARIABLE   NAMES = rubber, wheel, auto, tread.

/ GROUP      CODES(rubber, wheel, auto) = 1, 2, 3, 4.
             NAMES(wheel) = r_front, l_front,
                            r_rear, l_rear.
             NAMES(auto) = lemon, nogo, heap, junker.

/ TABULATE   VARIABLE = tread.
             GROUPING = rubber, auto, wheel.
             GPLOT.

/ PRINT      LINESIZE = 80.

/ END
```

Output 9D.2

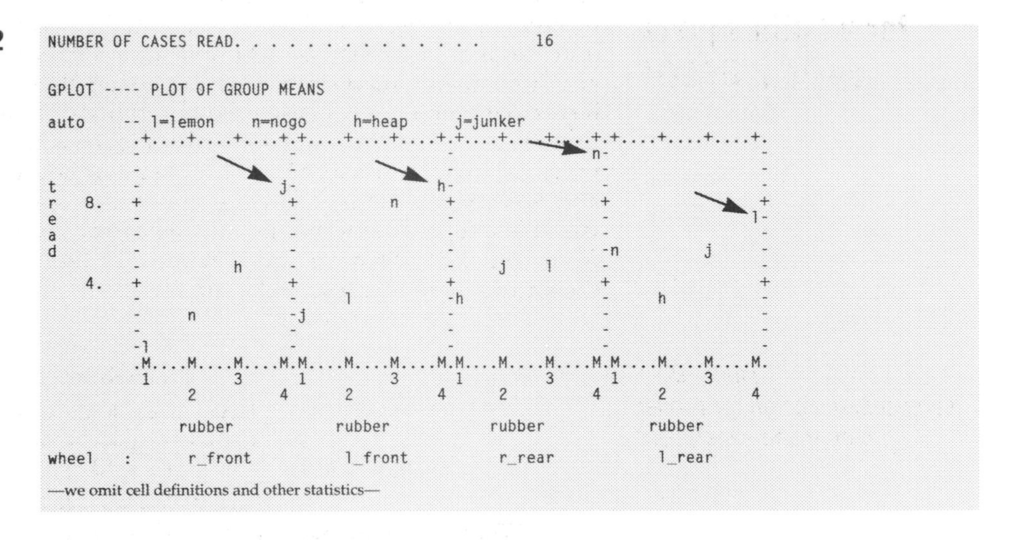

```
NUMBER OF CASES READ. . . . . . . . . . . . . .          16

GPLOT ---- PLOT OF GROUP MEANS
auto     -- l=lemon    n=nogo      h=heap      j=junker

     rubber            rubber            rubber            rubber
wheel   :    r_front           l_front           r_rear            l_rear
```

—we omit cell definitions and other statistics—

From the miniplots in Output 9D.2, we see that the fourth rubber compound always proves most durable (the most tread remains after testing), regardless of the auto or wheel position.

Example 9D.3 An incomplete block design with four factors: GPLOT

John (1971, p. 135) discusses an example of a partially confounded incomplete block design used for an agricultural experiment measuring potato crop yield. The data and the design are also described in Example 2V.4. The cells in the experiment are defined using four factors: three design factors and one blocking factor. The design factors are sulphate of ammonia, sulphate of potash, and nitrogen. Each factor has two levels (yes and no), so there are 8 possible combinations. The three factors were tested in eight locations (blocks), so a total of 64 measurements (cells) could be collected. This design is "incomplete" because only half of these experiments were done (see the figure accompanying Example 2V.4).

Each record in the file POTATO.DAT contains the potato yield in pounds along with four codes to indicate its block and the presence or absence of the three chemicals. The data are displayed in Example 2V.4.

The plot structure is defined in the TABULATE paragraph as follows:

y-axis	– YIELD (the VARIABLE)
x-axis	– the eight BLOCKs (BLOCK is first in the GROUP list)
symbols	– Yes or No for AMMONIA (second in the GROUP list)
frames	– four Yes/No combinations of POTASH and NITROGEN (these variables are last in the GROUP list)

Input 9D.3

```
/ INPUT      FILE = 'potato.dat'.
             FORMAT = FREE.
             VARIABLES = 5.

/ VARIABLE   NAMES = block, ammonia, potash, nitrogen,
                     yield.

/ GROUP      CODES(ammonia, potash, nitrogen) = 0, 1.
             NAMES(ammonia, potash, nitrogen) = no, yes.

/ TABULATE   GPLOT.
             VAR = yield.
             GROUPING = block, ammonia, potash, nitrogen.
/ END
```

Output 9D.3

```
NUMBER OF CASES READ. . . . . . . . . . . . . .        32

GPLOT ---- PLOT OF GROUP MEANS

ammonia -- n=no       y=yes
         .......+.......+........+.......+........+.......+........+.......+..
         -                -                -                -                -
         -                -                -                -  y y   y   n   -
y        -                -                -                -      n   n y   -
i        -                -y               -                -n               -
e   350  +                +      y       y +          y     +                +
l        -                -  n n   y n   n -    y         n -                -
d        -                -                -y n   n n     y -                -
         -                -                -                -                -
         -                -                -                -                -
    175  +                +                +                +                +
         -          y n   -                -                -                -
         -n y n y     y   -                -                -                -
         -        n       -                -                -                -
         .M.M.M.M.M.M...M..M.M.M.M.M.M.M.+..M.M.M.M.M.M.M.+..M.M.M.M.M.M...M..
                4                4                4                4
                        8                8                8                8

              block            block            block            block

potash  :      no               no               yes              yes
nitrogen:      no               yes              no               yes
```

— we omit the cell definitions, descriptive statistics for non-empty cells and the panel of tests —

Because the design of this experiment is incomplete, the miniplots displayed in Output 9D.3 include empty cells designated as missing by "M" on the plot frames. As we might expect, the potato yield is highest for cells with more nutrients added. The miniplots show little variation in yield among the eight blocks.

Example 9D.4 A repeated measures ANOVA design: RPLOT

The RPLOT option allows you to plot means when a single variable is measured two or more times for each subject. All repeated measures must be on one record. If the repeated measures are recorded on six lines in the file, your format statement must read all six lines as one record.

In this repeated measures ANOVA design, a number of animals have been measured six times (at 30, 60, 90, 120, 150, and 180 minutes). Animals are grouped by their prenatal exposure to alcohol (CONTROL or ETHANOL) and by exposure as an adult animal to a DRUG: VEHICLE (a control group) or MORPHINE. Each input record contains data for one animal: 6 responses plus codes for PRENATAL treatment and DRUG. The six responses are named Y_30, Y_60, Y_90, Y_120, Y_150, and Y_180. We do not show the data for this example.

In Input 9D.4, the GROUP paragraph defines and names the levels of the factors. In the TABULATE paragraph, RPLOT generates one position on the x-axis for each variable in the VARIABLE list: we specify Y_30 to Y_180 as the six repeated measurements. The first position is for the first variable in the list (Y_30); the second, for Y_60, etc. The value of the mean for each of these repeated measures is given on the y-axis. We use GROUPING to identify TREATMENT and DRUG as the grouping factors. RPLOT uses the group name of the **first** grouping variable, TREATMNT, to define plot symbols. RPLOT also generates frames using all combinations of the **second and remaining** variables in the list. Here, 9D produces two plots, one for each level of DRUG (VEHICLE and MORPHINE).

Input 9D.4

```
/ INPUT      VARIABLES = 8.
             FORMAT = FREE.

/ VARIABLE   NAMES = y_30, y_60, y_90, y_120, y_150,
                     y_180, treatmnt, drug.

/ GROUP      CODES(treatmnt) = 0, 1.
             NAMES(treatmnt) = control, ethanol.
             CODES(drug) = 0, 1.
             NAMES(drug) = vehicle, morphine.

/ TABULATE   RPLOT.
             VARIABLES = y_30 TO y_180.
             GROUPING = treatmnt, drug.
/ END
```

Output 9D.4

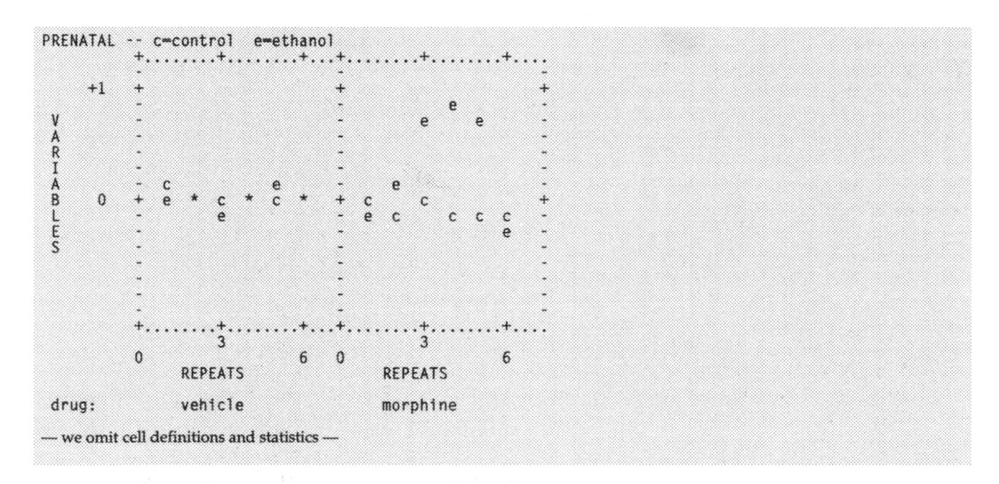

In Output 9D.4, RPLOT generates two miniplots for the two levels of DRUG (VEHICLE and MORPHINE). The cell means of the repeated measure (Y_30 to Y_180) are plotted on the *y*-axis against their order in the VARIABLE list on the *x*-axis. (Y_180, the sixth element in the list, is designated as 6 on the *x*-axis.) The plot symbols represent the two levels of TREATMNT (C for CONTROL, E for ETHANOL). 9D prints an asterisk when the plotting positions of the two groups coincide in the plot frame. We see that the animals exposed prenatally to ethanol (denoted by E's in the plot) exhibit a marked response to morphine that is not found for the control animals (C) or for either group in the VEHICLE experiment.

Example 9D.5 More than one variable per frame: VPLOT

VPLOTS are similar to GPLOTS, except that the cell means of all variables in the TABULATE VARIABLE list are plotted along the *y*-axis in each plot, and the symbols identify the variables instead of the categories within a variable. When you want to analyze multiple variables in a complex factorial design, VPLOT displays can help you examine the relationships between the variables for each subgroup in the design structure.

We generate plots for the survey data presented in Example 9D.1 showing three depression measures simultaneously for different groups. In the GROUP paragraph in Input 9D.5, we classify subjects by AGE and SEX (the three cutpoints on AGE produce four groups). In the TABULATE paragraph, we specify DEPRESS and CRY as the VARIABLES to be plotted. VPLOT uses all variables in the VARIABLE list to define the *y*-axis of the plots. The plot symbols will be the first characters of the variable names. We specify AGE and SEX as the two GROUPING factors. The order of the variables in the GROUPING list determines how the plots are formed. The first variable, AGE, defines the *x*-axis; the second and any remaining variables form the plot frames (two plot frames will be formed for the two levels of SEX). See Special Features for hints on how to generate one frame per case.

Input 9D.5

```
/ INPUT      VARIABLES = 37.
             FORMAT = FREE.
             FILE = 'survey.dat'.

/ VARIABLE   NAMES = id, sex, age, marital, educatn,
                     employ, income, religion,
                     (10)depress, (12)cry.

/ GROUP      CUTP(age) = 30, 50, 65.
             CODES(sex) = 1, 2.
             NAMES(sex) = male, female.

/ TABULATE   VAR = depress, cry.
             GROUPING = age, sex.
             VPLOT.
/ END
```

Output 9D.5

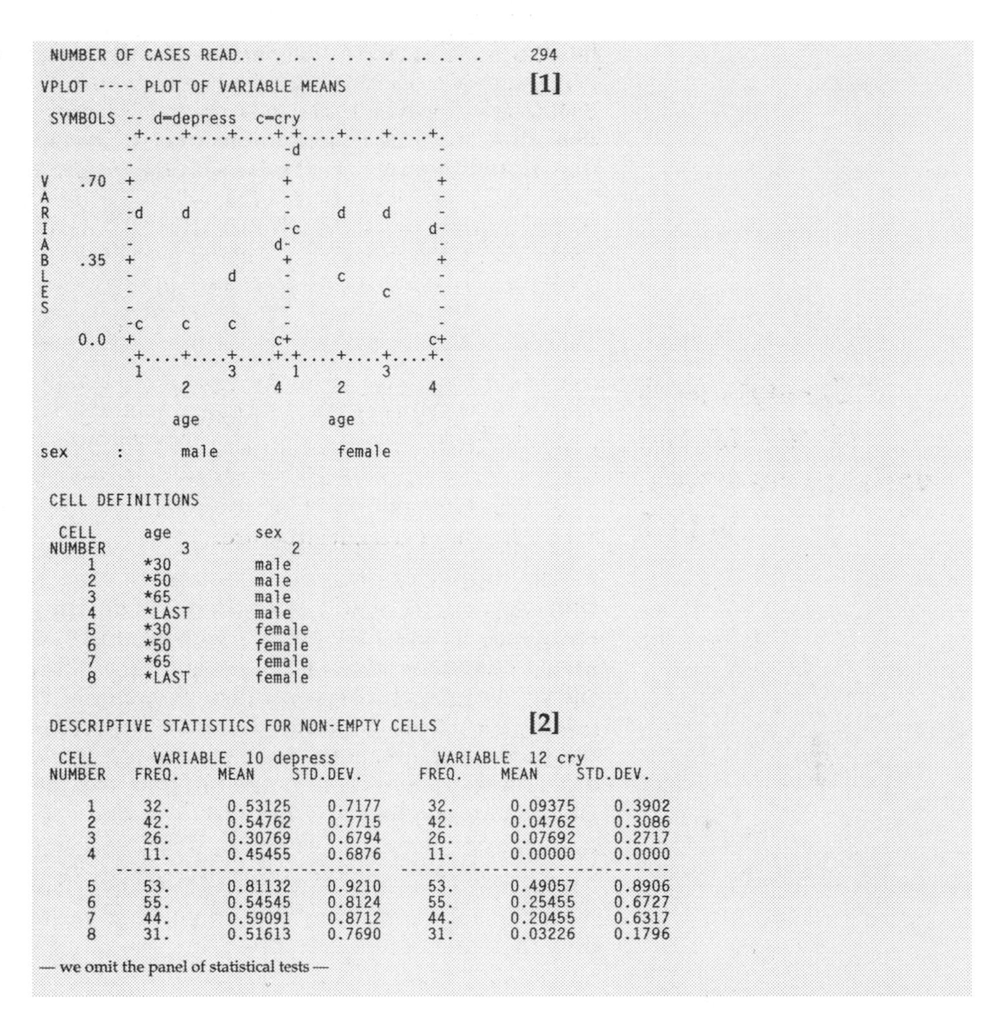

```
NUMBER OF CASES READ. . . . . . . . . . . . . .        294
VPLOT ---- PLOT OF VARIABLE MEANS                       [1]

 SYMBOLS -- d=depress  c=cry
         .+....+....+....+.+....+....+....+.
         -                -d               -
         -                -                -
V   .70  +                +                +
A        -                -                -
R        -d    d          -     d    d     -
I        -                -c              d-
A        -               d-                -
B   .35  +                +                +
L        -          d     -     c          -
E        -                -          c     -
S        -                -                -
         -c    c    c     -                -
    0.0  +               c+               c+
         .+....+....+....+.+....+....+....+.
          1         3      1         3
               2         4      2         4

              age              age

sex      :       male             female

 CELL DEFINITIONS

  CELL      age          sex
 NUMBER       3             2
     1     *30          male
     2     *50          male
     3     *65          male
     4     *LAST        male
     5     *30          female
     6     *50          female
     7     *65          female
     8     *LAST        female

 DESCRIPTIVE STATISTICS FOR NON-EMPTY CELLS             [2]

  CELL      VARIABLE  10 depress            VARIABLE  12 cry
 NUMBER   FREQ.     MEAN    STD.DEV.      FREQ.     MEAN    STD.DEV.

     1    32.     0.53125   0.7177        32.     0.09375   0.3902
     2    42.     0.54762   0.7715        42.     0.04762   0.3086
     3    26.     0.30769   0.6794        26.     0.07692   0.2717
     4    11.     0.45455   0.6876        11.     0.00000   0.0000
   ---------------------------        ---------------------------
     5    53.     0.81132   0.9210        53.     0.49057   0.8906
     6    55.     0.54545   0.8124        55.     0.25455   0.6727
     7    44.     0.59091   0.8712        44.     0.20455   0.6317
     8    31.     0.51613   0.7690        31.     0.03226   0.1796

— we omit the panel of statistical tests —
```

Bold numbers below correspond to those in Output 9D.5.

[1] VPLOT generates miniplots for the two levels of SEX (MALE and FEMALE). The cell means of the two variables specified are plotted along the y-axis, and the plot symbols are the first characters of the variable names. The four levels of AGE are plotted along the x-axis. (Note that group indices are used rather than actual AGE values.) We see that females tend to cry more and to be more depressed than males; this is especially true of the youngest female group.

[2] Panels of descriptive statistics for up to four variables are printed across the page (LINESIZE controls how many).

Example 9D.6 Plots of cell means and marginal means: PLOTMEANS and MARGIN

PLOTMEANS. The PLOTMEANS display is an alternative to GPLOTS which plots the same means in a different format. Panels for two or more variables can be displayed side-by-side. We again use the survey data described in Example 9D.1. INCOME, TOTAL, and HEALTHY are the variables to be plotted, and EDUCATN, SEX, and RELIGION are the GROUPING variables. In Input 9D.6, we specify two TABULATE paragraphs. In the first, we specify PLOTMEANS to generate plots showing the cell means for the 32 combinations of levels in EDUCATN, SEX, **and** RELIGION. Note that an END paragraph is included after each TABULATE paragraph. (You can also respecify group codes and cutpoints for each subproblem by inserting a new GROUP paragraph after each TABULATE paragraph.)

MARGIN. In the second TABULATE paragraph we specify a marginal subset, collapsing over religion. The MARGIN command is used with PLOTMEANS and has a special notation for defining marginal subsets. One character position represents each grouping factor. A letter indicates that the factor should be used; a period means "collapse over." All original cell means are plotted, but the cell means that belong to the same marginal subset are plotted on the same line. If there are two grouping factors, the statement

MARGIN = 'AB', 'A.','.B', '..'.

specifies how cells are defined.

'AB' – Two letters. All possible combinations of the two GROUPING variables are used (this is the preassigned value if MARGIN is not specified).

'A.' – A letter in the first position and a period in the second. Classification is by the first GROUPING variable only.

'.B' – A period in the first position and a letter in the second. Classification is by the second GROUPING variable only.

'..' – Two periods. There is no GROUPING variable by which to classify cases. Only one cell is formed.

The number of letters and periods between the apostrophes must be the same as the number of GROUPING variables in the TABULATE paragraph. The position of the letters and periods between the apostrophes determines how marginal subsets are classified. For example, MARGIN = 'X..Y.Z' directs 9D to form cells using all combinations (marginals) of the first, fourth, and sixth GROUPING variables. The mean and standard deviation are computed for each cell. In the plot of cell means, one line is used for each marginal subset; the values of the original means from the six-way combinations are plotted. To obtain a plot of the means for only the combinations of X, Y, and Z, specify a new TABULATE paragraph with GROUPING = X, Y, Z.

For our marginal subset, 'es.', 9D defines cells based on the 8 levels of EDUCATN and SEX only, ignoring RELIGION. You can specify more than one marginal subset in a single statement: e.g., MARGIN = 'es.', 'e.r', '.sr'. Note that only the order of letters and periods is significant; 'AB.' is equivalent to 'es.'.

Input 9D.6

```
/ INPUT     VARIABLES = 37.
            FORMAT = FREE.
            FILE = 'survey.dat'.

/ VARIABLE  NAMES = id, sex, age, marital, educatn,
                    employ, income, religion, (29)total,
                    (32)healthy.

/ GROUP     CUTP(age) = 35, 65.
            CODES(sex) = 1, 2.
            NAMES(sex) = male, female.
            CODES(religion) = 1 to 4.
            NAMES(religion) = protest, catholic, jewish,
                              none.
            CUTP(educatn) = 2, 3, 4.
            NAMEs(educatn) = no_grad, hs_grad, some_col,
                             col_grad.

/ TABULATE  VAR = income, total, healthy.
            GROUPING = educatn, sex, religion.
            PLOTMEANS.
/ END

TABULATE  MARGIN = 'es.'.    /
/ END
```

Output 9D.6

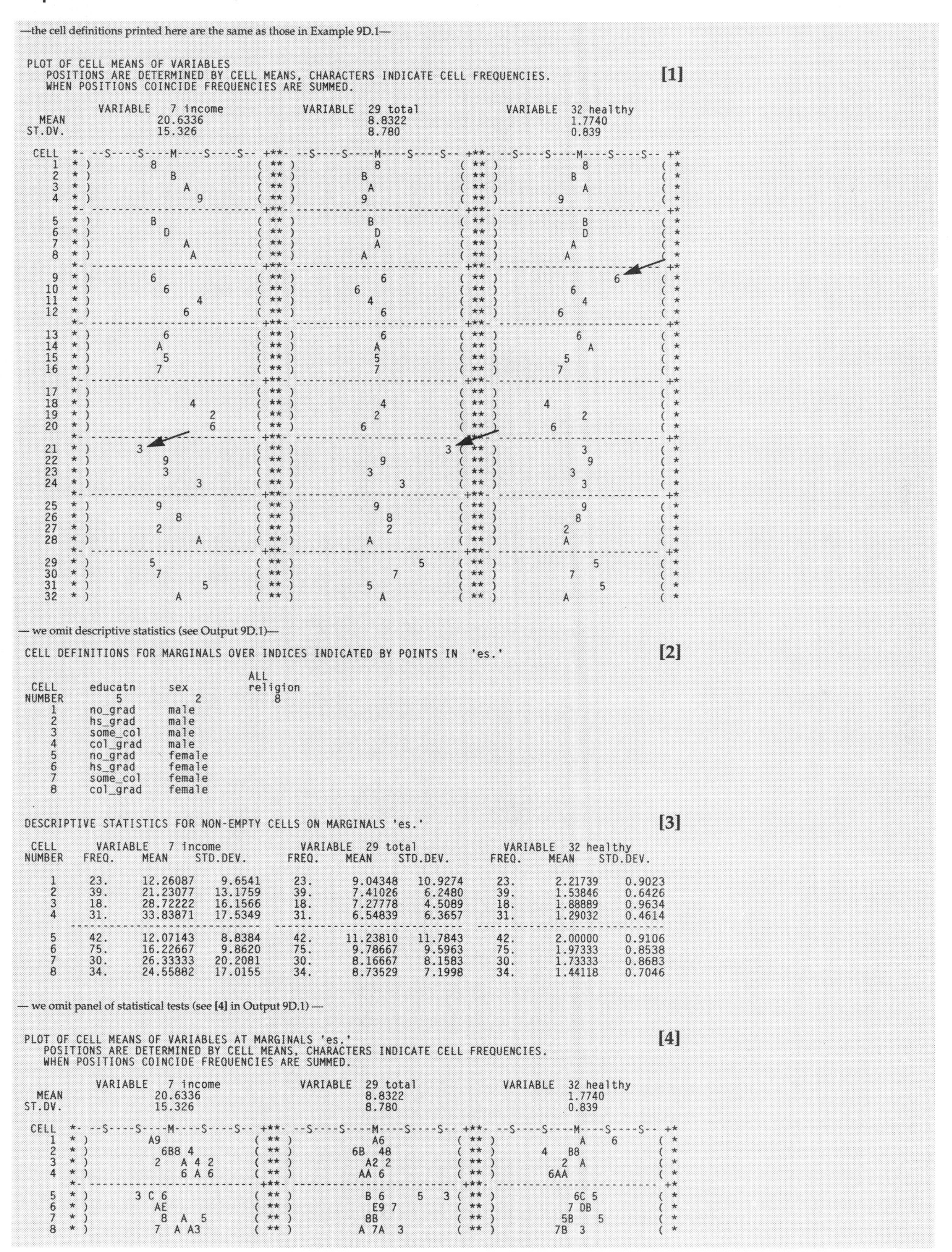

—the cell definitions printed here are the same as those in Example 9D.1—

```
PLOT OF CELL MEANS OF VARIABLES                                                    [1]
   POSITIONS ARE DETERMINED BY CELL MEANS, CHARACTERS INDICATE CELL FREQUENCIES.
   WHEN POSITIONS COINCIDE FREQUENCIES ARE SUMMED.

        VARIABLE   7 income          VARIABLE  29 total          VARIABLE  32 healthy
 MEAN            20.6336                       8.8322                       1.7740
ST.DV.           15.326                        8.780                        0.839

 CELL  *- --S----S----M----S----S-- +**- --S----S----M----S----S-- +**- --S----S----M----S----S-- +*
   1   * )          8              ( ** )           8              ( ** )           8             ( *
   2   * )             B           ( ** )          B               ( ** )          B              ( *
   3   * )               A         ( ** )           A              ( ** )             A           ( *
   4   * )                 9       ( ** )          9               ( ** )        9                ( *
       *- ------------------------ +**- ------------------------ +**- ------------------------ +*
   5   * )          B              ( ** )          B               ( ** )             B           ( *
   6   * )            D            ( ** )           D              ( ** )             D           ( *
   7   * )               A         ( ** )           A              ( ** )           A             ( *
   8   * )                A        ( ** )          A               ( ** )          A              ( *
       *- ------------------------ +**- ------------------------ +**- ------------------------ +*
   9   * )          6              ( ** )            6             ( ** )                 6       ( *
  10   * )            6            ( ** )         6                ( ** )           6             ( *
  11   * )                 4       ( ** )           4              ( ** )             4           ( *
  12   * )               6         ( ** )            6             ( ** )         6             ( *
       *- ------------------------ +**- ------------------------ +**- ------------------------ +*
  13   * )            6            ( ** )            6             ( ** )            6            ( *
  14   * )          A              ( ** )           A              ( ** )              A          ( *
  15   * )            5            ( ** )           5              ( ** )          5              ( *
  16   * )          7              ( ** )           7              ( ** )         7               ( *
       *- ------------------------ +**- ------------------------ +**- ------------------------ +*
  17   * )                         ( ** )                          ( ** )                         ( *
  18   * )                4        ( ** )            4             ( ** )      4                  ( *
  19   * )                   2     ( ** )           2              ( ** )             2           ( *
  20   * )                   6     ( ** )          6               ( ** )        6                ( *
       *- ------------------------ +**- ------------------------ +**- ------------------------ +*
  21   * )        3                ( ** )                        3 ( ** )             3           ( *
  22   * )            9            ( ** )            9             ( ** )              9          ( *
  23   * )            3            ( ** )          3               ( ** )           3             ( *
  24   * )                3        ( ** )              3           ( ** )            3            ( *
       *- ------------------------ +**- ------------------------ +**- ------------------------ +*
  25   * )           9             ( ** )           9              ( ** )            9            ( *
  26   * )              8          ( ** )             8            ( ** )           8             ( *
  27   * )           2             ( ** )             2            ( ** )         2               ( *
  28   * )                A        ( ** )          A               ( ** )         A               ( *
       *- ------------------------ +**- ------------------------ +**- ------------------------ +*
  29   * )          5              ( ** )                  5       ( ** )               5         ( *
  30   * )           7             ( ** )              7           ( ** )           7             ( *
  31   * )                  5      ( ** )          5               ( ** )                5        ( *
  32   * )             A           ( ** )            A             ( ** )         A               ( *
```

— we omit descriptive statistics (see Output 9D.1)—

```
CELL DEFINITIONS FOR MARGINALS OVER INDICES INDICATED BY POINTS IN  'es.'          [2]

                                      ALL
  CELL     educatn      sex          religion
 NUMBER       5            2            8
    1     no_grad      male
    2     hs_grad      male
    3     some_col     male
    4     col_grad     male
    5     no_grad      female
    6     hs_grad      female
    7     some_col     female
    8     col_grad     female
```

```
DESCRIPTIVE STATISTICS FOR NON-EMPTY CELLS ON MARGINALS 'es.'                       [3]
```

CELL NUMBER	VARIABLE 7 income FREQ.	MEAN	STD.DEV.	VARIABLE 29 total FREQ.	MEAN	STD.DEV.	VARIABLE 32 healthy FREQ.	MEAN	STD.DEV.
1	23.	12.26087	9.6541	23.	9.04348	10.9274	23.	2.21739	0.9023
2	39.	21.23077	13.1759	39.	7.41026	6.2480	39.	1.53846	0.6426
3	18.	28.72222	16.1566	18.	7.27778	4.5089	18.	1.88889	0.9634
4	31.	33.83871	17.5349	31.	6.54839	6.3657	31.	1.29032	0.4614
5	42.	12.07143	8.8384	42.	11.23810	11.7843	42.	2.00000	0.9106
6	75.	16.22667	9.8620	75.	9.78667	9.5963	75.	1.97333	0.8538
7	30.	26.33333	20.2081	30.	8.16667	8.1583	30.	1.73333	0.8683
8	34.	24.55882	17.0155	34.	8.73529	7.1998	34.	1.44118	0.7046

— we omit panel of statistical tests (see **[4]** in Output 9D.1) —

```
PLOT OF CELL MEANS OF VARIABLES AT MARGINALS 'es.'                                  [4]
   POSITIONS ARE DETERMINED BY CELL MEANS, CHARACTERS INDICATE CELL FREQUENCIES.
   WHEN POSITIONS COINCIDE FREQUENCIES ARE SUMMED.

        VARIABLE   7 income          VARIABLE  29 total          VARIABLE  32 healthy
 MEAN            20.6336                       8.8322                       1.7740
ST.DV.           15.326                        8.780                        0.839

 CELL  *- --S----S----M----S----S-- +**- --S----S----M----S----S-- +**- --S----S----M----S----S-- +*
   1   * )         A9              ( ** )           A6             ( ** )            A    6       ( *
   2   * )          6B8 4          ( ** )         6B  48           ( ** )       4   B8            ( *
   3   * )          2    A 4 2     ( ** )           A2 2           ( ** )           2  A          ( *
   4   * )              6 A 6      ( ** )           AA 6           ( ** )         6AA             ( *
       *- ------------------------ +**- ------------------------ +**- ------------------------ +*
   5   * )       3 C 6             ( ** )           B 6     5    3 ( ** )             6C 5        ( *
   6   * )           AE            ( ** )             E9 7         ( ** )             7 DB        ( *
   7   * )            8  A  5      ( ** )           8B             ( ** )            5B    5      ( *
   8   * )          7  A A3        ( ** )          A 7A  3         ( ** )           7B  3         ( *
```

[1] PLOTMEANS displays the 32 cells along the vertical axis. The horizontal axis contains an M for the overall mean of the VARIABLE being plotted (the first plot is for INCOME), so you can see the shift in individual cell means relative to the overall mean. Each S represents one pooled standard deviation from the overall mean. The width of each display is scaled to represent ± 2.5 standard deviations from the mean. The letters in the plots are codes for the number of observations in each cell. ("A" represents 10–19 cases, "B" represents 20–29 cases; see Commands for more information.)

For Protestant males (cells 1 to 4) and females (cells 5 to 8) average income increases with level of education. Compare this display with the plot frame for Protestants in **[1]** in Example 9D.1. Several outliers are also visible. The average HEALTH score for 6 male, Catholic, high school dropouts (cell 9) is far above average. Since HEALTH is scaled 1 = excellent, ..., 4 = poor, these subjects tend to have unusually poor health. There are no Jewish male high school dropouts (cell 17) and only three Jewish women who did not complete high school. The average income for these three women is among the lowest, and their average depression score (TOTAL) is among the highest.

[2] After collapsing over religion, 8 cells result. Their definitions appear here.

[3] For each cell identified in **[2]**, 9D prints descriptive statistics for income, total depression, and health.

[4] In this PLOTMEANS display, all of the original cell means are plotted, but the vertical axis is "collapsed" to represent the 8 EDUCATN by SEX levels only. Each line has four cell means, one for each level of RELIGION. The line for cell 1 displays the means from cells 1, 9, 17, and 25 in the first PLOTMEANS display. (When cell means coincide in their plotting position, their frequencies are summed, so there are not always four distinct values on a line.)

Example 9D.7 Interactive paragraphs

9D now allows you to transform your data, modify the grouping instructions, and request TABULATE paragraphs interactively. You can specify any number of GROUP, TABULATE, and TRANSFORM paragraphs after the INPUT, VARIABLE, and END paragraphs. In each case, the interactive paragraph should end with a slash, and the rest of the line after the slash should be blank. You can continue performing new transformations and analyses as long as you like; you only need to begin a new problem if you want to read in a different data set or change the defaults in the VARIABLE or SAVE, paragraphs. After the last interactive paragraph, type END/. Interactive mode allows you to perform the tests specified in one paragraph, view the results, and then decide whether to continue to the next paragraph or modify the last.

Imagine that the instructions for Example 9D.1 are stored in a file called 9D.INP. Enter 9D interactively (see Chapter 11). 9D will then print a banner and place you in the BMDP line editor at line 1. You may then type

```
GET 9D.INP
```

to GET your input file (see Chapter 11 for a list of line editor commands). After your file appears on the screen, type E for Execute to run the program. The output shown in Output 9D.1 will appear on your screen one portion at a time.

Use Insert mode in the line editor to add the interactive instructions shown following the first END paragraph in Input 9D.7. To exit insert mode, type //. When you execute this new set of instructions (E #, where # is the line number of the first new command), 9D will perform the analysis specified in each interactive paragraph. You can also run 9D from program control level, one paragraph at a time (see Chapter 11). **Note:** in the BMDP release for the PC, a full-screen editor replaces the line editor. See your *PC User's Guide* for details.

In Input 9D.7 we use interactive paragraphs after the initial analysis to specify cutpoints for a new grouping variable, AGE, which we then use in the TABU-

LATE paragraph. We omit the output for this example. Note that the old, noninteractive way of running subproblems in 9D (repeating TABULATE, GROUP, and END after the first END paragraph) is no longer available.

Input 9D.7

```
/ INPUT       VARIABLES = 37.
              FORMAT = FREE.
              FILE = 'survey.dat'.

/ VARIABLE    NAMES = id, sex, age, marital, educatn,
                      employ, income, religion.

/ GROUP       CODES(sex) = 1, 2.
              NAMES(sex) = male, female.
              CODES(religion) = 1 TO 4.
              NAMES(religion) = protest, catholic, jewish,
                                none.
              CUTP(educatn) = 2, 3, 4.
              NAMES(educatn) = no_grad, hs_grad, some_col,
                               col_grad.
/ TRANSFORM   USE = age LE 65.
/ TABULATE    VARIABLE = income.
              GROUPING = educatn, sex, religion.
              GPLOT.
/ END
 GROUP        CUTP(age) = 30, 50, 65.                /
 TABULATE     VARIABLE = income.
              GROUPING = educatn, sex, age.
              GPLOT.                                 /
 END /
```

Special Features

Organizing the data: structures for each type of plot

For the GPLOT and VPLOT options, each subject or animal is measured once for each variable. For VPLOT, two or more different variables are analyzed for each subject. RPLOT allows you to plot means when a single variable is measured two or more times for each subject or animal. The symbols in VPLOTS identify variables; symbols in GPLOTS and RPLOTS identify categories within variables.

Table 9D.1 provides an overview of the three miniplot commands. It will help you decide which option best fits your research design, and how to order the GROUPING variables in the TABULATE paragraph to produce the desired plots.

Table 9D.1
Organizing plot structures

		y-axis: Cell Means for	x-axis: Index of	Symbols	Frames Defined by Combinations of
GPLOT:	factorial designs	v_i (one var.)	g_1 levels	g_2	$g_3, g_4, \ldots$
RPLOT:	repeated measures	v_1 to v_p	variables v_1 to v_p	g_1	$g_2, g_3, \ldots$
VPLOT:	multiple variables	v_1 to v_p	g_1 levels	v_1 to v_p	$g_2, g_3, \ldots$

GROUPING = g_1, g_2, etc.
VARIABLE = $v_1, \ldots, v_p$

Obtaining one miniplot frame per subject

VPLOTS. It is possible to obtain a separate frame for each subject in your data by using an ID or case number variable as a grouping variable. For example, suppose each input record contains four variables: a subject ID, a week number, a drug dose level for the previous week, and an adverse drug response score. There is one record for each week's measurements, and each subject (ID) has more than one record. (Measurements for each subject need not be made at the same times, and subjects may have different numbers of records.)

Since all variables are plotted on a common scale, we divide dose by 10 to obtain better resolution for the other measure. The instructions are

```
/ TRANSFORM   dose = dose/10.
/ GROUP       CODES(id) = 1 TO 4.
/ TABULATE    VAR = dose, response.
              GROUPING = week, id.
              VPLOT.
```

The variables in the VARIABLE list are plotted in each frame, using the first character of each variable name as the plot symbol. WEEK defines the *x*-axis, and the remaining variable, ID, defines the plot frames. The plots are printed in the order that the ID numbers are listed in the CODES command of the GROUP paragraph. You can list as many ID numbers in the CODES command as you like, obtaining one plot for each subject listed. The values plotted in this example are not means, but individual data values (there is one observation per cell).

Figure 9D.2
Separate VPLOTS for each subject

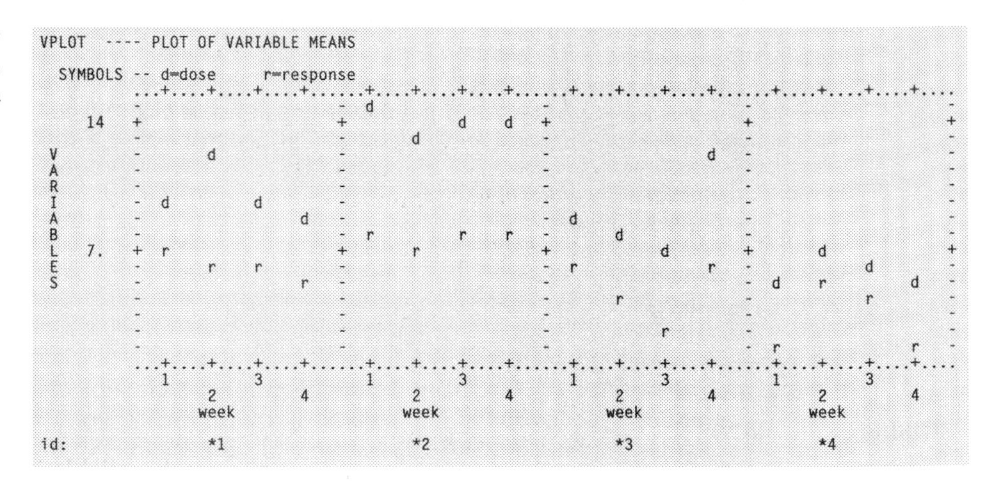

RPLOTS. Obtaining separate plots for each subject in a repeated measures design with all measurements on one record is slightly more complex. In the following example we use the Potthoff and Roy data described in program 2V. The data contain growth measures for 11 girls and 16 boys at ages 8, 10, 12, and 14 (we name the measures y_8, y_10, y_12, and y_14). We create an ID for each subject using KASE in the TRANSFORM paragraph (see Chapter 6). A GROUP paragraph specifies the ID CODES for which we want plots (you must specify CODES for any GROUPING variable with more than 10 values); we arbitrarily select the 5th, 10th, 15th, and 20th cases. We also create a dummy variable that provides the plot symbol. We do not show output; it is similar to VPLOTS in Figure 9D.2.

```
/ VARIABLE    NAMES = subject, sex, y_8, y_10, y_12, y_14.
/ GROUP       CODES(id) = 5, 10, 15, 20.
/ TRANSFORM   dummy = 1.0.
              id = kase.
/ TABULATE    GROUPING = dummy, id.
              VAR = y_8, y_10, y_12, y_14.
              RPLOT.
/ END
```

9D Commands

/ VARiable

The VARIABLE paragraph is optional. (See detailed description in Chapter 4.) Additional commands for 9D are FREQUENCY and CWEIGHT.

FREQuency = variable. `FREQ = COUNT.`

State the name or number of a variable containing the case frequency. A case frequency has the effect of repeating cases. Any record with a negative or zero case frequency (COUNT ≤ 0 in this example) is excluded from the calculations as though it were missing. See Chapter 4.

CWeight = variable. `CWEIGHT = SAMPWT.`

Specify a case weight variable. As with case frequencies, records with negative or zero case weight values are excluded from the computations. Case weights can be used to adjust for sampling from populations with unequal variances. Means and correlations are unaffected, but variances, standard deviations, and standard errors change. See Chapter 4.

/ TABULate

The TABULATE paragraph is required and can be repeated interactively to vary the configuration of the plots. See Order of Instructions.

GROUPing = list. `GROUP = AGE, SEX.`

Required to list the names or numbers of the variables used to classify cases into groups. Cells are formed from all possible combinations of the levels of the variables in the list; the levels for each variable are the intervals or codes assigned in the GROUP paragraph. See the definitions of GPLOT, RPLOT, and PLOT for how the order in the GROUPING list determines the miniplot structure and symbols. If a GROUPING variable has more than 10 distinct values, CODES or CUTPOINTS must be specified for it in the GROUP paragraph (see Chapter 5).

VARiable = list. `VARS = HEIGHT, WEIGHT.`

List the names or numbers of the variables to be analyzed. See GPLOT, RPLOT, and VPLOT definitions for additional information about the role of variables in this list. By default, all variables except GROUPING variables are analyzed.

GPLOT. `GPLOT.`

Requests miniplots of cell means for factorial structures. GPLOT generates one set of plots of the cell means for each variable in the TABULATE VAR list when there are two or more variables in the list. Up to eight plots per page are printed. The GROUPING and VARIABLE commands are essential to the plot structure.

y-axis	– cell means of one variable in the VARIABLE list
x-axis	– the levels of the first variable in the GROUPING list
symbols	– the first characters of the category names for the second variable in the GROUPING list
frames	– one frame for each unique combination of levels of the remaining variables in the GROUPING list (i.e., the third, fourth, etc.)

RPLOT. `RPLOT.`

Requests miniplots of cell means for repeated measures designs. RPLOT generates up to eight plots per page of the cell means specified in the TABULATE VARIABLE list. All variables in the list are treated as repeated measures of a single factor.

y-axis	– cell means of all variables in the VARIABLE list.
x-axis	– the order of "repeats." The program determines the order by the index of that variable in the VARIABLE list.
symbols	– the first characters of the category names of the first variable in the GROUPING list. Asterisks are used as plot symbols when

the plotting positions of two or more categories coincide.

frames – all combinations of levels of the second and any remaining variables in the GROUPING list.

VPLOT. `VPLOT.`

Requests miniplots of cell means for several variables simultaneously. VPLOTS are similar to GPLOTS, except that cell means of all variables in the TABULATE VARIABLE list are plotted along the *y*-axis in each plot, and the symbols identify the variables instead of categories within a variable.

y-axis – cell means of all variables in the VARIABLE list.

x-axis – the levels of the first variable in the GROUPING list.

symbols – the first characters of all variable names specified in the VARIABLE list. Asterisks are used as plot symbols when the plotting positions of two or more variables coincide.

frames – all combinations of levels of the second and any remaining variables in the GROUPING list.

SIZE = #1, #2. `SIZE = 10, 12.`

Specify the size of each plot frame as the number of characters for the horizontal axis (#1) and the number of lines for the vertical axis (#2). State the width first. The default plot frame is 15 characters wide, 12 lines high. The smallest plot frame allowed is 9 characters wide, 10 lines high.

PLOTmeans. `PLOT.`

Requests a vertical display of means that shows the shift in means from group to group, relative to the mean of all groups. Each group mean is represented in the plot by its frequency. If the frequency is more than nine, letters are used to represent the frequencies as follows:

Symbol	Frequency	Symbol	Frequency
A	10–19	J	100–199
B	20–29	K	200–299
C	30–39	L	300–399
D	40–49	M	400–499
E	50–59	N	500–599
F	60–69	O	600–699
G	70–79	P	700–799
H	80–89	Q	800–899
I	90–99	R	900–999
		S	> 1000

When marginal subsets are specified, cell means that belong to the same marginal subset are plotted together on the same line.

MARGINal = 'c_1','c_2',...

```
GROUP = SEX, INCOME.
MARGIN = 'SI', '.I', '..'.
```

Use with PLOTMEANS to identify marginal subsets, collapsing across one or more GROUPING variables. By default, cells are formed using all GROUPING variables. Each 'c_i' is a character string with one character for each GROUPING variable specified in the TABULATE paragraph. A period in place of a character tells the program to ignore that variable when defining cells. **Example:** The command above specifies three marginal subsets: 'SI' specifies cells to be formed using all combinations of SEX and INCOME, '.I' specifies classification by INCOME only, ignoring SEX, and '..' specifies one cell to be formed with no GROUPING classification.

Order of Instructions

■ indicates required paragraph

```
■ / INPUT
  / VARIABLE
  / GROUP
  / TRANSFORM
  / SAVE
  / PRINT
■ / TABULATE
■ / END
  data
  GROUP ... /
  TRANSFORM ... /
  TABULATE ... /
  END /
```

Repeat for additional problems. See Multiple Problems, Chapter 10. (applies to the whole block from / INPUT to END /)

May be repeated interactively (applies to GROUP ... /, TRANSFORM ... /, TABULATE ... /)

Summary Table for Commands Specific to 9D

	Paragraphs Commands	Defaults	Multiple Problems	See
	/ **VARiable**			
	FREQuency = variable.	no case frequency variable	–	Cmds
	CWeight = variable.	no case weights used	–	Cmds
■	/ **TABULate**			
■	GROUPing = list.	required	●	9D.1
▲	VARiable = list.	all variables used except GROUPING	●	9D.1
▲	GPLOT.	NO GPLOT.	●	9D.1
▲	RPLOT.	NO RPLOT.	●	9D.4
▲	VPLOT.	NO VPLOT.	●	9D.5
	SIZE = #1, #2.	15, 12	●	Cmds
	PLOTmeans.	NO PLOTMEANS.	●	9D.6
	MARGINal = 'c_1','c_2', ...	no statistics for marginals	●	9D.6

Key:
- ■ Required paragraph or command
- ▲ Frequently used command
- ● Value retained for multiple problems
- – Default reassigned

4F

Part 1
Forming Frequency Tables

Frequency tables (cross-tabulations) are used to summarize results for categorical data. Categories may be unordered (e.g., male-female, treatment group, answer to a multiple choice question), ordered (e.g., number of pregnancies, educational level, attitude on an ordered scale), or obtained by grouping continuous data into intervals (e.g., intervals of age, weight, IQ).

4F can form, analyze, and save frequency tables that are two-way (indexed by two categorical variables), multiway (by more than two categorical variables), or cross sections of a multiway table. The tabulation of EMPLOYMENT (private, government, none) by SMOKING status (yes, no) is an example of a two-way table. If such a tabulation is performed separately for each SEX, then each two-way table (one for male and one for female workers) is a cross section of a three-way table (EMPLOYMENT by SMOKING by SEX). Similarly, a four-way table can be separated into cross sections that are three-way or two-way tables.

4F can tabulate data read as cases or recorded as cell frequencies. Output from 4F includes tables of observed frequencies, descriptive statistics, a list of cases not included, and a table showing the reasons for excluding these cases. You can also obtain percentages of row totals, column totals, or of the total frequency for two-way and multiway tables. You can form marginal subtables from multiway tables by summing over one or more of the categorical variables. You can also stack variables by combining several categorical variables into a single set of categories.

Sometimes simply examining cross-tabulations and their percentages may be all you need to do. This chapter describes how to form and present such tables. At other times more sophisticated analyses are necessary in order to understand relationships between the categorical variables. Chapter 4F Part 2 describes how to analyze two-way frequency tables using more than two dozen tests and measures of association. Chapter 4F Part 3 describes how to fit log-linear models to multiway frequency tables.

Where to Find It

Examples

See also:

4F Part 2 Measures of Association and Tests of Influence for Two-way Tables
- Pearson χ^2 test
- Fisher's exact test
- odds ratio and Mantel-Haenszel statistic
- test of linear trend
- McNemar test of symmetry
- Kappa measure of reliability
- Kendall's tau
- measures of association
- other tests and measures

4F Part 3 Analysis of Log Linear Models
- screening effects and fitting models
- parameter estimates and deviates
- identification of extreme cells and strata
- structural zeros and tests of quasi-independence

Two-way Tables

The examples in this chapter use data on psychological and demographic measures from a survey of 465 women by L. S. Fidell and J. E. Prather (see Appendix D). We use six of the variables from the survey (some of the variable names have been changed slightly):

ESTEEM – a measure of self-esteem coded from 1 to 22 (grouped as low, medium, high)

HAP_STAT – happiness with current marital status on a scale from 1 to 48 (grouped as low, medium, high)

WOMENROL – attitude toward role of women on a scale ranging from 1 to 38 (grouped as conservative, moderate, liberal)

EDUCATN – education in number of years (grouped as less than 12 years, high school graduate, some college, college graduate)

WORKSTAT — work status at the time of the survey, coded as 0,1, 2 (grouped into two groups as paid work, homemaker)

MARITAL – marital status at the time of the survey: single, married

The data are organized as one record per woman, separating each measure with one or more blanks; for example, the data from the first two cases are

```
14  23  14  12  1  2
13  38  18  12  0  2
```

Some responses to the survey were omitted; asterisks in the data file mark the positions of missing values. The examples in this section show how to cross-classify two categorical variables to form two-way frequency tables. 4F forms the tables and applies the chi-square test of independence to each table created. Sometimes you will want additional analyses of your frequency tables: for example, other tests of independence, correlation, analysis of linear trend for ordered categories, etc. In Chapter 4F Part 2 we discuss the analysis of two-way tables.

Example 4F1.1 Basic setup for a two-way table

In the first example, we request a table in which ESTEEM is classified against EDUCATN. The INPUT and VARIABLE paragraphs are common to all programs and are described in Chapters 3 and 4. The data are in a file called FIDELL.DAT. See Chapter 2 for an explanation of the FILE or UNIT commands to describe the location of the data. (Alternatively, the data could follow the END paragraph.)

A CATEGORY (or GROUP) paragraph describes variables that are used to determine the categories in the frequency tables. The CATEGORY paragraph in Input 4F1.1 describes five variables, although for this example only the descriptions of ESTEEM and EDUCATN will be used. One of the variables, MARITAL status, has two levels and is described by CODE and NAME statements: the CODE statement describes all permissible values for the variable, and the NAME statement attaches labels to these values. There must be the same number of names as there are code values.

The other variables are described by CUTPOINT and NAME commands. The CUTPOINT list includes the upper endpoint of each interval except the last; the last interval ends at the maximum observed value (or the maximum permitted value when a MAXIMUM is specified in the VARIABLE paragraph). Therefore, there is one more interval than there are values, and the number of names must correspond to the number of intervals; i.e., one greater than the number of cutpoint values. For example, ESTEEM has two cutpoints, 11 and 16, but three names, LOW, MEDIUM, and HIGH. LOW corresponds to values less than or equal to 11, MEDIUM to values above 11 and less than or equal to 16, and HIGH to values greater than 16. CODES or CUTPOINTS need not be specified if there are 10 or fewer values to a variable and each value represents a different category. However, CODES and CUTPOINTS are desirable to assign NAMES to the categories or intervals, or when requesting a multiway table. See Chapter 5 for a detailed explanation of the CODE, CUTPOINT, and NAME instructions.

The frequency table is specified in the TABLE paragraph by setting the COLUMN variable equal to ESTEEM and the ROW variable to EDUCATN. The results are displayed in Output 4F1.1.

Input 4F1.1

```
/ INPUT       VARIABLES = 6.
              FORMAT IS FREE.
              FILE IS 'fidell.dat'.

/ VARIABLE    NAMES = esteem, hap_stat, womenrol, educatn,
                      workstat, marital.

/ CATEGORY    CODES (marital) = 1, 2.
              NAMES(marital) = single, married.
              CUTPOINTS(esteem) = 11, 16.
              NAMES(esteem) = low, medium, high.
              CUTPOINTS(educatn) = 11, 12, 15.
              NAMES(educatn) = '< 12', hs_grad, part_col,
                                 col_grad.
              CUTPOINTS(hap_stat) = 33, 41.
              NAMES(hap_stat) = low, medium, high.
              CUTPOINT(workstat) = 0.
              NAMES(workstat) = paidwork, homemakr.

/ TABLE       COLUMN = esteem.
              ROW = educatn.

/ END
```

Output 4F1.1

```
 CASE    1         2        3        4        5        6                              [1]
  NO. esteem   hap_stat womenrol educatn  workstat marital
----- -------- -------- -------- -------- -------- --------
    1 medium   low         14.00 hs_grad  homemakr married
    2 medium   medium      18.00 hs_grad  paidwork married
    3 low      medium      12.00 hs_grad  paidwork married
    4 medium   medium      25.00 < 12     homemakr married
    5 high     high        27.00 hs_grad  homemakr married
    6 medium   low         30.00 hs_grad  paidwork married
    7 medium   low         12.00 part_col homemakr married
    8 medium   medium       8.00 hs_grad  homemakr married
    9 high     high        24.00 col_grad homemakr married
   10 medium   low         32.00 hs_grad  paidwork single

NUMBER OF CASES READ. . . . . . . . . . . . . .     465

 VARIABLE       STATED VALUES FOR            GROUP CATEGORY    INTERVALS               [2]
 NO.   NAME   MINIMUM MAXIMUM MISSING   CODE INDEX   NAME      .GT.   .LE.
---- -------- ------- ------- -------  ------ ----- --------  ------ -------

  1  esteem                                      1  low               11.00
                                                 2  medium     11.00  16.00
                                                 3  high       16.00

  2  hap_stat                                    1  low               33.00
                                                 2  medium     33.00  41.00
                                                 3  high       41.00

  4  educatn                                     1  < 12              11.00
                                                 2  hs_grad    11.00  12.00

                                                 3  part_col   12.00  15.00
                                                 4  col_grad   15.00

  5  workstat                                    1  paidwork          0.000
                                                 2  homemakr   0.000

  6  marital                             1.000   1  single
                                         2.000   2  married
--------------------------------------------------------------------------------

DESCRIPTIVE STATISTICS OF DATA                                                         [3]

VARIABLE    TOTAL          STANDARD  ST.ERR   COEFF  SMALLEST LARGEST
NO. NAME    FREQ.   MEAN     DEV.   OF MEAN  OF VAR   VALUE    VALUE    RANGE

 1 esteem    465  14.166  3.9425  .18283  .27832  1.0000  22.000  21.000
 2 hap_stat  460  36.020  8.5534  .39880  .23747  1.0000  48.000  47.000
 3 womenrol  465  20.865  6.7582  .31340  .32391  1.0000  38.000  37.000
 4 educatn   462  13.251  2.2716  .10569  .17143  4.0000  24.000  20.000
 5 workstat  465  .76559  .87770  .04070  1.1464  0.0000  2.0000  2.0000
 6 marital   465  1.7785  .41571  .01928  .23374  1.0000  2.0000  1.0000

*** N O T E *** ONLY REQUESTED OPTIONS ARE PRINTED BELOW.
                TABLES ARE SEQUENCE NUMBERED ACROSS TABLE PARAGRAPHS.
```

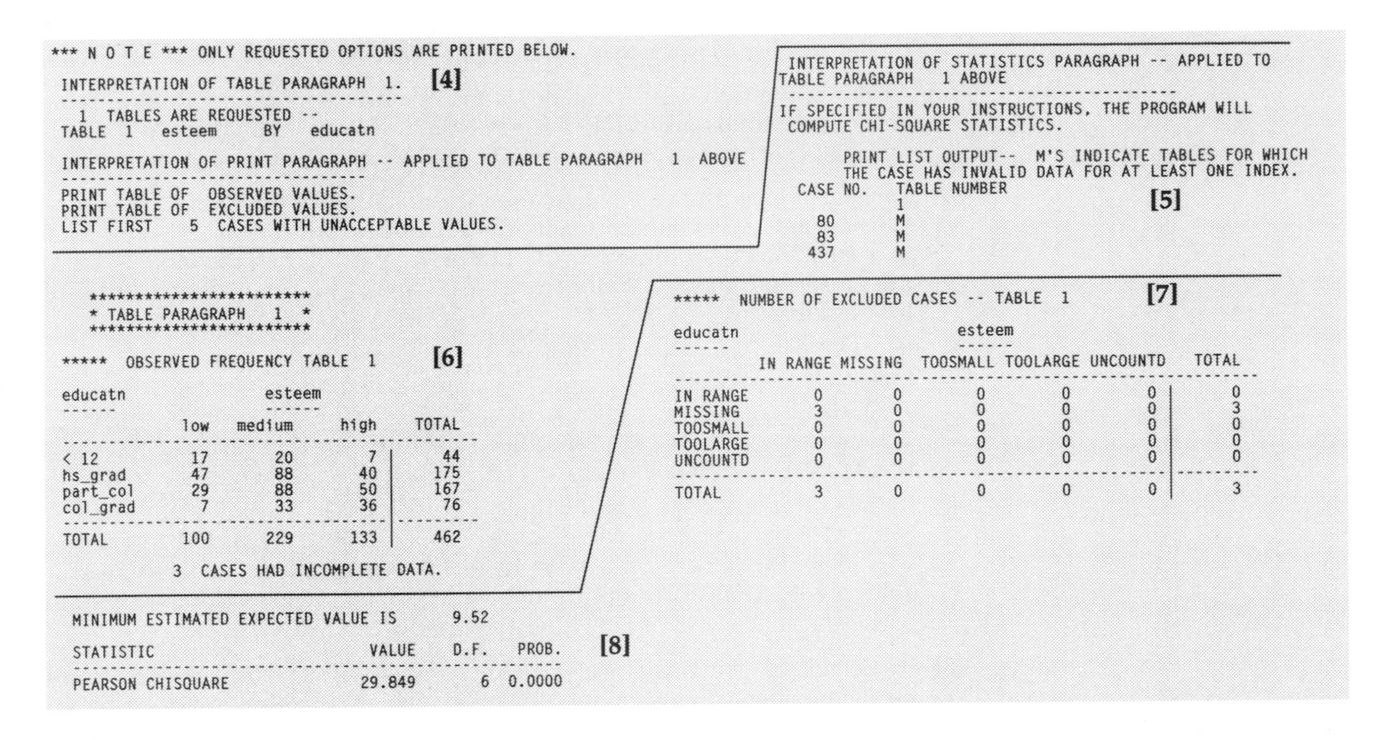

```
*** N O T E *** ONLY REQUESTED OPTIONS ARE PRINTED BELOW.
INTERPRETATION OF TABLE PARAGRAPH  1.      [4]
------------------------------------
  1  TABLES ARE REQUESTED --
TABLE  1    esteem      BY     educatn

INTERPRETATION OF PRINT PARAGRAPH -- APPLIED TO TABLE PARAGRAPH   1  ABOVE
------------------------------------
PRINT TABLE OF  OBSERVED VALUES.
PRINT TABLE OF  EXCLUDED VALUES.
LIST FIRST     5  CASES WITH UNACCEPTABLE VALUES.
```

```
INTERPRETATION OF STATISTICS PARAGRAPH -- APPLIED TO
TABLE PARAGRAPH    1 ABOVE
------------------------------------------
IF SPECIFIED IN YOUR INSTRUCTIONS, THE PROGRAM WILL
COMPUTE CHI-SQUARE STATISTICS.

      PRINT LIST OUTPUT--   M'S INDICATE TABLES FOR WHICH
      THE CASE HAS INVALID DATA FOR AT LEAST ONE INDEX.
 CASE NO.   TABLE NUMBER
               1                                   [5]
   80          M
   83          M
  437          M
```

```
************************
* TABLE PARAGRAPH   1  *
************************

*****  OBSERVED FREQUENCY TABLE  1        [6]

educatn              esteem
------               ------
              low   medium    high    TOTAL
-------------------------------------------
< 12           17      20        7  |    44
hs_grad        47      88       40  |   175
part_col       29      88       50  |   167
col_grad        7      33       36  |    76
-------------------------------------------
TOTAL         100     229      133  |   462

        3  CASES HAD INCOMPLETE DATA.
```

```
*****  NUMBER OF EXCLUDED CASES -- TABLE  1                    [7]

educatn                       esteem
------                        ------
            IN RANGE MISSING  TOOSMALL TOOLARGE UNCOUNTD   TOTAL
-----------------------------------------------------------------
IN RANGE       0        0        0        0        0   |     0
MISSING        3        0        0        0        0   |     3
TOOSMALL       0        0        0        0        0   |     0
TOOLARGE       0        0        0        0        0   |     0
UNCOUNTD       0        0        0        0        0   |     0
-----------------------------------------------------------------
TOTAL          3        0        0        0        0   |     3
```

```
MINIMUM ESTIMATED EXPECTED VALUE IS        9.52

STATISTIC                        VALUE    D.F.   PROB.    [8]
-------------------------------------------------------
PEARSON CHISQUARE               29.849       6  0.0000
```

[1] Prints the first ten cases of data. Grouping names are printed if specified.

[2} Prints grouping information with codes and names.

[3] For each categorical variable, 4F prints the following descriptive statistics:
- mean
- standard deviation
- minimum observed value within acceptable range limits
- maximum observed value within acceptable range limits
- total frequency of acceptable values
- number of cases with values defined as MISSING in the VARIABLE paragraph
- number of cases with values less than the lower limit (MINIMUM)
- number of cases with values greater than the upper limit (MAXIMUM)
- number of cases with values not equal to any of the stated CODES in the CATEGORY paragraph

Some of these will not be useful for qualitative data (e.g., MARITAL). However, when a categorical variable is created from ordered or continuous measures, such as the score for WOMENROL, the information may be useful. Only variables used as indices in at least one table are included in this list. We see here that the three cases excluded in [2] above had missing values for EDUCATN. Asterisks in the data file mark missing values (the default method); see Chapter 3 for more information about missing values.

[4] 4F prints an interpretation of your instructions. One table, ESTEEM by EDUCATN, will be printed, as requested in the first (and only) TABLE paragraph for this problem. Since we did not specify PRINT or STATISTICS paragraphs, 4F describes the default options.

[5] When data are excluded from one or more tables, 4F lists the cases for which data were omitted and prints an M under the affected table number(s). In this problem there are three cases (80, 83, and 437) that had unusable data for the EDUCATN by ESTEEM table. (The data in these three cases are present in the original data set and were deliberately altered to create the missing data.) Reasons for omitting these cases are given in the next panel. By default, 4F lists only the first five cases with unacceptable values; to increase that number, insert a PRINT paragraph with a LIST command into the instructions. The list of cases with unacceptable values is followed by the number of cases read by 4F for this problem.

[6] 4F prints the two-way frequency table of observed values. Row, column, and overall totals are printed around the border of the table. To omit the table of observed values, specify NO OBSERVED in the PRINT paragraph.

[7] Each table uses all available data for the variables in the TABLE paragraph. When some cases contain invalid data (missing, out of range, or not equal to a specified code) for either or both of the variables in the table, 4F compiles a second table to summarize these cases. The row/column headings for this table are

IN RANGE	–	valid for this categorical variable
MISSING	–	equal to a missing value code for this variable
TOOSMALL	–	below the specified minimum for this variable
TOOLARGE	–	above the specified maximum for this variable
UNCOUNTD	–	not equal to any code when category codes are specified

In the example we see that the three excluded cases had missing values for EDUCATN but had acceptable values (IN RANGE) for ESTEEM. For a two-way table, this table will always have a 0 in the cell for which both variables are IN RANGE. To omit the table of excluded cases from the output, specify NO EXCLUDE in the PRINT paragraph.

[8] By default 4F prints the Pearson chi-square statistic, which is used to test for independence between the row and column variables. We describe this statistic in Example 4F2.1 in Chapter 4F Part 2, along with many alternate statistics. In this example we reject the hypothesis of independence between ESTEEM and EDUCATN with a value less than 0.00005 (rounded in the output to 0.0000).

Example 4F1.2 Requesting several tables: the PAIR and CROSS commands

You can use the TABLE paragraph to form several tables by listing more than one row variable and more than one column variable. Also, the TABLE paragraph can be repeated before the END paragraph to create additional tables.

The instructions

```
/ TABLE    COLUMN = esteem, hap_stat.
           ROW = educatn, marital.
```

request two two-way tables by PAIRING (the default) the COLUMN and ROW variables. ESTEEM by EDUCATN is the first table requested, and HAP_STAT by MARITAL is the second one. When pairing row and column variables, there must be the same number of variables listed for both rows and columns.

In Input 4F1.2, CROSS is added to the TABLE paragraph. 4F will now create tables from all combinations of pairings of the ROW and COLUMN variables; the number of ROW and COLUMN variables need not be equal. For example:

```
/ TABLE    COLUMN = esteem, hap_stat.
           ROW = educatn, marital, esteem.
           CROSS.
```

requests the tables

1. ESTEEM by EDUCATN
2. ESTEEM by MARITAL
3. omitted
4. HAP_STAT by EDUCATN
5. HAP_STAT by MARITAL
6. HAP_STAT by ESTEEM

Tables that have the same variable for two indices are omitted; i.e., there is no table 3 (ESTEEM by ESTEEM).

Input 4F1.2
(replace the TABLE paragraph in Input 4F.1)

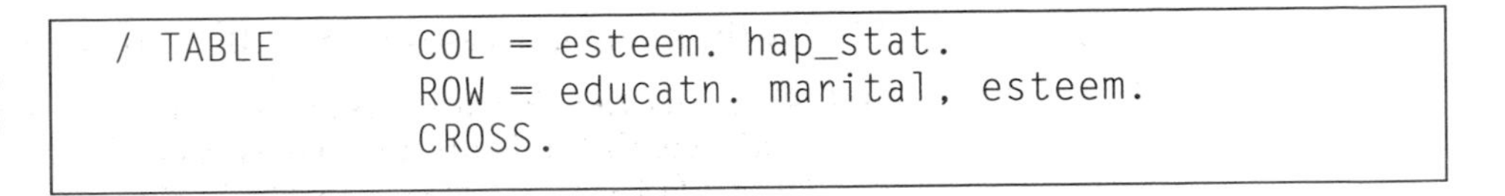

```
/ TABLE        COL = esteem. hap_stat.
               ROW = educatn. marital, esteem.
               CROSS.
```

Output 4F1.2

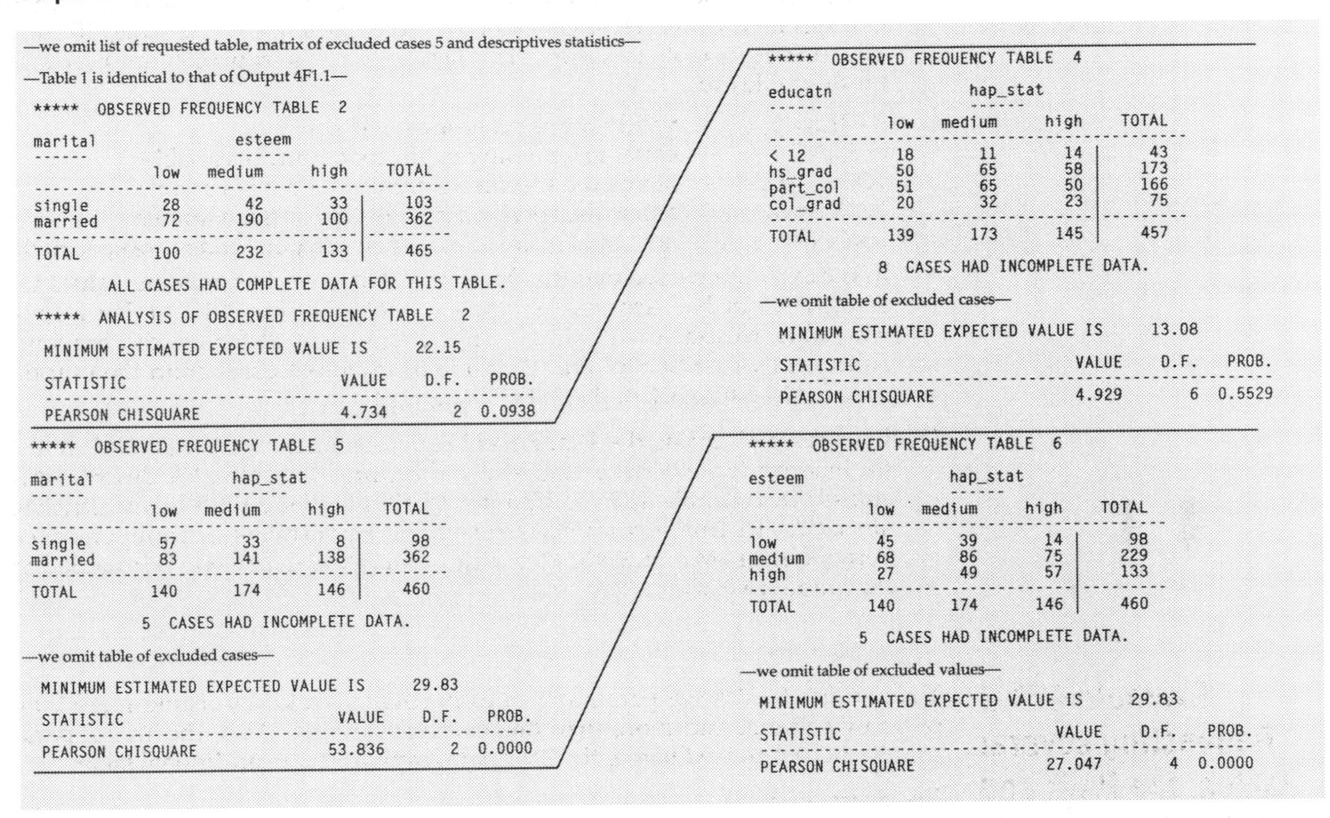

—we omit list of requested table, matrix of excluded cases 5 and descriptives statistics—

—Table 1 is identical to that of Output 4F1.1—

```
***** OBSERVED FREQUENCY TABLE  2

marital             esteem
------              ------
             low   medium      high    TOTAL
------------------------------------------------
single        28       42        33  |   103
married       72      190       100  |   362
---------------------------------------------
TOTAL        100      232       133  |   465

      ALL CASES HAD COMPLETE DATA FOR THIS TABLE.

***** ANALYSIS OF OBSERVED FREQUENCY TABLE   2

 MINIMUM ESTIMATED EXPECTED VALUE IS       22.15

 STATISTIC                         VALUE    D.F.   PROB.
 -------------------------------------------------------
 PEARSON CHISQUARE                 4.734       2  0.0938
```

```
***** OBSERVED FREQUENCY TABLE  4

educatn                hap_stat
------                 ------
                low   medium      high    TOTAL
---------------------------------------------------
< 12             18       11        14  |    43
hs_grad          50       65        58  |   173
part_col         51       65        50  |   166
col_grad         20       32        23  |    75
---------------------------------------------------
TOTAL           139      173       145  |   457

           8  CASES HAD INCOMPLETE DATA.
```

—we omit table of excluded cases—

```
 MINIMUM ESTIMATED EXPECTED VALUE IS          13.08

 STATISTIC                            VALUE    D.F.   PROB.
 ----------------------------------------------------------
 PEARSON CHISQUARE                    4.929       6  0.5529
```

```
***** OBSERVED FREQUENCY TABLE  5

marital             hap_stat
------              ------
             low   medium      high    TOTAL
------------------------------------------------
single        57       33         8  |    98
married       83      141       138  |   362
---------------------------------------------
TOTAL        140      174       146  |   460

     5  CASES HAD INCOMPLETE DATA.
```

—we omit table of excluded cases—

```
 MINIMUM ESTIMATED EXPECTED VALUE IS       29.83

 STATISTIC                         VALUE    D.F.   PROB.
 -------------------------------------------------------
 PEARSON CHISQUARE                53.836       2  0.0000
```

```
***** OBSERVED FREQUENCY TABLE  6

esteem                 hap_stat
------                 ------
                low   medium      high    TOTAL
---------------------------------------------------
low              45       39        14  |    98
medium           68       86        75  |   229
high             27       49        57  |   133
---------------------------------------------------
TOTAL           140      174       146  |   460

           5  CASES HAD INCOMPLETE DATA.
```

—we omit table of excluded values—

```
 MINIMUM ESTIMATED EXPECTED VALUE IS          29.83

 STATISTIC                            VALUE    D.F.   PROB.
 ----------------------------------------------------------
 PEARSON CHISQUARE                   27.047       4  0.0000
```

The resulting tables are displayed in Output 4F1.2. All available data are used in each table. For example, although table 1 excludes three observations that are missing values for EDUCATN, table 2 uses all 456 cases. Tables 5 and 6 each exclude five observations with no value for HAP_STAT, and table 4 excludes all eight observations that are missing values for either HAP_STAT or EDUCATN.

Example 4F1.3 Percentages

4F can print the observed frequencies as percentages of row totals, percentages of column totals, or percentages of total frequency. These percentages can help you interpret the frequency table. In Input 4F1.3 we request all three percentages. It is a rare occasion when you will want all three.

Input 4F1.3
(add to Input 4F1.1 after table)

```
/ PRINT        PERCENT = ROW, COLUMN, TOTAL.
```

Output 4F1.3

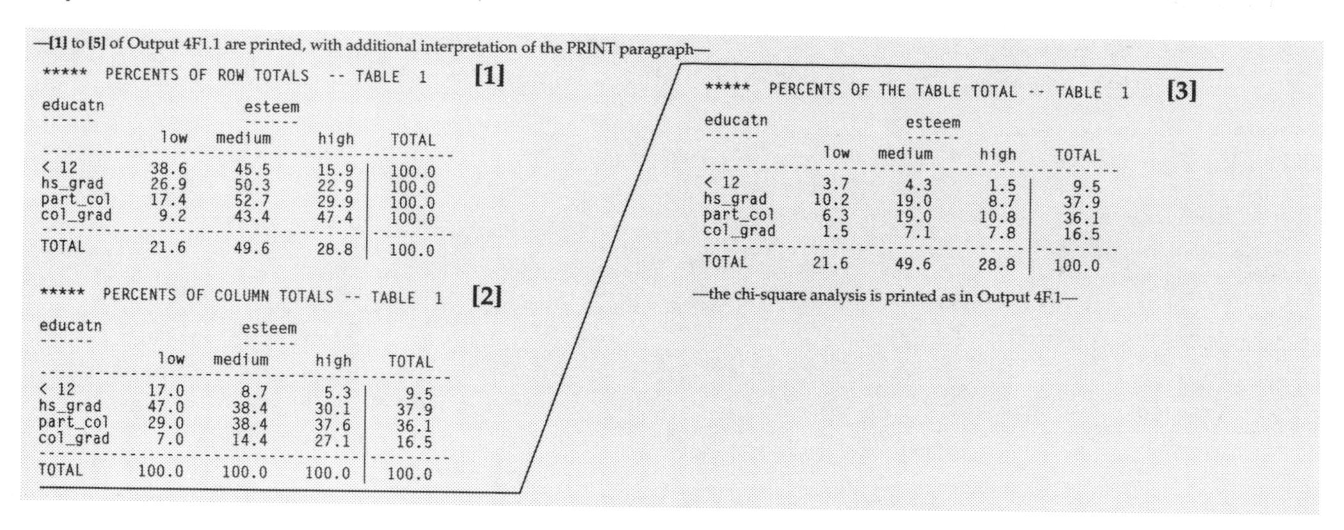
—[1] to [5] of Output 4F1.1 are printed, with additional interpretation of the PRINT paragraph—

***** PERCENTS OF ROW TOTALS -- TABLE 1 **[1]**

educatn	low	medium	high	TOTAL
< 12	38.6	45.5	15.9	100.0
hs_grad	26.9	50.3	22.9	100.0
part_col	17.4	52.7	29.9	100.0
col_grad	9.2	43.4	47.4	100.0
TOTAL	21.6	49.6	28.8	100.0

***** PERCENTS OF COLUMN TOTALS -- TABLE 1 **[2]**

educatn	low	medium	high	TOTAL
< 12	17.0	8.7	5.3	9.5
hs_grad	47.0	38.4	30.1	37.9
part_col	29.0	38.4	37.6	36.1
col_grad	7.0	14.4	27.1	16.5
TOTAL	100.0	100.0	100.0	100.0

***** PERCENTS OF THE TABLE TOTAL -- TABLE 1 **[3]**

educatn	low	medium	high	TOTAL
< 12	3.7	4.3	1.5	9.5
hs_grad	10.2	19.0	8.7	37.9
part_col	6.3	19.0	10.8	36.1
col_grad	1.5	7.1	7.8	16.5
TOTAL	21.6	49.6	28.8	100.0

—the chi-square analysis is printed as in Output 4F.1—

[1] 4F prints percentages of the total frequency in each row. For example, among the college graduates 36/76 = 47.4% have high self-esteem. Examination of this table shows the shift from low to high self-esteem as education increases, which may explain the highly significant chi-square test for independence between ESTEEM and EDUCATN.

[2] 4F prints percentages of the total frequency in each column. For example, among respondents with high self-esteem, 36/133 = 27.1% are college graduates. Since education is more likely to affect self-esteem than the converse, an examination of the row percentages **[1]** is more appropriate for this example.

[3] 4F prints percentages of the total table frequency.

Example 4F1.4 Conditioning: a separate table for each level of a third variable

Often the relationship between two variables depends on a third variable. In this example, we study the relationship between ESTEEM and EDUCATN, conditioning on work status to see if the relationship for women who work is the same as it is for those who are homemakers. We obtain separate frequency tables for the two levels of WORKSTAT.

Input 4F1.4
(replace the TABLE paragraph in 4F1.1 with the following)

```
/ TABLE        COL = esteem.
               ROW = educatn.
               CONDITION = workstat.
```

Output 4F1.4

```
***** OBSERVED FREQUENCY TABLE  1

USING LEVEL   paidwork   OF VARIABLE    5   workstat        [1]
              ********                      ********

educatn               esteem
------                ------
              low   medium     high     TOTAL
-----------------------------------------------
< 12            5       10        4  |     19
hs_grad        23       40       26  |     89
part_col       15       36       35  |     86
col_grad        6       19       26  |     51
-----------------------------------------------
TOTAL          49      105       91  |    245

        1 CASES HAD INCOMPLETE DATA.
```

—a table of excluded values similar to that in [1] of Output 4F1.1 is printed—

```
 STATISTIC                        VALUE     D.F.    PROB.
 ---------------------------------------------------------
 PEARSON CHISQUARE               10.569       6    0.1026
```

```
USING LEVEL   homemakr   OF VARIABLE    5   workstat        [2]
              ********                      ********
educatn               esteem
------                ------
              low   medium     high     TOTAL
-----------------------------------------------
< 12           12       10        3  |     25
hs_grad        24       48       14  |     86
part_col       14       52       15  |     81
col_grad        1       14       10  |     25
-----------------------------------------------
TOTAL          51      124       42  |    217

        2 CASES HAD INCOMPLETE DATA.
```

—we omit table of excluded cases—

```
 STATISTIC                        VALUE     D.F.    PROB.
 ---------------------------------------------------------
 PEARSON CHISQUARE               21.149       6    0.0017
```

[1] In Output 4F1.4, 4F prints a frequency table for women who work outside the home (PAIDWORK), which is the first level of the conditioning variable WORKSTAT. In contrast to the result presented in Output 4F1.1, the test of independence between ESTEEM and EDUCATN is not significant (p = 0.1026) for these women.

[2] 4F prints a frequency table for women who are homemakers, the second level of WORKSTAT. For these women the test of independence is highly significant (p = 0.0017). This indicates that the relationship between self-esteem and education is affected by whether or not the respondent is paid to work outside the home.

Multiway Tables

Many problems require you to study a relationship between three or more categorical variables: for example, survival by type of treatment by age and by sex. In this section we describe how to form multiway tables using 4F. 4F Part 3 describes how to fit a model to the frequencies in the cells in order to understand the relationships among the categorical variables.

Example 4F1.5 Basic setup for a multiway table

For this example we form a single four-way frequency table: EDUCATN by MARITAL by ESTEEM by WORKSTAT. In Input 4F1.5 we request the table by specifying the INDICES (names) of categorical variables in the TABLE paragraph. The TABLE paragraph may be repeated, but only one INDICES list may be specified per paragraph. The dimension of the table is the number of variables listed—i.e., this is a four-way table. 4F uses the first variable (EDUCATN) as the column index, the second (MARITAL) as the fastest changing row index, and so on, to the last (WORKSTAT) as the slowest changing row index.

Input 4F1.5
(replacement for the TABLE paragraph in Input 4F1.1)

```
/ TABLE    INDICES = educatn. marital, esteem. workstat.
```

The frequency table in Output 4F1.5 shows the totals for each row, summed over levels of the EDUCATN variable (a). 4F also prints partial column totals (b) for each level of EDUCATN, summed over levels of MARITAL (the fastest moving row variable), but not over levels of ESTEEM or WORKSTAT. Totals are also printed for each combination of levels of WORKSTAT and ESTEEM, summed over levels of MARITAL and EDUCATN (c). You can use STACK to change how these totals are computed (see Example 4F1.8).

Output 4F1.5

workstat	esteem	marital	< 12	hs_grad	part_col	col_grad	TOTAL	
paidwork	low	single	2	6	4	3	15	← a
		married	3	17	11	3	34	
		TOTAL	5 ← b	23	15	6	49	← c
	medium	single	4	12	9	7	32	
		married	6	28	27	12	73	
		TOTAL	10	40	36	19	105	
	high	single	1	8	13	8	30	
		married	3	18	22	18	61	
		TOTAL	4	26	35	26	91	
homemakr	low	single	6	2	4	1	13	
		married	6	22	10	0	38	
		TOTAL	12	24	14	1	51	
	medium	single	1	1	5	3	10	
		married	9	47	47	11	114	
		TOTAL	10	48	52	14	124	
	high	single	0	2	1	0	3	
		married	3	12	14	10	39	
		TOTAL	3	14	15	10	42	

TOTAL OF THE OBSERVED FREQUENCY TABLE IS 462

3 CASES HAD INCOMPLETE DATA.

The table of excluded values (not shown) is a two-way table classified by the first two variables listed in the INDICES statement. If the data are excluded due to a third or higher-level variable, the incomplete observation will show up in the cell for which both listed variables are IN RANGE. You can find the reason for excluding cases by studying the descriptive statistics or by switching the order of the indices in the table.

Example 4F1.6 Marginal subtables

A marginal subtable is an observed frequency table collapsed (summed) over one or more of the categorical variables (indices). When you collapse a table over a variable, that variable is eliminated as an index from the subtable. In Input 4F1.6, we use the MARGINAL instruction to request all two-way tables of the four-way table EDUCATN by MARITAL by ESTEEM by WORKSTAT. This instruction automatically produces all lower order tables as well—in this case all one-way tables. (For this four-way table, we could have requested three-way tables also. The four-way table of observed values is printed by default.)

Input 4F1.6
(we add to Input 4F1.5)

```
/ PRINT      MARGINALS = 2.
```

Output 4F1.6

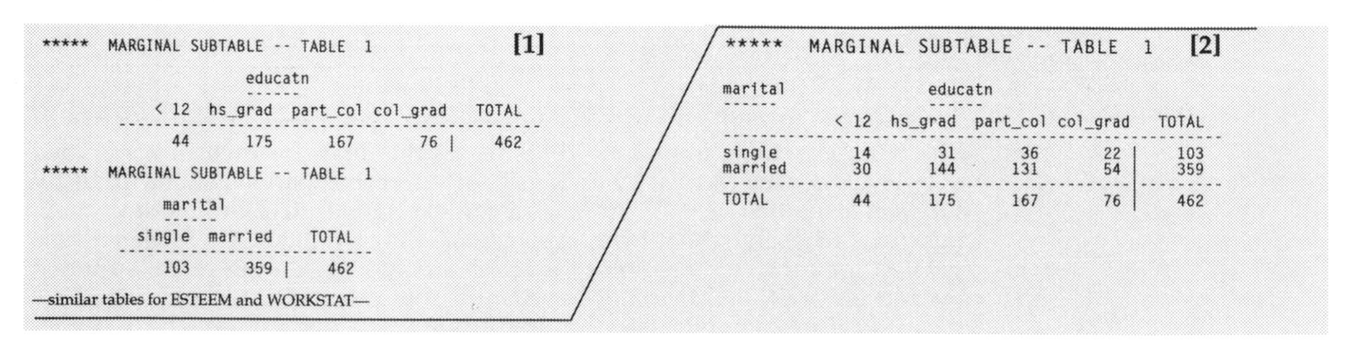

[1]

***** MARGINAL SUBTABLE -- TABLE 1

educatn	< 12	hs_grad	part_col	col_grad	TOTAL
	44	175	167	76	462

***** MARGINAL SUBTABLE -- TABLE 1

marital	single	married	TOTAL
	103	359	462

—similar tables for ESTEEM and WORKSTAT—

[2]

***** MARGINAL SUBTABLE -- TABLE 1

marital	< 12	hs_grad	part_col	col_grad	TOTAL
single	14	31	36	22	103
married	30	144	131	54	359
TOTAL	44	175	167	76	462

```
*****  MARGINAL SUBTABLE -- TABLE  1

esteem                  educatn
------                  ------
             < 12  hs_grad  part_col  col_grad    TOTAL
-----------------------------------------------------
low            17       47        29         7 |   100
medium         20       88        88        33 |   229
high            7       40        50        36 |   133
-----------------------------------------------------
TOTAL          44      175       167        76 |   462
*****  MARGINAL SUBTABLE -- TABLE  1
```

—we omit WORKSTAT by EDUCATN and ESTEEM by MARITAL—

```
workstat      marital
------        ------
           single  married    TOTAL
-------------------------------------
paidwork       77      168 |    245
homemakr       26      191 |    217
-------------------------------------
TOTAL         103      359 |    462

*****  MARGINAL SUBTABLE -- TABLE  1

workstat                esteem
------                  ------
              low   medium      high     TOTAL
-----------------------------------------------
paidwork       49      105        91 |    245
homemakr       51      124        42 |    217
-----------------------------------------------
TOTAL         100      229       133 |    462
```

Four one-way tables are printed in Output 4F1.6 (we only display results for EDUCATN and MARITAL). Six two-way tables follow (we show four).

[1] 4F prints the one-way marginal subtable formed by summing over all of the categorical variables except EDUCATN. This table, then, contains the total frequency for each level of EDUCATN. This is followed by the marginal table for MARITAL, and so on.

[2] 4F prints the two-way marginal subtable, EDUCATN by MARITAL. This is formed by summing over the remaining variables in the frequency table, ESTEEM and WORKSTAT. All combinations of levels of EDUCATN and MARITAL are shown in this subtable, as well as the row and column totals (which match the figures in the one-way tables above). The five remaining two-way marginal subtables follow.

Example 4F1.7 Percentages in a multiway table

4F can print percentages of the row total, the column total, or the total frequency of the table specified in the TABLE paragraph. ROW indicates the line in the table, and COLUMN a column in a panel. In order to obtain the percents that are relevant for a specific problem, the order in which the table factors are arranged can be controlled in the TABLE paragraph.

In Input 4F1.7 we request COLUMN percentages for the four-way table formed by EDUCATN, MARITAL, ESTEEM, and WORKSTAT. Since EDUCATN is the first variable listed in the list of INDICES, it is designated the column variable. Therefore percentages will be provided as a function of educational level attained.

Input 4F1.7
(add to Input 4F1.5)

```
/ PRINT       PERCENTS = COLUMN.
```

Output 4F1.7

—observed frequency table in Output 4F1.5 is printed—

```
***** PERCENTS OF COLUMN TOTALS -- TABLE  1

workstat esteem   marital                educatn
------   ------   ------                 ------
                           < 12  hs_grad  part_col  col_grad    TOTAL
------------------------------------------------------------------------

paidwork low      single   40.0    26.1     26.7     50.0   |   30.6
                  married  60.0    73.9     73.3     50.0   |   69.4
                  -----------------------------------------------------
                  TOTAL   100.0   100.0    100.0    100.0   |  100.0

         medium   single   40.0    30.0     25.0     36.8   |   30.5
                  married  60.0    70.0     75.0     63.2   |   69.5
                  -----------------------------------------------------
                  TOTAL   100.0   100.0    100.0    100.0   |  100.0

         high     single   25.0    30.8     37.1     30.8   |   33.0
                  married  75.0    69.2     62.9     69.2   |   67.0
                  -----------------------------------------------------
                  TOTAL   100.0   100.0    100.0    100.0   |  100.0

------------------------------------------------------------------------

homemakr low      single   50.0     8.3     28.6    100.0   |   25.5
                  married  50.0    91.7     71.4      0.0   |   74.5
                  -----------------------------------------------------
                  TOTAL   100.0   100.0    100.0    100.0   |  100.0

         medium   single   10.0     2.1      9.6     21.4   |    8.1
                  married  90.0    97.9     90.4     78.6   |   91.9
                  -----------------------------------------------------
                  TOTAL   100.0   100.0    100.0    100.0   |  100.0

         high     single    0.0    14.3      6.7      0.0   |    7.1
                  married 100.0    85.7     93.3    100.0   |   92.9
                  -----------------------------------------------------
                  TOTAL   100.0   100.0    100.0    100.0   |  100.0
```

In Output 4F1.7, 4F prints the percentages of column totals. These are percents of the total for a column in a panel of the table consisting of separate rows for married and unmarried women (see Output 4Fl.5). For each level of EDUCATN, there is a separate column total for each combination of WORKSTAT and ESTEEM. For example, among women who are college graduates, engage in paid work, and are low in self-esteem, half are married and half are single (first panel). Among women of the same educational level who are high in self-esteem and are homemakers, all are married (last panel). These percentages could be used to report the proportion married within each WORKSTAT by ESTEEM by EDUCATN category. To compute column percentages over all six rows in the PAIDWORK and all six rows in the HOMEMAKER strata, see Example 4F1.8.

Example 4F1.8 STACK: combining two variables into one

You can combine several original variables into a single new variable by stacking. Variables that are stacked together are treated as a single composite variable consisting of all possible combinations of the levels of the variables being stacked. This may be helpful if you want to condense the printing of your table without collapsing across any variables. Stacking variables also allows you to create tables of percentages that are especially interesting or that may fit your needs for reporting results. 4F can print more than one categorical variable horizontally (and/or vertically) by STACKING the indices. Note that in all analyses, variables STACKED together are treated as a single categorical variable and the dimension of the table is reduced accordingly.

In Input 4F1.8 we combine two categorical variables, creating a three-way table out of four categorical variables. For one of the row variables, we stack MARITAL and ESTEEM. The other row variable remains WORKSTAT, and the column variable remains EDUCATN, as in Example 4F1.5. The STACK command forms a new variable consisting of all possible categories of work status and self-esteem. Since MARITAL is a row index in the table (not the first index), the new STACKED variable is treated as the row index with 6 levels (2 levels of MARITAL times 3 levels of ESTEEM). The order of the variables in the TABLE instruction determines whether a STACKED variable will be a row or column variable.

We request percent of COLUMN totals in the PRINT paragraph in order to find percentages of women in the various categories of EDUCATN and WORKSTAT with differing marital status and self-esteem.

Input 4F1.8
(replace the table paragraph in Input 4F1.5 and add the following PRINT paragraph)

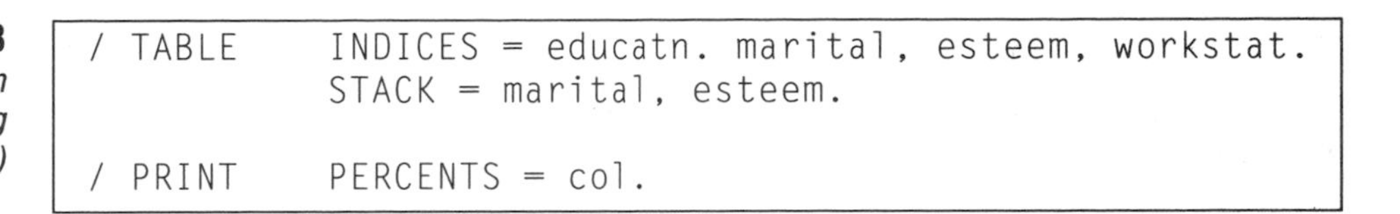

```
/ TABLE     INDICES = educatn. marital, esteem, workstat.
            STACK = marital, esteem.

/ PRINT     PERCENTS = col.
```

Output 4F1.8

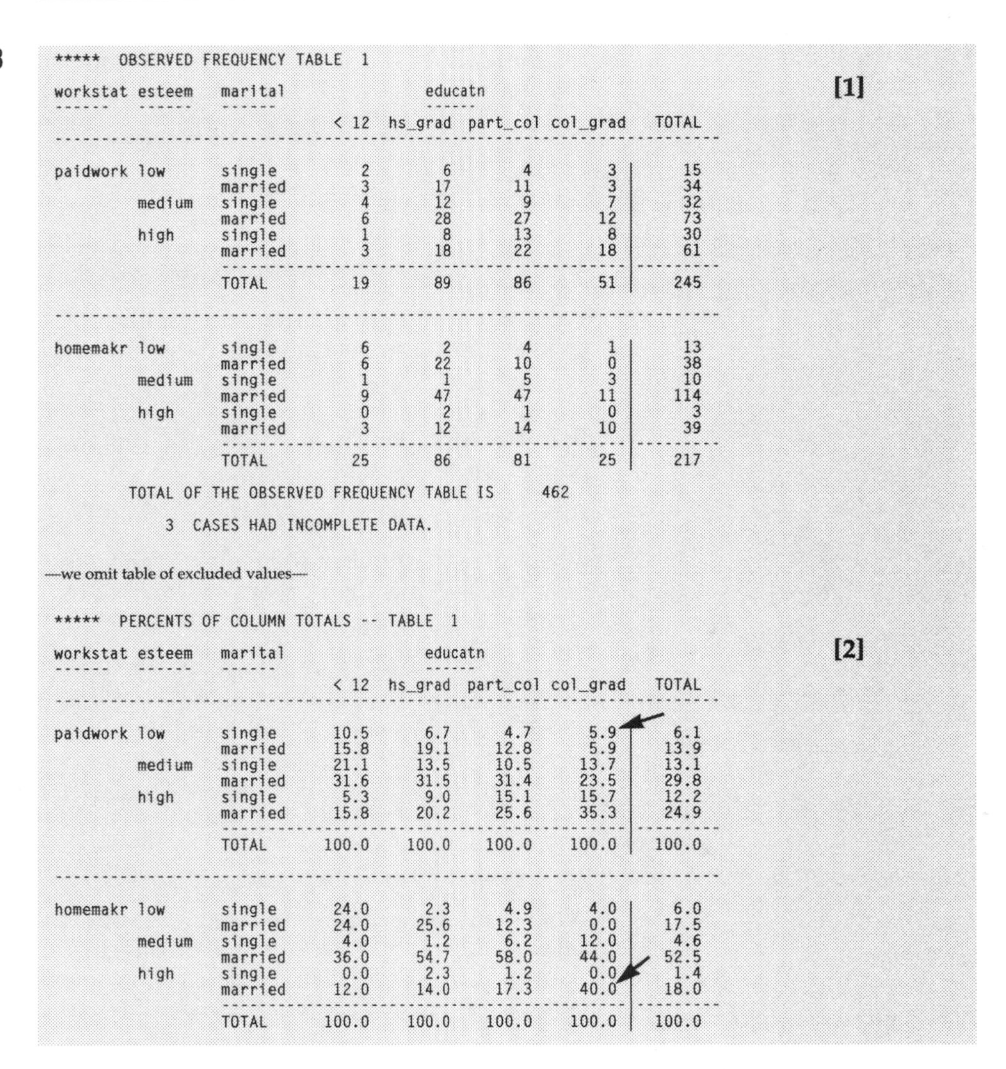

***** OBSERVED FREQUENCY TABLE 1 **[1]**

workstat	esteem	marital	< 12	hs_grad	part_col	col_grad	TOTAL
			educatn				
paidwork	low	single	2	6	4	3	15
		married	3	17	11	3	34
	medium	single	4	12	9	7	32
		married	6	28	27	12	73
	high	single	1	8	13	8	30
		married	3	18	22	18	61
		TOTAL	19	89	86	51	245
homemakr	low	single	6	2	4	1	13
		married	6	22	10	0	38
	medium	single	1	1	5	3	10
		married	9	47	47	11	114
	high	single	0	2	1	0	3
		married	3	12	14	10	39
		TOTAL	25	86	81	25	217

TOTAL OF THE OBSERVED FREQUENCY TABLE IS 462

3 CASES HAD INCOMPLETE DATA.

—we omit table of excluded values—

***** PERCENTS OF COLUMN TOTALS -- TABLE 1 **[2]**

workstat	esteem	marital	< 12	hs_grad	part_col	col_grad	TOTAL
			educatn				
paidwork	low	single	10.5	6.7	4.7	5.9	6.1
		married	15.8	19.1	12.8	5.9	13.9
	medium	single	21.1	13.5	10.5	13.7	13.1
		married	31.6	31.5	31.4	23.5	29.8
	high	single	5.3	9.0	15.1	15.7	12.2
		married	15.8	20.2	25.6	35.3	24.9
		TOTAL	100.0	100.0	100.0	100.0	100.0
homemakr	low	single	24.0	2.3	4.9	4.0	6.0
		married	24.0	25.6	12.3	0.0	17.5
	medium	single	4.0	1.2	6.2	12.0	4.6
		married	36.0	54.7	58.0	44.0	52.5
	high	single	0.0	2.3	1.2	0.0	1.4
		married	12.0	14.0	17.3	40.0	18.0
		TOTAL	100.0	100.0	100.0	100.0	100.0

[1] 4F prints the three-way observed frequency table. Row totals are based on combinations of WORKSTAT, ESTEEM, and MARITAL (summed over EDUCATN). For each combination of WORKSTAT and EDUCATN, the column totals are frequencies summed across the levels of ESTEEM and MARITAL. Compare this table with Output 4F1.5.

[2] 4F prints percents of column totals for each panel. There is a separate panel for each level of WORKSTAT. Here we see that single women with low self-esteem are 5.9% of the college graduates who are working (PAIDWORK). Among the 25 college graduates who are homemakers, 40% are married and report high self-esteem. Compare this output with Output 4F1.7; this takes up much less space and the descriptive statements are quite different. Your choice between the two will depend on the focus of your study.

Table Input and Saving Tables

4F accepts as input tabulated data, such as a table from a government or scientific report or a textbook. If the data are already in a frequency table, it may be more efficient to use the table as input than to return to the original data. The tabulated data can be entered in the form of a multiway frequency table (Example 4F1.9) or as a series of cases, each case containing a frequency and the indices of the appropriate cell (Example 4F1.10).

Example 4F1.11 shows how to SAVE a frequency table in a BMDP File, and Example 4F1.12 how to read a frequency table from a BMDP File.

Example 4F1.9 Using a multiway frequency table as input

Consider a frequency table that has two rows and three columns, such as

```
10  20  30
40  50  60
```

You can record only the frequencies, forming two lines with the frequencies as printed above. Or, you can record the frequencies on a single data line,

```
10 20 30 40 50 60
```

In either case, you must specify the dimensions of your TABLE in the INPUT paragraph, for example,

```
/ INPUT TABLE = 3, 2.
```

where 3 is the number of levels of the fastest moving index (columns in our example) and 2 is the number of levels of the index that changes more slowly (rows).

As a more complex example, we use the four-way frequency table presented in Output 4F1.5. The four categorical variables are EDUCATN, MARITAL, ESTEEM, and WORKSTAT. To use the table as input we specify

```
TABLE = 4, 2, 3, 2.
```

in the INPUT paragraph; 4 is the number of levels in the fastest moving index (the column index EDUCATN), 2 is the next fastest moving (MARITAL), 3 the next fastest (ESTEEM), and 2 is the slowest moving index (WORKSTAT).

When multiway tables are used as input, the number of VARIABLES in the INPUT paragraph is the number of categorical variables in the table (4 in our example). The format must describe all elements in the table to be read as input. 4F automatically assigns codes to each index; CODES and CUTPOINTS in the CATEGORY paragraph are ignored, and NAMES only need be specified. (NOTE: for all other types of input and in all other programs, CODES and CUTPOINTS are required whenever NAMES are specified.) When a multiway table is input, it is not possible to specify the same NAME for two different CODES for a single index; in all other types of input, all CODES with the same NAME are automatically combined into a single category.

Input 4F1.9

```
/ INPUT       VARIABLES = 4.
              FORMAT = FREE.
              TABLE = 4, 2, 3, 2.

/ VARIABLE    NAMES = educatn, marital, esteem, workstat.

/ CATEGORY    NAMES(educatn) = '< 12', hs_grad, part_col,
                                 col_grad.
              NAMES(esteem) = low, medium, high.
              NAMES(workstat) = paidwork, homemakr.
              NAMES(marital) = single, married.

/ TABLE       INDICES = educatn, marital, esteem, workstat.
```

Input 4F1.9 *(continued)*

```
/ TABLE      ROW = educatn. COL = esteem.

/ END

2   6   4   3
3  17  11   3
4  12   9   7
6  28  27  12
1   8  13   8
3  18  22  18
6   2   4   1
6  22  10   0
1   1   5   3
9  47  47  11
0   2   1   0
3  12  14  10
```

Note that we have specified two TABLE paragraphs in Input 4F1.9. The first will reproduce the original table that was read as input. If we had reordered the INDICES in the TABLE paragraph, then the order of the indices would have determined the format of the printed table; i.e., the first variable listed would be the column variable, etc. The second TABLE paragraph describes a two-way table, ESTEEM by EDUCATN, which is formed by summing the frequency table over the indices MARITAL and WORKSTAT. This second table is the same as that presented in Output 4F1.1, so we do not show it here.

Example 4F1.10 Using cell frequencies and indices as input

4F can also read input records that include frequency counts and cell indices; that is, each record contains the indices for a cell, and the count, or number of observations in that cell. The variable containing counts is identified by adding indices as a COUNT command to the TABLE paragraph as in Input 4F1.10. Similar to the usual input, the order of the variables is defined by the order of the NAMES in the VARIABLE paragraph; there is no need to order the records to correspond to the order of the cells in the table (although we have done so in the example to make it easier to read).

Because we are starting from a table, we have arbitrarily defined the levels of each variable used to index the table as 1, 2,.... If other values are used in their place, then those values would appear in the CODE lists. In the INPUT paragraph we have stated that we have five variables—four indices and one count variable (FREQ). In the TABLE paragraphs we request the same tables as in the previous example. We have added COUNT = FREQ to each TABLE paragraph to indicate that each record represents multiple observations; the number of observations is recorded in the variable FREQ. We do not show the output (the first table is the same as Output 4F1.5; the second, Output 4F1.1).

Input 4F1.10

```
/ INPUT      VARIABLES = 5.
             FORMAT IS FREE.

/ VARIABLE   NAMES = educatn, marital, esteem, workstat,
                     freq.

/ CATEGORY   CODES(educatn) = 1, 2, 3, 4.
             NAMES(educatn) = '< 12', hs_grad, part_col,
                              col_grad.
             CODES(esteem) = 1, 2, 3.
             NAMES(esteem) = low, medium, high.
```

Input 4F1.10
(continued)

```
                CODES(workstat) = 1 2.
                NAMES(workstat) = paidwork, homemakr.
                CODES(marital) = 1, 2.
                NAMES(marital) = single, married.

/ TABLE         INDICES = educatn, marital, esteem, workstat.
                COUNT = freq.
/ TABLE         ROW = educatn.
                COL = esteem.
                COUNT = freq.
/ END
1 1 1 1          2
2 1 1 1          6
3 1 1 1          4
4 1 1 1          3
1 2 1 1          3
2 2 1 1         17
3 2 1 1         11
4 2 1 1          3
. . . .
. . . .
. . . .
1 2 3 2          3
2 2 3 2         12
3 2 3 2         14
4 2 3 2         10
```

Example 4F1.11 Saving tables in a BMDP file

You may want to reanalyze a frequency table by requesting additional statistics, fitting different models, or looking at cross sections or marginal subtables. When there are many cases, it is time consuming to start each analysis from the original data. You may prefer to save the frequency tables in one analysis in order to reuse them. 4F can save frequency tables, as well as data, in a BMDP File.

4F saves tables in a data type format so that any BMDP program will be able to use the save file. To save a frequency table use the command

```
CONTENT = COUNT.
```

in the SAVE paragraph. Each table requested by the TABLE paragraph will be saved by the name COUNT1, COUNT2, etc. (instead of the old TABLE1, TABLE2). For each variable requested by the TABLE paragraph, the CODES and/or CUTPOINTS are saved instead of the original values (the last cutpoint is defined as two times the next to last cutpoint). FREQ, a variable containing the frequency counts, is added to the file. Thus each cell of the table is saved as a case with a case frequency.

In Input 4F1.11, we create a new BMDP File and store tabulated data in two subfiles of the BMDP File. We use two separate TABLE paragraphs to specify the tables. (You can create and SAVE several tables in a single TABLE paragraph. See Example 4F1.2.) We use a SAVE paragraph to specify what to store in the files and how to label them. This example demonstrates the VAX or IBM PC version of 4F. See Chapter 8 for saving BMDP Files on other computer systems. The two subfiles are written into the NEW BMDP File named FIDELL.SAV, CODED as FIDELL. You can access any table in the BMDP File by specifying its CONTENT (COUNT1, COUNT2, etc.). This use of CONTENT differs slightly from the usual BMDP File usage. Within 4F, you can save tables only as BMDP Files; you cannot write a table to a raw (ASCII) file.

Input 4F1.11

```
/ INPUT       VARIABLES = 6.
              FORMAT IS FREE.
              FILE IS 'fidell.dat'.

/ VARIABLE    NAMES = esteem, hap_stat, womenrol,educatn,
                      workstat, marital.

/ CATEGORY    CUTPOINTS(womenrol) = 17, 24.
              NAMES(womenrol) = conserv, moderate, liberal.
              CUTPOINTS(educatn) = 11, 12, 15.
              NAMES(educatn) = '< 12', hs_grad, part_col,
                               col_grad.
              CUTPOINTS(esteem) = 11, 16.
              NAMES(esteem) = low, medium, high.
              CUTPOINTS(hap_stat) = 33, 41.
              NAMES(hap_stat) = low, medium, high.
              CUTPOINT(workstat) = 0.
              NAMES(workstat) = paidwork, homemakr.
              CODES(marital) = 1, 2.
              NAMES(marital) = single, married.

/ TABLE       COLUMN = esteem.
              ROW = educatn.
/ TABLE       INDICES = educatn, marital, esteem, workstat.

/ SAVE        CONTENT = COUNT.
              FILE = 'fidell.sav'.
              CODE = fidell.
              NEW.
/ END
```

Output 4F1.11

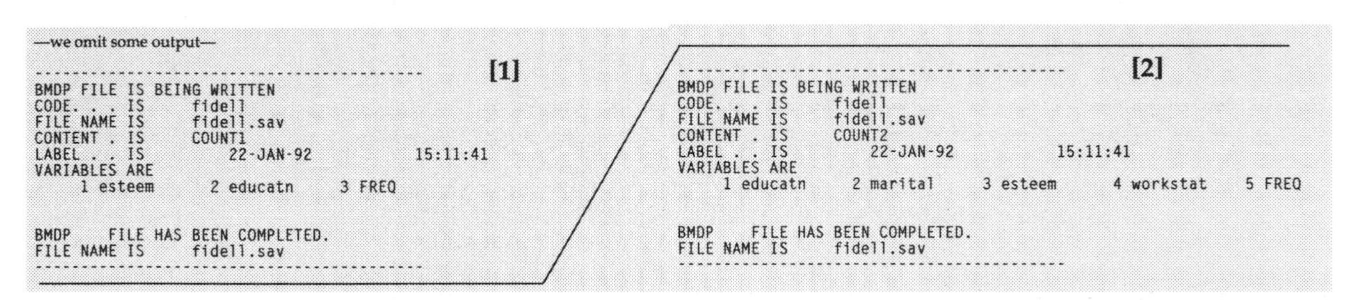

```
—we omit some output—

------------------------------------------                [1]
BMDP FILE IS BEING WRITTEN
CODE. . . IS      fidell
FILE NAME IS      fidell.sav
CONTENT . IS      COUNT1
LABEL . . IS          22-JAN-92            15:11:41
VARIABLES ARE
     1 esteem       2 educatn      3 FREQ


BMDP    FILE HAS BEEN COMPLETED.
FILE NAME IS      fidell.sav
------------------------------------------
```

```
------------------------------------------                [2]
BMDP FILE IS BEING WRITTEN
CODE. . . IS      fidell
FILE NAME IS      fidell.sav
CONTENT . IS      COUNT2
LABEL . . IS          22-JAN-92            15:11:41
VARIABLES ARE
     1 educatn      2 marital      3 esteem      4 workstat     5 FREQ


BMDP    FILE HAS BEEN COMPLETED.
FILE NAME IS      fidell.sav
------------------------------------------
```

[1] In Output 4F1.11, 4F describes each subfile and writes a message that the BMDP File has been saved. COUNT1 is saved as a two-way table, ESTEEM by EDUCATN, in the BMDP File named FIDELL.SAV and coded FIDELL.

[2] COUNT2 is saved as a four-way table, EDUCATN by MARITAL by ESTEEM by WORKSTAT. COUNT2 is added to BMDP File FIDELL.SAV, and the complete BMDP File is saved. 4F forms all of the subfiles in the BMDP Files before printing the frequency tables. Therefore if the run terminates before the analysis of the last tables, those tables may still be available in the BMDP File. CATEGORY, CODES and CUTPOINTS for the table are also saved in the BMDP File.

Example 4F1.12 Reading tables from a BMDP File

To enter a saved frequency table (i.e., CONTENT = COUNT) into any BMDP program, state CONTENT = COUNT1 or CONTENT = COUNT2, etc. in the INPUT paragraph. 4F can read and perform further analyses on tables that have been saved in a BMDP File. When there is only one table in the BMDP File, or when you want only the first table, you can omit the sequence number of the table; otherwise you must specify CONTENT. In this example, we request as input part of the second table (COUNT2) saved in Example 4F1.11. We omit variable names and category codes and names because they are stored in the files with the tables. The table in the BMDP File may be treated in the same manner as a table of cell frequencies read as input. Thus you can form and analyze a marginal subtable by specifying a subset of the variables (indices).

For 4F the command COUNT = FREQ must be added to the TABLE paragraph when the input save file is a table. You can use either variable names or numbers in the TABLE paragraph. The numbers refer to the order of the variables in the TABLE paragraph in the run that created the table, not the VARIABLE paragraph. (See the numbers assigned to the table variables in the BMDP File report in Output 4F1.11.) If you specify a CONDITION variable when you create a BMDP File, it will be saved after the last table factor and before FREQ.

In Input 4F1.12 the second table of counts (COUNT2) is read from the BMDP File FIDELL.SAV. It is the four-way table, EDUCATN by MARITAL by ESTEEM by WORKSTAT, created in the previous example. We then request a new table indexed by the first three indices of the original table; i.e., summed over WORKSTAT.

Input 4F1.12

```
/ INPUT      FILE IS 'fidell.sav'.
             CODE = fidell.
             CONTENT = COUNT2.

/ TABLE      COUNT = FREQ.
             INDICES = 1, 2, 3.

/ END
```

Output 4F1.12

```
INPUT BMDP FILE
CODE. . . IS      fidell
CONTENT . IS      DATA
LABEL . . IS
    22-JAN-92            15:11:41
VARIABLES
     1 educatn      2 marital      3 esteem      4 workstat     5 FREQ

VARIABLES TO BE USED
     1 educatn      2 marital      3 esteem      4 workstat     5 FREQ
```

—we omit intervening output—

```
    ************************
    * TABLE PARAGRAPH   2  *
    ************************

 *****  OBSERVED FREQUENCY TABLE  2

 VARIABLE     5   FREQ        USED AS COUNT VARIABLE.
                  ********

 esteem    marital               educatn
 ------    ------                ------
                      <12  hs_grad  part_col  col_grad    TOTAL
 ---------------------------------------------------------------
 low       single       8       8        8         4  |     28
           married      9      39       21         3  |     72
           ----------------------------------------------------
           TOTAL       17      47       29         7  |    100

 medium    single       5      13       14        10  |     42
           married     15      75       74        23  |    187
           ----------------------------------------------------
           TOTAL       20      88       88        33  |    229

 high      single       1      10       14         8  |     33
           married      6      30       36        28  |    100
           ----------------------------------------------------
           TOTAL        7      40       50        36  |    133

           TOTAL OF THE OBSERVED FREQUENCY TABLE IS 462
```

Special Features

Alternative commands for requesting multiway tables: CATVAR, ROW and COLUMN

In Example 4F1.5 we specify a multiway table by stating INDICES = educatn, marital, esteem, workstat. Two alternate ways of specifying this same table are:

```
/ TABLE COLUMN = educatn.  or    / TABLE CATVAR = educatn.
        ROW = marital.                   CATVAR = marital.
        CATVAR = esteem.                 CATVAR = esteem.
        CATVAR = workstat.               CATVAR = workstat.
```

Each CATVAR represents an additional dimension (level) for the table. The variable in the first CATVAR statement changes more rapidly than that in the second, etc. The advantage of these alternate representations is that many tables with the same number of levels may be requested in the same TABLE paragraph by specifying lists of variables in the same manner as described for two-way tables in Example 4F1.2. The option CROSS will apply to all possible combinations of variables across all the COLUMN, ROW, and CATVAR lists.

4F Part 1 Commands

Where to Find Commands

4F Part 1

To **form two-way tables of multiway tables**, see ROW, COLUMN, and INDICES in the TABLE paragraph and MARGINALS in the PRINT paragraph

To **input counts already in table form**, see TABLE in the INPUT paragraph or COUNT in the TABLE paragraph

4F Part 2

For **analysis of two-way tables**

4F Part 3

For **analysis of multiway tables: log-linear models**

4F Part 1 Commands

/ INPut

The INPUT paragraph is required. See Chapter 3 for INPUT options common to all programs. Commands specific to 4F include TABLE for inputting data already in tabulated form and CONTENT for BMDP File table input.

TABLE = # list. `TABLE = 4, 3, 2.`

Required if data are input as a multiway frequency table. The numbers (#) are the number of levels of each of the variables (indices) that define the table; their product is the number of cells in the table. The first number is the number of levels of the fastest varying index; the second number is the number of levels of the next fastest varying index, etc. The number of VARIABLES should be specified as the number of indices (categorical variables). The number of CASES is not used, and the FORMAT should describe the whole table. Example 4F1.9 demonstrates using a multiway frequency table as input. For data entered as cell indices and frequencies, see COUNT in the TABLE paragraph. Example: The table has 3 factors so we should also state VARIABLES = 3. The first index has 4 levels (i), the second has 3 levels (j), and the third has 2 (k). Letting f_{ijk} denote the frequency in cell (i,j,k), the order of the cell counts in the data is $f_{111}\ f_{211}\ f_{311}\ f_{411}\ f_{121} \cdots f_{132}\ f_{232}\ f_{332}\ f_{432}$.

CONTent = (*one only*)
DATA, COUNT#. `CONTENT = COUNT2.`

Instructs 4F to read raw DATA or a table of COUNTS from a BMDP File. By default, 4F assumes the file to be read contains DATA. # is the sequence number of the COUNT table in the BMDP File that was assigned when the file was created. The example instructs 4F to read the second table of counts in the BMDP File.

/ VARiable

The VARIABLE paragraph is optional. (See detailed description in Chapter 4.) Additional commands for 4F are FREQUENCY and MMV.

FREQuency = variable. `FREQ = COUNT.`

State the name or number of a variable containing the case frequency. FREQ works the same as COUNT in the TABLE paragraph. Either FREQ or COUNT is required when data are input as cell indices and frequencies, either in raw data form or as a BMDP File.

MMV = #. `MMV = 10.`

Specify the maximum number of missing values or values out of range that will be allowed before a case is excluded from the analysis. Default: use all acceptable values.

/ CATEGory (*or* GROUP)

CODES or CUTPOINTS are required for any categorical variable with more than 10 levels, and category NAMES are useful for labeling tables. The CATEGORY paragraph is described in detail in Chapter 5. 4F initially allocates space for all tables requested using the number of levels specified for each categorical variable or 10 (the default allocation). For multiway tables or many two-way tables, the use of 10 as the default allocation may cause 4F to exceed the space available for the data. For example, if one four-way table is requested and 10 levels are assumed for each index, 10^4 = 10,000 cells will be required. Therefore, we recommend that CODES or CUTPOINTS be specified for multiway tables or when multiple tables are requested.

/ TABLE

The TABLE paragraph is required and may be repeated. This paragraph defines tables to be formed. To form two-way tables, see ROW and COLUMN; for multiway tables see INDEX (or CATVAR).

COLumn = list. `COL = MARITAL.`

List the names or numbers of the variables that define column categories. 4F forms a separate table for each pair of variables in the ROW and COLUMN lists (see CROSS below for all combinations of ROW and COLUMN variables). If a COLUMN variable has more than ten distinct categories, you must specify its CODES or CUTPOINTS in the CATEGORY paragraph.

ROW = list. `ROW = WORKSTAT.`

List the names or numbers of the variables that define the row categories (see COLUMN, PAIR and CROSS). If a ROW variable has more than ten distinct categories, you must specify its CODES or CUTPOINTS in the CATEGORY paragraph. Examples 4F1.1 and 4F1.2 show how to form a two-way table using the ROW and COLUMN instructions.

INDices = list.

```
IND = WORKSTAT, MARITAL,
      RELIGION, ESTEEM.
```

List the names or numbers of the variables that define the multiway table. The first variable listed is the fastest moving index in the table and will form the columns of the table. (Note that this deviates from ANOVA programs such as 2V and 4V, in which the first listed variable is the slowest moving index.) See Example 4F1.5.

CATvar = variable.

```
CAT = WORKSTAT. CAT=MARITAL.
CAT = RELIGION. CAT=ESTEEM.
```

List the name or number for each variable used to define a multiway table. This is an alternative to the INDICES list. The CATVAR list may be repeated. If CAT-

VAR is repeated, the index in the first CATVAR statement varies more rapidly than that of the second CATVAR statement, etc. **Example:** The ordering of variables is the same in the above INDICES and CATVAR statement examples.

PAIR. *or* **CROSS.** `CROSS.`

If you state CROSS, 4F forms two-way tables from all possible combinations of the ROW and COLUMN (and/or CATVAR) variables. The first COLUMN variable is crossed with all the ROW variables in turn, then the second COLUMN variable, and so on. If PAIR (the default) is specified, the first ROW variable and the first COLUMN variable form the first table, the second ROW variable and the second COLUMN variable form the second table, and so on; there must be the same number of variables in each COLUMN and ROW (and/or CATVAR) statement. Example 4F1.2 shows how to request several two-way tables using the CROSS command.

CONDition = list. `COND = SEX.`

List the names or numbers of variables used to stratify the tables. The CONDITION list is used to form separate frequency tables for each category of the CONDITION variable. If you specify more than one CONDITION variable, each variable is used in turn. If the variable takes on more than ten distinct values, CODES or CUTPOINTS must be specified for it in the CATEGORY paragraph. See Example 4F1.4.

COUNT = list. `COUNT = FREQ.`

Required when data are input as cell indices and frequencies. List names or numbers of variable(s) that contain frequency counts. Tables are formed separately for each variable in the list. State COUNT = FREQ when reading a BMDP File containing a table of counts (see also INPUT CONTENT above). The COUNT command overrides the VARIABLE FREQUENCY command.

STACK = list. `STACK = MARITAL, WORKSTAT.`

List the names or numbers of variables used to form a new table factor that consists of all possible combinations of the levels of the variables. An index in one table cannot appear in more than one STACK list. Stacking only takes place when all the variables in the STACK list appear as indices in a table. If the STACK specification includes the column (first) categorical variable in the table, the new STACKED variable is printed horizontally. In all analyses the new index is treated as a single index. The STACK command may be repeated to create more than one stacked variable. STACK is demonstrated in Example 4F1.8.

Additional TABLE instructions

Chapter 4F Part 3 shows additional TABLE instructions to use when fitting log-linear models. The EMPTY and INITIAL instructions allow you to specify cells whose frequencies are to be set to structural zeros. The DELTA instruction adds a constant to each cell. The SYMBOLS instruction assigns symbols by which variables are referred to in model specifications.

/ PRINT

The PRINT paragraph is optional. It is used to specify the results to be printed. If this paragraph is omitted, 4F automatically prints OBSERVED and EXCLUDED tables and LISTS the first five excluded cases. Any other parameters must be specified to be printed. The PRINT paragraph may be repeated after each TABLE paragraph. Additional PRINT features are described in Chapter 4F Parts 2 and 3.

OBServed. `NO OBSERVED.`

Unless you specify NO OBSERVED, 4F prints tables of observed frequencies.

EXCLuded. `NO EXCL.`

Unless you specify NO EXCL, 4F prints a two-way table of frequencies of the excluded cases. This table follows each frequency table from which cases have been excluded and shows the reasons for exclusion. See Example 4F1.1 for a table of excluded cases.

PERCent = *(one or more).* `PERC = ROW, COL.`
NONe, ROW, COL, TOTal.

Specify optional tables of PERCENTS to be printed. Specify ROW for a table of percents of row totals, COL for a table of percents of the column totals, and TOTAL for a table of percents of the total frequency. Default: NONE of the tables are printed. Tables of percentages are shown in Examples 4F1.3, 4F1.7, and 4F1.8.

MARGinal = #. `MARG = 2.`

Use to request marginal subtables for all dimensions up to and including order #. The marginal table for the highest order is the observed table and is printed by default. For example, if you have a three-way table and request MARG = 2, 4F will print all possible marginal tables: the three-way table, three two-way tables, and three one-way tables. Example 4F1.6 describes marginal subtables.

LIST = #. `LIST = 20.`

LIST the first # of cases that were not included in one or more tables. # is the maximum number of cases to be printed in a case-by-table display. By default, 4F prints the first five excluded cases.

BAR = 'c'. `BAR = 'I'.`

Specify the character to be printed along vertical lines in two-way and multi-way tables. The default character is a vertical bar (|).

New Feature

CASE = #. `CASE = 50.`

Specify the number of cases for which data are printed. Values flagged by MISS, MAX, and MIN in the VARIABLE paragraph specifications are replaced by MISSING, GT MAX, and LT MIN. Default: 10.

New Feature

MINimum. `MIN.`

Prints data for cases in which at least one variable has a value less than its lower limit as specified in the VARIABLE paragraph. Default: such cases are not printed unless you specify DATA.

New Feature

MAXimum. `MAX.`

Prints data for cases in which at least one variable has a value greater than its upper limit as specified in the VARIABLE paragraph. Default: such cases are not printed unless you specify DATA.

New Feature

MISSing. `MISS.`

Prints data for cases that have at least one variable with a missing value. Missing values in the input may be identified in several ways: (1) an asterisk (the default missing value character), (2) a missing value code identified using MISS in the VARIABLE paragraph, (3) a character identified as MCHAR in the INPUT paragraph, (4) blanks (applies only to input read with FORTRAN fixed format). Default: such cases are not printed unless you specify DATA.

New Feature

METHod = VARiable *or* **CASE.** `METH = CASE.`

Use to determine how the printout is organized. VARIABLE is the default value.

VARIABLE – Prints for each case only as many variables as fill the print line (up to ten variables) and begins the next line with the same variables for the next case. Thus the data for each variable are in an easy-to-scan column. When the program finishes printing the first subset of variables for all cases, it continues with another subset.

CASE – Prints the values of all variables for one case (possibly filling several print lines) before printing data for the next case.

The following commands work only in the batch mode

New Feature

MEAN. `NO MEAN.`

Unless you specify NO MEAN, 4F prints the mean, standard deviation, and frequency of each used variable.

New Feature

EXTReme. `NO EXTR.`

Unless you state NO EXTR, 4F prints the minimum and maximum values for each variable used.

New Feature

EZSCores. `NO EZSC.`

Prints z-scores for the minimum and maximum values of each used variable.

New Feature

ECASe. `NO ECAS.`

Prints the case numbers containing the minimum and maximum values of each variable.

New Feature

SKewness. `SK.`

Prints skewness and kurtosis for each used variable.

Additional PRINT Commands

See 4F Part 2 for additional PRINT options that concern analysis of two-way frequency tables.

/ SAVE

The SAVE paragraph is required only if you want to create a BMDP File. See Chapter 8 for details about the SAVE paragraph and Example 4F1.11 for specifics about saving tables.

CONTent = *(one or both)* `CONTENT = COUNT.`
DATA, COUNT.

Instructs 4F to save raw data or tables of counts in a BMDP File. When tables of counts are saved, a CONTENT is generated for each of the tables requested. The content consists of the word COUNT followed by the sequence number of the table: for example, COUNT1 is the first table requested. The CONTENT is printed when the BMDP File is formed and must be used to recall a table other than the first. Example 4F1.11 shows how to save tables of counts in a BMDP File.

Order of Instructions

■ indicates required paragraph

■ / INPUT
/ VARIABLE
/ GROUP or CATEGORY
/ TRANSFORM
/ SAVE
■ / TABLE
/ FIT
/ STATISTICS
/ PRINT
■ / END
data

Repeat for subproblems (/ TABLE through / PRINT)

Repeat for additional problems. See Multiple Problems, Chapter 10. (/ INPUT through data)

Summary Table for Commands Specific to 4F, Part 1

Paragraphs Commands	Defaults	Multiple Problems	See
/ **INPut**			
TABLE = # list.	input is not a table	–	4F1.9
CONTent = (*one only*) DATA, COUNT#.	DATA	–	4F1.12
/ **VARiable**			
FREQuency = variable.	no case frequency variable	–	Cmds
MMV = #.	all acceptable values	–	Cmds
▲ / **CATEGory** (*or* **GROUP**)			
■ / **TABLE**			
▲ COLumn = list.	none	–	4F1.1
▲ ROW = list.	none	–	4F1.1
or INDices = list.	see ROW and COL	–	4F1.5
or CATvar = list.	see ROW and COL	–	SpecF
PAIR. *or* CROSS.	PAIR.	–	4F1.2
CONDition = list.	no condition variables	–	4F1.4
COUNT = list.	the variable stated in /VAR FREQ =	–	4F1.10
STACK = list.	no stacked variables	–	4F1.8
/ **PRINT**			
OBServed. *or* NO OBS.	OBSERVED.	●	Cmds
EXCLuded. *or* NO EXCL.	EXCLUDED.	●	Cmds
PERCent = (*one or more*) NONe, ROW, COL, TOT.	NONE	●	4F1.3
MARGinal = #.	no marginal subtables	●	4F1.6
LIST = #.	5	●	Cmds
BAR = 'c'.	vertical bar (I)	●	Cmds
CASE = #.	10	●	Cmds
MISSing.	NO MISS.	●	Cmds
MINimum.	NO MIN.	●	Cmds
MAXimum.	NO MAX.	●	Cmds
METHod = VARiables *or* CASE.	VARIABLES	–	Cmds
MEAN. *or* NO MEAN.	MEAN.	●	Cmds
EXTReme. *or* NO EXTR.	EXTReme. values printed	●	Cmds
EZSCores.	NO EZSC. (*z*-scores)	●	Cmds
ECASe.	NO ECAS. (Case #'s of Min and Max printed)	●	Cmds
SKewness.	NO SK. (skewness and kurtosis)	●	Cmds
/ **SAVE**			
CONTent = (one or both) DATA, COUNT.	DATA	–	4F1.11

Key:
- ■ Required command or paragraph
- ▲ Frequently used command or paragraph
- ● Value retained for multiple problems
- – Default reassigned

4F

Part 2
Analysis of Two-Way Tables

For two-way frequency tables, 4F can compute more than two dozen statistics that can be used as tests of independence or as measures of association and correlation. Not all measures are appropriate for a given table. We subdivide the examples into classes based on the structure of the table, the size of the table, and whether or not the categories of the table factors are ordered. In our notation, 2 x 2 is a table with two rows and two columns, R x C is a table of any size (including a 2 x 2 table), and R x R is a square table.

Categories for an index of the table may or may not be ordered. Examples of unordered categories are sex, religion, or occupation. Ordered categories may be formed naturally (e.g., number of pregnancies, educational level completed) or may be obtained by grouping continuous data into intervals (e.g., intervals of height, driving speed). Different statistics may be appropriate depending on your hypotheses about the variables and whether or not categories are ordered.

For all two-way tables, 4F can print expected frequencies under the hypothesis of independence between the row and column variables. 4F also can print the difference between observed and expected frequencies and functions of the difference, such as standardized deviates.

Chapter 4F Part 1 shows how to form two-way (and multiway) frequency tables. For analysis of multiway frequency tables using log-linear models, see Chapter 4F, Part 3. For further reading, see Bishop et al. (1975), Cox (1970), Fienberg (1977), Fleiss (1981), Greenland (1983).

Table 4F2.1 summarizes the 26 tests and measures produced by 4F for two-way tables. These tests are described in more detail in later tables. Formulas appear in Appendix B.11.

Table 4F2.1
Overview and usage of statistics and measures

Unordered Categories		
R x C	**2 x 2**	**R x R**
Pearson χ^2 test	χ^2 (Yates corrected χ^2)	McNemar's test of symmetry
G^2 likelihood ratio test	Fisher's exact test	Kappa measure of reliability
ϕ (phi)	ϕ (phi)	
Contingency coefficient, C	Yule's Q	
Cramér's V	Yule's Y	
Goodman and Kruskal τ	Cross-product (odds) ratio α	
Optimal prediction λ	Mantel-Haenszel	
Optimal prediction λ^*		
Uncertainty coefficient, U		
Ordered Categories		
R x C	**2 x 2**	**2 x C or R x 2**
Gamma, Γ	Tetrachoric correlation, r_t	Test of linear trend
Kendall's τ_b		
Stuart's τ_c		
Product-moment correlation, r		
Spearman rank correlation, r_s		
Somers' D		

Where to Find It

Examples

Where to Find It *(continued)*

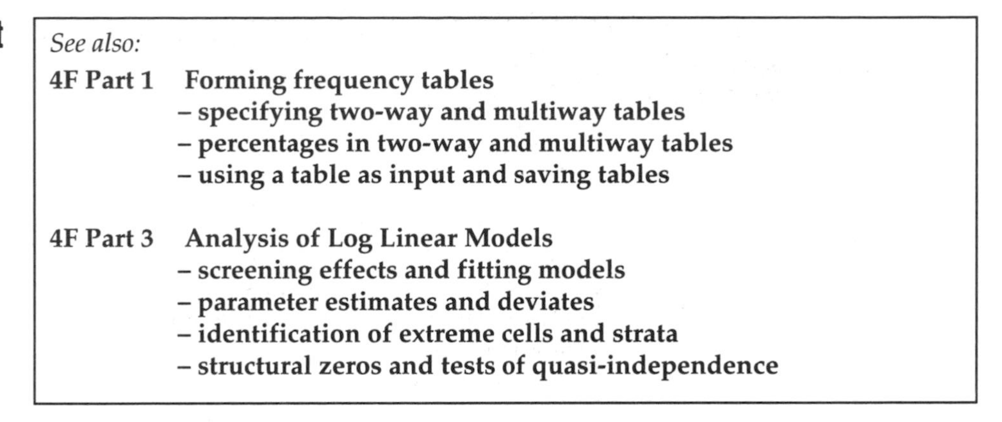

See also:

4F Part 1 Forming frequency tables
- **specifying two-way and multiway tables**
- **percentages in two-way and multiway tables**
- **using a table as input and saving tables**

4F Part 3 Analysis of Log Linear Models
- **screening effects and fitting models**
- **parameter estimates and deviates**
- **identification of extreme cells and strata**
- **structural zeros and tests of quasi-independence**

The R x C Table

For the examples in this section, we use the survey on demographic and psychological variables described in Chapter 4F, Part 1 (see Appendix D). These examples demonstrate the analysis of two-way tables in which neither of the variables necessarily has ordered categories. Much of the material in this section can apply to any table, including 2 x 2. If the only statistic you need is the usual Pearson χ^2, additional instructions are unnecessary. The basic setup plus a TABLE paragraph always produces a χ^2 test. This statistic is explained in Example 4F2.1 along with other tests.

Table 4F2.2: Tests and measures for R x C tables (unordered categories)

Test or Measure	Command	Range of Values	Test that Measure is Zero	Interpretation and Example
Pearson χ^2 test	default	0,∞	χ^2 with (R–1)(C–1)df	Tests of independence of rows and columns or equality of proportions between rows and columns (4F2.1)
G^2 likelihood ratio test	LRCHI	0,∞	"	
ϕ (phi)	CONT	0,m	use χ^2 test	Same as χ^2, but not dependent on sample size. Used to compare tables (4F2.1)
Contingency coeff., *C*	CONT	0,1	"	
Cramér's *V*	CONT	0,1	"	
Goodman and Kruskal τ	TAUS	0,1		Compare random proportional prediction of one index with conditional proportion prediction based on second index (4F2.3)
Optimal prediction λ	LAMB	0,1		Measure of predictive association—relative degree of success with which one index can be used to predict second index (4F2.3)
Optimal prediction λ*	LAMB	0,1		Similar to λ, but adjusted to have common row marginals (4F2.3)
Uncertainty coeff., *U*	UNC	0,1	use G^2 test	Relative reduction in uncertainty about one index when the second index is known (4F2.3)

Note: The notation is defined in Appendix B.11.

Example 4F2.1 Tests of independence: chi-square, phi, *C*, Cramér's *V*

In Input 4F2.1 we create the 3 x 4 table, ESTEEM by EDUCATN, formed in Example 4F1.1. In the STATISTICS paragraph we request tests of independence appropriate for an R x C table: Pearson chi-square (by default), likelihood ratio chi-square (LRCHI), and measures of CONTINGENCY. The data are stored in a system file called FIDELL.DAT. The instructions for the INPUT, VARIABLE, and CATEGORY paragraphs are described in Chapters 3–5.

Input 4F2.1

```
/ INPUT        VARIABLES = 6.
               FORMAT IS FREE.
               FILE IS 'fidell.dat'.

/ VARIABLE     NAMES = esteem, hap_stat, womenrol,
                       educatn, workstat, marital.

/ CATEGORY     CODES(marital) = 1, 2.
               NAMES(marital) = single, married.
               CUTPOINTS(esteem) = 11, 16.
               NAMES(esteem) = low, medium, high.
               CUTPOINTS(educatn) = 11, 12, 15.
               NAMES(educatn) = '< 12', hs_grad, part_col,
                                col_grad.
               CUTPOINTS(hap_stat) = 33, 41.
               NAMES(hap_stat) = low, medium, high.
               CUTPOINT(workstat) = 0.
               NAMES(workstat) = paidwork, homemakr.

/ TABLE        COLUMN = esteem.
               ROW = educatn.

/ STATISTICS LRCHI.
               CONTINGENCY.
/ END
```

Output 4F2.1

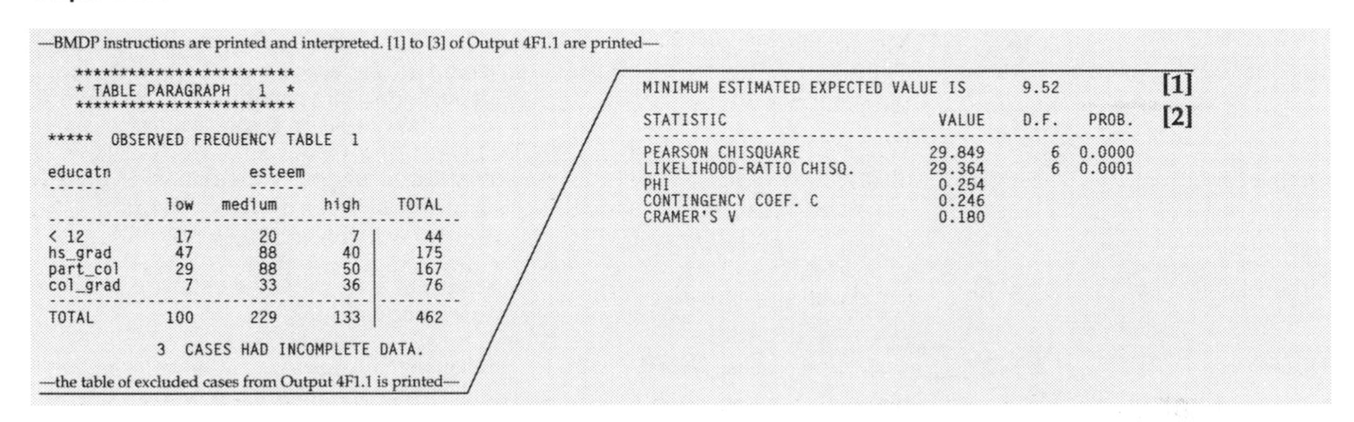

—BMDP instructions are printed and interpreted. [1] to [3] of Output 4F1.1 are printed—

```
    ***********************
    * TABLE PARAGRAPH   1  *
    ***********************

***** OBSERVED FREQUENCY TABLE  1

educatn              esteem
------               ------
            low   medium     high    TOTAL
----------------------------------------------
< 12         17       20        7 |     44
hs_grad      47       88       40 |    175
part_col     29       88       50 |    167
col_grad      7       33       36 |     76
----------------------------------------------
TOTAL       100      229      133 |    462

          3  CASES HAD INCOMPLETE DATA.
```

—the table of excluded cases from Output 4F1.1 is printed—

```
MINIMUM ESTIMATED EXPECTED VALUE IS      9.52          [1]

STATISTIC                    VALUE    D.F.   PROB.     [2]
------------------------------------------------------
PEARSON CHISQUARE           29.849       6  0.0000
LIKELIHOOD-RATIO CHISQ.     29.364       6  0.0001
PHI                          0.254
CONTINGENCY COEF. C          0.246
CRAMER'S V                   0.180
```

Bold numbers below correspond to those in Output 4F2.1.

[1] Under the hypothesis of independence between ROW and COLUMN variables, the expected frequency in a cell is its

(row total) x (column total)/(total frequency in table).

4F prints the smallest expected value. A very small expected value indicates that the chi-square statistic for testing independence may be poorly approximated by a chi-square distribution and the resulting p-value may be misleading. In Example 4F2.2 we discuss how small is too small. The smallest value in this example, 9.52, is acceptable. With very small values, you may want to consider combining categories of one or more of the variables (see Example 4F2.4). Rows or columns that are entirely zero are excluded from the computations of all statistics in a two-way frequency table.

[2] 4F automatically computes and prints the Pearson chi-square test of independence between rows and columns without requiring a STATISTICS paragraph. The Pearson chi-square statistic is defined as

$$\chi^2 = \frac{\sum_i \sum_j \left(f_{ij} - F_{ij}\right)^2}{F_{ij}}$$

where f_{ij} is the observed frequency in a cell (i,j) and F_{ij} is the expected frequency in that cell as in the formula in **[1]** above. 4F prints the value of the chi-square statistic, the degrees of freedom (D.F.), and the probability value. That value is the probability of getting a chi-square value larger than that observed when the null hypothesis of independence is true, i.e., its p-value. If you choose to test a hypothesis at a given level of significance (say .01), a PROBABILITY less than this level is a significant result. In the example, we can see that there is a significant relationship between self-esteem and education ($p < .00005$). Percentages may be useful in identifying associations of interest between the variables (as seen in Example 4F1.3).

The likelihood ratio chi-square statistic, G^2, is an alternative to the usual Pearson chi-square test. This test is based on maximum likelihood theory and is used more than any other test statistic in the analysis of multiway frequency tables. It is defined as

$$G^2 = 2\sum_i \sum_j f_{ij} \ln\left(f_{ij} / F_{ij}\right)$$

When the null hypothesis of independence is not appropriate (i.e., when the row and column variables are associated), the expected value of the chi-square statistic is proportional to the sample size, N. Therefore, chi-square results for tables based on different sample sizes are not comparable. Measures that are not dependent on N, such as the contingency coefficient C, phi, and Cramér's V, can be used to compare tables with different sample sizes. See Table 4F2.3 and Kendall and Stewart (1967).

When the table is 2 x 2, phi corresponds to Cramér's V. Otherwise V is always less than or equal to one, but phi can take on values greater than 1. The contingency coefficient C is normalized differently from phi and V. All three measures are zero if (and only if) the chi-square statistic is zero. Therefore, while the three measures are scaled differently, all three test the same null hypothesis of independence.

Example 4F2.2 Expected values and cell deviates

In a two-way frequency table, expected frequencies are of interest for two reasons:

1. Expected values can help you decide whether the distribution of your chi-square statistic is well approximated by the theoretical or tabulated chi-square distribution. The following guideline is commonly accepted. No cell may have an expected value less than 1, and not more than 20% of the cells should have expected values less than 5. When more cells have small expected frequencies than indicated by this rule, the distribution of the chi-square statistic can differ widely from the tabulated chi-square distribution. Example 4F2.4 shows how to combine categories with small expected frequencies when this rule is violated. Some investigators use a more

conservative rule suggested by Fisher—that no cell may have an expected value less than 5. We recommend Fisher's rule for 2 x 2 tables. When expected values are small, use Fisher's exact test (see Example 4F2.5).

2. Expected frequencies can be used to examine patterns in the data that contribute to a significant value for the chi-square test. For example, we can ask whether the counts in a few cells fall far below or far above those predicted from the pattern of the remainder of the table. In addition to the expected values, 4F can compute various cell deviates to help you assess how the observed counts deviate from those expected.

In Input 4F2.2 we request the expected frequencies for the table formed in Example 4F2.1 and three of the six types of deviates measuring the difference between observed and expected cell frequencies.

Input 4F2.2

(replace the STATISTICS paragraph in Input 4F2.1 with the following PRINT paragraph)

```
/ PRINT        EXPECTED.
               DIFFERENCE.
               ADJUSTED.
               CHISQ.
```

Output 4F.2

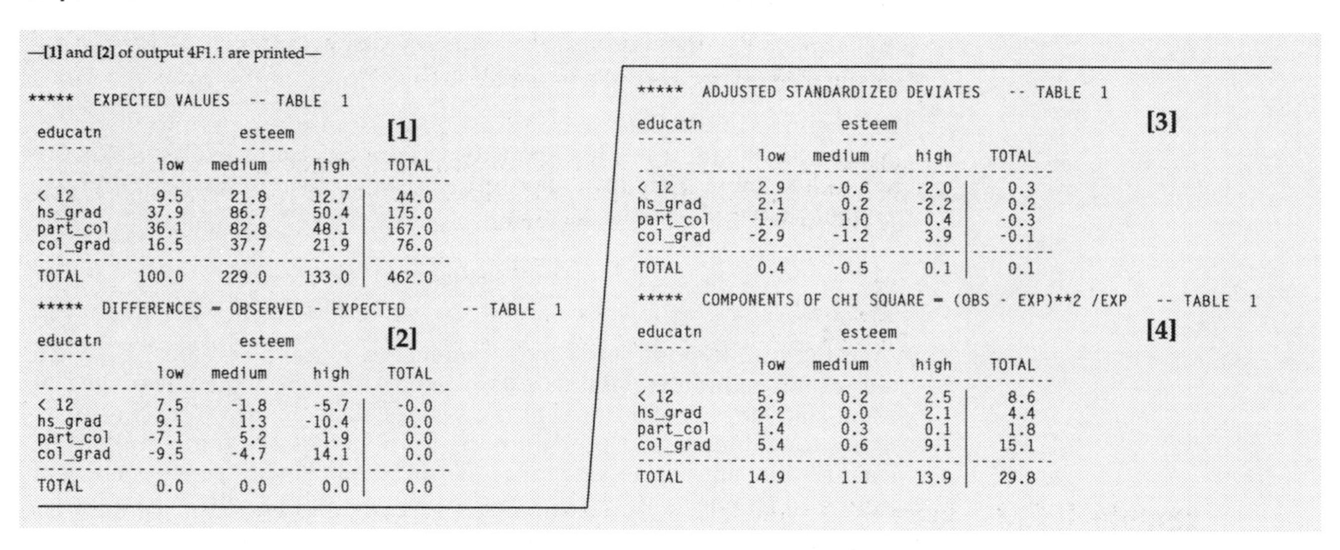

—[1] and [2] of output 4F1.1 are printed—

***** EXPECTED VALUES -- TABLE 1 [1]

educatn	esteem low	medium	high	TOTAL
< 12	9.5	21.8	12.7	44.0
hs_grad	37.9	86.7	50.4	175.0
part_col	36.1	82.8	48.1	167.0
col_grad	16.5	37.7	21.9	76.0
TOTAL	100.0	229.0	133.0	462.0

***** DIFFERENCES = OBSERVED - EXPECTED -- TABLE 1 [2]

educatn	esteem low	medium	high	TOTAL
< 12	7.5	-1.8	-5.7	-0.0
hs_grad	9.1	1.3	-10.4	0.0
part_col	-7.1	5.2	1.9	0.0
col_grad	-9.5	-4.7	14.1	0.0
TOTAL	0.0	0.0	0.0	0.0

***** ADJUSTED STANDARDIZED DEVIATES -- TABLE 1 [3]

educatn	esteem low	medium	high	TOTAL
< 12	2.9	-0.6	-2.0	0.3
hs_grad	2.1	0.2	-2.2	0.2
part_col	-1.7	1.0	0.4	-0.3
col_grad	-2.9	-1.2	3.9	-0.1
TOTAL	0.4	-0.5	0.1	0.1

***** COMPONENTS OF CHI SQUARE = (OBS - EXP)**2 /EXP -- TABLE 1 [4]

educatn	esteem low	medium	high	TOTAL
< 12	5.9	0.2	2.5	8.6
hs_grad	2.2	0.0	2.1	4.4
part_col	1.4	0.3	0.1	1.8
col_grad	5.4	0.6	9.1	15.1
TOTAL	14.9	1.1	13.9	29.8

[1] 4F prints the table of expected values under the hypothesis of independence between rows (EDUCATN) and columns (ESTEEM). For example, were there no association between ESTEEM and EDUCATN, you would expect 21.9 women to have high self-esteem and be college graduates (rather than the 36 who were in this cell, as seen in the observed frequency table in Example 4F2.1).

Since expected values are computed using the total frequency in the table, each cell frequency affects the computation of all the expected values. Also, the total of the observed frequencies must equal the total of the expected frequencies in each row and in each column. Therefore, when comparing observed and expected values (or any of the deviates described below) it is necessary to allow for the influence of a single divergent cell frequency (i.e., one with a large difference between observed and expected frequencies) on the calculation of all the other values, especially on those in the same row or column.

[2] For each cell 4F prints the table of differences between observed and expected values. High positive values indicate that more women had these combinations of attributes than expected, and high negative values indicate that too few women had the combination.

[3] 4F prints the table of adjusted standardized deviates. The differences between observed and expected values are normalized to have variance equal to one when the data are from a multinomial distribution (Haberman, 1973):

$$\frac{f_{ij} - F_{ij}}{\sqrt{F_{ij}(1 - r_i / N)(1 - c_j / N)}}$$

where r_i is the row total and c_j is the column total. These values are similar to z-scores. Extreme values, say greater than 3.0 in absolute value, indicate deviations that are unusually large when row and column variables are independent. The largest standardized discrepancy between observed and expected frequencies is for college graduates with high self-esteem. The positive sign indicates that there are more women than expected in this cell.

[4] 4F prints the table of components of the Pearson chi-square statistic:

$$(f_{ij} - F_{ij})^2 / F_{ij}$$

Of the total chi-square value of 29.8, 9.1 is contributed by one cell: college graduates with high self -esteem.

Other measures of deviation. Standardized deviates are the square roots of the components of Pearson's chi-square statistic,

$$(f_{ij} - F_{ij}) / F_{ij}^{1/2}$$

where f_{ij} and F_{ij} are the observed and expected frequencies, respectively, for each cell (i,j). Components of the likelihood ratio chi-square statistic are also available. They are of the form:

$$2 f_{ij} \ln(_{\text{fij}} / F_{ij})$$

Freeman-Tukey deviates are similar to z-scores when the data are from a Poisson distribution; they are defined as

$$(f_{ij})^{1/2} + (f_{ij} + 1)^{1/2} - (4F_{ij} + 1)^{1/2}$$

Example 4F2.3 Measures of prediction and uncertainty: Goodman and Kruskal's tau, lambda

Input 4F2.3 requests several measures of prediction and uncertainty for the 3 x 4 table formed in Example 4F2.1, ESTEEM by EDUCATN. These statistics measure the gain in predicting one categorical variable from another (or reduction in uncertainty) when the value of one variable is known, relative to when it is not known.

The conditions imposed on the frequencies differ for the various measures of uncertainty and prediction. Goodman and Kruskal (1954) discuss lambda, lambda-star, and Goodman and Kruskal's τ. Brown (1975) discusses the uncertainty coefficient. All four of these measures assume no ordering of the categories, and they are inherently asymmetric. That is, the gain in prediction is different depending on the direction of prediction. For example, the gain in prediction of ESTEEM from EDUCATN differs from the gain in prediction of EDUCATN from ESTEEM. Moreover, since self-esteem is more likely to be influenced by education than the converse, only the former direction is appropriate. 4F computes gain for both directions of prediction, and you must choose the one appropriate for your purposes. You will want to predict the variable considered "dependent" from the variable considered "independent." Both lambda and the uncertainty coefficient also have a symmetric form, which can be interpreted as an average gain in prediction with no direction specified.

Input 4F2.3
(replace the STATISTICS paragraph in 4F2.1 with the following)

```
/ STATISTICS   TAUS.
               LAMBDA.
               UNCERTAINTY.
```

Output 4F2.3

—[1] and [5] of Output 4F1.1 are printed—

```
MINIMUM ESTIMATED EXPECTED VALUE IS      9.52

STATISTIC                          VALUE    D.F.   PROB.
-------------------------------------------------------
PEARSON CHISQUARE                 29.849       6  0.0000

STATISTIC                          VALUE    ASE1 T-VALUE DEP.
------------------------------------------------------------
LAMBDA SYMMETRIC                   0.025   0.024
LAMBDA ASYMMETRIC                  0.013   0.035            1
LAMBDA ASYMMETRIC                  0.035   0.032            4
LAMBDA-STAR ASYMMETRIC             0.019   0.052            1
LAMBDA-STAR ASYMMETRIC             0.041   0.038            4
UNCERTAINTY COEF NORMED            0.028   0.010
UNCERTAINTY ASYMMETRIC             0.031   0.011            1
UNCERTAINTY ASYMMETRIC             0.025   0.009            4
TAU ASYMMETRIC                     0.027   0.010            1
TAU ASYMMETRIC                     0.018   0.007            4
```

For all of the measures in Output 4F2.3, an estimate of the asymptotic standard error under the alternative hypothesis (that the parameter is not equal to zero) is printed in the column labeled ASE1. This asymptotic standard error can be used in setting an approximate 95% confidence interval for the parameter being estimated—the limits are VALUE + 2(ASE1). The values in the column labeled *t*-value are not appropriate here. They are used for other tests (see, for example 4F2.5). Note that under independence between row and column variables, all these statistics are zero. However, all but the uncertainty coefficient can be zero when independence does not hold.

When the statistic is asymmetric in its treatment of the two categorical variables, the index of the dependent variable (the one to be predicted) is indicated in the column labeled DEP. In our example, the index for ESTEEM is 1 (it is the first variable on each data record) and the index for EDUCATN is 4.

Example 4F2.4 Combining rows and columns when there are small expected values

When there are too many small expected frequencies (see Example 4F2.2 for guidelines for minimum values), we recommend that you collapse categories to form new categories with larger frequencies. This is best done by examining the table of expected frequencies and modifying the names of the CODES or CUTPOINTS in the CATEGORY paragraph—CODES or CUTPOINTS with the same name are combined when 4F forms the table (see Example 4F2.5). Use your knowledge of the data to collapse categories where appropriate.

Alternatively, 4F can automatically collapse adjacent categories until the minimum expected value exceeds a specified MINIMUM value. Such a method may combine categories that are neighboring but dissimilar. Therefore, this option should be used with caution, especially when categories are not ordered. Depending on the problem, it may be more reasonable to run the program twice: once to determine the expected cell frequencies, and a second time to collapse categories by modifying cutpoints or names in the CATEGORY paragraph.

In Input 4F2.4 we form the frequency table ESTEEM by EDUCATN for each level of WORKSTAT, i.e., CONDITIONED on WORKSTAT. (See Example 4F1.4 for a discussion of conditioning.) Each of the two tables formed (women who are paid to work (PAIDWORK) and homemakers) contains expected frequencies less than five. To demonstrate automatic collapsing, we specify MINIMUM = 5 in the STATISTICS paragraph. This statement requests computing the chi-square test of independence after collapsing neighboring rows and/or columns so that the minimum expected frequency exceeds the specified value (5). We also request EXPECTED frequencies in the PRINT paragraph to determine how the categories were collapsed.

Input 4F2.4
(replace the TABLE and STATISTICS paragraph in Input 4F2.1 and add the following PRINT paragraph)

```
/ TABLE          COL = esteem.
                 ROW = educatn.
                 CONDITION = workstat.

/ STATISTICS     MINIMUM = 5.

/ PRINT          EXPECTED.
```

Output 4F2.4

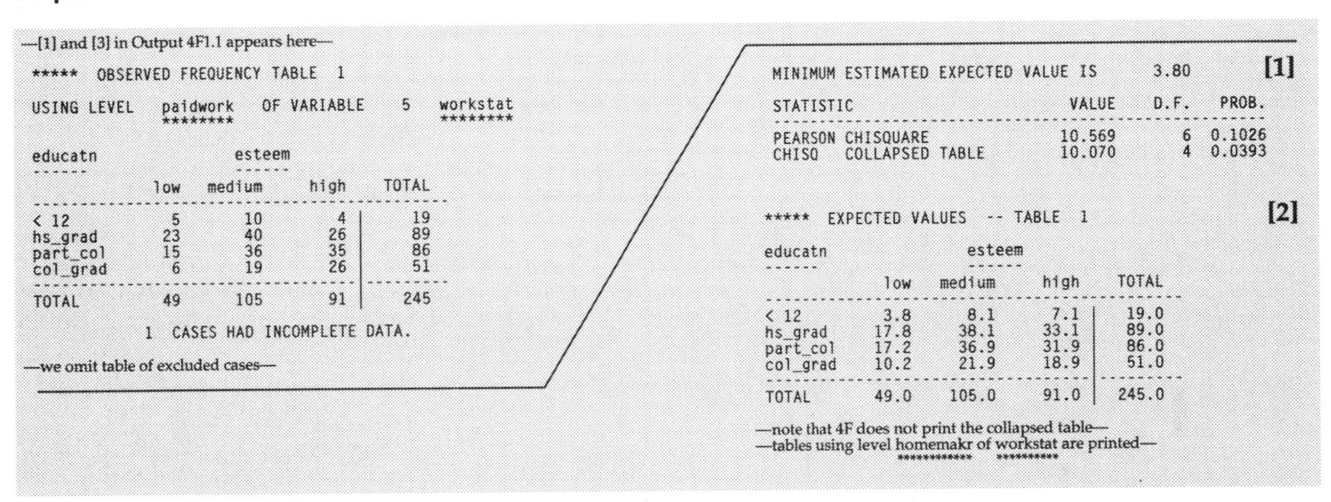

—[1] and [3] in Output 4F1.1 appears here—

```
***** OBSERVED FREQUENCY TABLE  1

USING LEVEL  paidwork   OF VARIABLE   5   workstat
             ********                     ********

educatn              esteem
------               ------
            low   medium      high     TOTAL
---------------------------------------------
< 12          5       10         4  |     19
hs_grad      23       40        26  |     89
part_col     15       36        35  |     86
col_grad      6       19        26  |     51
---------------------------------------------
TOTAL        49      105        91  |    245

         1  CASES HAD INCOMPLETE DATA.
```

—we omit table of excluded cases—

```
MINIMUM ESTIMATED EXPECTED VALUE IS        3.80        [1]

STATISTIC                          VALUE    D.F.   PROB.
--------------------------------------------------------
PEARSON CHISQUARE                 10.569      6   0.1026
CHISQ   COLLAPSED TABLE           10.070      4   0.0393


***** EXPECTED VALUES  -- TABLE  1                     [2]

educatn                 esteem
------                  ------
            low     medium      high     TOTAL
-----------------------------------------------
< 12        3.8        8.1       7.1  |    19.0
hs_grad    17.8       38.1      33.1  |    89.0
part_col   17.2       36.9      31.9  |    86.0
col_grad   10.2       21.9      18.9  |    51.0
-----------------------------------------------
TOTAL      49.0      105.0      91.0  |   245.0
```

—note that 4F does not print the collapsed table—
—tables using level homemakr of workstat are printed—

[1] In Output 4F2.4, the minimum expected value in the table for women who are paid to work is 3.80. Since this is less than the specified MINIMUM of 5, automatic collapsing is performed. First, 4F prints the chi-square value of the original observed table without collapsing (Pearson's chi-square = 10.569). Then the program collapses the table by combining adjacent rows and columns until the minimum expected value is greater than the MINIMUM specified. The program first scans the table of expected values for the minimum expected value. Then 4F tests whether the row total is smaller than the column total. If so, 4F combines rows (as in this example); if not, columns are combined. 4F combines this row (or column) with the adjacent row (or column) having the lower total. If both neighbors have the same total, 4F combines with the neighbor that has the higher index. The program then scans the modified table for the minimum expected value. If it is greater than MINIMUM, 4F stops collapsing; otherwise the process is repeated.

When the collapsing process is finished, 4F prints the Pearson chi-square statistic for the collapsed table (10.070 with 4 D.F.).

[2] The table of expected values contains one expected value less than 5. Since the row total (19) corresponding to this expected value is less than its column total (49), collapsing proceeds by combining this row with its neighbor. In the modified table there will be no cells with an expected frequency less than five and so the procedure for collapsing terminates. Since the collapsed table has three rows instead of four, the degrees of freedom for the collapsed chi-square is 4, and not 6 as in the original table. The collapsed table is not printed.

Note: When collapsing is performed, the statistics phi, the contingency coefficient C and Cramér's *V* are computed from the collapsed chi-square statistic. All other statistics are computed from the original observed table.

The 2 x 2 Table

Two of the variables in our data set (MARITAL and WORKSTAT) each have only two levels. We form 2 x 2 tables from these variables for each level of EDUCATN. The first of the examples in this section (4F2.5) illustrates a variety of statistics for testing independence, describing correlation, and computing odds ratios for 2 x 2 tables. The second example (4F2.6) shows how to compare 2 x 2 tables for different groups.

Test or Measure	Command	Range of Values	Test that Measure is Zero	Interpretation and Example
χ_y^2 (Yates corrected χ^2)	default	$0,\infty$	χ^2 with 1 df	Tests of independence of rows or equality of proportions between rows or columns (4F2.5)
Fisher's exact test	FISH	0,1	prob. of more extreme configuration (one- or two-tail)	
ϕ (phi)	CONT	–1,1	use χ^2 test	Same as χ^2 but not dependent on N. Used to compare tables (4F2.5)
Yule's Q	CONT	–1,1	"	Probability of similar (dissimilar) ranking on two ordered indices among cases with different values for both indices (4F2.5)
Yule's Y	CONT	–1,1	"	Similar to Q but less weighting for similar and dissimilar ranking (4F2.5)
Cross-product ratio α	CONT	$0,\infty$	to test α=1 use χ^2 test	Odds ratio—ratio of two proportions (4F2.5)
Mantel-Haenszel	CONT			Combines and compares odds

Example 4F2.5 The odds ratio and tests of independence: Fisher's exact test, Yates' chi-square, phi, *C*, Yule's *Q* and *Y*

In Input 4F2.5 we form the 2 x 2 table MARITAL by WORKSTAT at each level of EDUCATN. The results for each level of education appear in Output 4F2.5.

Input 4F2.5
(replace the TABLE and STATISTICS paragraphs in Input 4F2.1 with the following)

```
/ TABLE       COLUMN = marital.
              ROW = workstat.
              CONDITION = educatn.

/ STATISTICS FISHER.
              CONTINGENCY.
```

Output 4F2.5

—BMDP features shown in [1] to [3] of Output 4F1.1 are printed—

```
***** OBSERVED FREQUENCY TABLE  1

USING LEVEL   < 12        OF VARIABLE    4   educatn
              ********                       ********
workstat      marital
------        ------
         single  married     TOTAL
---------------------------------
paidwork      7       12  |    19
homemakr      7       18  |    25
--------------------------|------
TOTAL        14       30  |    44

     ALL CASES HAD COMPLETE DATA FOR THIS TABLE.

MINIMUM ESTIMATED EXPECTED VALUE IS        6.05

STATISTIC                       VALUE    D.F.   PROB.
------------------------------------------------------
PEARSON CHISQUARE               0.389       1  0.5328
FISHER EXACT TEST(1-TAIL)                      0.8290
FISHER EXACT TEST(2-TAIL)                      1.0000
YATES CORRECTED CHISQ.          0.088       1  0.7665
ROW RELATIVE SYMMETRY CHISQ     4.100       1  0.0429
COL RELATIVE SYMMETRY CHISQ     0.088       1  0.7665
PHI - CRAMER'S V                0.094
MAXIMUM VALUE FOR PHI           0.784
CONTINGENCY COEF. C             0.094
MAX.VALUE FOR CONTINGEN.        0.617

STATISTIC                       VALUE     ASE1  T-VALUE DEP.
------------------------------------------------------------
YULE'S Q                        0.200    0.313    0.612
CROSS-PRODUCT RATIO             1.500
YULE'S Y                        0.101    0.161    0.618
LN(CROSS-PRODUCT RATIO)         0.405    0.652    0.621
```

```
USING LEVEL   hs_grad     OF VARIABLE    4   educatn
              ********                       ********
workstat      marital
------        ------
         single  married     TOTAL
---------------------------------
paidwork     26       63  |    89
homemakr      5       81  |    86
--------------------------|------
TOTAL        31      144  |   175

     ALL CASES HAD COMPLETE DATA FOR THIS TABLE.

MINIMUM ESTIMATED EXPECTED VALUE IS       15.23

STATISTIC                       VALUE    D.F.   PROB.
------------------------------------------------------
PEARSON CHISQUARE              16.429       1  0.0001
FISHER EXACT TEST(1-TAIL)                      0.0000
FISHER EXACT TEST(2-TAIL)                      0.0000
YATES CORRECTED CHISQ.         14.863       1  0.0001
ROW RELATIVE SYMMETRY CHISQ    75.002       1  0.0000
COL RELATIVE SYMMETRY CHISQ     7.070       1  0.0078
PHI - CRAMER'S V                0.306
MAXIMUM VALUE FOR PHI           0.456
CONTINGENCY COEF. C             0.293
MAX.VALUE FOR CONTINGEN.        0.415

STATISTIC                       VALUE     ASE1  T-VALUE DEP.
------------------------------------------------------------
YULE'S Q                        0.740    0.117    3.736
CROSS-PRODUCT RATIO             6.686
YULE'S Y                        0.442    0.104    4.466
LN(CROSS-PRODUCT RATIO)         1.900    0.516    4.797
```

```
USING LEVEL   part_col    OF VARIABLE    4   educatn
              ********                       ********
workstat      marital
------        ------
         single  married     TOTAL
---------------------------------
paidwork     26       60  |    86
homemakr     10       71  |    81
--------------------------|------
TOTAL        36      131  |   167

     ALL CASES HAD COMPLETE DATA FOR THIS TABLE.

MINIMUM ESTIMATED EXPECTED VALUE IS       17.46

STATISTIC                       VALUE    D.F.   PROB.
------------------------------------------------------
PEARSON CHISQUARE               7.892       1  0.0050
FISHER EXACT TEST(1-TAIL)                      0.0040
FISHER EXACT TEST(2-TAIL)                      0.0078
YATES CORRECTED CHISQ.          6.870       1  0.0088
ROW RELATIVE SYMMETRY CHISQ    54.158       1  0.0000
COL RELATIVE SYMMETRY CHISQ     3.064       1  0.0800
PHI - CRAMER'S V                0.217
MAXIMUM VALUE FOR PHI           0.509
CONTINGENCY COEF. C             0.212
MAX.VALUE FOR CONTINGEN.        0.453

STATISTIC                       VALUE     ASE1  T-VALUE DEP.
------------------------------------------------------------
YULE'S Q                        0.509    0.152    2.706
CROSS-PRODUCT RATIO             3.077
YULE'S Y                        0.274    0.095    2.909
LN(CROSS-PRODUCT RATIO)         1.124    0.411    2.985
```

```
USING LEVEL   col_grad    OF VARIABLE    4   educatn
              ********                       ********
workstat      marital
------        ------
         single  married     TOTAL
---------------------------------
paidwork     18       33  |    51
homemakr      4       21  |    25
--------------------------|------
TOTAL        22       54  |    76

     ALL CASES HAD COMPLETE DATA FOR THIS TABLE.

MINIMUM ESTIMATED EXPECTED VALUE IS        7.24

STATISTIC                       VALUE    D.F.   PROB.
------------------------------------------------------
PEARSON CHISQUARE               3.036       1  0.0814
FISHER EXACT TEST(1-TAIL)                      0.0675
FISHER EXACT TEST(2-TAIL)                      0.1085
YATES CORRECTED CHISQ.          2.171       1  0.1407
ROW RELATIVE SYMMETRY CHISQ    14.040       1  0.0002
COL RELATIVE SYMMETRY CHISQ     9.877       1  0.0017
PHI - CRAMER'S V                0.200
MAXIMUM VALUE FOR PHI           0.447
CONTINGENCY COEF. C             0.196
MAX.VALUE FOR CONTINGEN.        0.408

STATISTIC                       VALUE     ASE1  T-VALUE DEP.
------------------------------------------------------------
YULE'S Q                        0.482    0.238    1.792
CROSS-PRODUCT RATIO             2.864
YULE'S Y                        0.257    0.145    1.910
LN(CROSS-PRODUCT RATIO)         1.052    0.619    1.954
```

—results for the Mantel-Haenszel statistics follow. They are explained in Example 4F2.6—

In Output 4F2.5, 4F prints the following statistics and measures:

Independence. For all 2 x 2 tables, 4F automatically prints the familiar Pearson chi-square test of independence, as well as the Yates' corrected chi-square (χ^2). Yates' correction is intended to improve the approximation to the chi-square distribution. Some studies indicate that Yates' test may be too conservative.

4F prints Fisher's exact test for a 2 x 2 table only when the minimum expected frequency is less than 20. (We conditioned on EDUCATN in order to obtain a minimum expected frequency less than 20.) When the minimum expected frequency is greater than 20, 4F prints a message to that effect instead of Fisher's test, since the Pearson χ^2 test can be safely used. Note that Fisher's exact test is recommended for a 2 x 2 table when the minimum expected frequency is less than 5. The two tail probability is the sum over all configurations (e.g., patterns of cell frequencies) whose probability, assuming independence, is less than or equal to the probability of the observed configuration. Each pattern must preserve the observed row and column totals. The one-tail probability is the sum over the configurations that have a probability less than or equal to that observed when the cell with minimum frequency is decremented to zero. If your hypothesis is one-sided, the configurations appropriate to test the hypothesis may differ from those used to compute the one-tail probability; if they differ from those used by 4F, the one-tail probability exceeds 0.5 and is therefore nonsignificant. **Note:** Unless a one-sided hypothesis is being tested, the two-tail probability for Fisher's exact test should be used.

In this example all of the tests of independence agree. They are all nonsignificant for this strata (education < 12 years), significant for high school graduates and partial college, and again nonsignificant for college graduates. Not surprisingly, the significant tests result because the likelihood of married women being homemakers is greater than that for single women in these two educational strata.

Relative Symmetry. The tests of row and column relative symmetry are included for each 2 x 2 model. Relative symmetry is calculated to determine whether changes in the rows and columns are in the process of change. This test does not assume that row totals are fixed. However, if the row totals are fixed, these tests of symmetry may be the most meaningful interpretation of the model. For more information on these tests, see Bishop, Fienberg, and Holland, (1975).

Contingency. The CONTINGENCY command also produces phi (which is equal to Cramér's *V* for a 2 x 2 table), the contingency coefficient *C*, Yule's *Q* and *Y*, and the cross-product, or odds ratio. These statistics (except for the cross-product ratio) are described in Example 4F2.1. In a 2 x 2 table, the phi coefficient can readily be interpreted by paying attention to its sign (positive or negative). In this case phi is positive; that is, the higher code for marital status, in this case married, which is coded 2, is associated with the higher code for work status (in this case "homemakr," which is also coded 2). The contingency coefficient *C* takes on only positive values.

If the row and column totals are held fixed, as they are in computation of expected cell frequencies, the totals determine the maximum value that these measures can attain. In the 2 x 2 table, 4F prints maximum values for phi and the contingency coefficient *C*. The maximum value for phi is the largest positive value of phi attainable when the observed phi is positive and the largest negative value of phi when the observed phi is negative. The maximum value for *C* is obtained from that of phi.

Association and correlation. Yule's *Q* and *Y* are measures of association. These measures are zero if (and only if) the chi-square statistic is zero. Therefore, the null hypothesis that a measure is zero can be tested by the chi-square test, or by the ratio of the statistic to its estimated standard error. 4F prints the *t*-values for these statistics. The column ASE1 provides an estimate of the asymptotic standard error under the alternative hypothesis that the statistic is not equal to zero. This asymptotic standard error can be used to set an approximate 95% confidence interval for the parameter being estimated–the limits are VALUE + 2(ASE1). See Brown and Benedetti (1977a) for a derivation of these tests. As explained for phi above, the coding of the two dichotomous variables determines the direction of association.

Cross-product (odds) ratio. In a 2 x 2 table the cross-product ratio is also called the odds ratio. This is the ratio of the major diagonal ($f_{11} * f_{22}$) to the minor diagonal ($f_{12} * f_{21}$). If there is no association between the two dichotomous categorical variables, the odds ratio will be one. The odds for married women in the lowest educational strata to be homemakers is 18/12. The odds for single women is 7/7. The ratio of these odds is

$$\frac{18/12}{7/7} = 1.5$$

In this educational strata (< 12) married women may be slightly more likely to be homemakers than single women, but the increase is not significant, as explained below. The odds ratio can be used as an approximation to relative risk when the event (single here) is rare. The use of the cross-product in retrospective studies is discussed by Mantel and Haenszel (1959) and by Fleiss (1981).

Since the sample cross-product ratio has a very skewed distribution, it is usually transformed by taking its natural logarithm for purposes of hypothesis testing and estimation. 4F prints both the cross-product and ln(cross-product), as well as the estimate of the asymptotic standard error and *t*-value of the latter. The asymptotic standard error can be used to set approximate confidence limits for ln(cross-product)—the 95% confidence limits are approximately VALUE + 2(ASE1). For the example, the 95% confidence limits for ln(cross-product ratio) are 0.405 + 2(0.652), or –0.899 to 1.709. Taking natural antilogs of these limits, the 95% confidence limits for the cross-product ratio are 0.41 and 5.52. Note that 1.0 falls within this interval. Therefore, we cannot conclude that married women are more likely to be homemakers than single women in this strata.

The *t*-value is the ratio of the statistic to its estimated asymptotic standard error under the null hypothesis that the true parameter is zero. Here the *t*-value is used to test the null hypothesis that the natural logarithm of the odds ratio is equal to zero (or odds ratio = 1). If the absolute value of *t* exceeds 2, this null hypothesis can be rejected. See Brown and Benedetti (1977a) for a derivation of these tests. (Do not use a *t*-value based on ASE1—it is for the alternative hypothesis).

The odds ratio for the remaining three strata are 6.69, 3.08, and 2.86. For a comparison of odds ratios across strata, see Example 4F2.6.

Example 4F2.6 The Mantel-Haenszel statistic and test of homogeneity of odds ratios

In our previous example, four 2 x 2 tables were formed (WORKSTAT by MARITAL for each level of EDUCATN) and statistics were computed for each table. In this section we describe how to test whether the odds-ratios computed for each statistic and test of level are similar (homogeneous) and how to estimate an odds-ratio common to all the levels and test whether it is unity. In Input 4F2.6 we use Input 4F2.5 again.

Input 4F2.6
(reuse the TABLE and STATISTICS paragraphs in Input 4F2.5)

```
/ TABLE        COLUMN = marital.
               ROW = workstat.
               CONDITION = educatn.

/ STATISTICS FISHER.
             CONTINGENCY.
```

Output 4F2.6

```
—we omit output shown in Output 4F2.5—
 --------------------------------------------------------------------
 MANTEL-HAENSZEL STATISTICS FROM COMBINING THE TWO-BY-TWO TABLES ABOVE
 RISK       3.42  CHI-SQUARE      23.58      TAIL-PROB.      0.0000

 RISK OBTAINED AS ANTILOG OF WEIGHTED COMBINATION OF LN(RISK)
 RISK       3.30       APPROX. 95% CONFIDENCE LIMITS      1.96 TO       5.57
 TEST FOR HOMOGENEITY - CHI-SQUARE       3.41  DF  3  PROB.  0.3320
 --------------------------------------------------------------------
```

In Output 4F2.5, the cross-product ratios for the four strata are 1.50, 6.69, 3.08, and 2.86. Here in Output 4F2.6, the test for homogeneity (last line in the output) is a test that all the cross-product ratios are equal. In this case the chi-square associated with this test is nonsignificant ($p = 0.33$). This is an illustration of a common phenomenon in statistics: several inferences from the same data may reach conflicting conclusions. Here we conclude that the cross-product ratios do not differ significantly. Yet from tests of each individual cross-product ratio, we conclude that there is a definite association between marital status and work status for high school graduates and for those completing partial college. For the other strata, we accept that there is no difference. These contradictions occur because accepting the null hypothesis simply means that there is insufficient evidence to reject it. The test for homogeneity has been shown to have poor power (Greenland, 1983).

Two different methods are used to combine the cross-product ratios across the four education strata. The Mantel-Haenszel statistic (called RISK) combines the numerators and denominators of the cross-products (after adjusting for the total frequency in each table). A chi-square test with one degree of freedom is used to test whether the RISK is equal to one. A second method of combining ratios is to combine them on a logarithmic scale. Confidence limits for the estimate can easily be obtained on the log scale and the limits transformed back to the original scale. In our example the Mantel-Haenszel risk is 3.42 and the second method produces an estimate of 3.30. The test of the Mantel-Haenszel statistic is highly significant ($p < .00005$), and the 95% confidence interval for the second estimate does not include unity. Therefore, both statistics lead to the same conclusion that there is a positive association between being married and being a homemaker.

Special Structures

In this section, we demonstrate statistics available for two-way tables that have special structures:

- measures of correlation when both variables have ordered categories (Example 4F2.7)
- a test of linear trend for a 2 x K or K x 2 table where the variable with K levels has ordered categories (Example 4F2.8)
- a test of reliability and a test of symmetry when the table is square; e.g., when the same subjects are measured twice on the same variable (Example 4F2.9)

For the first two examples, we use the Fidell data set from Example 4F2.1.

Table 4F2.4: Tests and measures for tables with Special structures

Test or Measure	Command	Range of Values is	Test that Measure Zero	Interpretation and Example
		R x R (*Unordered Categories*)		
McNemar's test of symmetry	MCN	0,∞	χ^2 with R(R – 1)/2 df*	Tests symmetry by comparing each pair of cells about diagonal (4F2.9) *see Appendix B
Kappa	MCN	-1,1	$t = \kappa / s_k$	Test of reliability for square two-way tables (4F2.9)
		R x C (*Ordered Categories*)		
Gamma, Γ	GAM	-1,1	$\Gamma / s_0(\Gamma)$	Probability of similar (dissimilar) ranking on two indices among cases with different values for both indices (4F2.7)
Kendall's τ_b	GAM	-1,1	$\tau_b / s_0(\tau_b)$	Numerator as in Γ, denominator ≥ that of Γ (4F2.7)
Stuart's τ_c	GAM	-1,1	$\tau_c / s_0(\tau_c)$	Numerator as in Γ, τ_b. Denominator ≥

Table 4F2.4 (continued)

Test or Measure	Command	Range of Values is	Test that Measure Zero	Interpretation and Example
Spearman rank corr., r_s	SPEAR	-1,1	$r_s/s_0(r_s)$	As in r except ranks of the indices are the values (4F2.7)
Product-moment correlation, r	CORR	-1,1	$r/s_0(r)$	Correlation using values of the indices with frequencies as number of replicates (4F2.7)
		***R* x *C* (Ordered Categories)**		
Somers' D	GAM	-1,1	$D/s_0(D)$	Probability of similar (dissimilar) ranking on two indices among cases with different values for row index (4F2.7)
		2 x 2		
Tetrachoric corr., r_t	TETRA	-1,1	$r_t/s_0(r_t)$	Correlation of bivariate normal with the probabilities $a/N, b/N, c/N, d/N$ in the four quadrants
		2 x *C* or *R* x 2 (Ordered Categories)		
Test of linear trend	LIN	0,∞	χ^2 with 1 df	Test of linear trend on the proportions of the first row or first column (4F2.8)

Example 4F2.7 The R x C table with ordered categories: correlations, gamma, Kendall's and Stuart's taus, Somers' *D*

Several measures of association are appropriate for *R* x *C* tables when both categorical variables have ordered categories. In Input 4F2.7 we request these statistics to assess the relationship between ESTEEM (self-esteem) and EDUCATN (educational level completed) for the Fidell data set.

Input 4F2.7
(replace the STATISTICS paragraph in Input 4F2.1)

```
/ STATISTICS   GAMMA.
               CORRELATION.
               SPEARMAN.
```

—the observed table printed in Output 4F2.1 is printed here—

```
MINIMUM ESTIMATED EXPECTED VALUE IS       9.52

STATISTIC                          VALUE    D.F.   PROB.
---------------------------------------------------------
PEARSON CHISQUARE                 29.849       6  0.0000

STATISTIC                          VALUE    ASE1  T-VALUE DEP.
---------------------------------------------------------------
GAMMA                              0.321   0.058   5.302
KENDALL TAU-B                      0.213   0.040   5.302
STUART TAU-C                       0.210   0.040   5.302
SOMERS D                           0.203   0.038   5.302    1
SOMERS D                           0.224   0.042   5.302    4
CORRELATION COEFFICIENT            0.244   0.044   5.222
SPEARMAN RANK CORR.                0.240   0.044   5.341
```

GAMMA in the STATISTICS paragraph requests several measures of association between self-esteem and education: gamma, Kendall's and Stuart's tau, and the two values of Somers' *D*, an asymmetric measure. Gamma and the two measures of tau differ only in how ties are treated (see Appendix B.11). Of the three, gamma is greater than or equal to Kendall's tau, which in turn is greater than or equal to Stuart's tau. The correlation coefficient and Spearman rank correlation are produced by CORRELATIONS and SPEARMAN, respectively, in the STATISTICS paragraph.

Tests that gamma, the tau statistics, and the correlations are zero are discussed by Brown and Benedetti (1977a). These tests use the ratio of the estimate of the measure to its approximate asymptotic standard error under the hypothesis that the true measure is zero. For sufficiently large samples, this ratio is approximately a t statistic (T-VALUE in the output). Also, for sufficiently large samples, confidence limits can be constructed for the measure by using the asymptotic standard error formula developed by Goodman and Kruskal (1972)—ASE1 in the printed output. The approximate 95% confidence limits, for example, would be VALUE + 2(ASE1).

In Output 4F2.7, all of the measures of association show that self-esteem and educational level are positively correlated, and that the correlation is statistically significant. For example, the t-value for gamma is 5.302. The positive correlation trend indicates that higher self-esteem is associated with higher levels of education.

Somers' D shows the two values of association depending on which variable is considered the dependent (DEP) variable. The value in the DEP column is the index of the variable in the VARIABLE list that is considered the dependent variable. For self-esteem (the first variable in the list) as the dependent variable, the association as measured by Somers' D is .203. Considering education as the dependent variable (fourth in the list), the association is .224. The more appropriate form should be used.

Example 4F2.8 The 2 x C or R x 2 table with ordered categories: test of linear trend

When one categorical variable is dichotomous and the other is ordered, Cochran (1954) proposes a test of linear trend of the proportions on an arbitrary set of weights. This is equivalent to testing the slope of a simple linear regression (or table with ordered testing linear trend in an analysis of variance) where the dependent variable is the proportion. Tests of linear trend are designed to reveal whether proportions increase or decrease across levels of an ordered categorical variable. In Input 4F2.8 we cross classify the dichotomous variable, WORKSTAT, with the ordered variable, EDUCATN, to see if the proportions of women who work outside the home for pay change with the level of education they have completed. When CUTPOINTS are specified, the independent variable is arbitrarily assigned the values 1, 2, 3, etc. Otherwise the code values of the index with more than two categories are used as the values of the independent variable.

We request LINEAR in the STATISTICS paragraph to produce the desired test. We also request percents based on column totals (educational levels) in the PRINT paragraph to help interpret the results.

Input 4F2.8
(replace the TABLE and STATISTICS paragraphs in Input 4F.2 and add the following PRINT paragraph)

```
/ TABLE          COLUMN IS educatn.
                 ROW IS workstat.

/ STATISTICS     LINEAR.

/ PRINT          PERCENT = COL.
```

Output 4F2.8

```
***** OBSERVED FREQUENCY TABLE  1

workstat              educatn
------                ------
           < 12  hs_grad  part_col col_grad    TOTAL
----------------------------------------------------
paidwork     19       89        86       51 |    245
homemakr     25       86        81       25 |    217
----------------------------------------------------
TOTAL        44      175       167       76 |    462

         3  CASES HAD INCOMPLETE DATA.
```

—we omit table of excluded cases—

```
***** PERCENTS OF COLUMN TOTALS -- TABLE  1

workstat              educatn
------                ------
           < 12  hs_grad  part_col col_grad    TOTAL
----------------------------------------------------
paidwork   43.2     50.9      51.5     67.1 |   53.0
homemakr   56.8     49.1      48.5     32.9 |   47.0
----------------------------------------------------
TOTAL     100.0    100.0     100.0    100.0 |  100.0
```

```
MINIMUM ESTIMATED EXPECTED VALUE IS     20.67

STATISTIC                         VALUE    D.F.    PROB.
--------------------------------------------------------
PEARSON CHISQUARE                 8.247      3   0.0412
TEST FOR LINEAR TREND             6.133      1   0.0133
```

In Output 4F2.8, the probability associated with the test of linear trend is .0133. If testing at a significance level of .05, this would indicate a significant linear change in proportions of women working outside the home as a function of educational level attained. From the percents of column totals, we can see that the higher the educational level, the greater the proportion of women working outside the home. While 43.2% of women with less than 12 years of education work outside the home, 67.1% of college graduates do.

Example 4F2.9 The R x R table: McNemar's test of symmetry and Kappa measure of reliability

McNemar's test of symmetry is used when the same subjects are measured on the same variable at two different times, or for matched pairs. The test measures whether a change in one direction (for example, yes to no) is equal to change in the other direction (no to yes). The MCNEMAR command in the STATISTICS paragraph produces both the McNemar test of symmetry and the KAPPA test of reliability. The McNemar statistic tests for change around the measure of reliability diagonal rather than for independence. It tests the equality of frequencies in all pairs of cells that are symmetric around the diagonal. Frequencies in the major diagonal (upper left cell to lower right cell) are ignored. In the 2 x 2 table, this reduces to a test of equality of the two off-diagonal cells. A significant value (say PROB less than .01) indicates lack of symmetry—greater change in one direction than the other. Kappa, on the other hand, focuses on the major diagonal, testing if the diagonal counts are significantly larger than those expected by chance alone.

In Input 4F2.9 we use a set of political survey data. The big political story of late 1986 was the "Iranscam" affair—the secret sale of arms to Iran and the subsequent diversion of funds to the anti-Communist rebels in Nicaragua. A typical survey asked 344 community members, "Do you agree or disagree with President Reagan's handling of foreign affairs?" Each respondent was queried twice, first in October, 1986 and then in January, 1987 (before and after "Iranscam" became public). Their responses are shown in tabulated form in the first table in Output 4F2.9. (See Example 4F1.9 for how to set up table input.) We use the McNemar statistic to test whether a significant shift has taken place across the diagonal. We want to determine whether public opinion changed significantly in one direction. Following the output we discuss the Kappa statistic.

Input 4F2.9

```
/ INPUT          VARIABLES = 2.
                 FORMAT = FREE.
                 TABLE = 3, 3.

/ VARIABLE       NAMES ARE after, before.
```

Input 4F2.9 (continued)

```
/ TABLE          COL = after.
                 ROW = before.

/ CATEGORY       NAMES(before, after) = agree, disagree,
                                        unsure.

/ STATISTICS     MCNEMAR.

/ PRINT          PERCENT = TOTAL.

/ END
 47  56  38  28  61  31  26  47  10
```

Output 4F.9

```
***** OBSERVED FREQUENCY TABLE  1              ***** PERCENTS OF THE TABLE TOTAL -- TABLE  1
before              after                      before              after
------              ------                     ------              ------
          agree disagree  unsure    TOTAL                agree disagree  unsure    TOTAL
------------------------------------------     -------------------------------------------
agree        47       56      38 |   141       agree      13.7     16.3    11.0 |   41.0
disagree     28       61      31 |   120       disagree    8.1     17.7     9.0 |   34.9
unsure       26       47      10 |    83       unsure      7.6     13.7     2.9 |   24.1
------------------------------------------     -------------------------------------------
TOTAL       101      164      79 |   344       TOTAL      29.4     47.7    23.0 |  100.0

***** ANALYSIS OF OBSERVED FREQUENCY TABLE   1

 MINIMUM ESTIMATED EXPECTED VALUE IS      19.06
 STATISTIC                         VALUE    D.F.    PROB.
 ---------------------------------------------------------
 PEARSON CHISQUARE                11.584      4   0.0207
 MCNEMAR TEST OF SYMMETRY         14.865      3   0.0019    [1]
 MARGINAL HOMOGENEITY             14.778      2   0.0006          [2]
 STATISTIC                         VALUE    ASE1 T-VALUE DEP.
 ------------------------------------------------------------
 KAPPA, MEAS. RELIABILITY          0.001   0.036   0.039          [3]
```

[1] In Output 4F2.9, McNemar's test is highly significant (PROB = .0019); this indicates a departure from symmetry in the table. Examining the table of percents we see, for example, that 16.3% of the respondents changed from "agree" to "disagree" between the two surveys, and only 8.1% changed from "disagree" to"agree."

[2] A test of marginal homogeneity (Everitt, 1977) indicates that the marginal proportions before and after differ significantly (PROB = .0006).

[3] Kappa is a form of correlation appropriate for a square table. It is a measure of reliability (consistency) of the two categorical variables. Kappa is frequently used when the same subject is being judged by two raters. Kappa then provides a measure of interrater reliability by focusing on the diagonal to see if it contains more counts than are expected by chance. Using the *T*-VALUE printed with KAPPA in Output 4F2.9 we are unable to say that the counts along the diagonal differ from those expected by chance.

As a better example, we could use kappa to measure how well two doctors agree in diagnosing the same patients. Figure 4F2.1 contains the results from two doctors diagnosing the same 77 psychiatric patients. If we analyze this table using 4F in a setup similar to the one above, we obtain a kappa of .61, with a *t*-value of 7.66. We see that there is good agreement between the two doctors, because we find that the counts along the diagonal differ significantly from those expected by chance. Fleiss (1981) suggests that values of kappa less than .40 reflect poor agreement. With kappa between .40 and .75, agreement is fair to good. Values of kappa above .75 indicate strong agreement.

Figure 4F2.1
Agreement between two physicians diagnosing psychiatric patients

		Doctor A		
		schizo	manic	other
	schizo	24	2	3
Doctor B	manic	5	16	6
	other	3	1	17

Special Features

Fitting log-linear models

4F can fit log linear models to two-way tables to obtain estimates of parameters and their standard errors. When a two-way table is specified and FIT is present, both the two-way and the requested log-linear analyses are performed. Log-linear analysis is discussed in Chapter 4F Part 3. Examples 4F3.1 through 4F3.4 show basic setups for screening and specifying models. Example 4F3.7 shows how to test all models in a two-way and three-way frequency table.

Structural zeros

4F can fit log-linear models to frequency tables in which some cells are defined as structural zeros. Structural zeros occur when some cells contain events with a known probability of zero (e.g., pregnancy rate in males). At times structural zeros are specified in order to assess the impact of one or more cells on a model by eliminating those cells from the calculations. When cells in a two-way table are identified as structural zeros (by one of the two methods shown in Examples 4F3.7 and 4F3.8), 4F computes a test of quasi-independence limited to that subset of the table which contains no structural zeros.

4F Part 2 Commands

Where to Find Commands

4F Part 1

For information on the following **table forming options:**

/ INPUT	TABLE =	CONTENT =		
/ TABLE	ROW =	INDEX =	PAIR.	CONDITION
	COUNT =	COLUMN =	CATVAR =	CROSS.
	STACK =			
/ PRINT	OBSERVED.	PERCENT =	LIST =	EXCLUDED.
	MARGIN =			
/ SAVE	CONTENT =			

4F Part 2

To compute **statistics and measures of association for two-way tables**, see commands in STATISTICS paragraph.

4F Part 3

/ INPut

The INPUT paragraph is required. See Chapter 3 for INPUT options common to all programs. 4F accepts data or tabulated input. Examples 4F1.9 and 4F1.12 show how to start from frequency tables as input. Example 4F1.10 shows how to use cell indices and frequency counts as input.

/ VARiable

The VARIABLE paragraph is optional. See Chapter 4 for VARIABLE options common to all programs. Chapter 4F Part 1 describes the / VAR FREQ command.

New Feature **MMV = #.** `MMV = 10.`

Specify the maximum number of missing values or values out of range that will be allowed before a case is excluded from the analysis. Default: use all acceptable values.

/ CATEGory (*or* **GROUP**)

CATEGORY CODES or CUTPOINTS are recommended. The CATEGORY paragraph is described in detail in Chapter 5. Variable(s) used as table indices with more than 10 levels must have CODES or CUTPOINTS.

/ TABLE

The TABLE paragraph is required and may be repeated. This paragraph defines tables to be formed.

4F Part 1. Examples 4F1.1 and 4F1.5 show how to use the ROW, COL, and INDICES commands. Use the optional CONDITION statement when you want to form separate two-way frequency tables for each level of another categorical variable (see Example 4F1.4). The COUNT command is required only when data are input as cell indices and frequency counts (see Examples 4F1.10 and 4F1.12). You can use the STACK command to treat two or more variables as a single composite variable (see Example 4F1.8).

4F Part 3. The EMPTY and INITIAL commands are optional and are used when you want to specify one or more cells as structural zeros to fit a log-linear model or perform a test of quasi-independence for a two-way table. Examples 4F3.7 and 4F3.8 show how to use EMPTY and INITIAL. SYMBOLS are used to define terms in models to be fitted. Usually the symbol for each variable is the first character of the variable name. If variable names are not specified for a two-way design, the letters A and B are used. A constant can be added to each cell frequency using the optional DELTA instruction.

/ STATistics `NO CHISQ. LRCHI.`

The STATISTICS paragraph is optional. Not all statistics are applicable to a single table. For example, the categories are either ordered or not. The question may be that of independence or prediction. You should select the statistics appropriate to your specific problem. Table 4F2.1 provides an overview of the usage of the statistics and measures available. The computations of all statistics except Fisher's exact test are affected by the addition of a constant DELTA to each cell frequency (see 4F Part 3). You can locate examples for many of the following commands in the Summary Table at the end of this chapter.

CHISQuare. – Pearson chi-square test and, when the table is 2 x 2, Yates' corrected chi-square (χ_y^2). These statistics are automatically computed unless you specify NO CHISQ.

LRCHI. – Likelihood ratio chi-square, G^2.

FISHer. – For a 2 x 2 table only, compute Fisher's exact probabilities, both 1-tail and 2-tail. Computed only when the minimum expected frequency is less than 20. This can be calculated as long as the sum of the table entries is less than 500.

CONTingency. – Contingency coefficient *C*, phi (ϕ), and Cramér's *V*. For a 2 x 2 table, also compute Yule's *Q* and *Y*, maximum value for *C*, maximum value for phi, and the odds-ratio. *C*, *V*, and phi are computed from the value of the chi-square test. When a series of 2 x 2 tables is formed by CONDITIONING the ROW and COLUMN variables on a third variable and CONTINGENCY is specified, 4F computes the Mantel -Haenszel statistic and a test of homogeneity of the cross -product ratios across the 2 x 2 tables. The computation of the contingency coefficient C, Cramér's V, and phi are affected when frequency tables are automatically collapsed.

TETRAchoric. – For a 2 x 2 table, compute the tetrachoric correlation r_t.

GAMma. – Gamma (Γ), Kendall's τ_b, Stuart's τ_c, and Somers' D.

CORRelation. – The Pearson product-moment correlation.

SPEARman. – Spearman rank correlation.

TAUS. – Goodman and Kruskal's τ.

LAMBda. – Goodman and Kruskal's λ and λ^*, both asymmetric and symmetric.

UNCertainty. – The uncertainty coefficients, both asymmetric and symmetric.

MCNemar. – For R x R tables, compute McNemar's test for symmetry and the kappa measure of reliability.

LINear. – For 2 x C or R x 2 tables, compute the linear trend of proportions.

NO ALL. – Do not print any statistics. NO ALL is overridden by any other requests in the STATISTICS paragraph. For example, NO ALL. LINEAR. requests that only LINEAR be computed.

MINimum = #. `MIN = 1.`

If the minimum expected frequency in a table is less than #, 4F combines adjacent rows or columns. Caution: Automatic collapsing can combine incompatible categories, especially when the categories are not ordered. Statistics affected by the collapsing are the chi-square test, phi, contingency coefficient C, and Cramér's V. The other statistics and all tables are not affected.

/ PRINT `NO EXCL.EXPECTED.ADJUSTED.`

The PRINT paragraph is optional. If this paragraph is omitted, 4F automatically prints OBSERVED and EXCLUDED tables and LISTS the first five excluded cases. Any other parameters must be specified to be printed. The PRINT paragraph may be repeated after each TABLE paragraph. Additional PRINT features, relevant for multiway tables, are described in Chapter 4F Part 3.

LIST = #. `LIST = 35.`

List the first # of cases that were not included in one or more tables. # is the maximum number of cases to be printed in a case-by-table display. Default: the first 5 cases excluded from any table will be printed.

BAR = 'c'. `BAR = 'I'.`

Specify the character to be printed along vertical lines in two-way and multiway tables. The default character is a vertical bar (I).

New Feature

METHod = VARiable *or* **CASE.** `METH = CASE.`

Use to determine how the printout is organized. VARIABLE is the default value.

VARIABLE – Prints for each case only as many variables as fill the print line (up to ten variables) and begins the next line with the same variables for the next case. Thus the data for each variable are in an easy-to-scan column. When the program finishes printing the first subset of variables for all cases, it continues with another subset.

CASE – Prints the values of all variables for one case (possibly filling several print lines) before printing data for the next case.

PERCent = (*one or more*). **NONe, ROW, COLumn, TOTal.** `PERC = TOT.`

Requests tables of PERCENTS. Specify ROW for a table of percents of the row totals, COLUMN for percents of the column totals, and TOTAL for percents of the total table frequency. Default: NONE of the tables of percents are printed.

OBServed. – Table of observed frequencies unless you specify NO OBS.

EXCluded. – Two-way table of frequencies of the excluded cases unless you specify NO EXCLUDED. This table follows each frequency table from which cases have been excluded and shows the reasons for exclusion.

EXPected. – Expected values of all cells under the null hypothesis of independence.

DIFference. – Differences between observed and expected values for all cells.

FReeman. – Freeman-Tukey deviates between observed and expected values for each cell.

STANdardized. – Print a table of the standardized deviates between observed and expected values for each cell.

ADJusted. – Adjusted standardized deviates between observed and expected values for each cell. See Haberman (1973) for a discussion of adjusted standardized deviates.

CHISQuare. – Components of the Pearson chi-square statistic for each cell.

LRCHI. – Components of the likelihood ratio chi-square statistic.

New Feature **CASE = #.** – Specify the number of cases for which data are printed. Values flagged by MISS, MAX, and MIN in the VARIABLE paragraph specifications are replaced by MISSING, GT MAX, and LT MIN. Default: 10.

New Feature **MINimum.** – Prints data for cases in which at least one variable has a value less than its lower limit as specified in the VARIABLE paragraph. Default: such cases are not printed.

New Feature **MAXimum.** – Prints data for cases in which at least one variable has a value greater than its upper limit as specified in the VARIABLE paragraph. Default: such cases are not printed.

New Feature **MISSing.** – Prints data for cases that have at least one variable with a missing value. Missing values in the input may be identified in several ways: (1) an asterisk (the default missing value character), (2) a missing value code identified using MISS in the VARIABLE paragraph, (3) a character identified as MCHAR in the INPUT paragraph, (4) blanks (applies only to input read with FORTRAN fixed format). Default: such cases are not printed unless you specify DATA.

The following commands work only in the batch mode

New Feature **MEAN.** – Unless you specify NO MEAN, 4F prints the mean, standard deviation, and frequency of each used variable.

New Feature **EXTReme** – Unless you state NO EXTR, 4F prints the minimum and maximum values for each variable used.

New Feature **EZSCores.** – Prints z-scores for the minimum and maximum values of each used variable.

New Feature **SKewness.** – Prints skewness and kurtosis for each used variable.

New Feature **ECASe.** – Prints the case numbers containing the minimum and maximum values of each variable.

We discuss additional PRINT commands, not relevant to two-way tables, in Chapter 4F, Parts 1 and 3.

/ SAVE

The SAVE paragraph is required to create a BMDP File. See Chapter 8 for details about the SAVE paragraph. See 4F Part 1 for information about saving tables in BMDP Files and reusing them in later analyses.

Order of Instructions

■ indicates required paragraph

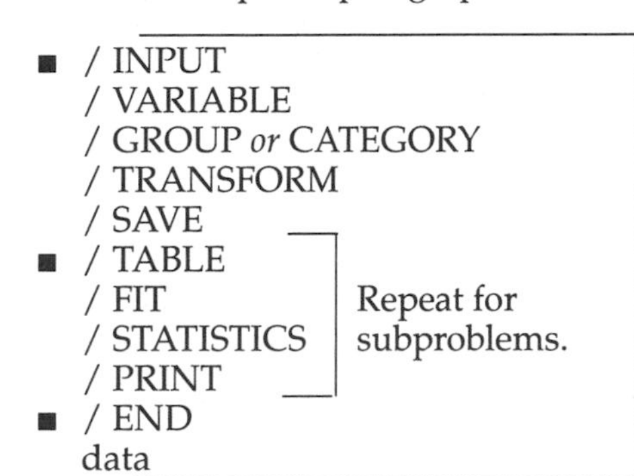

Repeat for additional problems. See Multiple Problems, Chapter 10.

Summary Table for Commands Specific to 4F, Parts 1 and 2

Paragraphs Commands	Defaults	Multiple Problems	See
■ / **INPut**			
TABLE = # list.	input is not a table	–	4F1.9
CONTent = (*one only*)	DATA	–	4F1.12
DATA, COUNT#.			
/ **VARiable**			
FREQuency = variable.	no case frequency variable	–	Cmds
MMV = #.	use all acceptable values	–	Cmds
▲ / **CATEGory** (*or* **GROUP**)			
■ / **TABLE**			
▲ COLumn = list.	none	–	4F1.1
▲ ROW = list.	none	–	4F1.1
or INDices = list.	see ROW and COL	–	4F1.5
or CATvar = list.	see ROW and COL	–	SpecF
PAIR. or CROSS.	PAIR.	–	4F1.2
CONDition = list.	no condition variables	–	4F1.4
COUNT = list.	the variable stated in /VAR FREQ =	–	4F1.10
STACK = list.	no stacking variables.	–	4F1.8
/ **STAtistics**			
CHISQuare. or NO CHISQ.	CHISQ.	●	4F2.1
LRCHI.	NO LRCHI.	●	4F2.1
FISHer.	NO FISH.	●	4F2.5
CONTingency.	NO CONT.	●	4F2.1
TETRAchoric.	NO TETRA.	●	Cmds

Summary Table for Commands Specific to 4F, Parts 1 and 2 *(continued)*

Paragraphs Commands	Defaults	Multiple Problems	See
GAMma.	NO GAM.	●	4F2.7
CORRelation.	NO CORR.	●	4F2.7
SPEARman.	NO SPEAR.	●	4F2.7
TAUS.	NO TAUS.	●	4F2.3
LAMBda.	NO LAMB.	●	4F2.3
UNCertainty.	NO UNC.	●	4F2.3
MCNemar.	NO MCN.	●	4F2.9
LINear.	NO LIN.	●	4F2.8
ALL. *or* NO ALL.	NO ALL. CHISQ.	●	Cmds
MINIMUM= #.	no automatic collapsing	●	4F2.4
/ **PRINT**			
LIST = #.	5	●	Cmds
BAR = 'c'.	vertical bar (I)	●	Cmds
METHod = VARiables *or* CASE.	VARIABLE	–	Cmds
PERCent = (*one or more*) NONe, ROW, COL, TOT.	NONE	●	4F1.3
OBServed. *or* NO OBS.	OBS.	●	Cmds
EXCLuded. *or* NO EXCL.	EXCL.	●	Cmds
EXPected.	NO EXP.	●	4F2.2
DIFference.	NO DIFF.	●	4F2.2
FReeman.	NO FR.	●	Cmds
STANdardized.	NO STAND.	●	Cmds
ADJusted.	NO ADJ.	●	4F2.2
CHISQuare.	NO CHISQ.	●	4F2.2
LRCHI.	NO LRCHI.	●	Cmds
MARGinal = #.	no marginal subtables	●	4F1.6
CASE = #.	10	●	Cmds
MISSing.	NO MISS.	●	Cmds
MINimum.	NO MIN.	●	Cmds
MAXimum.	NO MAX.	●	Cmds
MEAN. *or* NO MEAN.	MEAN.	●	Cmds
EXTReme. *or* NO EXTR.	EXTR. values printed	●	Cmds
EZSCores.	EZSC. (z-scores)	●	Cmds
ECASe.	ECAS. (Case #'s of Min and Max printed)	●	Cmds
SKewness.	SK. (skewness and kurtosis)	●	Cmds
/ **SAVE**			
CONTent = (*one or both*) DATA, COUNT.	DATA	–	4F1.11

Key:
- ■ Required paragraph or command
- ▲ Frequently used paragraph or command
- ● Value retained for multiple problems
- – Default reassigned

4F

Part 3
Analysis of Multiway Tables: Log-linear Models

When a frequency table is formed by two or more categorical variables, relationships among the variables may be modeled by a log-linear model fitted to the cell frequencies. The log-linear model represents the natural logarithm of the expected cell frequency as a linear combination of main effects and interactions in a manner similar to the usual analysis of variance model.

A test that a main effect of a factor is zero is equivalent to a test that the total frequencies are equal at all levels of the factor. For example, when gender is a factor, the test that the main effect of gender is zero is equivalent to testing whether the total number of males is equal to that of females.

A test that a second-order effect, an interaction between two factors, is zero is equivalent to testing whether the two factors are independent when there are at most two two-factor interactions in the log-linear model. When there are more than two two-factor interactions in the log-linear model, the test of a two-way (or higher-way) interaction is conditional on all the other interactions in the model and may not have a simple explanation.

Various models can be formed by including different combinations of effects. 4F fits and tests hierarchical models, those in which inclusion of higher order effects requires inclusion of all their component lower order effects. For example, if you include a sex by survival interaction in your model, the first-order effects of both sex and survival are also automatically included in the model.

4F provides several methods to screen and select log-linear models. By a single command it can compute all possible hierarchical models for two- or three-way tables. Again, by a single command, 4F can compute in multiway tables of any dimension a pair of tests for each interaction to determine the likely importance of the interaction in the model. You can specify hierarchical models; 4F will compute their tests of fit and estimate parameters for each model. Also, starting from a specified model, 4F can add or delete interactions in a stepwise manner to identify those terms needed or not needed in the model. With a single instruction, you can include certain second or higher order effects in all models to be tested.

For each model specified, 4F provides a test of fit of the model to the observed frequencies. You can request the expected cell frequencies that are estimated by the model, and estimates of the log-linear parameters with their standard errors. For the model specified, 4F can also print the differences between observed and expected frequencies, as well as functions of these differences, such as Freeman-Tukey deviates, standardized deviates, and components of the chi-square goodness-of-fit.

It is possible that, by the nature of the variables, some cells will always have zero frequency (structural zeros). For example, the cell for vaginal cancer in men must be zero. Or, you might want to test some hypothesis which excludes one or more cells, such as all cells on the diagonal. In both of these examples, models would be fitted to a subset of the table. 4F fits models, such as a model of quasi-independence (Goodman, 1968), to frequencies in a subset of the table. You can specify structural zeros or cells to be excluded from the model-fitting process; 4F fits the model to the data in the remaining cells and computes the appropriate test of fit. This allows you to fit models that apply only to some of the cells of the table.

The best fitting model may not be appropriate for all cells in the table even though the model is a good predictor of frequencies in most of the cells. 4F can identify, in a stepwise manner, cells whose observed frequencies deviate most from frequencies that would be expected under a given model. 4F screens for these divergent cells by finding the one with the largest Freeman-Tukey or standardized deviate, and then eliminates this cell by treating it as a structural zero. This process is repeated for a number of steps. In this way you can identify cells with unexpectedly large or small frequencies. This process can also help you find patterns or unusual features in your data.

The data in your frequency table may be fitted well by a particular model except for one of the levels of one of your categorical variables. For example, a model may fit well for people who are currently or previously married, but not those who have never been married. Or it may be a single level of one categorical variable that causes the interaction with another categorical variable. 4F can refit the model after eliminating, in turn, each one of the levels (strata) of an index. For example, the first analysis might exclude all those currently married, the second, all those never married, and so forth.

We recommend the texts by Bishop et al. (1975), Fienberg (1977), and Plackett (1974) for more detailed discussions of the log-linear model. Chapter 4F Part 1 shows how to form and input multiway frequency tables. Chapter 4F Part 2 shows how to analyze frequency tables with only two categorical variables.

Where to Find It

Examples

Where to Find It (continued)

Analysis by Log-linear Models

The purpose of analyzing a multiway table is to describe relationships among categorical variables cross-classified in the table. The analysis is based on fitting a hierarchical log-linear model to the cell frequencies. In this model, the natural logarithm of the expected cell frequency is written as an additive (linear) function of effects (first-order effects and higher order interactions among the variables).

Expected frequencies. In a two-way model, the expected frequency of a cell is a product of two terms, one depending on the row to which the cell belongs, representing a level of one categorical variable, and the other depending on the column to which the cell belongs, representing a level of the other categorical variable. For a two-way table with factors A and B, a model can be written as

$$\ln F_{ij} = \theta + \lambda_{A_i} + \lambda_{B_j} + \lambda_{AB_{ij}}$$

where the λ parameters sum to zero over each of their indices and θ is the zero-order or mean effect. The usual χ^2 test of independence tests that $\lambda_{AB_{ij}} = 0$ for all cells simultaneously.

In a table with more than two indices, the expected frequency of a cell may be a product of several terms, each representing a first-order effect or a higher order interaction. Since a logarithm of a product of terms is the sum of the logarithms of the terms, the logarithm of the expected frequency of a cell can be expressed as a linear or additive model—the log-linear model.

Take, for example, a three-way model in which the indices are A: gender, with i

levels; B: survival (yes, no), with j levels; and C: treatment, with k levels. We have an $I \times J \times K$ (or $2 \times 2 \times K$) frequency table. For a particular cell ijk, the natural logarithm of the expected frequency would be:

$$\ln F_{ijk} = \theta + \lambda_{A_i} + \lambda_{B_j} + \lambda_{C_k} + \lambda_{AB_{ij}} + \lambda_{AC_{ik}} + \lambda_{BC_{jk}} + \lambda_{ABC_{ijk}}$$

where $F_{ijk} = E(f_{ijk})$, the expected value of the observed cell frequency. The λs are the **effects**, with A, B, and C indicating the variable to which the effect refers. For example, λ_A means that the effect is due to variable A alone. In identifying an effect, the lower case subscript can be omitted. The order of the effect is the number of indices in the subscript. For example, λ_{AB} has two indices, A and B, so it is a second-order effect. The λs for each effect sum to zero when summed over the levels of each index.

The log-linear model written above is referred to as the saturated model, since it contains all possible effects. By excluding certain effects—setting them equal to zero—different models are formed. In a hierarchical model a higher order effect cannot be present unless all lower-order effects whose indices are subsets of the higher order effect are also included in the model. For example, if λ_{AB} is included in the model (i.e., is not zero), then λ_A and λ_B and θ must also be included in the model. 4F considers only hierarchical models, so models can be described by a minimal number of effects. For example, a full second-order model includes all the terms represented by the effects (θ, A, B, C, AB, BC, AC) with the three-way interaction set to zero. As a hierarchical model, this model can be described by the minimal set of effects (AB, AC, BC). If you include a higher order effect, you are automatically including the lower order effects contained by it. That is, (AB) implies (θ, A, B, AB).

Tests of goodness of fit. The goodness of fit of a log-linear model can be tested using either the usual Pearson goodness-of-fit chi-square statistic (described in Example 4F2.1) or the likelihood ratio statistic

$$G^2 = 2 \sum_{ijk} f_{ijk} \ln(f_{ijk} / F_{ijk})$$

where f_{ijk} is the observed frequency for that cell and F_{ijk} is the fitted frequency (expected frequency) for that cell as a result of fitting the log-linear model under consideration. Both the Pearson and likelihood ratio statistics are asymptotically distributed as chi-square, with $n - p$ degrees of freedom (df), where n is the number of cells and p is the number of independent parameters estimated. (When one or more cells either have an expected value of zero or are described as containing structural zeros, the number of degrees of freedom is reduced; see Brown and Fuchs, 1983.)

The likelihood ratio statistic (G^2) is *additive under partitioning* for nested models. Two models M1 and M2 are nested if all of the effects in M1 are a subset of the effects in M2. G^2 is "additive" because the statistic for M2 can be subtracted from the statistic for M1; that is, the difference in G^2 between the two models is a test of the additional effects in M2, given the effects in both models (i.e., the effects in M1). This difference in G^2 also has an asymptotic chi-square distribution, with degrees of freedom equal to the difference in the number of parameters (effects) fitted to the two models. The Pearson chi-square statistic is not additive under partitioning for nested models. Therefore, for some of the model tests, 4F computes only the likelihood ratio test (G^2).

Steps in log-linear modeling. The purpose of fitting the log-linear model is to aid in understanding the relationships among the categorical variables and to identify a parsimonious model. The analysis can usefully be divided into three stages:

1. screening for an appropriate model
2. testing, comparing, and understanding the models under consideration
3. examining cells or strata with large disparities between observed and expect ed frequencies under a chosen model.

In Examples 4F3.1 through 4F3.6, we start from data summarized as a four-way frequency table and take the analysis through the three stages in a straightforward fashion. In Example 4F3.1 we show tests of marginal and partial interaction to identify models that should be further tested or compared. In Example 4F3.2 we show how to test the models identified in the screening stage. In Example 4F3.3 we describe how to estimate log-linear parameters and their standard errors for the tested models. We show how to find the expected (fitted) values and deviations between observed and expected values for tested models in Example 4F3.4. In Examples 4F3.5 and 4F3.6 we show how to identify extreme cells and strata in a fitted model.

The remaining examples in this chapter deal with special features for log-linear modeling available through 4F. In Examples 4F3.7 and 4F3.8 we describe how to specify cells that contain structural zeros and analyze the resultant frequency table. This technique allows you to analyze tables in which observations in some cells cannot occur or are not of interest. Or, if the frequency table contains highly divergent cells, you may find that by eliminating these cells (specifying that they contain structural zeros) you can fit a simpler model—one which is appropriate for all of the cells except the one excluded. (It would then be necessary to describe how the eliminated cells deviate from the simpler model.)

Finally, we show how to test all possible hierarchical models in a three-way frequency table in Example 4F3.9. In Examples 4F3.10 and 4F3.11 we show how 4F can delete and add terms in a stepwise manner from a specified model.

Example 4F3.1 Using tests of marginal and partial association to screen effects

The examples in this chapter continue to analyze the data from the file FIDELL.DAT that was used in 4F Parts 1 and 2. The BMDP instructions to describe the data set in Input 4F3.1 are identical to those in Input 4F1.1 with one exception. We have renamed self-esteem (ESTEEM in Parts 1 and 2) as STEEM. The rationale for this change is that we will be specifying models in later examples in this chapter; at that time all factors in the table must have distinct first letters unless SYMBOLS is specified. Since education and esteem both began with an E, we have omitted the first letter from esteem in order to make it unique.

The four-way table formed in Input 4F3.1, education by marital status by self-esteem by work status, is identical to that in 4F1.5. We add

```
/ FIT ASSOCIATION = 3.
```

to request the automatic model screening feature in 4F. INDICES in the TABLE paragraph identifies the categorical variables to be cross-classified. The INPUT, VARIABLE, and CATEGORY paragraphs are described in detail in Chapters 3-5. The results are displayed in Output 4F3.1.

Input 4F3.1

```
/ INPUT      VARIABLES = 6.
             FORMAT IS FREE.
             FILE IS 'fidell.dat'.

/ VARIABLE   NAMES = steem, hap_stat, womenrol, educatn,
                     workstat, marital.

/ CATEGORY   CODES(marital) = 1, 2.
             NAMES(marital) = single, married.
             CUTPOINTS(steem) = 11, 16.
             NAMES(steem) = low, medium, high.
```

Input 4F3.1 *(continued)*

```
                CUTPOINTS(educatn) = 11, 12, 15.
                NAMES(educatn) = '< 12', hs_grad, part_col,
                                 col_grad.
                CUTPOINTS(workstat) = 0.
                NAMES(workstat) = paidwork, homemakr.

/ TABLE         INDICES = educatn, marital, steem, workstat.

/ FIT           ASSOCIATION = 3.

/ END
```

Output 4F3.1

```
*****  OBSERVED FREQUENCY TABLE  1                                [1]

workstat steem    marital              educatn
------   ------   ------               ------
                            < 12  hs_grad  part_col col_grad   TOTAL
-------------------------------------------------------------------------

paidwork low      single       2       6        4        3  |     15
                  married      3      17       11        3  |     34
                  ------------------------------------------|--------
                  TOTAL        5      23       15        6  |     49

         medium   single       4      12        9        7  |     32
                  married      6      28       27       12  |     73
                  ------------------------------------------|--------
                  TOTAL       10      40       36       19  |    105

         high     single       1       8       13        8  |     30
                  married      3      18       22       18  |     61
                  ------------------------------------------|--------
                  TOTAL        4      26       35       26  |     91
-------------------------------------------------------------------------

homemakr low      single       6       2        4        1  |     13
                  married      6      22       10        0  |     38
                  ------------------------------------------|--------
                  TOTAL       12      24       14        1  |     51

         medium   single       1       1        5        3  |     10
                  married      9      47       47       11  |    114
                  ------------------------------------------|--------
                  TOTAL       10      48       52       14  |    124

         high     single       0       2        1        0  |      3
                  married      3      12       14       10  |     39
                  ------------------------------------------|--------
                  TOTAL        3      14       15       10  |     42

        TOTAL OF THE OBSERVED FREQUENCY TABLE IS       462

           3  CASES HAD INCOMPLETE DATA.
```

—we omit table of excluded values—

```
*****  THE RESULTS OF FITTING ALL K-FACTOR MARGINALS.
       SIMULTANEOUS TEST THAT ALL K+1 AND HIGHER FACTOR INTERACTIONS ARE ZERO.        [2]

K-FACTOR     D.F.        LR CHISQ      PROB.        PEARSON CHISQ      PROB.         ITERATION
--------     ----        --------      -----        -------------      -----         ---------
0-MEAN        47          445.81     0.00000               546.00    0.00000
1             40          117.52     0.00000               124.79    0.00000            2
2             23           29.14     0.17567                26.16    0.29336            6
3              6            7.46     0.28076                 7.59    0.26933            7
4              0            0.       1.                      0.      1.

*****SIMULTANEOUS TEST THAT ALL K-FACTOR INTERACTIONS ARE SIMULTANEOUSLY ZERO.         [3]
       THE CHI-SQUARES ARE DIFFERENCES IN THE ABOVE TABLE.

K-FACTOR     D.F.        LR CHISQ      PROB.        PEARSON CHISQ      PROB.
--------     ----        --------      -----        -------------      -----
1              7          328.29     0.00000               421.21    0.00000
2             17           88.38     0.00000                98.63    0.00000
3             17           21.69     0.19711                18.57    0.35403
4              6            7.46     0.28076                 7.59    0.26933
```

[4]

***** ASSOCIATION OPTION SELECTED FOR ALL TERMS OF ORDER LESS THAN OR EQUAL TO 3

EFFECT	PARTIAL ASSOCIATION D.F.	CHISQUARE	PROB	ITER	MARGINAL ASSOCIATION D.F.	CHISQUARE	PROB	ITER
e.	3	120.04	0.0000					
m.	1	150.19	0.0000					
s.	2	56.37	0.0000					
w.	1	1.70	0.1926					
em.	3	5.94	0.1146	5	3	6.23	0.1011	2
es.	6	25.90	0.0002	5	6	29.36	0.0001	2
ew.	3	4.89	0.1804	6	3	8.40	0.0384	2
ms.	2	3.43	0.1801	5	2	4.42	0.1099	2
mw.	1	25.14	0.0000	5	1	26.18	0.0000	2
sw.	2	14.19	0.0008	5	2	18.41	0.0001	2
ems.	6	5.43	0.4897	6	6	7.74	0.2577	5
emw.	3	2.85	0.4160	5	3	3.61	0.3068	5

Bold numbers below correspond to those in Output 4F3.1

[1] 4F prints the observed frequency table, along with certain marginal totals for the cells formed by the first two categorical variables listed in the TABLE paragraph. See Example 4F1.5 for further details.

[2] 4F prints tests of models of full order. A test that all effects greater than order 2 are zero is performed by fitting the model that includes all zero, first, and second order effects—θ, *E*, *M*, *S*, *W*, *EM*, *ES*, *EW*, *MS*, *MW*, and *SW* in the notation of the example. (Zero-order effect refers to j in the log-linear equation in the overview above. The test that this effect is zero is equivalent to the test that all cells in the table contain equal frequencies.)

K-FACTOR refers to the highest order included in the model. For example, the line beginning with 2 (K-FACTOR = 2) gives the results of fitting a model with all first and second order effects—that is, one in which all third and higher order effect λs are set to zero. Lack of significance means that the model is a good fit to the observed data. Since the model on the line beginning with 2 is not significant, it may not be necessary to include any three-way interactions in the model. You should look for the lowest order model that produces a nonsignificant result. Given this output, then, you would confine further analysis to models that contain second order effects (and possibly one third order effect).

Comments: The test that all higher order terms are zero is a simultaneous test of multiple interactions; it may hide the marginal effect of one or more interactions among the more numerous nonsignificant interactions. Therefore, examining interactions even when the simultaneous test is non-significant may be of interest; however, the resulting inferences must be interpreted cautiously. Conversely, when a test for each interaction is performed separately at the 5% level of significance as in the tests of marginal and partial association described below, the probability of obtaining one or more significant results under the null hypothesis (i.e., the overall level of significance) greatly exceeds 5%. Therefore, there is a need for caution in interpreting "significant" effects that are found when screening many potential effects. (These comments are equally appropriate for examining effects in the analysis of variance, screening variables in a multiple regression, or performing multiple *t*-tests in order to choose variables whose means change between groups.)

[3] This section provides tests of nested models. Models of full order are nested within each other. The full second order model (*EM*,*ES*,*EW*,*MS*,*MW*,*SW*) is nested within the full third order model (*EMS*,*EMW*,*ESW*,*MSW*). We can, therefore, test the difference in fit between the two models to evaluate the hypothesis that effects of a given order are simultaneously zero. Since this is a test of the difference between nested models, only the likelihood ratio G^2 test, labeled LR CHISQ, is appropriate.

The line beginning with 3 (K-FACTOR = 3) is the difference between the lines beginning with 2 and 3 in **[2]**. This difference is not significant, so we again see that third-order effects are not likely to be necessary to provide a good fit to the observed table. (See comment above.) Since the test of second-order interactions (shown on the line beginning with 2) is significant, we see that at least some of the two-way interactions will significantly enhance the fit to the observed table. Therefore models to be evaluated should contain some second-order effects.

We conclude from **[2]** and **[3]** that second-order models should be investigated, but few third-order interactions are likely to be needed. Nonsignificance in **[2]** is what we are looking for, because it indicates adequate fit. Significance is what we are looking for in **[3]** because it indicates the necessity for including terms of that order.

[4] 4F prints tests of marginal and partial association in response to the ASSOCIATION = 3 command in the FIT paragraph.

Partial Association tests the significance of deleting a particular effect from a model. For example, to test the effect λ_{EM} (education by marital status), 4F computes the difference in G^2 between the full second order model (*EM,ES,EW,MS,MW,SW*) and the second order model from which λ_{EM} has been deleted (*ES,EW,MS,MW,SW*). A significant G^2 for an effect implies that the effect should be considered in models subjected to further analysis—deleting that effect makes a difference. Since this test is obtained as the difference between nested models, only the likelihood ratio test G^2 is appropriate.

Marginal Association tests the significance of deleting an effect from a model which contains all effects after summing over levels of categorical variables not included in the effect. For example, suppose you wanted to test the marginal effect of λ_{EM}. You would sum across levels of *S* and *W* to form a two-way *E* x *M* table, and then test for the necessity of including the *EM* term. The difference between the second-order (EM) model and the one with the *EM* term deleted (*E,M*) tests the *EM* effect. This is equivalent to the usual test of independence between *E* and *M*. A significant result again implies that the λ_{EM} effect makes a difference in the adequacy of fit and should be considered in models to be evaluated further.

The tests of partial and marginal association produce the same results for first and highest order effects. They can differ, sometimes substantially, for intermediate effects. If no marginals equal zero, the **degrees of freedom** associated with the chi-square tests of both marginal association and partial association of an effect *Z* are

$$(I-1)\,\delta^{ZA}\;(J-1)\delta^{ZB}\;(K-1)\delta^{ZC}\,(L-1)\delta^{ZD}$$

where $\delta^{ZA} = 1$ if *A* is part of *Z*, and zero otherwise; this is the same formula used for interactions in the analysis of variance. For example, the degrees of freedom for interaction *AB* are $(I-1)(J-1)$. *Note*: When there are observed cells with frequencies equal to zero, it is possible that one or more cells will have an expected value of zero. This will occur when a cell in a marginal subtable corresponding to one of the effects in the hierarchical model has a total frequency of zero. When one or more cells has an expected value of zero or is specified to contain a structural zero, the number of degrees of freedom is reduced; see Brown and Fuchs (1983).

The tests of marginal and partial association can be used simultaneously to screen the various interactions to determine whether they are necessary in the model for the data being used. If both partial and marginal tests are highly significant for an effect ($p < .01$), the effect should be retained in further models. If both tests are nonsignificant ($p > .05$), it is probably unnecessary to include the effect in further modeling. If one test is significant and the other is not, models with and without the questionable terms can be fit in a second pass of 4F to determine which terms are appropriate. This is explained further by Brown (1976).

Note: The analysis of a log-linear model is similar to that of an unbalanced analysis of variance. The tests of partial and marginal association are equivalent to the increase in the chi-square test-of-fit when the interaction is removed from a model (the model of full order if partial association is tested and the saturated model for a subtable if marginal association is tested). As with unbalanced analysis of variance, the effect of removing two interactions from a model may not be equal to the sum of the two chi-squares printed in the analysis.

From the output we see that by both partial and marginal criteria only one of the third order effects, λ_{MSW}, attains marginal significance, so the others are not needed in the final model. The tests of marginal and partial association are both highly significant for λ_{ES}, λ_{MW} and λ_{SW}; therefore these interactions belong in the model. For λ_{EW}, the test of marginal association is significant but the test for partial association is not; therefore it is doubtful whether λ_{EW} is needed in the model. Both tests are nonsignificant for λ_{EM} and λ_{MS}; therefore these are not needed in the model. We use these guidelines to choose which models to test. The models are tested in Example 4F3.2.

Example 4F3.2 Specifying log-linear models to be tested

Using the output of Example 4F3.1, we choose four models to examine further. These models include different combinations of terms indicated to be of interest by the tests of partial and marginal association. The second order effects, λ_{ES} and λ_{MW}, are included in all candidate models. The models are ordered from simple to more complex for ease in reading the output; terms less likely to be needed are added to models with terms more likely to be needed. The four models are specified in the FIT paragraph. A fifth model is also specified which is equivalent to the fourth model to illustrate the ability to describe hierarchical models by a reduced set of terms.

Each model is expressed in an abbreviated form so that, for example, the statement

```
MODEL = ES, MW.
```

specifies the hierarchical model

$$\theta + \lambda_E + \lambda_S + \lambda_M + \lambda_W + \lambda_{ES} + \lambda_{MW}$$

A model is expressed by a sequence of subscripts. You need not specify subscripts that are subsets of other subscripts. Therefore the first-order effects are not specified in the above model statement. For each model, 4F gives the two criteria for test of fit: Pearson chi-square and likelihood ratio chi-square. Associated with each test is the probability of obtaining a more extreme chi-square under the hypothesis that the model does fit.

Input 4F3.2
(replace the FIT paragraph in Input 4F3.1 with the following)

```
/ FIT      MODEL = es, mw.
           MODEL = es, mw, sw.
           MODEL = es, mw, sw, ms.
           MODEL = es, mw, sw, msw.
           MODEL = es, msw.
```

Output 4F3.2

—the observed table in Output 4F3.1 appears here—

```
                                 LIKELIHOOD-RATIO            PEARSON
MODEL                      D.F.  CHI-SQUARE   PROB         CHI-SQUARE   PROB      ITER.
-----                      ----  ----------   ----         ----------   ----      -----
es,mw.                      33     61.98     0.0017          71.37     0.0001       2
es,mw,sw.                   31     43.57     0.0664          51.01     0.0133       2
es,mw,sw,ms.                29     40.08     0.0827          41.13     0.0670       5
es,msw.                     27     33.37     0.1852          31.69     0.2439       2
es,msw.                     27     33.37     0.1852          31.69     0.2439       2
```

The tests of the models are presented in Output 4F3.2. Although the fourth model was specified as (*ES,MW,SW,MSW*), it is printed as (*ES,MSW*) which is its representation using the minimal number of hierarchical terms; note that its results are identical to those of the last model for which the minimal hierarchical model was specified. The test of lack of fit of the first model (*ES,MW*) is highly significant. The next model is significant for the Pearson chi-square but not for the likelihood-ratio test. Both tests are only approximate, so the choice between the two is a matter of preference. Since the second model appears to fit adequately, we could choose it as the final model. The model (*ES,MW,SW*) describes an association between education and self-esteem and between work status and both marital status and self-esteem. However, self-esteem is independent of marital status at each level of work status.

As indicated above, the test of lack-of-fit is a simultaneous test; there may still be interactions of interest embedded within the terms set to zero. One way of examining for additional interactions is to examine the differences in likelihood ratio chi-squares between models. For example, the difference between (*ES,MW,SW*) and (*ES,MW,SW,MS*) is 43.57 – 40.08 = 3.49 with 2 degrees of freedom. Since the chi-square of the difference is not significant, there is no reason to add MS to the previous model. The difference between (*ES,MW,SW*) and (*ES,MSW*) is 43.57 – 33.37 = 10.20 with 4 degrees of freedom ($p < .05$). This is a simultaneous test of two interactions λ_{MS} and λ_{MSW}, both of which are in the latter model but not in the former one. In this example the additional terms are only marginally significant, and since many tests of significance have been performed, we would probably choose the simpler model (*ES,MW,SW*) to represent the structure of the data.

Results After Fitting a Log-linear Model

Example 4F3.3 Estimates of log-linear model parameters and their standard errors

Relationships among the categorical variables can be better understood by examining the estimates of the parameters. In Input 4F3.3 we request the estimates of the log-linear parameters and of the multiplicative parameters β by specifying LAMBDA and BETA in the PRINT paragraph. 4F computes parameter estimates for the model that we chose as most satisfactory through examination of Output 4F3.2—a model containing all first-order effects and three of the six second-order effects.

Input 4F3.3
(add to input 4F3.1 after TABLE)

```
/ FIT          MODEL = es, mw, sw.

/ PRINT        LAMBDA.
               BETA.
```

Output 4F3.3

—the observed frequency in Output 4F3.1 appears here—

```
                                       LIKELIHOOD-RATIO          PEARSON
  MODEL                           D.F. CHI-SQUARE  PROB        CHI-SQUARE  PROB       ITER.
  -----                           ---- ----------  ----        ----------  ----       -----
  es,mw,sw.                         31     43.57   0.0664         51.01    0.0133        2

  ASYMPTOTIC STANDARD ERRORS OF THE PARAMETER ESTIMATES ARE COMPUTED   [1]
            BY THE DELTA METHOD.

THE ABOVE MODEL IS DIRECT.
```

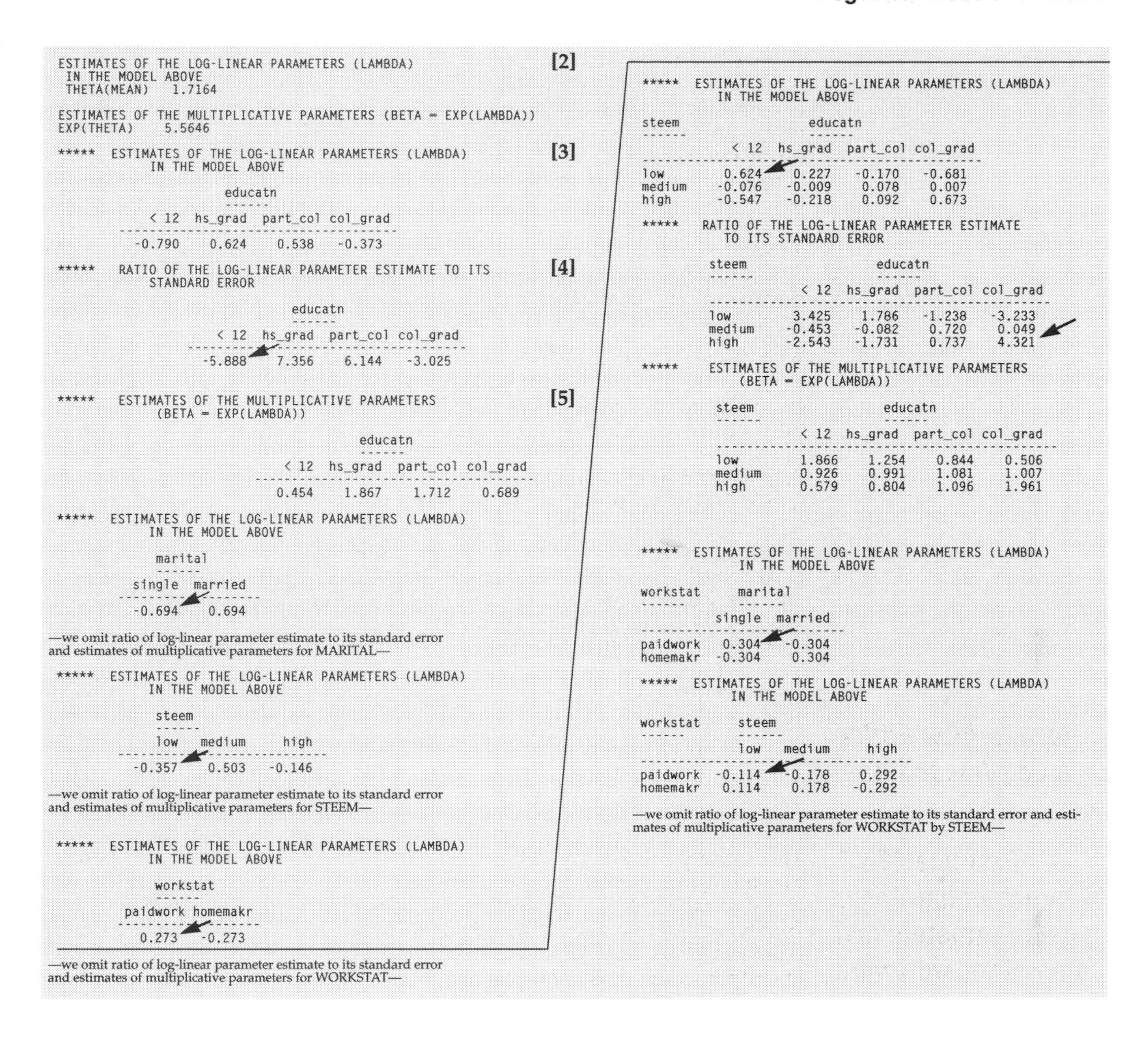

[2]

```
ESTIMATES OF THE LOG-LINEAR PARAMETERS (LAMBDA)
 IN THE MODEL ABOVE
 THETA(MEAN)    1.7164

ESTIMATES OF THE MULTIPLICATIVE PARAMETERS (BETA = EXP(LAMBDA))
EXP(THETA)     5.5646
```

[3]

```
*****  ESTIMATES OF THE LOG-LINEAR PARAMETERS (LAMBDA)
         IN THE MODEL ABOVE

                  educatn
                  ------

          < 12   hs_grad   part_col  col_grad
      -----------------------------------------
        -0.790    0.624     0.538    -0.373
```

[4]

```
*****   RATIO OF THE LOG-LINEAR PARAMETER ESTIMATE TO ITS
          STANDARD ERROR

                           educatn
                           ------

                  < 12   hs_grad   part_col  col_grad
              -----------------------------------------
                -5.888    7.356     6.144    -3.025
```

[5]

```
*****   ESTIMATES OF THE MULTIPLICATIVE PARAMETERS
          (BETA = EXP(LAMBDA))

                                   educatn
                                   ------

                          < 12   hs_grad   part_col  col_grad
                      -----------------------------------------
                         0.454    1.867     1.712     0.689

*****  ESTIMATES OF THE LOG-LINEAR PARAMETERS (LAMBDA)
         IN THE MODEL ABOVE

         marital
         ------

       single  married
     -------------------
      -0.694    0.694
```

—we omit ratio of log-linear parameter estimate to its standard error and estimates of multiplicative parameters for MARITAL—

```
*****  ESTIMATES OF THE LOG-LINEAR PARAMETERS (LAMBDA)
         IN THE MODEL ABOVE

         steem
         ------

         low    medium      high
     ----------------------------
      -0.357    0.503    -0.146
```

—we omit ratio of log-linear parameter estimate to its standard error and estimates of multiplicative parameters for STEEM—

```
*****  ESTIMATES OF THE LOG-LINEAR PARAMETERS (LAMBDA)
         IN THE MODEL ABOVE

         workstat
         ------

       paidwork  homemakr
     --------------------
        0.273    -0.273
```

—we omit ratio of log-linear parameter estimate to its standard error and estimates of multiplicative parameters for WORKSTAT—

```
*****  ESTIMATES OF THE LOG-LINEAR PARAMETERS (LAMBDA)
         IN THE MODEL ABOVE

steem              educatn
------             ------

              < 12   hs_grad   part_col  col_grad
----------------------------------------------------
low          0.624    0.227    -0.170    -0.681
medium      -0.076   -0.009     0.078     0.007
high        -0.547   -0.218     0.092     0.673

*****   RATIO OF THE LOG-LINEAR PARAMETER ESTIMATE
          TO ITS STANDARD ERROR

steem              educatn
------             ------

              < 12   hs_grad   part_col  col_grad
----------------------------------------------------
low          3.425    1.786    -1.238    -3.233
medium      -0.453   -0.082     0.720     0.049
high        -2.543   -1.731     0.737     4.321

*****   ESTIMATES OF THE MULTIPLICATIVE PARAMETERS
          (BETA = EXP(LAMBDA))

steem              educatn
------             ------

              < 12   hs_grad   part_col  col_grad
----------------------------------------------------
low          1.866    1.254     0.844     0.506
medium       0.926    0.991     1.081     1.007
high         0.579    0.804     1.096     1.961

*****  ESTIMATES OF THE LOG-LINEAR PARAMETERS (LAMBDA)
         IN THE MODEL ABOVE

workstat     marital
------       ------

           single  married
---------------------------
paidwork   0.304   -0.304
homemakr  -0.304    0.304

*****  ESTIMATES OF THE LOG-LINEAR PARAMETERS (LAMBDA)
         IN THE MODEL ABOVE

workstat     steem
------       ------

             low    medium     high
-------------------------------------
paidwork  -0.114   -0.178    0.292
homemakr   0.114    0.178   -0.292
```

—we omit ratio of log-linear parameter estimate to its standard error and estimates of multiplicative parameters for WORKSTAT by STEEM—

[1] In Output 4F3.3, this message describes the method used to compute the parameter estimates ($\hat{\lambda}$) and their asymptotic standard errors. When there are no structural zeros and the model is direct—i.e., the expected values can be expressed in closed form by a simple expression—the parameter estimates are obtained as described in the following paragraph and the standard errors are computed by the "delta" method (Lee, 1977). Otherwise, both the parameter estimates and their asymptotic standard errors are obtained by solving a set of linear equations involving $\ln F_{ijk}$ and then inverting the information matrix (see **[4]** below and Appendix B.13).

[2] The parameter estimate, θ, is the mean over all cells of the log of the expected frequencies. The multiplicative parameter estimate is the term e^{θ}.

[3] For the first effect in the model, EDUCATN, these are the log-linear parameter estimates ($\hat{\lambda}$s) at the four levels. The estimates are computed for **all** terms in the hierarchical model.

For those familiar with ANOVA parameters, if

$$x_{ijk} = \ln F_{ijk}$$

where F is the expected cell frequency, and not the observed frequency, then

$$\hat{\lambda}_{A_i} = \bar{x}_{i...} - \bar{x}_{....}$$

$$\hat{\lambda}_{AB_{ij}} = \bar{x}_{ij..} - \bar{x}_{i...} - \bar{x}_{.j..} - \bar{x}_{....}$$

$$\hat{\lambda}_{ABC_{ijk}} = \bar{x}_{ijk.} - \bar{x}_{ij..} - \bar{x}_{i.k.} - \bar{x}_{.jk.} - \bar{x}_{i...} + \bar{x}_{.j..} + \bar{x}_{..k.} - \bar{x}_{....}$$

where a period (.) indicates the sum over all levels of the omitted subscript. You can use parameter estimates to recalculate expected cell frequencies using the log-linear model. Take, for example, the cell representing working single women who have low self-esteem and did not complete high school (the first cell in Output 4F3.1). The observed frequency is 2. To find the expected frequency under the model (*ES,MW,SW*), we add parameter estimates:

$$\ln F_{1111} = 1.716 - .790 - .694 - .357 + .273 + .624 + .304 - .114 = .962$$

and take the antilog. The expected frequency under this model is:

$$e^{.962} = 2.62$$

However, the expected frequencies can be obtained directly by specifying EXPECTED in the PRINT paragraph (see next example).

[4] 4F prints the ratio of the parameter estimates to their asymptotic standard errors. With large samples, this ratio can be interpreted as a standard normal deviate (z). It can therefore be used as an indication of significance of the parameter estimate. For example, in the STEEM by EDUCATN interaction, we see that the estimated λ for college graduates with high self-esteem is large (4.321), implying a positive association between the two.

When the model is direct and no expected value is zero, the standard errors are computed by the delta method (Lee, 1977) unless VARIANCE is stated in the PRINT paragraph. In all other cases, the standard error is computed by inverting the information matrix (the only exception to this rule is described in the next paragraph). The latter method gives the correct asymptotic standard errors in all cases; the results of both methods are identical when the log-linear model is direct and there are no zero expected values.

When there is not enough space in computer memory to form the information matrix, the delta method is used to estimate the standard errors. When the model is not direct and there are no zero expected values, the standard errors produced by the delta method are conservative; i.e., they overestimate the asymptotic standard errors. When there are zero expected values, the standard errors produced by the delta method underestimate the asymptotic standard errors; the difference may be large when the parameter estimate has the same levels of the indices as those for a cell with a zero expected value. See Appendix B.13 for a more detailed explanation of the computation of the parameter estimates and their standard errors.

[5] Estimates of the multiplicative parameters, $\hat{\beta}$, where $\hat{\beta} = \exp(\hat{\lambda})$. Values of $\hat{\beta}$ greater than one indicate an increased probability of that combination of indices, while values of $\hat{\beta}$ less than one indicate a decreased probability. The null hypothesis would be that the βs are unity, which is equivalent to $\lambda = 0$ and can be tested by the ratios of the parameter estimates to their standard errors, **[4]** above.

Example 4F3.4 Expected (fitted) values and cell deviates

The chi-square test of fit provides an overall indication of how well the model fits the data, that is, how close the expected values under the model are to the observed frequencies. To better understand the nature of the fit, it is often useful to compare the observed and expected values for each cell, either directly or in terms of deviates based on the difference between the two. In Input 4F3.4, we take the model chosen as most satisfactory in Example 4F3.2 (*ES,MW,SW*) and request expected values and three types of available deviates in the PRINT paragraph.

Input 4F3.4
(add to Input 4F3.1 after TABLE)

```
/ FIT          MODEL = es, mw, sw.

/ PRINT        EXPECTED. STANDARDIZED. LRCHI.
```

Output 4F3.4

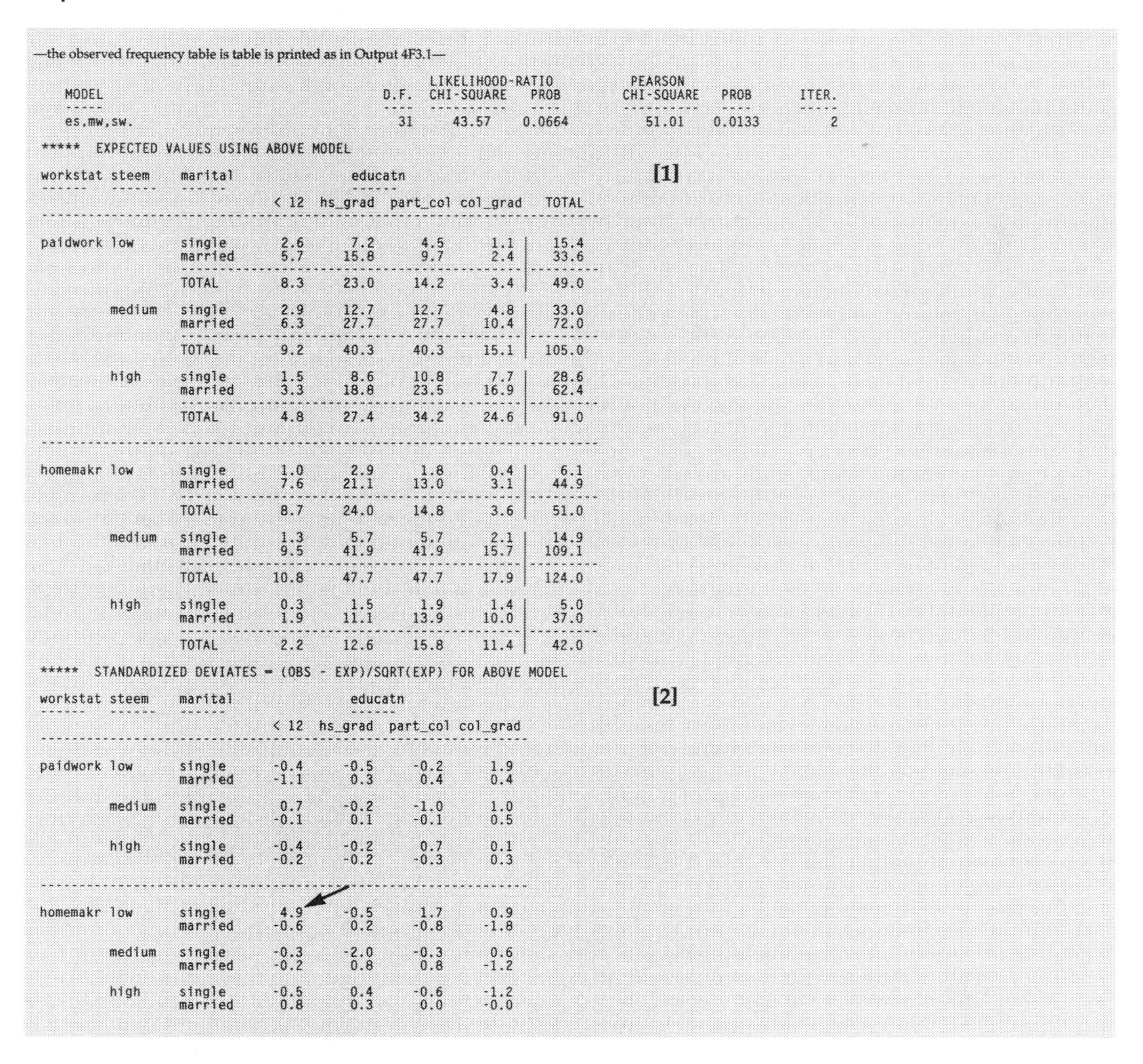

```
—the observed frequency table is table is printed as in Output 4F3.1—

                                    LIKELIHOOD-RATIO            PEARSON
  MODEL                        D.F.  CHI-SQUARE  PROB        CHI-SQUARE  PROB     ITER.
  -----                        ----  ----------  ----        ----------  ----     -----
  es,mw,sw.                     31      43.57   0.0664          51.01   0.0133       2

***** EXPECTED VALUES USING ABOVE MODEL

workstat steem    marital               educatn                              [1]
------   ------   ------                ------
                         < 12  hs_grad  part_col col_grad   TOTAL
----------------------------------------------------------------
paidwork low      single   2.6     7.2      4.5      1.1 |   15.4
                  married  5.7    15.8      9.7      2.4 |   33.6
                  ---------------------------------------|-------
                  TOTAL    8.3    23.0     14.2      3.4 |   49.0

         medium   single   2.9    12.7     12.7      4.8 |   33.0
                  married  6.3    27.7     27.7     10.4 |   72.0
                  ---------------------------------------|-------
                  TOTAL    9.2    40.3     40.3     15.1 |  105.0

         high     single   1.5     8.6     10.8      7.7 |   28.6
                  married  3.3    18.8     23.5     16.9 |   62.4
                  ---------------------------------------|-------
                  TOTAL    4.8    27.4     34.2     24.6 |   91.0
----------------------------------------------------------------
homemakr low      single   1.0     2.9      1.8      0.4 |    6.1
                  married  7.6    21.1     13.0      3.1 |   44.9
                  ---------------------------------------|-------
                  TOTAL    8.7    24.0     14.8      3.6 |   51.0

         medium   single   1.3     5.7      5.7      2.1 |   14.9
                  married  9.5    41.9     41.9     15.7 |  109.1
                  ---------------------------------------|-------
                  TOTAL   10.8    47.7     47.7     17.9 |  124.0

         high     single   0.3     1.5      1.9      1.4 |    5.0
                  married  1.9    11.1     13.9     10.0 |   37.0
                  ---------------------------------------|-------
                  TOTAL    2.2    12.6     15.8     11.4 |   42.0

***** STANDARDIZED DEVIATES = (OBS - EXP)/SQRT(EXP) FOR ABOVE MODEL

workstat steem    marital               educatn                              [2]
------   ------   ------                ------
                         < 12  hs_grad  part_col col_grad
---------------------------------------------------------
paidwork low      single  -0.4    -0.5     -0.2      1.9
                  married -1.1     0.3      0.4      0.4

         medium   single   0.7    -0.2     -1.0      1.0
                  married -0.1     0.1     -0.1      0.5

         high     single  -0.4    -0.2      0.7      0.1
                  married -0.2    -0.2     -0.3      0.3
---------------------------------------------------------
homemakr low      single   4.9    -0.5      1.7      0.9
                  married -0.6     0.2     -0.8     -1.8

         medium   single  -0.3    -2.0     -0.3      0.6
                  married -0.2     0.8      0.8     -1.2

         high     single  -0.5     0.4     -0.6     -1.2
                  married  0.8     0.3      0.0     -0.0
```

***** COMPONENTS OF LIKELIHOOD RATIO CHI-SQUARE = 2*OBS LN(OBS / EXP) FOR MODEL ABOVE **[3]**

workstat	steem	marital	educatn < 12	hs_grad	part_col	col_grad	TOTAL
paidwork	low	single	-1.1	-2.3	-0.9	6.1	1.9
		married	-3.9	2.5	2.7	1.5	2.8
		TOTAL	-4.9	0.3	1.8	7.6	4.7
	medium	single	2.6	-1.3	-6.2	5.4	0.5
		married	-0.6	0.7	-1.3	3.5	2.3
		TOTAL	2.1	-0.7	-7.5	8.9	2.8
	high	single	-0.8	-1.2	4.9	0.5	3.5
		married	-0.5	-1.5	-2.8	2.3	-2.6
		TOTAL	-1.4	-2.7	2.1	2.8	0.9
homemakr	low	single	21.0	-1.4	6.5	1.7	27.8
		married	-2.9	1.8	-5.3	0.0	-6.3
		TOTAL	18.2	0.4	1.2	1.7	21.5
	medium	single	-0.5	-3.5	-1.3	2.0	-3.3
		married	-1.0	10.7	10.7	-7.9	12.5
		TOTAL	-1.6	7.2	9.4	-5.8	9.2
	high	single	0.0	1.1	-1.3	0.0	-0.2
		married	2.6	1.8	0.2	-0.0	4.6
		TOTAL	2.6	2.9	-1.1	-0.0	4.5

[1] *Expected values* (F_{ijkl}) are obtained by fitting the log-linear model to the observed frequencies (f_{ijkl}) by Haberman's (1972) iterative proportional fitting algorithm. Compare this table with the table of observed frequencies in Output 4F3.1.

In 4F Part 1 we indicated that it was undesirable for more than 20% of the cells to have expected frequencies less than 5 and that none should be below 1. Both guidelines are violated in this example: 42% of the cells have expected values less than 5, and two cells have expected values less than 1. In multiway tables, these rules are often violated because the total number of collected data points is constrained by cost or availability of subjects, while the total number of cells increases with the dimension of the table.

Strict adherence to the preceding rule would restrict the analysis of many tables. A modified rule seems to be appropriate. The probability associated with the overall test of fit of the model may be affected by violating the preceding rule. However, a comparison between two similar models is less affected since the comparison is based on the totals in the marginal subtables corresponding to effects in the model. Therefore, when there are adequate frequencies in the marginal subtables, the probability associated with a test of the difference of two models remains appropriate. Hence, tests of lower order effects (two-way and possibly three-way interactions) should usually be less affected.

In assessing the effect of a sparse cell on the overall test of fit of a model, it is worth noting that the expected value appears in the denominator of the usual (Pearson) chi-square statistic. When the observed value in a cell exceeds its expected value, the contribution to the chi-square will be inversely proportional to the expected value and may be large for values near zero. However, when the observed value is zero, the contribution to the chi-square statistic is the expected value, which is small. Therefore, nonzero observed values tend to have inflated contributions in cells with expected values near zero.

In this example the two cells with the smallest expected values (0.3 and 0.4) do not contribute greatly to the chi-square statistic. The largest contribution to the likelihood ratio chi-square statistic (see **[3]** below) is 21.0 from a cell with an expected value of 1.0. In a later example (4F3.7) we will specify that this cell contains a structural zero and fit a model to the data in the remainder of the table.

[2] *Standardized deviates* are the square roots of the components of Pearson's chi-square statistics,

$$(f_{ijkl} - F_{ijkl}) / F_{ijkl}^{\ 1/2}$$

Since these expected values are similar to *z*-scores, any absolute value larger than 2.58 is not consistent with the model. (In a normal distribution approximately 1% of the values are larger than |2.58|.) Note the value 4.9—there are more homemaker women who are single with less than a high school education who have low self-esteem (6 women) than expected by the model (1 woman).

[3] *Components of the likelihood ratio chi-square statistic,*

$$2f_{ijkl} \ln(f_{ijkl} / F_{ijkl})$$

The likelihood ratio chi-square components show how each cell and stratum contributes to the overall likelihood ratio chi-square value for goodness-of-fit. Other measures of deviation are also available:

The *simple differences between observed and expected frequencies* can be obtained:

$$f_{ijkl} - F_{ijkl}$$

Freeman-Tukey deviates are defined as

$$F_{ijkl}^{\ 1/2} + (f_{ijkl} + 1)^2 - (4F_{ijkl} + 1)^{1/2}$$

Freeman-Tukey deviates provide an alternative to standardized deviates when the data are from a Poisson distribution. A table of *components of Pearson's chi-square* is available:

$$(f_{ijkl} - F_{ijkl})^2 / F_{ijkl}$$

These components show the contribution of cells to the overall Pearson chi-square value for goodness-of-fit.

See 4F COMMANDS for a list of all features available in the PRINT paragraph for analysis of multiway frequency tables.

Identifying Extreme Cells and Strata

Sometimes the fit of a model may suffer because of a few cells whose expected frequencies are inconsistent with the data in the remaining cells. One or a few cells, therefore, can be the cause of a significant chi-square test of fit for a model that would otherwise be adequate. These cells may well be the most interesting in the table. 4F can identify in a stepwise manner the cells whose frequencies deviate from the expected frequencies under a proposed model by finding cells with the largest standardized or Freeman-Tukey deviate. This is demonstrated in Example 4F3.5. After a cell is identified as having the largest deviate, it can be treated as a structural zero (Examples 4F3.7 and 4F3.8) and the model fit to the remaining cells. 4F can also delete, in turn, each level of a categorical variable to see the effect of that level on the chi-square test of fit (Example 4F3.6).

Example 4F3.5 Identifying extreme cells

In order to identify extreme cells, in Input 4F3.5 we request in the FIT paragraph that standardized deviates (CELL IS STANDARDIZED) be used to identify up to five extreme cells (STEP = 5). The model specified in the FIT paragraph is fitted to the data. If the test of fit of the model is significant, then standardized deviates are computed for all cells and the cell with the largest absolute deviate is identified. This cell is then declared to be a structural zero and the model is refitted to all the remaining cells. The stepping continues until the maximum number of steps is reached or the test of fit of the model becomes nonsignificant.

Input4F3.5
(replace the FIT paragraph in Input 4F3.1 with the following)

```
/ FIT        MODEL = es, mw, sw.
             MODEL = es, mw.
             CELL IS STANDARDIZED.
             STEP = 5.
```

Output 4F3.5

—the observed frequency table of Output 4F3.1 appears here—

```
                                      LIKELIHOOD-RATIO          PEARSON
 MODEL                          D.F.  CHI-SQUARE   PROB      CHI-SQUARE  PROB      ITER.
 -----                          ----  ----------   ----      ----------  ----      -----
 es,mw,sw.                       31      43.57     0.0664       51.01    0.0133        2

*** N O T E *** STEPWISE CELL DELETION IS NOT PERFORMED.  P-VALUE OF MODEL FIT       [1]
                ( 0.0664 ) EXCEEDS CRITERION PROBABILITY ( 0.0500 ).

                                      LIKELIHOOD-RATIO          PEARSON
 MODEL                          D.F.  CHI-SQUARE   PROB      CHI-SQUARE  PROB      ITER.
 -----                          ----  ----------   ----      ----------  ----      -----
 es,mw.                          33      61.98     0.0017       71.37    0.0001        2

 ***** CRITERION TO SELECT CELLS IS MAXIMUM STANDARDIZED DEVIATE = (OBS. - EXP.) /SQRT(EXP.).   [2]

                                  MAXIMUM        FOUND IN CELL
STEP  CHISQUARE  D.F.   PROB     DEVIATION  workstat steem    marital  educatn

  0     61.98     33   0.00166
                                   5.156    homemakr low      single   < 12
  1     47.10     32   0.04156
                                   2.362    homemakr low      single   part_col
  2     42.82     31   0.07686

*** N O T E *** P-VALUE EXCEEDS SPECIFIED PROBABILITY LEVEL.  STEPPING ENDS.
```

[1] In Output 4F3.5, the first model is the one selected as most satisfactory in Example 4F3.2. This model has a nonsignificant test of fit when compared with the level of CRITERION PROB, which is the alpha level for determining an adequate fit. In the example the criterion is set at .05 (the default value) but this criterion can be changed (see 4F COMMANDS). Because of the nonsignificant test of fit, indicating an adequate model, stepping is not performed.

[2] The second model (*ES,MW*) has a highly significant test of lack of fit (chi-square = 61.98, p = .00166). Standardized deviates are computed for all the cells. The stepwise cell deletion algorithm identifies the cell (homemaker, low esteem, single, less than high school education) which has the largest standardized deviate. This cell is then replaced by a structural zero and the log-linear model is refitted to the remaining cells in the table. The resulting chi-square at step 1 is 47.10 which is a reduction of 14.88 from the original model. This model still has a marginally significant test of lack of fit. Therefore, the process is repeated and a second cell (homemaker, low esteem, single, partial college education) is identified and replaced by a structural zero. The resulting chi-square is now 42.82, a reduction of 4.28 from the preceding step and the test of lack of fit is no longer significant.

The two cells identified as extreme share three common attributes: homemaker, single, and low self-esteem. In Examples 4F3.7 and 4F3.8, we eliminate all four cells with these attributes (by defining them as structural zeros) and fit models to the remaining data.

Example 4F3.6 Identifying extreme strata

In this example we identify strata (levels of an index or categorical variable) containing observed frequencies that differ greatly from the expected frequencies of the models specified in Example 4F3.5. We allow all indices to be candidates for identification of extreme strata by specifying STRATA = ALL. As seen in 4F COMMANDS, we could have listed a subset of categorical variables to be examined for extreme strata.

Input 4F3.6
(replace the FIT paragraph in Input 4F3.1 with the following)

```
/ FIT        MODEL = es, mw, sw.
             MODEL = es, mw.
             STRATA = ALL.
```

Output 4F3.6

—the observed frequency table in Output 4F3.1 appears here—

```
                                           LIKELIHOOD-RATIO          PEARSON
   MODEL                             D.F.  CHI-SQUARE   PROB       CHI-SQUARE   PROB      ITER.
   -----                             ----  ----------   ----       ----------   ----      -----
   es,mw,sw.                          31     43.57     0.0664         51.01     0.0133        2

   *****DELETION OF STRATA

VARIABLE  CATEGORY  CHISQUARE  D.F.     PROB.        [1]
steem     low         14.34     20    0.81305
steem     medium      29.37     20    0.08074
steem     high        37.93     20    0.00904
educatn   < 12        27.19     22    0.20424
educatn   hs_grad     34.00     22    0.04916
educatn   part_col    36.98     22    0.02380
educatn   col_grad    27.58     22    0.18993

                                           LIKELIHOOD-RATIO          PEARSON
   MODEL                             D.F.  CHI-SQUARE   PROB       CHI-SQUARE   PROB      ITER.
   -----                             ----  ----------   ----       ----------   ----      -----
   es,mw.                             33     61.98     0.0017         71.37     0.0001        2

   *****DELETION OF STRATA

VARIABLE  CATEGORY  CHISQUARE  D.F.     PROB.        [3]
steem     low         31.91     21    0.05979
steem     medium      38.35     21    0.01173
steem     high        38.20     21    0.01219
educatn   < 12        45.20     24    0.00551
educatn   hs_grad     47.83     24    0.00265
educatn   part_col    45.72     24    0.00478
educatn   col_grad    42.50     24    0.01134
```

[1] The first model is the one chosen as most satisfactory in the screening process (Example 4F3.2). 4F eliminates each category of each index in turn. In this example 4F first shows the results when eliminating each level of STEEM, then EDUCATN. If a variable has only two levels, such as MARITAL or WORKSTAT, its categories are not eliminated.

For each category, the output shows the results of fitting the specified model with only that category eliminated. For example, if the model (*ES,MW,SW*) were fit to the remaining cells after eliminating the LOW category for STEEM, the likelihood ratio test of fit to the model (G^2) would be 14.34 with df = 20 instead of 43.567 with df = 31, which was obtained when fitting the same model to the frequencies in the original table. That is, elimination of the low level of self-esteem (one-third of the cells) reduces the chi-square statistic by two-thirds. This may indicate that one or more cells at this level are not fitted well by the model. Cells at the low level of self-esteem were identified in the previous example as contributing to a significant lack of fit of a simpler model. Elimination of levels of education did not achieve such startling reductions in chi-square.

[2] The test of fit of the second model (*ES,MW*) is highly significant. It is therefore appropriate to ask if the significant lack of fit is primarily due to the frequencies in a single category.

[3] Elimination of all cells referring to low self-esteem leads to an almost 50% reduction in the chi-square statistic. The resulting test of lack of fit is marginally nonsignificant (p = .060). Elimination of no other single strata provides as large an effect on the chi-square.

Structural Zeros and Tests of Quasi-Independence

The term "structural zero" is used to refer to a cell whose probability of containing an outcome is known to be zero. Note that we are not referring to cells with zero frequencies because too few observations were collected, rather to cells that are naturally empty (e.g., males with vaginal cancer) or empty by design (the diagonal of a two-way table is not of interest). Models fit to data in tables containing such cells are fitted to all cells except those with structural zeros. If you specify a cell as a structural zero, 4F computes the expected value for that cell using all cells except those declared as structural zeros. Deviates, such as standardized deviates or Freeman-Tukey deviates, are computed from the observed and expected frequencies for all cells, including those specified as containing structural zeros.

You can also specify cells with positive frequency as structural zeros to see the effect of deleting cells from a model. If you do, 4F computes deviates as the difference between observed and expected frequencies where the expected frequency is computed from all cells except the cells designated as structural zeros. The test of quasi-independence in a frequency table is a test of independence limited to that part of the table with cells not designated as structural zeros.

There are two ways to describe structural zeros in 4F. In the following examples, we illustrate the two methods and describe the effect of structural zeros on the analysis.

Example 4F3.7 Method 1: Defining empty cells

In the TABLE paragraph of Input 4F3.7, we state the indices of the cells specified as structural zeros. In Example 4F3.5 we identified two cells as extreme with respect to the model (*ES,MW*). These cells had three attributes in common: homemaker, single, and low self-esteem. In this example and the next, we eliminate the four cells that have these attributes in common. They are the cells with indices (1,1,1,2), (2,1,1,2), (3,1,1,2), and (4,1,1,2), where the indices are ordered with the fastest changing index first (as in the INDICES statement in the TABLE paragraph). That is, EDUCATN is the first index and WORKSTAT is the last, while homemaker is the second level of WORKSTAT. The statement

```
EMPTY = 1,1,1,2,  2,1,1,2,  3,1,1,2,  4,1,1,2.
```

in the TABLE paragraph specifies the four cells that will be treated as structural zeros in all analyses that are requested for this table. If the number of indices in the EMPTY list is not a multiple of the number of indices in the table, 4F prints an error message and stops. Note that integers must be used to describe the levels of the indices; category names or codes cannot be used.

Note: If you specify a CONDITION variable (i.e., if the tables are cross sections of a higher way table), the number of indices in the table includes the CONDITION variable. That is, you must specify as the last index in each cell (the slowest moving variable) the level of the CONDITION variable. Example 4F1.4 shows how to condition tables.

Input 4F3.7
(replace the TABLE and FIT paragraphs in Input 4F3.1 with the following)

```
/ TABLE   INDICES ARE educatn, marital, steem, workstat.
          EMPTY = 1,1,1,2,  2,1,1,2,  3,1,1,2,  4,1,1,2.

/ FIT     MODEL = es, mw, sw.
          MODEL = es, mw.

/ PRINT   EXPECTED.
          STANDARDIZED.
```

Output 4F3.7

```
***** OBSERVED FREQUENCY TABLE  1
ASTERISK INDICATES MISSING VALUE                                        [1]
workstat steem    marital            educatn
------   ------   ------             ------
                           < 12  hs_grad  part_col col_grad    TOTAL
--------------------------------------------------------------------
paidwork low      single     2       6        4        3   |    15
                  married    3      17       11        3   |    34
                  ------------------------------------------|-------
                  TOTAL      5      23       15        6   |    49

         medium   single     4      12        9        7   |    32
                  married    6      28       27       12   |    73
                  ------------------------------------------|-------
                  TOTAL     10      40       36       19   |   105

         high     single     1       8       13        8   |    30
                  married    3      18       22       18   |    61
                  ------------------------------------------|-------
                  TOTAL      4      26       35       26   |    91
--------------------------------------------------------------------

homemakr low      single     6 *     2 *      4 *      1 *       0
                  married    6      22       10        0   |    38
                  ------------------------------------------|-------
                  TOTAL      6      22       10        0   |    38

         medium   single     1       1        5        3   |    10
                  married    9      47       47       11   |   114
                  ------------------------------------------|-------
                  TOTAL     10      48       52       14   |   124

         high     single     0       2        1        0   |     3
                  married    3      12       14       10   |    39
                  ------------------------------------------|-------
                  TOTAL      3      14       15       10   |    42

      TOTAL OF THE OBSERVED FREQUENCY TABLE IS      449
           SUMMED OVER   44  CELLS WITHOUT STRUCTURAL ZEROS             [2]

          3  CASES HAD INCOMPLETE DATA.
```

—we omit table of excluded cases—

```
                                          LIKELIHOOD-RATIO         PEARSON
  MODEL                              D.F.  CHI-SQUARE   PROB      CHI-SQUARE  PROB     ITER.   [3]
  -----                              ----  ----------   ----      ----------  ----     -----
  es,mw,sw.                           27     23.92     0.6350       21.18    0.7777      4
```

—we omit table of expected values and standardized deviates for the first model—

```
                                          LIKELIHOOD-RATIO         PEARSON
  MODEL                              D.F.  CHI-SQUARE   PROB      CHI-SQUARE  PROB     ITER.   [4]
  -----                              ----  ----------   ----      ----------  ----     -----
  es,mw.                              29     41.49     0.0624       38.62    0.1092      4
```

```
***** EXPECTED VALUES USING ABOVE MODEL
ASTERISK INDICATES MISSING VALUE
workstat steem    marital            educatn
------   ------   ------             ------                             [5]
                           < 12  hs_grad  part_col col_grad    TOTAL
--------------------------------------------------------------------
paidwork low      single    1.9     8.0      4.4      1.1  |   15.4
                  married   4.2    17.4      9.7      2.3  |   33.6
                  ------------------------------------------|-------
                  TOTAL     6.2    25.3     14.1      3.4  |   49.0

         medium   single    2.9    12.7     12.7      4.8  |   33.0
                  married   6.3    27.7     27.7     10.4  |   72.0
                  ------------------------------------------|-------
                  TOTAL     9.2    40.3     40.3     15.1  |  105.0

         high     single    1.5     8.6     10.8      7.7  |   28.6
                  married   3.3    18.8     23.5     16.9  |   62.4
                  ------------------------------------------|-------
                  TOTAL     4.8    27.4     34.2     24.6  |   91.0
--------------------------------------------------------------------

homemakr low      single    0.4 *   1.7 *    0.9 *    0.2 *     0.0
                  married   4.8    19.7     10.9      2.6  |   38.0
                  ------------------------------------------|-------
                  TOTAL     4.8    19.7     10.9      2.6  |   38.0

         medium   single    0.8     3.7      3.7      1.4  |    9.7
                  married  10.0    43.9     43.9     16.5  |  114.3
                  ------------------------------------------|-------
                  TOTAL    10.8    47.7     47.7     17.9  |  124.0

         high     single    0.2     1.0      1.2      0.9  |    3.3
                  married   2.0    11.6     14.6     10.5  |   38.7
                  ------------------------------------------|-------
                  TOTAL     2.2    12.6     15.8     11.4  |   42.0
```

```
***** STANDARDIZED DEVIATES = (OBS - EXP)/SQRT(EXP) FOR ABOVE MODEL
ASTERISK INDICATES MISSING VALUE
workstat steem    marital              educatn                          [6]
-------- ------   ------               ------
                           < 12  hs_grad  part_col  col_grad
-----------------------------------------------------------------
paidwork low      single    0.0    -0.7     -0.2      1.9
                  married  -0.6    -0.1      0.4      0.5

         medium   single    0.3    -0.8     -1.5      0.6
                  married  -0.5    -0.8     -1.0     -0.1

         high     single   -0.2     0.5      1.5      0.8
                  married   0.2     0.8      0.8      1.3
-----------------------------------------------------------------
homemakr low      single    8.7 *   0.3 *    3.2 *    1.6 *
                  married   0.5     0.5     -0.3     -1.6

         medium   single    0.3    -1.2      1.0      1.7
                  married   0.2     1.6      1.6     -0.8

         high     single   -0.5     0.5     -0.6     -1.1
                  married   0.0    -1.2     -1.5     -1.3
```

In Output 4F3.7 we observe:

[1] In the observed frequency table, note the asterisks beside the frequencies in the four cells identified above. The observed frequencies in these cells are ignored in all computations, including those involving log-linear models (except for the computation of deviates).

[2] 4F computes the marginal total frequencies in the observed table without those cells specified as structural zeros. Compare this output with Output 4F3.1. The total is now 449 instead of 462 and the sum is over 44 cells instead of the 48 cells in the unaltered observed frequency table.

[3] The test of lack of fit of each model is based on four fewer cells than in the unaltered table of observations. This reduces the degrees of freedom used to test the fit of a model. For example, the chi-square statistic for the model (*ES*,*MW*,*SW*) in this altered table is 23.92 with 27 degrees of freedom as compared to 43.57 with 31 degrees of freedom in the unaltered table. That is, elimination of four cells reduces the chi-square statistic by almost 50%. Since this model did not have a significant test of lack of fit in the original table, it is of interest to fit a model (*ES*,*MW*) that did not fit the original data well.

[4] The chi-square statistic for the model (*ES*,*MW*) fitted to the table with four structural zeros is 41.49 with 29 degrees of freedom as compared to a chi-square statistic of 61.98 with 33 degrees of freedom in the original table (see Output 4F3.2). The p-value has changed from highly significant ($p = .0017$) to marginally nonsignificant ($p = .0624$). Therefore, the four cells that were set to structural zeros contribute substantially to the lack of fit of the model.

[5] To understand how the observed frequencies in the cells set to structural zeros deviate from the model, the observed values can be compared to the expected values computed from the model fit to the remaining cells in the table. Alternatively, functions of the differences between observed and expected values, such as standardized deviates, can be examined. The expected values for all four cells that were replaced by structural zeros are less than the observed values in those cells. The largest differences are found in the cells corresponding to the lowest level of education and to partial college. These inferences are consistent with our inferences in Example 4F3.5 where these two cells were selected as contributing to a significant lack of fit of the model.

[6] Standardized deviates are computed for all cells. The standardized deviates for the cells with structural zeros compare the original observed frequencies to the expected frequencies estimated from the model in which these cells were treated as structural zeros; i.e., not fitted by the model. There are two very large deviates among the four cells with structural zeros. The cell

corresponding to the lowest educational level contributes greatly to the lack of fit of the model to the original table (see Examples 4F3.4 and 4F3.5).

Note: When there are either cells specified as structural zeros or cells with expected values equal to zero (which occurs if any marginal subtable corresponding to an effect in the model contains a cell with zero frequency), the domain of the model is affected since the model is fitted only to cells that are not structural zeros and that have a nonzero expected value. Two models can be compared when either: (1) one model is a direct subset of the other and the model is fitted to the same set of observed cell frequencies (i.e., have the same domain), or (2) the two models are the same (i.e., have identical nonzero terms) and one set of observed cell frequencies is contained within the other set, i.e., the domain of one model is contained within the domain of the other. A comparison of two models which violates both conditions can lead to incorrect inferences (Brown and Fuchs, 1983).

Example 4F3.8 Method 2: Specifying an initial FIT matrix

For the second method of declaring structural zeros, we specify the initial fitted matrix in Input 4F3.8 by

```
INITIAL = 24*1, 4*0, 20*1.
```

This specifies that the first 24 cells contain observed values, the next four contain zeros (i.e., structural zeros), followed by 20 cells containing observed values. To use this option you must know how the cells are ordered within the table. The cells are ordered so that the first index (EDUCATN in this example) is fastest moving, etc. **Note:** INITIAL must be zero or a positive number. You may specify any initial fit matrix—not only a matrix of zeros and ones; however, when the nonzero elements are not ones, the fitted values will be affected by the initial values.

Note that only the INITIAL statement differs from Example 4F3.7. The results of this example are identical to those shown in Output 4F3.7.

Input 4F3.8
(replace the TABLE and FIT paragraphs in Input 4F3.1 with the following)

```
/ TABLE    INDICES ARE educatn, marital, steem, workstat.
           INITIAL = 24*1, 4*0, 20*1.

/ FIT      MODEL = es, mw, sw.
           MODEL = es, mw.

/ PRINT    EXPECTED.
           STANDARDIZED.
```

Other Modeling Options

Example 4F3.9 Testing all models in a three-way table

For a two- or three-way frequency table, all hierarchical models can be requested by specifying

```
/ FIT  ALL.
```

in the FIT paragraph. To illustrate this option we have requested a three-way table, EDUCATN by STEEM by MARITAL, in Input 4F3.9.

Input 4F3.9
(replace the TABLE and FIT paragraphs in Input 4F3.1 with the following)

```
/ TABLE   INDICES ARE educatn, marital, steem, workstat.

/ FIT     ALL.
```

Output 4F3.9

```
***** OBSERVED FREQUENCY TABLE  1

marital  steem            educatn
------   ------           -------
                  < 12  hs_grad  part_col  col_grad    TOTAL
-------------------------------------------------------------

single   low         8       8        8         4  |      28
         medium      5      13       14        10  |      42
         high        1      10       14         8  |      33
         -------------------------------------------------------
         TOTAL      14      31       36        22  |     103

married  low         9      39       21         3  |      72
         medium     15      75       74        23  |     187
         high        6      30       36        28  |     100
         -------------------------------------------------------
         TOTAL      30     144      131        54  |     359

         TOTAL OF THE OBSERVED FREQUENCY TABLE IS       462

              3  CASES HAD INCOMPLETE DATA.
```

—we omit table of excluded cases—

```
***** ALL MODELS ARE REQUESTED--

MODEL          DF   LIKELIHOOD-   PROB.      PEARSON   PROB.    ITERATIONS
                    RATIO CHISQ              CHISQ
-----          --   -----------   ------     -------   -------  ----------
e.             20      254.13     0.0000     278.30    0.0000       1
s.             21      317.80     0.0000     353.19    0.0000       1
m.             22      223.98     0.0000     227.05    0.0000       1
e,s.           18      197.76     0.0000     195.20    0.0000       1
s,m.           20      167.61     0.0000     151.25    0.0000       1
m,e.           19      103.94     0.0000     101.16    0.0000       1
e,s,m.         17       47.57     0.0001      51.51    0.0000       1
es.            12      168.40     0.0000     155.55    0.0000       1
em.            16       97.71     0.0000      92.51    0.0000       1
sm.            18      163.20     0.0000     146.36    0.0000       1
e,sm.          15       43.16     0.0001      45.22    0.0001       1
s,em.          14       41.35     0.0002      41.77    0.0001       1
m,es.          11       18.21     0.0768      19.77    0.0486       1
es,em.          8       11.98     0.1520      12.27    0.1394       1
em,sm.         12       36.93     0.0002      36.81    0.0002       1
sm,es.          9       13.79     0.1299      14.48    0.1061       1
es,em,sm.       6        7.74     0.2577       7.59    0.2700       4
```

In Output 4F3.9 the model with the fewest terms having a non-significant test of lack of fit is (*M,ES*). This model states that there is an interaction between education and self-esteem, but that both are independent of marital status. All models except the one with three two-factor interactions have simple interpretations. For example, (*ES,EM*) describes a model in which self-esteem is independent of marital status at each level of education (i.e., when conditioned on education). After identifying one or two models of interest, our next step would be to obtain a detailed analysis for these models (parameter estimates, expected values, cell deviates, etc.).

Example 4F3.10 Stepwise model building by deleting terms

Introduction to stepwise model building. Benedetti and Brown (1978) describe several methods for choosing an appropriate model. One family of methods is stepwise model building starting from a specified model. Terms can be added to the model or deleted from it. The terms added or deleted can be either simple effects or multiple effects. By simple effect we mean a single effect in a log-linear model. For example the models

$$(A,B,C) \text{ and } (A,BC)$$

differ only by the effect λ_{BC}; therefore they differ by a simple effect. By multiple effects we mean the difference between two models whose expressions differ by the addition of an index to an effect already in the model. For example, the models

$$(A,B) \text{ and } (AC,B)$$

differ by two effects $\lambda_C + \lambda_{AC}$. However, since the expression for the models differs only in the addition of one index, *C*, to a term already in the model, this change is called a multiple effect.

Adding effects. In Example 4F3.11, we demonstrate stepwise model building by adding effects. Starting from each MODEL specified in the FIT paragraph, 4F adds in turn each simple or multiple effect. After fitting all possible new models, 4F identifies the best model as that for which the test of significance of the difference is most significant (has the smallest tail probability). Since the number of degrees of freedom associated with the difference depends on the additional effect(s) being fitted, the criterion for the best model is not equivalent to having the largest chi-square test for the difference. This process is repeated until the maximum number of steps specified is performed, no more effects can be added, or both the test-of-fit of the best model and the test of the difference are nonsignificant.

Deleting effects. Starting from each MODEL specified in the FIT paragraph, 4F deletes in turn each simple or multiple effect. After fitting all possible new models, 4F identifies the best model as that for which the test of significance of the difference is least significant (has the largest tail probability). Since the number of degrees of freedom associated with the difference depends on the effect(s) being deleted, the criterion for the best model is not equivalent to that having the smallest chi-square test for the difference. This process is repeated until the maximum number of steps specified is performed, no more effects can be deleted, or both the test-of-fit of the best model and the test of the difference are significant.

Benedetti and Brown indicate that deleting terms from an overspecified model (one that has too many effects in it) is a "safer" procedure than adding terms. When the deletion option is chosen, we recommend deleting simple rather than multiple effects since deleting a multiple term (for example a second order interaction and all its higher order relatives) may be too drastic. (Conversely, if you choose the addition option, we recommend adding multiple rather than simple effects.)

In Example 4F3.10 the

```
DELETE = SIMPLE.
```

process is illustrated using two models. The first model is that containing all two-way interactions (as suggested by the results in Example 4F3.1), and the second model is that of all three-way interactions (based on our more conservative approach of identifying all possible interactions of interest).

Input 4F3.10

(replace the FIT paragraph in Input 4F3.1 with the following)

```
/ TABLE   INDICES ARE educatn, marital, steem, workstat.

/ FIT     MODEL = em, es, ew, ms, mw, sw.
          DELETE = SIMPLE.
          STEP = 5.

/ FIT     MODEL = ems, emw, esw, msw.
          DELETE = SIMPLE.
          STEP = 5.
```

Output 4F3.10

```
****************
*   MODEL  1   *
****************
                                                                              [1]
                                         LIKELIHOOD-RATIO         PEARSON
    MODEL                          D.F.  CHI-SQUARE   PROB     CHI-SQUARE   PROB     ITER.
    -----                          ----  ----------   ----     ----------   ----     -----
    em,es,ew,ms,mw,sw.              23     29.14     0.1757      26.16     0.2934       6
```

```
MODELS FORMED BY DELETING TERMS FROM MODEL --
   em,es,ew,ms,mw,sw.

                                             LIKELIHOOD-RATIO           PEARSON
    MODEL                               D.F.  CHI-SQUARE   PROB       CHI-SQUARE   PROB      ITER.
    -----                               ----  ----------   ----       ----------   ----      -----
    es,ew,ms,mw,sw.                      26     35.08     0.1099        35.81     0.0953        5
      DIFF. DUE TO DELETING  em.          3      5.94     0.1146

    em,ew,ms,mw,sw.                      29     55.04     0.0025        54.04     0.0032        5
      DIFF. DUE TO DELETING  es.          6     25.90     0.0002

    em,es,ms,mw,sw.                      26     34.03     0.1344        31.65     0.2050        6
      DIFF. DUE TO DELETING  ew.          3      4.89     0.1804

    em,es,ew,mw,sw.                      25     32.57     0.1422        31.25     0.1810        5
      DIFF. DUE TO DELETING  ms.          2      3.43     0.1801

    em,es,ew,ms,sw.                      24     54.28     0.0004        47.75     0.0027        5
      DIFF. DUE TO DELETING  mw.          1     25.14     0.0000

    em,es,ew,ms,mw.                      25     43.33     0.0129        40.43     0.0263        5
      DIFF. DUE TO DELETING  sw.          2     14.19     0.0008

STEP  1.     BEST MODEL FOUND IS --
    em,es,ms,mw,sw.

    es,ms,mw,sw.                         29     40.08     0.0827        41.13     0.0670        5
      DIFF. DUE TO DELETING  em.          3      6.05     0.1091

    em,ms,mw,sw.                         32     63.22     0.0008        65.69     0.0004        5
      DIFF. DUE TO DELETING  es.          6     29.19     0.0001

    em,es,mw,sw.                         28     37.71     0.1040        38.05     0.0974        4
      DIFF. DUE TO DELETING  ms.          2      3.68     0.1586

    em,es,ms,sw.                         27     59.28     0.0003        51.88     0.0027        4
      DIFF. DUE TO DELETING  mw.          1     25.26     0.0000

    em,es,ms,mw.                         28     51.51     0.0044        49.71     0.0070        4
      DIFF. DUE TO DELETING  sw.          2     17.48     0.0002

STEP  2.     BEST MODEL FOUND IS --
    em,es,mw,sw.

    es,mw,sw.                            31     43.57     0.0664        51.01     0.0133        2
      DIFF. DUE TO DELETING  em.          3      5.86     0.1184

    em,mw,sw.                            34     66.71     0.0007        76.67     0.0000        2
      DIFF. DUE TO DELETING  es.          6     29.00     0.0001

    em,es,sw.                            29     63.52     0.0002        55.23     0.0023        2
      DIFF. DUE TO DELETING  mw.          1     25.81     0.0000

    em,es,mw.                            30     55.75     0.0029        57.41     0.0019        2
      DIFF. DUE TO DELETING  sw.          2     18.04     0.0001

STEP  3.     BEST MODEL FOUND IS --
    es,mw,sw.

    e,mw,sw.                             37     72.94     0.0004        96.15     0.0000        2
      DIFF. DUE TO DELETING  es.          6     29.36     0.0001

    es,m,sw.                             32     69.75     0.0001        64.00     0.0007        2
      DIFF. DUE TO DELETING  mw.          1     26.18     0.0000

    es,mw.                               33     61.98     0.0017        71.37     0.0001        2
      DIFF. DUE TO DELETING  sw.          2     18.41     0.0001

STEP  4.     BEST MODEL FOUND IS --
    es,mw.

 STEPPING STOPS DUE TO CRITERION PROBABILITY (      0.050).

****************
*   MODEL  2   *
****************

                                             LIKELIHOOD-RATIO           PEARSON
    MODEL                               D.F.  CHI-SQUARE   PROB       CHI-SQUARE   PROB      ITER.      [2]
    -----                               ----  ----------   ----       ----------   ----      -----
    ems,emw,esw,msw.                      6      7.46     0.2808         7.59     0.2693        7

MODELS FORMED BY DELETING TERMS FROM MODEL --
    ems,emw,esw,msw.
```

—we omit step 1 models—

```
STEP  1.     BEST MODEL FOUND IS --
    ems,emw,msw.
```

—we omit step 2 models—

```
STEP  2.     BEST MODEL FOUND IS --
    ems,ew,msw.
```

—we omit step 3 models—

```
STEP  3.     BEST MODEL FOUND IS --
    es,em,ew,msw.
```

—we omit step 4 models—

```
STEP  4.      BEST MODEL FOUND IS --
    es,em,msw.

    es,msw.                                   27     33.37    0.1852          31.69    0.2439          2
      DIFF. DUE TO DELETING  em.               3      6.05    0.1091

    em,msw.                                   30     56.51    0.0024          54.52    0.0040          2
      DIFF. DUE TO DELETING  es.               6     29.19    0.0001

    es,em,sw,mw,ms.                           26     34.03    0.1344          31.65    0.2050          5
      DIFF. DUE TO DELETING  msw.              2      6.71    0.0349

STEP  5.      BEST MODEL FOUND IS --
    es,msw.

 MAX. NUMBER OF STEPS ( 5) PERFORMED. STEPPING STOPS.
```

[1] In Output 4F3.10, the first model specified as the starting model for deleting terms is (*EM,ES,EW,MS,MW,SW*). 4F deletes each of the six interactions in turn to form all possible models with only five interactions each. Since terms are being deleted, the term chosen to be deleted will be the one that has the least effect on the model (the change with the highest *p*-value). At step 1 4F identifies the effect λ_{EW} with a *p*-value of .1804 having the least effect and chooses as best model the model excluding this interaction (*EM,ES,MS,MW,SW*). Similarly, at the second step 4F selects the interaction MS for exclusion and chooses the model (*EM,ES,MW,SW*). At the third step 4F identifies (*ES,MW,SW*). Since the test of lack of fit of this model is still nonsignificant, 4F continues one more step and identifies (*ES,MW*). Since the test of lack of fit is now significant, the stepping algorithm terminates. The algorithm continues one step beyond the appropriate model, since the model at the last step will have a significant test of lack of fit; however, the model chosen to represent the data should have a nonsignificant test of lack of fit.

Note: When an interaction is deleted, main effects or interactions contained as subsets of the term being deleted continue to be included in the model. For example, when λ_{ES} is eliminated, both λ_E and λ_S continue to be included. As long as λ_E and λ_S are parts of other interactions in the remaining model, they will not appear explicitly in the model. However, when either or both are not within another interaction, then the main effect will appear explicitly.

[2] Starting from the model containing all three-way interactions (*EMS,EMW,ESW,MSW*), the algorithm identifies the model (*ES,MSW*) at the fifth step. Stepping stops since this is the maximum number of steps specified.

Example 4F3.11 Stepwise model building by adding terms

In this example we describe the final results of the stepwise process of adding terms to the model of all main effects (*E,S,M,W*). We first use the method ADD = SIMPLE and then the method ADD = MULTIPLE. For a description of the process of model building by adding and deleting terms, see Example 4F3.10.

Input 4F3.11
(replace the FIT paragraph in Input 4F3.1 with the following)

```
/ TABLE    INDICES ARE educatn, marital, steem, workstat.

/ FIT      MODEL = E, M,.S, W.
           DELETE = SIMPLE.
           STEP = 5.

/ FIT      MODEL = E, M,.S, W.
           DELETE = MULTIPLE.
           STEP = 5.
```

Output 4F3.11

```
****************
*   MODEL  1   *
****************

                                           LIKELIHOOD-RATIO            PEARSON                          [1]
    MODEL                             D.F.  CHI-SQUARE   PROB       CHI-SQUARE   PROB       ITER.
    -----                             ----  ----------   ----       ----------   ----       -----
    e,m,s,w.                            40    117.52     0.0000       124.79     0.0000         2

 MODELS FORMED BY ADDING TERMS TO MODEL --
    e,m,s,w.

                                           LIKELIHOOD-RATIO            PEARSON
    MODEL                             D.F.  CHI-SQUARE   PROB       CHI-SQUARE   PROB       ITER.
    -----                             ----  ----------   ----       ----------   ----       -----
    em,s,w.                             37    111.29     0.0000       111.30     0.0000         2
      DIFF. DUE TO  ADDING   em.         3      6.23     0.1011

    es,m,w.                             34     88.16     0.0000        84.19     0.0000         2
      DIFF. DUE TO  ADDING   es.         6     29.36     0.0001

    ew,m,s.                             37    109.12     0.0000       109.34     0.0000         2
      DIFF. DUE TO  ADDING   ew.         3      8.40     0.0384

    e,ms,w.                             38    113.10     0.0000       114.89     0.0000         2
      DIFF. DUE TO  ADDING   ms.         2      4.42     0.1099

    e,mw,s.                             39     91.34     0.0000       123.00     0.0000         2
      DIFF. DUE TO  ADDING   mw.         1     26.18     0.0000

    e,m,sw.                             38     99.12     0.0000        98.74     0.0000         2
      DIFF. DUE TO  ADDING   sw.         2     18.41     0.0001

 STEP  1.     BEST MODEL FOUND IS --
    e,mw,s.
```

—we omit step 2 models—

```
 STEP  2.     BEST MODEL FOUND IS --
    es,mw.
```

—we omit some step 3 models—

```
    sw,es,mw.                           31     43.57     0.0664        51.01     0.0133         2
      DIFF. DUE TO  ADDING   sw.         2     18.41     0.0001

 STEP  3.     BEST MODEL FOUND IS --
    sw,es,mw.

    sw,em,es,mw.                        28     37.71     0.1040        38.05     0.0974         4
      DIFF. DUE TO  ADDING   em.         3      5.86     0.1184

    ew,sw,es,mw.                        28     38.57     0.0880        44.49     0.0248         5
      DIFF. DUE TO  ADDING   ew.         3      5.00     0.1719

    ms,sw,es,mw.                        29     40.08     0.0827        41.13     0.0670         5
      DIFF. DUE TO  ADDING   ms.         2      3.49     0.1742

 STEP  4.     BEST MODEL FOUND IS --
    sw,em,es,mw.

 STEPPING STOPS DUE TO CRITERION PROBABILITY (      0.050).

****************
*   MODEL  2   *
****************

                                           LIKELIHOOD-RATIO            PEARSON                          [2]
    MODEL                             D.F.  CHI-SQUARE   PROB       CHI-SQUARE   PROB       ITER.
    -----                             ----  ----------   ----       ----------   ----       -----
    e,m,s,w.                            40    117.52     0.0000       124.79     0.0000         2

 MODELS FORMED BY ADDING TERMS TO MODEL --
    e,m,s,w.
```

—results for Step 1 and Step 2 are the same for those for MODEL 1—

```
    em,es,mw.                           30     55.75     0.0029        57.41     0.0019         2
      DIFF. DUE TO  ADDING   em.         3      6.23     0.1011

    ew,es,mw.                           30     53.58     0.0051        60.24     0.0009         2
      DIFF. DUE TO  ADDING   ew.         3      8.40     0.0384

    ms,es,mw.                           31     57.56     0.0026        60.93     0.0010         2
      DIFF. DUE TO  ADDING   ms.         2      4.42     0.1099

    sw,es,mw.                           31     43.57     0.0664        51.01     0.0133         2
      DIFF. DUE TO  ADDING   sw.         2     18.41     0.0001

    ems,mw.                             22     43.77     0.0038        41.26     0.0077         2
      DIFF. DUE TO  ADDING   ems.       11     18.21     0.0768

    es,emw.                             24     43.96     0.0077        40.51     0.0188         2
      DIFF. DUE TO  ADDING   emw.        9     18.01     0.0350

    esw,mw.                             22     32.88     0.0636        39.49     0.0124         2
      DIFF. DUE TO  ADDING   esw.       11     29.10     0.0022

    es,msw.                             27     33.37     0.1852        31.69     0.2439         2
      DIFF. DUE TO  ADDING   msw.        6     28.61     0.0001

 STEP  3.     BEST MODEL FOUND IS --
    es,msw.
```

```
—we omit some step 4 models—
    es,emw,msw.                                      18     18.75    0.4076          19.34    0.3710         5
       DIFF. DUE TO   ADDING    emw.                  9     14.62    0.1018

STEP  4.     BEST MODEL FOUND IS --
    es,emw,msw.

 STEPPING STOPS DUE TO CRITERION PROBABILITY (      0.050).
```

[1] With ADD = SIMPLE, only a single effect or interaction is added at each step. In this example, the stepping algorithm adds the interaction λ_{MW} at step 1, λ_{ES} at step 2, λ_{SW} at step 3, and λ_{EM} at step 4. Although the model at step 3 (*SW,ES,MW*) has a nonsignificant test of lack of fit, stepping continues until both the model and the term added do not have significant *p*-values. Since the term added at step 3 (λ_{SW}) had a significant *p*-value (p = .0001), the algorithm continued to add an extra effect. Therefore, the preferred model may be at a step before the last.

[2] Using the option ADD = MULTIPLE, the stepping algorithm added λ_{MW} at the first step, λ_{ES} at the second, and λ_{MSW} at the third. Including λ_{MSW} at the third step has the effect of entering three effects, λ_{MW}, λ_{MS}, and λ_{SW} into the model. The simultaneous test of the three was selected as the most significant addition to the previous model. Therefore, the algorithm arrives at a final model (*ES,MSW*) which differs from the model in **[1]**.

Note: The option ADD = SIMPLE adds single terms at each step. Therefore, a three-factor interaction such as λ_{MSW} cannot be considered unless all effects whose indices are subsets of *MSW*, λ_{MS}, λ_{MW}, and λ_{SW}, are already in the model. For this reason, the model in **[1]** could not consider λ_{MSW} as a potential effect because the contribution of the interaction λ_{MS} is nonsignificant and will never be included in the model.

Special Features

Specifying interactions to be included

Under some conditions you may want to test only those models in which specified terms are included. In the design of an experiment, for example, the planned analysis or the results of a preliminary analysis can affect the choice of models to be fitted to the data.

If the number of observations in each of the cells of a subtable is fixed by design, then the interaction containing indices of that subtable should be included in all models. For example, if in a clinical trial the number of patients to be examined is stratified by age and sex and a prespecified number is to be admitted at each age-sex combination, then the interaction between age and sex should be included in all models fitted even when that interaction is not statistically significant.

When there is a dependent variable, such as survival, and several independent variables, such as treatment, age, and sex, the models of interest are those that include the interactions between the independent variables, i.e., the interaction treatment by sex by age as well as all lower order relatives. When the dependent variable has only two levels, a model that includes all interactions between the independent variables is equivalent to a logistic model. Note: Program LR can be used to fit a logistic regression model.

You can INCLUDE interactions in all models by specifying

```
INCLUDE = list.
```

in the FIT paragraph, where items in the list represent hierarchical effects and are specified in the same manner as in the MODEL statement. For example,

```
INCLUDE = MW.
```

includes the interaction λ_{MW} and main effects λ_M and λ_W in all models tested.

INCLUDE affects all models by adding the interactions in the INCLUDE statement to those in all MODEL statements. INCLUDE also affects the models fitted by SIMULTANEOUS and ASSOCIATION options in the FIT paragraph. Each model of full order (see Example 4F3.1) is augmented by the terms in the INCLUDE statement as is each pair of models used to compute tests of marginal and partial association. The only models not affected by the INCLUDE statement are those produced by requesting ALL models in the FIT paragraph for three-way tables.

Effect of stacking on log-linear models

STACKING, as described in Example 4F1.8, reduces the dimension of a multiway table by creating a single categorical variable out of two or more original categorical variables. The original variables are said to be STACKED. All log-linear models are applied to the modified table.

Convergence and sparse tables

Each log-linear model is fitted by an algorithm published by Haberman (1972) called iterative proportional fitting. To control the accuracy of the final fit and to limit the maximum number of iterations, the CONVERGENCE criterion and the maximum number of ITERATIONS can be modified. In general, preassigned values can be used.

Sparse tables. When there are few observations per cell (the table is sparse), the number of iterations required for convergence tends to increase. At the same time the validity of the chi-square approximation (i.e., the estimation of the tail probability) becomes questionable.

Recommendation for empty cells

We recommend that you add a constant to all cells (/TABLE DELTA = .5.) when you have empty cells in your data that are not to be treated as structural zeros. Otherwise, some of the analyses are not performed correctly.

4F Part 3 Commands

Where to Find Commands

4F Part 1

For **specifying or constructing your multiway table,** see

- TABLE INDEX
- if your data are already in table form (instead of one record per subject), see INPUT TABLE or TABLE COUNT

4F Part 2

For **analysis of two-way tables**

4F Part 3

For **model screening,** see

- FIT ASSOCIATION

For **model fitting, see**

- FIT MODEL

To **print expected values, deviates, and model parameters,** see

/ INPut

The INPUT paragraph is required. See Chapter 3 for INPUT options common to all programs. 4F accepts data or tabulated input. Examples 4F1.10 and 4F1.11 show how to start from frequency tables as input. Example 4F1.10 shows how to start from frequency tables as input.

/ VARiable

The VARIABLE paragraph is optional. See Chapter 4 for VARIABLE options common to all programs. Chapter 4F Part 1 describes the VARIABLE FREQUENCY command.

New Feature

MMV = #. `MMV = 3.`

You can use the MMV command to specify the maximum number of allowed missing values or values out of range before a case is excluded from the analysis. Default: Use all acceptable values.

/ CATEGory (*or* GROUP)

CATEGORY CODES or CUTPOINTS are recommended if you are forming multiway frequency tables. The CATEGORY paragraph is described in detail in Chapter 5. Chapter 4F Part 1 describes the need for the CATEGORY paragraph in 4F.

/ TABLE

The TABLE paragraph is required and may be repeated. This paragraph defines tables to be formed. Chapter 4F Part 1 shows how to use the INDICES, CATVAR, CONDITION, STACK, and COUNT instructions. The EMPTY or INITIAL statement is optional. One or the other is used when you want to specify one or more cells as structural zeros.

Options for models with structural zeros

If you specify EMPTY or INITIAL for any TABLE paragraph, then the usual analysis for two-way tables will not be performed for any two-way table defined in the same paragraph, whether or not that table contains structural zeros.

EMPTY(#) = #list. `EMPTY(3) = 2,1,1,2, 3,2,2,1.`

The # in parentheses indicates the table for which you wish to specify empty cells (structural zeros). The (#) is unnecessary if there is only one table listed or if you want to specify empty cells in the first table. The # list designates the indices of the cell(s) to be declared as structural zeros. If the table is four-way, the first set of four numbers indicates a cell by specifying the levels of the four indices appropriate to that cell. If a CONDITION variable is specified for the table, the CONDITION index is the slowest moving variable in the table. See Example 4F3.7.

Example: The first number (2) specifies the second level of the first and fastest moving variable. The second number (1) specifies the first level of the second index. The third number (1) specifies the first level of the third index. The fourth number (2) specifies the second level of the fourth and slowest moving variable. The second set of four numbers specifies the second cell to be declared empty.

INITial(#) = # list. `INIT = 1 ,0 , 70 *1.`

The # in parentheses indicates the table for which you want to specify structural zeros. If you have only one table or the table for which you are specifying structural zeros is the first table, the (#) may be omitted. The # list indicates cells for which you want to specify values. Specifying a 1 indicates that the cell contains the observed frequency. Specifying a 0 declares the cell to be a structural zero. An INITIAL value may be any positive number. You can specify any initial fit matrix, not just a matrix of ones and zeros. See Example 4F3.8. In counting cells to determine position, remember that the first categorical variable in the INDICES list is the fastest moving index. If you specify a CONDITION variable for the table, then the CONDITION index is the slowest moving index in the table. Example: The first cell (of a total of 72 cells) contains the observed frequency, the second cell is declared empty, and the remaining 70 cells contain observed frequencies.

Other TABLE options

SYMBols = list. `SYMB = T, S, A.`

Specify SYMBOLS if you have more than one variable with the same first letter. You must specify either variable names or SYMBOLS in order to state MODELS in the FIT paragraph. Each categorical INDEX variable is labeled by a SYMBOL used to describe terms in models to be fitted. The first SYMBOL is used for the first INDEX variable and so on. Normally, the symbol for each variable is the first character of the variable name. If variable names are not specified, the letters A, B, C, ... are used.

DELTA = #. `DELTA = .5.`

You can use DELTA to add a constant to each cell frequency. Some investigators prefer to add a constant (most commonly 0.5) to each cell prior to analysis when cell frequencies are small. This has the effect of making the chi-square test more conservative. The constant is added in an attempt to compensate for any inflation in chi-square that might occur due to small expected values. The following are printed before DELTA is added: the observed frequency table, marginal subtables, and tables of percents. All other tables and statistics are computed after DELTA is added. (See exceptions for two-way tables in 4F Part 2.)

/ PRINT

```
NO OBS.   NO EXCL.
EXP.   LAMB.   LRCHI.
```

The PRINT paragraph is optional. If this paragraph is omitted, 4F automatically prints OBSERVED and EXCLUDED tables and LISTS the first five excluded cases. Any other options must be specified to be printed. The PRINT paragraph may be repeated after each TABLE paragraph. Additional PRINT features, relevant for multiway tables, are described in Chapter 4F Part 3.

LIST = #. `LIST = 20.`

List the first # of cases that were not included in one or more tables. # is the maximum number of cases to be printed in a case-by-table display. If you do not list cases first 5 cases excluded from any table will be printed.

New Feature

BAR = 'c'. `BAR = 'I'.`

Specify the character to be printed along vertical lines in two-way and multiway tables. The default character is a vertical bar (|).

PERCent = (*one or more*) `PERC = ROW, COL.`
NONe, ROW, COL, TOTal.

4F can print tables of PERCENTS. Specify ROW for a table of percents of the row totals, COL for a table of percents of the column totals, and TOTAL for a table of percents of the total table frequency. If no PERCENT is specified for a particular table, the PERCENT table specified for the previous table, if any, will be printed. Otherwise, NONE of the tables of percents will be printed.

New Feature **METHod = VARiable *or* CASE.** `METH = CASE.`

Use to determine how the printout is organized. VARIABLE is the default value.

VARIABLE. – Prints for each case only as many variables as fill the print line (up to ten variables) and begins the next line with the same variables for the next case. Thus the data for each variable are in an easy-to-scan column. When the program finishes printing the first subset of variables for all cases, it continues with another subset.

CASE. – Prints the values of all variables for one case (possibly filling several print lines) before printing data for the next case.

MARGinal = #. `MARG = 2.`

Print marginal subtables for all dimensions up to and including order #. The marginal table for the highest order is printed automatically. Example: If you have a four-way table and request MARG = 2, 4F will print the four-way table, six two-way tables, and the four one-way tables. Example 4F1.7 describes marginal subtables.

EXPected. – Expected values of all cells under the null hypothesis of independence.

LAMBda. – Estimates of the parameters of the log-linear model, and the estimates divided by their standard error. Not available with /FIT ALL. See Example 4F3.3.

BETA. – Estimates of the multiplicative parameters $\beta = e^{\lambda}$. Not available with /FIT ALL. See Example 4F3.3.

VARiance. – Correlation and covariance matrices between the estimates of the parameters of the log-linear model. When VARIANCE is specified, the standard errors of the parameter estimates are obtained by inverting the information matrix. Not available with /FIT ALL.

DIFference. – Differences between observed and expected values for all cells.

FReeman. – Freeman-Tukey deviates between observed and expected values are printed for each model specified in the FIT paragraph. Formula is shown in example 4F3.4.

STANdardized. – Print a table of the standardized deviates between observed and expected values for each cell for each model specified in the FIT paragraph. See 4F3.4.

ADJusted. – Adjusted standardized deviates between observed and expected values for each cell. See Haberman (1973) for a discussion of adjusted standardized deviates. See 4F3.4.

CHISQuare. – Components of the Pearson chi-square statistic associated with each cell for each model specified in the FIT paragraph.

LRCHI. – Components of the likelihood ratio chi-square statistic (G^2) associated with each cell for each MODEL specified in the FIT paragraph. Formula is shown in Example 4F3.4.

New Feature **CASE = #.** – Specify the number of cases for which data are printed. Values flagged by MISS, MAX, and MIN in the VARIABLE paragraph specifications are replaced by MISSING, GT MAX, and LT MIN. Default: 10.

New Feature **MINimum.** – Prints data for cases in which at least one variable has a value less than its lower limit as specified in the VARIABLE paragraph. Default: such cases are not printed unless you specify DATA.

New Feature **MAXimum.** – Prints data for cases in which at least one variable has a value greater than its upper limit as specified in the VARIABLE paragraph. Default: such cases are not printed unless you specify DATA.

New Feature **MISSing.** – Prints data for cases that have at least one variable with a missing value. Missing values in the input may be identified in several ways: (1) an asterisk (the default missing value character), (2) a missing value code identified using MISS in the VARIABLE paragraph, (3) a character identified as MCHAR in the INPUT paragraph, (4) blanks (applies only to input read with FORTRAN fixed format). Default: such cases are not printed unless you specify DATA.

OBServed. – Table of observed frequencies unless you specify NO OBSERVED.

The following commands work only in the batch mode

New Feature **MEAN.** – Unless you specify NO MEAN, 4F prints the mean, standard deviation, and frequency of each used variable.

New Feature **EXTReme** – Unless you state NO EXTR, 4F prints the minimum and maximum values for each variable used.

New Feature **EXCluded.** – Two-way table of frequencies of the excluded cases unless you specify NO EXCLUDED. This table follows each frequency table from which cases have been excluded and shows the reasons for exclusion.

New Feature **EZSCores.** – Prints *z*-scores for the minimum and maximum values of each used variable.

New Feature **SKewness.** – Prints skewness and kurtosis for each used variable.

New Feature **ECASe.** – Prints the case numbers containing the minimum and maximum values of each variable.

/ FIT

The FIT paragraph specifies models to be fitted. This paragraph is required for fitting log-linear models. The FIT paragraph may be repeated, and applies to the previous TABLE paragraph.

ALL. `ALL.`

Fit ALL hierarchical models in a two or three way table. If the table has more than three categorical variables, ALL is not available. See Example 4F3.9.

SIMULtaneous. `SIMUL.`

Print the simultaneous tests of all effects of a given order. These tests are also printed if ASSOCIATION is specified.

ASSOCiation = #. `ASSOC = 3.`

Print the tests that the partial and marginal associations are zero for each effect having a number of factors less than or equal to #. Each test requires the fitting of a model by iteration, which can be time-consuming in high-dimensional tables (5-way or more). Default: zero. See Example 4F3.1.

MODEL = list. `MODEL = AC, B.`

The MODEL command specifies a hierarchical log-linear model to be tested. Each list item represents the subscripts of the effects that define the hierarchical model to be fitted to the data. Each subscript must contain only SYMBOLS specified in the TABLE paragraph or, when SYMBOLS are not specified but variable names are, the first letters of the names of the categorical variables. The subscripts are separated by commas and terminated by a period. More than one model can be specified in one FIT paragraph, but each model must be specified in a separate MODEL command. See Example 4F3.2. **Example:** Effects λ_A and λ_C are included, as well as λ_{AC} and λ_B since higher order effects (*AC*) include all lower order effects (*A* and *C*) in hierarchical models.

DELete = (*one only*) `DEL = SIMPLE.`
SIMPle, MULTiple.

Build a MODEL in a stepwise manner, starting from a specified MODEL, by deleting SIMPLE or MULTIPLE effects at each step. When ADDING or DELETING terms from a model, the new model may differ from the previous model by a single effect (SIMPLE) or by a combination of effects (MULTIPLE), with the restriction that the more complex model differs from the simpler model by a term containing only one more index.

ADD = (*one only*) `ADD = MULT.`
SIMPle, MULTiple.

Build a MODEL in a stepwise manner, starting from a specified MODEL, by adding SIMPLE or MULTIPLE effects at each step. See DELETE above.

CELL = (*one only*) `CELL = STAN.`
NONe, STANdardized, FReeman.

Identify extreme cells in a stepwise manner using either the maximum STANDARDIZED deviate or FREEMAN-Tukey deviate at each step. See Example 4F3.5.

STEP = #. `STEP = 4.`

is the maximum number of steps for ADDING or DELETING effects in a stepwise manner from a user-specified MODEL, or the maximum number of cells to be identified as extreme in a stepwise manner. Default: zero. See Examples 4F3.5 and 4F3.10.

INClude = list. `INC = AB, BC.`

List terms to be included in all log-linear models fitted. The listed terms (or interactions) are like those in the MODEL command. Each interaction listed here is added to all MODEL commands and affects the models fitted under the SIMULT and ASSOC options. The only option not affected is ALL.

PROBability = #. `PROB = .025.`

is the criterion probability (level) to determine whether a test of fit is statistically significant. Stepping stops if the *p*-value (tail probability) of the test-of-fit is greater than this value. Default: .05.

STRATA = variable list *or* **ALL.** `STRATA = SEX, AGE.`

Eliminate in turn each category (stratum) for each categorical variable (index) specified. Categories of variables with only two strata are not eliminated. See Example 4F3.6.

CONVergence = #, #. `CONV = .005, .00001.`

The first number is the maximum permitted absolute difference and the second number is the maximum permitted relative difference between every observed and fitted marginal total in the model. Default: .01 and .00001. Each log-linear model is fitted by an algorithm published by Haberman (1972) called iterative proportional fitting. To control the accuracy of the final fit and to limit the maximum number of iterations, you can specify the CONVERGENCE criterion and the maximum number of ITERATIONS. In general, you can use preassigned values. Note that when there are few observations per cell (the table is sparse) the number of iterations required for convergence tends to increase.

ITERation = #. `ITER = 10.`

is the maximum number of iterations used to fit any model. 4F prints an error message if this number is reached. Default: 20. See CONVERGENCE above.

/ SAVE

The SAVE paragraph is required to create a BMDP File. See Chapter 8 for details about the SAVE paragraph. See 4F Part 1 for information about saving tables in BMDP Files and reusing them in later analyses.

Order of Instructions

■ indicates required paragraph

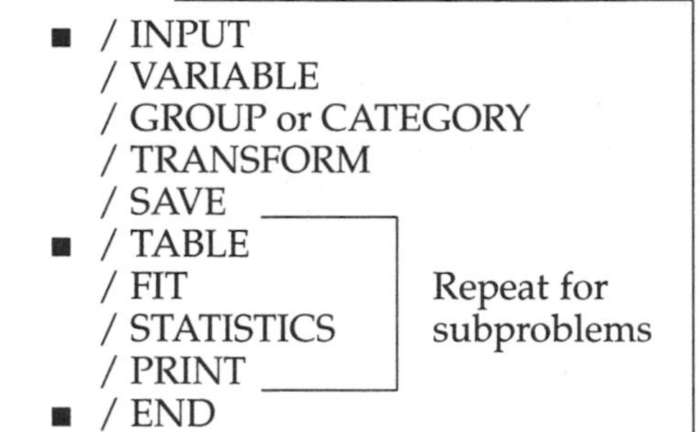

Repeat for additional problems.
See Multiple Problems, Chapter 10.

Summary Table for Commands Specific to 4F, Parts 1, 2 and 3

Paragraphs Commands	Defaults	Multiple Problems	See
■ / **INPut**			
TABLE = # list.	input is not a table	–	4F1.9
CONTent = (*one only*)	DATA	–	4F1.12
DATA, COUNT#.			
/ **VARiable**			
FREQuency = variable.	no case frequency variable	–	Cmds
MMV = #.	all acceptable values	–	Cmds
▲ / **CATEGory** (*or* **GROUP**)			
■ / **TABLE**			
▲ COLumn = list.	none	–	4F1.1
▲ ROW = list.	none	–	4F1.1
or INDices = list.	see ROW and COL	–	4F3.1
or CATvar = list.	see ROW and COL	–	4F1 Cmds
PAIR. *or* CROSS.	PAIR.	–	4F1.2
CONDition = list.	no condition variables	–	4F3.9
COUNT = list.	the variable stated in /VAR FREQ =	–	4F1.10
EMPTY(#) = # list.	no empty cells	–	4F3.7
or INITial(#) = # list.	all ones (no cells excluded)	–	4F3.8
SYMbols = list.	first letter of var. names	–	4F3 Cmds
STACK = list.	no stacked variables	–	4F1.8
DELTA = #.	0	–	4F3 Cmds
/ **STAtistics**			
CHISQuare. *or* NO CHISQ.	CHISQ.	●	4F2.1
LRCHI.	NO LRCHI.	●	4F2.1
FISHer.	NO FISH.	●	4F2.5
CONTingency.	NO CONT.	●	4F2.1
TETRAchoric.	NO TETRA.	●	Cmds

Summary Table (continued)

Paragraphs Commands	Defaults	Multiple Problems	See
GAMma.	NO GAM.	•	4F2.7
CORRelation.	NO CORR.	•	4F2.7
SPEARman.	NO SPEAR.	•	4F2.7
TAUS.	NO TAUS.	•	4F2.3
LAMBda.	NO LAMB.	•	4F2.3
UNCertainty.	NO UNC.	•	4F2.3
MINIMUM= #.	no automatic collapsing	•	4F2.4
MCNemar.	NO MCN.	•	4F2.9
LINear.	NO LIN.	•	4F2.8
ALL. *or* NO ALL.	NO ALL. CHISQ.	•	Cmds
/ PRINT			
EXPected.	NO EXP.	•	4F3.4
LAMBda.	NO LAMBDA.	•	4F3.3
BETA.	NO BETA	•	4F3.3
VARiance.	NO VARIANCE.	•	Cmds
STANdardized.	NO STAND.	•	4F3.4
FReeman.	NO FREEMAN.	•	4F3 Cmds
DIFference.	NO DIFF.	•	4F3 Cmds
ADJusted.	NO ADJ.	•	4F2.2
CHISQuare.	NO CHISQ.	•	4F3.4
LRCHI.	NO LRCHI.	•	Cmds
CASE = #.	10	•	Cmds
MISSing.	NO MISS.	•	Cmds
MINimum.	NO MIN.	•	Cmds
MAXimum.	NO MAX.	•	Cmds
LIST = #.	5	•	4F3 Cmds
PERCent = (*one or more*) NONe, ROW, COL, TOT.	NONE	•	4F3 Cmds
METHod = VARiables *or* CASE.	VARIABLES	–	Cmds
MARGinal = #.	no marginal subtables	•	4F3 Cmds
BAR = 'c'.	vertical bar (|)	•	4F1 Cmds
MEAN. *or* NO MEAN.	MEAN.	•	Cmds
EXTReme. *or* NO EXTR.	EXTReme. values printed	•	Cmds
EXCLuded. *or* NO EXCL.	EXCLUDED.	•	4F1.1
OBServed. *or* NO OBS.	OBSERVED.	•	4F1.1
EZSCores.	NO EZSC. (z-scores)	•	Cmds
ECASe.	NO ECASe. (Case #'s of Min and Max printed)	•	Cmds
SKewness.	NO SKewness. (skewness and kurtosis)	•	Cmds
■ **/ FIT**			
ASSOCiation = #.	0	–	4F3.1
MODEL = list.	no models tested	–	4F3.2
ALL.	NO ALL.	–	4F3.9
SIMULtaneous.	SIMUL. if ASSOC is used; otherwise NO SIMUL.	–	4F3.1
DELETE = (*one only*) SIMPle, MULTiple.	no effects deleted	–	4F3.10
ADD = (*one only*) SIMPle, MULTiple.	no effects added	–	4F3.11
CELL = (*one only*) NONe, STANdardized, FReeman.	NONE	–	4F3.5

Summary Table (continued)

Paragraphs Commands	Defaults	Multiple Problems	See
STEP = #.	0	–	4F3.5, 4F3.10
INCLude = list.	none	–	Cmds
PROBability = #.	.05	–	Cmds
STRATA = variable list *or* ALL.	no strata eliminated	–	4F3.6
CONVergence = #, #.	.01, .00001	–	Cmds
ITERations = #.	20	–	Cmds
/ **SAVE**			
CONTent = (*one or both*) DATA, COUNT.	DATA	–	4F1.11

Key:
- ■ Required command or paragraph
- ▲ Frequently used command or paragraph
- ● Value retained for multiple problems
- – Default reassigned

4M

Factor Analysis

Factor analysis is useful in exploratory data analysis. It has the following general objectives: to study the intercorrelations of a large number of variables by clustering them into common factors such that variables within each factor are highly correlated, to interpret each factor according to the variables with high loadings belonging to it, and to summarize many variables by a few factors or summary scores. Using a correlation or covariance matrix, factor analysis extracts the dimensions that account for the variability in a space defined by the variables, rotates the factors to obtain a simple interpretation (making the loadings for each factor either large or small, not intermediate), and provides factor scores for each case. The assumption is that the number of factors will be appreciably less than the original number of variables.

4M provides four methods of initial factor extraction from a correlation or covariance matrix, as well as several methods of rotation. Different methods of rotation are commonly used with different methods of extraction. You can specify the methods to be used, or 4M will use default options. Input can be data, a correlation or covariance matrix, factor loadings, or factor score coefficients. See BMDP Technical Report #8 for annotated computer output from 4M. A short introduction to factor analysis can be found in Afifi and Azen (1979); for an extensive account see Harman (1967).

Where to find it

Where to find it *(continued)*

Introduction

Factor analysis is a useful tool when a researcher wants a parsimonious description for a set of continuous, intercorrelated variables. For example, variables that measure interest in meeting new people, going to parties, and participating in team sports might be combined to form an "extroversion" factor. Body weight and height might be included in a "size" factor. Factor analysis can be useful during the pilot stages of questionnaire development to assess whether each item works as intended. Researchers also use factor analysis to verify that a subscale defined in the literature applies to their data; in other words, that the subscale variables load on the same factor.

Experts have known for many years that different methods of obtaining factors can yield different results. In general, several methods should be used and compared to see if they give similar results.

In the examples below, we do not assume any structure for the data. Rather, we use the data to suggest a structure. This is called *exploratory* factor analysis. Program 4M can also be used when the number of factors is predetermined and the primary goal is to confirm a hypothesis regarding the number of factors. However, to actually perform complete *confirmatory* factor analysis, where both the number of factors and the specific variables for each factor are predetermined, users should use the EQS Structural Equations program, available from BMDP. EQS is a separate program specifically developed for structural equation modeling and confirmatory factor analysis.

The data used throughout this chapter are from a survey of AIDS patients' reactions to their physicians conducted by Van Servellen based on a scale developed by Cope et. al. (1986). The 14 items in the survey questionnaire measure patient attitudes about physician personality, demeanor, competence and prescribed treatment using a Likert type scale of 1 to 5 for each item. Since seven of the 14 items were stated negatively, they have been recoded (reflected) so that 1 represents the most positive and 5 the least positive response on all items. The items and corresponding variable names are listed below (see Appendix D for more about the data).

Table 4M.1
Survey questionnaire items

Variable		Item
1. FRIENDLY	–	My doctor treats me in a friendly manner.
2. DOUBTABL	–	I have some doubts about the ability of my doctor.
3. COLD	–	My doctor seems cold and impersonal.
4. REASSURE	–	My doctor does his/her best to keep me from worrying.
5. CAREFUL	–	My doctor examines me as carefully as necessary.
6. NORSPECT	–	My doctor should treat me with more respect.
7. DOUBTTRT	–	I have some doubts about the treatment suggested by my doctor.
8. COMPETNT	–	My doctor seems very competent and well-trained.
9. INTEREST	–	My doctor seems to have a genuine interest in me as a person.

Table 4M.1 (continued)

10. UNANSQST	–	My doctor leaves me with many unanswered questions about my condition and its treatment.
11. JARGONMD	–	My doctor uses words that I do not understand.
12. CONFDNCE	–	I have a great deal of confidence in my doctor.
13. CONFIDE	–	I feel I can tell my doctor about very personal problems.
14. NOTFREE	–	I do not feel free to ask my doctor questions.

Our intentions are to investigate how the responses to the items relate to one another, how the responses for this sample agree with the subscales reported previously in the literature, and whether we can summarize each person's set of 14 scores by only a few factor scores.

New procedures have been added to 4M to provide the user with further information on scale evaluation. These include Cronbach's alpha for standardized and non-standardized responses to the items (a commonly used measure of reliability), and a second-order factor analysis to determine whether all the subscales can be considered to contribute to a single overall scale.

Example 4M.1 Basic factor analysis

For Example 4M.1, we perform a factor analysis using only default options; that is, extracting factors using principal components analysis and performing an analysis orthogonal rotation of *k* factors, where *k* is the number of eigenvalues greater than one. (Principal components is especially useful if the correlation matrix is not full rank, e.g., one variable is a linear combination of others.) In this first run, we hope to find out whether the default number of extracted factors accounts for most of the information contained in the original variables. We also check for (1) recording errors or outliers, (2) anomalies in the distributions, and (3) variables that are independent of the others and need not be included.

In Input 4M.1 there are no BMDP instructions specific to program 4M. The INPUT, VARIABLE, and END paragraphs are used by all programs. The data are stored on disk in a file called PATIENT.DAT. (The FILE command tells the program where to find the data and is used for systems like VAX and the IBM PC. For IBM mainframes, see UNIT in Chapter 3.) The MISSING command in the VARIABLE paragraph sets the value zero to missing in all 14 variables.

Input 4M.1

```
/ INPUT      VARIABLES = 14.
             FORMAT IS '14F1'.
             FILE IS 'patient.dat'.

/ VARIABLE   NAMES = friendly, doubtabl, cold, reassure,
                     careful, norspect, doubttrt, competnt,
                     interest, unansqst, jargonmd,
                     confdnce, confide, notfree.
             MISSING = 14*0.
/ END
```

The results of the analysis are presented in Output 4M.1. For information about running 4M interactively, see Chapter 11 and the Special Feature section in this chapter on Interactive factor analysis.

Output 4M.1

—the BMDP instructions read by 4M and printed and interpreted—

—we omit the default printing of the first ten cases— [1]

```
NUMBER OF CASES READ. . . . . . . . . . . . . .        68     [2]
  CASES WITH DATA MISSING OR BEYOND LIMITS . .          4
    REMAINING NUMBER OF CASES . . . . . . . .          64
```

—we omit the printing of missing value table—

DESCRIPTIVE STATISTICS OF DATA [3]

VARIABLE NO. NAME	TOTAL FREQ.	MEAN	STANDARD DEV.	ST.ERR OF MEAN	COEFF OF VAR	SMALLEST VALUE	SMALLEST Z-SCR	SMALLEST CASE	LARGEST VALUE	LARGEST Z-SCR	LARGEST CASE	RANGE
1 friendly	64	1.6563	.78110	.09764	.47161	1.0000	-0.84	1	5.0000	4.28	14	4.0000
2 doubtabl	64	1.9219	.99690	.12461	.51871	1.0000	-0.92	9	4.0000	2.08	4	3.0000
3 cold	64	1.7656	.86817	.10852	.49171	1.0000	-0.88	7	5.0000	3.73	54	4.0000
4 reassure	64	2.0938	1.0192	.12740	.48676	1.0000	-1.07	12	5.0000	2.85	7	4.0000
5 careful	64	1.9375	.85217	.10652	.43983	1.0000	-1.10	16	5.0000	3.59	25	4.0000
6 norspect	64	2.0156	.93422	.11678	.46349	1.0000	-1.09	7	5.0000	3.19	14	4.0000
7 doubttrt	64	2.2656	1.0426	.13033	.46019	1.0000	-1.21	7	4.0000	1.66	1	3.0000
8 competnt	64	1.6719	.75708	.09463	.45283	1.0000	-0.89	1	4.0000	3.08	4	3.0000
9 interest	64	1.9063	.95483	.11935	.50090	1.0000	-0.95	1	5.0000	3.24	14	4.0000
10 unansqst	64	2.2656	1.1442	.14303	.50504	1.0000	-1.11	12	5.0000	2.39	14	4.0000
11 jargonmd	64	2.4844	1.1408	.14260	.45917	1.0000	-1.30	12	5.0000	2.21	26	4.0000
12 confdnce	64	1.9375	.94070	.11759	.48552	1.0000	-1.00	1	5.0000	3.26	62	4.0000
13 confide	64	1.8750	.86373	.10797	.46066	1.0000	-1.01	2	5.0000	3.62	54	4.0000
14 notfree	64	1.8125	.94070	.11759	.51901	1.0000	-0.86	6	5.0000	3.39	8	4.0000

CORRELATION MATRIX [4]

		friendly 1	doubtabl 2	cold 3	reassure 4	careful 5	norspect 6	doubttrt 7	competnt 8	interest 9	unansqst 10	jargonmd 11	confdnce 12	confide 13
friendly	1	1.0000												
doubtabl	2	0.5561	1.0000											
cold	3	0.6283	0.5837	1.0000										
reassure	4	0.6393	0.4604	0.3481	1.0000									
careful	5	0.5157	0.4426	0.4948	0.5186	1.0000								
norspect	6	0.7035	0.5126	0.4939	0.5152	0.5396	1.0000							
doubttrt	7	0.4452	0.4784	0.2803	0.3347	0.3763	0.6312	1.0000						
competnt	8	0.6115	0.6806	0.4366	0.4314	0.5582	0.6358	0.4942	1.0000					
interest	9	0.7861	0.5758	0.6624	0.5474	0.6559	0.6423	0.3443	0.7033	1.0000				
unansqst	10	0.5655	0.6308	0.3992	0.5500	0.5382	0.5752	0.6451	0.6153	0.6188	1.0000			
jargonmd	11	0.3145	0.2711	0.3248	0.2061	0.1296	0.2609	0.2237	0.2421	0.1735	0.2525	1.0000		
confdnce	12	0.5536	0.7225	0.5065	0.4863	0.6287	0.6152	0.4703	0.7508	0.7002	0.6645	0.3097	1.0000	
confide	13	0.6882	0.5046	0.5953	0.5364	0.5068	0.7303	0.4429	0.4946	0.6592	0.5320	0.2396	0.6544	1.0000
notfree	14	0.6238	0.4242	0.3341	0.4656	0.3812	0.5814	0.5047	0.4917	0.5279	0.5631	0.2339	0.5067	0.5568

		notfree 14
notfree	14	1.0000

```
NUMBER OF VARIABLES TO BE USED. . . . . . . . .     14        [5]
UNROTATED FACTORS ARE PRINCIPAL COMPONENTS.
NUMBER OF FACTORS IS LIMITED TO THE NUMBER OF EIGENVALUES
    GREATER THAN      1.000
TOLERANCE LIMIT FOR MATRIX INVERSION. . . . . .     0.00010
VARIMAX ROTATION IS PERFORMED.
GAMMA . . . . . . . . . . . . . . . . . . . . .      1.0000
MAXIMUM NUMBER OF ITERATIONS FOR ROTATION . . .        50
CONVERGENCE CRITERION FOR ROTATION. . . . . . .   0.0000100
KAISERS NORMALIZATION . . . . . . . . . . . . .        YES
```

SQUARED MULTIPLE CORRELATIONS (SMC) OF EACH VARIABLE WITH ALL OTHER VARIABLES [6]

Variable	SMC
1 friendly	0.79437
2 doubtabl	0.69426
3 cold	0.65293
4 reassure	0.53863
5 careful	0.56487
6 norspect	0.72482
7 doubttrt	0.60191
8 competnt	0.72085
9 interest	0.81961
10 unansqst	0.68136
11 jargonmd	0.25326
12 confdnce	0.78381
13 confide	0.70377
14 notfree	0.50537

[7]

```
HISTOGRAM OF EIGENVALUES

  EIGENVALUE     HISTOGRAM

  1  7.75941     *****************************************************************************************************
  2  1.00688     *************
  3 0.968418     *************
  4 0.852149     ***********
  5 0.648881     ********
  6 0.541350     *******
  7 0.502360     ******
  8 0.425813     *****
  9 0.403107     *****
 10 0.301634     ****
 11 0.181310     **
 12 0.159318     **
 13 0.147750     **
 14 0.101616     *

CONDITION NUMBER =     76.36
```

[8]

```
COMMUNALITIES OBTAINED FROM  2 FACTORS AFTER    1 ITERATIONS.
-------------------------------------------------------------

THE COMMUNALITY OF A VARIABLE IS ITS SQUARED MULTIPLE
CORRELATION WITH THE FACTORS.

     1 friendly    0.7333
     2 doubtabl    0.5825
     3 cold        0.6916
     4 reassure    0.4700
     5 careful     0.5555
     6 norspect    0.6959
     7 doubttrt    0.7911
     8 competnt    0.6425
     9 interest    0.8238
    10 unansqst    0.7207
    11 jargonmd    0.1392
    12 confdnce    0.7010
    13 confide     0.6507
    14 notfree     0.5685
```

[9] [10]

```
FACTOR  VARIANCE      CUMULATIVE PROPORTION OF VARIANCE   CARMINES
        EXPLAINED     IN DATA SPACE   IN FACTOR SPACE       THETA
------  ---------    ----------------------------------    ------
   1     7.7594           0.5542          0.8851           0.9381
   2     1.0069           0.6262          1.0000
   3     0.9684           0.6953
   4     0.8521           0.7562
   5     0.6489           0.8026
   6     0.5414           0.8412
   7     0.5024           0.8771
   8     0.4258           0.9075
   9     0.4031           0.9363
  10     0.3016           0.9579
  11     0.1813           0.9708
  12     0.1593           0.9822
  13     0.1477           0.9927
  14     0.1016           1.0000

THE VARIANCE EXPLAINED BY EACH FACTOR IS THE EIGENVALUE FOR THAT FACTOR.

TOTAL VARIANCE IS DEFINED AS THE SUM OF THE POSITIVE EIGENVALUES OF THE
CORRELATION MATRIX.
```

[11]

```
UNROTATED FACTOR LOADINGS (PATTERN)
-----------------------------------
FOR PRINCIPAL COMPONENTS


                 FACTOR1    FACTOR2
                    1          2

friendly   1      0.841     -0.160
doubtabl   2      0.763      0.009
cold       3      0.684     -0.472
reassure   4      0.683     -0.063
careful    5      0.713     -0.216
norspect   6      0.824      0.133
doubttrt   7      0.640      0.618
competnt   8      0.799      0.061
interest   9      0.849     -0.321
unansqst  10      0.792      0.304
jargonmd  11      0.361      0.095
confdnce  12      0.837     -0.009
confide   13      0.796     -0.129
notfree   14      0.697      0.289

          VP      7.759      1.007

THE VP IS THE VARIANCE EXPLAINED BY THE FACTOR.
IT IS COMPUTED AS THE SUM OF SQUARES FOR THE
ELEMENTS OF THE FACTOR'S COLUMN IN THE FACTOR
LOADING MATRIX.
```

[12]

```
ORTHOGONAL ROTATION, GAMMA =      1.0000
ITERATION   SIMPLICITY
            CRITERION
    0       -0.506397
    1       -2.127010
```

```
  2        -2.127010
ROTATED FACTOR LOADINGS (PATTERN)
---------------------------------
```

[13]

		FACTOR1 1	FACTOR2 2
friendly	1	0.739	0.432
doubtabl	2	0.570	0.508
cold	3	0.826	0.093
reassure	4	0.556	0.401
careful	5	0.680	0.305
norspect	6	0.534	0.641
doubttrt	7	0.077	0.886
competnt	8	0.563	0.571
interest	9	0.851	0.315
unansqst	10	0.398	0.750
jargonmd	11	0.210	0.308
confdnce	12	0.638	0.543
confide	13	0.685	0.426
notfree	14	0.336	0.675
	VP	4.849	3.917

THE VP IS THE VARIANCE EXPLAINED BY THE FACTOR.
IT IS COMPUTED AS THE SUM OF SQUARES FOR THE
ELEMENTS OF THE FACTOR'S COLUMN IN THE FACTOR
LOADING MATRIX.

[14]

```
                    ROTATED FACTOR LOADINGS

          VARIABLES ARE DENOTED BY 1,...., 9, A,...., Z
          OVERLAPS ARE DENOTED BY AN ASTERISK.

         .+....+....+....+....+....+....+....+....+....+....+.
          -                                                   -
          -                                                   -
          -                                                   -
          -                                                   -
     1.0  +                                                   +
          -                                                   -
          -                           7                       -
          -                                                   -
          -                                   A               -
          -                                 E    6            -
          -                                       8 C         -
     .50  +                                       2           +
          -                                       4  D1       -
          -                                                   -
 F        -                              B           5   9    -
 A        -                                                   -
 C        -                                                   -
 T        -                                              3    -
 O   0.0  +                                                   +
 R        -                                                   -
 2        -                                                   -
          -                                                   -
          -                                                   -
          -                                                   -
          -                                                   -
    -.50  +                                                   +
          -                                                   -
          -                                                   -
          -                                                   -
          -                                                   -
          -                                                   -
    -1.0  +                                                   +
          -                                                   -
          -                                                   -
          -                                                   -
         .+....+....+....+....+....+....+....+....+....+....+.
        -1.0     -.60      -.20       .20       .60      1.0
             -.80      -.40       0.0       .40       .80

                              FACTOR1
```

```
SORTED ROTATED FACTOR LOADINGS (PATTERN)
----------------------------------------
```

[15]

		FACTOR1 1	FACTOR2 2
interest	9	0.851	0.315
cold	3	0.826	0.000
friendly	1	0.739	0.432
confide	13	0.685	0.426
careful	5	0.680	0.305
confdnce	12	0.638	0.543
doubtabl	2	0.570	0.508
reassure	4	0.556	0.401
doubttrt	7	0.000	0.886
unansqst	10	0.398	0.750
notfree	14	0.336	0.675
norspect	6	0.534	0.641
competnt	8	0.563	0.571
jargonmd	11	0.000	0.308
	VP	4.849	3.917

THE ABOVE FACTOR LOADING MATRIX HAS BEEN REARRANGED SO
THAT THE COLUMNS APPEAR IN DECREASING ORDER OF VARIANCE
EXPLAINED BY FACTORS. THE ROWS HAVE BEEN REARRANGED
SO THAT FOR EACH SUCCESSIVE FACTOR, LOADINGS GREATER
THAN 0.5000 APPEAR FIRST. LOADINGS LESS THAN 0.2500
HAVE BEEN REPLACED BY ZERO.

[16]

```
ABSOLUTE VALUES OF CORRELATIONS IN SORTED AND SHADED FORM
---------------------------------------------------------

    9 interest ■
    3 cold     ⊠■
    1 friendly ■⊠■
   13 confide  ⊠⊠■■
    5 careful  ⊠MMM■
   12 confdnce ■MM⊠⊠■
    2 doubtabl MMMMX■■
    4 reassure M+⊠MMXX■
    7 doubttrt +-XX+XX+■
   10 unansqst ⊠XMMM⊠⊠M⊠■
   14 notfree  M+⊠⊠+MXXMM■
    6 norspect ⊠M■■M⊠MM⊠MM■
    8 competnt ■X⊠MM■⊠XM⊠M⊠■
   11 jargonmd .++-.+-------■


THE ABSOLUTE VALUES OF
THE MATRIX ENTRIES HAVE BEEN PRINTED ABOVE IN SHADED FORM
ACCORDING TO THE FOLLOWING SCHEME


              LESS THAN OR EQUAL TO        0.098
      .      0.098  TO AND INCLUDING        0.197
      -      0.197  TO AND INCLUDING        0.295
      +      0.295  TO AND INCLUDING        0.393
      X      0.393  TO AND INCLUDING        0.491
      M      0.491  TO AND INCLUDING        0.590
      ⊠      0.590  TO AND INCLUDING        0.688
      ■                 GREATER THAN        0.688
```

[17]

```
FACTOR SCORE COVARIANCE (COMPUTED FROM FACTOR
STRUCTURE AND FACTOR SCORE COEFFICIENTS)
---------------------------------------------

THE DIAGONAL OF THE MATRIX BELOW CONTAINS THE SQUARED
MULTIPLE CORRELATIONS OF EACH FACTOR WITH THE VARIABLES.

               FACTOR1    FACTOR2
                     1          2

FACTOR1    1     1.000
FACTOR2    2    -0.000      1.000
```

[18]

```
ESTIMATED FACTOR SCORES AND MAHALANOBIS DISTANCES (CHI-SQUARE S) FROM
EACH CASE TO THE CENTROID OF ALL CASES FOR THE ORIGINAL DATA
( 14 D.F.) FACTOR SCORES (  2 D.F.) AND THEIR DIFFERENCE ( 12 D.F.).
EACH CHI-SQUARE HAS BEEN DIVIDED BY ITS DEGREES OF FREEDOM.
A MINUS SIGN AFTER THE CASE NUMBER DENOTES A CASE WITH NONPOSITIVE WEIGHT.

     CASE     CHISQ/DF CHISQ/DF CHISQ/DF FACTOR1   FACTOR2
  LABEL   NO.       14        2       12
           1     1.260    0.662    1.360   -0.864    0.760
           2     0.507    0.247    0.550   -0.099    0.696
           3     1.936    3.552    1.666    2.557   -0.753
           4     2.391    2.668    2.344    0.239    2.298
           5     0.145    0.103    0.151    0.370   -0.263
           6     1.596    0.329    1.807    0.585   -0.562
           7     1.841    0.083    2.133    0.187   -0.363
           8     1.731    0.189    1.989   -0.014    0.615
           9     0.741    0.271    0.819   -0.156   -0.720
          10     0.768    0.635    0.791    0.406    1.051
          11     0.654    0.306    0.712   -0.283    0.729
          12     1.736    5.391    1.127    1.814   -2.737
          13     0.145    0.103    0.151    0.370   -0.263
          14     2.631    3.887    2.421    2.078    1.859
          15     1.004    1.089    0.989   -0.291    1.447
          16     0.813    0.704    0.831   -1.168   -0.209
          17     0.393    0.672    0.347   -0.916   -0.711
          18     0.742    0.601    0.765   -0.222   -1.073
          20     0.579    1.039    0.502   -1.440   -0.066
```

—we omit cases 21 to 64—

```
          64     1.764    1.540    1.801    0.780    1.572
          65     0.256    0.949    0.141   -0.763   -1.147
          66     0.333    0.622    0.285   -0.787   -0.790
          67     0.817    0.471    0.875   -0.070   -0.968
          68     0.487    0.631    0.463   -0.646   -0.919
```

[19]

```
FACTOR SCORE COVARIANCE (COMPUTED FROM FACTOR SCORES)
-----------------------------------------------------

             FACTOR1  FACTOR2
                   1        2

FACTOR1    1   1.000
FACTOR2    2  -0.000    1.000
```

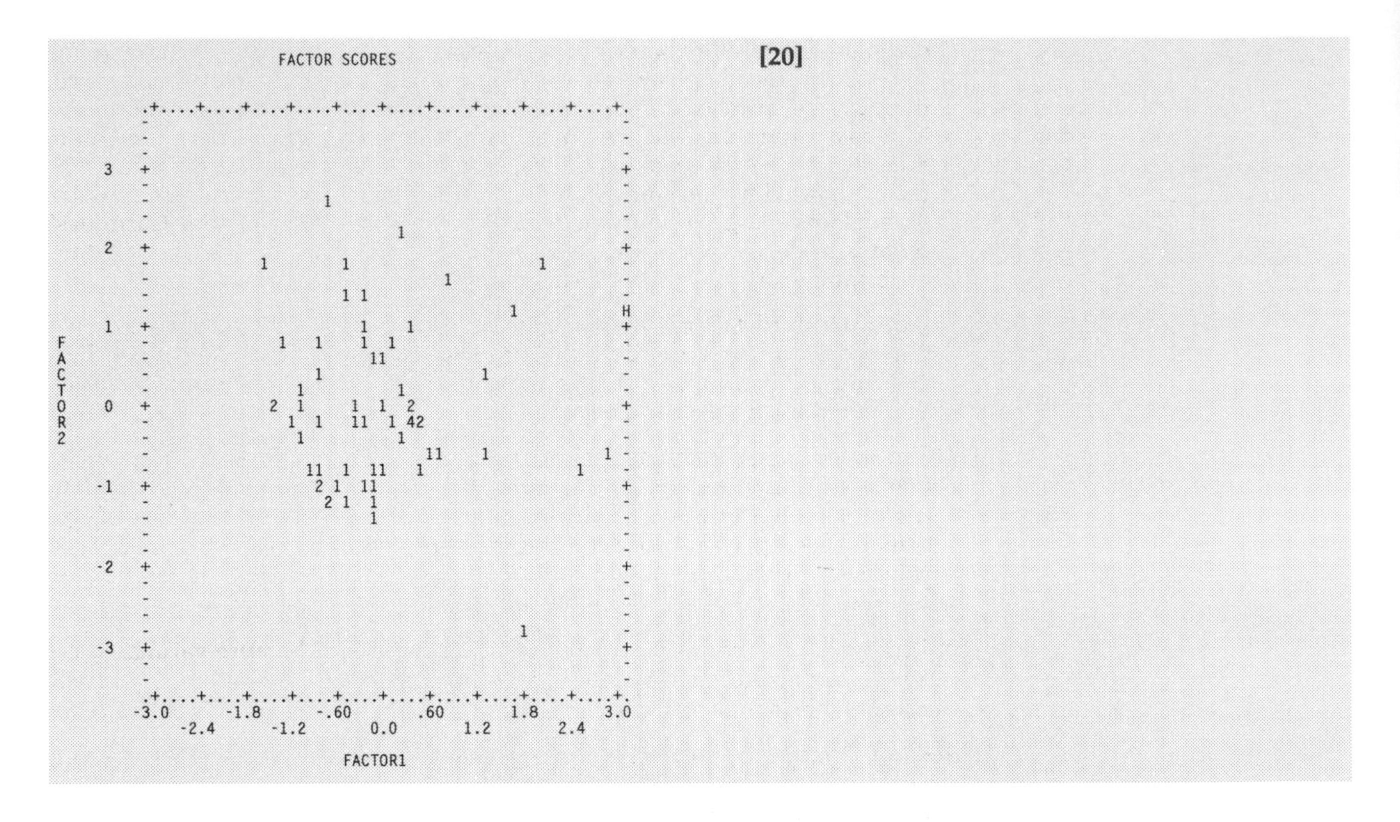

[1] We omit the default listing of data values for the first ten cases. You can specify the number of cases to print using CASE in the PRINT paragraph.

[2] Number of cases read. 4M uses only complete cases in all computations. That is, if the value of any variable in a case is missing or out of range, the case is omitted from all computations. If a USE statement is specified in the VARIABLE paragraph, only the values of variables in the USE statement are checked for missing value codes and values out of range.

[3] Univariate summary statistics. If narrow (LINESIZE = 80) output is requested. The CASE number for the SMALLEST and LARGEST values indicate first occurrences. This is less useful for this data set where there are multiple occurrences of the extreme scores. If case weights are specified, weighted means and standard deviations are computed.

[4] Correlation matrix. For an ordered, graphic version of this matrix, see **[16].**

[5] Summary table for factor analysis that will be performed on the data.

[6] Squared multiple correlation (SMC) of each variable with all other variables. A low SMC indicates that a variable is unrelated to the others, and may thus be unsuitable for inclusion in the factor scores. JARGONMD has the lowest SMC of the variables (0.25). We could consider removing such variables from subsequent factor analyses.

[7] Histogram of eigenvalues.

[8] The *condition number* of 76.36 is equal to the largest eigenvalue of the correlation matrix divided by the smallest (7.76/0.102), and is of interest to see how nearly singular the correlation matrix might be. Large condition numbers indicate greater linear dependence and fewer factors.

[9], [10] The **communality** of a variable is its squared multiple correlation with the factors extracted. It shows how much of each variable's total variance is accounted for by the factor. Note that only about 14% of JARGONMD's variance is explained by the two factors. The factor *eigenvalues* or factor variances are all listed under the heading "Variance Explained." The preassigned criterion for the number of factors is the number of factors with eigenvalues greater than one. Therefore, communalities are obtained for two factors. Ideally, when the user knows the number of factors in advance, they can be supplied to the program using the NUMBER instruction in the FACTOR paragraph. The *cumulative proportion* of the total variance in **[10]** is the sum of the variance explained (eigenvalues) up to and including the factor divided by the sum of all the eigenvalues. A successful factor analysis explains a large proportion of variance with very few factors. Carmines' theta is a special case of Cronbach's alpha, a measure of internal consistency ranging from 0.0 to 1.0 (Carmines and Zeller, 1979).

Since the third factor had an eigenvalue (variance) just under 1(0.97), we may wish to lower the default cutoff to include it (See Example 4M.2). Adding a third factor in this way would increase the total variance accounted for from 63% to about 70%.

[11] Unrotated factor loadings (pattern) for principal components. These loadings are the eigenvectors of the correlation matrix multiplied by the square roots of the corresponding eigenvalues. They are the correlations of the principal components with the original variables. The eigenvalues (VP) are printed at the bottom of each column. Note that the initial, unrotated factor solution may not give interpretable factors.

[12] Orthogonal rotation is performed. (Gamma is set to 1 because varimax rotation is performed. See Appendix B.14 for a discussion of gamma.) At each iteration, the simplicity criterion *G* (Afifi and Azen 1979, 335–337) is printed.

[13] Rotated factor loadings (pattern)—coefficients of the factors after rotation. The sums of squares of the coefficients are printed below each column (VP). When the rotation is orthogonal, as in this example, VP is the variance explained by the factor, and the rotated loadings are the correlations of the variables with the factors. A comparison of the rotated factor loadings with the unrotated ones shows that the larger loadings are larger than before and the smaller loadings are smaller. The object of rotation is to make each factor reflect information in only a few of the variables. In this example, the factors are not clearly separated. Better separation is achieved in Examples 4M.2 and 4M.3.

[14] Plots of the rotated factor loadings. The loadings of the variables for one factor are plotted against those of another factor. The sequence of numbers and letters used as plot symbols corresponds to the order of the variables in the VAR USE list (1, 2, 3, . . ., A, B, C, D, E), with FRIENDLY = 1 and NOTFREE = E. (In BMDP releases dated before 1988, asterisks are used as the plot symbol.) We discuss the interpretation of these plots in Example 4M.3.

[15] Sorted rotated factor loadings. The variables are reordered so the rotated factor loadings for each factor are grouped. By default, loadings less than 0.25 are set to zero. If a variable has two loadings greater than 0.5, it is ordered according to the larger of the two. The factors are sorted according to VP, the sum of the squared loadings for each factor. We use this display to name the factors, looking through the columns for variables with large loadings. We name the factors as follows:

1.*Empathy* – INTEREST,COLD, FRIENDLY, CONFIDE, CAREFUL,CONFDNCE, DOUBTABL, REASSURE, NORSPECT, COMPETNT

2. *Skill* – CONFDNCE, DOUBTABL, DOUBTTRT, UNANSOST, NOTFREE, NORSPECT, COMPETNT

Factor 1 is composed mostly of items that measure how the patients perceive their physician's personality and demeanor. Factor 2 is related to patient perception of the physician's skill and treatment ability. Four of the variables, CONFDNCE, DOUBTABL, NORSPECT, and COMPETNT, seem to load about equally on both factors. JARGONMD is little related to either factor and could form a third factor. In Examples 4M.2 and 4M.3, we see whether other estimation or rotation methods separate the variables more completely or produce the same factors given here.

[16] Correlations in shaded form. The variables are reordered according to the sorted rotated factor loadings. The values of the correlations are replaced by the symbols (or overprinted symbols) listed at the bottom of the display. On the terminal screen, numbers are used instead of symbols. To print only the first digit of each correlation see NCUT in the Commands section.

[17] Factor score covariances. The factor structure matrix is multiplied by the factor score coefficients to obtain the covariances of the factor scores. When the method of initial factor extraction is principal components (no communality estimates and only one iteration, as in our example), the factor score covariance matrix is the same as the factor correlation matrix. Otherwise, the diagonal of this covariance matrix contains the squared multiple correlations of each factor with the variables. In our example, this matrix is the identity matrix (i.e., all diagonal entries are 1.0) since the initial extraction is by principal components and the rotation is orthogonal.

[18] Mahalanobis distances and estimated factor scores. The three columns labeled CHISQ/DF show the Mahalanobis distances from each case to the centroid of all cases in the space defined by (1) the original variables, (2) the two factors, and (3) their difference. The factor scores for the two factors follow the distance measures. They should have a symmetric distribution about zero.

The calculation used in the first CHISQ/DF column uses the inverse of the correlation (or covariance) matrix and the standard scores. Its degrees of freedom are the number of variables (14). The second CHISQ/DF column is computed using the factor scores and the inverse of the factor score covariance matrix. Its degrees of freedom are the number of factors (2). The last CHISQ/DF column is a measure based on the "residuals" from regressing the variables on the factor scores. For large samples from a multivariate normal distribution, these three distances divided by their degrees of freedom are distributed approximately as chi-square divided by degrees of freedom. For a discussion of detecting outliers in multivariate data, see Hawkins (1974). Using the 99.9th percentile of the χ^2/df distribution as a cutoff, case 14 has larger values that could influence the formation of the factors. It may be useful to rerun the factor analysis using zero weight for case 14 to see whether the factor structure changes greatly. We do this in Example 4M.6.

Under the principal components method, the factor scores in the last two columns are obtained by multiplying standard scores for the original variables by the factor score coefficients. For each factor, the average score is zero and the expected standard deviation is 1.0.

[19] The factor score covariance matrix. The fact that this matrix is the same as the factor score covariance matrix computed from the factor structure and factor score coefficients is an indication of the numerical accuracy of the program.

[20] Plots of the factor scores. Since PCA and orthogonal rotation produce uncorrelated factors, the plot should show a cloud centered at the origin. The H on the right side on the plotting frame means that at least one case had a very high score on the Empathy factor (factor 1).

Example 4M.2 Factor extraction: maximum likelihood

Four methods of initial factor extraction are available: principal components; (PCA); iterated principal factor analysis (PFA), discussed in Harman (1967); maximum likelihood factor analysis (MLFA), discussed by Lawley and Maxwell maximum likelihood (1971); and Kaiser's (1970) Second Generation Little Jiffy (LJIFFY).

For a first analysis, we recommend the principal components method. If your correlation matrix happens to be singular, the analysis is still performed. This method is also used when the common factor model is not appropriate for the data. When the common factor model is appropriate and the correlation matrix is nonsingular, many statisticians prefer the maximum likelihood method, although this method may take considerable computation time when there are many variables. The use of the maximum likelihood method permits tests of hypotheses to be performed if a multivariate normal distribution can be assumed. Kaiser's Second Generation Little Jiffy also uses a nonsingular correlation matrix; it consists of image analysis followed by orthoblique rotation. Principal factor analysis is appropriate for data known to follow the common factor model and can also be used with a singular correlation matrix. However, principal components factor analysis is less popular than MLFA and LJIFFY. If a method is not specified, PCA is used (see Example 4M.1). When MLFA or PFA is specified, the maximum number of iterations can be specified (ITERATE). In addition, the convergence criterion (EPS) for PFA can be changed.

In Input 4M.2 we request maximum likelihood factor extraction (MLFA) by adding a METHOD command to the FACTOR paragraph. We also lower the cutoff eigenvalue for retaining factors from 1.0 to 0.9 via the CONSTANT command. As in example 4M.1, we use the default orthogonal rotation with gamma equal to 1.0.

Input 4M.2
(add a FACTOR paragraph to Input 4M.1 before END)

```
/ FACTOR     METHOD = MLFA.     CONSTANT = 0.9.
```

Output 4M.2

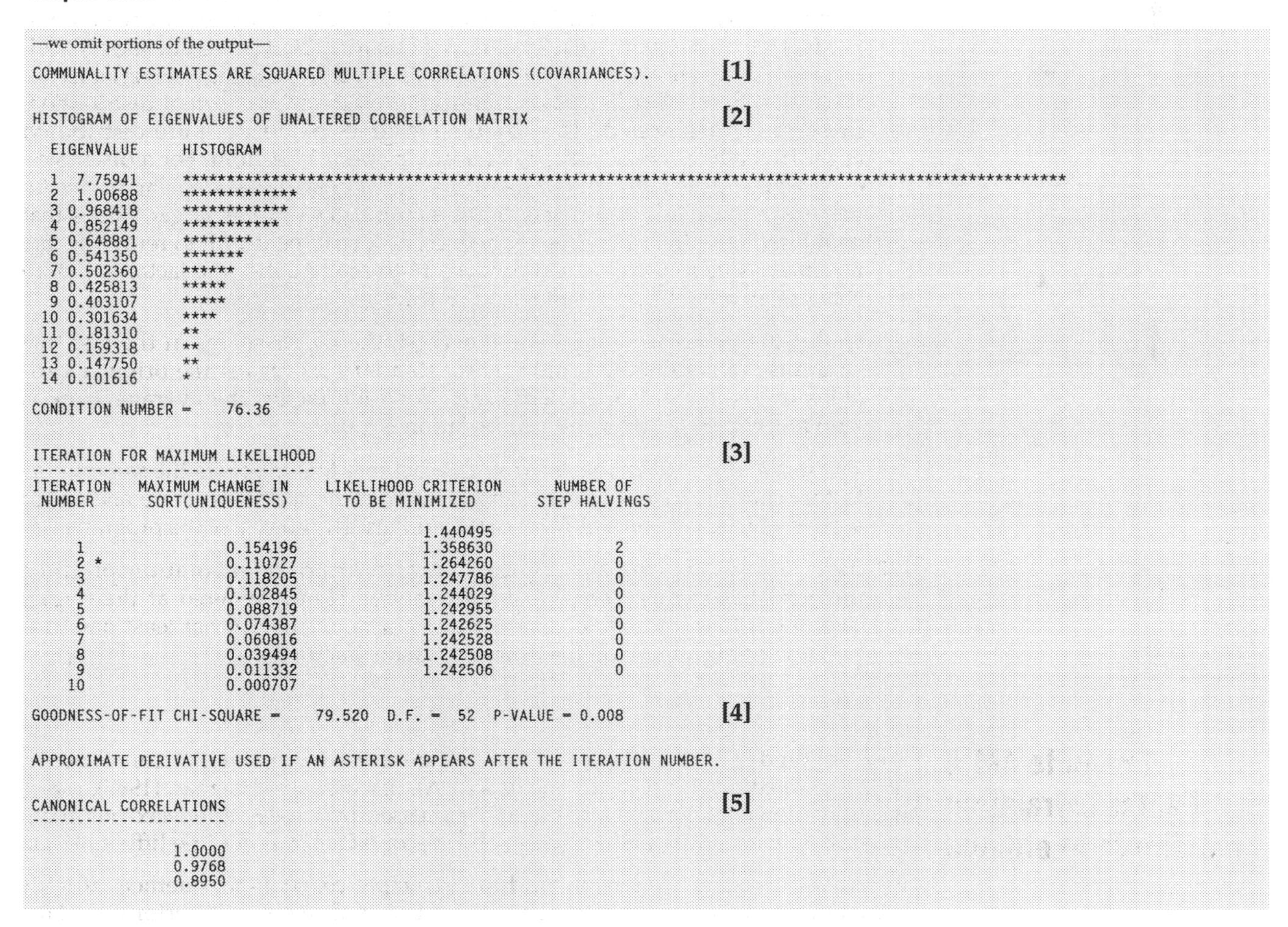

—we omit portions of the output—

```
COMMUNALITY ESTIMATES ARE SQUARED MULTIPLE CORRELATIONS (COVARIANCES).          [1]

HISTOGRAM OF EIGENVALUES OF UNALTERED CORRELATION MATRIX                         [2]

  EIGENVALUE     HISTOGRAM

  1  7.75941     ****************************************************************************************************
  2  1.00688     *************
  3 0.968418     ************
  4 0.852149     ***********
  5 0.648881     ********
  6 0.541350     *******
  7 0.502360     ******
  8 0.425813     *****
  9 0.403107     *****
 10 0.301634     ****
 11 0.181310     **
 12 0.159318     **
 13 0.147750     **
 14 0.101616     *

CONDITION NUMBER =    76.36

ITERATION FOR MAXIMUM LIKELIHOOD                                                 [3]
--------------------------------
ITERATION   MAXIMUM CHANGE IN    LIKELIHOOD CRITERION      NUMBER OF
 NUMBER      SQRT(UNIQUENESS)      TO BE MINIMIZED       STEP HALVINGS

                                       1.440495
     1               0.154196          1.358630                2
     2 *             0.110727          1.264260                0
     3               0.118205          1.247786                0
     4               0.102845          1.244029                0
     5               0.088719          1.242955                0
     6               0.074387          1.242625                0
     7               0.060816          1.242528                0
     8               0.039494          1.242508                0
     9               0.011332          1.242506                0
    10               0.000707

GOODNESS-OF-FIT CHI-SQUARE =    79.520   D.F. =   52   P-VALUE = 0.008          [4]

APPROXIMATE DERIVATIVE USED IF AN ASTERISK APPEARS AFTER THE ITERATION NUMBER.

CANONICAL CORRELATIONS                                                           [5]
----------------------

            1.0000
            0.9768
            0.8950
```

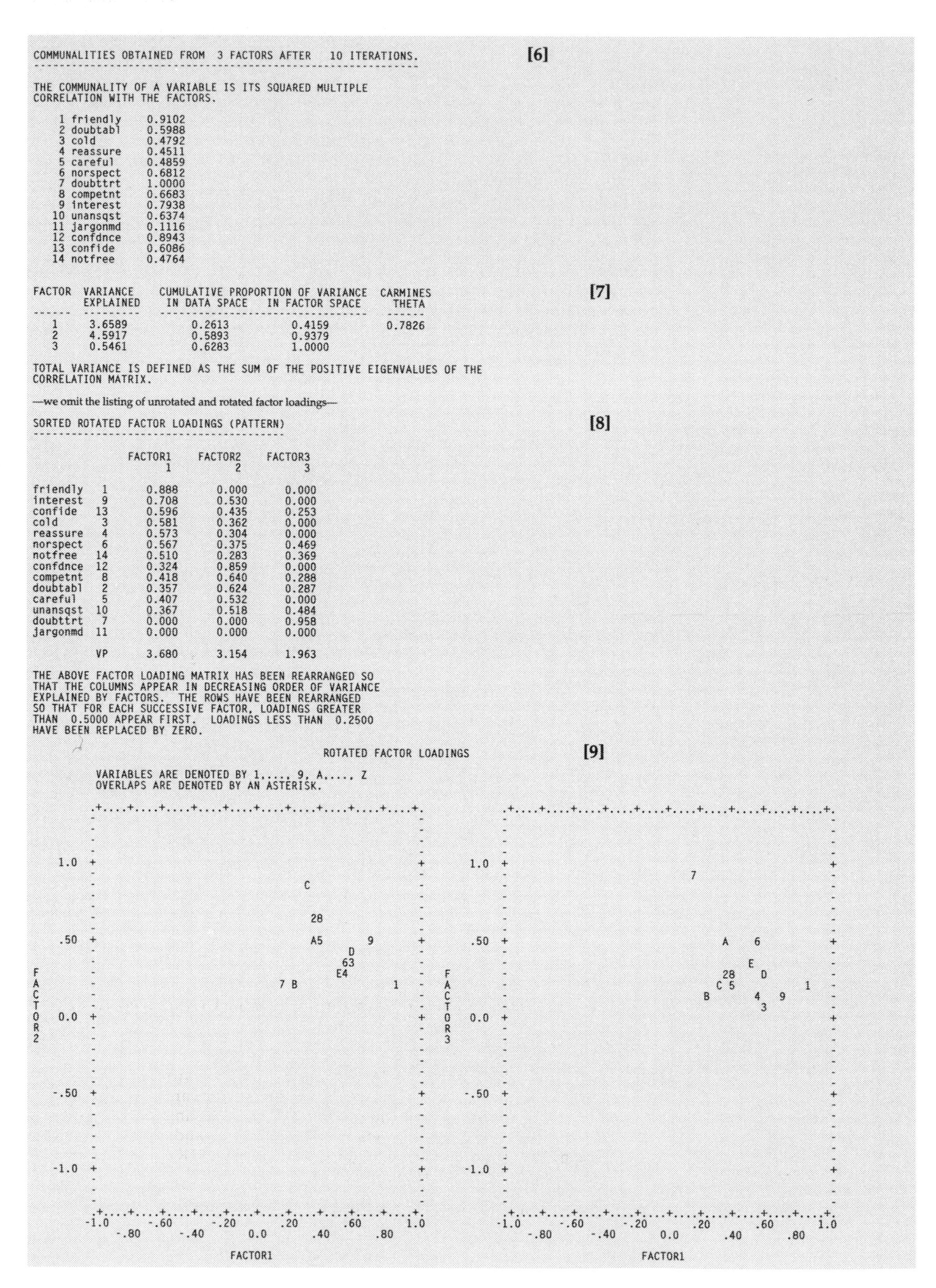

COMMUNALITIES OBTAINED FROM 3 FACTORS AFTER 10 ITERATIONS. **[6]**

THE COMMUNALITY OF A VARIABLE IS ITS SQUARED MULTIPLE CORRELATION WITH THE FACTORS.

1	friendly	0.9102
2	doubtabl	0.5988
3	cold	0.4792
4	reassure	0.4511
5	careful	0.4859
6	norspect	0.6812
7	doubttrt	1.0000
8	competnt	0.6683
9	interest	0.7938
10	unansqst	0.6374
11	jargonmd	0.1116
12	confdnce	0.8943
13	confide	0.6086
14	notfree	0.4764

[7]

FACTOR	VARIANCE EXPLAINED	CUMULATIVE PROPORTION OF VARIANCE IN DATA SPACE	CUMULATIVE PROPORTION OF VARIANCE IN FACTOR SPACE	CARMINES THETA
1	3.6589	0.2613	0.4159	0.7826
2	4.5917	0.5893	0.9379	
3	0.5461	0.6283	1.0000	

TOTAL VARIANCE IS DEFINED AS THE SUM OF THE POSITIVE EIGENVALUES OF THE CORRELATION MATRIX.

—we omit the listing of unrotated and rotated factor loadings—

SORTED ROTATED FACTOR LOADINGS (PATTERN) **[8]**

		FACTOR1 1	FACTOR2 2	FACTOR3 3
friendly	1	0.888	0.000	0.000
interest	9	0.708	0.530	0.000
confide	13	0.596	0.435	0.253
cold	3	0.581	0.362	0.000
reassure	4	0.573	0.304	0.000
norspect	6	0.567	0.375	0.469
notfree	14	0.510	0.283	0.369
confdnce	12	0.324	0.859	0.000
competnt	8	0.418	0.640	0.288
doubtabl	2	0.357	0.624	0.287
careful	5	0.407	0.532	0.000
unansqst	10	0.367	0.518	0.484
doubttrt	7	0.000	0.000	0.958
jargonmd	11	0.000	0.000	0.000
	VP	3.680	3.154	1.963

THE ABOVE FACTOR LOADING MATRIX HAS BEEN REARRANGED SO THAT THE COLUMNS APPEAR IN DECREASING ORDER OF VARIANCE EXPLAINED BY FACTORS. THE ROWS HAVE BEEN REARRANGED SO THAT FOR EACH SUCCESSIVE FACTOR, LOADINGS GREATER THAN 0.5000 APPEAR FIRST. LOADINGS LESS THAN 0.2500 HAVE BEEN REPLACED BY ZERO.

ROTATED FACTOR LOADINGS **[9]**

VARIABLES ARE DENOTED BY 1,...., 9, A...., Z
OVERLAPS ARE DENOTED BY AN ASTERISK.

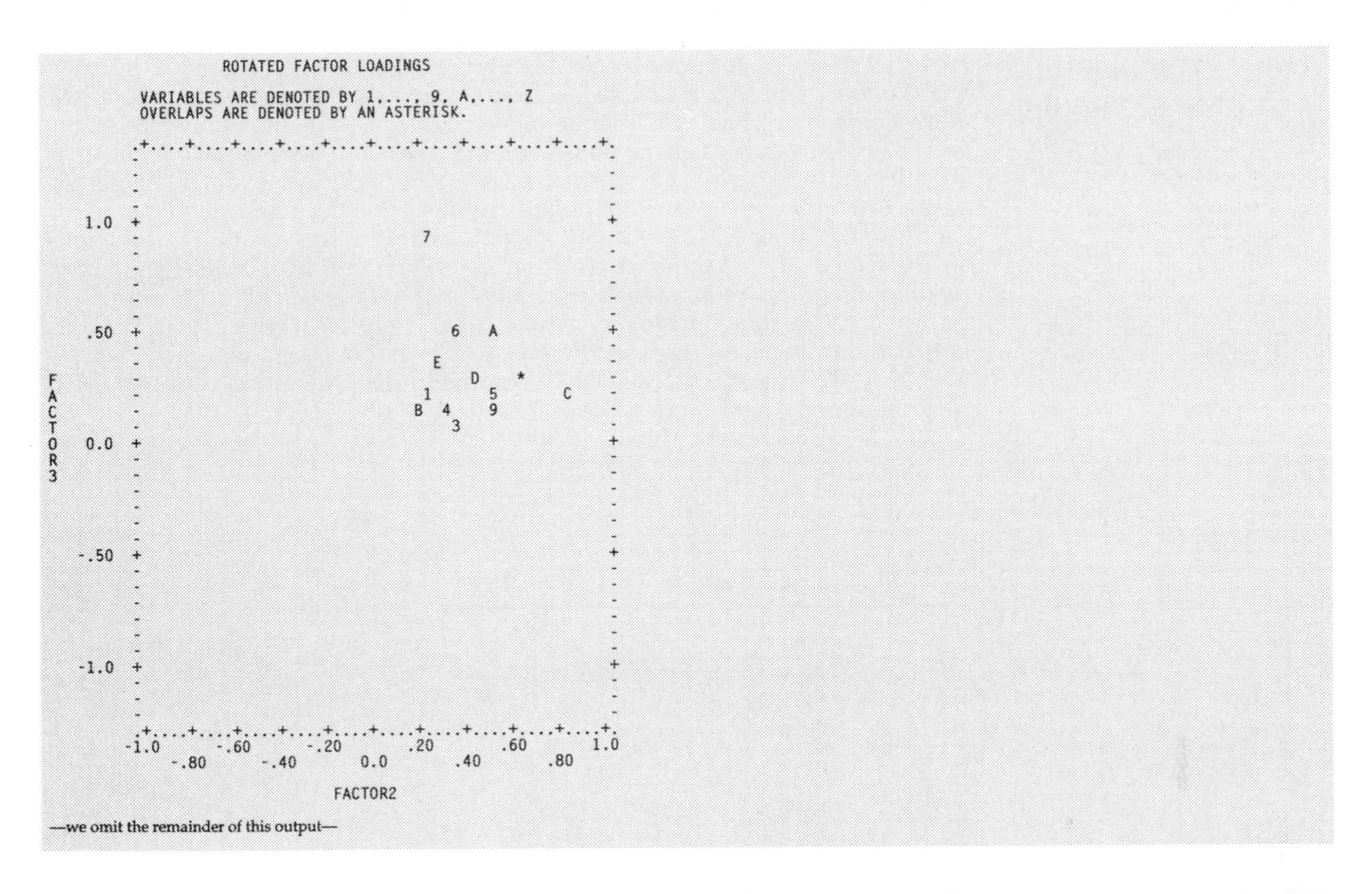

[1] The preassigned values for the communality estimates are squared multiple correlations when the method of initial factor extraction is maximum likelihood.

[2] Histogram of eigenvalues. The eigenvalues are computed from the original correlation matrix without communalities substituted for the diagonal elements. Three factors will be extracted since we changed CONSTANT to 0.9.

[3] Unlike PCA, iteration is required for maximum likelihood factor analysis. The third column reports the minus log likelihood at each step until convergence or the maximum number of iterations have been performed. The default maximum of 25 can be changed with the ITER command.

[4] The goodness of fit chi square is significant at 79.520.

[5] The canonical correlations are the multiple correlations of each factor with the original variables.

[6] Since the factors have been redefined somewhat, the communalities change. See Example 4M.1.

[7] Since the errors structure is different in MLFA compared to PCA (in Example 4M.1), the three factors account for only about 63% of the total variation of the original data.

[8] The loadings differ from PCA (Example 4M.1) because of the different method of factor extraction. However, the first factor can still be interpreted as reflecting Empathy and the second reflecting technical Skill, although there are more variables that load on both factor 1 and factor 2. The third factor mainly represents the single variable, DOUBTTRT. These results can be compared to oblique rotation in Example 4M.3.

[9] Three plots are produced since there are three pairs of factors.

Example 4M.3 Oblique rotation

Factors are rotated to obtain a simple interpretation. In other words, the goal is to make the loadings for each factor either large or small, not intermediate. 4M allows both orthogonal and oblique rotation. In orthogonal rotations, the factor scores for the various factors are uncorrelated. In oblique rotations, the factors are allowed to be correlated, which may result in "simpler" loadings than those for an orthogonal rotation. With oblique rotation there is a greater tendency for each variable to be associated with a single factor. In this example we demonstrate the results of Jennrich and Sampson's (1966) direct quartimin, the method of oblique rotation that we recommend. (See Special Features for more information on rotation options available.) In Input 4M.3, we add a ROTATE paragraph to the setup used in the previous example, and a PRINT paragraph with a LOLEV command. The LOLEV command determines the loadings to be replaced by zero in the Sorted Rotated Factor Loading panel.

Input 4M.3
(insert into Input 4M.2 before / END)

```
/ ROTATE        METHOD = DQUART.

/ PRINT         LOLEV = 0.35.
```

Output 4M.3

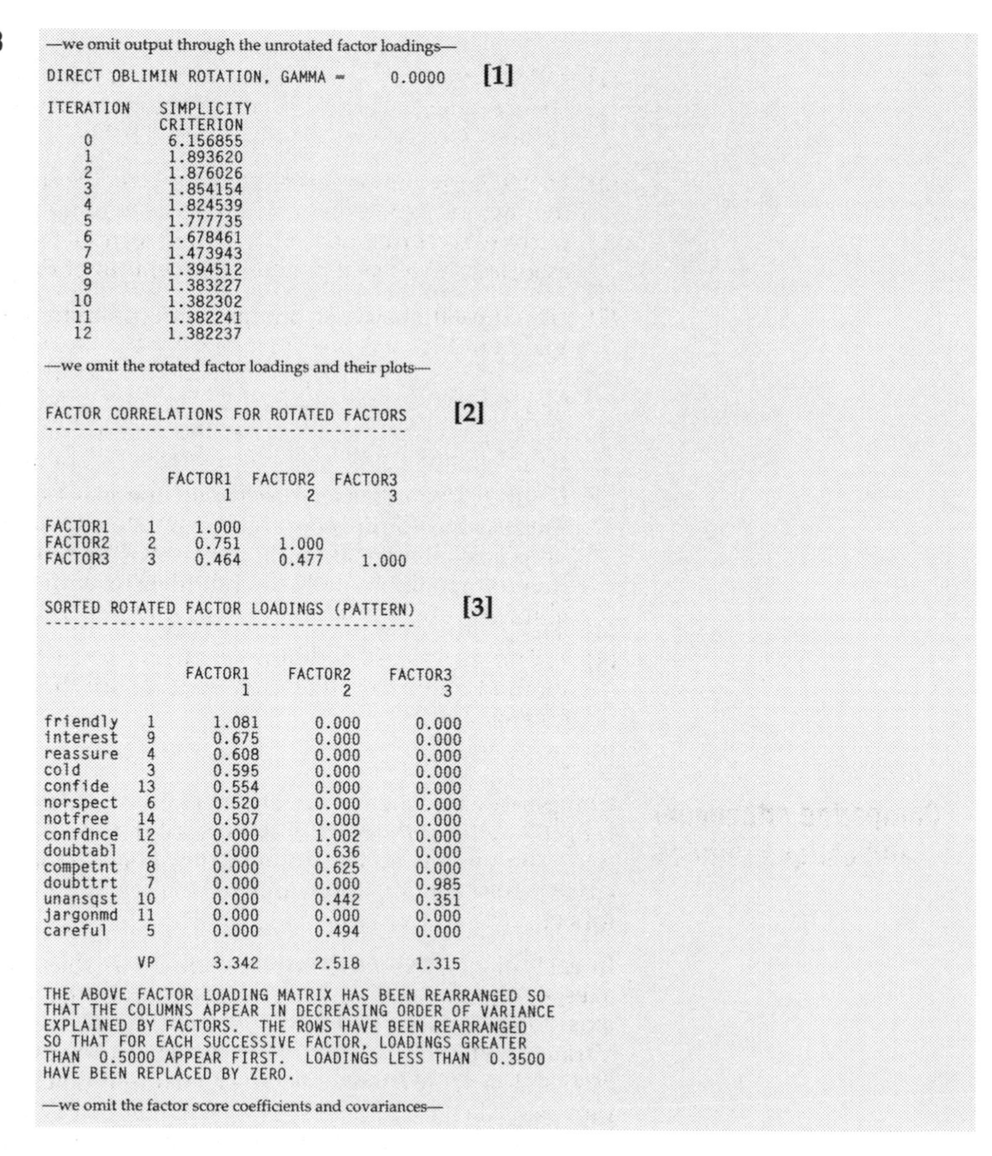

—we omit output through the unrotated factor loadings—

```
DIRECT OBLIMIN ROTATION, GAMMA =     0.0000    [1]

ITERATION   SIMPLICITY
             CRITERION
    0         6.156855
    1         1.893620
    2         1.876026
    3         1.854154
    4         1.824539
    5         1.777735
    6         1.678461
    7         1.473943
    8         1.394512
    9         1.383227
   10         1.382302
   11         1.382241
   12         1.382237
```

—we omit the rotated factor loadings and their plots—

```
FACTOR CORRELATIONS FOR ROTATED FACTORS         [2]
---------------------------------------

              FACTOR1  FACTOR2  FACTOR3
                 1        2        3

FACTOR1   1    1.000
FACTOR2   2    0.751    1.000
FACTOR3   3    0.464    0.477    1.000

SORTED ROTATED FACTOR LOADINGS (PATTERN)        [3]
----------------------------------------

                FACTOR1    FACTOR2    FACTOR3
                   1          2          3

friendly   1     1.081      0.000      0.000
interest   9     0.675      0.000      0.000
reassure   4     0.608      0.000      0.000
cold       3     0.595      0.000      0.000
confide   13     0.554      0.000      0.000
norspect   6     0.520      0.000      0.000
notfree   14     0.507      0.000      0.000
confdnce  12     0.000      1.002      0.000
doubtabl   2     0.000      0.636      0.000
competnt   8     0.000      0.625      0.000
doubttrt   7     0.000      0.000      0.985
unansqst  10     0.000      0.442      0.351
jargonmd  11     0.000      0.000      0.000
careful    5     0.000      0.494      0.000

          VP     3.342      2.518      1.315

THE ABOVE FACTOR LOADING MATRIX HAS BEEN REARRANGED SO
THAT THE COLUMNS APPEAR IN DECREASING ORDER OF VARIANCE
EXPLAINED BY FACTORS.  THE ROWS HAVE BEEN REARRANGED
SO THAT FOR EACH SUCCESSIVE FACTOR, LOADINGS GREATER
THAN  0.5000 APPEAR FIRST.  LOADINGS LESS THAN  0.3500
HAVE BEEN REPLACED BY ZERO.
```

—we omit the factor score coefficients and covariances—

[4]

```
ESTIMATED FACTOR SCORES AND MAHALANOBIS DISTANCES (CHI-SQUARE S) FROM
EACH CASE TO THE CENTROID OF ALL CASES FOR THE ORIGINAL DATA
( 14 D.F.) FACTOR SCORES (  3 D.F.) AND THEIR DIFFERENCE ( 11 D.F.).
EACH CHI-SQUARE HAS BEEN DIVIDED BY ITS DEGREES OF FREEDOM.
A MINUS SIGN AFTER THE CASE NUMBER DENOTES A CASE WITH NONPOSITIVE WEIGHT.

    CASE     CHISQ/DF CHISQ/DF CHISQ/DF FACTOR1  FACTOR2  FACTOR3
 LABEL   NO.      14        3       11
          1    1.260    2.208    1.002   -0.648   -0.743    1.713
          2    0.507    0.183    0.595    0.300    0.200    0.711
          3    1.936    2.119    1.886    1.195    1.927   -0.331
          4    2.391    2.507    2.359    0.816    2.113    1.600
          5    0.145    0.113    0.153    0.305    0.078   -0.257
          6    1.596    0.085    2.008    0.251    0.109   -0.259
          7    1.841    0.736    2.142    0.143   -0.195   -1.220
          8    1.731    0.199    2.149    0.453    0.108   -0.255
          9    0.741    0.329    0.853   -0.777   -0.254   -0.261
         10    0.768    0.638    0.804    0.649    1.226    0.667
         11    0.654    1.226    0.498    0.246    0.051    1.691
         12    1.736    1.417    1.824    0.328   -0.723   -1.190
         13    0.145    0.113    0.153    0.305    0.078   -0.257
         14    2.631    6.841    1.482    3.756    1.532    1.685
         15    1.004    1.112    0.974    0.482    0.243    1.686
         16    0.813    0.400    0.925   -0.960   -0.949   -0.230
         17    0.393    0.583    0.342   -0.984   -1.019   -1.201
         18    0.742    0.555    0.793   -0.768   -0.961   -1.200
         20    0.579    1.322    0.376   -1.077   -1.120    0.750

—we omit cases 21 to 64—

         65    0.256    0.615    0.159   -1.027   -1.101   -1.198
         66    0.333    0.574    0.267   -0.942   -1.013   -1.200
         67    0.817    0.764    0.832   -0.737   -0.214   -1.236
         68    0.487    0.565    0.466   -0.962   -0.955   -1.204

—we omit the rest of the output—
```

[1] 4M informs you that direct oblimin rotation is being used (DQUART is a special case of direct oblimin rotation where gamma = 0), and prints the value of GAMMA (see Harman 1967, p 324).

[2] Factor correlations for rotated factors. These are the correlations between the factors. When the rotation is orthogonal this matrix is not printed. The fairly high correlation of 0.751 between factors 1 and 2 indicates a positive association between patients' perception of Empathy and technical Skill.

[3] Sorted rotated factor loadings (pattern). Since the rotation is not orthogonal, VP is not the variance explained by the factor. However, large values of VP still indicate important factors, and small values indicate less important factors. Comparing the loadings here with those from the orthogonal rotation (Example 4M.2), we see that the goal of having each variable associated with a single factor is better met for the oblique rotation. In the orthogonal rotation, eight of the 14 variables had loadings greater than 0.35 for two or more factors. With the oblique rotation, only UNANSQST loads on more than one factor. In this MLFA run, the ordering of the factors is different, although they are similar to those in Output 4M.2. Factor 3 consists mostly of the variables DOUBTTRT and UNANSQST and may represent a "treatment" factor.

[4] Factor scores after oblique rotation. Even though a different estimation method and rotation is used, case 14 still has large CHISQ/DF values as in Output 4M.1.

Comparing orthogonal and oblique rotation

Comparisons of the two rotations can be made by examining the factor plots for MLFA with orthogonal rotation and the corresponding factor plots for MLFA with oblique rotation and identifying like factors. In Figure 4M.1, Factor 1 is associated with Empathy variables and Factor 2 is related to technical Skill factors (see **[14]** in Example 4M.1).

In each plot, we want to see clusters of variables at the ends of the sketched-in axes, close to an axis or at the intersection of the axes. Variables at the end of an axis define a factor; variables at the intersection are related to neither factor. Variables with extreme values on the 45 degree line are undesirable because their association with one or the other factors is unclear. The coordinates for these plots are taken from the ROTATED FACTOR LOADINGS (not shown). Note that the points are closer to the axes after oblique rotation. Variable 7 (DOUBTTRT) is at or near the origin, indicating that it has very low correlation (loading) with both factors 1 and 2. As noted in Output 4M.3, DOUBTTRT loads on Factor 3.

Figure 4M.1: Comparison of orthogonal vs. oblique rotation: Factor 1 (Empathy) vs. Factor 2 (Skill)

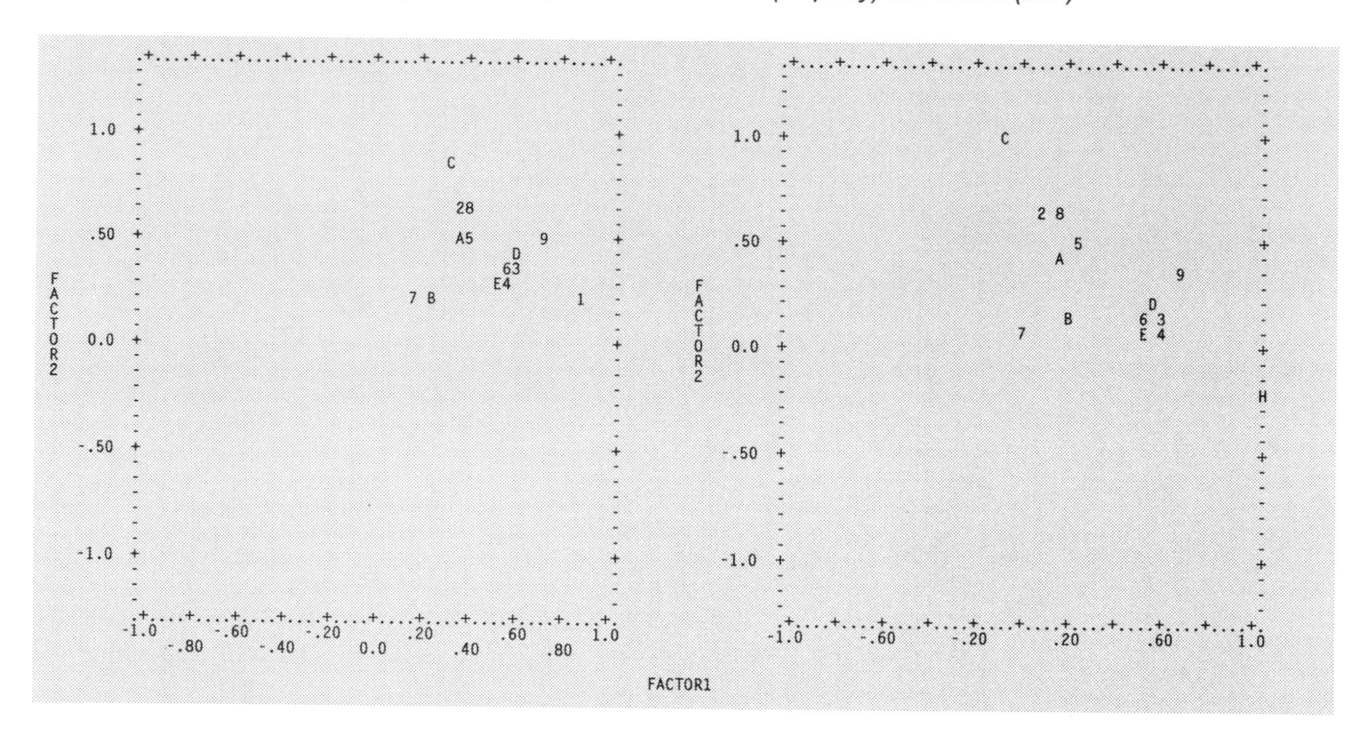

Example 4M.4 Printed Results

Many options are available in the PRINT paragraph for printing results. You can specify the number of cases to print from the original data matrix (five cases is the default). You can also request the factor score coefficients (FSCF) for each case, STANDARD scores, a COVARIANCE matrix, the INVERSE of the correlation (or covariance) matrix, PARTIAL correlations of each pair of variables after removing the effects of all other variables, a factor structure matrix (FSTR) when oblique rotation is performed, and the RESIDUAL correlation matrix. As you have seen in previous examples, the standard (default) output includes the correlation matrix and the shaded and sorted correlation matrix. Input 4M.4 includes four of the options. See 4M Commands for definitions of the others.

Input 4M.4
(replace the / PRINT paragraph of Input 4M.3)

```
/ PRINT       FSCF.
              STANDARD.
              INVERSE.
              PARTIAL.
```

Output 4M.4

—we omit summary statistics (same as 4M.1)

STANDARD SCORES   VARIABLE INDICES   **[1]**

LABEL	NO.	WEIGHT	1	2	3	4	5	6	7	8	9	10	11	12	13	14
	1	1.000	-0.8	0.1	0.3	0.9	0.1	0.0	1.7	-0.9	-0.9	-0.2	-0.4	-1.0	0.1	0.2
	2	1.000	0.4	1.1	0.3	0.9	0.1	0.0	0.7	0.4	0.1	0.6	0.5	0.1	-1.0	0.2
	3	1.000	0.4	0.1	1.4	0.9	2.4	2.1	-0.3	0.4	2.2	-0.2	-0.4	2.2	2.5	0.2
	4	1.000	0.4	2.1	0.3	-0.1	0.1	2.1	1.7	3.1	0.1	1.5	1.3	2.2	2.5	0.2
	5	1.000	0.4	0.1	0.3	-0.1	0.1	0.0	-0.3	0.4	0.1	-0.2	-0.4	0.1	0.1	0.2
	6	1.000	0.4	2.1	0.3	-0.1	0.1	0.0	-0.3	-0.9	0.1	-0.2	-0.4	0.1	0.1	-0.9
	7	1.000	0.4	0.1	-0.9	2.9	0.1	-1.1	-1.2	-0.9	-0.9	-0.2	1.3	0.1	0.1	0.2
	8	1.000	0.4	0.1	0.3	-0.1	0.1	0.0	-0.3	0.4	0.1	-0.2	-0.4	0.1	0.1	3.4
	9	1.000	-0.8	-0.9	0.3	-0.1	0.1	-1.1	-0.3	-0.9	-0.9	-0.2	-0.4	0.1	0.1	-0.9
	10	1.000	0.4	1.1	0.3	0.9	0.1	0.0	0.7	1.8	1.1	1.5	1.3	1.1	0.1	0.2

—we omit standard scores for cases 11 to 64—

LABEL	NO.	WEIGHT	1	2	3	4	5	6	7	8	9	10	11	12	13	14
	65	1.000	-0.8	-0.9	-0.9	-1.1	-1.1	-1.1	-1.2	-0.9	-0.9	-1.1	-1.3	-1.0	-1.0	-0.9
	66	1.000	-0.8	-0.9	-0.9	-0.1	-1.1	-1.1	-1.2	-0.9	-0.9	-0.2	-0.4	-1.0	-1.0	-0.9
	67	1.000	-0.8	0.1	-0.9	-0.1	0.1	0.0	-1.2	-0.9	-0.9	-1.1	-1.3	0.1	0.1	-0.9
	68	1.000	-0.8	-0.9	-0.9	-1.1	0.1	-1.1	-1.2	-0.9	-0.9	-0.2	-0.4	-1.0	-1.0	-0.9

—we omit the correlation matrix (same as 4M.1)—

INVERSE OF CORRELATION MATRIX **[2]**

		friendly 1	doubtabl 2	cold 3	reassure 4	careful 5	norspect 6	doubttrt 7	competnt 8
friendly	1	4.863034							
doubtabl	2	-0.358273	3.270729						
cold	3	-0.550444	-1.345295	2.881225					
reassure	4	-1.110046	-0.429560	0.586150	2.167440				
careful	5	0.336918	0.600691	-0.604524	-0.606147	2.298142			
norspect	6	-0.650941	0.264619	0.087738	-0.096660	-0.245070	3.634020		
doubttrt	7	-0.254247	-0.231558	-0.019224	0.339702	-0.182552	-1.183314	2.511991	
competnt	8	-0.493195	-0.991848	0.681864	0.372491	-0.214498	-0.789975	-0.215337	3.582339
interest	9	-2.311666	0.840923	-1.335196	0.131830	-0.596175	-0.179633	1.119958	-1.217118
unansqst	10	0.369978	-0.663233	0.400953	-0.528300	-0.188263	0.362593	-1.248274	0.081384
jargonmd	11	-0.515949	0.263532	-0.547029	-0.065457	0.226994	-0.073927	0.039315	-0.088020
confdnce	12	1.393027	-1.495483	0.562551	0.066096	-0.830344	0.089275	0.102561	-1.211711
confide	13	-0.596702	0.364490	-0.793740	-0.313858	0.264020	-1.357281	0.086441	0.822895
notfree	14	-0.800696	0.129969	0.296379	-0.013351	0.124765	-0.105683	-0.297045	-0.011484

		interest 9	unansqst 10	jargonmd 11	confdnce 12	confide 13	notfree 14
interest	9	5.543447					
unansqst	10	-1.118172	3.138315				
jargonmd	11	0.775620	-0.185304	1.339158			
confdnce	12	-1.197774	-0.297898	-0.618876	4.625621		
confide	13	-0.015857	-0.028612	0.238492	-1.470495	3.375805	
notfree	14	-0.027855	-0.353118	-0.034384	-0.224900	-0.296649	2.021711

—we omit the correlation matrix (same as 4M.1)—

PARTIAL CORRELATIONS **[3]**

		friendly 1	doubtabl 2	cold 3	reassure 4	careful 5	norspect 6	doubttrt 7	competnt 8	interest 9	unansqst 10	jargonmd 11	confdnce 12	confide 13
friendly	1	1.000												
doubtabl	2	0.090	1.000											
cold	3	0.147	0.438	1.000										
reassure	4	0.342	0.161	-0.235	1.000									
careful	5	-0.101	-0.219	0.235	0.272	1.000								
norspect	6	0.155	-0.077	-0.027	0.034	0.085	1.000							
doubttrt	7	0.073	0.081	0.007	-0.146	0.076	0.392	1.000						
competnt	8	0.118	0.290	-0.212	-0.134	0.075	0.219	0.072	1.000					
interest	9	0.445	-0.197	0.334	-0.038	0.167	0.040	-0.300	0.273	1.000				
unansqst	10	-0.095	0.207	-0.133	0.203	0.070	-0.107	0.445	-0.024	0.268	1.000			
jargonmd	11	0.202	-0.126	0.278	0.038	-0.129	0.034	-0.021	0.040	-0.285	0.090	1.000		
confdnce	12	-0.294	0.384	-0.154	-0.021	0.255	-0.022	-0.030	0.298	0.237	0.078	0.249	1.000	
confide	13	0.147	-0.110	0.255	0.116	-0.095	0.388	-0.030	-0.237	0.004	0.009	-0.112	0.372	1.000
notfree	14	0.255	-0.051	-0.123	0.006	-0.058	0.039	0.132	0.004	0.008	0.140	0.021	0.074	0.114

		notfree 14
notfree	14	1.000

THE ELEMENTS OF THIS MATRIX ARE THE PARTIAL CORRELATIONS OF EACH PAIR OF VARIABLES, PARTIALED ON ALL OTHER VARIABLES (I.E., HOLDING ALL OTHER VARIABLES FIXED).

—we omit the rest of the output up to the factor score coefficients (same as 4M.3)—

FACTOR SCORE COEFFICIENTS **[4]**

THESE COEFFICIENTS ARE FOR THE STANDARDIZED VARIABLES, MEAN ZERO AND STANDARD DEVIATION ONE.

		FACTOR1 1	FACTOR2 2	FACTOR3 3
friendly	1	0.55419	-0.07264	0.01434
doubtabl	2	0.02064	0.09838	-0.00449
cold	3	0.05544	0.02693	-0.00027
reassure	4	0.05294	0.01480	0.00029
careful	5	0.02698	0.06176	-0.00255
norspect	6	0.07966	0.03034	0.00003
doubttrt	7	-0.03118	-0.01404	1.01540
competnt	8	0.03557	0.11838	-0.00520
interest	9	0.16259	0.12124	-0.00288
unansqst	10	0.02680	0.07464	-0.00319
jargonmd	11	0.01064	0.00967	-0.00027
confdnce	12	0.01730	0.58127	-0.02858
confide	13	0.07010	0.04469	-0.00087
notfree	14	0.04637	0.00942	0.00043

—we omit the rest of the output—

Bold numbers below refer to Output 4M.4.

[1] Standard scores for each case. For each value, 4M computes and prints $z_{ij} = (x_{ij} - \bar{x}_j) / s_j$, where x_{ij} is the value of the jth variable in the ith case. Note the larger standard scores associated with case 14.

[2] Inverse of the correlation matrix.

[3] Partial correlations. These are the correlations between each pair of variables after removing the linear effects of all the other variables. This matrix is equivalent to the "anti-image" correlation matrix (except for the signs of the off-diagonal elements). It is often important to inspect the partial correlations to clearly determine variable redundancy, and to gain a clear view of which variables are intrinsically correlated.

[4] Factor score coefficients for factors from the MLFA analysis with an oblique rotation. The rotated factor loadings are multiplied by the factor correlations to obtain the factor structure (identical to factor loadings in the orthogonal case). The factor structure matrix contains the correlations of the factors with the variables. The factor structure is multiplied by the inverse of the correlation (or covariance) matrix of the original variables to obtain the factor score coefficients. These factor score coefficients are for the standardized variables.

Example 4M.5 Using a correlation matrix as input

4M can use a data file, a covariance or correlation matrix, factor loadings, or factor score coefficients as input. If you begin the analysis from a covariance or correlation matrix (instead of the data) you can save computing time.

Covariance or correlation matrix input may be (1) read from a BMDP File, or (2) entered directly. Correlation matrices formed using nonparametric methods, such as Spearman correlations, may also be used as input if we are concerned about outliers and asymmetric distributions. In this example, we use the correlation matrix from Example 4M.1 as input.

Reading matrices from a BMDP File. In addition to the INPUT paragraph CODE needed for BMDP Files (see Chapter 8), you may need to include CONTENT and TYPE. Use neither if the CONTENT was DATA when the File was written. Use TYPE to tell program 4M to treat other BMDP File matrices as correlations (or covariances). For example, if partial covariances were written to a BMDP File in program 6R using CONTENT IS COVAPART, TYPE IS COVA would be specified in order to use them as input to 4M. Note that the SHAPE command is not used with BMDP File input.

Reading matrices NOT in a BMDP File. In Input 4M.5, we enter the correlations following the END paragraph (alternatively, we could add FILE or UNIT to the INPUT paragraph and store the matrix in a separate file). These correlations are written in the form of a LOWER triangular matrix with one element (the diagonal) in the first row, two elements in the second, etc. We use the TYPE and SHAPE commands in the INPUT paragraph to specify that we are reading in a CORRELATION matrix with a LOWER triangular shape. (4M also accepts a SQUARE matrix containing all the rows and columns of a complete correlation matrix. Even when a square matrix is read as input, only the lower half is used in the analysis.)

When entering correlations directly and there are more elements than will fit on one line, the row is completed on the following line(s). However, each row of the input matrix should start on a new line. The number of variables in the INPUT paragraph should be equal to the number of variables in the matrix. The format reads the elements on one line. **You must use FORTRAN-type format.** FREE format is not available when using this type of input. The variable names appear in the same order as they occur in the rows of the matrix.

When a correlation or covariance matrix is used as input, the following are omitted from the output: the statistics for each of the variables, data for the first five cases, estimated factor scores and Mahalanobis distances, factor score plots, and the factor score covariance matrix that is computed from the factor scores. When the correlation (or covariance) matrix is read from a BMDP File **and** the data are on the same BMDP File as the matrix (with the same CODE), those statistics, scores, and plots are printed.

The analysis requested in Input 4M.5 is similar to that in Input 4M.1. It uses the same variables with the principal components method of factor extraction and standard defaults. We do not show output for this example.

Input 4M.5

```
/ INPUT     FORMAT = '8F9'.
            VAR = 14.
            TYPE = CORR.
            SHAPE = LOWER.
            CASE = 68.

/ VARIABLE NAMES ARE friendly, doubtabl, cold, reassure,
                     careful, norspect, doubttrt,
                     competnt, interest, unansqst,
                     jargonmd, confdnce, confide,
                     notfree.

/ END

 1.000
-0.556   1.000
-0.628   0.584   1.000
 0.639  -0.460  -0.348   1.000
 0.516  -0.443  -0.495   0.519   1.000
-0.704   0.513   0.494  -0.515  -0.540   1.000
-0.445   0.478   0.280  -0.335  -0.376   0.631   1.000
 0.611  -0.681  -0.437   0.431   0.558  -0.636  -0.494   1.000
 0.786  -0.576  -0.662   0.547   0.656  -0.642  -0.344   0.703
 1.000
-0.566   0.631   0.399  -0.550  -0.538   0.575   0.645  -0.615
-0.619   1.000
-0.315   0.271   0.325  -0.206  -0.130   0.261   0.224  -0.242
-0.174   0.253   1.000
 0.554  -0.723  -0.507   0.486   0.629  -0.615  -0.470   0.751
 0.700  -0.665  -0.310   1.000
 0.688  -0.505  -0.595   0.536   0.507  -0.730  -0.443   0.495
 0.659  -0.532  -0.240   0.654   1.000
-0.624   0.424   0.334  -0.466  -0.381   0.581   0.505  -0.492
-0.528   0.563   0.234  -0.507  -0.557   1.000
 / END
```

Example 4M.6 Case weights

The data in each case can be weighted by the values of a CWEIGHT or FREQ variable (see Chapter 4). CWEIGHT is used when the variance is not homogeneous over all cases (the weight is the inverse of the variance). FREQ is used to represent the frequencies of cases that are repeated in the data (the variable value is the frequency of the case). Either can be used to remove the influence of some cases (with weight = 0) while still obtaining their factor scores and Mahalanobis distances.

In this example, we remove from the computations a case that had extreme Mahalanobis distances in the space of the 14 variables. We want to see if its absence markedly alters the factor structure. Input 4M.6 shows the additional instructions needed to create case weights. We identify the variable WT_VAR as the case weight (CWEIGHT) in the VARIABLE paragraph. We create the case weight variable in the TRANSFORM paragraph. For every case but case 14, it will have the value 1.0.

Input 4M.6

```
/ INPUT       VARIABLES = 14.
              FORMAT is '14f1'.
              FILE IS 'patient.dat'.

/ VARIABLE    NAMES ARE friendly, doubtabl, cold, reassure,
                        careful, norspect, doubttrt,
                        competnt, interest, unansqst,
                        jargonmd, confdnce, confide,
                        notfree.
              MISS = 14*0.
              CWEIGHT = wt_var.

/ TRANSFORM wt_var = 1.
              IF (KASE EQ 14) THEN wt_var = 0.

/ FACTOR      METHOD = MLFA.   CONSTANT = 0.9.

/ ROTATE      METHOD = DQUART.

/ PRINT       LOLEV = 0.35.

/ END
```

Output 4M.6

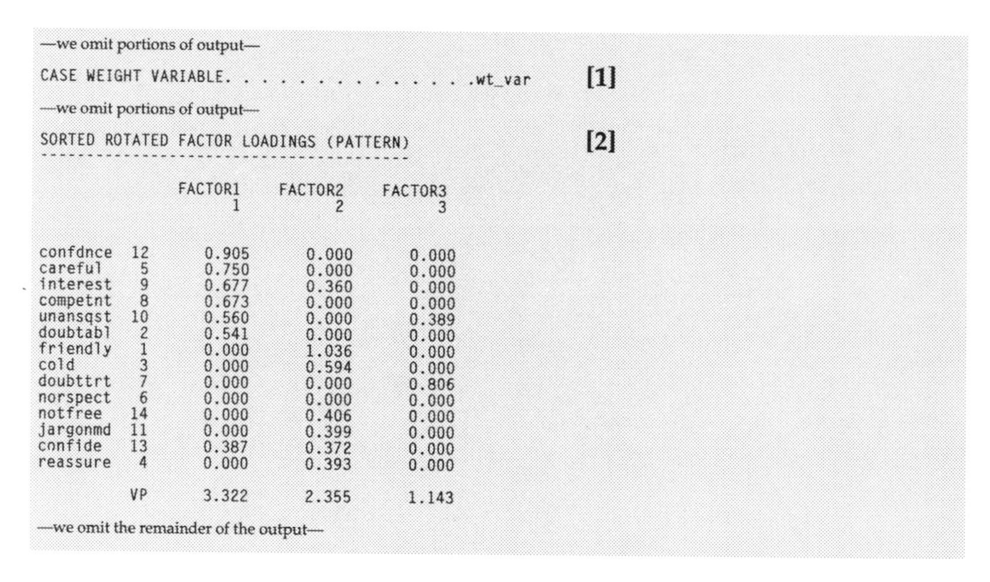

—we omit portions of output—

CASE WEIGHT VARIABLE. . . . . .wt_var **[1]**

—we omit portions of output—

SORTED ROTATED FACTOR LOADINGS (PATTERN) **[2]**

		FACTOR1 1	FACTOR2 2	FACTOR3 3
confdnce	12	0.905	0.000	0.000
careful	5	0.750	0.000	0.000
interest	9	0.677	0.360	0.000
competnt	8	0.673	0.000	0.000
unansqst	10	0.560	0.000	0.389
doubtabl	2	0.541	0.000	0.000
friendly	1	0.000	1.036	0.000
cold	3	0.000	0.594	0.000
doubttrt	7	0.000	0.000	0.806
norspect	6	0.000	0.000	0.000
notfree	14	0.000	0.406	0.000
jargonmd	11	0.000	0.399	0.000
confide	13	0.387	0.372	0.000
reassure	4	0.000	0.393	0.000
	VP	3.322	2.355	1.143

—we omit the remainder of the output—

[1] Identification of variable containing the case weights, WT_VAR. Cases with zero weight are not included in the univariate statistics panel (not shown).

[2] Sorted rotated factor loadings. The loadings differ only slightly from the ones shown in Output 4M.3. This may not be true when more cases are deleted.

We do not include the output for the Mahalanobis distances and factor scores (**[18]** in Output 4M.1). The distances and factor scores are computed for all cases, including those with zero weights.

Example 4M.7 Scale evaluation

The process of scale evaluation is done during the development of a new scale, the modification of an existing scale, or when adapting an existing scale to subjects from a different population. Ordinarily, researchers sum the resulting item scores and obtain a summated Likert scale. This summation only makes sense if the negatively worded items are reflected first.

In evaluating summated scales, three characteristics of summated scales are commonly assessed: validity, reliability, and number of dimensions. High validity and reliability are essential to the effectiveness of a scale (see Bollen, 1989 and Carmines & Zeller, 1979). In addition, scales that measure along a single dimension are easier to interpret (see McKennell, 1977 and Bourque & Clark, 1992). 4M now includes one of the most commonly used measures of reliability, Cronbach's alpha, and one method of determining dimensionality, second order factor analysis.

In this example, we again use the dataset presented in Example 4M.1. In this case we use the DOBLI command in the ROTATE paragraph to specify an oblique method of rotation. The use of one of the oblique rotation methods is essential for scale evaluation, since you do not want to force the factors to be independent. The CRON command in the print paragraph will produce both Cronbach's alpha and the second order factor results.

Input 4M.7

```
/ INPUT      VARIABLES = 14.
             FORMAT IS '14F1'.
             FILE IS 'patient.dat'.

/ VARIABLE   NAMES = friendly, doubtabl, cold, reassure,
                     careful, norspect, doubttrt, competnt,
                     interest, unansqst, jargonmd,
                     confdnce, confide, notfree.
             MISSING = 14*0.

/ ROTATE     METHOD = DOBLI.

/ PRINT      CRON.

/ END
```

Output 4M.7

—we omit portions of output which were shown in Output 4M.1—

DESCRIPTIVE STATISTICS OF DATA [1]

VARIABLE NO.	NAME	TOTAL FREQ.	MEAN	STANDARD DEV.	ST.ERR OF MEAN	COEFF OF VAR	SMALLEST VALUE	SMALLEST Z-SCR	SMALLEST CASE	LARGEST VALUE	LARGEST Z-SCR	LARGEST CASE	RANGE
1	friendly	64	1.6563	.78110	.09764	.47161	1.0000	-0.84	1	5.0000	4.28	14	4.0000
2	doubtabl	64	1.9219	.99690	.12461	.51871	1.0000	-0.92	9	4.0000	2.08	4	3.0000
3	cold	64	1.7656	.86817	.10852	.49171	1.0000	-0.88	7	5.0000	3.73	54	4.0000
4	reassure	64	2.0938	1.0192	.12740	.48676	1.0000	-1.07	12	5.0000	2.85	7	4.0000
5	careful	64	1.9375	.85217	.10652	.43983	1.0000	-1.10	16	5.0000	3.59	25	4.0000
6	norspect	64	2.0156	.93422	.11678	.46349	1.0000	-1.09	7	5.0000	3.19	14	4.0000
7	doubttrt	64	2.2656	1.0426	.13033	.46019	1.0000	-1.21	7	4.0000	1.66	1	3.0000
8	competnt	64	1.6719	.75708	.09463	.45283	1.0000	-0.89	1	4.0000	3.08	4	3.0000
9	interest	64	1.9063	.95483	.11935	.50090	1.0000	-0.95	1	5.0000	3.24	14	4.0000
10	unansqst	64	2.2656	1.1442	.14303	.50504	1.0000	-1.11	12	5.0000	2.39	14	4.0000
11	jargonmd	64	2.4844	1.1408	.14260	.45917	1.0000	-1.30	12	5.0000	2.21	26	4.0000
12	confdnce	64	1.9375	.94070	.11759	.48552	1.0000	-1.00	1	5.0000	3.26	62	4.0000
13	confide	64	1.8750	.86373	.10797	.46066	1.0000	-1.01	2	5.0000	3.62	54	4.0000
14	notfree	64	1.8125	.94070	.11759	.51901	1.0000	-0.86	6	5.0000	3.39	8	4.0000

—we omit portions of output—

```
SQUARED MULTIPLE CORRELATIONS (SMC) OF EACH VARIABLE WITH ALL          [2]
OTHER VARIABLES, AND CRONBACH'S ALPHA, WITH THAT VARIABLE REMOVED
-----------------------------------------------------------------

                  SMC      ALPHA

   1 friendly   0.79437   0.9264
   2 doubtabl   0.69426   0.9291
   3 cold       0.65293   0.9316
   4 reassure   0.53863   0.9317
   5 careful    0.56487   0.9309
   6 norspect   0.72482   0.9270
   7 doubttrt   0.60191   0.9329
   8 competnt   0.72085   0.9280
   9 interest   0.81961   0.9265
  10 unansqst   0.68136   0.9280
  11 jargonmd   0.25326   0.9405
  12 confdnce   0.78381   0.9266
  13 confide    0.70377   0.9281
  14 notfree    0.50537   0.9312

ALPHA FOR ALL VARIABLES = 0.9347          [3]

THIS IS CRONBACH'S STANDARDIZED ALPHA, COMPUTED FROM CORRELATIONS.
```

—we omit portions of the output—

```
FACTOR CORRELATIONS FOR ROTATED FACTORS                                [4]
---------------------------------------

             FACTOR1  FACTOR2
                   1        2

FACTOR1   1    1.000
FACTOR2   2    0.568    1.000

SORTED ROTATED FACTOR LOADINGS (PATTERN)
----------------------------------------

               FACTOR1     FACTOR2
                     1           2

cold       3     0.960      -0.286
interest   9     0.935       0.000
friendly   1     0.773       0.000
careful    5     0.733       0.000
confide   13     0.710       0.000
confdnce  12     0.624       0.306
reassure   4     0.563       0.000
doubtabl   2     0.552       0.300
competnt   8     0.528       0.374
doubttrt   7     0.000       0.956
unansqst  10     0.288       0.652
notfree   14     0.000       0.597
norspect   6     0.477       0.465
jargonmd  11     0.000       0.000

          VP     5.135       2.455

THE ABOVE FACTOR LOADING MATRIX HAS BEEN REARRANGED SO
THAT THE COLUMNS APPEAR IN DECREASING ORDER OF VARIANCE
EXPLAINED BY FACTORS.  THE ROWS HAVE BEEN REARRANGED
SO THAT FOR EACH SUCCESSIVE FACTOR, LOADINGS GREATER
THAN  0.5000 APPEAR FIRST.  LOADINGS LESS THAN  0.2500
HAVE BEEN REPLACED BY ZERO.

CRONBACH'S ALPHA IS THE STANDARDIZED ALPHA, COMPUTED FROM CORRELATIONS.      [5]
THE FIRST ALPHA IS CALCULATED USING ALL VARIABLES.  THE ALPHA FOR EACH
INDIVIDUAL FACTOR IS CALCULATED BY USING ONLY CERTAIN VARIABLES CHOSEN FOR
THEIR LOADINGS IN THE SORTED ROTATED FACTOR LOADING MATRIX.  FOR EACH FACTOR,
THE CALCULATION USES ONLY THE VARIABLES DISPLAYING A POSITIVE ROTATED FACTOR
LOADING ON THAT FACTOR, AS WELL AS A ZERO LOADING ON ALL OTHER FACTORS.
NOTE THAT ALPHA IS UNDEFINED IF ONLY ONE VARIABLE IS USED.

FACTOR  ALPHA   VARIABLES USED

        0.9347  - ALL -
     1  0.8847  interest  friendly  careful   confide   reassure
     2  0.6709  doubttrt  notfree
```

—we omit a portion of the output—

```
SCALE EVALUATION                                                       [6]
----------------

ROTATED SECOND-ORDER FACTORS CALCULATED FROM THE MATRIX OF FACTOR CORRELATIONS
------------------------------------------------------------------------------

               2ND-ORDR
                      1

FACTOR1   1       0.885
FACTOR2   2       0.885

          VP      1.568

THE VP IS THE VARIANCE EXPLAINED BY THE FACTOR.  A SINGLE
SECOND-ORDER FACTOR WITH A LARGE VP IMPLIES THAT THE FACTORS
SHARE COMMON VARIANCE.  THIS IS AN INDICATION OF OVERLAPPING
DIMENSIONS THAT CONTRIBUTE TO A SINGLE OVERALL DIMENSION.
FACTOR ROTATION IS DONE BY THE DIRECT QUARTIMIN METHOD.
```

Bold numbers below refer to Output 4M.7

[1] The descriptive statistics table should be examined to determine that the various items have similar standard deviations and data ranges. Since we have not specified the use of the covariance matrix, all the computations have been performed on the correlation matrix.

[2] Alpha is also computed with each item removed one at a time. The alpha statistic is reported along with the squared multiple correlation of each item with all other items. If the standardized alpha increases when an item is removed, then the item has a relatively low correlation with other items in the scale. Thus, if the investigator is examining the scale to determine possible items to be omitted, this item might be considered as a candidate. A low squared multiple correlation of an item with the other items is also an indication that an item is not measuring the same thing as other items. In this example, we have only one item, JARGONMD, that has not only a low multiple correlation with the other items, but also causes an increase in alpha when it is removed. In this case, the increase in alpha is minor.

Note that researchers normally use well-tested scales without revision. Therefore, they do not remove scale items unless there are compelling reasons to do so, or the revisions are suggested for specific target populations.

[3] The standardized alpha for all variables is 0.9347 using all the variables after they have been standardized. You can use this statistic when evaluating a summated scale of raw scores only if the standard deviations in **[1]** are similar. If not, you should recompute the factor analysis using the covariance matrix in the factor paragraph to obtain the raw score alpha. In this case, the standard deviations do not appear to be widely disparate and the level of alpha is high. Carmines and Zeller, (1979) recommend that scales should have an alpha of at least 0.80. The standardized alpha is computed as:

$$\text{Cronbach's Alpha} = \left(\frac{P}{P-1}\right)\left(1-\frac{\sum \text{Diagonal of matrix}}{\sum \text{all entries in matrix}}\right)$$

where P = the number of items, and the matrix is either the correlation matrix as used here, or the covariance matrix. Note that standardized alpha tends to increase with increasing P and increasing magnitude of the average correlation coefficient among items (see Carmines & Zeller, 1979).

[4] After an oblique rotation, the correlation between the rotated factors is 0.568, indicating that the factors are quite highly correlated. The sorted rotated factor loadings with all loadings less than 0.25 replaced by zero is given next.

[5] Cronbach's alpha is given for all items that have positive loadings on one factor and zero loadings on all other factors. The two alphas reported here are 0.8847 using five items in the first factor and 0.6709 from two items in the second factor. The purpose of these two alphas is that items which load high on a factor could be put into their own summated score scale for further analyses. These alphas predict the internal consistency of these scales.

[6] Two factors were identified in this example. The scale evaluation section calculates a second order factor to determine whether the two factors will load on a single factor. A single second order factor was obtained, indicating that the decision to sum all items into a single summated scale is reasonable. This could be further checked by using the 1M cluster analysis program where the variables are clustered.

Special Features

Specifying the number of factors

You can use the command NUMBER in the FACTOR paragraph to specify the maximum number of factors that you want computed. If you omit NUMBER, then the number of factors depends on the number of eigenvalues of the correlation matrix greater than the value of CONSTANT. The default value of CONSTANT is 1.0 if the unaltered correlation matrix is used, and 0 if you specify COMMUNALITY = SMCS or MAXROW. When the initial extraction method is PFA, you may wish to set the value of CONSTANT to the total variance explained divided by the number of variables.

Saving factor scores for further analysis

The SAVE paragraph makes it possible to store many of the results computed by 4M in a BMDP File (see Chapter 8 for details) for input to subsequent analyses. The results that you can save include the following:

DATA – The data and factor scores (both the data and the factor scores for each case are saved. The data are followed by the factor scores, which are named FACTOR1, FACTOR2, etc.).

COVA – The covariance matrix, variances, means, sample size, and sum of weights (this is equivalent to saving a correlation matrix; the correlation matrix can be used in an analysis that reads the covariance matrix as input and vice versa).

UFLD – Unrotated factor loadings.

RFLD – Rotated factor loadings.

FCOR – Factor correlations.

FSCF – Factor score coefficients.

To save any of the results listed above, use CONTENT in the SAVE paragraph to list their names. However, if you are saving only the data and factor scores, you can omit CONTENT, since the data (and scores) are saved by default. The results saved in a BMDP File can be used by any BMDP program.

Note that you can save the covariance matrix for analysis in other programs or for reanalysis in 4M—e.g., with a different linkage rule. The variables in the BMDP file will be numbered sequentially 1, 2, 3 ... When using CORR or COVA, if you have used fewer variables than you had available, the BMDP file variable numbers will not correspond to the previous variable numbers. It is important to keep that distinction in mind during future use of the BMDP file, when you should refer to variables by names rather than by number. Default: DATA.

Choice of matrix to be factored

You may use the FORM command in the FACTOR paragraph to factor the correlation matrix, covariance matrix, the correlation matrix about the origin (i.e., assuming variable means are zero), or the covariance matrix about the origin. If a choice is not specified, the correlation matrix is factored. Note that the FORM statement differs from the INPUT TYPE statement, which allows you to input a covariance matrix and request that the factor analysis be performed on the correlation matrix.

Initial communality estimates

Communality estimates can be specified as one of three BMDP instruction options or as a list of values, one for each variable. The three options are as follows (state COMMUNALITY = option in the FACTOR paragraph):

UNALT. – The diagonal of the correlation (or covariance) matrix remains unaltered.

SMCS. – The diagonal of the correlation (or covariance) matrix is replaced by the squared multiple correlations of each variable with the remaining variables. If the covariance matrix is factored, the squared multiple correlations are multiplied by the corresponding variances.

MAXROW. – The diagonal elements of the correlation (or covariance) matrix are replaced by the maximum absolute row values of the correlation matrix. If the covariance matrix is factored, these values are multiplied by the corresponding variances.

Computing factor scores when alternative input forms are used

When input is a correlation or covariance matrix from a BMDP File, you can obtain factor scores if the data are on the same BMDP File as the matrix and have the same CODE (see Chapter 8). To match up variables from the data with the variables in the correlation or covariance matrix, the SCORE statement must be given. This option avoids recomputing the covariance or correlation matrix without losing the ability to obtain factor scores when using several methods of factoring.

When factor loadings or factor score coefficients are input from a BMDP File, factor scores can be obtained if the data and covariance or correlation matrix are in the same BMDP File. This allows you to try several methods of rotation and to obtain factor scores without recomputing the covariance matrix or initial loadings. Moreover, by using this option, scores for one set of subjects can be obtained using factor score coefficients computed from another set of subjects. Factor scores can also be obtained for one set of variables using the factor score coefficients computed from another set of variables. For example, factor scores for posttreatment variables can be obtained using factor score coefficients from a pretreatment factor analysis of the same variables. To match variables from the covariance matrix with variables from the loadings or coefficients, the USE command can be used. To match variables from the data with variables in the loadings or coefficients matrix, the SCORE command must be given. Note that the covariance matrix is needed when SCORE is used, since the means and standard deviations required for computing standard scores for the factor scores are stored with the covariance matrix.

Tolerance

The inverse of the correlation or covariance matrix is computed by stepwise pivoting. The TOLERANCE limit is used to determine whether to pivot on a variable. If the TOLERANCE limit is not met for any variable, the computations continue when METHOD is PCA or PFA but terminate when METHOD is MLFA or LJIFFY.

Interactive factor analysis

4M now reads interactive FACTOR, ROTATE, GROUP, PRINT, SAVE, TRANS, CONTROL, END and FINISH paragraphs. In particular, you may do factorizations until you find one you that best fits your data. Then, you can do rotations, based on factorization, until you find one you like. You can use both general interactive commands and 4M-specific commands in interactive PLOT and PRINT paragraphs.

4M Commands

/ INPut

Required. See Chapter 3 for a full description of INPUT options. Additional options for 4M are CONTENT, TYPE, SHAPE, FACTOR, and CASE.

CONTent = *(one only)* `CONTENT = CORR.`
DATA, CORR, COVA, UFLD, RFLD, FCOR, *or* **FSCF.**

Use only if input is from a BMDP File. If input is not DATA, CONTENT must be specified and must be identical to that stated when the BMDP File was created (see SAVE below). Default: DATA.

TYPE = *(one only)* `TYPE = LOAD.`
DATA, CORR, COVA, LOAD, *or* **FSCF.**

If your input is not data, you must specify the TYPE of input unless it comes from a BMDP File and the CONTENT of the BMDP File is COVA or CORR. If input is factor loadings (LOAD), the factors are assumed to be orthogonal. If the input is FSCF or LOAD from a BMDP FIle, then the correlation or covariance matrix must be in the same file for 4M to perform rotation and calculate factor score coefficients. If the data are also on the file, 4M will calculate factor scores as well. Default: DATA.

SHAPE = SQUARE *or* **LOWER.** `SHAPE = LOWER.`

When TYPE is CORR or COVA, SHAPE describes the input matrix (matrix is not from a BMDP File). Each row of the matrix must begin in a new record. FORMAT describes the longest row. Example: A lower triangular matrix, one element (the diagonal) in the first row, two elements in the second, etc. Default: SQUARE is assumed.

FACTOR = #. `FACTOR = 6.`

Use when TYPE is LOAD or FSCF to specify the number of factors. The factors are assumed to be orthogonal when LOAD is used.

New Feature

CASE = #. `CASE = 25.`

Required when the input matrix is a correlation or covariance matrix. Specifies the number of cases used to create the matrix.

/ VARiable

The VARIABLE paragraph is optional. (See detailed description in Chapter 4.) Additional commands for 4M are FREQUENCY, CWEIGHT, SCORE.

FREQuency = variable. `FREQ = COUNT.`

Name or number of a variable containing the case frequencies. The value of the case frequency variable is the number of times that the case is to be repeated. Case frequencies must be zero or positive. If a case has zero frequency it is not used in computing loadings or factor score coefficients, but its scores and distances are computed. If omitted or set to zero, there are no case frequencies.

CWeight = variable. `CWEIGHT = VARINVSE.`

Name or number of a variable containing the case weight for unequal variances. Case weights must be zero or positive. If a case has zero CWEIGHT it is not used in computing loadings or factor score coefficients, but its scores and distances are computed. If omitted or set to zero, there are no case weights.

SCORE = list. `SCORES = 1 TO 12.`

Names or numbers of variables used for factor scores. This is used only when input is from a BMDP File and the data are **not** read as input. The SCORE command signals the program to compute factor scores and that the data is in a

BMDP File in the system file with the same CODE as the covariance or correlation matrix, factor loadings, or factor score coefficients read as input. Use names or numbers of variables to match variables from the data with variables in the correlation or covariance matrix.

/ FACTOR

Optional. Use to specify initial factor extraction.

METHod = *(one only)* `METHOD = MLFA.`
PCA, MLFA, LJIFFY, PFA.

Method used for initial factor extraction. Default: PCA.

PCA. – principal component analysis

MLFA. — maximum likelihood factor analysis

LJIFFY. – Kaiser's Second Generation Little Jiffy

PFA. – principal factor analysis

FORM = *(one only)* `FORM = COVA.`
CORR, COVA,
OCORR, OCOVA.

Matrix to be factored.

CORR. – the correlation matrix

COVA. – the covariance matrix

OCORR. – the correlation matrix about the origin

OCOVA. – the covariance matrix about the origin

You may choose to factor the correlation matrix, covariance matrix, the correlation matrix about the origin (i.e., assuming variable means are zero), or the covariance matrix about the origin. Default: CORR. Note that when a covariance matrix is being factored (COVA or OCOVA), the loadings printed may be those that relate to the raw data or those that relate to the standardized variables. State LOAD = CORR in the FACTOR paragraph to get the latter.

LOAD = CORR *or* **COVA.** `LOAD = CORR.`

State LOAD = CORR when FORM = COVA or OCOVA to get loadings for the standardized variables. Default: LOAD = COVA (loadings are for raw data).

NUMBer = #. `NUMBER = 3.`

Maximum number of factors obtained. Default: see CONSTANT below.

CONSTant = #. `CONSTANT = 0.9.`

The factors obtained are restricted to those with eigenvalues greater than CONSTANT. To be included, a factor must satisfy both NUMBER and CONSTANT criteria. For example, if you state NUMBER = 10 and CONSTANT = 0.9, but only eight eigenvalues are greater than 0.9, only eight factors are used. When the covariance matrix is factored, the limit is applied to the CONSTANT times the average variance of the variables. The number of factors for PCA, MLFA, and LJIFFY is determined from the eigenvalues of the unaltered correlation matrix. The number of factors for PFA is determined from the eigenvalues of the correlation matrix after the substitution of communality estimates at each iteration. Default: 1.0 if COMM = UNALT, 0 otherwise.

Note: A CONSTANT equal to the average communality is sometimes chosen for the PFA method, since the communalities are used instead of ones in the diagonal of the correlation matrix.

COMMunality = *(one only)* `COMM = MAXROW.`
UNALT, SMCS, MAXROW,
or **# list.** `COMM = .8, .95, .9,...`

Initial estimates of the communalities. If a Haywood case (communality > 1) occurs, 4M prints a warning and continues. Default: UNALT for METHOD = PCA; SMCS for other methods.

UNALT. – The diagonal of the correlation (or covariance) matrix remains unaltered.

SMCS. – The diagonal of the correlation (or covariance) matrix is replaced by the squared multiple correlations of each variable with the remaining variables. If the covariance matrix is factored, the squared multiple correlations are multiplied by the corresponding variances.

MAXROW. – The elements of the diagonal of the correlation (or covariance) matrix are replaced by the maximum absolute row values of the correlation matrix. If the covariance matrix is factored, these values are multiplied by the corresponding variances.

 # list. – Use one number for each variable used as the communality estimate. The first number is for the first variable in the USE list (if specified), the second number for the second variable, etc.

ITERate = #. `ITERATE = 10.`

Maximum number of iterations performed for factor extraction. Applies only to METHOD = MLFA or PFA. Default: 25.

EPS = #. `EPS = .01.`

Convergence criterion for iteration on communalities when METHOD = PFA. Iteration continues to number of times specified by ITERATE or until the estimated communalities from two successive iterations differ by EPS or less. Default: .001.

TOLerance = #. `TOL = .001.`

Tolerance limit. If the squared multiple correlation of any variable exceeds (1 – TOL), pivoting is not performed (the variable is considered a linear combination of the other variables). When METHOD is MLFA or LJIFFY, the program terminates if any variable does not pass the tolerance limit. Note that if a zero intercept model is used, then the R^2 is estimated under the assumption that all variables have zero means. Default: TOL = .0001.

/ ROTATE

Optional to specify factor rotation.

METHod = *(one only)* `METH = DQUART.`

VMAX. – varimax

DQUART. – direct quartimin rotation for simple loadings; recommended for oblique rotation

QRMAX. – quartimax

EQMAX. – equamax

ORTHOG. – orthogonal with gamma (see below)

DOBLI. – direct oblimin with gamma (see below)

ORTHOB. – orthoblique; this preassigned value for LJIFFY is only available with that factoring method

NONE. – no rotation

Default: VMAX unless METHOD = LJIFFY is specified in the FACTOR paragraph; then ORTHOB is used. To perform a simple principal components analysis, specify METHOD = PCA in the FACTOR paragraph and METHOD = NONE in the ROTATE paragraph.

GAMMA = #. `GAMMA = .75.`

Value of *G* used in simplicity criterion *G* for ORTHOG and DOBLI rotations. Default: 0.0 for DOBLI, 0.5 for NONE, 0.5 or 1.0 otherwise.

MAXIT = #. `MAXIT = 25.`

Maximum number of iterations for rotation. Default: 50.

CONSTant = #. `CONST = .0001.`

Convergence criterion for rotation. The convergence criterion is satisfied when the relative change in *G* between two successive iterations is less than CONSTANT. Default: .00001.

NORMal. `NO NORM.`

Kaiser's normalization is performed unless you specify NO NORMAL.

/ PRINT

Optional.

FIELDs = list. `FIELD = 3, 2*4, 9.1, -2.`

Controls the format of a data listing by specifying the number of character positions, decimal places, and whether a field is alphanumeric. A minus sign indicates an alphanumeric field. See Chapter 7 for a full explanation.

FSCORE = #. `FSCORE = 3.`

Number of factors for which factor scores are printed. If omitted, scores for all factors are printed. If no factor scores are desired, state FSCORE = 0.

Additional output

Select any or all of the following for additional output. The correlation matrix (CORR) and the shaded correlation matrix are printed automatically unless you specify otherwise.

STANDard. – standard scores for each case

COVAriance. – covariance matrix

CORRelation. – correlation matrix (printed unless you state NO CORR)

INVerse. – inverse of correlation or covariance matrix

FSCF. – factor score coefficients

PARTial. – partial correlations of each pair of variables are printed after removing the linear effects of all other variables; these partial correlations are the negatives of the "anti- image" correlations

FSTR. – factor structure matrix printed when oblique rotation is performed

RESI. – residual correlations, which are **R – AA′**, where **R** is the correlation matrix, **A** the initial loadings matrix, and **A′** the transpose of **A**

SHADE. – the correlation matrix is printed in shaded form after the variables are sorted according to the factor loadings; to omit state NO SHADE (see NCUT, below)

HILEV = #. `HILEV = .55.`

Value considered as a high factor loading. Variables are sorted for displays according to loadings greater than HILEV. If omitted, .5 is assumed.

LOLEV = #. `LOLEV = .20.`

Value considered as a low factor loading. Loadings less than LOLEV are replaced by zero in the Sorted Rotated Factor Loading display. If omitted, .25 is assumed.

NCUT = 5, 8, *or* **11.** `NCUT = 11.`

Controls printing of the shaded correlation matrix. If overprinting is not available, specify NCUT = 5. Specifying NCUT = 11 prints the first digit of each correlation and truncates the remaining digits (e.g., .2659 appears as 2). If you run 4M in interactive mode (see Chapter 11), NCUT = 11 is used; if you use noninteractive mode, NCUT = 8 (overprinting) is used.

LINEsize = #. `LINESIZE = 80.`

Number of columns for printing. If omitted, 132 is assumed for batch runs, 80 for interactive. Varying the number of columns may affect the output; for example, data printouts and matrices may vary depending on the LINESIZE entered.

New Feature

CRONbach. `CRON.`

Prints Cronbach's alpha.

New Feature

ITERation. `NO ITER.`

Unless you state NO ITER, 4M prints the log likelihood at each iteration.

New Feature

CASE = #. `CASE = 50.`

Specify the number of cases for which data are printed. Values flagged by MISS, MAX, and MIN in the VARIABLE paragraph specifications are replaced by MISSING, GT MAX, and LT MIN. Default: 10.

New Feature

MINimum. `MIN.`

Prints data for cases in which at least one variable has a value less than its lower limit as specified in the VARIABLE paragraph. Default: such cases are not printed unless you specify MIN.

New Feature

MAXimum. `MAX.`

Prints data for cases in which at least one variable has a value greater than its upper limit as specified in the VARIABLE paragraph. Default: such cases are not printed unless you specify MAX.

New Feature

MISSing. `MISS.`

Prints data for cases that have at least one variable with a missing value. Missing values in the input may be identified in several ways: (1) an asterisk (the default missing value character), (2) a missing value code identified using MISS in the VARIABLE paragraph, (3) a character identified as MCHAR in the INPUT paragraph, (4) blanks (applies only to input read with FORTRAN fixed format). Default: such cases are not printed unless you specify MISS.

New Feature

METHod = VARiable *or* **CASE.** `METH = CASE.`

Use to determine how the printout is organized. VARIABLE is the default value.

VARIABLE – Prints for each case only as many variables as fill the print line (up to ten variables) and begins the next line with the same variables for the next case. Thus the data for each variable are in an easy-to-scan column. When the program finishes printing the first subset of variables for all cases, it continues with another subset.

CASE – Prints the values of all variables for one case (possibly filling several print lines) before printing data for the next case.

The following commands work only in the batch mode

New Feature

MEAN. `NO MEAN.`

Unless you specify NO MEAN, 4M prints the mean, standard deviation, and frequency of each used variable.

New Feature

EZSCores. `NO EZSC.`

Unless you state NO EZSC, 4M prints *z*-scores for the minimum and maximum values of each used variable.

New Feature

ECASe. `NO ECAS.`

Unless you state NO ECAS, 4M prints the case numbers containing the minimum and maximum values of each variable.

New Feature

EXTReme. `NO EXTR.`

Unless you state NO EXTR, 4M prints the minimum and maximum values for each variable used.

New Feature

SKewness. `SK.`

Prints skewness and kurtosis for each used variable.

/ PLOT

Optional.

INITIAL = #.
FINAL = #.
FSCORE = #.

`FINAL = 5.  FSCORE = 5.`

The number of factors displayed in plots can be controlled. If omitted, 0 unrotated factors, 4 rotated factors, and scores for 3 factors are plotted.

INITIAL = #. – Number of unrotated factors for which factor loadings are plotted. For each pair of factors, the loadings of one factor are plotted against those of another.

FINAL = #. – Number of rotated factors for which factor loadings are plotted. For each pair of factors the loadings of one factor are plotted against those of the other. To omit, state FINAL = 0.

FSCORE = #. – Number of factors for which factor scores are plotted. For each pair of factors the scores of one factor are plotted against those of the other. To omit, state FSCORE = 0.

SIZE = #1, #2. `SIZE = 40, 25.`

Specify the size of the plot as the number of characters for the horizontal axis (#l) and the number of lines for the vertical axis (#2). State the width first. The default plot is 50 characters wide, 35 lines high.

/ SAVE

Required only if a BMDP File is to be created. See description of CODE, UNIT, and NEW options in Chapter 8. An additional command for 4M is CONTENT.

CONTent = *(one or more)*
DATA, COVA, UFLD,
RFLD, FCOR, FSCF.

`CONTENT = DATA, RFLD, FSCF.`

Specify the data or matrices to be saved in a BMDP File. If omitted, the data and factor scores are saved. The factor scores appear after the data in each case and are named FACTOR1, FACTOR2, etc.

DATA. – data and factor scores (both the data and the factor scores for each case are saved; the data are followed by the factor scores, which are named FACTOR1, FACTOR2, etc.)

COVA. – covariance matrix, variances, means, standard deviation, sample size, and sum of weights

UFLD. – unrotated factor loadings

RFLD. – rotated factor loadings

FCOR. – factor correlations

FSCF. – factor score coefficients

Note: Only CONTENT = DATA files may be written as an ASCII (FORMAT = F or G) file. The others may be saved as BMDP files only.

Order of Instructions

■ indicates required paragraph

```
■ / INPUT
  / VARIABLE
  / TRANSFORM
  / FACTOR
  / ROTATE
  / PRINT
  / PLOT
  / SAVE
■ / END
  data
  FACTOR ... /
  ROTATE ... /
  PLOT ... /
  PRINT ... /
  SAVE ... /
  TRANS ... /
  CONTROL ... /
  END /
  FINISH /
```

(FACTOR ... / through END /: May be repeated interactively.)

(/ INPUT through END /: Repeat for additional problems. See Multiple Problems, Chapter 10.)

Summary Table for Commands Specific to 4M

Paragraphs / Commands	Defaults	Multiple Problems	See
■ / **INPut**			
CONTent = DATA, CORR, COVA, UFLD, RFLD, FCOR, *or* FSCF.	DATA	–	4M.5
TYPE = DATA, CORR, COVA, LOAD, *or* FSCF.	DATA	–	4M.5
SHAPE = SQUARE *or* LOWER.	SQUARE	•	4M.5
FACTOR = #.	none	–	Cmds
CASE = #.	none	–	Cmds
/ **VARiable**			
FREQuency = variable.	no case frequency variable	•	4M.6
CWeight = variable.	no case weights	•	4M.6
SCORE = list.	none	–	SpecF
▲ / **FACTOR**			
METHod = *(one only)* PCA, MLFA, LJIFFY, PFA.	PCA	•	4M.2
FORM = *(one only)* CORR, COVA, OCORR, OCOVA.	CORR	•	Cmds, SpecF
LOAD = CORR *or* COVA.	COVA	•	Cmds
NUMBer = #.	# of factors determined by CONST.	–	SpecF
CONSTant = #.	1.0 if COMM = UNALT; 0 otherwise	•	4M.2
COMMunality = *(one only)* UNALT, SMCS, MAXROW, *or* # list.	UNALT for METH = PCA; SMCS otherwise	–	SpecF
ITERate = #.	25	•	4M.2
EPS = #.	.001	•	4M.2
TOLerance = #.	.0001	•	SpecF
▲ / **ROTATE**			
METHod = *(one only)* VMAX, DQUART, QRMAX, EQMAX, ORTHOG, DOBLI, ORTHOB, NONE.	VMAX (ORTHOB if FACTOR METH = LJIFFY).	•	4M.3, Cmds
GAMMA = #.	0.0 for DOBLI; 0.5 for NONE; 0.5 or 1.0 otherwise	–	Cmds
MAXIT = #.	50	•	Cmds
CONSTant = #.	.00001	•	Cmds
NORMal. *or* NO NORM.	NORM.	•	Cmds
/ **PRINT**			
FIELDs = list.	FORMAT if specified; otherwise 8.2 for numeric, -4 for label variables	•	Cmds
FSCORE = #.	scores for all factors printed	–	Cmds
STANDard.	NO STAND.	•	4M.4
COVAriance.	NO COVA.	•	Cmds
CORRelation.	CORR.	•	4M.1
INVerse.	NO INV.	•	4M.4

Summary Table (continued)

Paragraphs Commands	Defaults	Multiple Problems	See
FSCF.	NO FSCF.	•	4M.4
PARTial.	NO PART.	•	4M.4
FSTR.	NO FSTR.	•	Cmds
RESI.	NO RESI.	•	Cmds
SHADE.	SHADE.	•	4M.1
HILEV = #.	.5	•	Cmds
LOLEV = #.	.25	•	4M.3, Cmds
NCUT = 5, 8, *or* 11.	8 for noninteractive; 11 for interactive	•	Cmds
LINEsize = #.	132 for noninteractive; 80 for interactive	–	4M.1, Cmds
CRONbach.	NO CRONbach alpha.	•	Cmds
ITERation. *or* NO ITER.	ITERation.	•	Cmds
CASE = #.	10	•	Cmds
MISSing.	NO MISS.	•	Cmds
MINimum.	NO MIN.	•	Cmds
MAXimum.	NO MAX.	•	Cmds
METHod = VARiables *or* CASE.	VARIABLES	–	Cmds
MEAN. *or* NO MEAN.	MEAN.	•	Cmds
EZSCores. *or* NO EZSC.	EZSC. (z-scores)	•	Cmds
ECASe. *or* NO ECAS.	ECASe. (Case #'s of Min and Max printed)	•	Cmds
EXTReme. *or* NO EXTR.	EXTReme. values printed	•	Cmds
SKewness.	NO SK. (skewness and kurtosis)	•	Cmds
/ **PLOT**			
INITIAL = #.	0	–	Cmds
FINAL = #.	4	–	Cmds
FSCORE = #.	3	–	Cmds
SIZE = #1, #2.	50, 35	•	Cmds
/ **SAVE**			
CONTent = (*one or more*) DATA, COVA, UFLD, RFLD, FCOR, FSCF, CORR.	DATA.	–	SpecF

Key:
- ■ Required paragraph or command
- ▲ Frequently used command
- ● Value retained for multiple problems
- – Default reassigned

7M

Stepwise Discriminant Analysis

7M performs a discriminant analysis between two or more groups. The program finds the combination of variables that best predicts the category or group to which a case belongs. The group identification must be known for each case used in the analysis. The combination of predictor variables is called a classification function. This function can then be used to classify new cases whose group membership is unknown.

7M chooses the variables used in computing the linear classification functions in a stepwise manner. You can specify either forward or backward selection of variables; at each step the variable that adds the most to the separation of the groups is entered into (or the variable that adds the least is removed from) the discriminant function. You can also specify important group differences as contrasts. These contrasts guide the selection of variables. In addition, you can direct 7M interactively at the terminal, specifying which variables to enter or remove at each step.

7M evaluates the number of cases correctly classified into each group, performs a jackknife-validation procedure to reduce the bias in this evaluation, and presents a summary classification table. 7M can also classify cases which were not used in the computations. If there are only two groups, you can compute the discriminant function with 2R or 9R by coding the grouping variables with the values $n_1/(n_1 + n_2)$ and $-n_2/(n_1 + n_2)$, where n_1 and n_2 are the respective sample sizes; see also LR for two groups coded 0 and 1. The discriminant analysis in program 7M is one approach to one-way multivariate analysis of variance. For more discussion see Afifi and Azen (1979, p. 348). For a general introduction to discriminant analysis, see Tabachnick and Fidell (1983) or Afifi and Clark (1984).

Where to Find It

Example7M.1 Stepwise discriminant function analysis

We use the same data set that Fisher used in the paper in which discriminant function analysis was developed (see Appendix D). Sepal length and width and petal length and width were measured on samples of 50 irises from each of 3 species. Each data record contains the four measurements for one flower, plus a code for iris type—setosa, versicolor, or virginica. We want to determine which linear combination of these variables best allows discrimination between irises from different species and how much overlap between different species remains after these characteristics have been used for classification. The instructions in Input 7M.1 are common to all BMDP programs and are described in Chapters 3 through 5; we use no instructions specific to 7M for this run.

Input 7M.1

```
/ INPUT       FILE IS 'fisher.dat'.
              VARIABLES = 5.
              FORMAT IS FREE.

/ VARIABLE    NAMES ARE sepal_l, sepal_w, petal_l, petal_w,
                        iristype.

/ GROUP       VARIABLE = iristype.
              CODES(iristype) = 1 TO 3.
              NAMES(iristype) = setosa, versicol, virginic.
/ END
```

Output 7M.1

```
 CASE    1        2        3        4        5                       [1]
  NO. sepal_l  sepal_w  petal_l  petal_w  iristype
----- -------- -------- -------- -------- --------
    1   50.00    33.00    14.00     2.00 setosa
    2   64.00    28.00    56.00    22.00 virginic
    3   65.00    28.00    46.00    15.00 versicol
    4   67.00    31.00    56.00    24.00 virginic
    5   63.00    28.00    51.00    15.00 virginic
    6   46.00    34.00    14.00     3.00 setosa
    7   69.00    31.00    51.00    23.00 virginic
    8   62.00    22.00    45.00    15.00 versicol
    9   59.00    32.00    48.00    18.00 versicol
   10   46.00    36.00    10.00     2.00 setosa

NUMBER OF CASES READ. . . . . . . . . . . . . .       150        [2]

   VARIABLE         STATED VALUES FOR          GROUP CATEGORY    INTERVALS          [3]
   NO.   NAME    MINIMUM MAXIMUM MISSING   CODE  INDEX   NAME     .GT.   .LE.
   ---- --------  ------- ------- -------  ------ ----- --------  ------ ------

   5  iristype                              1.000    1  setosa
                                            2.000    2  versicol
                                            3.000    3  virginic
--------------------------------------------------------------------------------


GROUPING VARIABLE. . . iristype
                             CATEGORY    FREQUENCY
                             --------    ---------
                             setosa         50
                             versicol       50
                             virginic       50


DESCRIPTIVE STATISTICS OF DATA              [4]
----------- ---------- -- ----

 VARIABLE     TOTAL          STANDARD  ST.ERR    COEFF      S M A L L E S T      L A R G E S T
 NO. NAME      FREQ.  MEAN     DEV.   OF MEAN   OF VAR  VALUE  Z-SCR  CASE  VALUE  Z-SCR  CASE  RANGE

   1 sepal_l    150  58.433  8.2807  .67611   .14171  43.000 -1.86     59  79.000  2.48    39  36.000
   2 sepal_w    150  30.573  4.3587  .35588   .14256  20.000 -2.43     98  44.000  3.08    72  24.000
   3 petal_l    150  37.580  17.653  1.4414   .46974  10.000 -1.56     10  69.000  1.78    90  59.000
   4 petal_w    150  11.993  7.6224  .62236   .63555  1.0000 -1.44     37  25.000  1.71    23  24.000


TOLERANCE. . . . . . . . .        0.010               [5]
F-TO-ENTER . . . . . . . .        4.000      4.000
F-TO-REMOVE. . . . . . . .        3.996      3.996
METHOD . . . . . . . . . .            1
MAXIMUM FORCED LEVEL . . .            0
MAXIMUM NUMBER OF STEPS. .           10
GROUPING VARIABLE. . . . .            5
NUMBER OF GROUPS . . . . .            3
PRIOR PROBABILITIES. . . .      0.33333    0.33333    0.33333


    MEANS

       GROUP =    setosa        versicol        virginic        ALL GPS.       [6]
VARIABLE
  1 sepal_l      50.06000       59.36000        65.88000        58.43333
  2 sepal_w      34.28000       27.70000        29.74000        30.57333
  3 petal_l      14.62000       42.60000        55.52000        37.58000
  4 petal_w       2.46000       13.26000        20.26000        11.99333

COUNTS              50.           50.             50.            150.
```

—we omit panels showing standard deviations and coefficients of variation—

```
******************************************************************************

STEP NUMBER    0                                                          [7]

  VARIABLE        F TO    FORCE TOLERNCE *   VARIABLE       F TO    FORCE TOLERNCE
                 REMOVE   LEVEL          *                 ENTER    LEVEL
        DF =   2  148                    *          DF =   2  147
                                         *   1 sepal_l    119.26      1    1.00000
                                         *   2 sepal_w     49.16      1    1.00000
                                         *   3 petal_l   1180.16      1    1.00000
                                         *   4 petal_w    960.01      1    1.00000

******************************************************************************

STEP NUMBER    1                                                          [8]
VARIABLE ENTERED    3 petal_l

  VARIABLE        F TO    FORCE TOLERNCE *   VARIABLE       F TO    FORCE TOLERNCE
                 REMOVE   LEVEL          *                 ENTER    LEVEL
        DF =   2  147                    *          DF =   2  146
  3 petal_l    1180.16      1    1.00000 *   1 sepal_l     34.32      1    0.42822
                                         *   2 sepal_w     43.04      1    0.85718
                                         *   4 petal_w     24.77      1    0.76530


U-STATISTIC(WILKS' LAMBDA) 0.0586283    DEGREES OF FREEDOM    1    2    147
APPROXIMATE F-STATISTIC      1180.161   DEGREES OF FREEDOM      2.00    147.00

 F - MATRIX          DEGREES OF FREEDOM =      1   147

            setosa    versicol
versicol  1056.87
virginic  2258.26     225.35
```

```
CLASSIFICATION FUNCTIONS

           GROUP =    setosa         versicol        virginic
VARIABLE
  3 petal_1          0.78947         2.30037         2.99804

CONSTANT            -6.86963       -50.09645       -84.32415

************************************************************
```

—we omit for steps 2 and 3—

```
************************************************************

STEP NUMBER    4
VARIABLE ENTERED    1 sepal_1

  VARIABLE      F TO    FORCE TOLERNCE *   VARIABLE      F TO    FORCE TOLERNCE
              REMOVE   LEVEL           *               ENTER    LEVEL
        DF =  2  144                   *          DF =  2  143
  1 sepal_1     4.72     1    0.34799 *
  2 sepal_w    21.94     1    0.60886 *
  3 petal_1    35.59     1    0.36513 *
  4 petal_w    24.90     1    0.64931 *

U-STATISTIC(WILKS' LAMBDA) 0.0234386   DEGREES OF FREEDOM    4   2    147
APPROXIMATE F-STATISTIC      199.145   DEGREES OF FREEDOM    8.00    288.00

 F - MATRIX          DEGREES OF FREEDOM =    4  144

             setosa    versicol
versicol    550.19
virginic   1098.27     105.31

CLASSIFICATION FUNCTIONS

           GROUP =    setosa         versicol        virginic
VARIABLE
  1 sepal_1          2.35442         1.56982         1.24458
  2 sepal_w          2.35879         0.70725         0.36853
  3 petal_1         -1.64306         0.52115         1.27665
  4 petal_w         -1.73984         0.64342         2.10791

CONSTANT           -86.30846       -72.85261      -104.36832

CLASSIFICATION MATRIX                                                        [9]

GROUP      PERCENT    NUMBER OF CASES CLASSIFIED INTO GROUP -
           CORRECT
                      setosa    versicol  virginic
 setosa    100.0      50          0          0
 versicol   96.0       0         48          2
 virginic   98.0       0          1         49

TOTAL       98.0      50         49         51

JACKKNIFED CLASSIFICATION

GROUP      PERCENT    NUMBER OF CASES CLASSIFIED INTO GROUP -
           CORRECT
                      setosa    versicol  virginic
 setosa    100.0      50          0          0
 versicol   96.0       0         48          2
 virginic   98.0       0          1         49

TOTAL       98.0      50         49         51

  SUMMARY TABLE                                                              [10]

         VARIABLE     F VALUE TO   NO. OF                                DEGREES
STEP     ENTERED      ENTER        VARIAB.                APPROXIMATE      OF
NO.        REMOVED      REMOVE     INCLUDED  U-STATISTIC  F-STATISTIC    FREEDOM
---  -------------  -------------  --------  -----------  -----------  -----------
  1   3 petal_1        1180.161        1        0.0586      1180.161    2.0  147.0
  2   2 sepal_w          43.035        2        0.0369       307.105    4.0  292.0
  3   4 petal_w          34.569        3        0.0250       257.503    6.0  290.0
  4   1 sepal_1           4.721        4        0.0234       199.145    8.0  288.0


            INCORRECT                  MAHALANOBIS D-SQUARE FROM AND          [11]
         CLASSIFICATIONS               POSTERIOR PROBABILITY FOR GROUP -

GROUP  setosa               setosa          versicol          virginic

  CASE
   1                      0.2 1.000       90.7 0.000       181.6 0.000
   6                      1.3 1.000       84.0 0.000       170.1 0.000
  10                      2.3 1.000      113.7 0.000       210.0 0.000
  18                      2.8 1.000       67.5 0.000       145.7 0.000
  26                      4.0 1.000      113.2 0.000       210.2 0.000
```

—we omit similar statistics for the remaining SETOSA cases—

```
GROUP  versicol             setosa          versicol          virginic

  CASE
   3                    105.3 0.000        2.2 0.996        13.1 0.004
   8                    131.7 0.000        8.4 0.960        14.8 0.040
   9           virginic 130.9 0.000        8.7 0.253         6.5 0.747
  11                     99.2 0.000        1.3 0.998        13.8 0.002
  12           virginic 149.0 0.000        8.4 0.143         4.9 0.857
```

—we omit similar statistics for the remaining VERISCOL cases—

```
GROUP  virginic              setosa          versicol         virginic

  CASE
   2                     208.6 0.000      27.3 0.000       1.9 1.000
   4                     207.9 0.000      31.7 0.000       4.5 1.000
   5           versicol  133.1 0.000       5.3 0.729       7.2 0.271
   7                     173.2 0.000      26.6 0.000      11.0 1.000
  13                     159.0 0.000      12.8 0.003       1.2 0.997
```

—we omit plot of group means—

[12]

```
EIGENVALUES

               32.19193      0.28539

CUMULATIVE PROPORTION OF TOTAL DISPERSION

                0.99121      1.00000
```

[13]

```
                   MULTIVARIATE TESTS
                                                          F APPROXIMATION
    STATISTICS                    VALUE              F        D.F.   D.F.  P-VALUE

WILKS' LAMBDA                    0.02344       199.14536        8     288   0.0000
   D.F.      4      2    147

PILLAI'S TRACE                   1.19190        53.46649        8     290   0.0000

HOTELLING-LAWLEY TRACE          32.47732       584.59180        8     144   0.0000

ROY'S MAXIMUM ROOT              32.19193

    TABLES ARE AVAILABLE FOR ROY'S TEST:  MORRISON (1976)


CANONICAL CORRELATIONS

               0.98482       0.47120

AVERAGE SQUARED CANONICAL CORRELATION              0.59595

VARIABLE       COEFFICIENTS FOR CANONICAL VARIABLES

  1 sepal_l      0.08294      0.00241
  2 sepal_w      0.15345      0.21645
  3 petal_l     -0.22012     -0.09319
  4 petal_w     -0.28105      0.28392

               STANDARDIZED  (BY POOLED WITHIN-GROUP VARIANCES)
VARIABLE       COEFFICIENTS FOR CANONICAL VARIABLES

  1 sepal_l      0.42695      0.01241
  2 sepal_w      0.52124      0.73526
  3 petal_l     -0.94726     -0.40104
  4 petal_w     -0.57516      0.58104

CONSTANT         2.10511     -6.66147

GROUP         CANONICAL VARIABLES EVALUATED AT GROUP MEANS

   setosa        7.60760      0.21513
   versicol     -1.82505     -0.72790
   virginic     -5.78255      0.51277
```

—we omit plots of group means— [14]

```
GROUP setosa

   CASE    X      Y        CASE    X      Y        CASE    X      Y

     1   7.67  -0.13        64   7.13  -0.79        126   7.78   0.58
     6   7.21   0.36        65   8.06   0.30        135   7.22  -0.11
    10   8.68   0.88        68   6.82   0.46        136   7.33  -1.07
    18   6.25   0.44        69   7.19  -0.36        137   5.66  -1.93
    26   8.61   0.40        72   9.16   2.74        139   8.08   0.97
    31   6.76  -0.76        73   8.13   0.51        140   8.02   1.14
    36   7.99   0.09        79   7.37   0.57        144   7.50  -0.19
    37   8.33   0.23        80   9.13   1.22        145   7.59   1.21
    40   7.24  -0.27        88   9.47   1.83        146   7.92   0.21
    42   6.41   1.25        89   7.06  -0.66        150   8.31   0.64
```

—we omit the coordinates for the remaining SETOSA cases and all of the VERISCOL and VIRGINIC cases—

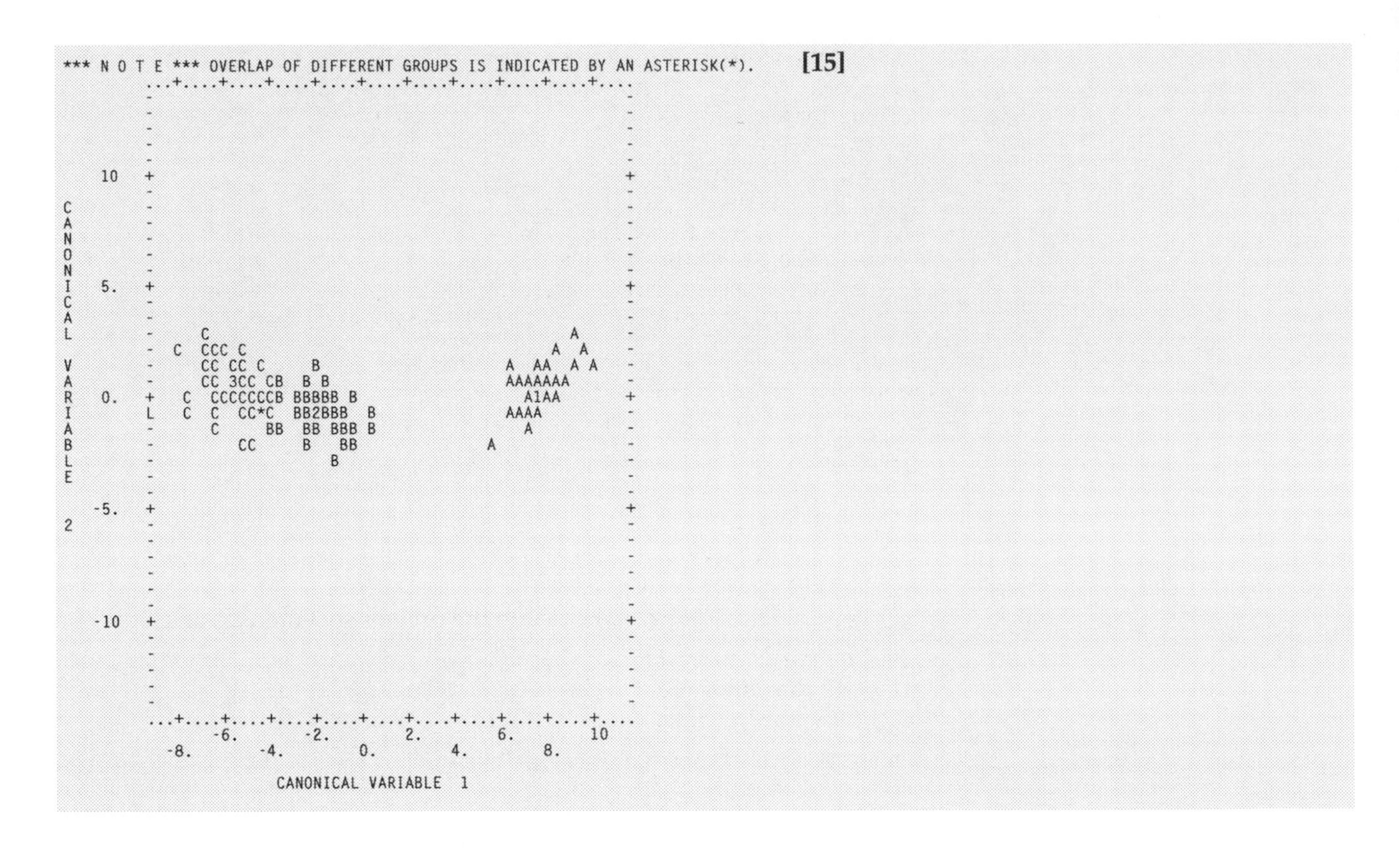

[1] Prints the first ten cases of data. Group names are printed if specified.

[2] Number of cases read. 7M uses only complete cases in all computations. That is, if the value of any variable in a case is missing or out of range, 7M omits the case from all computations. If you specify a USE list in the VARIABLE paragraph, only the variables in the USE list are considered in determining the completeness of each case. When you specify CODES for the GROUPING variables, 7M uses only cases with one of the specified codes.

[3] The grouping variable codes and names as interpreted from instructions.

[4] This table gives all descriptive statistics divided by group. This table includes:

- Mean
- Standard Deviation
- Standard Error of the Mean
- Coefficient of Variation
- Smallest value with case number and *z*-score for that case
- Largest value with case number and *z*-score for that case

[5] Interpretation of instructions specific to this program. The grouping variable is IRISTYPE, the fifth variable.

[6] Additional descriptive statistics for each variable in each group and for all groups combined (we show means only). The mean values for each group are especially useful in interpreting the differences among the groups. The pooled (all groups) standard deviation s_p is the square root of the variance

$$s_p^2 = \frac{\sum (N_k - 1)s_k^2}{\sum (N_k - 1)}$$

where s_k^2 is the variance of group k and N_k is the sample size of group k.

[7] Step 0 is the step before any variable is entered into the discriminant function. 7M prints the *F*-to-enter for each variable. At step 0 the *F*-to-enter for a variable corresponds to the *F* statistic computed from a one-way analysis of variance (ANOVA) of that variable for the groups used in the analysis. The degrees of freedom correspond to those of the one-way ANOVA; i.e., $(g - 1)$ and $\sum (N_k - 1)$, where g is the number of groups. When multiple variables are being considered in a stepwise manner, it is not clear how to attach *p*-values to the *F*s. Within a step, *F*s may be used to assess relative importance among the candidate variables. See Special Features for a discussion of LEVEL, and Commands for a discussion of tolerance.

[8] The variable with the highest *F*-to-enter at step 0 (PETAL_L—petal length) enters into the discriminant functions; this is the variable that discriminates best between groups. At step 1, 7M computes the following:

- *F-to-remove for the variable in the equation.* This is equal to the *F*-to-enter at step 0.
- *F-to-enter for each variable not in the equation.* This is equal to the *F* statistic corresponding to the one-way ANOVA on the residuals of the variable; i.e., at each step 7M computes the *F*-to-enter from a one-way analysis of covariance where the covariates are the previously entered variables.
- *Wilks' lambda or U statistic.* This is a multivariate analysis of variance statistic that tests the equality of group means for the variable(s) in the discriminant function.
- *Approximate F statistic.* This is a transformation of Wilks' lambda that can be compared with the *F* distribution; at step 1 the *F* statistic is the same as that in the one-way ANOVA between the group means for the variable entered.
- *F matrix.* This contains *F* values computed from the Mahalanobis D^2 statistics that test the equality of group means for each pair of groups. The test is based only on the variables currently in the discriminant functions; these values are proportional to distance measures.
- *Classification functions.* For each group, the corresponding function is the linear combination of variables that best discriminates that group from the rest of the cases. For example, after step 4 the classification function for the setosa group is

```
y = 2.35442*SEPAL_L+2.35879*SEPAL_W -1.64306*PETAL_L
    -1.73984*PETAL_W - 86.30816
```

7M evaluates this function and the functions for versicolor and virginica for each case. Then it assigns the case to the group for which the classification function has the largest value. These functions can also be used to classify new cases as shown in Example 7M.2.

[9] At the last step, in addition to the results printed at each step, 7M prints the

- *classification matrix.* 7M classifies each case into the group with the highest posterior probability and prints the number of cases classified into each group and the percent of correct classifications. In this example classification is exceptionally good; only three cases are misclassified.
- *jackknifed classification matrix.* The classification matrix above may provide an overly optimistic estimate of the probability of misclassification. Therefore 7M also classifies each case into the group with the highest posterior probability according to classification functions computed from all the data except the case being classified. This is a special case of the general cross-validation method in which the classification functions are computed on a subset of cases, and the probability of misclassification is estimated from the remaining cases. When each case is left out in turn, the method is known as the jackknife. See Lachenbruch and Mickey (1968).

[10] A summary table. A one-line summary of each step, including the *F*-to-enter (or remove) for the variable entered (or removed), the Wilks' lambda *U* statistic, and the approximate *F* statistic.

[11] The classification of each case. For each case, 7M computes the Mahalanobis distance to each group mean and the posterior probabilities of belonging to each group. The posterior probability for the distance of a case from a group is the ratio of $\exp(D^2)$ for the group over the sum of $\exp(D^2)$ for all groups. Prior probabilities, if assigned, affect these computations (see Appendix B.15). Unusual cases are those with large Mahalanobis distances, that is, cases distant from their group means. Large distances may be due to data errors, incorrect group specification, or chance. For large samples from a multivariate normal distribution, the distance from a case to its group mean is approximately distributed as a chi-square with degrees of freedom equal to the number of variables selected.

7M notes each incorrectly classified case in the output (cases 5, 9, and 12). You can save a variable indicating predicted group membership in a BMDP File (see Special Features).

[12] Eigenvalues of the matrix $\mathbf{W}^{-1/2}\ \mathbf{B}\mathbf{W}^{-1/2}$ are computed, where $\mathbf{B}$ is the between-groups sums of cross products and $\mathbf{W}$ is the pooled (within-groups) sum of squares (see Appendix B.15). 7M also prints the canonical correlations between the variables entered and the dummy variables representing the groups, and the coefficients for the canonical variables. The first canonical variable is the linear combination of variables that best discriminates among the groups; the second canonical variable is the next best linear combination orthogonal to the first one, etc. 7M adjusts the canonical variables so that the (pooled) within-group variances are 1 and their overall mean is zero. 7M evaluates the canonical variables at the group means. In this example, the first canonical variable accounts for most of the discrimination between groups (.99121 of the total dispersion). This canonical variable is closely related to the difference between sepal size and petal size.

[13] Multivariate tests are performed on the model. This table includes:

- Wilks' Lambda
- Pillai's Trace statistic
- Hotelling-Lawley Trace statistic
- Roy's Maximum Root statistic

[14] 7M plots the group means in a scatterplot. The axes are the first two canonical variables. (We do not show this plot.)

[15] 7M plots the group means and all cases in a scatterplot (here shown slightly edited from its full size). The axes are the first two canonical variables. If the data followed a multivariate normal distribution with equal variances and covariances among groups, then each scatter of points would be ellipsoidal with approximately the same size and direction of tilt. The dissimilar ellipses in this figure suggest a departure from assumptions, but the equations developed still divide the groups quite sharply. Group A (SETOSA) is well differentiated from the other two groups; group B (VERSICOL) slightly overlaps group C (VIRGINIC). (The program assigns the letters A, B, C because the group names do not have unique first letters.) Numbers 1, 2, 3 denote group means. The "L" printed on the left border of the plot frame indicates a point off the scale of the *x*-axis. Note that here most of the discrimination is provided by the first canonical variable. If there is only one canonical variable, 7M prints a histogram. The presence of an outlier can severely affect a discriminant analysis. Such an outlier can sometimes be identified in this plot (the case number can be obtained from the coordinate listing above the plot).

Running 7M interactively

If you run 7M interactively, you can specify the variables to be entered or removed from the discriminant equation at each step. For example, suppose that the instructions for Example 7M.1 are stored in a file called 7M.INP (you may also enter the instructions directly, using the BMDP line editor). Enter 7M interactively (see Chapter 11). 7M will then print a banner and place you in the BMDP Line Editor at line 1. You may then type

```
1 COM> GET 7M.INP <Return>
```

to "Get" your input file (user responses are in bold type). See Chapter 11 for a list of line editor commands. After the 11 lines of the file appear on the screen, you then type E to Execute the instructions:

```
12 COM> E <Return>
```

The program output shown in Output 7M.1 will appear on your screen in sections. After printing Step 0, 7M will ask you which variable you want to move in or out of the equation at each step; you can follow the default order shown in Output 7M.1 or specify the variables you prefer in any order. At Step 1 you could enter PETAL_W into the equation instead of PETAL_L. For example, at step 0, 7M asks the following question:

```
ENTER VARIABLE TO MOVE NEXT:
Press <RETURN> for: petal_l--->petal_w <Return>
```

Instead of the suggested variable, we chose to move PETAL_W into the equation. All of the usual information for step 1 is now printed. Then, at the next step, 7M suggests SEPAL_W.

```
ENTER VARIABLE TO MOVE NEXT:
Press <RETURN> for: sepal w---><Return>
```

This time we follow the program's suggestion by pressing the <Return> key. You can choose to move any variable either into or out of the equation. If, at this step, we again specified PETAL_W, that variable would be removed from the equation.

Example 7M.2 Cross-validating with random subsamples and classifying new cases

7M can calculate the classification functions based on a subset of the data and then use these functions to classify the remaining cases. This feature allows you to cross-validate your results without having to collect new data. You randomly subdivide the cases in each group into two subgroups, use the first subgroup to estimate the classification function, and then use the function to classify the new cases in the second subgroup. By observing the proportion of correct classifications for the second subgroup, you have an empirical measure of the success of the discrimination (Lachenbruch and Mickey, 1968).

In Input 7M.2 we randomly select about 80% of the cases from each of the original groups to compute the classification functions, and we use the other 20% to estimate the percent of cases misclassified. We use a TRANSFORM paragraph to add the constant 3 to the IRISTYPE code for approximately 20% of the cases, forming new codes 4, 5, and 6. That is, for each case a uniform random number is generated and, when it is less than 0.2, IRISTYPE is set to IRISTYPE + 3 (group 1 becomes group 4, etc.). The integer 7832 is a seed number for the random number generator (seeds should be large integers between 1 and 30000). The original group names are SETOSA, VERSICOL, and VIRGINIC; we name the smaller subgroups NEWSET, NEWVERS, NEWVIRG.

Only those groups identified in the GROUP paragraph USE list are used to calculate the classification functions; however, cases in other groups are still classified. Here we specify that only groups 1, 2, 3 (the larger subsets) are to be used in estimating the function. The cases in groups 4, 5, and 6 are not included in the GROUP USE list, but are classified into groups 1, 2, or 3. Note: 7M is the only program that allows a GROUP USE list.

Input 7M.2

```
 / INPUT       FILE IS 'fisher.dat'.
               VARIABLES = 5.
               FORMAT IS FREE.

 / VARIABLE    NAMES ARE sepal_l, sepal_w, petal_l,
                         petal_w, iristype.

 / TRANSFORM   IF (RNDU(7832) LE .2)
                  THEN iristype = iristype + 3.

 / GROUP       VARIABLE = iristype.
               CODES(iristype) = 1 TO 3.
               NAMES(iristype) = setosa, versicol, virginic,
                                 newset, newvers, newvirg.
               USE = setosa TO virginic.

/ END
```

Output 7M.2

—we omit the tables and information shown in Output 7M.1—

```
NUMBER OF CASES READ. . . . . . . . . . . . .      150


  VARIABLE        STATED VALUES FOR         GROUP CATEGORY    INTERVALS
  NO.   NAME    MINIMUM MAXIMUM MISSING   CODE  INDEX   NAME     .GT.   .LE.
  ---- --------  ------- ------- -------  ------ -----  --------  ------ -------

   5  iristype                            1.000    1  setosa
                                          2.000    2  versicol
                                          3.000    3  virginic
                                          4.000    4  newset
                                          5.000    5  newvers
                                          6.000    6  newvirg
------------------------------------------------------------------------------

GROUPING VARIABLE. . . iristype
                                CATEGORY    FREQUENCY
                                --------    ---------
                                setosa         41
                                versicol       41
                                virginic       41
                                newset          9
                                newvers         9
                                newvirg         9

    MEANS

       GROUP =    setosa        versicol      virginic      newset        newvers
VARIABLE
  1 sepal_l      50.07317      59.02439      65.95122      50.00000      60.88889
  2 sepal_w      34.26829      28.04878      30.02439      34.33333      26.11111
  3 petal_l      14.65854      42.43903      55.73171      14.44444      43.33333
  4 petal_w       2.43902      13.19512      20.36585       2.55556      13.55556

COUNTS              41.           41.           41.            9.            9.

       GROUP =    newvirg       GPS.USED
VARIABLE
  1 sepal_l      65.55556      58.34959
  2 sepal_w      28.44444      30.78049
  3 petal_l      54.55556      37.60976
  4 petal_w      19.77778      12.00000
```

[1]

—we omit standard deviations and coefficients of variation for the six groups, and steps 0, 1, 2—

```
******************************************************************************

STEP NUMBER   3
VARIABLE ENTERED    4 petal_w

  VARIABLE      F TO    FORCE TOLERNCE *   VARIABLE      F TO   FORCE TOLERNCE
              REMOVE   LEVEL          *                ENTER   LEVEL
        DF =  2  118                   *          DF =  2  117
  2 sepal_w     46.71     1   0.72129 *   1 sepal_l      3.03     1   0.35225
  3 petal_l     35.53     1   0.72461 *
  4 petal_w     27.89     1   0.66884 *

U-STATISTIC(WILKS' LAMBDA) 0.0246684   DEGREES OF FREEDOM   3   2    120
APPROXIMATE F-STATISTIC      211.099   DEGREES OF FREEDOM     6.00   236.00

 F - MATRIX            DEGREES OF FREEDOM =    3  118

              setosa    versicol virginic newset     newvers
versicol    569.44
virginic   1161.50     116.19
newset        0.03     205.87    418.86
newvers     237.60       1.64     33.02    145.44
newvirg     415.16      39.21      0.53    253.60      17.95
```

Output 7M.2

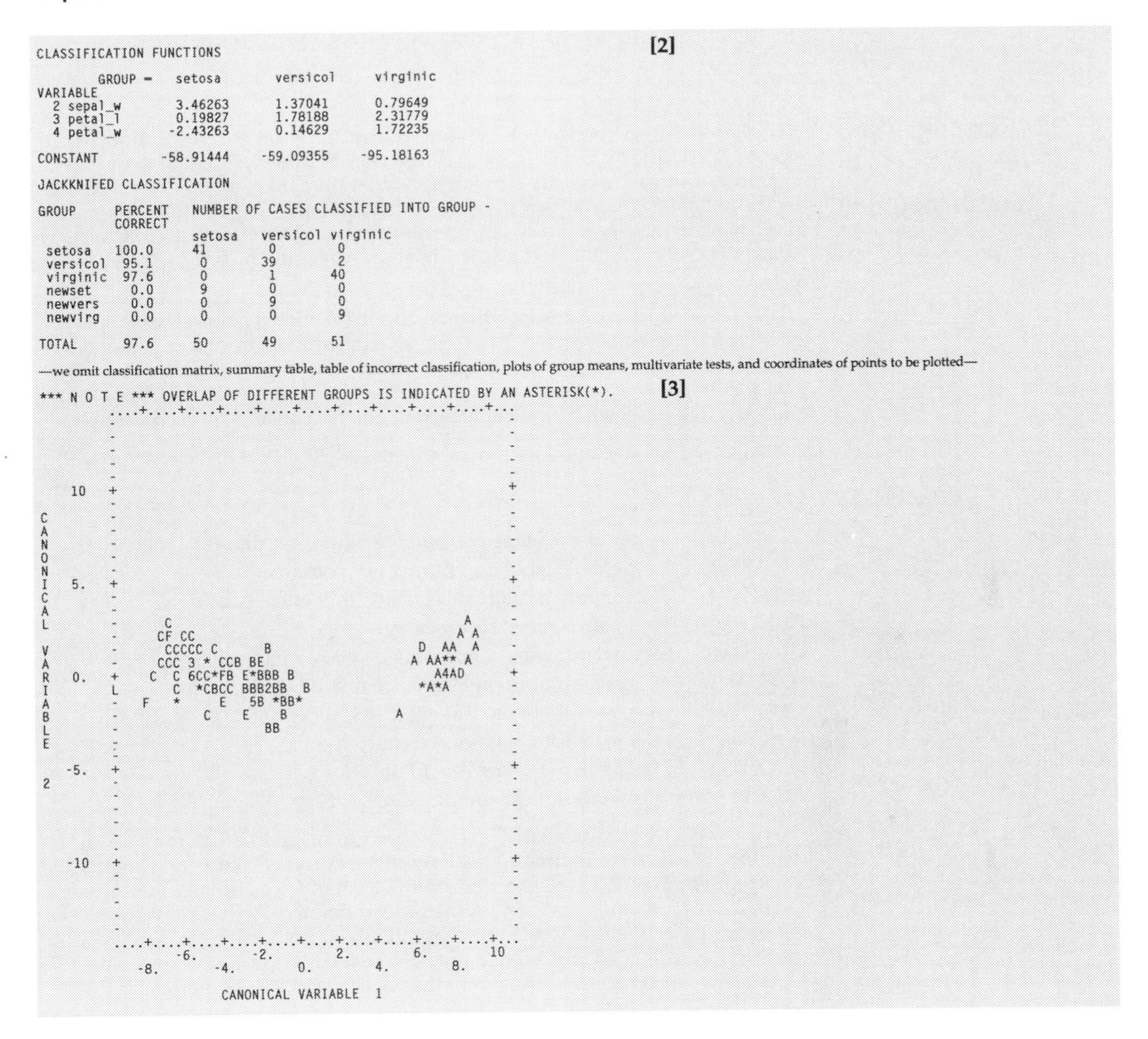

```
CLASSIFICATION FUNCTIONS                                          [2]

        GROUP =    setosa        versicol      virginic
VARIABLE
  2 sepal_w        3.46263        1.37041       0.79649
  3 petal_l        0.19827        1.78188       2.31779
  4 petal_w       -2.43263        0.14629       1.72235

CONSTANT         -58.91444      -59.09355     -95.18163

JACKKNIFED CLASSIFICATION

GROUP      PERCENT   NUMBER OF CASES CLASSIFIED INTO GROUP -
           CORRECT
                     setosa   versicol virginic
 setosa    100.0     41         0        0
 versicol   95.1      0        39        2
 virginic   97.6      0         1       40
 newset      0.0      9         0        0
 newvers     0.0      0         9        0
 newvirg     0.0      0         0        9

TOTAL       97.6     50        49       51
```

—we omit classification matrix, summary table, table of incorrect classification, plots of group means, multivariate tests, and coordinates of points to be plotted—

```
*** N O T E *** OVERLAP OF DIFFERENT GROUPS IS INDICATED BY AN ASTERISK(*).     [3]
          ....+....+....+....+....+....+....+....+....+....+...
          -                                                    -
          -                                                    -
          -                                                    -
          -                                                    -
   10     +                                                    +
C         -                                                    -
A         -                                                    -
N         -                                                    -
O         -                                                    -
N         -                                                    -
I   5.    +                                                    +
C         -                                                    -
A         -                                                    -
L         -      C                                       A     -
          -     CF CC                                  A A     -
V         -      CCCCC C      B                   D  AA  A     -
A         -     CCC 3 * CCB BE                   A AA** A      -
R   0.    +    C  C 6CC*FB E*BBB B                  A4AD       +
I         L       C  *CBCC BBB2BB  B              *A*A         -
A         -   F   *     E   5B *BB*                            -
B         -          C    E    B              A                -
L         -                   BB                               -
E         -                                                    -
          -                                                    -
  -5.     +                                                    +
2         -                                                    -
          -                                                    -
          -                                                    -
          -                                                    -
  -10     +                                                    +
          -                                                    -
          -                                                    -
          -                                                    -
          -                                                    -
          ....+....+....+....+....+....+....+....+....+....+...
                   -6.       -2.        2.         6.        10
              -8.       -4.        0.         4.        8.

                        CANONICAL VARIABLE  1
```

[1] In Output 7M.2, our randomly selected subsamples (NEWSET, NEWVERS, and NEWVIRG) contain approximately 20% of the cases from the original groups. Note that the last column of means is based, not on all groups, but on the groups used in the classification functions (SETOSA, VERSICOL, and VIRGINIC).

[2] The classification functions and jackknifed classification matrix. Note that the computed classification functions differ from those computed from all cases. Only three groups (USE = 1 TO 3 in the GROUP paragraph) are used to estimate the classification functions, but cases in all six groups are classified according to the functions. 7M classifies all the cases in NEWSET correctly into SETOSA; all cases in NEWVIRG and NEWVERS are also classified correctly. In data where the separation is less complete, a different random selection of cases might yield considerably different results.

[3] The canonical variable plot (here shown slightly edited from its full size) now contains six symbols. Group D (NEWSET) overlaps group A (SETOSA);

group E (NEWVERS) overlaps group B (VERSICOL); and group F (NEWVIRG) overlaps group C (VIRGINIC). There are now six numbers denoting group means. 7M prints an asterisk where groups overlap.

Example 7M.3 Using contrasts to direct stepping

7M allows you to specify one or more contrasts to direct the stepping when you have specific hypotheses about differences between particular groups. For example, if three groups are structured so that they may be considered as low, medium, and high, you can perform a discriminant analysis that maximizes the distance between the two extreme groups. (The effect of contrasts on computations is described in greater detail in Appendix B.15.)

In Example 7M.3 we use 7M to examine a data set stored in a file named ULCER.DAT which contains information about 157 ulcer patients (see Appendix D). The 157 patients entered a two-phase clinical trial. In phase 1, they were treated with one of two active drugs. Patients whose ulcer healed entered into phase 2. In phase 2, patients received no treatment but were followed to determine the time at which their ulcer recurred. The file contains thirteen variables:

TREATMNT	– one of two active treatments used in study's healing phase
AGE	– age of patient
SEX	– sex of patient
FAM_HIST	– number of family members with ulcer disease
CIGARETS	– cigarette smoking (number of packs per day)
NITEPAIN	– presence of pain at night at study entry (yes/no)
PAINFREQ	– a pain score at study entry
OUTCOME	– final outcome at end of two-phase trial
SQRTTIME	– square root of months since first ulcer diagnosis
LOG_TIME	– logarithm of months since first ulcer diagnosis
B_ACID	– square root of basal acid output
PEAKACID	– square root of peak acid output
TROUBLE	– previous ulcer complications (yes/no)

We perform a discriminant analysis between groups defined by the OUTCOME variable, which records time to healing and recurrence status at various followup times. This variable is coded from 1 to 9, but we are only interested in codes 2, 3, 4, 5, 8, and 9. We NAME the groups defined by these codes to make three groups: NO_ HEAL (patients whose ulcer did not heal in phase 1), RECUR (patients who healed but whose ulcer recurred within 6 months), and OK (patients with no recurrence in 6 months). See Chapter 5 for more information about the GROUP paragraph.

In this example we want to choose variables that will maximize the discrimination between the first (NO_HEAL) and third (OK) groups, so we state

```
CONTRAST = -1, 0, 1.
```

in the DISCRIMINANT paragraph. Note that these coefficients should sum to zero. (In a previous run with no contrasts specified, only the variables SQRTTIME and B ACID entered the discriminant function, and 12 subjects were misclassified from NO_HEAL into OK or vice versa.) We also state NO STEP in the PRINT paragraph so that only the first and last steps will be printed. We state CONTRAST in the PLOT paragraph to obtain a canonical variable plot using the contrasts specified in the DISCRIMINANT paragraph.

Input 7M.3

```
/ INPUT        VARIABLES = 13.
               FORMAT IS FREE.
               FILE IS 'ulcer.dat'.

/ VARIABLE     NAMES = treatmnt, age, sex, fam_hist,
                       cigarets, nitepain, outcome,
                       sqrttime, b_acid, peakacid, trouble,
                       log_time, painfreq.
               MISS = (age)99, 9, 9, 99, 9.

/ GROUP        VARIABLE = outcome.
               NAME(outcome) = no_heal, recur, recur,
                               ok, ok, ok.
               CODE(outcome) = 2 , 3, 4, 5, ,8 ,9.

/ DISCRIM      CONTRAST = -1, 0, 1.

/ PRINT        NO STEP.

/ PLOT         CONTRAST.

/ END
```

Output 7M.3

—we omit standard 7M output shown in Output 7M.1—

```
 VARIABLE          STATED VALUES FOR               GROUP  CATEGORY    INTERVALS
 NO.   NAME    MINIMUM MAXIMUM MISSING   CODE  INDEX    NAME      .GT.   .LE.
 ---- -------- ------- ------- -------  ------ -----  --------   ------ -------

  2  age                        99.00
  3  sex                        9.000
  4  fam_hist                   9.000
  5  cigarets                   99.00
  6  nitepain                   9.000
  7  outcome                             2.000    1   no_heal
                                         3.000    2   recur
                                         4.000    2   recur
                                         5.000    3   ok
                                         8.000    3   ok
                                         9.000    3   ok
-------------------------------------------------------------------------------

  MEANS

     GROUP =    no_heal        recur          ok          ALL GPS.
VARIABLE
  1 treatmnt    1.36842      1.53333      1.52632      1.50000
  2 age        49.36842     46.91111     44.13158     46.33333
  3 sex         1.05263      1.13333      1.28947      1.17647
  4 fam_hist    0.36842      0.33333      0.18421      0.28431
  5 cigarets    0.84211      0.68889      0.55263      0.66667
  6 nitepain    0.73684      0.57778      0.52632      0.58824
  8 sqrttime   11.03179      8.77391      5.11055      7.82972
  9 b_acid      2.47637      2.14382      1.69663      2.03917
 10 peakacid    6.07600      6.06720      5.29976      5.78293
 11 trouble     0.47368      0.73333      0.76316      0.69608
 12 log_time    1.76547      1.66373      1.10918      1.47609
 13 painfreq    5.94737      4.46667      3.76316      4.48039

COUNTS             19.          45.          38.         102.
```

—we omit tables of information and further descriptive statistics—

```
****************************************************************************
STEP NUMBER    0

  VARIABLE      F TO   FORCE TOLERNCE *   VARIABLE      F TO   FORCE TOLERNCE
               REMOVE  LEVEL          *                ENTER   LEVEL
        DF =   1  100                 *          DF =  1   99
                                      *   1 treatmnt     1.25     1    1.00000
                                      *   2 age          1.49     1    1.00000
                                      *   3 sex          5.04     1    1.00000
                                      *   4 fam_hist     1.25     1    1.00000
                                      *   5 cigarets     2.53     1    1.00000
                                      *   6 nitepain     2.30     1    1.00000
                                      *   8 sqrttime    15.69     1    1.00000
                                      *   9 b_acid       8.26     1    1.00000
                                      *  10 peakacid     5.15     1    1.00000
                                      *  11 trouble      5.15     1    1.00000
                                      *  12 log_time     8.93     1    1.00000
                                      *  13 painfreq     4.01     1    1.00000
****************************************************************************

STEP NUMBER    4
VARIABLE ENTERED    3 sex

  VARIABLE      F TO   FORCE TOLERNCE *   VARIABLE      F TO   FORCE TOLERNCE
               REMOVE  LEVEL          *                ENTER   LEVEL
        DF =   1   96                 *          DF =  1   95
   3 sex         5.33     1   0.97602 *   1 treatmnt     1.31     1    0.96754
   8 sqrttime   14.10     1   0.98902 *   2 age          0.05     1    0.92785
   9 b_acid      9.00     1   0.96508 *   4 fam_hist     0.00     1    0.96521
  11 trouble     6.37     1   0.95363 *   5 cigarets     1.19     1    0.98052
                                      *   6 nitepain     1.51     1    0.97893
                                      *  10 peakacid     0.26     1    0.66286
                                      *  12 log_time     2.45     1    0.16480
                                      *  13 painfreq     3.87     1    0.92144

U-STATISTIC(WILKS' LAMBDA) 0.7129242   DEGREES OF FREEDOM   4    1     99
APPROXIMATE F-STATISTIC        9.664   DEGREES OF FREEDOM    4.00     96.00

 F - MATRIX        DEGREES OF FREEDOM =    4   96

              no_heal   recur
recur           2.59
ok              9.66     4.78

CLASSIFICATION FUNCTIONS

            GROUP =   no_heal        recur          ok
VARIABLE
  3 sex               7.90475        8.72316        9.95224
  8 sqrttime          0.39502        0.31181        0.17108
  9 b_acid            2.81200        2.32386        1.78962
 11 trouble           2.39978        3.87590        4.26219

CONSTANT            -11.48802      -11.32179      -11.09687

CLASSIFICATION MATRIX

GROUP      PERCENT    NUMBER OF CASES CLASSIFIED INTO GROUP -
           CORRECT
                      no_heal   recur      ok
 no_heal    57.9       11         5         3
 recur      44.4       12        20        13
 ok         78.9        1         7        30

TOTAL       59.8       24        32        46

JACKKNIFED CLASSIFICATION

GROUP      PERCENT    NUMBER OF CASES CLASSIFIED INTO GROUP -
           CORRECT
                      no_heal   recur      ok
 no_heal    47.4        9         7         3
 recur      37.8       15        17        13
 ok         73.7        3         7        28

TOTAL       52.9       27        31        44
```

[1]

—we omit summary table, table of incorrect classification, eigenvalues, multivariate test, etc—

```
VARIABLE       COEFFICIENTS FOR CANONICAL VARIABLES

  3 sex           1.15414
  8 sqrttime     -0.12623
  9 b_acid       -0.57630
 11 trouble       1.04981
               STANDARDIZED  (BY POOLED WITHIN-GROUP VARIANCES)
VARIABLE       COEFFICIENTS FOR CANONICAL VARIABLES

  3 sex           0.43344
  8 sqrttime     -0.67166
  9 b_acid       -0.55634
 11 trouble       0.47661

CONSTANT          0.07495

GROUP         CANONICAL VARIABLES EVALUATED AT GROUP MEANS

    no_heal      -1.03256
    recur        -0.19017
    ok            0.74148
```

—we omit table of points to be plotted—

[2]

```
HISTOGRAM OF CANONICAL VARIABLE

                                                   o
                                              r    r             o
                                    o         r    r           o o
                                  ror       o r    r      o    o o          oo
            r n                   rrr  onrror r    r     oro o o ro  o      oo      o    o
     n      nnnr     nr rnor nrr  nrn rrnnronnr  ronnr rrrrrnoroonororro ro oro     r    r  oo
.+....+....+....+....+....+....+..1.+....+....+.2..+....+....+.3..+....+....+....+....+....+....+....+....+..
          -2.4      -1.8      -1.2      -.60      0.0        .60      1.2       1.8       2.4       3.0
     -2.7      -2.1      -1.5      -.90      -.30       .30       .90      1.5       2.1       2.7
```

[1] From the classification matrix in Output 7M.3, we observe that only four subjects were classified incorrectly from NO_HEAL into OK or vice versa. 7M has been fairly successful in discriminating between these two groups. However, fewer than half of the cases from the intermediate (RECUR) group were classified correctly. Jackknifing indicates a still lower apparent success of the classification.

[2] Since only one contrast is specified for this example, only one canonical variable is estimated. Instead of a bivariate canonical plot, the histogram of the canonical variable is printed. The means for the three groups are marked on the histogram scale by 1, 2, and 3. Note that in this example the points are labeled N, R, and O, because each group name starts with a unique letter.

Special Features

Forward and backward stepping (ENTER and REMOVE)

The discriminant analysis begins with no variables in the classification function. Variables are entered or removed one at a time according to the criterion used (see METHOD option). After all significant variables are entered, they can be removed from the classification function according to a second set of *F*-to-enter and *F*-to-remove limits. The commands ENTER and REMOVE in 7M's DISCRIMINANT paragraph operate the same as in program 2R. See 2R Special Features for a discussion of forward and backward stepping.

Stepping criteria

Controlling the order in which variables enter the model (LEVEL and FORCE). The LEVEL and FORCE statements in 7M are the same as in program 2R. You can use the LEVEL statement in the DISCRIMINANT paragraph to specify a complete or partial ordering for variable entry by dividing the variables into levels containing one (complete ordering) or more (partial ordering) variables. You can use the FORCE statement with the LEVEL statement to ensure that a variable or group of variables remains in the classification function regardless of the influence of other variables.

In the LEVEL statement you assign positive integers or zeros to each of the variables. Variables assigned zeros are excluded from the analysis. 7M begins stepping with the lowest positive value of LEVEL and enters all possible variables with that LEVEL (provided they pass the *F*-to-enter and tolerance limit) before considering variables at the next higher level. Variables not assigned any value for LEVEL are considered for entry only after all other variables have entered or have failed the *F*-to-enter test.

Variables with a LEVEL less than or equal to FORCE are forced into the classification function regardless of the *F*-to-enter limit and are not removed. As an example of **complete ordering** for three of the variables, we could specify the following LEVEL and FORCE commands for the ulcer data in Example 7M.3:

```
LEVEL = 1, 3, 2.
FORCE = 3.
```

By assigning LEVELS to the first three variables (TREATMNT, AGE, and SEX), we ensure that TREATMNT will enter the discriminant function first (its LEVEL is 1), followed by SEX (LEVEL 2) and AGE (LEVEL 3). The FORCE command ensures that all three variables will remain in the function, regardless of their F-to-enter values. The remaining variables are then free to enter in the usual manner.

As an example of **partial ordering**, we use the following instructions with the ulcer data. The tabbing feature is used to assign levels to the nonconsecutive variables:

```
LEVEL = 1, 3, 2, (NITEPAIN)4, (PAINFREQ)4,
        (B_ACID)5, (PEAKACID)5.
FORCE = 3.
```

The first three variables are forced in as above. The pain variables, NITEPAIN and PAINFREQ, are considered next, followed by the acid output variables, B_ACID and PEAKACID. The remaining variables are automatically assigned a LEVEL of 6. Variables with the same LEVEL value compete for entry in the usual manner.

When variables are eliminated by setting LEVEL = 0, 7M prints the *F*-to-enter statistic for these variables at each step. (When you eliminate variables from the analysis by omitting them from the USE list in the VARIABLE paragraph, 7M does not print their *F*-to-enter values.)

Forcing in all variables. By default, all variables except the GROUPING variable have LEVEL = 1. Therefore, specifying FORCE = 2 without a LEVEL instruction forces all variables into the model.

Removing variables from the classification function (METHOD). You can choose either of two criteria to remove variables from the classification function (variables always enter according to *F* to-enter):

- A variable is removed if its *F*-to-remove is less than the *F*-to-remove limit (METHOD = 1). This is the default.
- A variable is removed to obtain a smaller Wilks' lambda than was obtained at an earlier step with the same number of variables in the function (METHOD = 2).

Limiting the number of steps (STEPS). You can specify a maximum number of steps. Stepping ends if this number is reached, even if *F*-to-enter and *F*-to-remove limits indicate an extended analysis.

Prior probabilities

Unless you state otherwise, a case is assumed to have equal probability of being in any group. If you specify prior probabilities, they affect the constant term of the classification function and consequently the computation of the posterior probabilities and the classification table.

Using jackknifing to compute Mahalanobis' D^2 and posterior probabilities

After the final step of the discriminant analysis, 7M automatically computes the Mahalanobis distance from each group mean to each case and the posterior probabilities for each case. The case is used in the computation of both the group mean and the cross-products matrices. See **[8]** in Example 7M.1.

When you specify JACKKNIFE in the DISCRIMINANT paragraph, 7M eliminates each case in turn from the computations. Thus 7M computes the Mahalanobis distance as the distance from the eliminated case to the groups formed by the remaining cases (Lachenbruch and Mickey, 1968). In the output the distances and posterior probabilities are labeled "jackknifed distances and posterior probabilities."

Plots

7M automatically prints the canonical variable plot (Example 7M.1) when there are two or more nonzero canonical variables. Only one nonzero canonical variable exists when one of the following occurs:

- only two groups are analyzed
- only one variable enters the classification function
- a single contrast is supplied
- all data points are on a line

7M prints a histogram in these situations. You can suppress either display by stating NO CANONICAL in the PLOT paragraph. When group memberships overlap extensively, you can limit the groups displayed in each plot frame by using GROUP statements in the PLOT paragraph (e.g., for six groups, one GROUP statement could request the first three groups, another GROUP statement the last three groups). See PLOT GROUP in 7M Commands. See Example 7M.3 for the use of CONTRASTS in the canonical variable plot.

Saving the canonical variable scores and an indicator of predicted group membership

When a BMDP File is created, 7M saves the data, the canonical variables, and the index of the group into which each case is classified. The canonical variables are named CNVR1, CNVR2, etc., and placed immediately after the data for each case. The index of the group into which the case is classified comes last and is named PREDICTD. The BMDP File can then be read as input to other BMDP programs (see Chapter 8).

7M Commands

/ INPut

Required. See Chapter 3 for a description of available INPUT commands. An additional, rarely used command specific to 7M is GROUP.

GROUP = #. `GROUP = 3.`

Specify the number of groups to be formed. This option is used only when neither CODES nor CUTPOINTS are specified. You must specify a number equal to the number of distinct codes in the data set for the GROUPING variable. Each code defines a group. The groups will be named *1, *2, etc., in the output.

/ VARiable

See Chapter 4 for a description of VARIABLE commands.

/ GROUP

Required. Used to specify the name or number of the variable used to classify cases into groups. The earlier method of specifying a grouping variable with the GROUPING command in the VARIABLE paragraph can still be used, although this will only work for batch jobs and will not work interactively. See the 1990 *BMDP Manual* for details. To make the printed results easier to read, we suggest that you specify group NAMES. Define NAMES (and CUTPOINTS or CODES) in the GROUP paragraph (see Chapter 5).

New Syntax

VARiable = variable. `VARIABLE = IRISTYPE.`

List name or number of the GROUPING variable (between subject factors). Only one grouping factor is allowed. You must specify CODES or CUTPOINTS in the GROUP paragraph if the variable takes on more than 10 distinct values. Groups must have unique first letters in order for the names to appear in the analysis of variance table; if not, the groups are labeled G, H,..., etc. in the table. **Example:** The variable IRISTYPE contains values for separating the cases into cells (groups).

USE = list. `USE = 1, 2.`

List the names or numbers of groups used in computing the cross-products matrix and estimating the classification functions. Cases in groups that are not listed here, but that have acceptable values for the GROUPING variable will still be classified. By default, 7M uses all groups in the computations. **Example:** Assuming that three groups were specified, only groups 1 and 2 are used in computing classification functions; cases from all three groups are classified.

PRIOR = # list. `PRIOR = .25, .75.`

Specify the prior probabilities for groups used in analysis. The probabilities should add to unity. The first number is the prior probability for the first group, the second for the second group, etc. The order is that specified in the CODES or CUTPOINTS, or that of input (not that of the USE statement in the GROUP paragraph). Unequal prior probabilities affect the computation of the constant term in the classification function and thus computation of the posterior probabilities and classification table (see Appendix B.15); they do not affect the *F* statistics, Wilks' lambda, or the selection of variables. 7M assumes equal priors. **Example:** Priors for the first and second groups are .25 and .75, respectively.

/ DISCriminant

Optional. The parameters listed below (except JACKKNIFE) can influence the stepping (number and order of variables entering the classification functions). By default, variables enter in a forward manner using $F = 4.0$ for entering, $F = 3.996$ for removal.

ENTER = #l, #2. `ENTER = 1.0, 5.0.`
REMOVE = #1, #2. `REMOVE = 0.0, 4.9.`

Specify *F*-to-enter and *F*-to-remove limits. ENTER #1 and REMOVE #1 are used during forward stepping to control the entry or removal of variables. ENTER and REMOVE #2 are used for backward stepping. Entering variables must have *F*s greater than the ENTER limit. (For removal, *F*s less than the REMOVE limit.) REMOVE must be less than ENTER. Default: ENTER = 4.0, 4.0.; REMOVE = 3.996, 3.996. (no backward stepping). See discussion of *F* values in Example 7M.1. **Example:** During forward stepping, variables are entered if their $F > 1.0$ and none are removed (REMOVE *F* is 0.0). Then, during backward stepping, variables are removed whose $F < 4.9$, and they are not reentered unless *F*-to-enter is greater than or equal to 5.0.

JACKknife. `JACK.`

Requests posterior probabilities and Mahalanobis' D^2 for each case, without using the case in the computation of the group mean and pooled within-group covariance matrix. By default, jackknifing is not used to compute these statistics. (However, the jackknifed classification matrix is printed.)

LEVEL = # list. `LEVEL = 1, 3, 2.`

Specify the order in which variables should enter the classification function, using one number for each variable. The first number is the level assigned to the first variable read as input, the second to the second variable read as input, etc. Stepping begins at the lowest positive LEVEL and enters all possible variables (those that pass the *F*-to-enter and TOLERANCE criteria) at that level before considering variables at the next higher level. Variables at the same level compete for entry according to METHOD. 7M prints *F*-to-enter for variables with level zero, but they are not entered into the equation. Default: all variables are assigned a level of one. **Example:** Variable 1 is considered first, then variable 3, and then variable 2. Since no levels are assigned for the remaining variables, they are assigned level 4 and are considered last. (See Special Features.)

FORCE = #. `FORCE = 1.`

Variables with LEVEL greater than zero and less than or equal to FORCE (if FORCE is greater than zero) enter and are not removed. (ENTER and REMOVE limits are ignored when FORCE is specified.) By default, zero is assumed, and there is no forcing. **Example:** Variables with LEVEL = 1 are included in the classification function. See Special Features.

METHod = 1 *or* **2.** `METH = 2.`

Method 1 removes variables from the classification function using *F*-to-remove limits; Method 2 removes variables to reduce Wilks' lambda. Default is Method 1. See Special Features.

CONTrast = # list. `CONTRAST = - 1, 0, 1.`

List the coefficients of a contrast used to direct the stepping. The first number is the coefficient of the first group, the second number of the second group, etc. (according to CODE or CUTPOINT order—not USE in the GROUP paragraph). These should sum to zero. You can repeat this instruction to specify additional contrasts. When you specify CONTRASTS, the computation of the *F*-to-enter, *F*-to-remove, and Wilks' lambda are affected. The coefficients of the classification functions, the classification matrix, and Mahalanobis distances are not directly affected, since they depend only on the set of variables selected and not on the values of the contrasts. If you specify more than one contrast, all the specified contrasts are used simultaneously to direct the stepping. See Appendix B.15. **Example:** The contrast directs 7M to maximize the difference between the first and third groups.

STEP = #. `STEP = 4.`

Specify the maximum number of steps in the analysis. The default number of steps is based on ENTER and REMOVE (or LEVEL, FORCE, METHOD, CONTRAST), and the maximum number is 2*(1 + number of variables used).

TOLerance = #. `TOL = .001.`

Specify a tolerance limit. No variable enters the classification function whose squared multiple correlation (R^2) with already entered variables exceeds (1-TOL), or whose entry will cause the tolerance of already entered variables to be unacceptable. Default: .01.

/ PRINT

The PRINT paragraph is optional. When you use the PRINT paragraph as an interactive paragraph, you can use the following 7M-specific commands as well as the general interactive PRINT paragraph commands.

CORRelation. `CORR.`

Prints the within-groups correlation matrix.

WITHin. `WITH.`

Prints the within-groups covariance matrix.

STEP. `NO STEP.`

Unless you state NO STEP, 7M prints results at every step. If you specify NO STEP, they are printed only at step 0 and the last step.

CLASSification = # list. `CLASS = 1, 5, 6, 7.`

List the steps at which the classification matrix and jackknifed classification matrix are printed. They are printed automatically after the last step.

POSTerior. `NO POST.`

Unless you specify NO POST, 7M prints Mahalanobis' D^2 and posterior probabilities for each case at the last step.

POINT. `NO POINT.`

Unless you specify NO POINT, 7M prints the values of the canonical variables used in the plot.

New Feature

CASE = #. `CASE = 50.`

Specify the number of cases for which data are printed. Values flagged by MISS, MAX, and MIN in the VARIABLE paragraph specifications are replaced by MISSING, GT MAX, and LT MIN. Default: 10.

New Feature

MINimum. `MIN.`

Prints data for cases in which at least one variable has a value less than its lower limit as specified in the VARIABLE paragraph. Default: such cases are not printed unless you specify DATA.

New Feature

MAXimum. `MAX.`

Prints data for cases in which at least one variable has a value greater than its upper limit as specified in the VARIABLE paragraph. Default: such cases are not printed unless you specify DATA.

New Feature

MISSing. `MISS.`

Prints data for cases that have at least one variable with a missing value. Missing values in the input may be identified in several ways: (1) an asterisk (the default missing value character), (2) a missing value code identified using MISS in the VARIABLE paragraph, (3) a character identified as MCHAR in the INPUT paragraph, (4) blanks (applies only to input read with FORTRAN fixed format). Default: such cases are not printed unless you specify DATA.

New Feature

METHod = VARiable *or* **CASE.** `METH = CASE.`

Use to determine how the printout is organized. VARIABLE is the default value.

VARIABLE – Prints for each case only as many variables as fill the print line (up to ten variables) and begins the next line with the same variables for the next case. Thus the data for each variable are in an easy-to-scan column. When the program finishes printing the first subset of variables for all cases, it continues with another subset.

CASE – Prints the values of all variables for one case (possibly filling several print lines) before printing data for the next case.

The following commands work only in batch mode

New Feature

MEAN. `NO MEAN.`

Unless you specify NO MEAN, 7M prints the mean, standard deviation, and frequency of each used variable.

New Feature

EZSCores. `NO EZSC.`

Unless you state NO EZSC, 7M prints z-scores for the minimum and maximum values of each used variable.

New Feature

ECASe. `NO ECAS.`

Unless you state NO ECAS, 7M prints the case numbers containing the minimum and maximum values of each variable.

New Feature

EXTReme. `NO EXTR.`

Unless you state NO EXTR, 7M prints the minimum and maximum values for each variable used.

New Feature

SKewness. `SK.`

Prints skewness and kurtosis for each used variable. Default: no skewness printed.

/ PLOT

The PLOT paragraph is optional. When you use the PLOT paragraph as an interactive paragraph, you can use the following 7M-specific commands as well as the general interactive PLOT paragraph commands.

CONTRast. `CONTRAST.`

Directs 7M to use contrasts specified in the DISCRIMINANT paragraph to compute the canonical variables for the plots. By default, contrasts are not used.

CANONical. `NO CANON.`

Unless you state NO CANON, 7M prints a plot of the first two canonical variables. If there is only one canonical variable, 7M prints a histogram.

GROUP = list.

```
GROUP = 1, 2.
GROUP = 3, 4.
```

List names or numbers of groups to be plotted in the canonical variable plot. Each GROUP command defines a separate plot. This instruction may be repeated. If you specify GROUPS, the canonical plot is printed even if NO CANON is stated. By default, 7M plots cases from all groups. **Example:** Two plots are printed: cases from groups 1 and 2 in the first, from groups 3 and 4 in the second.

/ SAVE

Required only if you wish to create a BMDP File. When you use the SAVE paragraph as an interactive paragraph, you can use the following 7M-specific commands as well as the general interactive SAVE paragraph commands. After the data and canonical variable scores, 7M automatically saves the variable PREDICTD, which indicates the group into which each case is classified, and several canonical variables, the number depending on the NCAN option below. See Chapter 8 for description of the CODE, UNIT, and NEW commands.

NCAN = #. `NCAN = 2.`

Specify the number of canonical variables for which scores are saved in a BMDP File. By default, NCAN is one less than the number of groups used in the analysis, or NCAN is equal to the number of CONTRASTS specified in the DISCRIMINANT paragraph, if any. If you specify a value of NCAN greater than the number of contrasts, the first set of canonical variable(s) contains the canonical variables evaluated for the contrasts, ordered by the size of the eigenvalues. The remaining set contains the residual variables (variables orthogonal to those corresponding to the set of contrasts).

MAHALanobis. `MAHAL.`

Saves the Mahalanobis distances from the group means for each case in a BMDP File. The Mahalanobis distances are appended to each case after the data, canonical variable scores, and PREDICTD, and are named MAHAL1, MAHAL2, etc.

POSTerior. `POST.`

Saves the posterior probabilities of group membership for each case in a BMDP File. The posterior probabilities are appended to each case after the data, canonical variable scores, PREDICTD, and Mahalanobis distances, and are named POSTP1, POSTP2, etc.

Order of Instructions

■ indicates required paragraph

```
■ / INPUT
  / VARIABLE
  / TRANSFORM
■ / GROUP
  / SAVE
  / DISCRIMINANT
  / PRINT          Repeat for additional problems.
  / PLOT           See Multiple Problems, Chapter 10.
■ / END
  data
  TRANSFORM ...
  DISCRIMINANT ... /   May be
  GROUP ... /          repeated
  PRINT ... /          interactively.
  PLOT ... /
  SAVE ... /
  CONTROL ... /
  END /
  FINISH /
```

Summary Table for Commands Specific to 7M

Paragraphs / Commands	Defaults	Multiple Problems	See
■ / **INPut**			
GROUP = #.	use only when neither CODE nor CUTPOINT is used	–	Cmds
■ / **GROUP**	required to specify grouping variables		
■ VARiable = variable.	none	●	7M.1, 7M.3
USE = list.	all groups used	●	7M.2
PRIOR = # list.	equal priors assumed	–	SpecF
/ **DISCriminant**			
JACKknife.	NO JACK.	●	SpecF
ENTER = #1, #2.	4.0, 4.0	●	SpecF
REMOVE = #1, #2.	3.996, 3.996	●	SpecF
METHod = 1 *or* 2.	METHOD = 1.	●	SpecF
LEVEL = # list.	equal levels	–	SpecF
FORCE = #.	no forcing	●	SpecF
CONTrast = # list.	NO CONTRAST.	●	7M.3
STEP = #.	2*(number of variables after transformations)	●	SpecF
TOLerance= #.	.01	●	Cmds
/ **PRINT**			
WITHin.	NO WITHIN.	●	Cmds
CORRelation.	NO CORR.	●	Cmds
COVAriance.	NO COVA.	●	Cmds
STEP. *or* NO STEP.	STEP.	●	7M.3
CLASS = # list.	classification matrices printed only after last step	●	Cmds
POSTerior. *or* NO POST.	POST.	●	Cmds
POINT. *or* NO POINT.	POINT.	●	Cmds
MEAN. *or* NO MEAN.	MEAN.	●	Cmds
EXTReme. *or* NO EXTR.	EXTReme. values printed	●	Cmds

Summary Table for Command Specific to 7M

Paragraphs Commands	Defaults	Multiple Problems	See
EZSCores. *or* NO EZSC.	EZSC. (z-scores)	●	Cmds
ECASe. *or* NO ECASe.	ECASe. (Case #'s of Min and Max printed)	●	Cmds
SKewness *or* NO SK.	NO SK. (skewness and kurtosis)	●	Cmds
CASE = #.	10	●	Cmds
MISSing.	NO MISS.	●	Cmds
MINimum.	NO MIN.	●	Cmds
MAXimum.	NO MAX.	●	Cmds
METHod = VARiables *or* CASE.	VARIABLES	–	Cmds
/ PLOT			
CONTRast.	NO CONTR.	●	7M.3
CANONical. *or* NO CANON.	CANON.	●	Cmds
GROUP = list.	cases from all groups plotted	●	SpecF
/ SAVE			
NCAN = #.	(# of groups –1) or (# of contrasts)	●	Cmds
MAHALanobis.	NO MAHAL.	●	Cmds
POSTerior.	NO POST.	●	Cmds

Key:
- ■ Required paragraph or command
- ▲ Frequently used command or paragraph
- ● Value retained for multiple problems
- – Default reassigned

2R

Stepwise Regression

2R fits a multiple linear regression equation in a stepwise manner by entering or removing one variable at a time from a list of potential predictors. You can request forward stepping (beginning with no predictors) and backward stepping (beginning with all predictors). You can direct 2R interactively at the terminal, specifying which variables to enter or remove at each step and which variables to use in diagnostic plots. You can force variables into the equation, and you can define sets of variables to enter or remove in a single step. You can also request additional analyses, plots, and diagnostics using interactive paragraphs.

The regression model fitted to the data is

$$y = \beta_0 + \beta_1 x_1 + \beta_2 x_2 + \ldots + \beta_p x_p + \varepsilon$$

where

- y is the dependent (outcome) variable,
- x_i is the ith independent variable,
- β_i is the ith regression coefficient,
- β_0 is the intercept,
- p is the number of independent variables, and
- ε is the error with mean zero.

2R offers extensive diagnostics to assist you in checking for outliers, cases with unusual influence on the model, and violations of assumptions. Many options are available for printing and plotting the diagnostics. Caseplots display three diagnostics side-by-side for each case: a measure of leverage, a measure of influence, and a standardized residual. You may request added variable plots and miniplots, half-normal probability plots of deleted standardized residuals (with simulation envelopes), and an error variance plot for identifying departures from homogeneity of variance. You can also create your own bivariate plot using any of the diagnostics or the variables in the model.

BMDP Technical Report #48 contains annotated output of stepwise regression using 2R and 9R. For initial data screening, we suggest first using program 2D to assess the distribution of each variable; then use 6D to plot each potential independent variable against the dependent variable. For a general introduction to regression, see Afifi and Clark (1984). More advanced material is contained in Draper and Smith (1981). Regression diagnostics are discussed in Belsley, Kuh, and Welsch (1980).

Where to Find It

Examples

Special Features

Example 2R.1 Simple regression

For our first example, we use the Werner blood chemistry data (Appendix D) to illustrate simple linear regression. The model is

$$y = a + bx + \varepsilon$$

where y is weight and x is HEIGHT. We want to see the relationship between the two and estimate the intercept a and slope b.

In the REGRESS paragraph in Input 2R.1, we specify WEIGHT as the dependent variable and HEIGHT as the independent or predictor variable. (The word "independent" is historical and not to be taken literally, as the two variables are not statistically independent.) Although there are nine variables in the data, we only give NAMES to the first four. This does not affect the results.

Input 2R.1

```
/ INPUT     VARIABLES = 9.
            FORMAT IS FREE.
            FILE IS 'werner.dat'.

/ VARIABLE  NAMES = id, age, height, weight.
            USE = height, weight.

/ REGRESS   DEPEND =  weight.
            INDEP = height.
/ END
```

Output 2R.1

```
 CASE    3        4          [1]
  NO. height   weight
----- -------- --------
    1    67.00   144.00
    2    64.00   160.00
    3    62.00   128.00
    4    68.00   150.00
    5    64.00   125.00
    6    67.00   130.00
    7    64.00   118.00
    8    65.00   119.00
    9    60.00   107.00
   10    65.00   135.00

NUMBER OF CASES READ. . . . . . . . . . . . . .     188          [2]
   CASES WITH DATA MISSING OR BEYOND LIMITS . .       4
      REMAINING NUMBER OF CASES . . . . . . . .     184

DESCRIPTIVE STATISTICS OF DATA          [3]
----------- ---------- -- ----

 VARIABLE    TOTAL            STANDARD  SKEW-                 SMALLEST        LARGEST
 NO. NAME     FREQ.  MEAN      DEV.     NESS   KURTOSIS  VALUE  Z-SCR  VALUE  Z-SCR  RANGE

   3 height     184  64.484  2.4827   -0.120   -0.071   57.000 -3.01   71.000  2.62  14.000
   4 weight     184  131.37  20.475    1.018    1.826   94.000 -1.83   215.00  4.08  121.00

*** N O T E *** KURTOSIS VALUES GREATER THAN ZERO INDICATE A DISTRIBUTION WITH HEAVIER TAILS THAN NORMAL DISTRIBUTION.

REGRESSION INTERCEPT. . . . . . . . . . . . . . .NON ZERO        [4]
REGRESSION WEIGHT VARIABLE. . . . . . . . . . . .
PRINT COVARIANCE MATRIX . . . . . . . . . . . .        NO
PRINT CORRELATION MATRIX. . . . . . . . . . . .        NO
PRINT ANOVA AT EACH STEP. . . . . . . . . . . .       YES
PRINT STEP OUTPUT . . . . . . . . . . . . . . .       YES
PRINT REGRESSION COEFFICIENT SUMMARY TABLE. . .       YES
PRINT PARTIAL CORRELATION SUMMARY TABLE . . . .        NO
PRINT F-RATIO SUMMARY TABLE . . . . . . . . . .        NO
PRINT SUMMARY TABLE . . . . . . . . . . . . . .       YES
PRINT DATA OR DIAGNOSTICS . . . . . . . . . . .        NO
PRINT CORRELATION OF REGRESSION COEFFICIENTS. .        NO
PRINT NORMAL PROBABILITY PLOT . . . . . . . . .        NO
PRINT DETRENDED NORMAL PROBABILITY PLOT . . . .        NO
PRINT PLOTS FOR XVAR AND YVAR . . . . . . . . .        NO
PRINT PLOTS AND DATA. . . . . . . . . . . . . .        NO
PRINT PLOTS WITH STATISTICS . . . . . . . . . .        NO
PRINT DIAGNOSTIC PLOT(S). . . . . . . . . . . .        NO
PRINT CASE-BY-STATISTIC PLOTS . . . . . . . . .        NO
PRINT ADDED VARIABLE MINIPLOTS. . . . . . . . .        NO

STEPPING ALGORITHM. . . . . . . . . . . . . . . .F
MAXIMUM NUMBER OF STEPS . . . . . . . . . . . .        18
DEPENDENT VARIABLE. . . . . . . . . . . . . . .         4 weight
MINIMUM ACCEPTABLE F-TO-ENTER . . . . . . . . .    4.000,    4.000
MAXIMUM ACCEPTABLE F-TO-REMOVE. . . . . . . . .    3.900,    3.900
MINIMUM ACCEPTABLE TOLERANCE. . . . . . . . . . 0.01000
SUBSCRIPTS OF THE INDEPENDENT VARIABLES . . . .       3
```

— we omit step 0 —

```
 STEP NO.    1
 ---------------
 VARIABLE ENTERED     3 height
 MULTIPLE R              0.4759
 MULTIPLE R-SQUARE       0.2265        [5]
 ADJUSTED R-SQUARE       0.2222
 STD. ERROR OF EST.     18.0571

 ANALYSIS OF VARIANCE
                SUM OF SQUARES     DF    MEAN SQUARE      F RATIO
       REGRESSION  17375.953        1    17375.95          53.29            [6]
       RESIDUAL    59342.883      182    326.0598

                        VARIABLES IN EQUATION FOR weight                  .                 VARIABLES NOT IN EQUATION

                                STD. ERROR  STD REG                  F         .               PARTIAL                   F
     VARIABLE        COEFFICIENT  OF COEFF    COEFF    TOLERANCE  TO REMOVE  LEVEL.    VARIABLE       CORR.  TOLERANCE  TO ENTER   LEVEL
(Y-INTERCEPT       -121.72305 )
height       3        3.92491       0.5377    0.476   1.00000      53.29       1 .  [7]

***** F LEVELS(  4.000,  3.900) OR TOLERANCE INSUFFICIENT FOR FURTHER STEPPING


          STEPWISE REGRESSION COEFFICIENTS         [8]

  VARIABLES   0 Y-INTCPT   3 height
STEP
   0          131.3696*     3.9249
   1         -121.7231*     3.9249*
```

— we omit the remainder of the output —

[1] Prints the first ten cases of data.

[2] Number of cases read. 2R uses only complete cases in all computations. That is, if the value of any variable in a case is missing or out of range, 2R omits the case from all computations. If you specify a USE list in the VARIABLE paragraph, only the variables in the USE list are considered in determining the completeness of each case.

[3] This table gives all descriptive statistics divided by group. This table includes:
- Mean
- Standard Deviation
- Skewness
- Kurtosis
- Smallest value with case number and z-score for that case
- Largest value with case number and z-score for that case

[4] Interpretation of regression instructions specific to this program.

[5] The MULTIPLE R is the same as the simple correlation between HEIGHT and WEIGHT since there is only one independent variable.

[6] In the ANALYSIS OF VARIANCE table, the SUM OF SQUARES ratio 17376/(17376 + 59343) is the same as the MULTIPLE R SQUARE and is the proportion of the total variation in WEIGHT accounted for by HEIGHT. The F RATIO is the ratio of the two mean squares (17376/326) and will be large when any of the predictor variables help explain the variation in the dependent variable. The F RATIO will be small only when all of the regression coefficients are close to zero compared to their standard errors.

[7] The estimated coefficient for HEIGHT (b) is about 3.92, with a standard error of about 0.54. This means that for every inch in height, the average increase in weight is approximately 3.9 pounds. If we had first standardized HEIGHT and WEIGHT to have mean zero and unit variance, the estimated regression coefficient would be 0.476. The equation relating WEIGHT to HEIGHT is

$$\text{predicted WEIGHT} = -121.7 + 3.92 * \text{HEIGHT}$$

Since this equation was obtained only for HEIGHTS between 57 and 71 inches (see top output panel), extrapolating far outside this range is not valid.

[8] Prints the regression coefficients at each step.

Example 2R.2 Stepwise regression

In this example we build a multiple regression model to predict mortality, using data from 60 U.S. Standard Metropolitan Statistical Areas. For each city, the data include an age-adjusted mortality rate; socioeconomic factors (% non-white, education, and population density); a climatological factor (rainfall); and two air pollution measures (SO_2 and NO_x). (McDonald and Schwing, 1973; Henderson and Velleman, 1981) We use a modified version of the data, omitting some of the original variables. Our goal is to select a good set of predictor variables from these candidate variables—that is, to separate the more important variables from those that may not be necessary.

2R enters and removes variables in a stepwise manner. By default, 2R does forward stepping by entering the variable whose addition would increase R^2 most at each step (the variable with the largest F-to-enter). 2R also checks variables already in the equation to ensure that they are still adding markedly to the strength of prediction (F-to-remove). To do backward stepping (beginning with all predictors), see discussion in Special Features. You may use the default values for F-to-enter and F-to-remove or specify cutoff values for a particular problem. These statistics are explained in the discussion of Output 2R.2.

Note: When running 2R interactively, you can select the variable to enter or remove at each step. See Special Features.

Sometimes the nature of your problem may demand that you control the order in which candidate variables enter the equation. For example, in building this model to predict mortality, you might want to allow for all sociological and environmental factors first and then see if the addition of any air pollution measures enhances the prediction of mortality. For more about controlling variable entry, see Special Features.

In Input 2R.2, only the REGRESS paragraph is specific to 2R. We specify MORTALTY as the dependent variable. The INPUT and VARIABLE paragraphs are common to all programs and are described in Chapters 3 and 4. The TRANSFORM paragraph is discussed in Chapter 6. Because of concern about asymmetric univariate distributions, we create two new variables by taking the logarithm of the SO_2 and NO_x measures and allow the program to select the better version of each measure. In this default run, we read the data stored in a file named AIRPOLL.DAT (see Appendix D). In addition to the variables discussed above, each record contains the name of the city. Since the names can contain up to eight characters, we split them into two parts and read them as two LABEL variables. These will be used as case labels in Example 2R.3. Output 2R.2 is printed in wide format (132 columns); for interactive use the default output is narrow (80 columns).

Input 2R.2

```
/ INPUT       VARIABLES = 9.
              FORMAT IS FREE.
              FILE IS 'airpoll.dat'.

/ VARIABLE    NAMES = name1, name2, rain, educatn, pop_den,
                      nonwhite, nox, so2, mortalty.
              LABELS ARE name1, name2.

/ TRANSFORM   log_so2 = LOG(so2).
              log_nox = LOG(nox).

/ REGRESS     DEPENDENT = mortalty.

/ END
```

Output 2R.2

—we omit the printing of the first ten cases—

```
 NUMBER OF CASES READ. . . . . . . . . . . . . .        60
                                                                                  [1]
DESCRIPTIVE STATISTICS OF DATA
----------- ---------- -- ----

 VARIABLE    TOTAL           STANDARD  SKEW-                SMALLEST       LARGEST
 NO. NAME     FREQ.   MEAN     DEV.    NESS  KURTOSIS   VALUE  Z-SCR   VALUE  Z-SCR   RANGE

   3 rain        60  37.367  9.9847  -0.763    0.946  10.000 -2.74  60.000  2.27  50.000
   4 educatn     60  10.973  .84530  -0.214   -0.862  9.0000 -2.33  12.300  1.57  3.3000
   5 pop_den     60  3866.1  1464.5   1.282    2.892  1441.0 -1.66  9699.0  3.98  8258.0
   6 nonwhite    60  11.870  8.9211   1.075    0.637  .80000 -1.24  38.500  2.99  37.700
   7 nox         60  22.650  46.333   4.910   26.687  1.0000 -0.47  319.00  6.40  318.00
   8 so2         60  53.767  63.390   1.817    2.960  1.0000 -0.83  278.00  3.54  277.00
   9 mortalty    60  940.38  62.212   0.091   -0.055  790.70 -2.41  1113.0  2.77  322.30
  10 log_so2     60  1.3885  .65040  -0.679   -0.180  0.0000 -2.13  2.4440  1.62  2.4440
  11 log_nox     60  1.0091  .51438   0.353    0.163  0.0000 -1.96  2.5038  2.91  2.5038

 ***  NOTE    *** KURTOSIS VALUES GREATER THAN ZERO INDICATE A DISTRIBUTION
                  WITH HEAVIER TAILS THAN NORMAL DISTRIBUTION.
```

—we omit portion of the output—

```
STEPPING ALGORITHM. . . . . . . . . . . . . . .F                                         [2]
MAXIMUM NUMBER OF STEPS . . . . . . . . . . . .      22
DEPENDENT VARIABLE. . . . . . . . . . . . . . .       9 mortalty
MINIMUM ACCEPTABLE F TO ENTER . . . . . . . . .   4.000,   4.000
MAXIMUM ACCEPTABLE F TO REMOVE. . . . . . . . .   3.900,   3.900
MINIMUM ACCEPTABLE TOLERANCE. . . . . . . . . . 0.01000
SUBSCRIPTS OF THE INDEPENDENT VARIABLES . . . .     3    4    5    6    7
                                                    8   10   11

STEP NO.    0                            [3]
---------------

STD. ERROR OF EST.     62.2124

ANALYSIS OF VARIANCE
                    SUM OF SQUARES      DF     MEAN SQUARE
       RESIDUAL     228352.78           59    3870.386

                        VARIABLES IN EQUATION FOR mortalty              .             VARIABLES NOT IN EQUATION

                                   STD. ERROR  STD REG                F        .               PARTIAL                   F
     VARIABLE      COEFFICIENT    OF COEFF    COEFF     TOLERANCE  TO REMOVE  LEVEL.   VARIABLE       CORR.  TOLERANCE   TO ENTER   LEVEL
(Y-INTERCEPT        940.38165 )                                                   .
                                                                                  . rain        3  0.50942  1.00000      20.33        1
                                                                                  . educatn     4 -0.51032  1.00000      20.42        1
                                                                                  . pop_den     5  0.26138  1.00000       4.25        1
                                                                                  . nonwhite    6  0.64355  1.00000      41.00        1
                                                                                  . nox         7 -0.07752  1.00000       0.35        1
                                                                                  . so2         8  0.42598  1.00000      12.86        1
                                                                                  . log_so2    10  0.40341  1.00000      11.27        1
                                                                                  . log_nox    11  0.29194  1.00000       5.40        1
STEP NO.    1                            [4]
---------------
VARIABLE ENTERED     6 nonwhite

MULTIPLE R                0.6435
MULTIPLE R-SQUARE         0.4142
ADJUSTED R-SQUARE         0.4041

STD. ERROR OF EST.     48.0264
                                                                                              [5]
ANALYSIS OF VARIANCE
                    SUM OF SQUARES      DF     MEAN SQUARE          F RATIO
       REGRESSION   94573.805            1    94573.80               41.00
       RESIDUAL     133778.98           58    2306.534

                        VARIABLES IN EQUATION FOR mortalty          [6]    .      VARIABLES NOT IN EQUATION  [7]

                                   STD. ERROR  STD REG                F        .               PARTIAL                   F
     VARIABLE      COEFFICIENT    OF COEFF    COEFF     TOLERANCE  TO REMOVE  LEVEL.   VARIABLE       CORR.  TOLERANCE   TO ENTER   LEVEL
(Y-INTERCEPT        887.11084 )                                                   .
nonwhite    6         4.48785      0.7009     0.644     1.00000      41.00       1 . rain        3  0.34935  0.82926       7.92        1
                                                                                  . educatn     4 -0.50226  0.95641      19.23        1
                                                                                  . pop_den     5  0.35223  0.99984       8.07        1
                                                                                  . nox         7 -0.11676  0.99966       0.79        1
                                                                                  . so2         8  0.42808  0.97463      12.79        1
                                                                                  . log_so2    10  0.48366  0.99725      17.41        1
                                                                                  . log_nox    11  0.22606  0.96403       3.07        1

STEP NO.    2                            [8]
---------------
VARIABLE ENTERED     4 educatn

MULTIPLE R                0.7496
MULTIPLE R-SQUARE         0.5619
ADJUSTED R-SQUARE         0.5466

STD. ERROR OF EST.     41.8918

ANALYSIS OF VARIANCE
                    SUM OF SQUARES      DF     MEAN SQUARE          F RATIO
       REGRESSION   128322.10            2    64161.05               36.56
       RESIDUAL     100030.69           57    1754.924

                        VARIABLES IN EQUATION FOR mortalty                 .          VARIABLES NOT IN EQUATION

                                   STD. ERROR  STD REG                F        .               PARTIAL                   F
     VARIABLE      COEFFICIENT    OF COEFF    COEFF     TOLERANCE  TO REMOVE  LEVEL.   VARIABLE       CORR.  TOLERANCE   TO ENTER   LEVEL
(Y-INTERCEPT       1211.37598 )                                                   .
educatn     4       -28.93121      6.5974    -0.393     0.95641      19.23       1 . rain        3  0.15756  0.65847       1.43        1
nonwhite    6         3.91554      0.6251     0.561     0.95641      39.23       1 . pop_den     5  0.27441  0.94079       4.56        1
                                                                                  . nox         7  0.00057  0.94519       0.00        1
                                                                                  . so2         8  0.38246  0.93235       9.60        1
                                                                                  . log_so2    10  0.42717  0.93438      12.50        1
                                                                                  . log_nox    11  0.29679  0.96056       5.41        1

STEP NO.    3                            [9]
---------------
VARIABLE ENTERED    10 log_so2

MULTIPLE R                0.8012
MULTIPLE R-SQUARE         0.6419
ADJUSTED R-SQUARE         0.6227

STD. ERROR OF EST.     38.2141

ANALYSIS OF VARIANCE
                    SUM OF SQUARES      DF     MEAN SQUARE          F RATIO
       REGRESSION   146575.00            3    48858.33               33.46
       RESIDUAL     81777.789           56    1460.318
```

```
                    VARIABLES IN EQUATION FOR mortalty                              .                VARIABLES NOT IN EQUATION

                                    STD. ERROR  STD REG                     F          .                        PARTIAL                         F
   VARIABLE         COEFFICIENT    OF COEFF     COEFF      TOLERANCE    TO REMOVE  LEVEL.    VARIABLE          CORR.    TOLERANCE      TO ENTER    LEVEL
(Y-INTERCEPT        1111.93799 )                                                        .
educatn     4        -23.41198        6.2173    -0.318     0.89611        14.18        1 . rain          3   0.33992   0.59347          7.19          1
nonwhite    6          3.91783        0.5702     0.562     0.95641        47.20        1 . pop_den       5   0.10780   0.76098          0.65          1
log_so2    10         27.97665        7.9132     0.292     0.93438        12.50        1 . nox           7  -0.23234   0.76022          3.14          1
                                                                                       . so2           8   0.09567   0.38685          0.51          1
                                                                                       . log_nox      11  -0.06310   0.37825          0.22          1
```

[10]

```
STEP NO.    4
---------------
VARIABLE ENTERED       3 rain

MULTIPLE R                0.8266
MULTIPLE R-SQUARE         0.6833
ADJUSTED R-SQUARE         0.6602

STD. ERROR OF EST.       36.2639

ANALYSIS OF VARIANCE
                       SUM OF SQUARES       DF      MEAN SQUARE          F RATIO
          REGRESSION   156023.98             4      39006.00             29.66
          RESIDUAL      72328.805           55       1315.069

                    VARIABLES IN EQUATION FOR mortalty                              .                VARIABLES NOT IN EQUATION

                                    STD. ERROR  STD REG                     F          .                        PARTIAL                         F
   VARIABLE         COEFFICIENT    OF COEFF     COEFF      TOLERANCE    TO REMOVE  LEVEL.    VARIABLE          CORR.    TOLERANCE      TO ENTER    LEVEL
(Y-INTERCEPT         943.76917 )                                                        .
rain        3          1.64526        0.6138     0.264     0.59347         7.19        1 . pop_den       5   0.11239   0.76095          0.69          1
educatn     4        -13.88546        6.8878    -0.189     0.65753         4.06        1 . nox           7  -0.10150   0.61966          0.56          1
nonwhite    6          3.31995        0.5853     0.476     0.81752        32.17        1 . so2           8   0.15573   0.37873          1.34          1
log_so2    10         34.63850        7.9100     0.362     0.84214        19.18        1 . log_nox      11   0.12165   0.29082          0.81          1

***** F LEVELS(  4.000,   3.900) OR TOLERANCE INSUFFICIENT FOR FURTHER STEPPING


         STEPWISE REGRESSION COEFFICIENTS

  VARIABLES  0 Y-INTCPT  3 rain       4 educatn    5 pop_den    6 nonwhite  7 nox         8 so2         10 log_so2  11 log_nox
STEP
   0          940.3817*     3.1741     -37.5586       0.0111       4.4879      -0.1041       0.4181        38.5874      35.3093
   1          887.1108*     1.8296     -28.9312       0.0115       4.4879*     -0.1200       0.3257        35.4586      21.3136
   2         1211.3760*     0.8007     -28.9312*      0.0080       3.9155*      0.0005       0.2573        27.9767      24.2406
   3         1111.9380*     1.6453     -23.4120*      0.0031       3.9178*     -0.2141       0.0903        27.9767*     -7.4259
   4          943.7692*     1.6453*    -13.8855*      0.0031       3.3200*     -0.0974       0.1398        34.6385*     15.3551
```

[11]

```
***   NOTE    *** 1) REGRESSION COEFFICIENTS FOR VARIABLES IN
                     THE EQUATION ARE INDICATED BY AN ASTERISK.
                  2) THE REMAINING COEFFICIENTS ARE THOSE WHICH WOULD BE
                     OBTAINED IF THAT VARIABLE WERE TO ENTER IN THE NEXT STEP

 SUMMARY TABLE

STEP          VARIABLE                    MULTIPLE      CHANGE    F TO     F TO    NO.OF VAR.
 NO.    ENTERED        REMOVED            R     RSQ     IN RSQ    ENTER   REMOVE   INCLUDED
  1    6 nonwhite                     0.6435 0.4142    0.4142     41.00                1
  2    4 educatn                      0.7496 0.5619    0.1478     19.23                2
  3   10 log_so2                      0.8012 0.6419    0.0799     12.50                3
  4    3 rain                         0.8266 0.6833    0.0414      7.19                4
```

[12]

2R uses only complete cases. Cases with missing values or values outside specified range limits (see MIN and MAX in the VARIABLE paragraph) are excluded from the analysis. All input variables are checked for invalid values unless you specify a USE list in the VARIABLE paragraph. If only the dependent variable is missing, 2R computes its predicted value.

[1] Descriptive statistics for the original and derived variables (using complete cases). Note that the high skewness and kurtosis values for the original air pollution measures (and their largest standard scores) are reduced after log transformation. Skewness, kurtosis, and standard scores are omitted if output is narrow (LINESIZE = 80).

[2] Interpretation of the REGRESS paragraph. Since we have specified only the DEPENDENT variable, the remainder of the options assume their default values. The stepping algorithm is "F"; that is, the entry or removal of variables from the equation is based on *F*-to-enter or *F*-to-remove limits (explained in **[6]** below).

[3] Step 0 includes the simple (Pearson) correlation of each independent variable with the dependent variable (listed in the column labeled PARTIAL CORR). At Step 0 the *F*-to-enter value for each independent variable is the *F* statistic appropriate for testing a single correlation, i.e., the significance of the simple linear regression of the dependent variable on that independent variable. See **[7]** for more about *F*-to-enter.

[4] 2R prints results at each step. We describe the results at Step 1, where NONWHITE is the first variable to enter the equation (it has the largest *F*-to-enter at Step 0). 2R prints the multiple correlation, *R* (the correlation of the dependent variable *y* with its predicted value), as well as

– *multiple* R^2

– *adjusted* R^2: $R^2 - p(1 - R^2)/(N - p')$,

p = no. of independent vars
$p' = p + 1$ if intercept is present
p if intercept is set to 0
N = number of cases

– *standard error of the estimate:*

$$\sqrt{\sum w_j\left(y_j - \hat{y}_j\right)^2 / (N - p')}$$

In this sample, NONWHITE accounts for approximately 40% (.4142) of the variability of the mortality rate.

[5] The analysis of variance table for the regression, containing

– *regression sum of squares:* $\sum w_j(\hat{y}_j - \bar{y})^2$

– *residual sum of squares:* $\sum w_j(y_j - \hat{y}_j)^2$

– *F ratio:* regression mean square divided by residual mean square

The *F* ratio is an overall test of the significance of the coefficients of the independent variables in the regression equation. Since 2R selects variables for inclusion in the equation in a manner that maximizes the *F* ratio, the level of significance of the *F* ratio cannot be obtained from the standard *F* distribution (see Forsythe et al., 1973); that is, it is not clear what probability is associated with each *F* statistic. When important variables have not yet been included in the equation, the denominator (the residual mean square) of the *F* ratio is inflated and the size of the *F* may be falsely small.

[6] At Step 1 the model is the equation of a straight line: MORTALTY = 887.11 + 4.488 NONWHITE. For independent variables in the equation, 2R prints

– *estimated regression coefficient:* b_i

– *estimated standard error of the coefficient:* $s(b_i)$

– *estimated standardized regression coefficient:* $b_i s_i / s_y$ The regression coefficient for standardized variables—omitted if output is narrow (LINESIZE = 80).

– *tolerance:* A check that the variable is not too highly correlated (collinear) with one or more of the other variables already in the equation. The computations lose numerical accuracy when TOLERANCE values are close to zero. See Special Features.

– *F-to-remove: F*-to-remove tests the regression coefficient to determine the relative importance of variables already in the equation. It is equal to $[b_i/s(b_i)]^2$. It is also the ratio

$$\frac{\mathrm{SS(removed)} - \mathrm{SS(residuals)}}{\mathrm{SS(residuals)}/(N - p')}$$

where SS(removed) is the sum of squares of residuals after removing the independent variable in question. If the *F*-to-remove of a variable is less than the acceptable limit (3.9 in this default run) and less than the value for the other variables entered, the variable is removed from the equation. The *F*-to-remove values are helpful in determining the relative importance of the selected variables at a given step. The usual tabled *F* values should not be used to test the need to remove a variable from the model (but see comment below concerning *F*-to-enter). Note that the *F*-to-remove value for EDUCATN in Step 2 is equal to its *F*-to-enter in Step 1.

- *level:* You can use LEVEL to order the entry of variables (see Special Features).

[7] For each independent variable not yet in the equation, 2R prints

- *partial correlation:* The correlation of each independent variable with the dependent variable, removing the linear effect of variables already in the equation (see Appendix B.16).
- *tolerance* (see **[6]** above)
- *F-to-enter:* F-to-enter tests the regression coefficient for each remaining independent variable as if it were entered separately into the equation at the next step. Let SS(residuals) be the sum of squares of residuals at this step and SS(next) be the residual sum of squares after the potential independent variable is entered. Then F-to-enter is the ratio

$$\frac{\text{SS(residuals)} - \text{SS(next)}}{\text{SS(next)}/(N - p' - 1)}$$

The F-to-enter value indicates how much the independent variable would enhance the prediction of the dependent variable by entering the equation next. The usual tabled F values (percentiles of the F distribution) should not be used to test the need to include a variable in the model. The distribution of the largest F-to-enter is affected by the number of variables available for selection, their correlation structure, and the sample size. When the independent variables are correlated, the critical value for the largest F can be much larger than that for testing one preselected variable. That is, using the usual table can yield a predictive equation with so many variables that the equation performs poorly on new data. See BMDP Technical Report #48 for further discussion.

In Step 1, EDUCATN has the largest F-to-enter value (19.23). Since this value exceeds the default F-to-enter limit of 4.0, EDUCATN will be entered in the next step.

- *level* (see **[6]** above)

[8] At Step 2 EDUCATN enters. The adjusted R^2 increases to .55.

[9] The air pollution measure LOG_SO2 enters, increasing the adjusted R^2 to .62, an increase of almost 8%.

[10] The last variable to enter is RAIN, bringing the adjusted R^2 to .66. The remaining variables have F-to-enter values less than 4.0 (the default value), so the stepping stops. The F-to-remove values of the entered variables are all greater than the remove limit of 3.9, so no variables are removed. The final model from this default run is

MORTALTY = 943.77 + 1.65 RAIN – 13.89 EDUCATN + 3.32 NONWHITE
+ 34.64 LOG_SO2

Note that the F-to-remove for EDUCATN is now marginal (4.06). Some data analysts might want to consider a simplified equation without this variable. The model resulting from a default run is often satisfactory, but you should take care to ensure that no outliers or gross recording errors exert undue influence on the correlations. In later examples, we find that data values for several cities are very unusual. Note that when the independent variables are intercorrelated, two or more subsets of variables may perform almost equally well. For more information see program 9R.

[11] Stepwise regression coefficients. This table shows the values of the regression coefficients of the variables in the equation (those with asterisks) at each step, and the coefficients of the other variables as if each had been entered separately into the regression equation.

[12] A summary table reports the variable entered, the multiple R, R^2, change in R^2, and the F-to-enter at each step.

Example 2R.3 Printing data, residuals, predicted values, correlations, and summary tables

Commands in the PRINT paragraph allow you to request additional results. For a full list of available options, see 2R Commands. Some results are printed automatically:

- ANOVA – analysis of variance table
- STEP – results at each step
- COEF – summary table of regression coefficients
- SUM – summary table of R, R^2, F-to-enter, and F-to-remove at each step

Each of these panels of output may be suppressed by stating NO ANOVA, NO STEP, etc.

In Input 2R.3, we specify DATA in the PRINT paragraph to request a printout of data, residuals, and predicted values for each case. We also specify CORRELATION, PARTIAL, and FRATIO to obtain the correlation matrix of the variables, the partial correlations, and a summary table of F-to-enter and F-to-remove values at each step. Regression diagnostics are also available for printing; see Example 2R.6 and Special Features: Printing data and diagnostics.

Input 2R.3
(add to Input 2R.2 before END)

```
/ PRINT   DATA.
          CORRELATION.
          PARTIAL.
          FRATIO.
```

Output 2R.3

```
CORRELATION MATRIX

                rain       educatn    pop_den    nonwhite   nox        so2        mortalty   log_so2    log_nox        [1]

                     3          4          5          6          7          8          9         10         11
rain       3    1.0000
educatn    4   -0.4904     1.0000
pop_den    5   -0.0088    -0.2350     1.0000
nonwhite   6    0.4132    -0.2088    -0.0127     1.0000
nox        7   -0.4873     0.2244     0.1654     0.0184     1.0000
so2        8   -0.1069    -0.2343     0.4264     0.1593     0.4094     1.0000
mortalty   9    0.5094    -0.5103     0.2614     0.6435    -0.0775     0.4260     1.0000
log_so2   10   -0.1212    -0.2562     0.4702     0.0524     0.3582     0.7738     0.4034     1.0000
log_nox   11   -0.3683     0.0180     0.3466     0.1897     0.7054     0.6905     0.2919     0.7328     1.0000
```

— we omit the regression steps —

```
         F-TO-ENTER OR F-TO-REMOVE OF EACH VARIABLE AT EACH STEP

  VARIABLES  3 rain      4 educatn   5 pop_den   6 nonwhite  7 nox       8 so2      10 log_so2  11 log_nox        [2]
 STEP
   0         20.3259     20.4236      4.2530     41.0026      0.3507     12.8581     11.2738      5.4040
   1          7.9235     19.2306      8.0735     41.0026*     0.7878     12.7891     17.4054      3.0697
   2          1.4257     19.2306*     4.5603     39.2342*     0.0000      9.5952     12.4993      5.4093
   3          7.1852     14.1797*     0.6466     47.2043*     3.1384      0.5081     12.4993*     0.2199
   4          7.1852*     4.0641*     0.6908     32.1742*     0.5621      1.3422     19.1765*     0.8112


         PARTIAL CORRELATIONS

  VARIABLES  3 rain      4 educatn   5 pop_den   6 nonwhite  7 nox       8 so2      10 log_so2  11 log_nox        [3]
 STEP
   0          0.5094     -0.5103      0.2614      0.6435     -0.0775      0.4260      0.4034      0.2919
   1          0.3493     -0.5023      0.3522      0.6435*    -0.1168      0.4281      0.4837      0.2261
   2          0.1576     -0.5023*     0.2744      0.6385*     0.0006      0.3825      0.4272      0.2968
   3          0.3399     -0.4495*     0.1078      0.6763*    -0.2323      0.0957      0.4272*    -0.0631
   4          0.3399*    -0.2623*     0.1124      0.6075*    -0.1015      0.1557      0.5085*     0.1217

 SUMMARY TABLE

STEP        VARIABLE                 MULTIPLE    CHANGE    F TO    F TO    NO.OF VAR.          [4]
 NO.   ENTERED       REMOVED         R      RSQ   IN RSQ    ENTER   REMOVE   INCLUDED
  1   6 nonwhite                   0.6435 0.4142 0.4142    41.00                 1
  2   4 educatn                    0.7496 0.5619 0.1478    19.23                 2
  3  10 log_so2                    0.8012 0.6419 0.0799    12.50                 3
  4   3 rain                       0.8266 0.6833 0.0414     7.19                 4


SERIAL CORRELATION  0.0457                                                                 [5]
DURBIN-WATSON STATISTIC  1.9062 BASED ON   60 CASES


 LIST OF PREDICTED VALUES, RESIDUALS, AND VARIABLES
 - CASES WITH MISSING VALUES ARE MARKED WITH A MINUS SIGN
   BETWEEN THE CASE NUMBER AND CASE LABEL.
 - ASTERISKS (UP tO 3) TO THE RIGHT OF A RESIDUAL INDICATE THAT
   THE RESIDUAL DEVIATES FROM THE MEAN BY MORE THAN THAT NUMBER                            [6]
   OF STANDARD DEVIATIONS.
 - MISSING VALUES AND VALUES OUT OF RANGE ARE DENOTED BY
   VALUES GREATER THAN OR EQUAL TO 2.12676E+37 IN ABSOLUTE VALUE.
```

```
  CASE                                      9          3          4          5           6         7          8
NO. LABEL      PREDICTED  RESIDUAL   WEIGHT  mortalty   rain       educatn  pop_den     nonwhite   nox        so2
  1 akronOH     935.2594  -13.3594    1.000   921.9000   36.0000   11.4000  3243.0000    8.8000   15.0000    59.0000
  2 albanyNY    915.3450   82.5550**  1.000   997.9000   35.0000   11.0000  4281.0000    3.5000   10.0000    39.0000
  3 allenPA     935.3380   27.0621    1.000   962.4000   44.0000    9.8000  4260.0000    0.8000    6.0000    33.0000
  4 atlantGA   1004.7467  -22.4467    1.000   982.3000   47.0000   11.1000  3125.0000   27.1000    8.0000    24.0000
  5 baltimMD   1042.3705   28.6295    1.000  1071.0000   43.0000    9.6000  6441.0000   24.4000   38.0000   206.0000
  6 birmhmAL   1081.4894  -51.4894*   1.000  1030.0000   53.0000   10.2000  3325.0000   38.5000   32.0000    72.0000
```

— we omit results for cases 7 through 26 —

```
 27 kansasMO    897.4135   22.2866    1.000   919.7000   35.0000   12.0000  3262.0000   12.6000    4.0000     4.0000
 28 lancasPA    944.3673 -100.2673**  1.000   844.1000   43.0000    9.5000  3214.0000    2.9000    7.0000    32.0000
 29 losangCA    892.9724  -31.1724    1.000   861.8000   11.0000   12.1000  4700.0000    7.8000  319.0000   130.0000
 30 louisvKY    978.3204   10.9796    1.000   989.3000   30.0000    9.9000  4474.0000   13.1000   37.0000   193.0000
 31 memphsTN   1056.5135  -50.5135*   1.000  1006.0000   50.0000   10.4000  3497.0000   36.7000   18.0000    34.0000
 32 miamiFL     927.6211  -66.2211*   1.000   861.4000   60.0000   11.5000  4657.0000   13.5000    1.0000     1.0000
 33 milwauWI    930.8878   -1.6878    1.000   929.2000   30.0000   11.1000  2934.0000    5.8000   23.0000   125.0000
 34 minnplMN    872.5390  -14.9390    1.000   857.6000   25.0000   12.1000  2095.0000    2.0000   11.0000    26.0000
 35 nashvlTN   1012.8209  -51.8209*   1.000   961.0000   45.0000   10.1000  2082.0000   21.0000   14.0000    78.0000
 36 newhvnCT    923.0425    0.1575    1.000   923.2000   46.0000   11.3000  3327.0000    8.8000    3.0000     8.0000
 37 neworlLA   1002.1705  110.8295*** 1.000  1113.0000   54.0000    9.7000  3172.0000   31.4000   17.0000     1.0000
 38 newyrkNY    972.2457   22.3542    1.000   994.6000   42.0000   10.7000  7462.0000   11.3000   26.0000   108.0000
 39 philadPA   1001.6129   13.3871    1.000  1015.0000   42.0000   10.5000  6092.0000   17.5000   32.0000   161.0000
```

— we omit the printout for the remaining cities and the panel of results for LOG SO2 and LOG NOX —

[1] Correlation matrix of the variables.

[2] Summary table of the *F*-to-enter and *F*-to-remove values at each step. Variables included in the equation are marked with asterisks.

[3] Summary table of partial correlations.

[4] Summary table of regression results. Table includes, *R*, R^2, and the change in the statistics as more variables are entered. In addition, the *F*-to-Enter and *F*-to-Remove stepping criteria are shown, with the variable maintaining the largest *F*-to-Enter value entered into the model first (nonwhite), followed in order by the remainder of the variables in descending order of *F*-to-Enter values (educatn, log_so2, and rain).

[5] The serial correlation is calculated and printed along with the Durbin-Watson test.

[6] 2R prints the residual, predicted value, and data for each case. Asterisks after each residual indicate the number of standard deviations that lie between the value and the overall mean; three asterisks represent three or more standard deviations. New Orleans has the most extreme residual.

Example 2R.4 Plotting predicted values, observed values, and residuals

In Input 2R.4 we use commands in the PLOT paragraph to request a number of standard plots for predicted values, observed values, and residuals:

RESIDUALS – plots of residuals and residuals squared against predicted values

NORMAL – normal probability plot of residuals

DNORMAL – detrended normal probability plot

VARIABLE – predicted and observed values of the dependent variable against the independent variable RAIN, and residuals against RAIN

PREP – partial residual plot for LOG_SO2

Both VARIABLE and PREP allow you to specify more than one variable to be plotted. We also specify a plot SIZE 44 characters wide and 12 lines high (smaller than the default size of 50, 35).

Input 2R.4
(add to Input 2R.2 before END)

```
/ PLOT  RESIDUALS.
        VARIABLE = rain.
        NORMAL.
        DNORMAL.
        PREP = log_so2.
        SIZE = 44, 12.
```

Output 2R.4

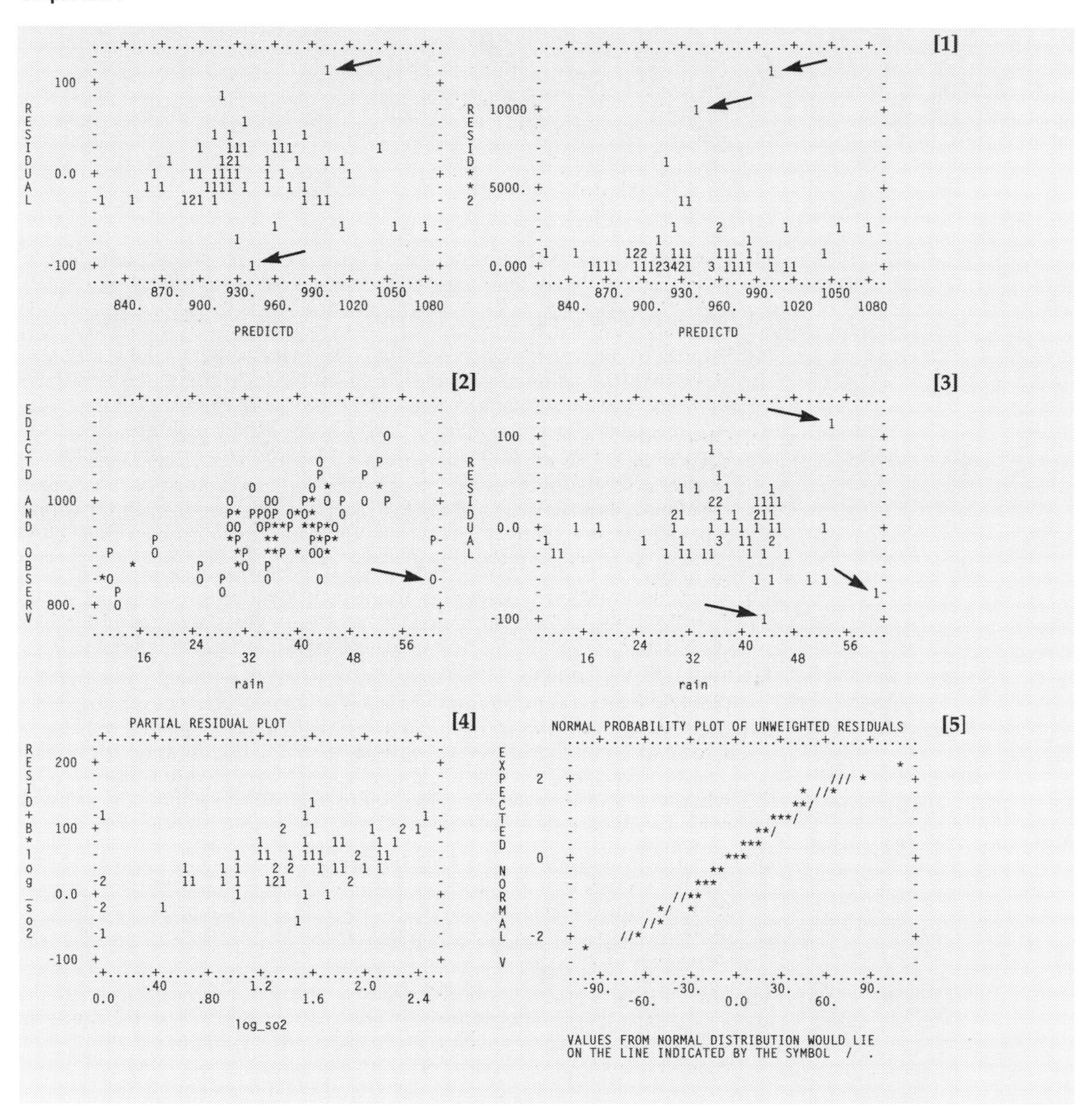

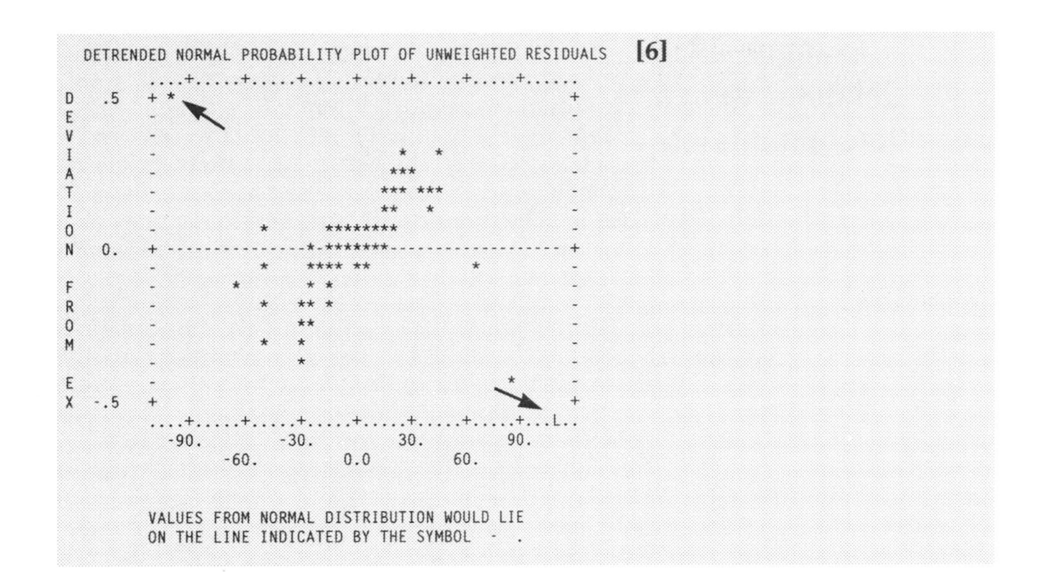

[1] Plots of residuals and residuals squared against the predicted values. 2R prints the number of points at each position. From the printout of residuals in Output 2R.3, we identify the largest positive residual as New Orleans and the largest negative residual as Lancaster. That is, the model predicts a lower mortality rate for New Orleans and a higher rate for Lancaster than is observed. (We add the dashed line for clarity.) The squared residuals for these two cities stand out at the top of the second plot.

[2] Observed (O) and predicted (P) values of the dependent variable plotted against the values of the independent variable RAIN. An asterisk represents an overlap between observed and predicted values. Note the pair of values for Miami plotted on the right side of the plot. Because this point is widely separated from the others, it could have a disproportionate leverage effect on the regression. For more about leverage, see Example 2R.6.

[3] The residuals plotted against RAIN (we add the dashed line). The largest positive residual belongs to New Orleans; the largest negative residual belongs to Lancaster. The point at the edge of the plot is again Miami.

[4] The partial residual plot for EDUCATN (we add the dashed line). This is a plot of the residual plus the contribution of the independent variable to the regression against the observed values of that variable: $[(y - \hat{y}) + bx]$ against x for each case. Systematic deviations from the line bx can reveal what transformation f to use on x so that $f(x)$ is linearly related to y. For example, when y is linearly related to $\log(x)$, the plot will resemble a logarithmic curve. See Larsen and McCleary (1972).

[5] A normal probability plot of the residuals. The residual is plotted against the expected normal deviate corresponding to its rank. If the residuals are from a normal distribution, they will fall along the line indicated by slashes. Because of the small plot size used here, several slashes sometimes fall on a single line.

[6] The detrended normal probability plot is similar to the normal probability plot except that the linear trend is removed and the line in **[5]** is rotated until it is horizontal. This plot has higher resolution than the normal plot. Residuals from a normal distribution should cluster around zero with little apparent pattern. Here there is a lack of random scatter about the line, and there may be a problem with outliers. The L on the bottom right corner of the plot frame indicates that the residual for New Orleans falls outside the plot. The point in the top left corner of the plot represents Lancaster.

Example 2R.5 Obtaining added variable miniplots

In this example, we use data from the air pollution study to illustrate use of the added variable miniplot commands. We want to plot on the vertical axis the residual of the mortality rate (the dependent variable) obtained when the variable(s) already in the regression are used. In this case the first variable to enter is LOG_NOX, so the vertical axis plots the mortality residual when LOG_NOX is the independent variable. The horizontal axis plots the residual in the next variable being considered for entry, when it is predicted by the independent variable(s) already entered. Miniplots are ordered by *F*-to-enter values (highest *F*-to-enter on the left, lowest on the right).

In Input 2R.5, we add a command to the TRANSFORM paragraph in Input 2R.2 and add a PLOT and PRINT paragraph. The TRANSFORM paragraph command restricts the USE of RAIN variable values and certain cases. The PLOT paragraph includes three commands for added variable miniplots. The NADD command activates the option and specifies that up to four (4) miniplots be printed at each step of the regression. The SADD command requests that miniplots be printed for the first (1) and the last step (LAST). The EXTREME command lists the five extreme cases for each plot. The PRINT paragraph sets the number of characters printed per line to 80.

Input 2R.5
(add TRANSFORM command and PLOT and PRINT to Input 2R.2)

```
/ TRANSFORM  USE = rain GE 20 AND KASE NE 28 AND KASE NE
             59 AND KASE NE 29 AND KASE NE 47 AND KASE
             NE 48 AND KASE NE 49.

/ PLOT       NADD = 4.
             SADD = 1, LAST.
             EXTREME = 5.

/ PRINT      LINESIZE = 80.
```

Output 2R.5

– we omit preliminary output –

```
STEP NO.    1
--------------
VARIABLE ENTERED   11 LOG_NOX                        [1]

MULTIPLE R              0.6501
MULTIPLE R-SQUARE       0.4227
ADJUSTED R-SQUARE       0.4114

STD. ERROR OF EST.     43.6472

ANALYSIS OF VARIANCE
                    SUM OF SQUARES     DF    MEAN SQUARE      F RATIO
        REGRESSION  71134.563           1   71134.56            37.34
        RESIDUAL    97158.813          51   1905.075

      VARIABLES IN EQUATION                  VARIABLES NOT IN EQUATION
--------------------------------------------  ------------------------------
                   STD.ERR           F                    PARTIAL         F
VARIABLE   COEFF.  OF COEFF  TOL.  REMOVE(L)  VARIABLE  CORR.   TOL.   ENTER(L)
--------------------------------------------  ------------------------------
(CONSTANT 874.0245)
LOG_NOX   81.6287  13.3585 1.0000   37.34(1)  RAIN      0.5119 0.9980   17.75(1)
                                              EDUCATN  -0.5353 0.9273   20.09(1)
                                              POP_DEN   0.0827 0.7540    0.34(1)
                                              NONWHITE  0.5942 0.9112   27.28(1)
                                              NOX      -0.1208 0.2326    0.74(1)
                                              SO2      -0.0997 0.3785    0.50(1)
                                              LOG_SO2  -0.1959 0.2914    1.99(1)

ADDED VARIABLE PLOTS FOR VARIABLES NOT IN EQUATION.
(PLOTS ARE ORDERED BY F-TO-ENTER VALUES).
ZEROES INDICATE CASES OMITTED DUE TO ZERO WEIGHT, ZERO FREQUENCY, OR TRANSFORM     [2]
USE=0. TO EXCLUDE "ZERO" CASES FROM PLOTS, SET USE=-1.
```

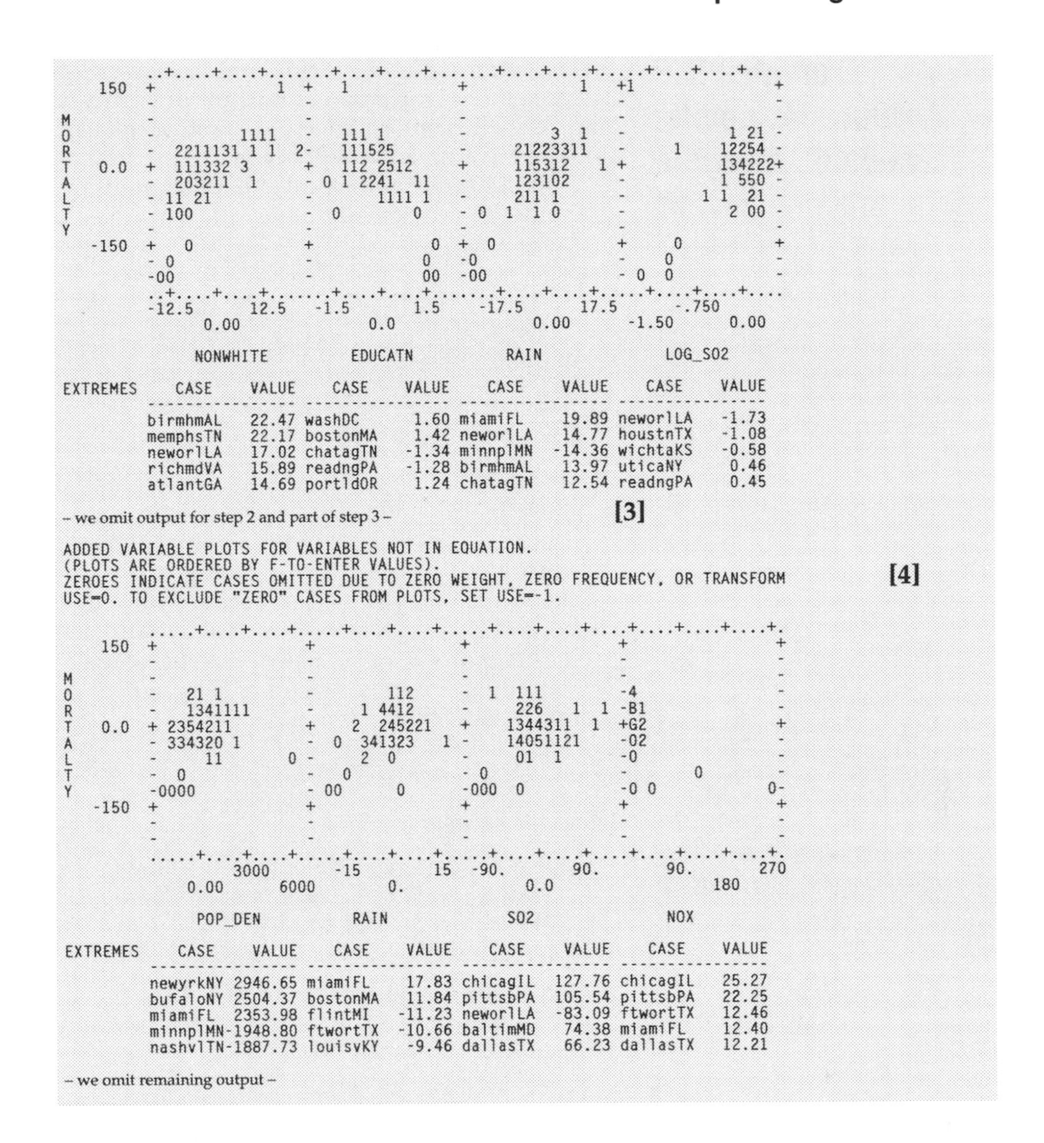

```
              NONWHITE              EDUCATN               RAIN                LOG_SO2

EXTREMES    CASE     VALUE      CASE     VALUE      CASE     VALUE      CASE     VALUE
          ---------------- ---------------- ---------------- ----------------
          birmhmAL   22.47  washDC      1.60  miamiFL    19.89  neworlLA   -1.73
          memphsTN   22.17  bostonMA    1.42  neworlLA   14.77  houstnTX   -1.08
          neworlLA   17.02  chatagTN   -1.34  minnplMN  -14.36  wichtaKS   -0.58
          richmdVA   15.89  readngPA   -1.28  birmhmAL   13.97  uticaNY     0.46
          atlantGA   14.69  portldOR    1.24  chatagTN   12.54  readngPA    0.45
```

– we omit output for step 2 and part of step 3 – **[3]**

```
ADDED VARIABLE PLOTS FOR VARIABLES NOT IN EQUATION.
(PLOTS ARE ORDERED BY F-TO-ENTER VALUES).
ZEROES INDICATE CASES OMITTED DUE TO ZERO WEIGHT, ZERO FREQUENCY, OR TRANSFORM
USE=0. TO EXCLUDE "ZERO" CASES FROM PLOTS, SET USE=-1.
```

[4]

```
              POP_DEN               RAIN                 SO2                  NOX

EXTREMES    CASE     VALUE      CASE     VALUE      CASE     VALUE      CASE     VALUE
          ---------------- ---------------- ---------------- ----------------
          newyrkNY 2946.65  miamiFL    17.83  chicagIL  127.76  chicagIL   25.27
          bufaloNY 2504.37  bostonMA   11.84  pittsbPA  105.54  pittsbPA   22.25
          miamiFL  2353.98  flintMI   -11.23  neworlLA  -83.09  ftwortTX   12.46
          minnplMN-1948.80  ftwortTX  -10.66  baltimMD   74.38  miamiFL    12.40
          nashvlTN-1887.73  louisvKY   -9.46  dallasTX   66.23  dallasTX   12.21
```

– we omit remaining output –

[1] 2R prints results at each step. In step 1, LOG_NOX is the first variable to enter the equation. 2R prints the multiple correlation, R^2, the adjusted R^2, and the standard error of the estimate for LOG_NOX. Statistics for variables not entered into the equation include partial correlation and *F*-to-Enter values.

[2] Added variable miniplots for step 1 of the regression. 2R prints plots for the first four variables (NADD = 4) not included in the equation. Unentered variables with the highest *F*-to-Enter values are chosen for the plots. Plots are also ordered according to *F*-to-Enter values; the variable with the highest value, NONWHITE (*F*-to-Enter = 27.28), is printed first. Numbers in plots indicate frequency.

[3] We omit results for step 2 of the regression since we did not specify miniplots for that step. In step 3 preliminary results, three variables are included: EDUCATN, NONWHITE, and LOG_NOX. EDUCATN enters first. Five variables do not enter: RAIN, POP_DEN, NOX, SO2, and LOG_SO2.

[4] Added variable miniplots for the third (LAST) step of the regression. Five variables have not yet entered the equation. 2R plots four of those variables (ranked by *F*-to-enter values). If we had specified NADD = 5, 2R would have printed all five of the remaining variables.

Example 2R.6 Advanced example: caseplots, bivariate plots, and diagnostics

Before interpreting the results of a regression analysis, it is a good idea to find out whether your data contain observations that violate important assumptions. When there are several independent variables, one or more multivariate outliers may distort the analysis. 2R offers a variety of diagnostic statistics to assist in identification of unusual observations (see Table 2R.1). The diagnostics fall roughly into three groups:

- *Residual statistics* identify potential outliers in the dependent variable space (*Y* space).
- *Leverage statistics* identify potential outliers in the independent variable space (*X* space).
- *Influence statistics* combine the effects of leverage and residuals to identify influential cases.

Observations having either extreme leverage or large residuals are said to be **potentially** influential. Influence statistics confirm whether or not the potentially influential cases exert undue force on the regression. Most influence statistics measure how deleting a case changes the estimates of the regression coefficients, while some statistics measure how deleting a case changes the coefficients and their variance. It is usually sufficient to consider one statistic from each of the three groups. The diagnostic statistics are available for both printing and plotting, either at the end of the analysis or at each step. You can also save them in a BMDP File.

Table 2R.1
Diagnostics: measures of leverage, residual, and influence

Leverage Measures	
HATDIAG[1] Diagonal elements of the "hat," or projection, matrix (h_{ii}). The h_{ii} vary between $1/n$ and 1, with an average value of p/n. **X** is the (n x p) matrix of centered independent variables; $\mathbf{x}_i^t$ is the ith row of **X**.	$\frac{1}{n}+x_i^t(X^tX)^{-1}x_i$
RHAT[2] A monotonic function of HATDIAG that emphasizes outlying values in the independent variable space more than HATDIAG.	$\frac{h_{ii}}{1-h_{ii}}$
MAHAL *Mahalanobis distance* : the distance of each case to the mean of all cases used in the regression equation. Computed using only the independent variables.	$(n-1)\left(h_{ii}-\frac{1}{n}\right)$
AP2[3] The 2nd factor of the Andrews-Pregibon statistic in one dimension. Small values are considered extreme.	$1-h_{ii}$
Residual Measures	
RESIDUAL The usual residual, e_i.	$y_i-\hat{y}_i$
STRESID[1] The *standardized residual*, r_i, is the usual residual divided by an estimate of its standard error.	$\frac{e_i}{\hat{\sigma}\sqrt{1-h_{ii}}}$
DELRESID[1] The *deleted residual*, $e_{(i)}$, for the ith case is its residual from the regression equation computed with the ith case deleted.	$\frac{e_i}{1-h_{ii}}$
DSTRESID[1] The *deleted standardized residual*, t_i, is the usual residual divided by an independent estimate of its standard error (the same as DELRESID divided by SE(DELRESID)). We use the notation t because DSTRESID follows a t-distribution with $(n-p-1)$ df.	$\frac{e_i}{\hat{\sigma}_{(i)}\sqrt{1-h_{ii}}}$
AP1[3] The 1st factor of the Andrews-Pregibon statistic in one dimension. Small values are considered extreme.	$1-\frac{r_i^2}{n-p}$
Q1[3] $Q1_i$ is the amount of decrease in the residual sum of squares when the ith case is deleted.	$\frac{e_i^2}{1-h_{ii}}$

Table 2R.1 (continued)

U	U_i is proportional to the contribution of the ith residual to the residual sum of squares.	$\frac{ne_i^2}{\hat{\sigma}^2\ (n-p)}$
WRESIDUL	If you specified a WEIGHT variable in the VARIABLE paragraph and c_i is the value stored in WEIGHT, then w_i is the case weight divided by the average of all the case weights ($\Sigma c_i / n$).	$\sqrt{w_i}\ e_i$
Influence Measures		
COOK[1]	Cook's distance is a measure of the influence of the ith case on the estimated regression coefficients.	$\frac{e_{(i)}^2 h_{ii}}{p\hat{\sigma}^2}$
MODCOOK[4,2]	The modified Cook's distance is a measure of the influence of the ith case on the regression coefficients and the variance of the coefficients.	$\left[\frac{(n-p)h_{ii}}{p(1-h_{ii})}\right]^{1/2} \lvert t_i \rvert$
DFFITS[4,5]	A measure of influence similar to MODCOOK. In addition to being scaled differently, the sign of the residual is preserved.	$\left(\frac{h_{ii}}{1-h_{ii}}\right)^{1/2} t_i$
AP[3]	The Andrews-Pregibon statistic in one dimension. Used to identify single influential observations.	$AP2 \cdot AP1$

Note: 2R allows case weights (c_i), but for simplicity the c_i are omitted from these formulas.)
[1]Cook and Weisberg (1982) [2]Atkinson (1985) [3]Draper and John (1981)
[4]Welsch and Kuh (1977) [5]Belsley, Kuh, and Welsch (1980)

Instead of using cutoffs or calibration points to identify extreme values for each diagnostic, many users prefer to examine plots of the diagnostics for points that stand apart from the others. Several options available for plotting diagnostics are summarized in Table 2R.2.

Table 2R.2
Summary of options for diagnostic plots

1. **Caseplots.** The CASEPLOT command displays three diagnostics for each case: a standardized residual, a leverage measure, and an influence measure. The diagnostics appear in side-by-side miniplots with case numbers and labels running down the left side. Since all three diagnostics are simultaneously visible, these plots provide a wealth of information at a glance.
2. **BMDP's STANDARD selection.** The command DIAGNOSTICS = STANDARD requests five plots:
   - added variable plot (or partial regression leverage plot)
   - STRESID versus a scaled version of HATDIAG (letters mark cases with extreme values of MODCOOK)
   - STRESID versus MODCOOK (letters mark cases with extreme values of HATDIAG)
   - half-normal probability plot of DSTRESID
   - error variance plot
   - transformation plot
3. **Your own bivariate scatterplots.** Use XVAR and YVAR lists to create plots of your design, choosing from the 16 diagnostics in Table 2R.1, the dependent variable, and the independent variables. Also available are the predicted value (PREDICTD) and four values used in the standard plots (XRESIDUL, YRESIDUL, STRES**2, and AP2*PRED).

In Input 2R.6, we request the caseplots and bivariate plots. We use the bivariate plot option to make our own caseplot for the influence measure DFFITS. In the TRANSFORM paragraph we create a variable containing the case number for each case (the BMDP case sequence number KASE is defined in Chapter 6). We then plot the case number against DFFITS:

```
YVAR = DFFITS.    XVAR = 'case #'.
```

We request a plot SIZE of 60, 18 so that all 60 cases will be uniquely identified on the x-axis. The caseplots, bivariate plots, and summary panels are printed at each step if you add the command STEP to the PLOT paragraph.

We also use a PRINT paragraph to request a printout of four diagnostics for each case. You can request a printout of ALL diagnostics, a STANDARD set (STRESIDL, DSTRESID, HATDIAG, and MODCOOK), or a list of just the diagnostics of interest to you. For a summary of printout options, see Special Features: Printing Data and Diagnostics.

Input 2R.6
(add to Input 2R.2 before END)

```
/ TRANSFORM   'case #' = KASE.

/ PLOT        CASEPLOTS.
              XVAR = 'case #'.
              YVAR = DFFITS.
              SIZE = 60, 18.

/ PRINT       DIAGNOSTICS = STRESID, HATDIAG, MODCOOK,
                            DFFITS.
```

Output 2R.6

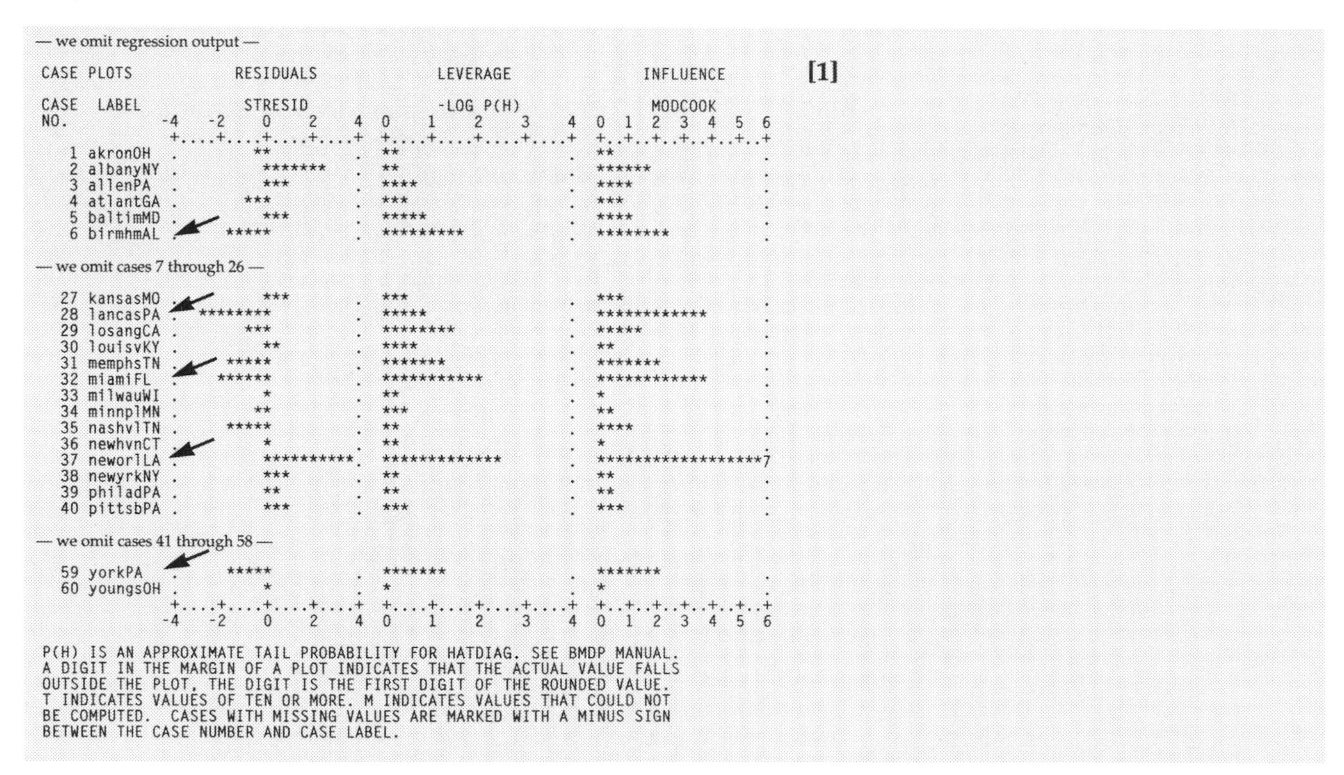

```
— we omit regression output —
CASE PLOTS            RESIDUALS                LEVERAGE               INFLUENCE          [1]

CASE  LABEL            STRESID                 -LOG P(H)               MODCOOK
NO.           -4   -2    0    2    4  0    1    2    3    4  0  1  2  3  4  5  6
               +....+....+....+....+  +....+....+....+....+  +..+..+..+..+..+..+
    1 akronOH  .        **         .  **                  .  **                .
    2 albanyNY .         *******   .  **                  .  ******            .
    3 allenPA  .         ***       .  ****                .  ****              .
    4 atlantGA .       ***         .  ***                 .  ***               .
    5 baltimMD .         ***       .  *****               .  ****              .
    6 birmhmAL .     *****         .  *********           .  ********          .
— we omit cases 7 through 26 —
   27 kansasMO .         ***       .  ***                 .  ***               .
   28 lancasPA .  ********         .  *****               .  ************      .
   29 losangCA .       ***         .  ********            .  *****             .
   30 louisvKY .         **        .  ****                .  **                .
   31 memphsTN .     *****         .  *******             .  *******           .
   32 miamiFL  .    ******         .  ***********         .  ************      .
   33 milwauWI .         *         .  **                  .  *                 .
   34 minnplMN .        **         .  ***                 .  **                .
   35 nashvlTN .     *****         .  **                  .  ****              .
   36 newhvnCT .         *         .  **                  .  *                 .
   37 neworlLA .         **********.  *************       .  *****************7
   38 newyrkNY .         ***       .  **                  .  **                .
   39 philadPA .         **        .  **                  .  **                .
   40 pittsbPA .         ***       .  ***                 .  ***               .
— we omit cases 41 through 58 —
   59 yorkPA   .     *****         .  *******             .  *******           .
   60 youngsOH .         *         .  *                   .  *                 .
               +....+....+....+....+  +....+....+....+....+  +..+..+..+..+..+..+
              -4   -2    0    2    4  0    1    2    3    4  0  1  2  3  4  5  6

P(H) IS AN APPROXIMATE TAIL PROBABILITY FOR HATDIAG. SEE BMDP MANUAL.
A DIGIT IN THE MARGIN OF A PLOT INDICATES THAT THE ACTUAL VALUE FALLS
OUTSIDE THE PLOT, THE DIGIT IS THE FIRST DIGIT OF THE ROUNDED VALUE.
T INDICATES VALUES OF TEN OR MORE. M INDICATES VALUES THAT COULD NOT
BE COMPUTED.  CASES WITH MISSING VALUES ARE MARKED WITH A MINUS SIGN
BETWEEN THE CASE NUMBER AND CASE LABEL.
```

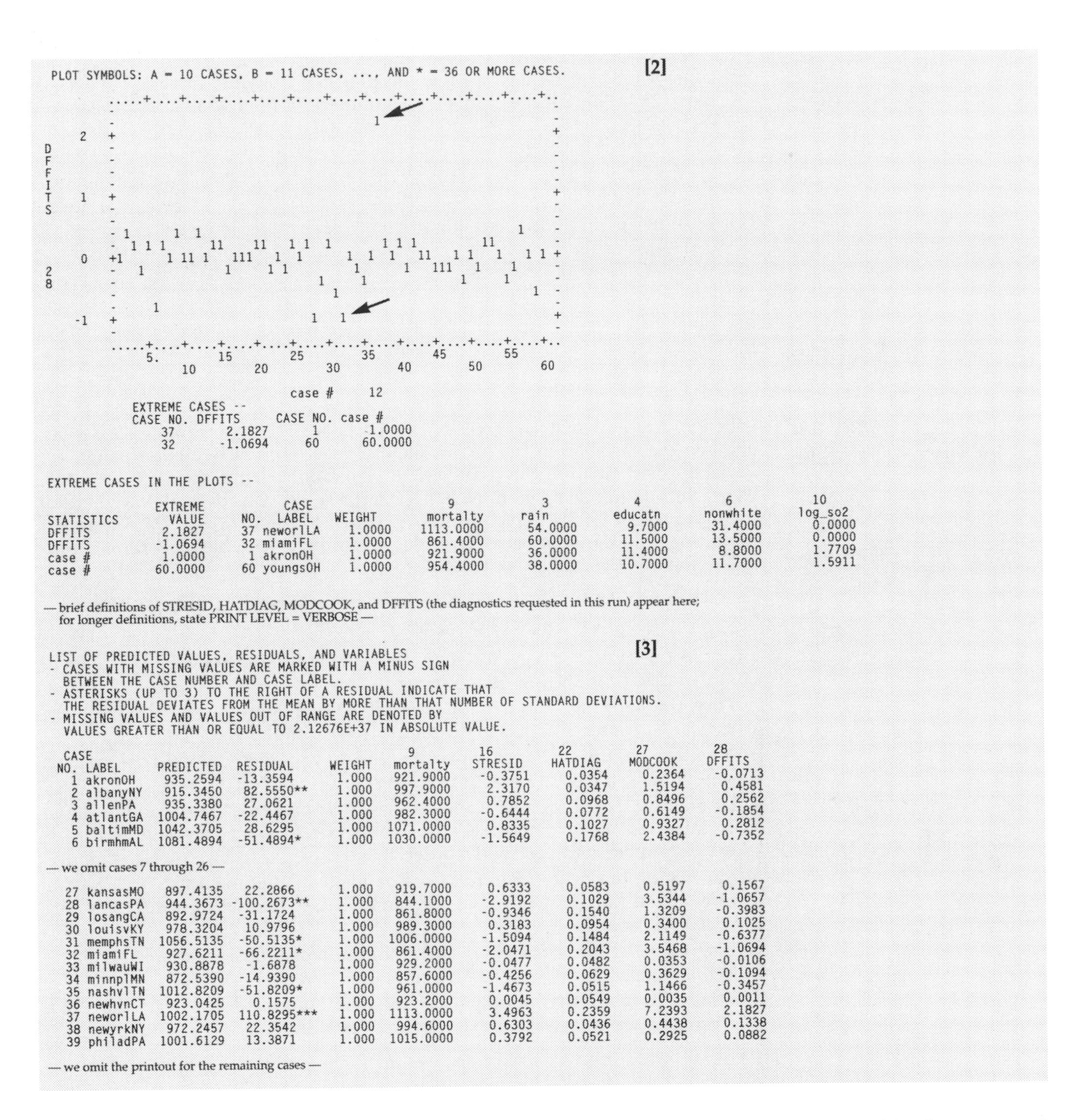


```
EXTREME CASES --
CASE NO. DFFITS     CASE NO. case #
   37      2.1827      1       1.0000
   32     -1.0694     60      60.0000
```

EXTREME CASES IN THE PLOTS --

STATISTICS	EXTREME VALUE	CASE NO.	CASE LABEL	WEIGHT	9 mortalty	3 rain	4 educatn	6 nonwhite	10 log_so2
DFFITS	2.1827	37	neworlLA	1.0000	1113.0000	54.0000	9.7000	31.4000	0.0000
DFFITS	-1.0694	32	miamiFL	1.0000	861.4000	60.0000	11.5000	13.5000	0.0000
case #	1.0000	1	akronOH	1.0000	921.9000	36.0000	11.4000	8.8000	1.7709
case #	60.0000	60	youngsOH	1.0000	954.4000	38.0000	10.7000	11.7000	1.5911

— brief definitions of STRESID, HATDIAG, MODCOOK, and DFFITS (the diagnostics requested in this run) appear here; for longer definitions, state PRINT LEVEL = VERBOSE —

[3]

```
LIST OF PREDICTED VALUES, RESIDUALS, AND VARIABLES
- CASES WITH MISSING VALUES ARE MARKED WITH A MINUS SIGN
  BETWEEN THE CASE NUMBER AND CASE LABEL.
- ASTERISKS (UP TO 3) TO THE RIGHT OF A RESIDUAL INDICATE THAT
  THE RESIDUAL DEVIATES FROM THE MEAN BY MORE THAN THAT NUMBER OF STANDARD DEVIATIONS.
- MISSING VALUES AND VALUES OUT OF RANGE ARE DENOTED BY
  VALUES GREATER THAN OR EQUAL TO 2.12676E+37 IN ABSOLUTE VALUE.
```

CASE NO.	LABEL	PREDICTED	RESIDUAL	WEIGHT	9 mortalty	16 STRESID	22 HATDIAG	27 MODCOOK	28 DFFITS
1	akronOH	935.2594	-13.3594	1.000	921.9000	-0.3751	0.0354	0.2364	-0.0713
2	albanyNY	915.3450	82.5550**	1.000	997.9000	2.3170	0.0347	1.5194	0.4581
3	allenPA	935.3380	27.0621	1.000	962.4000	0.7852	0.0968	0.8496	0.2562
4	atlantGA	1004.7467	-22.4467	1.000	982.3000	-0.6444	0.0772	0.6149	-0.1854
5	baltimMD	1042.3705	28.6295	1.000	1071.0000	0.8335	0.1027	0.9327	0.2812
6	birmhmAL	1081.4894	-51.4894*	1.000	1030.0000	-1.5649	0.1768	2.4384	-0.7352

— we omit cases 7 through 26 —

CASE NO.	LABEL	PREDICTED	RESIDUAL	WEIGHT	9 mortalty	16 STRESID	22 HATDIAG	27 MODCOOK	28 DFFITS
27	kansasMO	897.4135	22.2866	1.000	919.7000	0.6333	0.0583	0.5197	0.1567
28	lancasPA	944.3673	-100.2673**	1.000	844.1000	-2.9192	0.1029	3.5344	-1.0657
29	losangCA	892.9724	-31.1724	1.000	861.8000	-0.9346	0.1540	1.3209	-0.3983
30	louisvKY	978.3204	10.9796	1.000	989.3000	0.3183	0.0954	0.3400	0.1025
31	memphsTN	1056.5135	-50.5135*	1.000	1006.0000	-1.5094	0.1484	2.1149	-0.6377
32	miamiFL	927.6211	-66.2211*	1.000	861.4000	-2.0471	0.2043	3.5468	-1.0694
33	milwauWI	930.8878	-1.6878	1.000	929.2000	-0.0477	0.0482	0.0353	-0.0106
34	minnplMN	872.5390	-14.9390	1.000	857.6000	-0.4256	0.0629	0.3629	-0.1094
35	nashvlTN	1012.8209	-51.8209*	1.000	961.0000	-1.4673	0.0515	1.1466	-0.3457
36	newhvnCT	923.0425	0.1575	1.000	923.2000	0.0045	0.0549	0.0035	0.0011
37	neworlLA	1002.1705	110.8295***	1.000	1113.0000	3.4963	0.2359	7.2393	2.1827
38	newyrkNY	972.2457	22.3542	1.000	994.6000	0.6303	0.0436	0.4438	0.1338
39	philadPA	1001.6129	13.3871	1.000	1015.0000	0.3792	0.0521	0.2925	0.0882

— we omit the printout for the remaining cases —

[1] The CASEPLOTS for the default run, after the four variables have entered the model. The residual and influence measures are STRESID and MODCOOK. The leverage measure is a scaled version of HATDIAG: the negative log of the *p*-values associated with a function of h_{ii} (e.g., the *p*-values .1, .01, .001 become 1, 2, and 3, respectively). This function of h_{ii} has an F distribution when the independent variables have a multivariate normal distribution. We see that five cases stand out: Birmingham, Lancaster, Miami, New Orleans, and York. The mortality rate for Lancaster is considerably lower than expected, while that for New Orleans is markedly higher. Other diagnostic plots are available to show in more detail why these cases are unusual.

[2] DFFITS is a measure of influence which some prefer to MODCOOK. MODCOOK uses the absolute value of the residuals, while DFFITS preserves the sign. When you request a bivariate plot like this one, or the STANDARD plots, 2R automatically prints (1) a summary of the two extreme values on each axis under each plot, and (2) a summary panel of extreme cases from all the plots for a step, along with the value of the dependent variable and independent variables currently in the model. You may use EXTREME to request that more than two such values be identified and printed.

[3] Printout of the requested statistics for each case. If you specify PRINT DATA, the values for all the variables are inserted in this panel between MORTALTY and STRESID.

Special Features

Controlling the entry and removal of variables

When you specify an INDEPENDENT list in the REGRESS paragraph, 2R considers only the variables in the list for inclusion. 2R reports *F*-to-enter values for variables omitted from the INDEPENDENT list but part of the VARIABLE USE list. If you do not specify an INDEPENDENT list, 2R considers all variables on the input record (or USE list), plus those created in the TRANSFORM paragraph, excluding only the dependent variable, case weight variable, and LABEL variables, if any.

Note that if the TOLERANCE check for any of these candidate variables fails, the variable will not be used (see TOLERANCE).

2R provides several options for controlling the entry and removal of independent variables from the regression model:

- **Forward stepping.** The default method. At each step the variable with the largest *F*-to-enter is added to the model (see Example 2R.2).
- **Backward stepping.** Begin with all candidate variables in the model. At each step, remove the variable with the lowest *F*-to-remove.
- **Interactive execution.** At each step you respond to a query about which variable to enter or remove. Thus, you can do forward stepping and backward stepping, and control the order of entry.
- **Specifying the order of entry.** Commands available include SEQUENTIAL, LEVEL, and the options for sets of variables (see Special Features, Stepping Sets of Variables).
- **Forcing variables to remain in the model.** You can force a variable to remain in the model regardless of its *F*-to-enter or *F*-to-remove values.
- **Limiting the number of steps.** Use the STEP command to specify the maximum number of forward and backward steps allowed.

Forward stepping. A variable can enter if its *F*-to-enter is larger than the default cutoff value of 4.0. Once entered, a variable can be removed if its *F*-to-remove is less than the default cutoff of 3.996. The preassigned values for ENTER and REMOVE may allow the inclusion of too many variables in the model. To include only those variables whose *F*-to-enter values exceed, for example, 8.0, add these commands to the REGRESS paragraph:

```
ENTER = 8.0.
REMOVE = 7.99.
```

Backward Stepping. You can do backward stepping by forcing all independent variables to enter the equation before removing any. When the independent variables are highly intercorrelated, the final set of variables may differ from that selected by forward stepping. You may want to examine both solutions. To request backward stepping, assign two values to each ENTER and REMOVE limit. The first value for each limit controls the forward stepping, and the second controls backward stepping. 2R uses the first set of limits to step in the variables—when the forward stepping is completed, the second set is invoked and backward stepping begins. For example, you could add the following instructions to the REGRESS paragraph in Example 2R.2:

```
ENTER = 0.0, 5.00.
REMOVE = 0.0, 4.99.
```

Because of the very low values for the first pair of limits, all variables are entered into the equation. They are then removed one by one until only those variables with *F*-to-remove values greater than the second REMOVE limit (4.99) are left. If you do not want to see all forward steps, you can add a PRINT paragraph directing the program to omit output for intermediate steps (NO STEP).

Specifying the order of entry. 2R offers two commands to control the order of entry of variables: SEQUENTIAL and LEVEL. SEQUENTIAL forces variables (or sets of variables) to enter the equation in order from the INDEPENDENT list. The ENTER and REMOVE default limits are changed so that all variables in the list are forced into the equation. However, if you specify ENTER and REMOVE limits, they will not be overridden.

In the LEVEL statement you assign a positive number or zero to each variable on the input record, plus those added through transformation. Variables assigned a level of zero are excluded from the analysis, but their *F*-to-enter values are printed at each step. Variables with LEVEL 1 are considered for entry before variables with LEVEL 2, etc. Variables with the same value compete for entry. ENTER, REMOVE, and TOLERANCE checks are in effect.

For the air pollution data, you might want to consider socioeconomic variables first (LEVEL 1), climatological variables next (LEVEL 2), and finally air pollution measures (LEVEL 3). Thus you would state

```
      LEVEL = 0, 0, 2, 1, 1, 1, 3, 3, 0, 3, 3.
or    LEVEL =       (3)2, 3*1.
```

where the variables, in order, are NAME_1, NAME_2, RAIN, EDUCATN, POP_DEN, NONWHITE, NOX, SO2, MORTALTY, LOG_SO2, and LOG_NOX. The second LEVEL statement is an abbreviation for the first. 2R knows that the first two variables are case LABEL variables and that MORTALTY is the dependent variable, so the zeros for these variables need not be specified. We did not identify the level 3 variables in the second command because variables for which no level is specified are automatically assigned the next higher number. After examining many steps in an exploratory interactive run, you may want to obtain hard copy from a noninteractive run by specifying LEVELS for all of the variables.

Forcing variables to remain in the model. Any variable can be FORCED to stay in the equation regardless of its *F*-to-enter and *F*-to-remove values. To do this, use the FORCE command with LEVEL. Variables with a LEVEL less than or equal to FORCE enter the equation regardless of the *F*-to-enter limit and are not removed. (FORCE = 2 forces all variables with LEVEL 1 or 2 into the equation.) See also SEQUENTIAL above.

Interactive variable selection. 2R is one of four stepwise model building programs (2L, 7M, 2R, and LR) that allow you to enter or remove variables from the model interactively. Decisions about moving variables are based on knowledge of your problem and information from the supporting statistics and diagnostics in the output. See Chapter 11 for information about running BMDP interactively. On a VAX or IBM PC, invoke the program in interactive mode by typing BMDP 2R without specifying input and output files. (On the VAX, press RETURN when the program asks you for input and output files.) The interactive run begins in the BMDP line editor at line 1 with the following prompt:

```
1 Com>
```

You can either type the input with the BMDP line editor or use the Get command to bring in the input from a preexisting file. To execute the instructions, type E and press Return:

```
25 Com>E <Return>
Lines   1 through  15 are executing ...
```

The program output will appear on your screen. At the end of each screenful of output, 2R shows the following prompt. To continue viewing output a screen at a time, simply press <Return>.

```
Press <RETURN> to continue <Return>
```

After printing descriptive statistics (some statistics are omitted in interactive mode so that the output will fit on the computer screen), 2R queries you at each step about which variable you want to move into or out of the regression equation:

```
ENTER VARIABLE TO MOVE NEXT:
PRESS RETURN TO MOVE: nonwhite
<Return>
```

You can specify the variables that you prefer in any order by typing their names or numbers and pressing <RETURN>. After each step, 2R queries you about the variables you want to use in the added variable and error variance plots. You can press Return to choose the default plots or specify a variable of your choice.

Stepping sets of variables

2R allows you to group the independent variables into sets and request that they be entered and removed as blocks. The INPVAL and OUTPVAL commands substitute for ENTER and REMOVE when using sets. All variables in a set enter together and leave together, regardless of individual *F*-to-enter and *F*-to-remove values. Sets have three uses: (1) you can group together variables that have meaning when used together, as in the following example; (2) you can generate design variables for a categorical variable (see also program LR); and (3) you can perform analysis of variance using regression software by generating sets of design variables for the factors.

We use sets of variables to analyze data from the Berkeley Guidance Study (see Weisberg, 1985 and Appendix D), monitoring a group of 32 girls born in Berkeley, California, between January 1928 and June 1929. The variables are:

ID	– identification number
WT_2, WT_9, WT_18	– weight (kg) at ages 2, 9, and 18
HT_2, HT_9, HT_18	– height (cm) at ages 2, 9, and 18
LEG_9, LEG_18	– leg circumference at ages 9 and 18
STRONG_9, STRONG_18	– composite measure of strength at ages 9 and 18
FATNESS	– measure of fatness on a 7-point scale (1 = slender, 7 = fat), determined from a photograph taken at age 18

We want to discover what variables at earlier ages predict FATNESS at age 18. We use sets to group together size variables at ages 2, 9, and 18.

To do this, we specify SETNAMES = SIZE_2, SIZE_9, SIZE_18. We then define the variables within these sets: SIZE_2 includes weight and height at age 2, SIZE_9 includes the same variables at age 9, etc. The set variables are then specified in the INDEP list. Note that we define a set of size variables for age 18 even though we do not specify it in the INDEP list. This allows us to obtain *F* values and *p*-values for SIZE_18 without including it in the regression model. For backward stepping with sets, see INPVAL and OUTPVAL options.

Figure 2R.1
Input for stepping sets of variables

```
/ INPUT       FILE = 'fatness.dat'.
              VARIABLES = 12.
              FORMAT = FREE.

/ VARIABLE    NAMES = id, wt_2,  ht_2,
                      wt_9,  ht_9,  leg_9,  strong_9,
                      wt_18, ht_18, leg_18, strong18,
                      fatness.
              USE = wt_2, ht_2, wt_9, ht_9, wt_18, ht_18,
                    fatness.

/ REGRESS     DEPENDENT = fatness.
              SETNAMES = size_2, size_9, size_18.
              size_2  = wt_2,  ht_2.
              size_9  = wt_9,  ht_9.
              size_18 = wt_18, ht_18.
              INDEP   = size_2, size_9.

/ END
```

Figure 2R.2: Output for stepping sets of variables (we show only step 2 from the analysis)

```
STEP NO.    2
---------------
VARIABLE ENTERED  SET size_2
                    2 wt_2
                    3 ht_2

MULTIPLE R             0.8374
MULTIPLE R-SQUARE      0.7012
ADJUSTED R-SQUARE      0.6570

STD. ERROR OF EST.     0.5139

ANALYSIS OF VARIANCE
                  SUM OF SQUARES       DF    MEAN SQUARE       F RATIO
        REGRESSION  16.736685           4   4.184171           15.84
        RESIDUAL     7.1305099         27   0.2640930

                   VARIABLES IN EQUATION FOR FATNESS                     .              VARIABLES NOT IN EQUATION

                                   STD. ERROR  STD REG                F (P-VAL.)     .              PARTIAL                    F (P-VAL.)
    VARIABLE       COEFFICIENT     OF COEFF    COEFF     TOLERANCE    TO REMOVE  LEVEL.   VARIABLE     CORR.   TOLERANCE    TO ENTER   LEVEL
(Y-INTERCEPT        13.94846 )                                                        .

SET size_2                                                            5.16(0.013)   1 . SET size_18   0.66952              10.16(0.001)    0
wt_2         2     -0.09552         0.0936   -0.161    0.44603                      1 . wt_18      8  0.65728  0.45051                      0
ht_2         3     -0.08402         0.0413   -0.336    0.40666                      1 . ht_18      9  0.04129  0.16526                      0

SET size_9                                                           26.09(0.000)   1 .
wt_9         4      0.17594         0.0268    1.091    0.40077                      1 .
ht_9         5     -0.04519         0.0276   -0.293    0.34591                      1 .

***** P-VALUES( 0.050, 0.051) OR TOLERANCE INSUFFICIENT FOR FURTHER STEPPING
```

The F-to-enter is defined as in **[7]** in Example 2R.2, except that SS(next) is the residual sum of squares after the potential set of variables is entered. When a set enters the equation, you get regression coefficients for each variable and the F-to-remove (plus p-value) for the set. Our SIZE_9 variables enter in Step 1, followed by the SIZE_2 variables in Step 2, yielding the equation

$$\text{FATNESS} = 13.95 - 0.10\ \text{WT\_2} - 0.08\ \text{HT\_2} + 0.18\ \text{WT\_9} - 0.05\ \text{HT\_9}$$

The two sets together account for 70% of the variability of FATNESS (multiple RSQ = .7012).

Interactive paragraphs

2R allows you to transform your data and request REGRESS, PLOT and PRINT paragraphs interactively. You can specify any number of REGRESS, PLOT, PRINT, and TRANSFORM paragraphs after you specify INPUT, VARIABLE, and END paragraphs. You must always state REGRESS as one of the interactive paragraphs even if you do not specify REGRESS instructions (original REGRESS commands default for the interactive analysis). Interactive paragraphs should end with a slash, and the rest of the line after the slash should be blank.

In Figure 2R.3 we use interactive paragraphs after the initial analysis to specify a printout of data, residuals, and predicted values along with plots of residuals and residuals squared versus predicted values. This is an interactive version of Example 2R.2. For more information about running BMDP interactively, see Chapter 10.

Figure 2R.3
Specifying interactive PLOT, PRINT, and REGRESS paragraphs

```
/ INPUT        VARIABLES = 9.
               FORMAT = FREE.
               FILE = 'airpoll.dat'.

/ VARIABLE     NAMES = name1, name2, rain, educatn,
                       pop_den, nonwhite, nox, so2,
                       mortalty.
               LABELS = name1, name2.

/ TRANSFORM    log_so2 = LOG(so2).
               log_nox = LOG(nox).

/ REGRESS      DEPENDENT = mortalty.

/ END

PRINT              DATA./

PLOT         RESIDUALS./

REGRESS            /

END  /
```

Printing data and diagnostics

2R includes a variety of options to print data and diagnostics for each case or for extreme cases. The interaction of these options is summarized in Table 2R.3. PLOT paragraph commands, by default, produce output at the last step. You can change this default be specifying steps to obtain output using the STEP command. PRINT commands produce output at the end only.

Table 2R.3
Printing data and diagnostics

Commands		Printout Contains
For each case		
/ PLOT	**DATA.**	**data** from variables in XVAR and YVAR **diagnostics** from XVAR, YVAR, and PLOT DIAG = STAND.
/ PRINT	**DATA.**	variables included in the model, predicted value, and residual (see **[4]**, Output 2R.3)
	DIAG* =	diagnostics, dependent variable, predicted value, and residual (see **[3]**, Output 2R.5)
	DATA. **DIAG* =**	diagnostics follow the data

*DIAG = list, STAND, or ALL.

Table 2R.3 (continued)

For extreme cases only (*use EXTREME to specify no. of cases reported*)		
/ PLOT	XVAR = YVAR =	extreme diagnostics, variables listed in the XVAR and YVAR lists, and variables in the model (see [2], Output 2R.5)
	DIAG =	extreme diagnostics from the standard plots and variables in the model
	XVAR = YVAR = DIAG =	the two panels described above are combined

LINESIZE: narrow output for terminal viewing

Unless you are using 2R in interactive mode, to obtain narrow output for the computer terminal you must specify LINESIZE in the PRINT paragraph. LINESIZE = 80 is the default for interactive mode. If you specify any number less than 120, 2R uses a maximum of 80 columns to print panels of statistics and results at each step. Narrow output omits some statistics: skewness, kurtosis, smallest and largest standard scores, and, from each step in the regression, standardized regression coefficients and variable numbers. Data printouts and correlation matrices respond to finer control than other portions of the output. You can use LINESIZE to control the number of columns printed per panel.

Missing data

2R uses only complete cases. That is, unless you specify a VARIABLE USE list, 2R checks all variables for missing values and values out of range.

Cases with missing values or values out of range are excluded from the analysis, whether or not those variables are used in the regression. 2R will, however, compute a predicted value for cases where the selected independent variables are complete and the dependent variable is missing. We recommend that you use a VARIABLE USE list to ensure that the program does not exclude more cases than necessary from the analysis. The INDEPENDENT variable list in the REGRESSION paragraph is **not** used to identify complete cases. If you specify several REGRESS paragraphs within the same problem, your USE statement should include all variables under consideration. When you want to use several sets of independent variables, and different cases have missing or extreme values for these variables, you can maximize the number of cases used in each analysis by specifying a separate problem for each dependent variable (rather than separate REGRESS paragraphs within a single problem). Each problem can have its own USE list. See Multiple Problems and the FOR % notation, Chapter 10. See program AM for description and estimation of missing data; 2R will accept data, covariance, or correlation matrices from AM.

Using case deletion variables

It is possible to delete cases interactively in 2R by creating case deletion variables in the TRANSFORM paragraph. The effect of entering one of these indicator variables into the equation is to remove the case from the analysis. This allows you to assess the interplay of deleting influential cases and adding important predictors to the model. (See BMDP Technical Report #48, pp. 33–35, and Mickey et al, 1967, pp. 105–111.) Each case deletion variable is coded 0 for all cases except the case to be deleted, which is coded 1. Instead of adding case deletion variables for each of the 60 cities, you could do so for the extreme cases detected in Example 2R.6. This is done as follows:

```
/ TRANSFORM  BIRMHAM  = KASE EQ 6.
             LANCASTR = KASE EQ 28.
                     etc.
```

For an explanation of the logical operator EQ, see the TRANSFORM paragraph, Chapter 6. *F*-to-enter for a case deletion variable represents how the removal of the case will reduce the residual sum of squares and does not incorporate information about high leverage values. Thus we want to look at diagnostic plots at each step to understand the influence of outlying cases.

> ***Caution:*** *We do not suggest that a valid analysis will result from deleting unusual cases; decisions to delete cases should be based on knowledge of the subject matter.*

After determining which cases you want to delete in an interactive run, you may want to rerun the entire example with the cases deleted (see TRANSFORM OMIT), since case deletion variables affect degrees of freedom, etc. When case deletion variables enter the regression equation, some of the plots are affected, and the half-normal plot is not printed. The added variable plot remains useful, but the tolerance is distorted. The added variable plot for a case deletion variable is meaningless. In the STRESID versus MODCOOK and STRESID versus –LOG P(H) plots, cases that have been removed using case deletion variables appear at the bottom left corner of the plot.

The score test accompanying the error variance plot is distorted, but the distribution of points in the plot remains roughly the same. The error variance plot for a case deletion variable (e.g., AP2*YORK) is meaningless.

Table 2R.4 shows a condensed summary of an interactive run using case deletion variables, with the *F*-to-enter of the variable entered at each step, plus the *F*-to-enter at each step of LOG_SO2. We have underlined the case deletion variables used. Thus, you can see that the LANCASTR dummy variable enters at step 3 (the data for Lancaster are not used).

Table 2R.4
Summary of interactive stepping

Step No.	Variable Entered	R^2	*F*-to-enter	*F*-to-enter of LOG_SO2
0				11.3
1	6 NONWHITE	0.41	41.0	17.4
2	4 EDUCATN	0.56	19.2	12.5
3	13 LANCASTR	0.61	7.6	13.3
4	15 YORK	0.64	4.1	13.8
5	5 POP_DEN	0.71	12.6 (*3.5 before YORK enters*)	5.1
6	14 NEWORLNS	0.73	4.3	12.7
7	12 MIAMI	0.76	6.3	7.5
8	3 RAIN	0.81	13.4 (*3.5 before MIAMI enters*)	15.8
9	10 LOG_SO2	0.86	15.8	

Note that the deletion of certain cases (underlined in Table 2R.4) strongly affects the significance of variables in the regression. For example, the case York hides the importance of POP DEN, and Miami masks the effect of RAIN. In the displays, New Orleans often appears unusual; it ranks high in NONWHITE, RAIN, and MORTALTY and low in SO2 and EDUCATN. When we remove New Orleans by entering its case deletion variable, the *F*-to-enter of LOG_SO2 jumps to 12.7. LOG_SO2 enters last, increasing the R^2 to .86 (compared to .68 in the default run of Example 2R.1). After accounting for the sociological and climatological factors, the addition of LOG_SO2 into the model increases R^2 by 5%. These four cases clearly have a strong effect on the modeling.

Regression coefficients for case deletion variables are meaningless, so we rerun the analysis after deleting the cases (see OMIT in the TRANSFORM paragraph). Running the analysis without these four cities, we obtain the model

$$\text{MORTALTY} = 923.28 + 2.20\ \text{RAIN} - 15.08\ \text{EDUCATN} + .0095\ \text{POP\_DEN} + 2.23\ \text{NONWHITE} + 29.24\ \text{LOG\_SO2}$$

The multiple R^2 is .82. Compare this model to the one obtained in the default run of Example 2R.2.

Regression weights

In ordinary least squares regression, we assume that the random errors for each observation are independent with a common variance σ^2. A generalization of this approach is to allow the errors to have different variances. When this is true, accepted procedure is to weight according to the inverse of the variances. This technique is called weighted least squares regression.

For example, suppose we know that the error variance is inversely proportional to the variable ERRWT. To perform weighted least squares regression, we can specify

```
/ REGRESSION WEIGHT = errwt.
```

This WEIGHT option in the REGRESSION paragraph may be specified either alone or in addition to case frequencies and/or case weights in the VARIABLE paragraph. So the same analysis may have more than one kind of weighting to adjust for case frequency, sampling design, and random error structure. As before, if a case has a zero or negative regression weight, it is excluded from the regression calculations, although it will be included in the descriptive statistics if its case frequency and case weights are positive. To estimate predicted values for cases not used to compute the regression equation, set their case weights to zero (see Appendix C).

Tolerance

If the squared multiple correlation (R^2) of any independent variable with the other independent variables in the regression equation is too high, the computations lose numerical accuracy. For example, if a variable has an R^2 value greater than .99 with the other independent variables, its TOLERANCE value ($1 - R^2$) is below .01 and it cannot enter the equation. Note that the dependent variable is not considered here; our concern is with collinearity among the independent variables. To assign a value larger than the default of .01, use the TOLERANCE command in the REGRESS paragraph. Use 9R if you need a lower tolerance limit. If you have highly intercorrelated independent variables, you may want to consider 4R for regression on principal components or ridge regression.

Using a correlation or covariance matrix as input

2R reads data only, unless a BMDP File is used as input. DATA, a COVARIANCE matrix, or a CORRELATION matrix may be input from a BMDP File. For exploratory data analysis you may want to use program AM to estimate a correlation or covariance matrix from data with missing values and then use the matrix as input to 2R. When the input is not DATA, the CONTENT of the BMDP File must be specified. Several BMDP programs create BMDP Files that are not DATA, COVA, or CORR. One example would be FCOR in 4M. You can use these files as input in 2R, but you must specify with the INPUT CONTENT statement the parameter used in the SAVE paragraph where the file was created. You must also use the INPUT TYPE statement to specify how they are to be analyzed (as DATA, COVA, or CORR). See INPUT commands in this chapter.

Saving regression results and diagnostics

You can use a SAVE paragraph to make a BMDP File of the predicted values, residuals, and regression diagnostics along with the DATA. For more information about the diagnostics available for saving, see SAVE commands below. You can also save the COVARIANCE matrix, allowing subsequent analyses to start from the computed means and covariances (this may result in a substantial reduction in computing time). For example, you might add the following instructions to Input 2R.2:

```
/ SAVE   FILE = 'airpoll.sav'.
         CODE = airpoll.
         CONTENT = DATA, COVA.
         NEW.
```

Two files are saved: (1) data plus PREDICTD and RESIDUL and (2) covariance matrix. (See Chapter 8 for definitions of CODE, NEW, FILE, and CONTENT.) 2R prints messages in the output informing you that two files are saved. To save diagnostics along with the data, specify, for example,

```
/ SAVE   FILE = 'airpoll.sav'.
         CODE = diag.
         NEW.
         DIAGNOSTICS = DSTRESID, HATDIAG, MODCOOK.
```

The output contains all input data, followed by any new transformed variables (e.g., LOG_SO2), PREDICTD and RESIDUL (automatically saved by default), and the three listed diagnostic statistics for each case. You can use the DIAGNOSTICS command to list any of the statistics named in Table 2R.1, or you can state DIAGNOSTICS = ALL. To output a raw data file of data and diagnostics for a printed report or a high-resolution graphics program, see Chapter 8.

Regression models with no intercept

The regression equation usually contains an intercept (a constant term), e.g., in

$$y = \beta_0 + \beta_1 x_1 + \beta_2 x_2 + \dots + \beta_p x_p$$

β_0 is the intercept. In some problems the intercept is defined as zero; this can be specified by

```
TYPE = ZERO.
```

in the REGRESS paragraph. When this option is specified, the covariances are the sums of products of observed values instead of the sums of products of deviations from the sample mean.

2R Commands

/ INPut

The INPUT paragraph is required. See Chapter 3 for INPUT commands common to all programs. 2R accepts data, a covariance matrix, or a correlation matrix as input. If the input is not data, it must be from a BMDP File.

BMDP File Input

When the input is not data, you must specify the CONTENT of the BMDP File. When the CONTENT is DATA, COVA, or CORR, 2R recognizes how the input is to be analyzed. Several BMDP programs create BMDP Files with a different CONTENT that is not analyzed by 2R. You can use other types of BMDP Files as input to 2R, but you must use a TYPE statement to specify how they are to be analyzed (i.e., as a correlation or covariance matrix, or as data).

CONTent = parameter. `CONTENT = COVAPART.`

Specify parameter for BMDP File input. The parameter must be the same as that stated in the SAVE paragraph when the file was created. You must specify CONTENT if input is not DATA (the default), COVA, or CORR. Use TYPE when you want to change the way the BMDP File is analyzed. **Example:** A partial covariance matrix file from 6R is read in.

TYPe = *(one only)* `TYPE = CORR.`
DATA, COVA, *or* **CORR.**

Use to determine how input from a BMDP File should be analyzed. Rarely used. **Example:** When CONTENT = COVAPART, 2R assumes that the input should be analyzed as a covariance matrix (the program reads the first four letters of the CONTENT name, e.g., COVA for COVAPART). To analyze the partial covariance matrix as a correlation matrix, specify TYPE = CORR.

/ VARiable

The VARIABLE paragraph is optional and is described in detail in Chapter 4. Additional commands for 2R are CWEIGHT and FREQUENCY.

CWeight = variable. `CWEIGHT = SAMPWT.`

Specify a case weight variable. As with case frequencies, records with negative or zero case weight values are excluded from the computations. Case weights can be used to adjust for sampling from populations with unequal variances. Means and correlations are unaffected, but variances, standard deviations, and standard errors change. See Chapter 4.

FREQuency = variable. `FREQ = COUNT.`

State the name or number of a variable containing the case frequency. A case frequency has the effect of repeating cases. Any record with a negative or zero case frequency (COUNT ≤ 0 in this example) is excluded from the calculations as though it were missing. See Chapter 4.

/ REGRess

The REGRESS paragraph is required and may be repeated before the END paragraph to specify additional analyses of the same data. You must specify the DEPENDENT variable in the first REGRESS paragraph. Options describe the role of variables in the equation, the method of entry or removal of variables, and the method of treating the intercept.

DEPendent = variable. `DEPEND = mortalty.`

Required to specify the name or number of the dependent (predicted) variable.

INDEPendent = list. `INDEP = nonwhite, educatn.`

List the names or numbers of the independent (predictor) variables. By default, 2R uses all variables except the DEPENDENT variable, the case WEIGHT, and LABEL variables, if any. If a USE statement was included in the VARIABLE paragraph, then only those specified variables will be available to enter the regression equation. Even when a USE = LIST statement is included in the VARIABLE paragraph, the output will include all variables named in the VARIABLE paragraph. However, those not included in the USE statement are set to LEVEL = 0 and will not be included in the regression equation.

ENTER = #1, #2. `ENTER = 0.5, 6.00.`
REMOVE = #1, #2. `REMOVE = 0.0, 5.99.`

Use to define F-to-enter and F-to-remove limits. ENTER #1 and REMOVE #1 are used during forward stepping to control the entry or removal of variables. ENTER #2 and REMOVE #2 are used for backward stepping. Entering variables must have Fs greater than the ENTER limit. (For removal, Fs less than the REMOVE limit.) REMOVE must be less than ENTER. Default: ENTER = 4.0, 4.0.; REMOVE = 3.9, 3.9. (no backward stepping). **Example:** During forward stepping, variables are entered if the $F > 0.5$ and none are removed (REMOVE F is 0.0). During backward stepping, variables are removed whose F is less than or equal to 5.99 and are not reentered unless their F is greater than or equal to 6.00.

TOLerance = #. `TOL = .05.`

TOLERANCE guards against singularity. Use a TOLERANCE number between 0.01 and 1.0 to prevent loss in accuracy due to inclusion of numerically redundant variables. An independent variable does not enter the regression equation if it does not pass the TOLERANCE limit, i.e., if its squared multiple correlation (R^2) with independent variables already in the equation exceeds (1 – TOL), or if its entry will cause the R^2 of any previously entered variable with the independent variables in the equation to exceed 1 – TOL. See Appendix B.16. Note that if you use a zero intercept model, then 2R estimates the R^2 under the assumption that all variables have zero means. The default TOLERANCE is .01. Use 9R if you need a lower tolerance limit than .01.

STEP = #. `STEP = 4.`

State the maximum number of forward and backward steps allowed. Default: 2 * the number of variables.

WEIGHT = variable. `WEIGHT = errwt.`

Specify a regression weight variable for weighted least squares regression (see Special Features). By default, no regression weights are used.

TITLE = 'text'. `TITLE = 'RUN 2-BACKWARD'.`

Specify a problem title to be printed on the output. A title too long to fit in the output will be truncated. Default: no title.

Controlling the order of entry of variables

SEQuential. `SEQ.`

If you specify SEQ, variables or sets of variables (see SETNAMES option) are forced to enter in the order specified in the INDEPENDENT list rather than according to their *F*-to-enter values. Note that the ENTER and REMOVE (or INP and OUTP) default limits are changed so that all variables are forced into the equation. You can specify ENTER and REMOVE (or INP and OUTP) limits to ensure that 2R excludes extraneous variables.

LEVEL = # list. `LEVEL = 2, 1, 1, 1, (6)2.`

Define the order in which variables enter the equation. Lower numbers are considered first. Use one number for each input variable (not for each variable in the USE list). Variables with LEVEL = 0 are not used, but 2R prints *F*-to-enter statistics for them at each step. Default: LEVEL = 1 for each variable. DEPEND and WEIGHT variables are automatically used, regardless of any assigned level. **Example:** Variables 2, 3, and 4 are considered first for entry into the equation; 1 and 6 are considered next. No level is specified for variable 5, so it is considered last (its level is 3, the highest level specified plus 1).

FORCE = #. `FORCE = 1.`

Variables with LEVEL less than or equal to FORCE (if FORCE is greater than zero) enter the equation and are not removed. That is, ENTER and REMOVE limits are ignored when FORCE is specified. Default: FORCE = 0 (no forcing). Only the TOLERANCE test (see TOLERANCE option below) can be used to stop a forced variable from entering the equation. **Example:** All variables with LEVEL = 1 are included in the equation.

Stepping sets of variables

SETNames = 'name1', 'name2', ... `SETN = SIZE_2, SIZE_9.`
'name1' = list. `SIZE_2 = WT_2, HT_2.`
'name2' = list. ... `SIZE_9 = WT_9, HT_9.`

Define sets of independent variables so that all variables in a set are entered into or removed from the regression equation in the same step (e.g., sets of design variables for building analysis of variance models). By default, variables are not grouped into sets. For each subset of variables, you must specify a name (the example shows SIZE_2 and SIZE_9) and a list of the names or numbers of the variables in the set. The command SETNAMES is used to list the set names. The set names may be used in the REGRESS INDEPENDENT list. However, if one of the variables in the set does not pass the analysis tolerance limits, it will automatically be eliminated from the set. ***Caution:*** Avoid choosing set names that are close in spelling to other REGRESS commands such as TOLERANCE and FORCE, or an error message may result.

INPval = #1, #2. `INP = .30, .01.`
OUTPval = #1, #2. `OUTP = .99, .011.`

Use only for stepping sets of variables. Specify the pseudo-p-value-to-enter and p-value-to-remove limits when stepping in sets of variables. #1 is used for forward stepping, #2 is used for backward stepping. By default, INP = 0.05, 0.05. and OUTP = 0.051, 0.051. (no backward stepping). **Example:** During forward stepping, variables or sets of variables will be entered with INP less than .30 (they will not be removed unless OUTP is greater than .99). During backward stepping, variables are removed whose OUTP is greater than .011 and are not reentered unless INP is less than .01.

Rarely used commands

TYPE = *(one only)* `TYPE = ZERO.`
NONzero, ZERO, FLOAT.

Use to specify the method for treating the intercept. Default: NONZERO. The intercept, β_0, is usually part of a regression equation; however, in a few problems the intercept must equal zero because the regression plane passes through the origin. The intercept can be treated in three ways:

NONZERO. – intercept is always in the equation

ZERO. – intercept is never in the equation (rarely used)

FLOAT. – intercept starts out in the equation but may be removed if its F-to-remove is very small (rarely used). Avoid this option unless you are experienced in regression analysis.

2R prints the adjusted R^2 only when there is an intercept in the regression equation. If you specify ZERO or FLOAT, R^2 is computed relative to deviations of the dependent variable from zero rather than from the mean of the dependent variable. Thus, in some applications R^2 may not be useful. Evaluation of the regression should then be made only with respect to the residual mean square. If you repeat the REGRESS paragraph in one problem, you cannot change TYPE after the first REGRESS paragraph.

METHod = *(one only)* `METH = R.`
F, FSWAP, R, RSWAP.

Use to determine the methods for entering or removing variables at each step. F-to-enter values must exceed ENTER for a variable to enter the equation; F-to-remove must be less than REMOVE for a variable to be removed (see ENTER and REMOVE options above). FSWAP, R, and RSWAP are rarely used. Default: F.

F. – Remove the variable with the smallest F-to-remove. If no variable qualifies, enter the variable with the largest F-to-enter.

FSWAP. – If no variable meets the removal limit, exchange a variable in the equation with one not yet entered to increase multiple R. If no exchange increases multiple R, variables enter as in F.

R. – Remove the variable with the smallest F-to-remove if removal results in larger R than that for the previous smaller set. If it does not, variables enter as in F.

RSWAP. – Remove variable as in R. If no variable meets criterion, exchange a variable in the equation with a variable not yet entered to increase R. If no exchange increases R, variables enter as in F.

Method F requires the fewest computations, R the next fewest. FSWAP and RSWAP require many more computations; you may want to use program 9R instead.

/ PRINT `DATA. NO STEP.`

The PRINT paragraph is optional. Use it to specify the results to be printed (or omitted). Note that the interactive PRINT paragraph accepts only 2R PRINT commands, plus LINE, PAGE, LEVEL, DEBUG, CASE and VAR. If the paragraph is omitted, 2R automatically prints ANOVA, STEP, COEF, and SUM. You must specify the remaining parameters.

New Feature

METHod = VARiable *or* **CASE.** `METH = CASE.`

Use to determine how the printout is organized. VARIABLE is the default value.

VARIABLE – Prints for each case only as many variables as fill the print line (up to ten variables) and begins the next line with the same variables for the next case. Thus the data for each variable are in an easy-to-scan column. When the program finishes printing the first subset of variables for all cases, it continues with another subset.

CASE – Prints the values of all variables for one case (possibly filling several print lines) before printing data for the next case.

DIAGnostics = list, ALL, *or* **STANDard.** `DIAG = STAND.`

Prints one or more of the regression diagnostics for each case after stepping stops, along with the dependent variable, predicted value, and residual. When you request a printout of the diagnostics, and PRINT LEVEL = NORMAL, 2R prints a short description of each requested diagnostic. When PRINT LEVEL = VERBOSE, a long description is printed. If you specify DATA and DIAG, one panel contains both data and diagnostics. Default: no diagnostics.

list – select names from the 21 diagnostics listed in PLOT paragraph

ALL – all 21 diagnostics

STAND – STRESID, DSTRESID, HATDIAG, and MODCOOK

LINEsize = #. `LINE = 80.`

Specify the number of columns of output width. Use a number less than 120 to obtain narrow output (80 columns). Default: 132 for noninteractive jobs, 80 for interactive. Note that narrow output eliminates some statistics.

ANOVA. – Unless you specify NO ANOVA, 2R prints an analysis of variance table for each step. If you specify NO ANOVA, the table is printed only at the end, and NO STEP also takes effect.

STEP. – Unless you specify NO STEP (or NO ANOVA), 2R prints the results at every step.

COEF. – Unless you specify NO COEF, 2R prints a summary table of regression coefficients.

SUM. – Unless you specify NO SUM, 2R prints a summary table with R, R^2, F-to-enter, and F-to-remove for each step.

COVA. – Prints a covariance matrix of the variables.

CORR. – Prints a correlation matrix of the variables.

DATA. – Prints the input data, residuals, and predicted values. See Output 2R.2 for more information.

PART. – Prints a summary table of partial correlations at each step. The partial correlation of a variable in the equation is its correlation with the dependent variable after removing the linear effect of the remaining variables in the equation.

FRATio. – Prints a summary table of F-to-enter and F-to-remove values.

RREG. – Prints the correlations between regression coefficients.

New Feature **CASE = #.** – Specify the number of cases for which data are printed. Values flagged by MISS, MAX, and MIN in the VARIABLE paragraph specifications are replaced by MISSING, GT MAX, and LT MIN. Default: 10.

New Feature **MINimum.** – Prints data for cases in which at least one variable has a value less than its lower limit as specified in the VARIABLE paragraph. Default: such cases are not printed unless you specify DATA.

New Feature **MAXimum.** – Prints data for cases in which at least one variable has a value greater than its upper limit as specified in the VARIABLE paragraph. Default: such cases are not printed unless you specify DATA.

New Feature **MISSing.** – Prints data for cases that have at least one variable with a missing value. Missing values in the input may be identified in several ways: (1) an asterisk (the default missing value character), (2) a missing value code identified using MISS in the VARIABLE paragraph, (3) a character identified as MCHAR in the INPUT paragraph, (4) blanks (applies only to input read with FORTRAN FIXED format). Default: such cases are not printed unless you specify DATA.

The following commands work only in the batch mode

New Feature **MEAN.** – Unless you specify NO MEAN, 2R prints the mean, standard deviation, and frequency of each used variable.

New Feature **EZSCores.** – Unless you state NO EZSC, 2R prints *z*-scores for the minimum and maximum values of each used variable.

New Feature **EXTReme.** – Unless you state NO EXTR, 2R prints the minimum and maximum values for each variable used.

New Feature **SKewness.** – Unless you state NO SK, 2R Prints skewness and kurtosis for each used variable.

New Feature **ECASe.** – Prints the case numbers containing the minimum and maximum values of each variable.

/ PLOT `RESIDUAL. NORMAL.`

Optional. No plots are printed unless specified. Note that the interactive PLOT paragraph in 2R will read only the 2R PLOT commands and not those of the general interactive PLOT. Standard plots available in several BMDP regression programs are listed first below.

RESIDual. – Plots residuals and residuals squared against predicted values.

NORMal. – Prints a normal probability plot of residuals.

DNORMal. – Prints a detrended normal probability plot of the residuals.

VARiable = list. `VAR = rain, educatn.`

2R prints two plots for each variable specified (by name or number): (1) the observed values and predicted values of the dependent variable against the specified variable, and (2) the residuals against the specified variable.

PREP = list. `PREP = log_so2, educatn.`

List the names or numbers of independent variables to be plotted in partial residual plots as follows: residual + (coefficient * observed value of the independent variable) against the same observed value, i.e., $[(y - \hat{y}) + bx]$ against x.

SIZE = #1, #2. `SIZE = 60, 80.`

Specify the plot size as the number of characters along the x-axis (#1) and number of lines on the y-axis (#2). Default size is 50 columns wide, 35 lines high.

Regression Diagnostics

The following 21 statistics can be plotted and printed. The first 17 can also be saved in a BMDP File. See Table 2R.1 for formulas and Table 2R.2 for an overview of plots. You may include any diagnostics in the XVAR and YVAR lists, and you may state DIAG = STAND. State CASEPLOTS to obtain case-by-statistics plots. Their BMDP names, in the order in which they are added to the data for each case, are as follows (VT = number of variables after transformations):

PREDICTD	– predicted value of the dependent variable	(VT + 1)
Residual measures (outliers in *y*)		
RESIDUAL	– residual	(VT + 2)
WRESIDUL	– weighted residual (use with VARIABLE WEIGHT)	(VT + 3)
STRESID	– standardized residual	(VT + 4)
DELRESID	– deleted residual	(VT + 5)
DSTRESID	– deleted standardized residual	(VT + 6)
AP1	– Andrews-Pregibon statistic factor 1	(VT + 7)
Q1	– decrease in residual sum of squares when the *i*th case is deleted	(VT + 8)
U	– proportional to the contribution of the *i*th residual to the residual sum of squares	(VT + 9)
Leverage measures (outliers in the *x*-space)		
HATDIAG	– diagonal element of "hat," or projection, matrix	(VT + 10)
RHAT	– ratio of HATDIAG to 1 – HATDIAG	(VT + 11)
AP2	– Andrews-Pregibon statistic factor 2	(VT + 12)
MAHAL	– Mahalanobis distance	(VT + 13)
Influence measures (influence of the *i*th case on coefficients)		
COOK	– Cook's distance	(VT + 14)
MODCOOK	– modified Cook's distance	(VT + 15)
DFFITS	– measure of influence of the *i*th case on the estimated regression coefficients	(VT + 16)
AP	– Andrews-Pregibon statistic in one dimension	(VT + 17)
Additional diagnostics available for PLOT and PRINT		
'STRES2'**	– squared standardized residual	
'AP2*PRED'	– an independent or predicted variable, scaled by AP2	
XRESIDUL	– residual from regressing x_q on $x_1,\ldots,x_k$	
YRESIDUL	– residual from regressing y on $x_1,\ldots,x_k$	

CASEplots. `CASE.`

Requests a three-panel graphic display with diagnostics for each case: a residual measure (STRESID), a leverage measure (scaled HATDIAG), and an influence measure (MODCOOK). Caseplots are the graphic equivalent of case by diagnostic listings; they plot the case number on one axis and a diagnostic measure on the other. To print CASEPLOTS at each step, see STEP.

XVAR = list. `XVAR = predictd, rain.`
YVAR = list. `YVAR = stresid, modcook.`

Specify the names or numbers of variables or diagnostics to plot on the *x*-axis and *y*-axis, respectively. The first YVAR is plotted against the first XVAR, the second against the second, etc. **Example:** The standardized residual is plotted against the predicted values, and the modified Cook's distance is plotted versus the independent variable RAIN.

DIAGnostics = plot name *or* **STANdard.** `DIAG = ADDVAR.`

Prints up to six plots designed to highlight data that violate the assumptions of a least squares regression analysis. When running interactively, 2R will query you about variables to use in the first two plots. You can specify DIAG = STANDARD to obtain all six plots, or choose any one or combination of the following commands:

ERRVAR.	– error variance plot (STRES**2 vs. AP2*x_q)
ADDVAR.	– added variable plot (YRESIDUL vs. XRESIDUL)
DSTRESID.	– half–normal probability plot of DSTRESID
LEVRES.	– STRESID versus MODCOOK (Letters mark cases with extreme values of HATDIAG)
INFLRES.	– STRESID versus a scaled version of HATDIAG (letters mark cases with extreme values of MODCOOK)
YTRNS.	– transformation plot. (See YTRN command).

Other associated options

STEP = #.
EXTREME = #.
SIZE = #1, #2.
DATA.
STATISTICS.

ENVELOPE. SIMULATION = #. SEED = #.	for half-normal plot
RANGE = RESIDUAL or ORIGINAL.	for added variable miniplots

STEP = # *or* **ALL.** `STEP = 1,3.`

Specify the steps for which you want to obtain diagnostic plots. Default: plots are printed for the last step only. **Example:** Plots are printed for the first and third steps.

RANGe = RESIDual. `RANGE = ORIG.`
or **ORIGinal.**

Use for the added variable plot (DIAG = STAND) to scale the plots based on the RESIDUAL values or based on the values of the ORIGINAL variables. Default: RESIDUAL.

ENVElope. `ENVELOPE.`

Prints simulation envelopes in the half-normal probability plots. When 39 sets of residuals are simulated, the probability that the observed *n*th ordered residual lies above its envelope is .05. See SIMULATION and SEED to change the number of simulations and the seed for the random number generator.

SIMUlation = #. `SIMULATION = 10.`

State the number of simulations to perform in computing ENVELOPE. Default: 39.

SEED = #. `SEED = 27341.`

Specify a seed number (no greater than 30,000) to generate a sequence of random numbers for computing the SIMULATION envelopes. Default: 14381.

YTRN. `YTRN.`

Prints Atkinson's estimate of λ for the suggested Box-Cox power transformation of the dependent variable Y. The suggested transform is $(y^{\lambda}-1)/\lambda$. The printed score statistic and *p*–value test the null hypothesis $p = 1$ (no transformation is needed). Specifying YTRN also produces an added variable diagnostic plot of XRESIDUAL versus YRESIDUAL where XRESIDUAL and YRESIDUAL are the dependent and predictor variable residuals after removing the linear effect of all other predictor variables in the model. The YTRN command yields the same plot as DIAG = YTRN.

EXTReme = #. `EXTREME = 5.`

Specify the number of extreme values in diagnostic plots and added variable miniplots for which 2R prints: (1) below each plot for each extreme value, the case number and statistic, (2) after the last plot (at the specified step, if STEP is used), a summary of extreme values in all plots. Applies to plots specified in DIAG = STAND, XVAR and YVAR, YTRN, and NADD—not to the CASEPLOTS. Labels for label variables are substituted for case numbers. You can obtain at the most up to ten EXTREME values. Default: 2. Maximum: 10.

STATistics. `STAT.`

Prints summary statistics for each diagnostic plot: correlation, the equation of the least squares regression line for the points in the plot, etc.

DATA. `DATA.`

Prints for each case the values displayed in the plots (when XVAR and YVAR or DIAG = STAND are specified). If independent variables are specified in XVAR or YVAR, they are printed too. If STEP is specified, results appear at each step.

NADD = number. `NADD = 3.`

Specify the number of added variable miniplots to plot at each requested step. (You request steps in SADD.) The maximum number of miniplots allowed is the number of variables not entered into the equation. 2R changes NADD to the number of excluded variables if you specify more. **Example:** three miniplots are printed at each step.

SADD = #, LAST, ALL. `SADD = 1,3,4.`

Provide the steps at which you want 2R to print miniplots. You can list numbers (for specific steps), LAST (for the last step), or ALL (for all steps). Default: miniplots printed at ALL steps. **Example:** added variable miniplots are printed at the first, third, and fourth steps.

ASIZE = #1, #2. `ASIZE = 20, 18.`

Specify miniplot size as the number of characters along the x-axis (#1) and number of lines on the y-axis (#2). Minimum values are nine characters wide, ten lines high. Default values are 15 characters wide, 12 lines high. **Example:** each added variable miniplot is 20 characters wide and 18 lines high.

/ SAVE

The SAVE paragraph is required only if you want to create a BMDP File. See CODE, UNIT, and NEW in Chapter 8. Additional 2R commands are listed below.

CONTent = *(one or both)* **DATA, COVA.** `CONTENT = DATA, COVA.`

Specify the matrices to be saved in a BMDP File. When DATA is specified, predicted values and residuals are saved for each case after the data. When the REGRESS paragraph is repeated, only the values from the **first** regression are saved. Use separate problems (rather than separate REGRESS paragraphs) to save values from more than one regression analysis. Default: DATA.

The predicted values are saved as PREDICTD (VT + 1), and the residuals are saved as RESIDUAL (VT + 2), where VT is the total number of variables (after transformation, if any). The PREDICTD and the RESIDUAL are saved no matter what is stated in the DIAG option.

DIAGnostics = list, STANDard, *or* **ALL.** `DIAG = MODCOOK, DFFITS.`

When CONTENT = DATA, state DIAGNOSTICS to add regression diagnostics to the data for each case (see PLOT commands). Default: no diagnostics are added.

list – use the names of the first 17 diagnostics listed in the PLOT paragraph. For example, use DSTRESID to save the deleted standardized residuals.

STAND – STRESID, DSTRESID, HATDIAG, and MODCOOK are added to the data for each case.

ALL – The first 17 diagnostics listed in the PLOT paragraph are added to the data for each case. The last four diagnostics cannot be saved.

Order of Instructions

■ indicates required paragraph

```
■ / INPUT
  / VARIABLE
  / TRANSFORM
  / SAVE
■ / REGRESS ] Repeat for subproblems
  / PRINT
  / PLOT                                  Repeat for additional problems.
■ / END                                   See Multiple Problems, Chapter 10.
  data
  TRANSFORM ... /
  GROUP ... /
  PRINT ... /
  PLOT ... /          May be repeated
  REGRESS ... /       interactively
  SAVE ... /
  CONTROL ... /
  END /
  FINISH /
```

Summary Table for Commands Specific to 2R

Paragraphs Commands	Defaults	Multiple Problems	See
■ / **INPut**			
CONTent = parameter.	DATA	–	Cmds
TYPe = DATA, COVA, *or* CORR.	parameter specified in CONTENT	–	Cmds
▲ / **VARiable**			
CWeight = variable.	no case weights used	–	Cmds
FREQuency = variable.	no case frequency variable	–	Cmds
■ / **REGRess**			
■ DEPendent = variable.	required for first problem	●	2R.1
▲ INDEPendent = list.	all variables except DEPEND, WEIGHT, and LABEL	●	2R.1
ENTER = #1, #2.	ENTER = 4.0, 4.0.	●	SpecF
REMOVE = #1, #2.	REMOVE = 3.9, 3.9.	●	SpecF
TOLerance = #.	.01	●	SpecF
STEP = #.	2 * no. of var.	●	Cmds
WEIGHT = variable.	no regression weights	–	SpecF
TITLE = 'text'.	no title printed	–	Cmds
SEQuential.	no sequential forcing	–	SpecF
LEVEL = # list.	LEVEL = 1 for each variable	–	SpecF
FORCE = #.	no forcing (FORCE = 0.)	●	SpecF

Summary Table (continued)

Paragraphs Commands	Defaults	Multiple Problems	See
SETNames = list. 'name1' = list. 'name2' = list. ...	variables not grouped in sets	•	SpecF
INPval = #1, #2.	INP = 0.05, 0.05.	•	Cmds
OUTPval = #1, #2.	OUTP = 0.051, 0.051.	•	Cmds
TYPE = *(one only)* NONzero, ZERO, FLOAT.	NONZERO	•	SpecF
METHod = *(one only)* F, FSWAP, R, RSWAP.	F	•	Cmds
/ PRINT			
ANOVA. *or* NO ANOVa.	ANOVA.	•	2R.3
STEP. *or* NO STEP.	STEP.	•	2R.3
COEF. *or* NO COEF.	COEF.	•	2R.3
SUM. *or* NO SUM.	SUM.	•	2R.3
MEAN. *or* NO MEAN.	MEAN.	•	Cmds
EXTReme. *or* NO EXTR.	EXTReme values printed	•	Cmds
EZSC. *or* NO EZSC.	EZSC. (*z*-scores)	•	Cmds
SKewness.	SK. (skewness and kurtosis)	•	Cmds
ECASe.	No ECASe. (Case #'s of Min and Max printed)	•	Cmds
COVA.	NO COVA.	•	Cmds
CORR.	NO CORR.	•	Cmds
DATA.	NO DATA.	•	Cmds
PART.	NO PART.	•	Cmds
FRATio.	NO FRAT.	•	Cmds
RREG.	NO RREG.	•	Cmds
CASE = #.	10	•	Cmds
MISSing.	NO MISS.	•	Cmds
MINimum.	NO MIN.	•	Cmds
MAXimum.	NO MAX.	•	Cmds
METHod = VARiables *or* CASE.	VARIABLES	–	Cmds
DIAGnostics = list, STAND, *or* ALL.	no diagnostics	–	2R.6
LINEsize = #.	132 batch; 80 interactive	•	SpecF
/ PLOT			
RESIDual.	NO RESID.	•	2R.4
NORMal.	NO NORM.	•	2R.4
DNORMal.	NO DNORM.	•	2R.4
VARiable = list.	no plots printed	•	2R.4
PREP = list.	NO PREP.	•	2R.4
SIZE = #1, #2.	50, 35.	•	2R.2
CASEplots.	no caseplots	–	2R.6
XVAR = list.	no variables or diagnostics	–	2R.6
YVAR = list.	plotted		2R.6
DIAGnostics = ADDVAR, ERRVAR, DSTRESID, INFLRES, LEVRES, YTRNS, STANdard.	no plots printed	–	Cmds
STEP = # *or* ALL.	plots at last step only	–	Cmds
RANGe = RESIDual *or* ORIGinal.	RESID	•	Cmds
ENVElope.	envelope not printed	–	2R.6

Summary Table (continued)

Paragraphs Commands	Defaults	Multiple Problems	See
SIMUlation = #.	39	–	Cmds
SEED = #.	14,381	–	Cmds
YTRN.	no YTRN.	–	Cmds
EXTReme = #.	2	–	2R.5
STATistics.	no statistics with plots	–	Cmds
DATA.	NO DATA.	–	Cmds
NADD = #.	number of excluded variables	–	2R.5
SADD = #, LAST, ALL.	ALL.	–	2R.5
ASIZE = #1, #2.	15, 12	●	Cmds
/ SAVE			
CONTent = DATA *and/or* COVA.	DATA	–	Cmds
DIAGnostics = list, STANdard, *or* ALL.	no diagnostics saved	–	SpecF

Key:
- ■ Required paragraph or command
- ▲ Frequently used command
- ● Value retained for multiple problems
- – Default reassigned

Derivative-Free Nonlinear Regression

AR can estimate the parameters for a wide variety of nonlinear functions by least squares using a pseudo-Gauss-Newton iterative algorithm (see Ralston and Jennrich, 1978). Because AR does not require you to specify derivatives, you can use it with functions that are nonlinear in the parameters and for which derivatives are difficult or impossible to specify. You can specify the regression function for AR either by using one of seven built-in functions or by including transformation statements in a FUNCTION paragraph. (If you are using a mainframe computer, you can also specify functions in a FORTRAN subroutine.) You can fit the value of a parameter, or you can impose upper and lower limits on individual parameters or on arbitrary linear combinations of the parameters. AR can also estimate functions of the parameters and their standard errors. You can specify weights for each case and obtain iteratively reweighted and maximum likelihood parameter estimates. You can also obtain ridged estimates, the serial correlation, the Durbin-Watson statistic, and the runs test for the residuals. (See 3R in Volume 2 for more on maximum likelihood estimates, ridged estimates, and estimates incorporating independent prior statistical knowledge of one or more parameters.) You can save the predicted values with the data for further analysis by other programs. The DIFEQ and DIFIN paragraphs enable you to handle differential equations without writing a FORTRAN subroutine. For details, see BMDP Technical Report #85 (Ralston et al., 1987).

Five other BMDP programs perform nonlinear regression: 3R, LR, PR, 2L, and 4F. 3R is similar to AR, but 3R uses derivatives in the Gauss-Newton algorithm and in the asymptotic covariance matrix, while AR uses a secant approximation to those derivatives. The other programs fit particular nonlinear models: use LR for stepwise logistic regression, PR for logistic regression with more than two outcome groups, 2L for survival analysis with covariates, and 4F for log-linear models. 4F is described in this volume; 3R, LR, and 2L, in Volume 2. Readers unfamiliar with nonlinear regression may want to read Draper and Smith (1981), Chapter 10. See Ralston and Jennrich (1978) for technical details about derivative-free nonlinear regression.

Where to Find It

Examples

Example AR.1 Using a built-in function to specify an exponential model

We analyze data (shown below in Data Set AR.1) from a study on the emptying of a radioactively labeled meal from the stomach of a normal subject. A simple exponential function frequently fits gastric emptying data:

$$f = p_1 e^{p_2 t}$$

where t is the time after ingestion, p_1 is the percent of the meal in the stomach at $t = 0$, and p_2 is the emptying rate. The emptying rate is generally reported in terms of the half-time, $t_{1/2}$. Half-time is the time until half the original meal remains in the stomach: $t_{1/2} = \ln(.5)/p_2$.

Data set AR.1 Percent of meal remaining in the stomach at time t (from Elashoff, Reedy, and Meyer, 1982)

Time (minutes) after ingestion	Percent of meal remaining
2.6	94.5
4.9	95.5
7.5	92.6
10.7	92.1
13.3	90.2
20.5	80.0
30.5	69.2
40.5	54.8
50.5	48.2
60.5	44.8
80.5	32.6
90.5	25.8
100.5	19.8
118.5	12.4
141.5	5.2
151.5	4.9

The instructions in the INPUT and VARIABLE paragraphs shown in Input AR.1 are common to all BMDP programs (see Chapters 3 and 4). The data are stored on disk in a file named GASTRIC.DAT. The FILE command tells the program where to find the data and is used for systems like VAX and IBM PC. (For IBM mainframes, see UNIT, Chapter 3.)

AR requires two additional paragraphs: REGRESS and PARAMETER. In the REGRESS paragraph you must specify the DEPENDENT variable and the number of PARAMETERS you want to estimate. In the PARAMETER paragraph you must specify INITIAL values for the parameters.

In Input AR.1 we fit a simple exponential function for the gastric emptying data by specifying Function 1 (NUMBER = 1) and two PARAMETERS in the REGRESS paragraph (see Table AR.1 below). When you use a built-in function other than Function 6 or 7, you must name the INDEPENDENT variable in the REGRESS paragraph. Otherwise (except for Function 6 or 7) AR uses the first variable named in the VARIABLE paragraph as the independent variable. For Functions 6 and 7, AR uses as independent variables all variables not named in the DEPENDENT statement of the REGRESS paragraph. If you specify a USE list in the VARIABLE paragraph, only the variables in your list are used.

Table AR.1: The seven built-in functions

Number	Function	Comments
1	$f_1 = \begin{cases} p_1 e^{p_2 x} + p_3 e^{p_4 x} + \ldots + p_{m-1} e^{p_m x} & \text{(if } m \text{ is even)} \\ p_1 e^{p_2 x} + p_3 e^{p_4 x} + \ldots + p_m & \text{(if } m \text{ is odd)} \end{cases}$	Sum of exponentials. Particularly useful for compartmental analysis (see Jacquez, 1985).
2	$f_2 = 1 / f_1$	One/(sum of exponentials)
3	$f_3 = p_1 + p_2 e^{p_3 x} + (p_4 + p_5 e^{p_6 x}) \sin (p_7 + p_8 x) \quad (m = 8)$	Exponentially damped sine wave. Useful for nonstationary time series analyses.
4	$f_4 = \begin{cases} \dfrac{p_1 + p_3 x + p_5 x^2 + \ldots + p_m x^{(m-1)/2}}{1 + p_2 x + p_4 x^2 + \ldots + p_{m-1} x^{(m-1)/2}} & \text{(if } m \text{ is odd)} \\ \dfrac{p_1 + p_3 x + p_5 x^2 + \ldots + p_{m-1} x^{(m/2)-1}}{1 + p_2 x + p_4 x^2 + \ldots + p_m x^{(m/2)}} & \text{(if } m \text{ is even)} \end{cases}$	Rational polynomials. A flexible family of curves which has been used to approximate functions (see Hastings, 1955).
5	$f_5 = p_1 x^{p_2} e^{p_3 x^{p_4}} + p_5 x^{p_6} e^{p_7 x^{p_8}} + \ldots$	Growth functions. If m is not a multiple of four, the additional parameters are treated as zero. This includes the Gompertz functions (see Stevens, 1951).
6	$f_6 = p_1 x_1 + p_2 x_2 + p_3 x_3 + \ldots + p_m x_m$	Constrained linear regression. To estimate an intercept, one of the x variables must be a vector of ones.
7	$f_7 = p_1 + p_2 e^{(p_3 x_1 + p_4 x_2 + \ldots + p_m x_{m-2})}$	Compare with function number one above.

where m = number of parameters
x = value of the independent variable

Good initial estimates of the parameters are important to ensure convergence and reduce the number of iterations required. For a simple function, initial estimates can often be read from a plot of the data. In our example it seems reasonable to assume that at time 0, 100% of the meal would be in the stomach. This yields an initial estimate of $p_1 = 100$. We note that half emptying ($f = 50$) occurs around $t = 45$. Since

$$f = p_1 e^{p_2 t}$$

and we set initial $p_1 = 100$, we have

$$50 = 100e^{45p_2}$$

Solving for p_2, we get ln(.5)/45 = –.0154. Note that the initial estimates must be listed in the same order as the parameters. When initial estimates cannot be obtained from a plot, use values from a similar analysis or from an analysis by a different method.

Input AR.1

```
/ INPUT        FILE = 'gastric.dat'.
               VARIABLES = 2.
               FORMAT = FREE.

/ VARIABLE     NAMES = time, percent.

/ REGRESS      DEPENDENT = percent.
               INDEPENDENT = time.
               NUMBER = 1.
               PARAMETERS = 2.

/ PARAMETER    INITIAL = 100, -.0154.

/ END
```

Output AR.1

—we omit a portion of the output—

```
  DEPENDENT VARIABLE. . . . . . . . . . . . . . .   percent                [1]
  WEIGHTING VARIABLE. . . . . . . . . . . . . . .
  NUMBER OF PARAMETERS. . . . . . . . . . . . . .         2
  NUMBER OF CONSTRAINTS . . . . . . . . . . . . .         0
  TOLERANCE FOR PIVOTING. . . . . . . . . . . . . 1.0E-08
  TOLERANCE FOR CONVERGENCE BASED ON RESIDUAL . . 1.0E-05
  TOLERANCE FOR CONVERGENCE BASED ON PARAMETERS . 1.0E-04
  MAXIMUM NUMBER OF ITERATIONS. . . . . . . . . .        50
  MAXIMUM NUMBER OF INCREMENT HALVINGS. . . . . .         5

PARAMETERS TO BE ESTIMATED                                                 [2]

                 P1                P2
MINIMUM     -0.212676E+38     -0.212676E+38
MAXIMUM      0.212676E+38      0.212676E+38
INITIAL        100.000000         -0.015400


  USING THE ABOVE SPECIFICATIONS THIS PROGRAM COULD USE UP TO 1490 CASES.

NUMBER OF CASES READ. . . . . . . . . . . . .           16

  VARIABLE                          STANDARD                                [3]
  NO. NAME            MEAN          DEVIATION       MINIMUM       MAXIMUM

   1  time          57.781250       50.298031       2.600000    151.500000
   2  percent       53.912495       34.240322       4.900000     95.500000

ITER. INCR.   RESIDUAL SUM      PARAMETERS                                  [4]
 NO.  HALV.    OF SQUARES          P1              P2

  0    0      620.53096147     100.000000      -0.016940
  0    0      372.90740309     100.000000      -0.015400
  0    0      353.46753571     110.000000      -0.015400
  1    0      216.43200882     105.604545      -0.015644
  2    0      216.41991251     105.695335      -0.015671
  3    0      216.41561810     105.672041      -0.015661
  4    0      216.41561326     105.675109      -0.015659
```

[5]

```
THE RESIDUAL SUM OF SQUARES ( =      216.416      ) WAS SMALLEST WITH THE
FOLLOWING PARAMETER VALUES

PARAMETER        ESTIMATE              ASYMPTOTIC            COEFFICIENT
                                  STANDARD DEVIATION       OF VARIATION

P1              105.675109              2.146332              0.020311
P2               -0.015659              0.000717             -0.045760


ESTIMATE OF ASYMPTOTIC CORRELATION MATRIX


               P1          P2
                    1           2

P1          1   1.0000
P2          2  -0.6432     1.0000
```

[6]

[7]

```
                                         ESTIMATED MEAN      PSEUDO *
VARIABLE       MEAN         VARIANCE     SQUARED ERROR       R-SQUARE

percent   53.912495   1172.399536          15.458259          0.9877
                           DF=   15       DF=   14

*** N O T E *** PSEUDO R-SQUARE=1.0 MINUS THE RATIO OF WEIGHTED RESIDUAL SS TO (N-1) TIMES WEIGHTED VARIANCE.  WHEN THE NONLINEAR
                MODEL FITS THE DATA LESS WELL THAN THE MEAN, THE PSEUDO R-SQUARE WILL BE NEGATIVE.
```

[8]

CASE NO.	NAME	RESIDUAL	OBSERVED 2 percent	PREDICTED 2 percent	STD. DEV. PREDICTED	COOK DISTANCE	WEIGHT * DIAG HAT MAT	STANDARDIZED RESIDUAL	1 time
1		-6.959101	94.500000	101.459101	1.944640	0.671613	0.244635	-2.036546	2.600000
2		-2.369982	95.500000	97.869982	1.786503	0.059568	0.206465	-0.676678	4.900000
3		-1.365369	92.599998	93.965367	1.630583	0.015128	0.171999	-0.381640	7.500000
4		2.727123	92.099998	89.372875	1.471747	0.045588	0.140122	0.748008	10.700000
5		4.392738	90.199997	85.807259	1.369118	0.098013	0.121261	1.191859	13.300000
6		3.341673	80.000000	76.658327	1.199517	0.040874	0.093079	0.892481	20.500000
7		3.652998	69.199997	65.546999	1.171729	0.046173	0.088817	0.973344	30.500000
8		-1.246216	54.799999	56.046216	1.249238	0.006274	0.100955	-0.334289	40.500000
9		0.277467	48.200001	47.922534	1.335461	0.000367	0.115372	0.075033	50.500000
10		3.823651	44.799999	40.976348	1.394863	0.077895	0.125864	1.040180	60.500000
11		2.641489	32.599998	29.958510	1.414806	0.038565	0.129489	0.720081	80.500000
12		0.183859	25.799999	25.616140	1.384084	0.000177	0.123927	0.049961	90.500000
13		-2.103180	19.799999	21.903180	1.334265	0.021046	0.115166	-0.568677	100.500000
14		-4.123248	12.400000	16.523248	1.213389	0.063983	0.095244	-1.102538	118.500000
15		-6.326048	5.200000	11.526048	1.032404	0.102960	0.068951	-1.667500	141.500000
16		-4.955392	4.900000	9.855392	0.952200	0.052573	0.058654	-1.299042	151.500000

Bold numbers below correspond to those in Output AR.1.

[1] AR lists the values we specified for NUMBER (the built-in function number), DEPENDENT (the name of the dependent variable), and PARAMETERS (the number of parameters) in the REGRESS paragraph. The other values reported were preassigned by the program.

[2] Because we did not specify lower (MINIMUM) and upper (MAXIMUM) limits for each parameter, AR assigns extremely low minimum and extremely high maximum values. The INITIAL parameter values are those we set in our input.

[3] AR uses only complete cases (in this example, 16 cases) for all computations. That is, if the value for any variable in a case is missing or out of range, the case is omitted from computation. If you include a USE list in the VARIABLE paragraph, only the values of variables in the USE list are checked for missing values and values out of range. (See Chapter 4 for more on USE.)

[4] At each iteration AR prints:

- the number of the iteration. Note that the program calculates $m + 1$ values at iteration zero to form its estimates of derivatives, where m is the number of parameters. In this example we have two parameters, and AR calculates three values at iteration zero.
- the number of increment halvings (see Appendix B.17)
- the calculation of weighted sum of squared residuals RSS = $\Sigma w_j(y_j - \hat{y}_j)^2$ where y_j is the observed value of the dependent variable in case j, $\hat{y}_j$ is the predicted value from the function, and w_j is the weight. Because we specified no weighting variable, all $w_j = 1$ in this example. The summation is over all cases used in the analysis.

[5] The minimum sum of squared residuals and estimates of the parameters. The final parameter estimate of p_1 is 105.7%. Of course it is impossible for more than 100% of the meal to be present at time 0. All later computations in this example use the parameter estimates printed in this panel. (See Example AR.4 for ways of restricting parameter values.)

AR also prints the asymptotic standard deviations (standard errors) of the parameter estimates. These asymptotic standard deviations are estimates, and their accuracy is highly dependent on sample size, degree of nonlinearity, measurement error variance, and the correct specification of weights. Note that any asymptotic standard deviation with the precise value of zero indicates a problem with the estimate. Either the estimate lies on a boundary of the permissible parameter space (equal to its specified upper or lower limit), or there are extremely high correlations among the estimated parameters, or the estimate is so poor that if it is changed the value of the function is not affected. In such cases AR fixes the value of the parameter and estimates the remaining parameters conditional upon this fixed value. If an estimate lies on a boundary, AR prints a message describing the boundary.

The coefficient of variation, a dimensionless measure of the relative precision of the estimate, is the ratio of the standard deviation to the estimate.

[6] The asymptotic correlation matrix of the parameter estimates. Here the parameter estimates are not highly correlated; a correlation near ±1.0 suggests that the solution may be unstable. The matrix computation is described in Appendix B.17.

[7] The mean square error is equal to the weighted sum of squared residuals divided by $N - m$ degrees of freedom, where N is the number of cases with nonzero weights and m is usually the number of independent parameters in the function. (When linear constraints have been specified, or other degeneracies occur, m may be less than the number of parameters.) In this example, the mean square error is 216.416/14 = 15.46, and the pseudo R-square value is 0.9877.

[8] For each case AR prints

- the residual $y - \hat{y}$, where y is the observed value of the dependent variable and $\hat{y}$ is the predicted value of the function at that value of the independent variable
- the observed value y (in this example, PERCENT of meal remaining)
- $\hat{y}$, the predicted value of y, given the independent variable
- the standard deviation of the predicted value
- Cook's distance
- weight*diagonal hat matrix value
- the standardized residual
- the value of the independent variable (in this example, TIME)

The three diagnostic statistics—the standardized residual, weight*diagonal hat matrix, and Cook's distance—are computed by formulas similar to those used for linear regression; to the degree to which asymptotic (local) linearity in the parameters can be assumed, the interpretation should be similar. The standardized residual (the residual divided by an estimate of its standard error) is large when predicted and observed values of the y variable are far apart. The weight*diagonal hat matrix value is large for cases where partial derivatives of the regression function with respect to the parameters are large. Cook's distance is large for cases that may have a great influence on the parameter estimates. See 2R for further details about these diagnostic statistics in linear regression. (Note that in an interactive run, panel **[8]** is divided into two panels.)

Example AR.2 Using the FUNCTION and PLOT paragraphs with a Michaelis-Menten model

If the nonlinear function you want to fit is not one of the built-in functions, you can specify it in a FUNCTION paragraph. Rules for the FUNCTION paragraph are similar to those for the TRANSFORM paragraph (see Chapter 6), except that if you have specified a VARIABLE USE list, statements in the FUNCTION paragraph can refer only to the variables included in your list. When your input includes a FUNCTION paragraph, you do not need a NUMBER statement and you need not specify the INDEPENDENT variable in the REGRESS paragraph.

To determine hepatic dissociation and binding capacities, Sugiyama, Yamada, and Kaplowitz (1982) collected the data shown in Data Set AR.2. Such data are often fit with a Michaelis-Menten model, which states that the velocity of an enzyme-catalyzed reaction is $Y = Vx/(K + x)$, where x is the concentration of substrate (the independent variable), V is the maximum velocity approached as $x \rightarrow \infty$, and K is the substrate concentration necessary for half-maximum velocity. (For more on the Michaelis-Menten model, see Colquhoun, 1971, pp. 257-263.)

Data set AR.2 Spectrophotometric binding responses of Rose Bengal by pooled rat Z protein

Concentration (μM) Of Rose Bengal (dose)	Spectrophotometric response
.027	12.7
.044	16.0
.073	20.4
.102	22.3
.175	26.0
.257	28.8
.483	29.6
.670	31.4

In Input AR.2 we define the Michaelis-Menten function in the FUNCTION paragraph:

```
F = MAXRESP*DOSE/(HALFDOSE + DOSE)
```

where MAXRESP and HALFDOSE are the parameters V and K, and DOSE is the independent variable x. In the PARAMETER paragraph we must specify the initial values of the parameters. A plot of DOSE versus RESPONSE (not shown) indicates that the maximum observed response is in the low 30s, so we set the initial value of MAXRESP to 33. We estimate the dose that gives half the maximum response (33/2) to be about .05, so we use .05 as the initial value for HALFDOSE. Note that the parameter names given here must match those in the FUNCTION paragraph. To plot the predicted and observed values of the dependent variable (RESPONSE) against the independent variable (DOSE), we identify DOSE in the PLOT paragraph.

Input AR.2

```
/ INPUT          FILE = 'hepatic.dat'.
                 VAR = 2.
                 FORM = FREE.

/ VARIABLE       NAMES = dose, response.

/ REGRESS        PARAM = 2.
                 DEPEND = response.

/ FUNCTION       F = maxresp*dose/(halfdose + dose).

/ PARAMETER      NAME = maxresp, halfdose.
                 INIT = 33, .05.
```

Input AR.2 *(continued)*

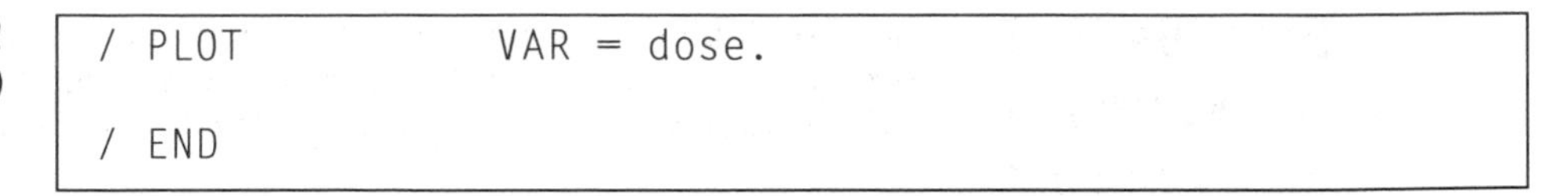

```
/ PLOT          VAR = dose.

/ END
```

Output AR.2

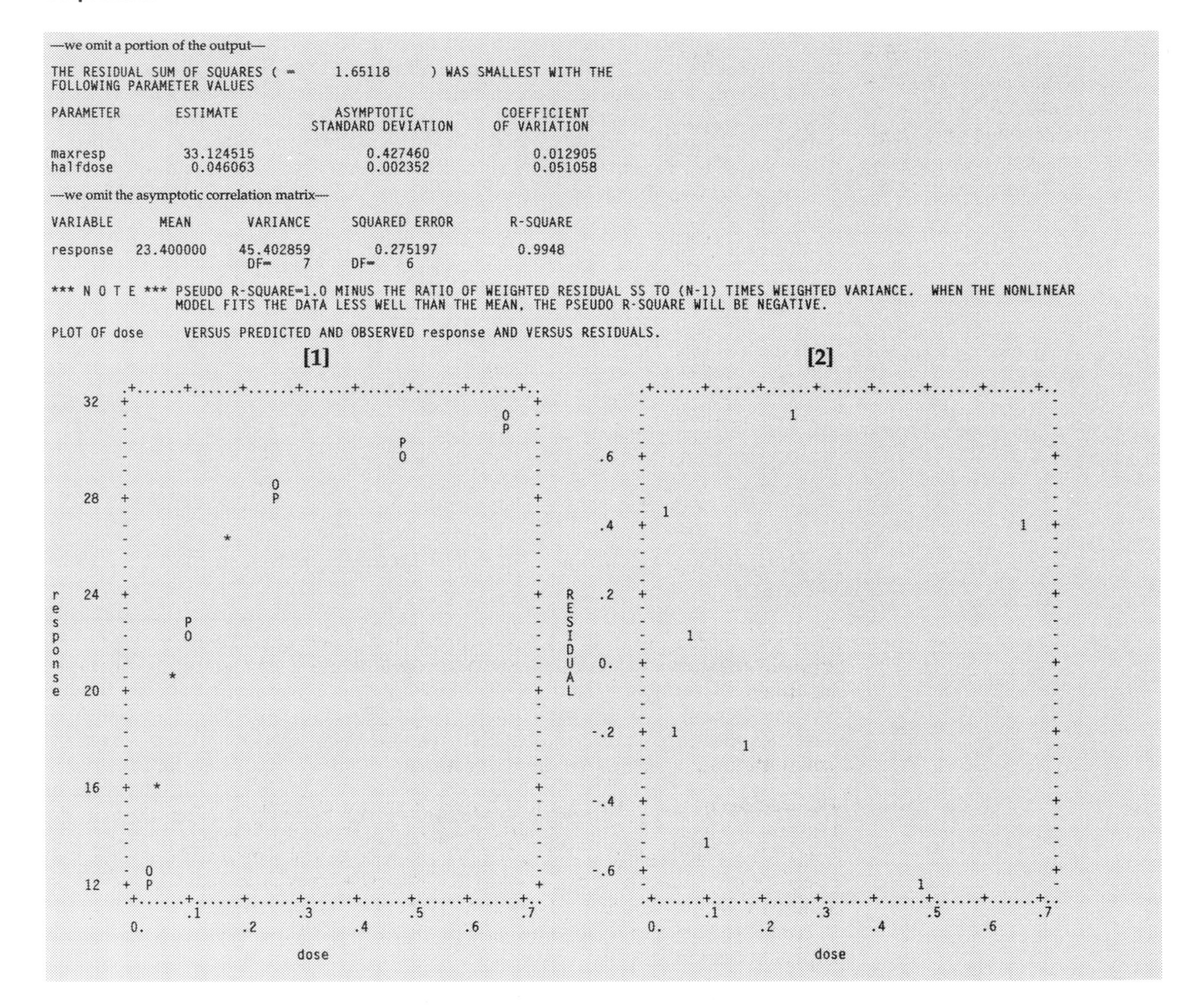

—we omit a portion of the output—

```
THE RESIDUAL SUM OF SQUARES ( =     1.65118     ) WAS SMALLEST WITH THE
FOLLOWING PARAMETER VALUES

PARAMETER        ESTIMATE              ASYMPTOTIC          COEFFICIENT
                                   STANDARD DEVIATION     OF VARIATION

maxresp          33.124515             0.427460            0.012905
halfdose          0.046063             0.002352            0.051058
```

—we omit the asymptotic correlation matrix—

```
VARIABLE       MEAN        VARIANCE     SQUARED ERROR        R-SQUARE

response   23.400000     45.402859        0.275197           0.9948
                          DF=     7     DF=     6

*** N O T E *** PSEUDO R-SQUARE=1.0 MINUS THE RATIO OF WEIGHTED RESIDUAL SS TO (N-1) TIMES WEIGHTED VARIANCE.  WHEN THE NONLINEAR
                MODEL FITS THE DATA LESS WELL THAN THE MEAN, THE PSEUDO R-SQUARE WILL BE NEGATIVE.

PLOT OF dose     VERSUS PREDICTED AND OBSERVED response AND VERSUS RESIDUALS.
```


[1] AR plots the predicted (P) and observed (O) values of the dependent variable RESPONSE versus the independent variable DOSE. Where the predicted and observed values are so close that P and O overlap, an asterisk (*) is printed. The fit here shows no consistent over- or under-prediction.

[2] AR plots the residual (observed minus predicted value of the dependent variable, $y - \hat{y}$) at each value of the independent variable.

Example AR.3 Using weights and requesting functions of parameters

Data Set AR.3 shows radioactivity counts in the blood of a baboon sampled at specific times after a single injection containing radioactive sulphate. We want to fit the data to a two-compartment model (the sum of two exponentials):

$$f = p_1 e^{p_2 t} + p_3 e^{p_4 t}$$

Dataset AR.3 Radioactivity in the blood of a baboon named Brunhilda (personal communication from Shires, as reported by Jennrich and Bright, 1976)

Counts $\times 10^{-4}$	Time
15.1117	2
11.3601	4
9.7652	6
9.0935	8
8.4820	10
7.6891	15
7.3342	20
7.0593	25
6.7041	30
6.4313	40
6.1554	50
5.9940	60
5.7698	70
5.6440	80
5.3195	90
5.0938	110
4.8717	130
4.5996	150
4.4968	160
4.3602	170
4.2668	180

Jennrich and Bright list the weights (not shown) used to adjust for Poisson variation in the counts. The quantity minimized is the weighted sum of squared residuals

$$\Sigma\, w_j (y_j - \hat{y}_j)^2$$

where the sum is taken over all cases read.

We decide not to enter Jennrich and Bright's computed weights as data, because we want AR to compute weights by the same formula Jennrich and Bright used, i.e., weight = $1/\text{COUNT}^2$. To accomplish this in Input AR.3, we specify the WT formula in the TRANSFORM paragraph and include a WEIGHT statement in the REGRESS paragraph. (Note that we use the observed response to compute the weights. Example AR.5 shows how to use weights based on the expected response.)

To obtain a sum of two exponential terms, we specify NUMBER and PARAMETERS in the REGRESS paragraph. We arrive at INITIAL values for the parameters by the process described in Examples AR.1 and AR.2.

We also want to obtain estimates (and their standard errors) for two functions of the parameters: the sum of the starting counts P1 + P3 at time zero, and the ratio of the two rate parameters, P2/P4. We include an FPARM statement in the REGRESS paragraph to give the number of functions, and we specify the functions in an FPARM paragraph:

```
/ FPARM     G1 = P1 + P3.
            G2 = P2/P4.
```

(In pre-1988 versions of AR, G1 and G2 are the only allowable names of the functions; the 1988 and later versions allow you to choose names.) AR evaluates these functions at each iteration, estimates their standard deviations, and includes them in the correlation matrix of the parameters (see Landaw et al., 1982).

To assess fit and detect outliers, we need a plot of residuals and of residuals squared against predicted COUNT values, so we include RESIDUAL in the PLOT paragraph. We also specify NORM and DNORM to obtain normal and detrended normal probability plots of the residuals. (For an explanation of NORM and DNORM, see 2R.)

Input AR.3

```
/ INPUT          FILE = 'baboon.dat'.
                 VARIABLES = 2.
                 FORMAT = FREE.

/ VARIABLE       NAMES = count, time.

/ TRANSFORM      wt = 1/count**2.

/ REGRESS        DEPENDENT = count.
                 INDEPENDENT = time.
                 WEIGHT = wt.
                 PARAMETERS = 4.
                 NUMBER = 1.
                 FPARM = 2.

/ PARAMETER      INITIAL = 10, -.1, 5, -.01.

/ FPARM          G1 = P1 + P3.
                 G2 = P2/P4.

/ PLOT           VAR = time.
                 RESID.
                 NORM.
                 DNORM.
/ END
```

Output AR.3

—we omit a portion of this output—

```
ITER.  INCR.   RESIDUAL SUM      PARAMETERS                                                                          [1]
 NO.   HALV.    OF SQUARES          P1             P2              P3             P4              G1              G2

  0     0       6.11006135       10.000000      -0.100000       5.000000      -0.011000       15.000000        9.090909
  0     0       5.67399033       10.000000      -0.110000       5.000000      -0.010000       15.000000       11.000000
  0     0       5.58621404       10.000000      -0.100000       5.000000      -0.010000       15.000000       10.000000
  0     0       5.55493937       11.000000      -0.100000       5.000000      -0.010000       16.000000       10.000000
  0     0       4.98566387       10.000000      -0.100000       5.500000      -0.010000       15.500000       10.000000
  1     1       0.89957465       10.099602      -0.100230       5.018603       0.001328       15.118206      -75.471809
  2     0       0.25715543        7.585836      -0.166169       8.631417      -0.004068       16.217253       40.850507
```

—we omit similar statistics for iteration 3 through 19—

```
 20     0       0.01381511       11.324207      -0.228190       7.370006      -0.003174       18.694213       71.891918
 21     1       0.01381510       11.324534      -0.228186       7.369672      -0.003174       18.694206       71.900502

THE RESIDUAL SUM OF SQUARES ( =    1.381510E-02 ) WAS SMALLEST WITH THE
FOLLOWING PARAMETER VALUES

PARAMETER        ESTIMATE           ASYMPTOTIC          COEFFICIENT
                                STANDARD DEVIATION     OF VARIATION

P1               11.324534           0.891301             0.078705
P2               -0.228186           0.019297            -0.084567
P3                7.369672           0.110819             0.015037
P4               -0.003174           0.000142            -0.044660
G1               18.694206           0.913812             0.048882
G2               71.900502           5.565673             0.077408

                                      ESTIMATED MEAN       PSEUDO *
VARIABLE      MEAN        VARIANCE     SQUARED ERROR        R-SQUARE

COUNT       5.745782      2.929567        0.000813          0.9944
                          DF=   20     DF=   17

*** N O T E *** PSEUDO R-SQUARE=1.0 MINUS THE RATIO OF WEIGHTED RESIDUAL SS TO (N-1) TIMES WEIGHTED VARIANCE.  WHEN THE NONLINEAR
                MODEL FITS THE DATA LESS WELL THAN THE MEAN, THE PSEUDO R-SQUARE WILL BE NEGATIVE.
```

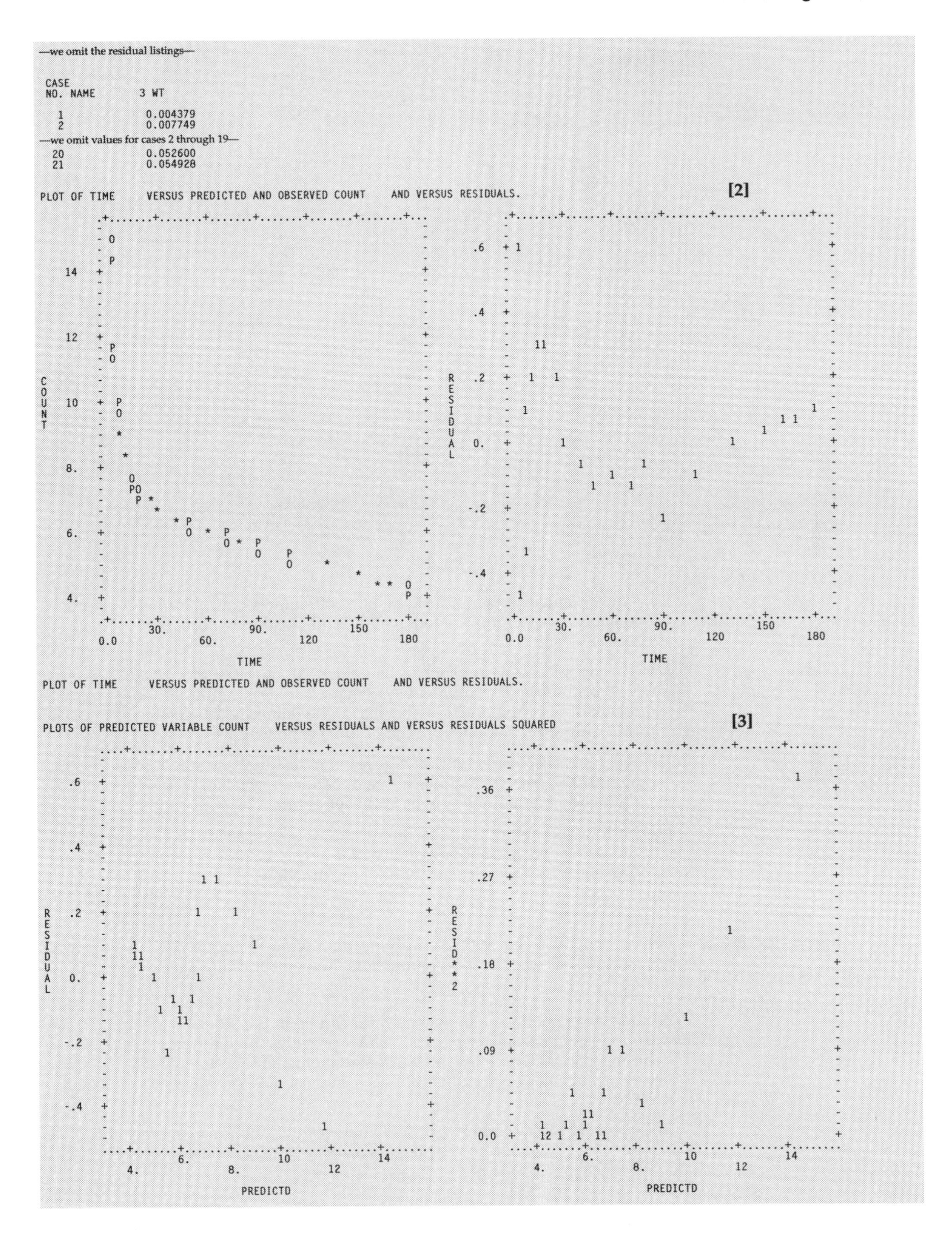

—we omit the residual listings—

```
 CASE
 NO. NAME       3 WT

   1            0.004379
   2            0.007749
```

—we omit values for cases 2 through 19—

```
  20            0.052600
  21            0.054928
```

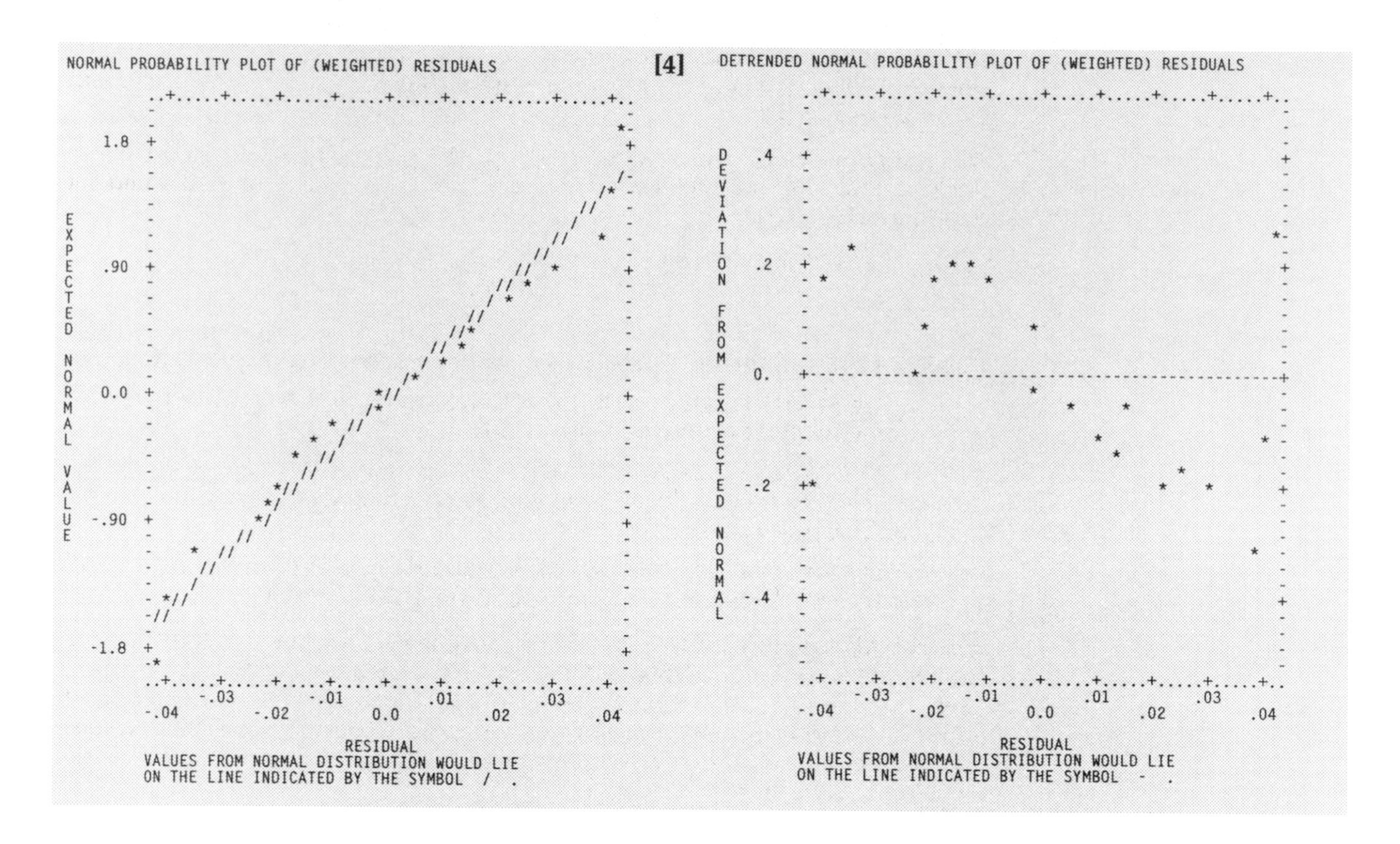

[1] The increment halving, residual sum of squares, and parameter estimates at successive iterations. Note that G1 and G2, the two functions of parameters, are included.

[2] The right-hand figure shows the RESIDUAL by TIME plot obtained by the VAR command. The apparent systematic variation in the residuals suggests a lack of fit of the function and that these data might be better fit by some other function, possibly the sum of three exponentials.

[3] AR plots the residuals (left figure) and residuals squared (right figure) against the predicted values of the dependent variable, COUNT. Note again the patterning of the residuals in the left figure.

[4] Both the normal probability plot of the weighted residuals (left figure) and the detrended normal probability plot of the weighted residuals (right figure) show a systematic lack of fit of the function.

Example AR.4 Imposing linear inequality constraints

Let us reanalyze the gastric emptying data from Example AR.1 to illustrate how to impose constraints on parameters. Remember that we used Function 1:

$$f = p_1 e^{p_2 t}$$

Because no more than 100% of the meal can be in the stomach at time = 0, we want to restrict parameter p_1 to ≤ 100. AR provides three different ways of constraining parameters: MIN and MAX statements, FIXED statements, and CONSTRAINT statements. We illustrate each in turn. (We NAME p_1 START and p_2 RATE.)

MIN and MAX statements. AR lets you set minimum and/or maximum values for each parameter. To restrict p_1 to ≤ 100, you would state:

```
/ PARAM   NAME = START, RATE.
          MAX = 100.
          INIT = 99, -.0126.
```

MIN and MAX statements follow the same rules as those used in the VARIABLE paragraph. The INITIAL value must be within (but not equal to) the upper and lower limits you set.

FIXED statements. If you want to insist that the parameter START stay fixed at 100 (and that only RATE be estimated), you can fix the value of START and set its initial value to 100:

```
/ PARAM      NAME = START, RATE.
             FIXED = START.
             INIT = 100, -.0126.
```

CONSTRAINT statements. To specify inequality constraints (denoted as ICs), use:

- a CONSTRAINT statement in the REGRESS paragraph to specify the number of inequality constraints you want.
- a CONSTRAINT statement in the PARAM paragraph to specify the coefficients (c_i) for each linear combination of parameters you want to constrain: $c_1p_1 + c_2p_2 + ... = IC$.
- a LIMIT statement in the PARAM paragraph to specify the lower (#1) and upper (#2) limits for each parameter combination: $\#1 \le IC \le \#2$.

In Input AR.4 we use a CONSTRAINT statement to constrain $p_1 \le 100$. First we specify one inequality constraint (CONSTRAINT = 1) in the REGRESS paragraph. Then we identify CONSTRAINT coefficients $c_1 = 1$ and $c_2 = 0$ with lower and upper LIMITS of 0 and 100 in the PARAM paragraph. Because the INITIAL values cannot be the same as the lower and upper limits, we change the initial value for p_1 from 100 to 99.

Input AR.4

```
/ INPUT      FILE = 'gastric.dat'.
             VARIABLES = 2.
             FORMAT = FREE.

/ VARIABLE   NAME = time, percent.

/ REGRESS    DEPENDENT = percent.
             INDEPENDENT = time.
             NUMBER = 1.
             PARAMETERS = 2.
             CONSTRAINT = 1.

/ PARAM      NAME = start, rate.
             CONSTRAINT = 1, 0.
             LIMIT = 0, 100.
             INITIAL = 99, -.0126.

/ END
```

Output AR.4

—we omit a portion of the output—

```
ITER. INCR.   RESIDUAL SUM      PARAMETERS                              [1]
 NO.  HALV.    OF SQUARES          start           rate

  0    0      578.74274661       99.500000      -0.012600
  0    0      578.15889390       99.000000      -0.012600

  5    0      330.42390584      100.000000      -0.014523

THE RESIDUAL SUM OF SQUARES ( =     330.424      ) WAS SMALLEST WITH THE
FOLLOWING PARAMETER VALUES

PARAMETER        ESTIMATE           ASYMPTOTIC          COEFFICIENT     [2]
                                STANDARD DEVIATION      OF VARIATION

start           100.000000          0.000000 *            0.000000
rate             -0.014523          0.000637             -0.043840

 AN * INDICATES THAT THE PARAMETER ESTIMATE IS (LOCALLY) HIGHLY CORRELATED
 WITH OTHER PARAMETERS OR THAT THE PARAMETER REACHED THE BOUNDARY LIMIT.
```

```
—we omit a portion of the output—

                                        ESTIMATED MEAN     PSEUDO *
VARIABLE      MEAN        VARIANCE      SQUARED ERROR      R-SQUARE

percent    53.912495   1172.399536        22.028261         0.9812
                           DF=   15     DF=   15
```

[1] The values of P1, START, begin at the initial value we specified (99) and go to the limit we specified (100).

[2] Because the estimated value of START lies on a boundary, the asymptotic standard deviation for START is undefined and is set to 0. The estimate and standard error of RATE are conditional on START equaling this boundary value.

Example AR.5 Using maximum likelihood estimation in a quantal bioassay

You can use AR to obtain maximum likelihood estimates for data sampled from a distribution in the exponential family (which includes the normal, binomial, Poisson, and gamma), using an iterative reweighting method described by Jennrich and Moore (1975). This method provides a meaningful residual analysis and standard errors for the parameter estimates. A method appropriate for maximum likelihood estimation for data from any distribution is described in Volume 2 of this manual.

Here we use the iterative reweighting method to estimate relative potency in a quantal bioassay. A bioassay is a study designed to estimate the relative potency of an unknown (test) substance with respect to a standard preparation, by administering fixed doses of the unknown and standard preparations and then comparing biological responses. In quantal bioassay the biological response is all-or-none; e.g., induction of a tumor in a mouse, death of an insect following spraying, or germination of a spore.

The response data for a particular dose can be summarized by reporting the number of subjects tested and the proportion responding. When the proportion responding is plotted against logdose for each test preparation, the dose response curves are typically sigmoidal in shape. To estimate the relative potency of a test preparation with respect to a standard, one assumes that the test preparation can be regarded as a dilution of the standard preparation and thus that the dose-response curves for the two preparations are *parallel*, or similar in shape (see Figure AR.1). The relative potency of the test preparation with respect to the standard is estimated by finding the ratio between equivalent doses of the two preparations; that is, between doses with equal mean responses.

In the past, for computational reasons, the sigmoidal dose-response curves resulting from quantal bioassays were usually transformed to produce linearity, using either the probit or the logit transformation. The two transformations produce nearly identical results in most cases (see Finney, 1978). In this example we fit the logistic function directly to obtain the maximum likelihood estimate of relative potency and its confidence interval. We use data (shown in Data Set AR.5) from an assay of insulin by the mouse convulsion method, analyzed in detail in Chapters 17 and 18 of Finney (1978). The columns of Data Set AR.5 represent drug group (0 for the standard and 1 for the test insulin preparation), insulin dose (x .001 IU), number of mice tested (N), and number of mice convulsing (r). In Figure AR.1 we plot these data; logdose is on the x-axis, and r/N (the proportion of mice convulsing at each logdose) is on the y-axis. Our goal is to fit two logistic curves, one for the s (standard) data points and one for the t (test) data points. (See the curves in Figure AR.1.)

Data set AR.5
Insulin study for a quantal bioassay

Drug	Dose	N	r
0	3.4	33	0
0	5.2	32	5
0	7.0	38	11
0	8.5	37	14
0	10.5	40	18
0	13.0	37	21
0	18.0	31	23
0	21.0	37	30
0	28.0	30	27
1	6.5	40	2
1	10.0	30	10
1	14.0	40	18
1	21.5	35	21
1	29.0	37	27

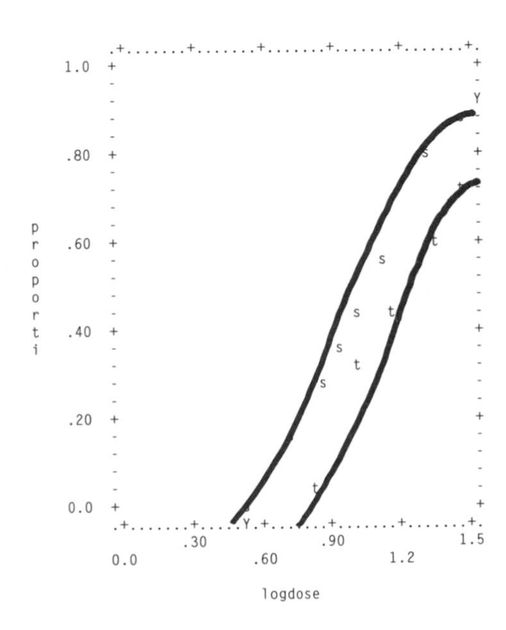

Figure AR.1
Logistic curves fit to Data Set AR.5

The logistic function. We assume that the outcome variable for each case (r_{ij}, the number of subjects responding) has a binomial distribution with parameters p_{ij} and N_{ij}, where p_{ij} (the probability of a response to dose j in group i) is given by the logistic model

$$p_{ij} = \frac{1}{1+e^{-(a_i+bx_{ij})}} = \frac{e^{(a_i+bx_{ij})}}{1+e^{(a_i+bx_{ij})}}$$

where x = logdose. We assume that the slope parameter β is the same for both drugs, but that the two curves may be shifted horizontally from each other.

We wish to estimate log relative potency, $M = (\alpha_t - \alpha_s)/\beta$, where t and s refer to the test and standard preparations. Therefore, we define the three parameters to be estimated as $p_1 = \alpha_s$, $p_2 = \alpha_t - \alpha_s$, and $p_3 = \beta$. For each case, we let $F = N_p$ be the expected value of r (the observed number of animals responding) where N is the number of animals tested and p is the probability of a response given by the logistic function. Since the variable DRUG is 0 for the standard and 1 for the test preparation, the quantity $p_1 + p_2$DRUG will equal α_s when DRUG = 0 and α_t when DRUG = 1. We can now write the equation for F as

$$F = \frac{Ne^{(p_1+p_2\text{drug}+p_3\text{logdose})}}{1+e^{(p_1+p_2\text{drug}+p_3\text{logdose})}}$$

In Input AR.5 we define this function in the FUNCTION paragraph. (If you are using a mainframe computer, you may specify the function in a FORTRAN subroutine; see Special Features.)

Iterative reweighting for maximum likelihood estimation. Since the outcome variable for each case (r_{ij}, the number of subjects responding) is assumed to have a binomial distribution with parameters p_{ij} and N_{ij},

$$\mathrm{Var}(r_{ij}) = p_{ij}\,(1 - p_{ij})N_{ij}$$

Therefore, we define weights $w_{ij} = 1/(\text{est Var}(r_{ij}))$ by replacing p_{ij} with F_{ij}/N_{ij} in the variance formula

$$w_{ij} = N_{ij}/[F_{ij}(N_{ij} - F_{ij})]$$

In Input AR.5 we specify the formula for the weights at the end of the FUNCTION paragraph. Note that weights must be *initialized;* we initialize the weights to 1.0 in the TRANSFORM paragraph.

To obtain maximum likelihood estimates of the parameters and their standard errors, we turn off the convergence criterion by setting HALVING = 0 and CONVERGENCE = -1 in the REGRESS paragraph. Since the convergence criterion has been turned off, the program runs for the default 50 iterations, which is usually more than enough. The maximum number of iterations can be controlled with the ITER command in the REGRESS paragraph. In addition, because the weights equal 1 divided by the estimated measurement variance, we rescale the mean square error to 1 (MEANSQUARE = 1), also in the REGRESS paragraph.

Note: For more information on maximum likelihood estimation, see BMDP Technical Report #9 (Jennrich and Moore, 1975) and Volume 2 of this manual.

Estimating log potency. To construct a confidence interval for the log potency, M, we need an estimate of its standard deviation. Thus, we define a function (G1) of the parameters by specifying FPARM = 1 in the REGRESS paragraph and setting G1 = P2/P3 in an FPARM paragraph. We must specify INITIAL values for each parameter in the PARAMETER paragraph. Starting with the assumption of equal relative potency, we have an initial value of zero for p_2. Noting that for logdose = 1 (dose = 10) and logdose = 1.3 (dose = 20) the proportions responding are about 0.4 and 0.7 respectively, we find initial values of -4.6 for p_1 and 4.2 for p_3 by substitution in the formula.

Input AR.5

```
/ INPUT        FILE = 'bioassay.dat'.
               VARIABLES = 4.
               FORMAT = FREE.

/ VARIABLES    NAMES = drug, dose, N, r.

/ REGRESS      DEPENDENT = proportn.
               PARAM = 3.
               FPARM = 1.
               WEIGHT = wt.
               MEANSQUARE = 1.
               HALVING = 0.
               CONVERGENCE = -1.

/ TRANSFORM    wt = 1. proportn = r/N.
               logdose = LOG(dose).

/ FUNCTION     xnumer = EXP(P1 + P2*drug + P3*logdose).
               denom = 1.0 + xnumer.
               f = xnumer/denom. wt = N*denom/f.
```

Input AR.5 *(continued)*

```
/ PARAMETER        INITIAL = -4.6, 0, 4.2.

/ FPARM            G1 = P2/P3.

/ PLOT             VAR = logdose.

/ END
```

Output AR.5

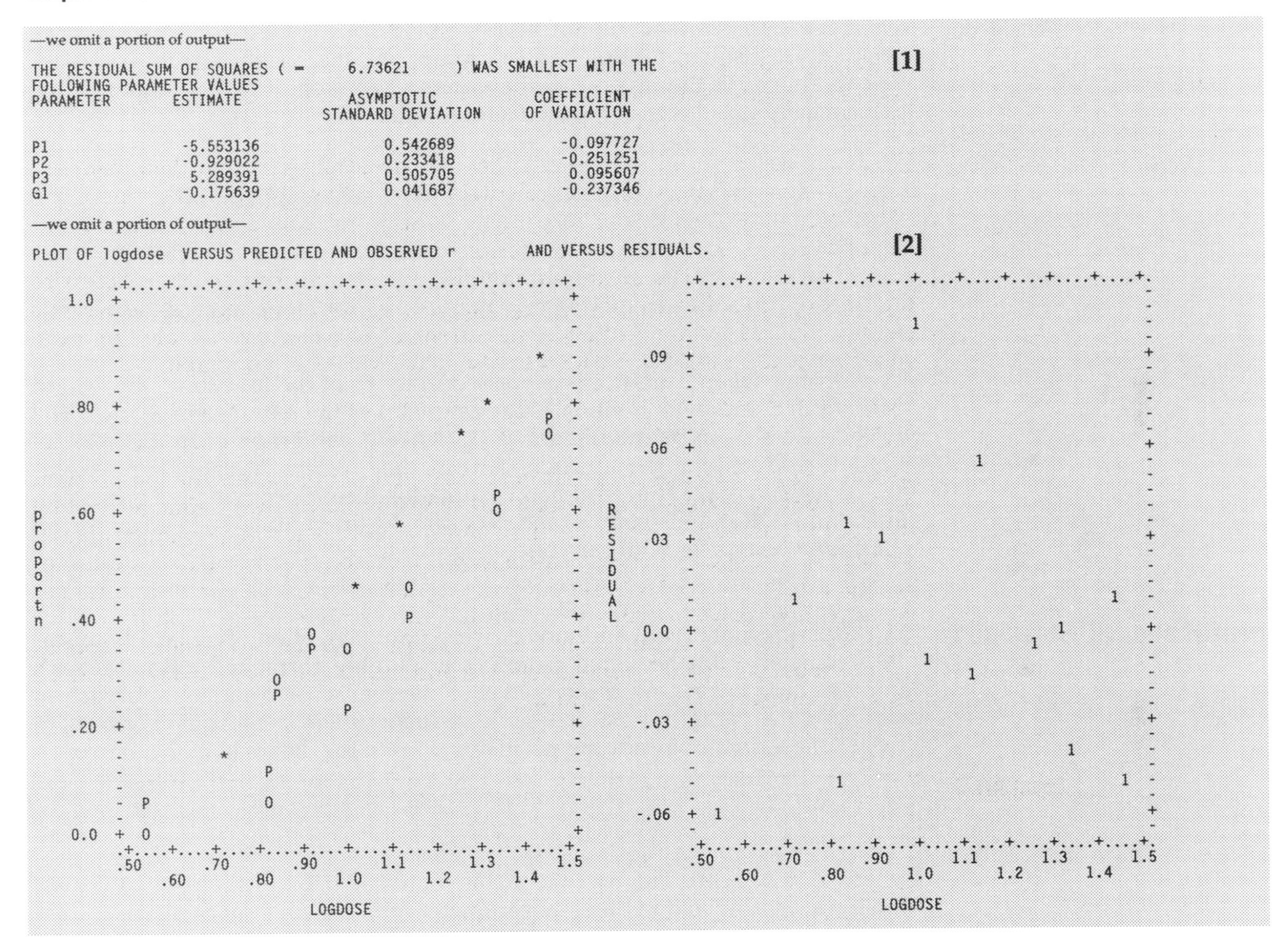

—we omit a portion of output—

THE RESIDUAL SUM OF SQUARES (= 6.73621) WAS SMALLEST WITH THE FOLLOWING PARAMETER VALUES [1]

PARAMETER	ESTIMATE	ASYMPTOTIC STANDARD DEVIATION	COEFFICIENT OF VARIATION
P1	-5.553136	0.542689	-0.097727
P2	-0.929022	0.233418	-0.251251
P3	5.289391	0.505705	0.095607
G1	-0.175639	0.041687	-0.237346

—we omit a portion of output—

PLOT OF logdose VERSUS PREDICTED AND OBSERVED r AND VERSUS RESIDUALS. [2]

We show only the relevant portion of output. The rest of the output gives fitted values and residuals.

[1] Estimating relative potency. The estimated value –.1756 for M (the log potency, M = G1) agrees closely with the value Finney obtained. The estimated relative potency $R = 20(10^M) = 13.347$. The usual R is multiplied by 20 in this case because doses of the test preparation were given in units of 20 (x .001) IU, whereas doses of the standard preparation were given in units of 1 (x .001) IU. The estimated asymptotic standard deviation for M is .041687, and the calculation –.1756 ± 2(.041687) gives an approximate 95% confidence interval for log potency ranging from –.2590 to –.0922. The confidence interval for R ranges from 11.02 to 16.17—that is, $20(10^{-.259})$ to $20(10^{-.0922})$—values close to the fiducial limits obtained by Finney using either probits or logits; see his Table 17.14.1. Thus, it takes about 13 times as much of the standard preparation to produce the same response as the test preparation.

[2] The PLOT paragraph produces a plot of the fitted curves so that we can assess the goodness of fit of the logistic function and the degree to which the curves actually appear parallel. An O denotes an observed value, and a P denotes a predicted value. The plot shows no evidence of lack of fit. In the right hand plot, the y-axis represents the residual.

Special Features

Weights

When the error variance is not homogeneous but varies from case to case, weighted least squares estimation is appropriate; the weight (w_j) for each case is proportional to the inverse of the variance. In weighted least squares, the quantity minimized is the weighted sum of squared residuals:

$$\Sigma w_j (y_j - \hat{y}_j)^2$$

where the sum is taken over all cases read. Weighted residuals are

$$\sqrt{w_j} \cdot e_j$$

where e_j are the residuals. Unlike 2R and lR, AR does not normalize the weights.

For problems where variance is expressed in terms of the function f, the weight varies from iteration to iteration as the estimates of the parameters change. The weight variable can be modified in a FUNCTION paragraph or in a FORTRAN subroutine describing the function. This method, called "iteratively reweighting," is used in maximum likelihood estimation. See "Specifying the regression function in FORTRAN" (below), and program 3R (in Volume 2). You can assign a plot symbol to cases with zero case weight.

PRIOR and RIDGE values

Poor experimental design and model overparameterization often lead to unstable parameter estimates. Such estimates are highly correlated estimates with large standard errors and difficulties in convergence. A common approach to stabilizing the parameter estimates and reducing their mean square error is to "ridge" the estimates by adding positive values to the diagonal of the matrix

$$\left(\frac{\partial f'}{\partial \theta} \mathbf{W} \frac{\partial f}{\partial \theta} \right)$$

(where $\mathbf{W}$ = weight). The inverse of this matrix is used in the pseudo-Gauss-Newton algorithm and in the final parameter estimate covariance matrix.

When independent knowledge provides prior estimates and standard errors of some parameters, ridging can be given a statistical justification, and the resulting "shrinkage" of the parameter estimates can be directed toward the prior values of those parameters (Landaw et al., 1982). AR and 3R allow you to use PRIOR and RIDGE instructions in the PARAMETER paragraph to incorporate such prior knowledge. If you use either of these commands, you must use both. For each parameter, supply a known PRIOR value and a RIDGE value equal to σ^2/σ_i^2, where σ^2 is an estimate of the (weighted) residual mean square error and σ_i is the standard error of the ith parameter's prior value.

As a simple example of using PRIOR and RIDGE, let us suppose that data from a colleague's laboratory or from a review of the literature provide an independent estimate of 0.050 for the half-saturation constant (HALFDOSE) in Example AR.2, with a standard error of 0.003 μM. We wish to incorporate this prior knowledge into our analysis of Data Set AR.2. The value of σ_i^2 for this parameter is $(0.003)^2$, and Output AR.2 provides an estimate for the residual meansquare error of $\sigma^2 = 0.2752$. Therefore, the RIDGE value for this parameter is $0.2752/(0.003)^2 = 30577.8$. We add

```
PRIOR = (2) 0.050.
RIDGE = (2) 30577.8.
```

to the PARAMETER paragraph in Input AR.2. The (2) indicates that the prior information is for parameter p_2.

When we rerun the program, the new estimate of HALFDOSE is 0.047614, with a standard error of 0.001888. This standard error is smaller than the 0.002356 obtained without prior knowledge in Output AR.2, and the estimate itself has shifted slightly (from 0.046064 to 0.047614) toward the prior value of 0.050. The standard error of the estimate of MAXRESP also shows a reduction (from 0.427681 to 0.374375). The correlation between the estimates of MAXRESP and HALFDOSE is 0.7292, rather than the 0.8053 obtained without prior knowledge.

Use of appropriate prior information for unidentifiable or poorly identifiable models can yield much improved stabilization of parameter estimates. For more details on PRIOR and RIDGE, see Landaw et al. (1982) and Chapter 3R in Volume 2 of this manual.

LOSS Function

You can replace the least square criterion with your own loss function, such as minus the logarithm of the likelihood (for maximum likelihood estimation). You must specify LOSS in the REGRESS paragraph and then compute XLOSS in the FUNCTION paragraph or in a PARFUN FORTRAN subroutine (see below). AR then sums XLOSS over all cases and uses it instead of the residual sum of squares in the convergence criterion. LOSS is especially useful for maximum likelihood estimation. For more details see Chapter 3R in Volume 2 of this manual.

ıe FUNCTION paragraph

If the nonlinear function you want to fit is not one of the built-in functions, you can specify it in a FUNCTION paragraph. Rules for the FUNCTION paragraph are similar to those for the TRANSFORM paragraph (see Chapter 6), with the following exceptions:

- If you specify a VARIABLE USE list, statements in the FUNCTION paragraph can refer only to variables included in your list.
- You must use F to refer to the function.
- If you supply parameter NAMES in the PARAMETER paragraph, you must use those names in the FUNCTION paragraph; otherwise, you must use P1, P2, etc., for the parameters.

When your input includes a FUNCTION paragraph, you do not need a NUMBER statement and you need not specify the INDEPENDENT variable in the REGRESS paragraph. For examples of input that include FUNCTION paragraphs, see Examples AR.2 and AR.5.

To perform maximum likelihood estimation by iterative reweighting, specify the formula for the weights in the FUNCTION paragraph and initialize the weights in the TRANSFORM paragraph (see Example AR.5). To replace the least squares criterion with your own loss function, specify the XLOSS function in the FUNCTION paragraph and state LOSS in the REGRESS paragraph (see "LOSS function," above). For other commands often used in the FUNCTION paragraph, see AR Commands and "Specifying the regression function in FORTRAN," below.

Specifying the regression function in FORTRAN

If you are using a mainframe computer, you may define a nonlinear regression function in a FORTRAN subroutine, which temporarily becomes part of the AR program. The FORTRAN subroutine is called once for each case during each iteration (or halving). The calling sequence is

```
SUBROUTINE PARFUN(F,P,X,N,KASE,NVAR,NPAR,IPASS,XLOSS,ITER)
IMPLICIT REAL*8 (A-H,O-Z)
DIMENSION P(NPAR),X(NVAR)
```

The rules of FORTRAN (not the rules of BMDP instructions) apply in the subroutine. The subroutine **must** evaluate F, the value of the function for the case.

The second statement, IMPLICIT REAL*8, specifies that all real values are in double precision. You must include all real variables in a double precision statement of this sort. All FORTRAN functions can be called. Use double precision versions of the functions. Failure to preclude division by 0 in a FORTRAN function may cause AR to stop executing without an error message.

The following values should not be changed:

NPAR	– The number of parameters in the function.
NVAR	– The total number of variables (not the number of independent variables), equal to the sum of the variables read and those added by transformations. If you specify USE in the VARIABLE paragraph, NVAR is the largest subscript of any variable in the USE statement.
KASE	– The sequence number of the case for which the function is being computed.
IPASS	– The number of times the data pass to the subroutine. Used primarily in maximum likelihood estimation.
P(1),...,P(NPAR)	– The current values of the parameters.

The following values are only changed for special uses:

X(1),...,X(NVAR)	– The observed values of all variables, in the order of their appearance in the data matrix. Note that you may not refer to variables by their names.
N	– The regression function NUMBER specified in the REGRESS paragraph. It can be used to select functions in your FORTRAN subroutine.
XLOSS	– A utility or loss function to replace that of least squares as a criterion for convergence. XLOSS is most useful for maximum likelihood estimation.
ITER	– The current iteration number. Use to make statements like IF (ITER.EQ.25) WRITE(6,...

For iterative reweighting in Example AR.5, we can specify the function and weights in this FORTRAN subroutine:

```
XNUMER = DEXP(P(1) + P(2)*X(1) + P(3)*X(6))
DENOM = 1.0 + XNUMER
F = X(3)*XNUMER/DENOM
X(5) = DENOM/F
```

where *X*(1) is the first variable DRUG, *X*(6) is LOGDOSE, *X*(3) is *N*, and *X*(5) is WT. (See your system consultant or installation instructions for information on linking FORTRAN subroutines with input files for your computer.)

Solving systems of differential equations

You can use the FUNCTION paragraph together with the DIFEQ and DIFIN paragraphs to fit a system of differential equations (e.g., a compartmental model) to data. For instructions, see BMDP Technical Report #85 (Ralston et al., 1987). In earlier versions of AR you must use a FORTRAN transformation for this purpose.

Saving the data and predicted values

You can save the predicted values and their standard errors (but not the residuals) in either a BMDP File or a raw data file. You can save predicted values and standard errors from more than one analysis of the same data by specifying in the SAVE paragraph the NUMBER of analyses for which the predicted values are to be saved. Otherwise, only the predicted values from the first analysis are saved (see AR Commands). If you alter the value of any variable (such as the WEIGHT variable) in a TRANSFORM paragraph or a FORTRAN subroutine, the value of that variable is also computed and saved. For more information on the SAVE paragraph, see AR Commands and Chapter 8.

Replicate error

When replicates exist, you may request an ANOVA test of lack of fit against replicate error. Descriptive statistics are available for each value of the independent variable, along with a Box-Cox diagnostic plot (see 7D) for selecting a variance-stabilizing transformation.

Plot variables

XVAR and YVAR commands (like those in 2R) added to the PLOT paragraph allow you to plot variables in the VAR USE list, predicted values, residuals (including standardized and weighted residuals), and logs of the absolute values of the residuals or of the predicted values.

AR Commands

To specify the regression function, use

- one of AR's seven built-in functions (see Table AR.1) by NUMBER in the REGRESS paragraph
- your own function in a FUNCTION paragraph (see Example AR.2)
- your own function in a FORTRAN subroutine (see Special Features)

/ REGRess

The REGRESS paragraph is required and may be repeated after the data to specify additional analyses of the same data. You must specify the DEPENDENT variable and number of parameters in the first REGRESS paragraph.

DEPENDent = variable. `DEPEND = COUNT.`

Required in the first REGRESS paragraph to specify the name or number of the dependent (predicted) variable.

INDEPendent = variable. `INDEP = TIME.`

Identify one independent variable for a built-in function. If you use Function 1, 2, 3, 4, or 5 and you do not specify an independent variable, AR assigns the first variable named in the VARIABLE paragraph as the independent variable for regression analysis. For Functions 6 and 7, AR uses as independent variables all variables not named in the DEPENDENT statement of the REGRESS paragraph; if you specify a VARIABLE USE list, only the variables in your list are used. If you use a FUNCTION paragraph or a FORTRAN subroutine, the function definition specifies the independent variable directly.

PARAMeter = #. `PARAM = 2.`

Required in the first REGRESS paragraph to specify the number of parameters in the regression function.

RCONVergence = #. `RCONV = .001.`

The RCONV command sets the convergence criterion based on the Sum of Squared Residuals. To assure that estimates from the last iteration are used in the summary results, state RCONVERGENCE = –1 and PCONVERGENCE = –1. Default: .00001. This values may not coincide with the iteration of the minimum residual sum or squares. If the minimum residual sum of squares does not occur in the final iteration, a message is printed that the program must be rerun to obtain the estimated covariance matrix corresponding to the parameter estimates. Note that this message does not appear if RCONV = –1 and PCONV = –1.

PCONVergence = #. `PCONV = .001.`

PCONV allows you to set the convergence criterion based on parameter values. Both RCONV and PCONV must be satisfied for AR to declare convergence. Default: .0001.

TITLE = 'c'. `TITLE = 'GASTRIC EMPTYING'.`

Assign a title of up to 160 characters (80 characters in interactive mode) to label output. By default, no title is printed.

NUMBer = #. `NUMB = 1.`

Specify one of the seven built-in functions (Table AR.1) by its code number. If you use FORTRAN subroutines, AR passes NUMBER to the subroutine that evaluates the function; you can then use NUMBER as an index in the subroutine.

PRINT = list. `PRINT = TIME.`

Designate the names or numbers of variables whose observed values are to be printed for each case. In addition to the observed value, AR prints the predicted value and its standard error, the value of the dependent variable and its residual, Cook's distance, WEIGHT*diagonal of the hat matrix, and the standardized residual value for each case. To suppress printing, specify PRINT = 0. By default, values are printed for all variables.

WEIGHT = variable. `WEIGHT = WT.`

Identify the name or number of the variable containing weights. This command provides a case weight rather than a regression weight. The case weight is normalized so that the average weight is 1. By default, all cases have the same weight of 1.0.

ITERations = #. `ITER = 25.`

Specify the maximum number of iterations. If convergence has not occurred by this number of iterations, the algorithm is terminated. Default: 50.

HALVing = #. `HALV = 0.`

Set the maximum number of increment halvings, as described in Special Features. Default: 5.

FPARM = #. `FPARM = 2.`

Required when the FPARM paragraph is used, to designate the number of functions of parameters.

TOLerance = #. `TOL = .001.`

Defined as one minus the squared multiple correlation of a given independent variable with all other independent variables in the function, where each "independent variable" in this context is the partial derivative of the regression function with respect to a particular parameter. AR computes the tolerance by using secant approximations to partial derivatives, and stops if the tolerance value becomes too low, indicating near-singularity in the covariance matrix of independent variables. Default: 10^{-8}.

CONSTraint = #. `CONST = 1.`

Required when the CONSTRAINT command in the PARAMETER paragraph is used, to identify the number of linear constraints on the parameters.

WITHIN = list. `WITHIN = DOSE, SEX.`

List the set of independent variables to be checked for replicate response (dependent) values. When the WITHIN instruction is used, the goodness-of-fit sum of squares is reported. Cases with replicate values for the WITHIN variables must be adjacent in the input data set. For information about analysis of replicate data, see Draper and Smith, (1981).

Options for special situations, such as maximum likelihood estimation

PASS = #. `PASS = 2.`

Set the number of times AR passes data to a FORTRAN subroutine. Default: 1.

MEANSQuare = #. `MEANSQ = 1.`

Specify to obtain information theory standard errors (e.g., for maximum likelihood estimation) or to assign a value to the mean square error other than the (weighted) sum of squared residuals divided by the degrees of freedom. AR uses the specified value to compute standard errors, Cook's distances, and standardized residuals. To delete the MEANSQUARE value you assigned in a previous problem, state a negative value.

LOSS. `LOSS.`

Replace the default least squares criterion with the desired loss function specified as XLOSS in a FUNCTION paragraph or a FORTRAN subroutine (see Special Features). The convergence criterion is applied to the sum of XLOSS.

Commands used for systems of Ordinary Differential Equations (ODEs)

In addition to the usual commands for AR, the commands below are specific for systems of differential equations. See BMDP Technical Report #85.

Commonly used options

NEQN = #. `NEQN = 5.`

Number of Ordinary Differential Equations in the system. Required for Ordinary Differential Equations. Must be less than or equal to ten.

ITIME = #. `ITIME = 4.`

Specify the number (not the name) of the time variable. Required for Ordinary Differential Equations. When solving Ordinary Differential Equations, the INDEPENDENT command is ignored and the ITIME command is expected.

Less commonly used options

MAXC = #. `MAXC = 5000.`

Maximum number of times the "DIFEQ" (differential equation) expressions are evaluated as used in the Runge-Kutta algorithm. Default: 10,000. This variable is called MXCNT in the FORTRAN subroutine.

RE = #. `RE = 0.00001.`

Relative error of integration tolerance. Default value is 10^{-6}.

AE = #. `AE = 0.001.`

Absolute error of integration tolerance. Default value is 10^{-4}.

RIDGE = 1, 2, ***or*** **0.** `RIDGE = 2.`

1 – ridge option

2 – marquardt option

0 or any other value – no ridging or marquardting.

Note: This command is not the same as the RIDGE command in the PARAMETER paragraph. See BMDP Technical Report #85. Initial standard deviation values for marquardting are provided by the dependent variable values of dummy cases whose time values are the negative of the corresponding parameter's subscript.

For example, if 0.1 is the initial standard deviation value for parameter 3, a dummy case with dependent variable equal to 0.1 and time variable equal to –3 is added.

IY = #. `IY = 3.`

Specify the number (not the name) of the dependent (Y) variable. Only needed when RIDGE options are used (marquardt or ridge regression). When the RIDGE option is used (RIDGE = 1 or RIDGE = 2), the DEPEND instruction is ignored and the IY instruction is expected.

/ PARAMeter

Required to specify initial values for the parameters. May be repeated.

INITial = # list. `INIT = 99, -014.`

Required. Specify initial values for parameter estimates, one per parameter. Values must be within (but not equal to) the specified lower and upper limits for the parameters and for the specified linear combinations of the parameters. Good initial estimates are important to avoid saddle-points or unstable solutions of the sum of squared residuals.

Note: If the final residual sum of squares seems too large, if there appear to be unusual features in estimates of parameters or residuals, or if system error messages (such as underflow, overflow, or divide check) are printed, check whether: (1) the function evaluated using the initial, largest, and smallest values of the parameters is extremely large or small; (2) the FORTRAN statements (if used) are correct; if you do not find an error, try different initial values.

MAXimum = # list. `MAX = 100, -.01.`

Specify the upper limit for each parameter. Note that MIN < INITIAL < MAX. Default: no upper limit.

MINimum = # list. `MIN = 98, -.02.`

Specify the lower limit for each parameter. Note that MIN < INITIAL < MAX. Default: no lower limit.

NAME = list. `NAME = start, rate.`

Assign a name (up to eight characters) to each parameter. The names you specify replace the default parameter names P1, P2, etc.

FIXED = parameter names *or* numbers. `FIXED = start.`

List parameters to be held fixed at their initial values. If you specified parameter NAMES, you must use those names in the FIXED statement. Otherwise, use P1, P2, etc. By default, no parameters are held fixed.

DELTA = # list. `DELTA = .9, -.009.`

Specify the increment for the starting values of each parameter. Rarely used. Default: .1* INIT(I) when INIT(I) ≠ 0, where (I) is the index of parameters; .01 otherwise.

CONSTraint = # list. `CONST = 1, 0.`

Specify the coefficients of inequality constraints of the parameters in one constraint, $\#1 \leq c_1p_1 + c_2p_2 + \ldots \leq \#2$. The first coefficient (c_1) is that of the first parameter, etc. You must have specified the number of constraints in the REGRESS paragraph, and you must use CONSTRAINT in conjunction with LIMIT (below). To constrain the sum of the first and third parameters, state CONST = 1, (3)1. The CON-

STRAINT statement can be repeated in the same PARAMETER paragraph to specify additional constraints, and each CONSTRAINT statement cancels all constraints specified in previous PARAMETER paragraphs. To delete constraints set in a previous problem, state CONSTRAINT = 0. Default: no constraints.

LIMIT = #1, #2. `LIMIT = 0, 1.`

Specify lower (#1) and upper (#2) LIMITS for the linear combination of parameters specified in the preceding CONSTRAINT statement (see above). If only one number is specified, it is used as a lower limit. An upper limit without a lower limit is specified as LIMIT = (2)#. Note that #1 < INITIAL < #2. Default: a lower limit of $-\infty$ and an upper limit of ∞.

PRIOR = # list. `PRIOR = (2) 0.2.`

Supply a known PRIOR estimate for each parameter. Must be used in conjunction with RIDGE (below). Default: no PRIOR estimates. See Special Features.

RIDGE = #list. `RIDGE = (2) 2500.`

Specify one RIDGE value per parameter. RIDGE values correspond to, and must be used with PRIOR estimates (above). Default: no RIDGE values. See Special Features.

/ FUNction

Use the rules of the TRANSFORM paragraph to specify the function for which parameters are to be estimated. If you specify a VARIABLE USE list, statements in the FUNCTION paragraph can refer only to the variables included in your list. If you supply parameter NAMES in the PARAMETER paragraph, you must use those names in the FUNCTION paragraph; otherwise, you must use P1, P2, etc., for the parameters. See Special Features, Examples AR.2 and AR.5, and Chapter 6.

F = function.

```
F = maxresp*dose/
      (halfdose + dose).
```

Required to specify the function for which parameters are to be estimated.

XLOSS = function. `XLOSS = -log(likelihd).`

Replace the least squares criterion with your own loss function for maximum likelihood estimation. State LOSS in the REGRESS paragraph. The convergence criterion is applied to the sum of XLOSS.

NEW = 0. `IF(KASE EQ 7) THEN NEW = 0.`

The variable NEW is available in the FUNCTION paragraph and is set to zero whenever any derivative (given in the DIFEQ paragraph) has a discontinuity (i.e., a change in functional form).

/ FPARM

Use the rules of the TRANSFORM paragraph to define functions of parameters for which estimates and their standard errors are to be computed. Note that you cannot use input variables. That is because variables have different values for different cases, but parameters do not. For syntax rules, see Chapter 6, Example AR.3, and the FPARM command in the REGRESS paragraph.

Variable = Transformation.

```
G1 = P1 + P3.
G2 = P2/P4.
```

/ DIFIN

A DIFIN paragraph is used to specify initial values for each differential equation and for the starting time. See BMDP Technical Report #85.

Z1 = expression. `Z1 = 0.`
Z2 = expression. `Z2 = 0.`
...

Zi sets the initial value of the *i*th Ordinary Differential Equation, where "expression" can be a numeric constant or an algebraic expression. If an algebraic expression is used, it must conform to the rules of the TRANSFORM paragraph.

T = expression. `T = 5.`

T is an internal variable that sets the initial time. "Expression" may be a numeric constant or an algebraic expression. If an algebraic expression is used, it must conform to the rules of the TRANSFORM paragraph. Default value is zero.

/ DIFEQ

A DIFEQ paragraph is used to specify the differential equations. See BMDP Technical Report #85.

DZ1 = expression. `DZ1 = -(K1 + K2)*Z1 + K3*Z2.`
DZ2 = expression. `DZ2 = K1*Z1 - K3*Z2.`

DZi = specifies the *i*th Ordinary Differential Equation, where "expression" is an algebraic expression that follows the rules of the TRANSFORM paragraph.

/ PRINT

Optional.

ALLP. *or* **NO ALLP.** `ALLP.`

ALLP requests printing of all parameters at the completion of each iteration. NO ALLP requests printing of as many parameters as will fit across the page (or screen) at each iteration; any remaining parameters are printed after the iterations have finished. Default: ALLP for interactive runs, NO ALLP for noninteractive runs.

SERCOR = KASE, OBSERVED, PREDICTD, *or* **variable.** `SERCOR = KASE.`

Prints the serial correlation of the residuals, with cases ordered by case number (KASE), the observed value of the dependent variable, the predicted value of the dependent variable, or the values of an independent variable. For cases with duplicate values of the ordering variable, the average of the residuals with the duplicate values is used as a single value.

RUN = KASE, OBSERVED, PREDICTD, *or* **variable.** `RUN = KASE.`

Prints the runs statistic for the residuals, with cases ordered by case number (KASE), the observed value of the dependent variable, the predicted value of the dependent variable, or the values of an independent variable. For cases with duplicate values of the ordering variable, the average of the residuals with the duplicate values is used as a single value.

FORMAT = F *or* **E.** `FORMAT = E.`

Specify FORMAT = E to obtain the iteration panel, parameter estimate summary, and case summary panel in scientific (E) notation. Default: F.

/ PLOT

VARiable = list. `VAR = TIME.`

For each variable specified (by name or number), the program plots observed and predicted values of the dependent variable against the variable specified, and the residuals ($y - \hat{y}$) against the specified predictor variable.

RESIDual. `RESID.`

Plots residuals ($y - \hat{y}$) and residuals squared against the predicted values ($\hat{y}$).

NORMal. `NORM.`

Prints normal probability plots of the residuals. If you specify weights, AR plots the weighted residuals.

DNORMal. `DNORM.`

Prints detrended normal probability plots of the residuals. If you specify weights, AR plots the detrended weighted residuals.

XVAR = list. `XVAR = PREDICTD, HEIGHT.`
YVAR = list. `YVAR = RESIDUAL, WEIGHT.`

Specify plots using lists of *x*-variables and *y*-variables. The list elements may be names or numbers of variables in the USE list, or selections from the list below. The first variable in XVAR is paired with the first variable in YVAR, the second with the second, etc. In addition to variables, the lists may include:

PREDICTD	– predicted value
RESIDUAL	– residual, which is observed – predicted
STRESIDL	– standardized residual
WRESIDUL	– weighted residual, which is (sqrt(case weight) * residual)
LOGRES	– ln(absolute value of residual)
LOGPRED	– ln(absolute value of predicted)
LOGWRES	– ln(absolute value of weighted residual)

ZEROWT = symbol. `ZEROWT = '*'.`

Specify a symbol to be assigned to cases with zero case weights in plots. The symbol may be any special character. By default, cases with zero weight are omitted from the plots (ZEROWT = ' '.).

SIZE = #1, #2. `SIZE = 45, 30.`

Specify plot size as the number of characters along the *x*-axis (#1) and the number of lines on the *y*-axis (#2). Default: 40, 25 in interactive mode; 50, 35 in non-interactive mode.

/ SAVE

Use to create a BMDP File containing predicted values and their standard errors (but not residuals), along with data. See Chapter 8 for an explanation of CODE, FILE, and NEW.

NUMBer = #. `NUMB = 2.`

Specify the number of analyses (subproblems) from which predicted values and their standard errors are to be saved in a BMDP File (along with data). AR saves the predicted values from the first analysis as variable VT+1 with the name PREDICT1, and saves the standard errors as VT+2 with the name STDPRED1 (where VT is the total number of variables read and transformed). Values from the second analysis are stored as VT+3 and VT+4 and are named PREDICT2 and STDPRED2. Note that VT+3 and VT+4 do not have the same definitions in AR as in 2R. By default, the results from the first analysis are saved.

Order of Instructions

■ indicates required paragraph

■ / INPUT
/ VARIABLE
/ TRANSFORM
/ SAVE
■ / REGRESS
■ / PARAMETER
/ DIFIN
/ DIFEQ
/ FUNCTION
/ FPARM
/ PRINT
/ PLOT
■ / END
data
/ REGRESS
/ PARAMETER
/ DIFIN
/ DIFEQ
/ FUNCTION
/ END

Repeat for additional analyses of same data. (applies to the second REGRESS through END)

Repeat for additional problems. See Multiple Problems, Chapter 10. (applies to INPUT through the final END)

Summary Table for Commands Specific to AR

	Paragraphs Commands	Defaults	Multiple Problems	See
■	/ **REGRess**			
■	DEPENDent = variable.	none	●	AR.1
■	PARAMeters = #.	none	●	AR.1
▲	INDEPendent = variable.	1st variable named in VARIABLE	●	AR.1
		paragraph (exception: Function 6)	●	AR.1
	NUMBer = #.	no built-in function used	●	AR.1
	PRINT = list.	all observed and predicted variables		
	or PRINT = 0.	and diagnostics	●	Cmds
	WEIGHT = var.	all cases have same weight	●	AR.3
	CONSTraint = #.	none	●	AR.4
	RCONVergence = #.	10^{-5}	●	Cmds
	PCONVergence = #.	10^{-4}	●	Cmds
	ITERations = #.	50	●	Cmds
	HALVings = #.	5	●	AR.5
	FPARM = #.	none	–	AR.3
	TOLerance = #.	10^{-8}	●	Cmds
	TITLE = 'text'.	no title printed	–	Cmds
	WITHIN =list.	none	–	Cmds
	PASS = #.	1	●	Cmds
	MEANSQuare = #.	none	●	AR.5
	LOSS.	NO LOSS.	●	SpecF
	NEQN = #.	none	●	Cmds
	ITIME = #.	none	●	Cmds
	MAXC = #.	10,000	●	Cmds
	RE = #.	10^{-6}	●	Cmds
	AE = #.	10^{-4}	●	Cmds
	RIDGE = #.	no ridging or Marquardting	●	Cmds
	IY = #.	none	●	Cmds

Summary Table (continued)

	Paragraphs / Commands	Defaults	Multiple Problems	See
■	/ **PARAMeter**			AR.1
■	INITial = # list.	none	●	AR.1
	MAXimum = # list.	none	●	AR.4
	MINimum = #list.	none	●	AR.4
	NAME = list.	P1, P2, etc.	●	AR.4
	DELTA = # list.	1* INIT when INIT = 0; .01 otherwise	●	Cmds
	FIXED = list.	none	–	AR.4
	CONSTraint = # list.	no constraint	●	AR.4
	LIMITs = #1, #2.	$\pm \infty$	●	AR.4
	PRIOR = # list.	none	–	SpecF
	RIDGE = # list.	none	–	SpecF
	/ **DIFIN**			
	Z1 = expression.	none	–	Cmds
	Z2 = expression. *etc.*			
	T = expression.	0	–	Cmds
	/ **DIFEQ**			
	DZ1 = expression.	none	–	Cmds
	DZ2 = expression. *etc.*			
▲	/ **FUNction**			AR.2
▲	F = function.	none	–	AR.2
	XLOSS = function.	none	–	SpecF
	NEW = 0.	none	–	Cmds
	/ **FPARM**			AR.3
	variable = Transformation.	none	–	AR.3
	/ **PRINT**			
	ALLP. *or* NO ALLP.	ALLP interactive; NO ALLP noninteractive	●	Cmds
	SERCOR = variable.	none	●	Cmds
	RUN = variable.	none	●	Cmds
	FORMAT = F *or* E.	F	●	Cmds
	/ **PLOT**			
	VARiable = list.	no plots involving indep. variables	●	AR.2
	RESIDual.	NO RESID.	●	AR.3
	NORMal.	NO NORM.	●	AR.3
	DNORMal.	NO DNORM.	●	AR.3
	XVAR = list.	none	–	Cmds
	YVAR = list.	none	–	Cmds
	ZEROWT = symbol.	no ZEROWT plotted	–	Cmds
	SIZE = #1, #2.	40, 25 interactive; 50, 35 noninteractive	●	Cmds
	/ **SAVE**			
	NUMber= #.	1	–	SpecF

Key:
- ■ Required paragraph or command
- ▲ Frequently used command or paragraph
- ● Value retained for multiple problems
- – Default reassigned

3S

Nonparametric Statistics

Nonparametric statistics involve less demanding assumptions about the distribution of the data than do standard statistical tests; in particular, nonparametric tests do not rely on assumptions of normality. Many nonparametric tests use ranks rather than values; when this is the case, 3S automatically converts any scores or measurements into ranks. The statistics computed by 3S are appropriate for four different problems:

- **Two or more independent groups.** The Mann-Whitney rank-sum test and the Kruskal-Wallis one-way analysis of variance provide tests of the null hypothesis that independent samples from two or more groups come from identical populations. Multiple comparisons are available for the Kruskal-Wallis test.
- **Paired observations.** The sign test and Wilcoxon signed-rank test both test the hypothesis of no difference between paired observations.
- **Randomized blocks.** The Friedman two-way analysis of variance is the nonparametric equivalent of a two-way ANOVA with one observation per cell or a repeated measures design with a single group. Multiple comparisons are available for the Friedman test. Kendall's coefficient of concordance is a normalization of the Friedman statistic.
- **Rank correlations.** The Kendall and Spearman rank correlations estimate the correlation between two variables based on the ranks of the observations.

These statistics are discussed in many texts, including Siegel (1956), Hollander and Wolfe (1973), Conover (1980), and Lehmann (1975). Each of these nonparametric statistics has a parallel parametric test. The Kruskal-Wallis test corresponds to a one-way analysis of variance (see 7D or 1V). The sign and Wilcoxon tests correspond to the paired *t* test, and the Mann-Whitney test corresponds to the pooled variance two-sample *t* test (see also 3D for sign, Wilcoxon, and Mann-Whitney tests in addition to *t* tests). The rank correlations have a parallel in the usual Pearson product-moment correlation coefficient computed by 3D. Several nonparametric tests and measures, including the Kendall and Spearman rank correlation coefficients, are also available in 4F.

Except for dropping the assumption that the data are normally distributed, the nonparametric test statistics have assumptions similar to their parametric counterparts. For example, the Mann-Whitney test and the pooled two-sample *t* both assume that the samples are obtained from distributions that are identical (have the same shape when plotted) under the null hypothesis. In the Mann-Whitney test and the two-sample *t* test, the actual probability of rejecting the null hypothesis when it is true depends on the ratio of the variances of the two groups (Pratt, 1964).

Where to find it

Examples

Example 3S.1 The Mann-Whitney rank-sum test

The Mann-Whitney (Wilcoxon) rank-sum test is a nonparametric analog of the two-sample *t* test for independent samples. The Mann-Whitney statistic, which is also reported by 3D, is computed whenever the Kruskal-Wallis test is requested for data with two groups. The Kruskal-Wallis test for more than two groups is discussed in Example 3S.2. These statistics are explained in Appendix B.18.

In Example 3S.1 we analyze the Exercise data described in Chapter 2 and Example 3D.2. The data are stored on disk in a file named EXERCISE.DAT. We test whether PULSE_2, pulse rate after exercise, differs significantly between smokers and nonsmokers. The data file has a case whose PULSE_2 is erroneous (265 instead of 165). The impact of this outlier is lessened when we use this method, which is based on ranks and not on exact values.

The INPUT, VARIABLE, and GROUP paragraphs in Input 3S.1 are common to all BMDP programs (see Chapters 3, 4, and 5). The FILE command tells the program where to find the data and is used for systems like IBM PC and VAX. (For IBM mainframes, see UNIT, Chapter 3.)

A GROUPING variable must be specified for the Mann-Whitney test. In this example we specify our grouping variable (SMOKE) in the GROUP paragraph, and then use CODES and NAMES to identify the values of the variable. The TEST paragraph is required in 3S; here we use it to request the Kruskal-Wallis test (KRUSKAL). As noted above, we will also get results for the Mann-Whitney test.

Input 3S.1

```
/ INPUT      FILE IS 'exercise.dat'.
             VARIABLES = 6.
             FORMAT IS FREE.

/ VARIABLE   NAMES = id, sex, smoke, age, pulse_1, pulse_2.

/ GROUP      VARIABLE = smoke.
             LABEL = id.
             CODES(smoke) = 1, 2.
             NAMES(smoke) = smoke, nosmoke.
```

Input 3S.1 *(continued)*

```
/ TEST        VARIABLE = pulse_2.
              KRUSKAL.
/ END
```

Output 3S.1

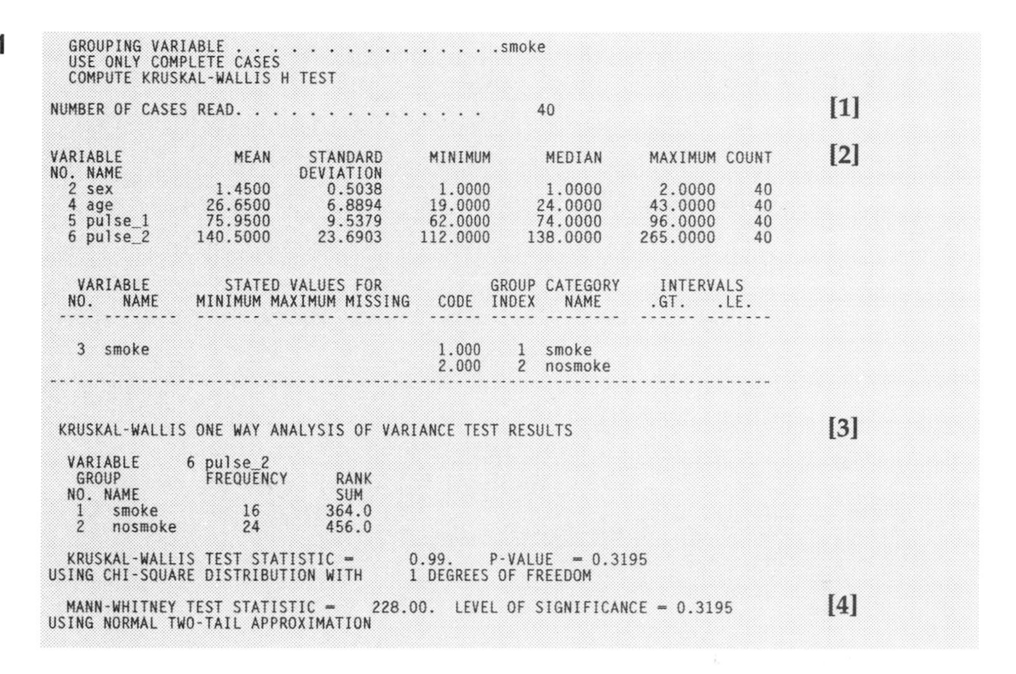

```
  GROUPING VARIABLE . . . . . . . . . . . . . . . .smoke
  USE ONLY COMPLETE CASES
  COMPUTE KRUSKAL-WALLIS H TEST

NUMBER OF CASES READ. . . . . . . . . . . . .        40                          [1]

VARIABLE            MEAN    STANDARD     MINIMUM      MEDIAN      MAXIMUM COUNT   [2]
NO. NAME                    DEVIATION
  2 sex           1.4500      0.5038      1.0000      1.0000       2.0000    40
  4 age          26.6500      6.8894     19.0000     24.0000      43.0000    40
  5 pulse_1      75.9500      9.5379     62.0000     74.0000      96.0000    40
  6 pulse_2     140.5000     23.6903    112.0000    138.0000     265.0000    40

   VARIABLE        STATED VALUES FOR          GROUP CATEGORY     INTERVALS
  NO.   NAME    MINIMUM MAXIMUM MISSING   CODE  INDEX   NAME      .GT.    .LE.
  ---- --------  ------- ------- -------  ------ -----  --------  ------ -------

   3  smoke                               1.000    1  smoke
                                          2.000    2  nosmoke
 ------------------------------------------------------------------------------

 KRUSKAL-WALLIS ONE WAY ANALYSIS OF VARIANCE TEST RESULTS                         [3]

  VARIABLE      6 pulse_2
   GROUP          FREQUENCY      RANK
  NO. NAME                        SUM
   1   smoke         16         364.0
   2   nosmoke       24         456.0

  KRUSKAL-WALLIS TEST STATISTIC =       0.99.    P-VALUE  = 0.3195
USING CHI-SQUARE DISTRIBUTION WITH      1 DEGREES OF FREEDOM

  MANN-WHITNEY TEST STATISTIC =     228.00.  LEVEL OF SIGNIFICANCE = 0.3195       [4]
USING NORMAL TWO-TAIL APPROXIMATION
```

[1] 3S reads 40 cases. Only complete cases are used in the computations; i.e., cases that have no values missing or out of range. All variables are checked for acceptable values unless you specify a USE list in the VARIABLE paragraph, in which case only the variables in the USE list are checked. See Example 3S.2 for how to use all available data for each test request.

[2] 3S prints descriptive statistics for each variable except the designated LABEL variable:

- mean
- standard deviation
- minimum observed value (not out of range)
- maximum observed value (not out of range)

[3] For each variable specified, 3S reports the sample size (frequency) and sum of ranks by subgroup, the Kruskal-Wallis test statistic, and the level of significance. If VARIABLE is not specified in the TEST paragraph, then the results are shown for all variables other than the LABEL and GROUPING variables. Here there is no significant difference in post-exercise pulse values between smokers and nonsmokers.

[4] When there are two groups, 3S computes the Mann-Whitney (Wilcoxon) rank-sum test statistic and its level of significance, which coincides with that of the Kruskal-Wallis test statistic. See Appendix B.18 for more about significance levels.

Example 3S.2 The Kruskal-Wallis test and multiple comparisons

We use the Werner blood chemistry data (Appendix D) to illustrate the Kruskal-Wallis statistic for more than two groups. We are testing whether cholesterol values for women in four different age groups come from identical populations. To classify the data into four AGE groups, we use the GROUP paragraph with CODES and CUTPOINTS specified.

We specify COMPARE in the TEST paragraph to request multiple comparisons; that is, 3S will compare every possible pair of groups. The Werner data consist of 188 cases; cholesterol and age values are present for every case, but there are missing values (marked by asterisks in the file) for height, weight, albumin, uric acid, and calcium. 3S normally deletes cases with missing values from all computations. We specify NO DELCASE to use all available data for each comparison; that is, all cases with acceptable values for AGE and CHOLSTRL. See Special Features for more about NO DELCASE. Because we obtain statistics based on ranks rather than actual values, we do not specify range limits for cholesterol.

Input 3S.2

```
/ INPUT      FILE = 'werner.dat'.
             VARIABLES = 9.
             FORMAT = FREE.
             NO DELCASE.

/ VARIABLE NAMES = id, age, height, weight, brthpill,
                   cholstrl, albumin, calcium, uricacid.
             LABEL = id.

/ GROUP      VARIABLE = age.
             CUTPOINTS(age) = 25, 35, 45.
             NAMES(age) = under_26, _26to35, _36to45,
                          over_45.

/ TEST       VARIABLE = cholstrl.
             COMPARE.
             KRUSKAL.
/ END
```

Output 3S.2

```
THE LONGEST RECORD MAY HAVE UP TO   80 CHARACTERS.
  GROUPING VARIABLE . . . . . . . . . . . . . . .age
  DO NOT DELETE INCOMPLETE CASES
  COMPUTE KRUSKAL-WALLIS H TEST
  COMPUTE MULTIPLE COMPARISONS FOR KRUSKAL-WALLIS TEST

NUMBER OF CASES READ. . . . . . . . . . . . .        188                [1]

—descriptive statistics are printed as in Example 3S.1—

 KRUSKAL-WALLIS ONE WAY ANALYSIS OF VARIANCE TEST RESULTS               [2]

  VARIABLE     6 cholstrl
   GROUP         FREQUENCY       RANK
  NO. NAME                        SUM
   1   under_26      52         4049.5
   2   _26to35       60         4816.0
   3   _36to45       42         4655.0
   4   over_45       34         4245.5

  KRUSKAL-WALLIS TEST STATISTIC =        23.35.   P-VALUE = 0.0000
USING CHI-SQUARE DISTRIBUTION WITH       3 DEGREES OF FREEDOM

         MULTIPLE COMPARISONS                                           [3]

THE NULL HYPOTHESIS IS REJECTED IF ZSTAT IS LARGER THAN
THE CRITICAL VALUE ZC, WHERE 1-PHI(ZC)= ALPHA/(K(K-1)),
PHI IS THE CUMULATIVE STANDARD NORMAL DISTRIBUTION FUNCTION,
ALPHA IS THE DESIRED OVERALL SIGNIFICANCE LEVEL, AND
K IS THE NUMBER OF GROUPS COMPARED.

WITH   4  GROUPS , THE CRITICAL Z VALUES ARE:
      2.39  FOR OVERALL ALPHA OF .10 (*)
      2.64  FOR OVERALL ALPHA OF .05 (**)

  COMPARISONS            ZSTAT       DIF        SE
under_26 - _26to35       0.23      -2.39      10.31
under_26 - _36to45       2.92**   -32.96      11.29
under_26 - over_45       3.92**   -46.99      12.00
_26to35  - _36to45       2.79**   -30.57      10.94
_26to35  - over_45       3.82**   -44.60      11.68
_36to45  - over_45       1.12     -14.03      12.55
```

[1] 3S reads 188 cases. In this example, 3S eliminates cases only when data needed for the particular test (i.e., AGE and CHOLSTRL) are missing or out of range. If NO DELCASE were omitted, seven cases would be eliminated

because of missing values for HEIGHT, WEIGHT, ALBUMIN, URICACID, and CALCIUM. All calculations would use the remaining 181 complete cases.

[2] For each variable specified, 3S reports the frequency and sum of ranks by subgroup, the Kruskal-Wallis test statistic, and the level of significance. If no variables are specified, then the results are shown for all variables other than the LABEL and GROUPING variables. The differences in cholesterol values for the four groups are significant ($p < .0005$).

[3] The COMPARE command generates multiple comparisons of all possible pairs of groups and reports the z values for a two-tailed test. In this example, the under_26 group differs significantly at the .05 level from both the 36to45 and the over_45 groups (ZSTAT values of 2.92 and 3.92 are greater than the critical value of 2.64). See Appendix B.18 for multiple comparison formulas.

Example 3S.3 The sign test and Wilcoxon signed-rank test

We use 3S to repeat the SIGN test and Wilcoxon signed-rank test performed in Example 3D.5 for the Exercise data. The hypothesis for both tests is that there is no difference between matched variables or paired observations. In Input 3S.3, the matched variables are PULSE_1 and PULSE_2, pulse rate before and after exercise. We test whether PULSE_1 differs significantly from PULSE_2. 3S automatically converts differences between PULSE_1 and PULSE_2 into ranks for the Wilcoxon test.

Input 3S.3

```
/ INPUT      FILE IS 'exercise.dat'.
             VARIABLES = 6.
             FORMAT IS FREE.

/ VARIABLE   NAMES = id, sex, smoke, age, pulse_1, pulse_2.

/ TEST       VARIABLES = pulse_1, pulse_2.
             SIGN.
             WILCOXON.
/ END
```

Output 3S.3

```
—we omit the descriptive statistics as shown in Example 3S.1—

SIGN TEST RESULTS                                                   [1]
NUMBER OF NON-ZERO DIFFERENCES

                    pulse_1    pulse_2
                          5          6

pulse_1    5              0
pulse_2    6             40          0

NUMBER OF POSITIVE (COLUMN VARIABLE)-(ROW VARIABLE) DIFFERENCES

                    pulse_1    pulse_2
                          5          6

pulse_1    5              0
pulse_2    6              0          0

TWO-TAIL P-VALUES OF SIGN TEST USING EXACT CALCULATION FOR 100
OR FEWER NONZERO DIFFERENCES, AND NORMAL APPROXIMATION OTHERWISE

                    pulse_1    pulse_2
                          5          6

pulse_1    5         1.0000
pulse_2    6         0.0000     1.0000
```

```
WILCOXON SIGNED RANKS TEST RESULTS                                  [2]
NUMBER OF NON-ZERO DIFFERENCES

              pulse_1    pulse_2
                    5          6

pulse_1    5        0
pulse_2    6       40          0


SMALLER SUM OF LIKE-SIGNED RANKS

              pulse_1    pulse_2
                    5          6

pulse_1    5     0.00
pulse_2    6     0.00       0.00

TWO-TAIL P-VALUES OF WILCOXON SIGNED RANKS TEST
USING EXACT CALCULATION WHEN
  - THERE ARE AT MOST 8 NONZERO DIFFERENCES OR
  - THE SMALLER SUM OF RANKS IS AT MOST 2.5.
THE NORMAL APPROXIMATION IS USED OTHERWISE.

              pulse_1    pulse_2
                    5          6

pulse_1    5   1.0000
pulse_2    6   0.0000     1.0000
```

[1] 3S computes the sign test for each pair of variables in the TEST paragraph VARIABLES list. We look at the flagged values. For each case, the total number of nonzero differences between paired PULSE_1 and PULSE_2 values (40 here) is printed in the first panel of results for the sign test. The number of positive differences appears in the second panel. Here all 40 cases showed an increase in pulse rate after exercise, so there are no positive differences. The third panel reports the level of significance of the sign test corresponding to a two-sided test of the hypothesis that the + and – signs of the differences are equally probable (each sign has probability 0.5). See Appendix B.18 for more information on significance levels.

[2] 3S computes the Wilcoxon signed-rank-test for each pair of variables. The first panel of results lists the number of nonzero differences. The second panel gives the value of the smaller of the sum of ranks for positive differences and the sum of ranks for negative differences. In the third panel 3S reports the level of significance of the Wilcoxon signed rank test for a two-sided test of the hypothesis that the populations have the same location parameter. See Appendix B.18.

Example 3S.4 Friedman's two-way analysis of variance and Kendall's coefficient of concordance

We analyze corrected data from Siegel (1956, p. 233; see Data Set 3S.1), using Friedman's two-way analysis of variance and the Kendall coefficient of concordance. The Friedman test is an extension of the sign test to more than two matched or paired variables. This arrangement of data is known as a randomized block design. The rows are the blocks, and the columns are the treatments. Blocks are formed using matched samples or repeated measures (as here). The null hypothesis is that of no treatment differences (the alternative hypotheses relate to differences in location).

In this example, the data in each row are the relative ranks (from 1 to 20) assigned by staff psychologists and speech therapists to 20 mothers based on effectiveness of child rearing. If the data were scores, 3S would convert them to ranks. We test whether there is no difference among the ranks of the mothers.

Data Set 3S.1
Siegel data

		Mothers																			
		1	2	3	4	5	6	7	8	9	10	11	12	13	14	15	16	17	18	19	20
	1	1	2	3	4	5	6	7	8	9	10	11	12	13	14	15	16	17	18	19	20
J	2	5	1	16	8	9	2	6	10	4	3	11	13	7	12	17	18	19	15	14	20
	3	3	2	7	5	14	9	15	16	6	11	8	10	1	4	19	12	20	13	17	18
u	4	8	3	10	11	4	2	5	13	9	1	14	7	6	15	16	12	19	17	18	20
	5	2	1	16	8	15	4	6	9	7	10	11.5	5	3	17	11.5	14	19	18	13	20
d	6	16	17	5	13	15	11	7	4	9	2	18	3	6	1	19	12	10	8	14	20
	7	12	9	14	6	7	2	3	10	5	4	17	8	1	15	13	16	18	11	20	19
g	8	11	2	13	10	7	3	4	14	6	5	17	9	1	12	8	16	20	15	18	19
	9	9.5	2	15	6	5	7	8	11	9.5	3	13	4	1	14	12	15	20	19	17	18
e	10	2	4	16	3	10	6	14	17	15	7	19	9	1	8	5	13	11	18	12	20
	11	11	14	12	8	7	2	5	10	3	4	13	9	1	18	6	15	19	16	17	20
s	12	8	1	13	3	5	2	14	9	6	10	15	11	19	4	7	12	18	17	16	20
	13	5	3	13	2	8	1	9	12	4	6	14	10	11	7	15	18	16	17	19	20

Note: Each case contains the rankings of one judge for all twenty mothers. The identification number of the judge is not recorded in the file.

Input 3S.4

```
/ INPUT       FILE = 'siegel.dat'.
              VARIABLES = 20.
              FORMAT = FREE.

/ TEST        FRIEDMAN.

/ END
```

Output 3S.4

—descriptive statistics, as in Example 3S.1 are printed here—

```
FRIEDMAN TWO-WAY ANALYSIS OF VARIANCE TEST RESULTS                    [1]

  VARIABLE          RANK
  NO. NAME          SUM
   1   X(1)         93.5
   2   X(2)         61.0
   3   X(3)        153.5
```

—we omit results for variables X(4) through X(18)

```
  19   X(19)       214.0
  20   X(20)       254.0


  FRIEDMAN TEST STATISTIC =     146.61.  LEVEL OF SIGNIFICANCE = 0.0000   [2]
ASSUMING CHI-SQUARE DISTRIBUTION WITH     19 DEGREES OF FREEDOM

  KENDALL COEFFICIENT OF CONCORDANCE =   0.5936                            [3]
```

[1] For each case 3S ranks the observations or scores for each variable (mother). A case corresponds to a judge or test. For each variable 3S prints the sum of the ranks. Since variable names were not included in the input, the variables are labeled X(1) through X(20).

[2] 3S next reports the value of the Friedman test statistic and its level of significance. Here the Friedman statistic is significant, suggesting consistent differences in child rearing effectiveness between mothers. We could use COMPARE to determine which pairs of mothers differ significantly. See Appendix B.18.

[3] The Kendall coefficient of concordance is a normalization of the Friedman statistic and has the same level of significance. The Kendall coefficient can range from 0 to 1.

Example 3S.5 Kendall and Spearman rank correlations

The Kendall and Spearman correlations estimate the association between two variables based on the ranks of the observations. They are appropriate for data whose observations can be ranked, whether or not an exact numerical value can be assigned. The two correlations are equally powerful, but are scaled differently. The Spearman correlation coefficient and level of significance for matched data are also provided by 3D (see Example 3D.5). When the variables are categorical you may use 4F, which also computes standard errors for the correlations.

We use the Werner blood chemistry data (Appendix D) to illustrate the Kendall and Spearman rank correlations.

Input 3S.5

```
/ INPUT      FILE = 'werner.dat'.
             VARIABLES = 9.
             FORMAT = FREE.

/ VARIABLE   NAMES = id, age, height, weight, brthpill,
                     cholstrl, albumin, calcium, uricacid.
             MAXIMUM = (cholstrl)400.
             MINIMUM = (cholstrl)150.
             LABEL = id.

/ TEST       KENDALL.
             SPEARMAN.

/ END
```

Output 3S.5

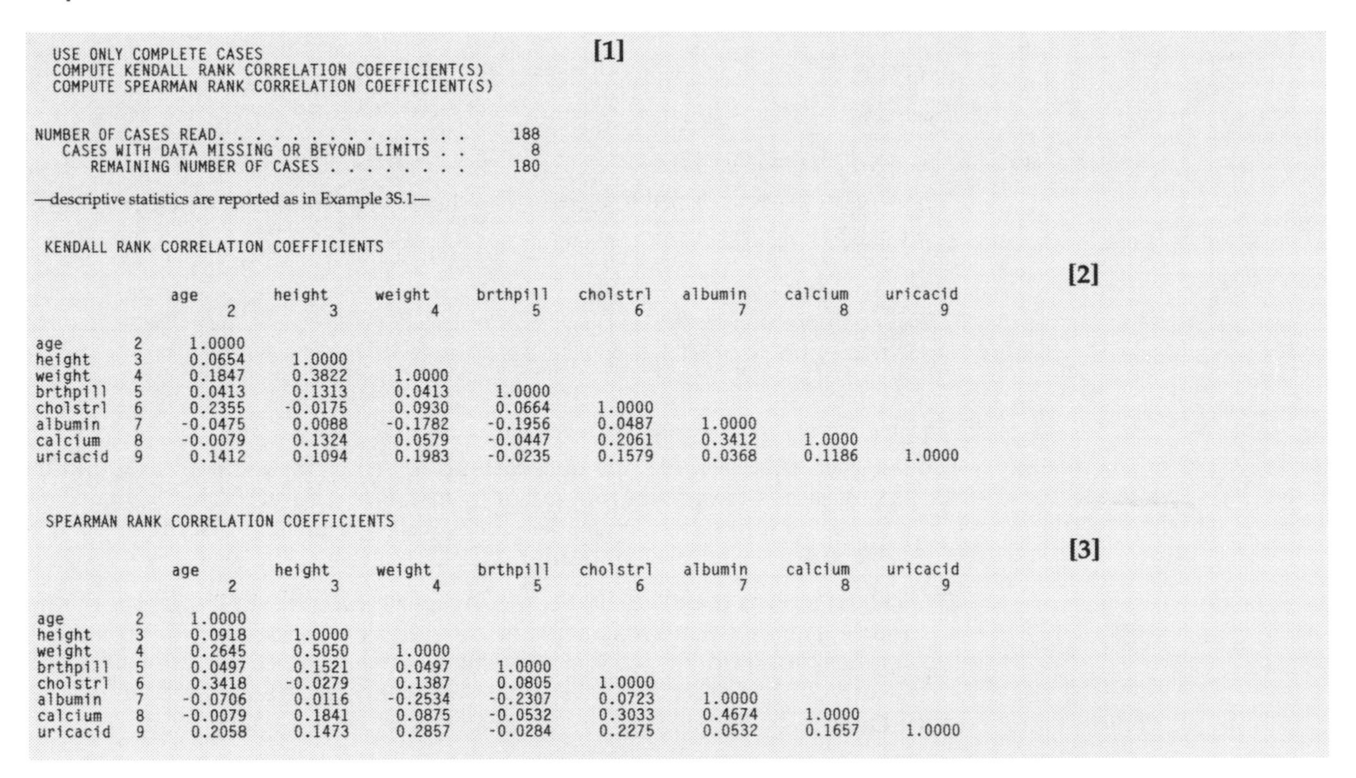

[1]

USE ONLY COMPLETE CASES
COMPUTE KENDALL RANK CORRELATION COEFFICIENT(S)
COMPUTE SPEARMAN RANK CORRELATION COEFFICIENT(S)

NUMBER OF CASES READ. 188
CASES WITH DATA MISSING OR BEYOND LIMITS . . 8
REMAINING NUMBER OF CASES 180

—descriptive statistics are reported as in Example 3S.1—

KENDALL RANK CORRELATION COEFFICIENTS [2]

		age 2	height 3	weight 4	brthpill 5	cholstrl 6	albumin 7	calcium 8	uricacid 9
age	2	1.0000							
height	3	0.0654	1.0000						
weight	4	0.1847	0.3822	1.0000					
brthpill	5	0.0413	0.1313	0.0413	1.0000				
cholstrl	6	0.2355	-0.0175	0.0930	0.0664	1.0000			
albumin	7	-0.0475	0.0088	-0.1782	-0.1956	0.0487	1.0000		
calcium	8	-0.0079	0.1324	0.0579	-0.0447	0.2061	0.3412	1.0000	
uricacid	9	0.1412	0.1094	0.1983	-0.0235	0.1579	0.0368	0.1186	1.0000

SPEARMAN RANK CORRELATION COEFFICIENTS [3]

		age 2	height 3	weight 4	brthpill 5	cholstrl 6	albumin 7	calcium 8	uricacid 9
age	2	1.0000							
height	3	0.0918	1.0000						
weight	4	0.2645	0.5050	1.0000					
brthpill	5	0.0497	0.1521	0.0497	1.0000				
cholstrl	6	0.3418	-0.0279	0.1387	0.0805	1.0000			
albumin	7	-0.0706	0.0116	-0.2534	-0.2307	0.0723	1.0000		
calcium	8	-0.0079	0.1841	0.0875	-0.0532	0.3033	0.4674	1.0000	
uricacid	9	0.2058	0.1473	0.2857	-0.0284	0.2275	0.0532	0.1657	1.0000

[1] 3S reads 188 cases, of which 180 are complete. All correlations are based on those 180 pairs. To use all data for each pair, state NO DELCASE. If there is a considerable amount of missing data, see program AM, which has several options for analyzing incomplete data.

[2] 3S calculates the Kendall rank correlation coefficients for all possible pairs of variables and reports the coefficients in matrix format.

[3] The Spearman rank correlation coefficients are printed in the same format as the Kendall statistics.

Special Features

Using all available data: NO DELCASE

When NO DELCASE is in effect, each test eliminates only cases with missing or out of range values for variables needed for that test. When you are performing tests on a number of variables, NO DELCASE maximizes the number of cases available for each test, but this means that all tests may not be based on the same cases. This differs from a VARIABLE USE list: a USE list includes all variables being tested in a problem, and only uses cases with acceptable values for all these variables. If NO DELCASE is used with KENDALL and SPEARMAN correlations, the correlations may be based on varying numbers and combinations of pairs of variables, reducing your ability to compare levels of correlation.

3S Commands

/ INPut

The INPUT paragraph is required for the first problem in each run. It is described in detail in Chapter 3. An additional command for 3S is DELCASE.

DELCASE. `NO DELCASE.`

State NO DELCASE if you want to use all available data for each variable tested. By default, 3S uses only cases complete for all variables (no values are missing or outside any specified range limits).

/ GROUP

See Chapter 5 for a description of GROUP commands.

New Syntax

VARiable = variable. `VAR = BRTHPILL.`

Required when KRUSKAL is specified in the TEST paragraph. State the name or number of a variable used to classify the cases into groups. If you prefer, you can still specify a grouping variable with the GROUPING command in the VARIABLE paragraph as described in the 1990 *BMDP Manual*. If the grouping variable takes on more than ten distinct values or codes, CODES or CUTPOINTS must be specified in the GROUP paragraph (see Chapter 5).

/ TEST

The TEST paragraph is required to specify the statistics to compute. It may be repeated after END for additional analyses of the same data.

WILcoxon. – Wilcoxon signed-rank test

KRUskal. – Kruskal-Wallis one-way analysis of variance and Mann-Whitney rank-sum test

SIGN. – Sign test

FRIEDman. – Friedman's two-way analysis of variance and Kendall's coefficient of concordance

KENDall. – Kendall rank correlation, τ_b

SPEARman. – Spearman rank correlation, r_s

New Feature

KS. – Kolmogorov-Smirnov test

Direct 3S to compute the specified tests. When KRUSKAL is specified, the other statistics are not computed, even if specified.

VARiables = list. `VAR = CHOLSTRL, AGE.`

List the names or numbers of variables to be included in the analysis. By default, 3S uses all variables except the GROUP and LABEL variables.

COMPare. `COMP.`

Use COMPARE with KRUSKAL or FRIEDMAN to obtain multiple comparisons for the Kruskal-Wallis and Friedman tests. 3S will compare every possible pair of groups.

TITLE = 'text'. `TITLE = 'PRE VERSUS POST'.`

Specify a title to print at the top of each output page. By default, no title is printed.

Order of Instructions

■ indicates required paragraph

```
■ / INPUT
  / VARIABLE
  / GROUP
  / TRANSFORM
  / SAVE
■ / TEST
  / PRINT
■ / END
  data
  / TEST    Repeat for
  / END     subproblems.
```

Repeat for additional problems.
See Multiple Problems, Chapter 10.

Summary Table for Commands Specific to 3S

Paragraphs Commands	Defaults	Multiple Problems	See
■ / **INPut**			
NO DELCASE.	DELCASE.	●	3S.1
▲ / **GROUP**			
▲ VARiable = variable.	no grouping variable; required for KRUSKAL	●	3S.1
■ / **TEST**			
▲ VARiables = list.	no tests performed	●	3S.1
KRUskal.	NO KRU.	●	3S.1
COMPare.	NO COMP.	●	3S.2
SIGN.	NO SIGN.	●	3S.3
WILcoxon.	NO WIL.	●	3S.3
FRIEDman.	NO FRIED.	●	3S.4
KENDall.	NO KEND.	●	3S.5
SPEARman.	NO SPEAR.	●	3S.5
KS.	NO KS.	●	Cmds
TITLE = 'text'.	no title printed	–	Cmds

Key:
■ Required paragraph
▲ Frequently used paragraph or command
● Value retained for multiple problems
– Default reassigned

2T

Box-Jenkins Time Series Analysis

2T is a versatile program for building and using the ARIMA time series and transfer function models of Box and Jenkins (1976). A time series is a sequence of observations on a single entity reported or measured at regular time intervals. Examples of time series include yearly gross national product, monthly traffic fatalities, and weekly weights of a dieter.

The *a priori* assumption of serial independence, common in other types of statistical analysis, is inappropriate for time series analysis. Correlation among observations is acceptable if the correlation depends on the time interval separating the observations; for example, if observations closer together in time are more highly correlated than those farther apart. In addition, 2T allows for the presence of drift (random change) or trend (deterministic change).

In contrast with the frequency domain (spectral) approach of program 1T, the time domain approach of 2T builds an explicit parametric model to account for the serial dependence and trend (or drift) in a series. Modeling provides not only a refined description of a time series, but also an important step toward predicting or controlling the series. For example, the government uses a model to predict unemployment in future months, and may take actions based on the model to reduce unemployment.

The Box-Jenkins Autoregressive-Integrated Moving Average (ARIMA) modeling method used by 2T enables you to identify a tentative model, estimate model parameters, and perform diagnostic checking or residual analysis. ARIMA models include autoregressive terms (which predict the series from its previous values), moving average terms (which predict the series from previous random errors, or "shocks"), and differencing (which accounts for either drift or trend in the series). Once you have reached a satisfactory model, you may forecast future values of the time series and analyze the effects of interventions or transfer functions.

2T provides a variety of powerful techniques for analyzing ARIMA models:

- time series plots
  - for the series (with or without differencing)
  - for transformations of the series
  - for residuals from any model selected

- identification
  - autocorrelation function (ACF)
  - partial autocorrelation function (PACF)
  - cross-correlation function (CCF)
- modeling of seasonal components, including seasonal differencing
- estimation
  - parameter estimates
  - t tests of the differences between the value of the estimates and zero
  - correlations among estimates
  - estimates of missing values
  - use of missing-value estimates in further analysis
- residual diagnostics
  - ACF and PACF of residuals
  - Ljung-Box portmanteau Q statistic
- forecasting
- intervention analysis
- multiple-input transfer function models
  - "prewhitening" (filtering to remove autocorrelation)
  - forecasts based on transfer function models

The definitive reference for ARIMA time series modeling is Box and Jenkins (1976). For a more accessible introduction to ARIMA models, Chatfield (1984), McCleary and Hay (1980), and Pankratz (1983) are readable and require only algebra as background. McCleary and Hay is geared to the social sciences; Pankratz discusses economic and forecasting applications. For an explanation of Box-Jenkins notation, which we use throughout this writeup, see "Defining an ARIMA model" below.

Where to Find It

Where to Find It (continued)

Special Features

Introduction

In this section we introduce the basic elements of the ARIMA model using ordinary algebra. We also summarize the ARIMA notation using the backshift (B) operator and del (∇) operator used in the ARIMA literature. In the next section we describe 2T features used in model building. Experienced users may want to skip to Example 2T.1.

Defining an ARIMA model

The Autoregressive (AR) Model. The simplest form of the ARIMA model is called the autoregressive model and is similar to a linear regression model. Let z_t stand for the value of a stationary time series at time t, that is, a time series that has no trend, but fluctuates about a constant value referred to as the *level* of the series. (We deal with trends below.) By autoregressive, we assume that current z_t values depend on *past* values from the same series. In symbols, at any t

$$z_t = C + \Phi_1 z_{t-1} + \Phi_2 z_{t-2} + \ldots + \Phi_p z_{t-p} + a_t$$

where C is the constant level, $z_{t-1}, z_{t-2} \ldots z_{t-p}$ are past series values (lags), the Φs are coefficients (similar to regression coefficients) to be estimated, and a_t is a random variable with mean zero and constant variance. The a_ts are assumed to be independent and represent random error or random shocks. Just as in regression, some of the Φ coefficients may be zero. If z_{t-p} is the furthest lag with a nonzero coefficient, the AR model is said to be of *order* p, denoted AR(p).

The Moving Average (MA) Model. One can imagine an extreme autoregressive model where the current observation depends on **all** past observations. If this were the true model, we would have too many parameters (Φs) to make estimation possible. However, if the Φs themselves are given by a few parameters, then estimation becomes feasible. For example, if $\Phi_i = -\Theta^i$ so that

$$z_t = -\Theta z_{t-1} - \Theta^2 z_{t-2} - \Theta^3 z_{t-3} - \ldots \quad + a_t$$

(we assume C is zero without loss of generality)

solving for a_t gives

$$a_t = z_t + \Theta z_{t-1} + \Theta^2 z_{t-2} + \Theta^3 z_{t-3} + \ldots$$

Multiplying the expression for a_{t-1} by θ, we obtain

$$\Theta a_{t-1} = \Theta z_{t-1} + \Theta^2 z_{t-2} + \Theta^3 z_{t-3} + \ldots$$

so by subtraction

$$a_t - \Theta a_{t-1} = z_t$$

This example shows that z_t as a linear function of all past lags is equivalent to z_t as a linear function of only a few past shocks. We call this type of model a moving average (MA) model since z_t is a weighted average of the uncorrelated a_t's. In general, we can model z_t as

$$z_t = a_t - \Theta_1 a_{t-1} - \Theta_2 a_{t-2} - \ldots - \Theta_q a_{t-q}$$

Such a model is called a moving average model of order q or MA(q).

The ARMA Model. We can combine the AR and MA models for stationary series to account for both past values and past shocks. Such a model is called an ARMA(p, q) model with p order AR terms and q order MA terms. For example, one ARMA(4, 2) model is written as

$$z_t = C + \Phi_1 z_{t-1} + \Phi_2 z_{t-2} + \Phi_4 z_{t-4} + a_t - \Theta_1 a_{t-1} - \Theta_2 a_{t-2}$$

Note that Φ_3, the coefficient of z_{t-3}, is zero in this example. By allowing AR and MA terms in one model, we often need only a few parameters to describe a complicated stationary series.

Trend, cycles, and ARIMA models. But how can we apply an ARMA model to a nonstationary series? Most real series show a trend, an average increase or decrease over time. For example, inflation generally increases over time, and deaths due to measles have decreased over time. Series also show cyclic behavior. For example, influenza deaths rise in winter and dip in summer.

We can remove trends and cycles from a series through differencing. For example, suppose t is in months and Y_t is a series with a linear trend. That is, every month the series increases on average (in expectation) by some constant amount C. Since $Y_t = C + Y_{t-1} + N_t$ where N_t is a random "noise" component with expectation zero, then the differences, $Y_t - Y_{t-1}$ are equal to $C + N_t$. Thus, $N_t + C$ (or just N_t) is a stationary series with the linear trend removed. We could now apply the ARMA model to $z_t = C + N_t$, even though it is not appropriate to do so to Y_t directly. By analogy, higher order polynomial trends, (i.e., quadratic trends, cubic trends) can be removed by differencing more than once.

Cycles can be removed by differencing at different lags. For example, a yearly cycle in a monthly series of stock prices can often be removed by forming the difference $z_t = Y_t - Y_{t-12}$, a 12th order (lag) difference. By differencing several times and/or at different lags, most series can be converted to a stationary series.

Once an ARMA model for z_t is known, we could reverse the differencing to form the original Y_t from the z_t. We term this process integration (although the word summation might seem more appropriate). The combined model for the original series Y_t, which involves autoregression, moving average, and integration (I) elements is termed the ARIMA(p, d, q) model (model of orders p, d, and q) with p AR terms, d differences, and q MA terms. The ARIMA model is often a parsimonious description of the behavior of a series.

The ARIMA model using backshift and del notation. To simplify notation, the ARIMA literature introduces the "backshift" operator, B. B operates on the observation Y_t by shifting it one point back in time. Thus

$$B(Y_t) = Y_{t-1}$$

B may be exponentiated in this manner:

$$B^2 = B[B(Y_t)] = B(Y_{t-1}) = Y_{t-2}$$

In general,

$$B^k(Y_t) = Y_{t-k}$$

For example, in backshift notation, differencing once is

$$z_t = (1 - B)Y_t$$

For the general case, involving differencing d times,

$$z_t = (1 - B)^d Y_t = \nabla^d Y_t$$

where the ∇ operator is substituted for $(1 - B)$ and serves as a differencing operator. Using backshift notation, the ARIMA model equation

$$z_t = C + (\Phi_1 z_{t-1} + \ldots + \Phi_p z_{t-p}) - (\Theta_1 a_{t-1} + \ldots + \Theta_q a_{t-q}) + a_t$$

becomes

$$z_t = C + (\Phi_1 B + \ldots + \Phi_p B^p) z_t + (1 - \Theta_1 B - \ldots - \Theta_q B^q) a_t$$

which is rearranged (substituting $(1- B)^d Y_t$ for z_t) as

$$(1 - \Phi_1 B - \ldots - \Phi_p B^p)(1 - B)^d Y_t = C + (1 - \Theta_1 B - \ldots \Theta_q B^q) a_t$$

To simplify notation further, the autoregressive polynomial

$$(1 - \Phi_1 B - \ldots - \Phi_p B^p)$$

is often abbreviated as $\Phi(B)$, and the moving average polynomial

$$(1 - \Theta_1 B - \ldots - \Theta_q B^q)$$

is abbreviated as $\Theta(B)$. Substituting, the model is compactly written as

$$\Theta(B) \nabla^d Y_t = C + \Theta(B) a_t$$

or

$$\nabla^d Y_t = \frac{C}{\Phi(B)} + \frac{\Theta(B)}{\Phi(B)} a_t$$

Note that the constant depends upon the ARIMA model (see "Interpreting the Constant," in Special Features in this chapter). Table 2T.1 in the Special Features section summarizes the correspondence between the ARIMA model written in backshift notation and the BMDP instructions. Note also that the del operator, ∇, is sometimes written $\nabla(B)$ to emphasize its equivalence with a backshift operator polynomial.

Additive versus multiplicative models—adjustment for known seasonality. Consider a time series z_t that we suspect depends on previous values at lag 1 and lag 12. An additive AR model for this series could be

$$(1 - \Phi_1 B - \Phi_{12} B^{12}) z_t = a_t$$

This model says that z_t depends on error plus the additive effects of z at lag 1 and lag 12. It is additive in the same sense that a two-way analysis of variance model with only row and column effects is additive.

This model could be modified by adding an "interaction" term to give

$$(1 - \Phi_1 B - \Phi_{12} B^{12} - \Phi_{13} B^{13}) z_t = a_t$$

We put the interaction at lag 13 because this allows us to formally rewrite the model as

$$(1 - \Phi_1 B)(1 - \Phi_{12} B^{12}) z_t = a_t$$

provided that $\Phi_{13} = \Phi_1 \Phi_{12}$. The latter model is a *multiplicative* model using lag 1 and 12 as opposed to the additive model. The multiplicative model allows an extra term to be added without additional parameters, often making for a better fit to real data. The two separate terms often have a simple interpretation. In our example, the $(1 - \Phi_1 B)$ term reflects a local dependence of z_t on the previous period's value, whereas the $(1 - \Phi_{12} B^{12})$ term represents a seasonal relationship. Often this seasonal lag can be determined from the context in which the data arise. The seasonal lag is usually 7 for daily data with weekly cycles and 12 for monthly data with yearly cycles. Both AR and MA elements of the ARIMA model may be re-expressed as a product of two or more multiplicative terms or factors.

The shorthand notation ARIMA$(p, d, q)(P, D, Q)_s$ is often used when an ARIMA model is factored into "local" and "seasonal" multiplicative terms where the seasonal period s is known. The lowercase p, d, and q are for the local effects factor and the uppercase P, D, and Q are for the global or seasonal term.

Building an ARIMA model

Building an ARIMA(p, d, q) model requires us to first determine the differencing (d orders), the AR terms needed (AR orders, p) and the MA terms needed (MA orders, q). This process is called model identification. We try to use the **fewest** number of these terms as possible. Once they have been identified, we can estimate the AR and MA parameters and check that the model fits adequately. When a reasonably fitting model has been derived, it can be used to generate forecasts, test for interventions, and predict the values of and explore the relationship with other series.

Series plots and stationarity. Examination of a time series plot allows us to evaluate some necessary conditions for fitting an ARIMA model. The most important of these is *stationarity* as discussed above. If we discover that our series is *nonstationary*, we may be able to render it stationary by differencing.

Another aspect of stationarity is that the variance of the series should remain constant. In some series the variance changes as a function of the level of the series, either because of the nature of the measurement or because of floor or ceiling effects. If the variance changes, we should consider transforming the series before analyzing it (see Box and Jenkins, 1976, for a complete discussion of transformations). However, care must be taken when using transformed series for forecasting (see Pankratz, 1983, pp. 256–258) or testing interventions (McCleary and Hay, 1980, pp. 172–175).

Autocorrelations and partial autocorrelations. We use the autocorrelation and partial autocorrelation function to determine p orders and q orders and confirm the differencing that was suggested by the plot. The autocorrelation at lag k, ACF(k), is the (linear) Pearson correlation between observations k time periods (lags) apart. If the ACF(k) differs significantly from zero, the serial dependence among the observations must be included in the ARIMA model. Like the ACF(k), the partial autocorrelation at lag k, or PACF(k), measures the correlation among observations k lags apart. However, the PACF(k) removes, or "partials out," all intervening lags. In general, for *autoregressive* processes [ARIMA(p, 0, 0)]

- the ACF declines exponentially
- the PACF spikes on the first p lag

By contrast, for *moving average* processes [ARIMA(0, 0, q)]

- the ACF spikes on the first q lags
- the PACF declines exponentially

Mixed processes [ARIMA(p, d, q)] decline on both ACF and the PACF. Finally, if the ACF or PACF declines slowly (i.e., has autocorrelations with t values > 1.6 for more than five or six consecutive lags), the process is probably not stationary and therefore should be differenced. Usually, a large number of significant autocorrelations or partial autocorrelations indicates nonstationarity and a need for further differencing. Figure 2T.1 shows theoretical plots of the ACF and PACF for each of these situations.

The ozone data. In Examples 2T.1, 2T.2 and 2T.7, we examine ozone data from a study in downtown Los Angeles (Box and Tiao, 1975). The data are the monthly averages of hourly readings of ozone (pphm) from January 1955 to December 1972 and are given in Input 2T.1. This series was subject to major interventions in 1960 and 1966. In 1960, two events occurred which might have reduced the ozone pollution level in downtown Los Angeles: (1) traffic was diverted when the Golden State Freeway was opened, and (2) a new law (rule 63) came into effect that reduced the allowable proportion of reactive hydrocarbons in the gasoline sold locally. In 1966, regulations were adopted requiring engine design changes that were expected to reduce the production of ozone in new cars.

In Examples 2T.1 and 2T.2, we ignore the interventions in 1960 and 1966 and build an ARIMA model for the series. We then treat the data more fully in the intervention analysis (Example 2T.7). To build a suitable ARIMA model for the ozone data, we proceed through the three steps recommended by Box and Jenkins (1976). First we identify a tentative model by examining a time series plot and plots of the autocorrelation and partial autocorrelation functions (Example 2T.1). Next we estimate the parameters of our tentative model (Example 2T.2). Finally, we check the adequacy of our model (Example 2T.2).

Figure 2T.1
Theoretical ACF and PACF for common ARIMA models

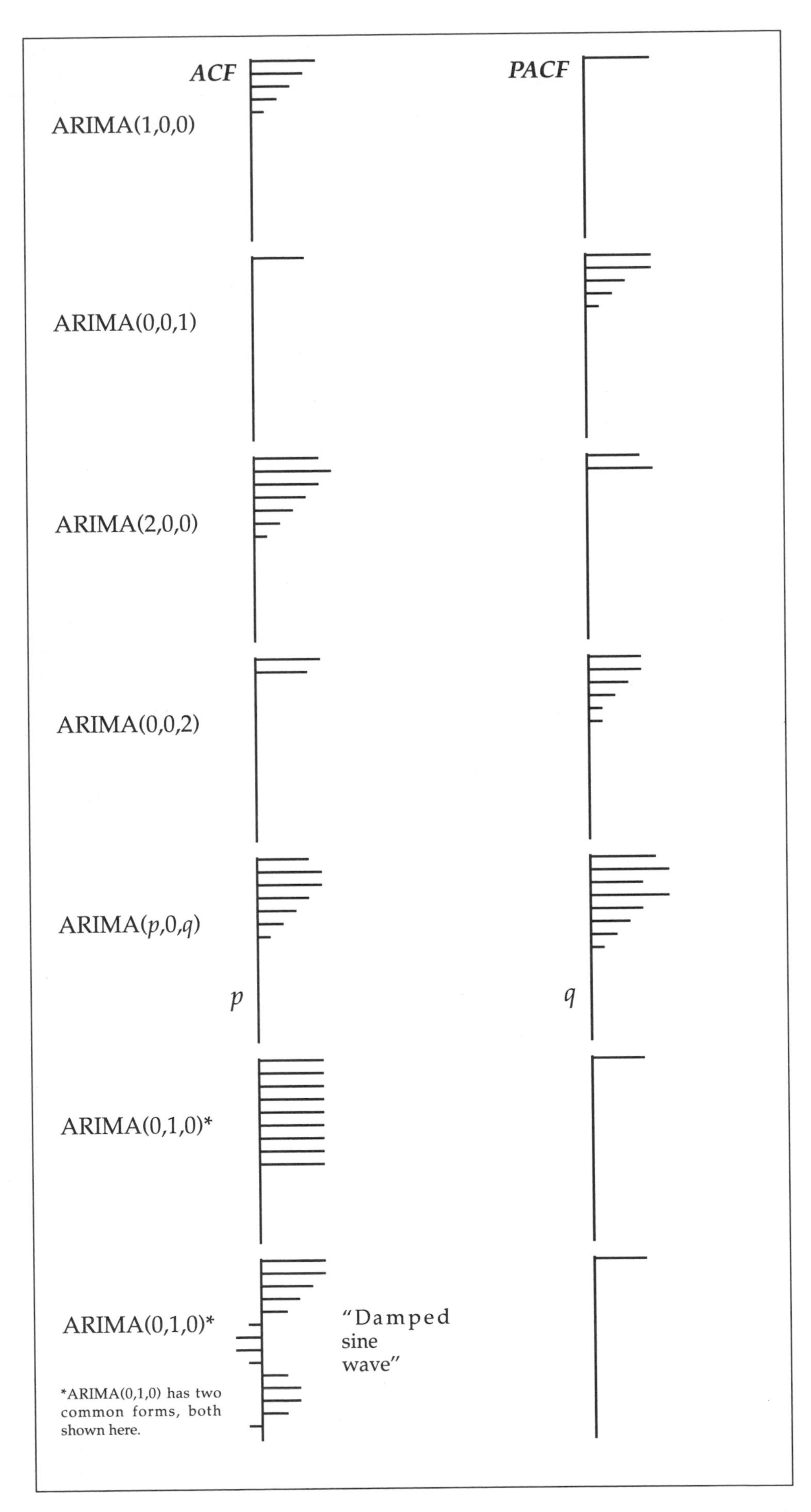

Univariate ARIMA models

In this example we begin to identify a tentative ARIMA model (the first step of model building) by plotting the ozone series and examining its autocorrelations and partial autocorrelations. Ordinarily, model identification is an exploratory process and the analyses to be performed are based upon previous results. So Input 2T.1 should be viewed as if each paragraph had been executed interactively with the result suggesting the next action.

Example 2T.1 ARIMA model identification

In Input 2T.1, the INPUT and VARIABLE paragraphs are common to all BMDP programs and define a single variable, OZONE, that will be read in STREAM format from the data. A full description of these paragraphs is given in Chapters 3 and 4. Unlike the paragraphs specific to 2T, the INPUT and VARIABLE paragraphs begin with a slash. The data are read using STREAM format with each line containing one year of monthly data.

The TPLOT, ACF, and PACF paragraphs are specific to program 2T. The VARIABLE instructions in the TPLOT, ACF and PACF paragraphs give the name or number of the time series variable to analyze. The SIZE statement in the TPLOT paragraph sets the horizontal plot width (Y_t) to 60 characters per line. SYMBOL is used to specify the plotting symbols to be used. When there are more data than symbols, the symbols are used repeatedly. In this example each symbol corresponds to a month. **A** is January, **B** is February, and so on. The MAXLAG sentence in the ACF and PACF paragraphs determines the number of autocorrelations or partial autocorrelations to be calculated. The appropriate number depends upon the length (n) of the series. Box and Jenkins (1976) recommend $n/4$ as an upper limit; here, we specify MAXLAG = 24 in order to save space in the presentation. When a time series is seasonal, the autocorrelations should cover three or four full annual cycles, if possible. DFORDERS in the last two paragraphs specifies that the computations be performed on the lag 12 differences in ozone ($z_t = Y_t - Y_{t-12}$) instead of on the original series. Note that there is no END instruction at the bottom of Input 2T.1. This is because we assume that the input continues in Example 2T.2.

The TPLOT, ACF, PACF and all other paragraphs unique to 2T **end** with a slash indicating that they are executed **interactively**. When 2T encounters a slash, it immediately executes the preceding paragraph. Therefore, instructions for the next paragraph must begin on a new line, not on the same line as the slash. Thus, when 2T encounters the slash at the end of the TPLOT paragraph, it prints the requested time series plot before processing the subsequent ACF paragraph.

See the Special Features section for a discussion on how to switch between the BMDP editor and fully interactive processing.

Caution: *For ARIMA modeling and estimation, the data must contain no gaps; a single missing datum divides the set into two series. When data are missing, 2T divides the series into BLOCKS of complete data, with one block for each set of contiguous observations. In the present case, no observations are missing. For more information, see Example 2T.4.*

Input 2T.1

```
 / INPUT    VARIABLE IS 1.
            FORMAT IS STREAM.
            TITLE IS 'Los Angeles ozone data: 1955-1972'.

/ VARIABLE  NAME IS ozone.

/ END
  2.7  2.0  3.6  5.0  6.5  6.1  5.9  5.0  6.4  7.4  8.2  3.9
  4.1  4.5  5.5  3.8  4.8  5.6  6.3  5.9  8.7  5.3  5.7  5.7
  3.0  3.4  4.9  4.5  4.0  5.7  6.3  7.1  8.0  5.2  5.0  4.7
  3.7  3.1  2.5  4.0  4.1  4.6  4.4  4.2  5.1  4.6  4.4  4.0
  2.9  2.4  4.7  5.1  4.0  7.5  7.7  6.3  5.3  5.7  4.8  2.7
  1.7  2.0  3.4  4.0  4.3  5.0  5.5  5.0  5.4  3.8  2.4  2.0
  2.2  2.5  2.6  3.3  2.9  4.3  4.2  4.2  3.9  3.9  2.5  2.2
  2.4  1.9  2.1  4.5  3.3  3.4  4.1  5.7  4.8  5.0  2.8  2.9
  1.7  3.2  2.7  3.0  3.4  3.8  5.0  4.8  4.9  3.5  2.5  2.4
```

Input 2T.1
(continued)

```
  1.6  2.3  2.5  3.1  3.5  4.5  5.7  5.0  4.6  4.8  2.1  1.4
  2.1  2.9  2.7  4.2  3.9  4.1  4.6  5.8  4.4  6.1  3.5  1.9
  1.8  1.9  3.7  4.4  3.8  5.6  5.7  5.1  5.6  4.8  2.5  1.5
  1.8  2.5  2.6  1.8  3.7  3.7  4.9  5.1  3.7  5.4  3.0  1.8
  2.1  2.6  2.8  3.2  3.5  3.5  4.9  4.2  4.7  3.7  3.2  1.8
  2.0  1.7  2.8  3.2  4.4  3.4  3.9  5.5  3.8  3.2  2.3  2.2
  1.3  2.3  2.7  3.3  3.7  3.0  3.8  4.7  4.6  2.9  1.7  1.3
  1.8  2.0  2.2  3.0  2.4  3.5  3.5  3.3  2.7  2.5  1.6  1.2
  1.5  2.0  3.1  3.0  3.5  3.4  4.0  3.8  3.1  2.1  1.6  1.3
 / END

TPLOT    VARIABLE IS ozone.     SIZE IS 60.
         SYMBOLS ARE a, b, c, d, e, f, g, h, i, j, k, l.     /

ACF      VARIABLE IS ozone.                                  /

ACF      VARIABLE IS ozone.  DFORDER IS 12.  MAXLAG IS 24.   /

PACF     VARIABLE IS ozone.  DFORDER IS 12.  MAXLAG IS 24.   /
```

(input continues in Input 2T.2)

Output 2T.1

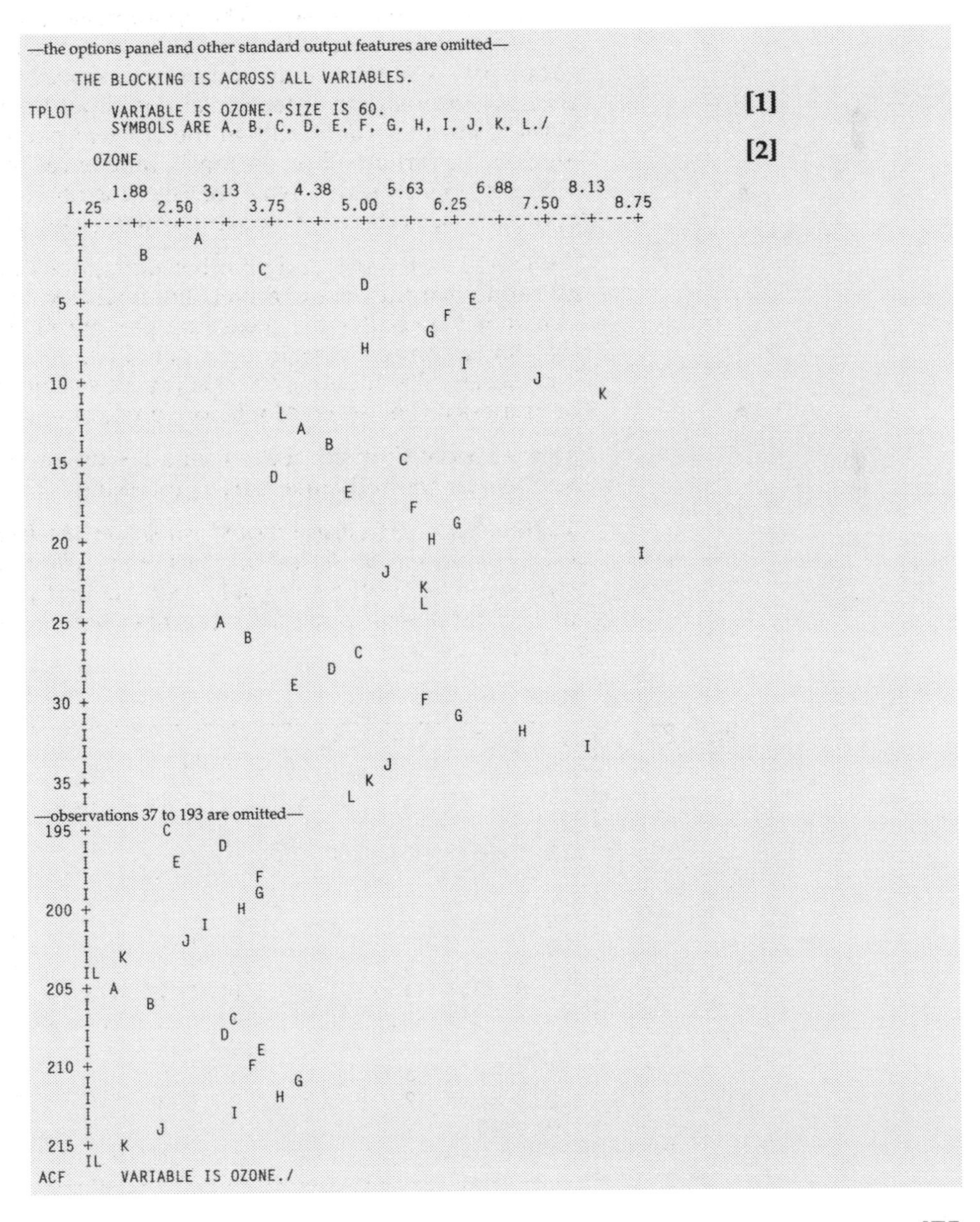

```
FIRST CASE NUMBER TO BE USED      -           1            [3]
LAST CASE NUMBER TO BE USED       -         216
NO. OF OBS. AFTER DIFFERENCING    -         216
MEAN OF THE (DIFFERENCED) SERIES  -      3.7727
STANDARD ERROR OF THE MEAN        -      0.1015
T-VALUE OF MEAN (AGAINST ZERO)    -     37.1719

AUTOCORRELATIONS                                               [4]

  1- 12     .73  .50  .28  .08 -.09 -.12 -.12  .01  .23  .46  .61  .67
  ST.E.     .07  .10  .11  .11  .11  .11  .11  .11  .11  .12  .12  .14

 13- 24     .57  .37  .14 -.04 -.18 -.20 -.15 -.03  .19  .41  .52  .57
  ST.E.     .15  .16  .16  .17  .17  .17  .17  .17  .17  .17  .17  .18

 25- 36     .44  .26  .08 -.13 -.25 -.24 -.19 -.08  .14  .32  .43  .48
  ST.E.     .19  .19  .20  .20  .20  .20  .20  .20  .20  .20  .20  .21

PLOT OF AUTOCORRELATIONS

          -1.0 -0.8 -0.6 -0.4 -0.2  0.0  0.2  0.4  0.6  0.8  1.0
LAG   CORR. +----+----+----+----+----+----+----+----+----+----+
                                     I
  1   0.729                       +  IXX+XXXXXXXXXXXXXXX       [5]
  2   0.495                     +    IXXXX+XXXXXXX
  3   0.276                     +    IXXXX+XX
  4   0.076                     +    IXX  +
  5  -0.088                     +  XXI    +
  6  -0.121                    +  XXXI     +
  7  -0.116                    +  XXXI     +
  8   0.013                    +     I     +
  9   0.233                    +     IXXXXXX
 10   0.462                    +     IXXXXX+XXXXXX
 11   0.607                    +     IXXXXX+XXXXXXXXX
 12   0.668                   +      IXXXXXX+XXXXXXXXXX
 13   0.567                   +      IXXXXXX+XXXXXXX
 14   0.372                  +       IXXXXXXX+X
 15   0.145                  +       IXXXX    +
 16  -0.043                  +      XI       +
 17  -0.180                  +   XXXXI       +
 18  -0.201                  +  XXXXXI       +
 19  -0.148                  +   XXXXI       +
 20  -0.032                  +      XI       +
 21   0.190                  +       IXXXXX  +
 22   0.409                  +       IXXXXXXX+XX
 23   0.523                 +        IXXXXXXXX+XXXX
 24   0.567                 +        IXXXXXXXX+XXXXX
 25   0.443                 +        IXXXXXXXX+XX
 26   0.262                 +        IXXXXXXX +
 27   0.076                +         IXX       +
 28  -0.127                +      XXXI         +
 29  -0.254                +   XXXXXXI         +
 30  -0.241                +   XXXXXXI         +
 31  -0.189                +    XXXXXI         +
 32  -0.078                +       XXI         +
 33   0.136                +         IXXX      +
 34   0.322                +         IXXXXXXXX +
 35   0.427                +         IXXXXXXXXX+X
 36   0.479                +         IXXXXXXXXX+XX
ACF      VARIABLE IS OZONE. DFORDER IS 12. MAXLAG IS 24./

FIRST CASE NUMBER TO BE USED      -           1
LAST CASE NUMBER TO BE USED       -         216
NO. OF OBS. AFTER DIFFERENCING    -         204            [6]
MEAN OF THE (DIFFERENCED) SERIES  -     -0.1485
STANDARD ERROR OF THE MEAN        -      0.0737
T-VALUE OF MEAN (AGAINST ZERO)    -     -2.0161

AUTOCORRELATIONS

  1- 12     .29  .10  .17  .18  .10  .08 -.03 -.10 -.08 -.09 -.11 -.41
  ST.E.     .07  .08  .08  .08  .08  .08  .08  .08  .08  .08  .08  .08

 13- 24    -.06 -.05 -.17 -.05  0.0 -.04  .03  .01 -.02  .08  .02 -.03
  ST.E.     .09  .09  .09  .09  .09  .09  .09  .09  .09  .09  .10  .10

PLOT OF AUTOCORRELATIONS

          -1.0 -0.8 -0.6 -0.4 -0.2  0.0  0.2  0.4  0.6  0.8  1.0
LAG   CORR. +----+----+----+----+----+----+----+----+----+----+
                                     I
  1   0.292                       +  IXX+XXXX
  2   0.104                      +   IXXX+
  3   0.175                      +   IXXXX
  4   0.180                      +   IXXXX
  5   0.098                      +   IXX +
  6   0.078                      +   IXX +
  7  -0.035                      +  XI   +
  8  -0.104                      +XXXI   +
  9  -0.083                      + XXI   +
 10  -0.086                      + XXI   +
 11  -0.106                      +XXXI   +
 12  -0.407                XXXXXX+XXXI   +
 13  -0.063                     +  XXI    +
 14  -0.054                     +   XI    +
 15  -0.167                     +XXXXI    +
 16  -0.053                     +   XI    +
 17  -0.002                     +    I    +
 18  -0.036                     +   XI    +
 19   0.028                     +    IX   +
 20   0.008                     +    I    +
 21  -0.017                     +    I    +
 22   0.077                     +    IXX  +
 23   0.024                     +    IX   +
 24  -0.031                     +   XI    +
```

```
PACF     VARIABLE IS OZONE. DFORDER IS 12. MAXLAG IS 24./
FIRST CASE NUMBER TO BE USED     =              1
LAST CASE NUMBER TO BE USED      =            216
NO. OF OBS. AFTER DIFFERENCING   =            204       [7]
MEAN OF THE (DIFFERENCED) SERIES =        -0.1485
STANDARD ERROR OF THE MEAN       =         0.0737
T-VALUE OF MEAN (AGAINST ZERO)   =        -2.0161

PARTIAL AUTOCORRELATIONS
  1- 12      .29  .02  .15  .10  .01  .02 -.11 -.11 -.06 -.05 -.03 -.37
  ST.E.      .07  .07  .07  .07  .07  .07  .07  .07  .07  .07  .07  .07

 13- 24      .23 -.06 -.05  .12 -.02  .04 -.01 -.09 -.01  .02 -.04 -.24
  ST.E.      .07  .07  .07  .07  .07  .07  .07  .07  .07  .07  .07  .07

PLOT OF PARTIAL AUTOCORRELATIONS
          -1.0 -0.8 -0.6 -0.4 -0.2  0.0  0.2  0.4  0.6  0.8  1.0
LAG   CORR. +----+----+----+----+----+----+----+----+----+----+
                                     I
  1   0.292                       +  IXX+XXXX
  2   0.020                       +  I  +
  3   0.152                       +  IXX+X
  4   0.099                       +  IXX+
  5   0.013                       +  I  +
  6   0.023                       +  IX +
  7  -0.112                       XXXI  +
  8  -0.114                       XXXI  +
  9  -0.058                       + XI  +
 10  -0.051                       + XI  +
 11  -0.032                       + XI  +
 12  -0.372                 XXXXXX+XXI  +
 13   0.229                       +  IXX+XXX
 14  -0.062                       +XXI  +
 15  -0.046                       + XI  +
 16   0.124                       +  IXXX
 17  -0.016                       +  I  +
 18   0.044                       +  IX +
 19  -0.014                       +  I  +
 20  -0.087                       +XXI  +
 21  -0.008                       +  I  +
 22   0.020                       +  IX +
 23  -0.037                       + XI  +
 24  -0.244                    XXX+XXI  +
```

Bold numbers correspond to those in Output 2T.1

[1] The BMDP instructions are "echoed" back as 2T executes them.

[2] The plot appears right after the TPLOT instruction, and shows a 12 month periodicity (seasonality) in the data. This implies that lag 12 differencing should be used.

[3] The ACF and PACF paragraphs first print a brief summary of the series including the number of observations and the series mean, standard error, and corresponding t score. This mean is a useful measure of the overall series level only when the series is stationary. Both the plot and the cyclic pattern and slow decays in the ACF below show that this original series is nonstationary due to the 12 month seasonality.

[4] The sample autocorrelations and their standard errors are printed with the ACFs and PACFs by default. This can be suppressed by stating NO PRINT.

[5] The ACF and PACF paragraphs also plot the sample autocorrelations and their approximate 95% confidence intervals which are denoted by the + symbol on both sides of the vertical axis. The autocorrelation point estimates are printed to the left of the plot. This can be suppressed by stating NO PLOT.

[6] The autocorrelations of the differenced series are much smaller overall now that the seasonality has been removed. Significant values occur only at lags 1 and 12. The plot implies that the differenced series is stationary and may follow a multiplicative MA model with coefficients at lags 1 and 12. Since the mean of the series is not very large compared to its standard error, in the next example we will tentatively fit a model without a constant term of the form

$$(1 - B^{12})Y_t = z_t = (1 - \Theta_1 B)(1 - \Theta_{12} B^{12})a_t$$

where B is the backshift operator defined in the introduction. The left hand side of this model corresponds to differencing at lag 12, and the right hand side is a multiplicative MA at lags 1 and 12.

[7] The PACF of the differenced series helps confirm our tentative choice of a pure MA model. There are several significant partial correlations implying that an AR model for this series would require more than two parameters. Thus the proposed MA model is more parsimonious.

Example 2T.2 Estimating parameters and diagnostic checking

Estimating parameters. After a tentative ARIMA model has been identified, we estimate its parameters using the ARIMA paragraph to specify the model and the ESTIMATE paragraph to actually perform the estimation. In 2T, an ARIMA and diagnostic paragraph must be specified before an ESTIMATE paragraph is given.

The backshift representation of the model is the basis for the ARIMA paragraph instructions. In representing the AR and MA polynomials, we drop all notation inside the parentheses except the order of the parameters. For example, in Input 2T.2, the MA polynomial for the ozone data, $(1 - \Theta_1 B)(1 - \Theta_{12} B^{12})$, is denoted by '(1),(12)'. Note that the parentheses are included and the entire expression is enclosed in apostrophes. [If the model had been additive instead of multiplicative, we would write '(1,12)' instead of '(1),(12)'.]

In the ARIMA paragraph, we use ARORDER, MAORDER, and DFORDER to specify the AR, MA, and difference polynomial parameter orders respectively. The VARIABLE instruction is used to specify the time series variable to be modeled.

In Input 2T.2, the CHECK paragraph directs 2T to print the MODEL currently held in memory, so that we can confirm that this is the model we want. (In interactive use of 2T, it is easy to lose track of the current model.)

The ESTIMATION paragraph gives instructions for estimating the current model's parameters for the series specified in the VARIABLE sentence of the ARIMA paragraph. RESIDUAL requests that residuals be stored in the variable ROZONE for diagnostic checking. PCORRELATION requests that the intercorrelations among the parameter estimates be printed. High intercorrelations indicate that the parameter estimates are unstable.

2T has two iterative methods of estimating model parameters: CLS (conditional least squares) and BACKCASTING (back forecasting). Unless you specify one of these methods in the ESTIMATION paragraph, 2T uses both methods. The CLS method is easier to compute. The backcasting method uses the model to "back forecast," or backcast, estimates of residuals preceding the first residual available when CLS is used. By default, the CLS estimates are used as initial estimates for the backcasting method. The backcasting method is more precise (especially for the early observations) but requires more processing time.

Diagnostic checking. Diagnostic checking consists of evaluating the adequacy of the estimated model. An adequate model satisfies four criteria:

1. Estimates of all parameters must differ significantly from zero. This is checked in the output from the ARIMA paragraph.

2. All AR parameter estimates must lie within the "bounds of stationarity." This guarantees that the model is stationary about its mean. For simple AR models, the requirements of the bounds of stationarity are

   AR(1)models: $|\Phi_1| < 1$

   AR(2) models: $|\Phi_2| < 1$

   $\Phi_2 - \Phi_1 < 1$

   $\Phi_2 + \Phi_1 < 1$

   In the AR(1) case, stationarity for Φ_1 is the same as saying that a correlation must be between –1 and 1. For higher order models, the criteria are more complex. See Box and Jenkins (1976).

3. All MA parameter estimates must lie within the "bounds of invertibility." This is the MA analog to stationarity in AR models. When the MA model is reexpressed as an infinite series of AR terms, (see Defining the ARIMA model, above) invertibility guarantees that this series converges. For simple MA models, the requirements of the bounds of invertibility are

   MA(1) models $|\Theta_1| < 1$

   MA(2) models: $|\Theta_2| < 1$

   $\Theta_2 - \Theta_1 < 1$

   $\Theta_2 + \Theta_1 < 1$

For models of the same order (i.e., AR(i) and MA(i)), the bounds of stationarity place limits on Φ_i that are identical to those placed on Θ_i by the bounds of invertibility. For models of order higher than 2, the bounds of stationarity and invertibility become very complex. See Box and Jenkins (1976) for a full explanation.

4. Residuals must not differ significantly from a series of pure random errors ("white noise") with mean zero. For white noise, the theoretical ACF and PACF are both zero at all lags. We pay particular attention to the low lags (1, 2, and 3). For residuals, the calculated standard errors tend to overestimate the true standard error (see Box and Pierce, 1970). Thus, when examining residuals, we adopt a stricter criterion than the usual ± 2 standard error bounds used in model identification. Pankratz (1983) suggests 1.25 standard errors for low lags and seasonal lags, and 1.6 standard errors elsewhere.

   Two other tests of residuals are also widely used. The first of these is the Ljung-Box portmanteau Q statistic. Q is a test of the null hypothesis that the ACF does not differ from zero, up to lag k. It is evaluated as a chi-square with $k - m$ degrees of freedom, where k is the number of lags examined and m is the number of parameters estimated ($p + q + 1$ if the model includes a CONSTANT, $p + q$ if not).

   The second test is to examine an ACF plot of the first differences of the residuals. Theoretically, this plot should consist of a single negative spike of –0.5 at lag 1. This corresponds to an MA(1) process with Θ_1 equal to 1.00, which is the result of taking the first differences of a white noise process.

   In Input 2T.2, we ask for the ACF and PACF of the residuals, ROZONE. The LBQ instruction in the ACF paragraph requests the Ljung-Box Q statistic.

Input 2T.2
(add to the end of Input 2T.1)

```
ARIMA       VARIABLE IS ozone. DFORDER IS 12.                  /
            MAORDER = '(1), (12)'.

CHECK       MODEL.

ESTIMATION  RESIDUALS = rozone    PCORRELATION.                /

ACF         VARIABLE IS rozone. MAXLAG IS 24. LBQ.             /

PACF        VARIABLE IS rozone. MAXLAG IS 24.                  /

END  /
```

Output 2T.2

```
ARIMA      VARIABLE IS OZONE.  DFORDER IS 12.  MAORDER = '(1),(12)'./
THE COMPONENT HAS BEEN ADDED TO THE MODEL.

THE CURRENT MODEL HAS                                          [1]
OUTPUT VARIABLE = OZONE
INPUT  VARIABLE = NOISE


CHECK      MODEL. /


SUMMARY OF THE MODEL

OUTPUT VARIABLE -- OZONE
INPUT VARIABLES -- NOISE

VARIABLE  VAR. TYPE  MEAN       TIME      DIFFERENCES           [2]
                                                 12
OZONE      RANDOM              1- 216   (1-B   )


PARAMETER VARIABLE    TYPE  FACTOR  ORDER ESTIMATE
       1  OZONE        MA      1       1     0.1000
       2  OZONE        MA      2      12     0.1000


       THE PSI-WEIGHTS HAVE NOT BEEN COMPUTED

ESTIMATION RESIDUALS = ROZONE.    PCORRELATION.    /
```

```
FIRST CASE NUMBER TO BE USED     -             1
LAST CASE NUMBER TO BE USED      -           216

ESTIMATION BY CONDITIONAL LEAST SQUARES METHOD

RELATIVE CHANGE IN EACH ESTIMATE LESS THAN 0.1000E-03

CORRELATION MATRIX OF PARAMETER ESTIMATES
      1         2                                                        [3]
 1  1.000
 2 -0.095    1.000

SUMMARY OF THE MODEL

OUTPUT VARIABLE -- OZONE
INPUT VARIABLES -- NOISE

VARIABLE  VAR. TYPE  MEAN      TIME      DIFFERENCES
                                                12
OZONE      RANDOM            1- 216   (1-B   )

PARAMETER VARIABLE   TYPE  FACTOR  ORDER ESTIMATE       ST. ERR.  T-RATIO   [4]
     1  OZONE         MA     1       1    -0.3624        0.0658    -5.51
     2  OZONE         MA     2      12     0.4882        0.0623     7.84

RESIDUAL SUM OF SQUARES   -        166.393616
DEGREES OF FREEDOM        -               202                               [5]
RESIDUAL MEAN SQUARE      -          0.823731

ESTIMATION BY BACKCASTING METHOD

RELATIVE CHANGE IN RESIDUAL SUM OF SQUARES LESS THAN 0.5000E-04

CORRELATION MATRIX OF PARAMETER ESTIMATES
      1         2
 1  1.000
 2 -0.070    1.000

SUMMARY OF THE MODEL                                                         [6]

OUTPUT VARIABLE -- OZONE
INPUT VARIABLES -- NOISE

VARIABLE  VAR. TYPE  MEAN      TIME      DIFFERENCES
                                                12
OZONE      RANDOM            1- 216   (1-B   )

PARAMETER VARIABLE   TYPE  FACTOR  ORDER ESTIMATE       ST. ERR.  T-RATIO
     1  OZONE         MA     1       1    -0.3685        0.0640    -5.76
     2  OZONE         MA     2      12     0.5832        0.0516    11.30

RESIDUAL SUM OF SQUARES   -        150.009171
DEGREES OF FREEDOM        -               202                               [7]
RESIDUAL MEAN SQUARE      -          0.742620
( BACKCASTS EXCLUDED )


ACF         VARIABLE IS ROZONE.   MAXLAG IS 24.   LBQ.

FIRST CASE NUMBER TO BE USED     -              13
LAST CASE NUMBER TO BE USED      -             216
NO. OF OBS. AFTER DIFFERENCING   -             204                          [8]
MEAN OF THE (DIFFERENCED) SERIES -         -0.2374
STANDARD ERROR OF THE MEAN       -          0.0578
T-VALUE OF MEAN (AGAINST ZERO)   -         -4.1046

AUTOCORRELATIONS

  1- 12      -.02  .15  .09  .18  .02  .13  0.0 -.01  .02 -.01  .08 -.07
  ST.E.       .07  .07  .07  .07  .07  .07  .08  .08  .08  .08  .08  .08
L.-B. Q       .10  4.7  6.3  13.  13.  17.  17.  17.  17.  17.  18.  19.

 13- 24       .07  .01 -.08  .04 -.01  .03  .05 -.03 -.05  .11 -.01 -.01
  ST.E.       .08  .08  .08  .08  .08  .08  .08  .08  .08  .08  .08  .08
L.-B. Q       20.  21.  22.  22.  22.  22.  23.  23.  24.  26.  26.  26.

PLOT OF AUTOCORRELATIONS

          -1.0 -0.8 -0.6 -0.4 -0.2  0.0  0.2  0.4  0.6  0.8  1.0
LAG   CORR. +----+----+----+----+----+----+----+----+----+----+
                                     I
  1  -0.017                      +   I  +
  2   0.150                      +   IXX+X
  3   0.085                     +    IXX +
  4   0.177                     +    IXXXX
  5   0.021                     +    IX  +
  6   0.131                     +    IXXX+
  7  -0.005                     +    I   +
  8  -0.008                     +    I   +
  9   0.015                     +    I   +
 10  -0.008                     +    I   +
 11   0.079                     +    IXX +
 12  -0.075                     +  XXI   +
 13   0.074                     +    IXX +
 14   0.013                     +    I   +
 15  -0.077                     +  XXI   +
 16   0.042                     +    IX  +
 17  -0.014                     +    I   +
 18   0.031                     +    IX  +
 19   0.052                     +    IX  +
 20  -0.029                     +   XI   +
 21  -0.048                     +   XI   +
 22   0.106                     +    IXXX+
 23  -0.006                     +    I   +
 24  -0.010                     +    I   +
```

—the PACF output is omitted—

[1] The model is defined by the ARIMA paragraph. In ARIMA models, the input series is always white noise (or random shock, a_t).

[2] The CHECK paragraph displays the current model as an MA model with two terms at lags 1 and 12. The parameter estimates are from the last ESTIMATION paragraph. Since an ESTIMATION paragraph has not yet been given, the initial values of 0.1 are displayed. They should be ignored.

[3] Results from PCORRELATION in the ESTIMATION paragraph give the correlation between the two parameter estimates as -0.095. Since this is far from 1.0 in absolute value, the parameter estimates are stable.

[4] The results of the CLS estimation are printed. The estimate of Φ_1 is –0.3624 with a standard error of 0.0658. The estimate of Φ_{12} is 0.4882 with a standard error of 0.0623. The *t*-ratios indicate that both estimates are several standard errors away from zero.

[5] The residual mean square from the CLS method is based on 202 degrees of freedom even though the original series had 216 observations. Twelve observations were lost by specifying 12th lag differences, leaving only 204 observations. Since there are 2 parameters, 202 degrees of freedom remain.

[6] The backcasting estimates are similar to the CLS estimates. Based on the backcasting estimates, the model for the ozone data is

$$(1 - B^{12})Y_t = (1 + 0.3685\,B)(1 - 0.5832\,B^{12})a_t.$$

The standard errors are somewhat smaller than those produced by CLS.

[7] The residual mean square is smaller using backcasting.

[8] The ACF of the residual series, ROZONE, is computed. We find that the mean of the residual series is almost four standard errors from zero and that there are significant correlations at lags 2 and 4, although they are not very large. This is confirmed by the Ljung-Box Q statistic which is equal to 13 by lag 4, which is significant for a chi-square variate with 4 – 2 = 2 degrees of freedom. Therefore the model obtained so far is not adequate. We have yet to take into account the 1960 and 1966 changes that might affect ozone. Further analysis is continued in Example 2T.7, where we include intervention effects.

Example 2T.3 Forecasting

Here we examine a series of 70 monthly measures of commercial bank real estate loan volume, in billions of dollars, from January 1973 to October 1978 taken from Pankratz (1983). In this example, the goal is to predict loan volume over the next few months after October 1978. The data are shown in Input 2T.3 below. The model building process (not shown but given in Pankratz, pp. 411–424) reveals that an ARIMA(0, 2, 1) model fits the data adequately. The twice differencing removes a quadratic trend yielding a stationary MA(1) series. In Input 2T.3 we begin forecasting at observation 65 and generate 18 future values. Thus we predict the series' last six observations and the future monthly values for one year.

A FORECAST paragraph is used to generate the 18 forecast values. The START sentence specifies the "forecast origin," the point at which forecasting is to begin. If you do not specify the START, 2T begins forecasting after the last observed value of the series. CASES specifies the number of forecasted values we want (18 in this example). RMS specifies the residual mean square of the series to be forecast. We specify RMS in order to use the residual mean square previously found from the backcasting method. If RMS is not specified, the CLS method residual mean square is used.

Note: If you do not specify RMS, 2T computes it by the CLS method. The number of the START case must be larger than either the number of parameter estimates (when RMS is not specified in the FORECAST paragraph) or the maximum order of the backshift operator (when RMS is specified).

Input 2T.3

```
/ INPUT          VARIABLE = 1.
                 FORMAT IS STREAM.

/ VARIABLE       NAME IS loan.

/ END
  46.5 47.0 47.5 48.3 49.1 50.1 51.1 52.0 53.2 53.9 54.5 55.2
  55.6 55.7 56.1 56.8 57.5 58.3 58.9 59.4 59.8 60.0 60.0 60.3
  60.1 59.7 59.5 59.4 59.3 59.2 59.1 59.0 59.3 59.5 59.5 59.5
  59.7 59.7 60.5 60.7 61.3 61.4 61.8 62.4 62.4 62.9 63.2 63.4
  63.9 64.5 65.0 65.4 66.3 67.7 69.0 70.0 71.4 72.5 73.4 74.6
  75.2 75.9 76.8 77.9 79.2 80.5 82.6 84.4 85.9 87.6
/ END

  ARIMA     VAR IS loan.  DFORDERS = 1, 1.  MAORDER IS '(1)'. /

  ESTIMATION                                                   /

  FORECAST START IS 65.     CASES ARE 18.
           RMS = .082115.                                      /

END   /
```

Output 2T.3 gives the forecasts, their standard errors, the observed values of the series (where available), and the residuals (i.e., the errors in prediction: $Y_t - \hat{Y}_t$). As we expect, the standard errors of prediction increase with distance from the actual data set. The magnitude of the residuals increases among the last six observations in the series. Examination of the TPLOT (not shown) reveals that the series is relatively flat for approximately the first two-thirds of the observations and then sharply increases during the last third. The series appears to be unstable or to have undergone some change (e.g., rapid inflation) that has biased our predictions from the beginning. In either case, we are alerted to using the present model with caution, especially for long-term forecasting.

Output 2T.3

—the output from the ARIMA and ESTIMATIO N paragraphs has been omitted—

FORECAST ON VARIABLE LOAN   FROM TIME PERIOD 65

PERIOD	FORECASTS	ST. ERR.	ACTUAL	RESIDUAL
65	78.90143	0.28656	79.20000	-0.29857
66	79.90285	0.54627	80.50000	-0.59715
67	80.90428	0.84418	82.60000	-1.69572
68	81.90571	1.17835	84.40000	-2.49429
69	82.90714	1.54590	85.90000	-2.99287
70	83.90856	1.94428	87.60000	-3.69144
71	84.90999	2.37140		
72	85.91142	2.82550		
73	86.91284	3.30511		
74	87.91427	3.80895		
75	88.91570	4.33593		
76	89.91712	4.88509		
77	90.91855	5.45557		
78	91.91998	6.04660		
79	92.92140	6.65751		
80	93.92283	7.28765		
81	94.92426	7.93648		
82	95.92568	8.60346		

STANDARD ERROR = 0.286557

The first six forecasts can be compared to the observed series to get an idea of the quality of the forecast. Not surprisingly, the residuals increase with time. This is why ARIMA models are usually only appropriate for short range forecasts. The forecast standard errors also increase with time.

Example 2T.4 Identifying and replacing (estimating) missing values

If a time series is not complete, 2T divides the series into blocks, with one block for each set of contiguous observations. You can use the BLOCK paragraph to specify how the blocks are to be arranged and which block is to be analyzed; if your instructions do not include a BLOCK paragraph, 2T analyzes the largest block. You can also use the BLOCK and FORECAST paragraphs to generate estimates of missing values and add these estimates to the data for analysis of the full series.

In this example, we use the loan data of Example 2T.3 to illustrate a procedure for identifying and replacing missing data. Although the loan data are complete, we create a missing value in Input 2T.4 by substituting the global missing code * for observation 40. We then use the BLOCK paragraph to show how 2T identifies missing values. The LIST sentence requests that the LOAN variable be examined for missing values. INDICATE generates a new variable, MLOAN. MLOAN is binary, coded 1 if the observation is present for all variables in the LIST, or 0 if the observation is missing for any variable. Variables specified in INDICATE may be used in future steps without having been added in the TRANSFORM paragraph. We then PRINT a few cases to locate the missing datum. The one missing datum (at observation 40) divides the data into two BLOCKS.

Having identified the missing datum, we fit a known ARIMA model to the first block of data and use the model to FORECAST the missing value and JOIN it to the series, recreating an unbroken series. In the interest of parsimony we use only the first block of data to identify the model for forecasting the missing value. The usual process of model identification (not shown) indicates that an ARIMA(1, 0, 0) is adequate for the first block. That is, the current loan volume in the first block can be modeled as a function of last month's volume plus random noise. The complete series more closely follows the ARIMA(0, 2, 1), but we would not know this until after we fill in the missing datum and analyze the entire series. (The quadratic trend is not apparent in the first block.) When we refine the model in later analyses, we can fit it to the entire data set.

The ARIMA paragraph defines the model ARIMA(1, 0, 0). In the BLOCK paragraph we specify the first block of data for analysis. The FORECAST paragraph generates one forecast, for the observation immediately following the first block (i.e., the missing observation). The JOIN sentence then joins the forecasted value(s) to the data set for future analyses. In the PRINT paragraph we request output for the CASES and VARIABLES listed.

Output 2T.4 shows that the missing datum has been estimated and entered into the data set. The value of the series at time 40 is forecast as 60.87, which is very close to the actual value of 60.7. MLOAN is not affected by the JOIN statement. Thus, it retains a record of which values of the series were measured and which were estimated.

If we now want to fit a model to the complete data set, we can add interactive paragraphs similar to those shown in Examples 2T.1, 2T.2, and 2T.3. If we do not want to perform further analysis at this point, but prefer to save the complete data set for future analyses, we add a SAVE paragraph (see Chapter 8).

Input 2T.4

```
/ INPUT          VARIABLE 1.
                 FORMAT IS STREAM.

/ VARIABLE       NAME IS loan.

/ END
46.5 47.0 47.5 48.3 49.1 50.1 51.1 52.0 53.2 53.9 54.5 55.2
55.6 55.7 56.1 56.8 57.5 58.3 58.9 59.4 59.8 60.0 60.0 60.3
60.1 59.7 59.5 59.4 59.3 59.2 59.1 59.0 59.3 59.5 59.5 59.5
59.7 59.7 60.5  *   61.3 61.4 61.8 62.4 62.4 62.9 63.2 63.4
63.9 64.5 65.0 65.4 66.3 67.7 69.0 70.0 71.4 72.5 73.4 74.6
75.2 75.9 76.8 77.9 79.2 80.5 82.6 84.4 85.9 87.6
/ END
```

Input 2T.4
(continued)

```
BLOCK        LIST = loan.           INDICATE = mloan.       /
PRINT        VAR = loan, mloan.     CASES = 35 TO 45.       /
ARIMA        VAR IS loan.           ARORDER IS '(1)'.       /
BLOCK        FIRST.                                         /
ESTIMATION   METHOD IS CLS.                                 /
FORECAST     CASE = 1.              JOIN.                   /
PRINT        VAR = loan, mloan.     CASES = 35 TO 45 .      /
END    /
```

Output 2T.4

```
     THE RANGE OF CASE NUMBERS FOR BLOCKS
     OF MISSING AND COMPLETE OBSERVATIONS.

     BLOCK          COMPLETE         MISSING
       1           1     39        40
       2          41     70

INDICATOR VARIABLE MLOAN     HAS BEEN CREATED.

PRINT          VAR = LOAN, MLOAN.  CASES = 35 TO 45.     /
 CASE     1            2
  NO.  LOAN        MLOAN
----- --------  --------
   35     59.50       1.00
   36     59.50       1.00
   37     59.70       1.00
   38     59.70       1.00
   39     60.50       1.00
   40   MISSING       0.00
   41     61.30       1.00
   42     61.40       1.00
   43     61.80       1.00
   44     62.40       1.00
   45     62.40       1.00
```

—we omit a portion of output—

```
FORECAST       CASE = 1.             JOIN.               /

 FORECAST ON VARIABLE LOAN      FROM TIME PERIOD    40

 PERIOD             FORECASTS       ST. ERR.      ACTUAL          RESIDUAL
   40              60.87125        0.42739

STANDARD ERROR = 0.427392          BY CONDITIONAL METHOD
```

—we omit a portion of output—

```
 CASE     1            2
  NO.  LOAN        MLOAN
----- --------  --------
   35     59.50       1.00
   36     59.50       1.00
   37     59.70       1.00
   38     59.70       1.00
   39     60.50       1.00
   40     60.87       0.00
   41     61.30       1.00
   42     61.40       1.00
   43     61.80       1.00
   44     62.40       1.00
   45     62.40       1.00
```

Intervention Analysis

You can also use 2T to model and test the effects of exogenous influences upon a time series. If these influences are present, they must be modeled so that their effects will not compromise the accuracy of forecasts. On the other hand, time series data may be collected and analyzed specifically to assess the effect of some intervention or change upon the series. Campbell and Stanley (1966) called this design the "interrupted time series quasi-experiment." While the methods of analysis originally suggested by Campbell and Stanley are invalid (because of the bias of variance estimates produced by serial dependence), the ARIMA model is statistically valid for the assessment of the significance of these effects. This work was developed by Box and Tiao (1975); Glass, Willson, and Gottman (1975); and McCleary and Hay (1980).

In Examples 2T.5 through 2T.7, we model interventions. In 2T.5, we lay the groundwork by examining the preintervention series of a patient before he is put on a drug. In 2T.6, we examine the effects of the drug, modeling it as a single, permanent intervention. In Example 2T.7, we return to the ozone data and consider two types of interventions, a single permanent change and a constant downward decline.

For intervention analysis, the ARIMA model becomes

$$\nabla^d Y_t = f(I_t) + N_t$$

where ∇ is the differencing polynomial (sometimes written $\nabla(B)$); I_t is a binary variable coded 1 during the effect and 0 at all other times; $f(I_t)$ is the intervention component we seek to model and test; and N_t may be any noise process (including a full ARIMA model).

Intervention effects may be either permanent or temporary. Typically, I_t is coded as a step function; that is, as 0 before the (hypothesized) effect and 1 after. When I_t is a step function (as the rest of our discussion assumes), permanent effects are modeled by I_t; temporary "pulse" effects, by $(1 - B)I_t$. The binary form of I_t and $(1 - B)I_t$ is

$\boldsymbol{I_t}$	0	0	0	1	1	1	1
$\boldsymbol{(1-B)I_t}$	0	0	0	1	0	0	0
time	n–3	n–2	n–1	n	n+1	n+2	n+3

Note that **you** must add I_t to the model; 2T does not automatically generate it. The TRANSFORM paragraphs of Input 2T.5, 2T.6, and 2T.7 give examples.

The general form of the intervention component is

$$f(I_t)\begin{cases} \dfrac{U(B)}{S(B)} I_t & \text{for permanent effects} \\[2ex] \dfrac{U(B)}{S(B)} (1-B) I_t & \text{for temporary effects} \end{cases}$$

where

$$\frac{U(B)}{S(B)} = \frac{\left(U_0 + U_1 B + \ldots + U_s B^s\right)}{1 - S_1 B - S_2 B^2 - \ldots - S_r B^r}$$

The expression $(U_0 + U_1B + U_2B^s + \ldots + U_sB^s)$ is termed the U polynomial and is often represented by the Greek letter omega (e.g., McCleary and Hay, 1980). It reflects the change in the level of the postintervention series.

The expression $(1 - S_1B - S_2B^2 - S_rB^r)$ is termed the S polynomial and is often represented by the Greek letter delta (again see McCleary and Hay, 1980). It expresses the rate at which the series approaches its asymptotic postintervention level; small values of S_i indicate rapid stabilization, while large values indicate that many observations will be necessary for the asymptotic level to be reached. The coefficients (S_i) in the S polynomial must satisfy limits referred to as the "bounds of system stability," which ensure that the postintervention series is stationary about its mean. These limits are identical to the limits placed on Φ_i by the bounds of stationarity:

for $d = 1$, $\quad |S_1| < 1$
for $d = 2$, $\quad |S_2| < 1$
$\quad S_2 - S_1 < 1$
$\quad S_2 + S_1 < 1$

While the intervention components $U(B)$ and $S(B)$ may be of arbitrary complexity, the principle of parsimony in model building suggests that the intervention component should be as simple as possible. In practice, one U parameter (for example, U_0) and one S parameter (for example, $1 - S_1B$) suffice to fit most intervention effects.

Intervention components are also either gradual or sudden in achieving their full effect. The dimensions temporary/permanent and sudden/gradual are independent of one another, and each has a characteristic form and specification (see Figure 2T.2).

When the form of an intervention cannot be specified *a priori*, based on theory, McCleary and Hay recommend a three-stage "blind" analysis using the simple intervention component $U_0/(1 - S_jB_j)$. For $j = 1$, the three steps are:

1. Attempt to fit a sudden, temporary effect:

$$[U_0/(1-S_1B)](1-B)I_t$$

If S_1 is too large (greater than or very near 1.00), rule out a temporary effect.

Figure 2T.2: Typical forms of the intervention components (from McCleary and Hay, 1980)

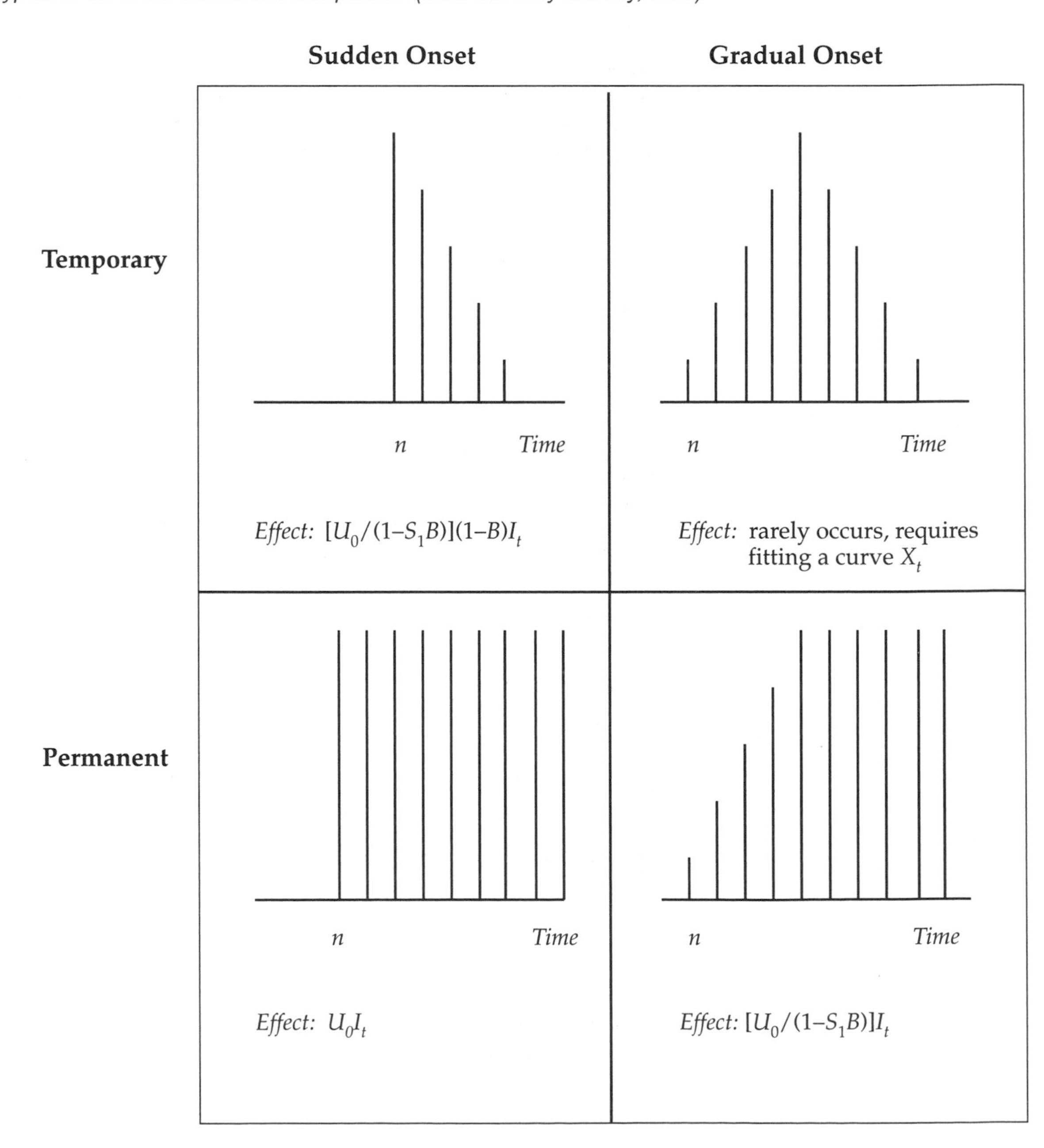

2. Attempt to fit a gradual, permanent effect:

$$[U_0/(1-S_1B)]I_t$$

If S_1 is too small (not significantly different from 0), rule out a gradual effect. If $S_1 = 1$, the effect does not level out, but indicates a linear trend (Mccleary and Hay, pp. 158–159).

3. Finally, try a sudden, permanent effect:

$$U_0I_t$$

If this is not significant, the null hypothesis of no intervention effect may not be rejected.

A model is acceptable only if all its parameter estimates (intervention components **and** ARIMA parameters for N_t) are significant. One common error is to accept an intervention model in which S_1 is significantly different from 0, but U_0 is not significant. This leads to the nonsensical interpretation that the postintervention series does not significantly differ in level from the preintervention series, and that it achieves that non-different level at a significant rate.

In 2T the INDEPENDENT paragraph specifies intervention components, which are composed of three elements: the U polynomial, the S polynomial, and a difference polynomial. One sentence of the INDEPENDENT paragraph corresponds to each element. For example, the model

$$Y_t \frac{U_o}{(1-S_1B)}(1-B)I_t + N_t$$

would be specified as

```
INDEPENDENT      VAR IS 1.              DFORDER = 1.
                 TYPE IS BINARY.        UPORDERS = '(0)'.
                                        SPORDERS = '(1)'.
```

The VAR sentence identifies the variable containing the intervention component. TYPE and DFORDER correspond to $(1 - B)I_t$. "UPORDERS" indicates U polynomial orders; "SPORDERS," S polynomial orders. Keeping these mnemonics in mind will help you to avoid syntax errors. TYPE = BINARY specifies a binary (1, 0) effect I_t. You may also model more complex interventions. For example, if you know that an intervention will occur over three months, you can set up the intervention variable X_t:

X_t	0	0	0	0	.333	.667	1.00	1.00
Time	–3	–2	–1	0	1	2	3	4

Specify NONRANDOM instead of BINARY for this example. For more information on TYPE, see 2T Commands.

Differencing Y_t complicates the model specification. McCleary and Hay consider $\nabla(B)$, the difference polynomial for Y_t, to be part of the noise component N_t. For example, if $\nabla(B)$ is $(1- B)(1 - B^{12})$—i.e.,

$$z_t = (1 - B)(1 - B^{12})Y_t$$

then McCleary and Hay's intervention model takes the form

$$Y_t \frac{U(B)}{S(B)}(1-B)I_t + \frac{\Theta(B)}{\Phi(B)(1-B)(1-B^{12})}a_t$$

2T treats $\nabla(B)$ as a separate element. Therefore, we must rewrite the model by multiplying both sides by $(1 - B)(1 - B^{12})$ to get

$$(1-B)(1-B^{12})Y_t = \frac{U(B)}{S(B)}(1-B)(1-B)(1-B^{12})I_t + \frac{\Theta(B)}{\Phi(B)}a_t$$

This has the effect of including $\nabla(B)$ in the difference polynomial of the intervention component. For this example, the INDEPENDENT paragraph would state DFORDERS = 1, 1, 12. (**not** DFORDERS = 1). One differencing is for the pulse intervention, the other first- and twelfth-order differencing are a result of differencing Y_t. All **must** be included in the INDEPENDENT paragraph. For more information on DFORDERS, see a "Specifying a model" in Special Features.

In Examples 2T.5 and 2T.6 we perform an intervention analysis of a series of daily measures of perceptual speed for a schizophrenic subject (from Glass et al., 1975). Beginning the 61st day, the subject was put on a daily chlorpromazine regimen.

Example 2T.5 Constructing a preintervention model

In Input 2T5 we read the data on perceptual speed (described above) as the series SPEED and then use the TRANSFORM paragraph to create a binary intervention variable CPZ (for chlorpromazine), which has a value of 0 before the intervention and a value of 1 after the intervention. Thus, we are modeling a permanent intervention. We then examine a TPLOT for evidence of drift or trend. The BLOCK paragraph selects the preintervention observations; i.e., the first 60 days (see 2T Commands) for building the ARIMA model. Since we have daily observations, we choose a MAXLAG of 15 for the ACF and PACF paragraphs, to expose any weekly periodicity. The ACF and PACF plots (not shown) lead us to choose an ARIMA(0, 1, 1) model. We then define the ARIMA model, ESTIMATE its parameters, and examine the ACF and PACF of the residuals for compliance with the diagnostic criteria. The output (not shown) confirms an ARIMA(0, 1, 1) model, that is, an MA(1) model after differences were taken. The constant of the differenced series (not shown) is not significant, indicating that the nonstationarity of the series is due to drift rather than trend (see "Interpreting the constant" in Special Features). For more information on the commands used in Input 2T.5, see 2T Commands and Examples 2T.1 through 2T.4.

Input 2T.5

Input data and construct binary intervention series

```
 / INPUT        VAR IS 1.
                FORMAT IS STREAM.

 / VARIABLE     NAME IS speed.

 / TRANSFORM    cpz = 0.
                IF (KASE GT 60) THEN cpz = 1.
 / END
55 56 48 46 56 46 59 60 53 58 73 69 72 51
72 69 68 69 79 77 53 63 80 65 78 64 72 77
82 77 35 79 71 73 77 76 83 73 78 91 70 88
88 85 77 63 91 94 72 83 88 78 84 78 75 75
86 79 76 87 66 73 62 27 52 47 65 59 77 47
51 47 49 54 58 56 50 54 45 66 39 51 39 27
39 37 43 41 27 29 27 26 29 31 28 38 37 26
31 45 38 33 33 25 24 29 37 35 32 31 28 40
31 37 34 43 38 33 28 35
 / END
```

Input 2T.5 *(continued)*

Check for stationarity	`TPLOT VAR IS speed. SIZE IS 60. /`
Select prevention observations	`BLOCK RANGE IS 1,60. /`

Identify and define preliminary preintervention model	`ACF VAR IS speed. MAXLAG IS 15. /`
	`PACF VAR IS speed. MAXLAG IS 15. /`
	`ACF VAR IS speed. MAXLAG IS 15. DFORDER IS 1. /`
	`ARIMA VAR IS speed. DFORDER IS 1. MAORDER IS '(1)'. /`

Estimate model parameters and store residuals	`ESTIMATION RESID IS rspeed3. /`

Check model adequacy	`ACF VAR IS rspeed3. MAXLAG IS 15. LBQ. /`
	`PACF VAR IS rspeed3. MAXLAG IS 15. /`
	`ACF VAR IS rspeed3. DFORDER IS 1. MAXLAG IS 15. /`

```
TPLOT       VAR IS speed.          SIZE IS 60.        /
BLOCK       RANGE IS 1,60.                            /
ACF         VAR IS speed.        MAXLAG IS 15.        /
PACF        VAR IS speed.        MAXLAG IS 15.        /
ACF         VAR IS speed.        MAXLAG IS 15.
            DFORDER IS 1.                             /
ARIMA       VAR IS speed.        DFORDER IS 1.
            MAORDER IS '(1)'.                         /
ESTIMATION RESID IS rspeed3.                          /
ACF         VAR IS rspeed3.      MAXLAG IS 15.
            LBQ.                                      /
PACF        VAR IS rspeed3.      MAXLAG IS 15.        /
ACF         VAR IS rspeed3.      DFORDER IS 1.
            MAXLAG IS 15.                             /
```

—no output is given for this example—

Example 2T.6 Identifying the form of the intervention and estimating intervention parameters

The next step in our analysis is to identify whether the effect of the intervention is temporary or permanent and sudden or gradual. Because we want to change our model to include the intervention, we must ERASE the preintervention model. Thus, in Input 2T.6 we begin by erasing the preintervention model and widening the BLOCK range to include the entire series. The ARIMA paragraph is the same one we used in Input 2T.5; however, we now add an INDEPENDENT paragraph to identify CPZ as a binary intervention variable whose effect we initially assume (as McCleary and Hay recommend) to be sudden and temporary.

This effect has three elements, as shown in Figure 2T.3:

U_0	specified as	UPORDERS = '(0)'.
$(1 - S^1 B)$	specified as	SPORDERS = '(1)'.
$(1 - B)I_t$	specified as	DFORDERS = 1, 1.

The sentence DFORDERS = 1, 1 includes 1 for the pulse intervention and 1 for the differencing of Y_t in the ARIMA paragraph.

Upon estimating the parameters for this model, we note that the estimate for S_1 is 1.01, which is outside the bounds of system stability. We therefore reject a sudden, temporary effect and try to fit a gradual, permanent effect. To prevent any adverse effects due to starting estimation from the extreme value, we ERASE the model. We then respecify the ARIMA paragraph and add a new INDEPENDENT paragraph that changes DFORDERS = 1, 1 to DFORDERS = 1; i.e., changes $(1 - B)I_t$ to I_t. When we estimate the new parameters, we find that all parameters differ significantly from zero and that S_1 lies within the bounds of system stability. Finally, we check the residuals to confirm that they resemble white noise.

Input 2T.6 *(add to the end of Input 2T.5)*

Erase preintervention model	`ERASE MODEL. /`
Select entire series	`BLOCK RANGE IS 1,120. /`
Redefine preliminary model	`ARIMA VAR IS speed . DFORDER IS 1. / MAORDER IS'(1)'. /`
Specify intervention components for a sudden, temporary effect	`INDEPENDENT VAR IS cpz. TYPE IS BINARY. DFORDERS ARE 1,1. UPORDERS ARE '(0)'. SPORDERS ARE '(1)'. /`
Estimate model parameters	`ESTIMATION RESID = rspeed4. PCOR. /`
Erase model	`ERASE MODEL. /`
Redefine model for a gradual, permanent effect	`ARIMA VAR IS speed. DFORDER IS 1. MAORDER IS '(1)'. / INDEPENDENT VAR IS cpz. TYPE IS BINARY. DFORDER IS 1. UPORDER IS '(0)'. SPORDER IS '(1)'. /`
Estimate model parameters	`ESTIMATION RESID = rspeed5. PCOR. /`
Check model adequacy	`ACF VAR IS rspeed5. MAXLAG IS 15. LBQ. /` `PACF VAR IS rspeed5. MAXLAG IS 15. /`
Check model adequacy	`ACF VAR IS rspeed5. DFORDER IS 1. MAXLAG IS 15. /`

```
ERASE          MODEL.                                     /

BLOCK          RANGE IS 1,120.                            /

ARIMA          VAR IS speed .    DFORDER IS 1.    /
                                 MAORDER IS'(1)'.   /

INDEPENDENT    VAR IS cpz.       DFORDERS ARE 1,1.
               TYPE IS BINARY.   UPORDERS ARE '(0)'.
                                 SPORDERS ARE '(1)'.     /

ESTIMATION     RESID = rspeed4.  PCOR.        /

ERASE MODEL.                                  /

ARIMA          VAR IS speed.     DFORDER IS 1.
                                 MAORDER IS '(1)'. /
INDEPENDENT    VAR IS cpz.       DFORDER IS 1.
               TYPE IS BINARY.   UPORDER IS '(0)'.
                                 SPORDER IS '(1)'. /

ESTIMATION     RESID = rspeed5.  PCOR.        /

ACF            VAR IS rspeed5.   MAXLAG IS 15.
               LBQ.                                /

PACF           VAR IS rspeed5.   MAXLAG IS 15.     /

ACF            VAR IS rspeed5.
               DFORDER IS 1.
               MAXLAG IS 15.                       /
```

Output 2T.6

—we omit a portion of the output—

```
ESTIMATION  RESID = RSPEED4.    PCOR./

ESTIMATION BY BACKCASTING METHOD

RELATIVE CHANGE IN RESIDUAL SUM OF SQUARES LESS THAN 0.5000E-04

CORRELATION MATRIX OF PARAMETER ESTIMATES
       1        2        3
 1  1.000
 2  0.028    1.000
 3  0.058    0.326    1.000

SUMMARY OF THE MODEL

OUTPUT VARIABLE -- SPEED
INPUT VARIABLES -- NOISE     CPZ

VARIABLE  VAR. TYPE  MEAN      TIME     DIFFERENCES
                                               1
SPEED       RANDOM            1- 120  (1-B  )
                                               1      1
CPZ         BINARY            1- 120  (1-B  ) (1-B  )

PARAMETER VARIABLE   TYPE  FACTOR  ORDER ESTIMATE        ST. ERR.  T-RATIO
      1  SPEED        MA     1       1     0.7844        0.0556    14.10
      2  CPZ          UP     1       0    -22.23         5.9399    -3.74
      3  CPZ          SP     1       1     1.010         0.0055    183.64     [1]

RESIDUAL SUM OF SQUARES   =      10829.032227
DEGREES OF FREEDOM        =               115
RESIDUAL MEAN SQUARE      =         94.165497
( BACKCASTS EXCLUDED )

ERASE   MODEL.                  /                                           [2]
UNIVARIATE TIME SERIES MODEL ERASED
```

```
ESTIMATION  RESID = RSPEED5.   PCOR./

—we omit a portion of the output—

PARAMETER VARIABLE   TYPE  FACTOR  ORDER ESTIMATE        ST. ERR.  T-RATIO      [3]
    1  SPEED          MA      1       1    0.7701         0.0578    13.33
    2  CPZ            UP      1       0   -13.28          6.1357    -2.16
    3  CPZ            SP      1       1    0.5414         0.2278     2.38

RESIDUAL SUM OF SQUARES  =     10710.915039
DEGREES OF FREEDOM       =              116
RESIDUAL MEAN SQUARE     =        92.335472          [4]

—we omit the remainder of the output—
```

[1] The S_1 estimate of 1.01 lies outside the bounds of system stability. We therefore rule out a temporary effect and try to fit a gradual, permanent effect.

[2] Whenever you obtain out-of-bounds estimates, erase the model to prevent adverse effects due to starting estimation from the extreme value. Erasing a model resets initial values to 0.1; otherwise, the latest estimate is used.

[3] All parameter estimates are significant, and the S_1 estimate lies within the bounds of system stability.

[4] The residual mean square error of 92.34 is smaller than the 101.82 obtained for the preintervention model (not shown), indicating that we have improved the fit of the model by adding the intervention component.

Analysis of residuals (not shown) confirms that the residuals do not differ from white noise.

Example 2T.7 Modeling several interventions

We now return to the ozone data of Examples 2T.1 and 2T.2. In Example 2T.2 we found that the ARIMA model alone does not fit the data well. Therefore we need to consider the interventions that occurred in 1960 and 1966. We will denote as I1 the freeway opening and law change that occurred in 1960 and as I2 the 1966 introduction of automobiles with redesigned engines.

Rather than rely on the "blind" three-step method of McCleary and Hay for determining the form of the intervention, it is important to use our knowledge of the subject matter as a guide. We may also plot fitted values and residuals, fit alternate models, and compare forecasts with the actual series during the intervention period (with a forecasting origin just prior to the intervention), although we do not present these steps in this example.

For the ozone data, Box and Tiao (1975) argue that I1 might be expected to produce a relatively sudden and permanent step change in the ozone level, because opening the freeway and changing the law should affect the ozone level immediately and permanently. The I2 intervention in 1966 might represent a constant (hopefully downward) change from year to year as an increasing proportion of newly designed vehicles make up the car population. Because of the summer-winter differential in the atmosphere temperature inversion and the difference in sunlight density, the net effect of reducing ozone by engine changes might differ by season; winter oxidant pollution is low, but in summer it is high. Thus I2 should be represented by two intervention terms in the model. With I1, this requires a total of three intervention terms.

In Input 2T.7, three indicator variables are created in the TRANSFORM paragraph for the three intervention terms:

$$\text{I1960} = \begin{cases} 0 & t < \text{January 1960 (i.e., } t < \text{case 60)} \\ 1 & t \geq \text{January 1960} \end{cases}$$

$$\text{Summer} = \begin{cases} 1 & \text{“Summer” months June–October beginning 1966 (i.e., beginning with case 132)} \\ 0 & \text{Otherwise} \end{cases}$$

$$\text{Winter} = \begin{cases} 1 & \text{“Winter” months November–May beginning 1966 (i.e., beginning with case 132)} \\ 0 & \text{Otherwise} \end{cases}$$

Following the principles of McCleary and Hay, since the 1960 intervention is thought to be sudden and permanent, we will use a term of the form $U_0 I_t$ in the model. Since the summer or winter intervention is thought to represent a continuing change from year to year, two terms of the form $U_0 I_t / (1 - B^{12})$ will be used. This latter form is the same as that for a gradual, permanent change with S equal to 1. When $S = 1$, the change does not "level off," but increases (or decreases) linearly from year to year. Box and Tiao use $S = 1$ since (at least up through the end of the series in 1972), the effect of the new vehicles is thought to be continuing. Recalling the ARIMA model for the noise component in Examples 2T.1 and 2T.2, the full model is

$$Y_t = U_{10}\text{I1960} + \frac{U_{20}\text{Summer}}{(1 - B^{12})} + \frac{U_{30}\text{Winter}}{(1 - B^{12})} + \frac{(1 - \Theta_1 B)(1 - \Theta_{12}B^{12})a_t}{(1 - B^{12})}$$

or, using the form required by 2T:

$$(1 - B^{12})Y_t = U_{10}\ \text{I1960}\ (1 - B^{12}) + U_{20}\ \text{Summer} + U_{30}\ \text{Winter} + (1 - \Theta_1 B)\ (1 - \Theta_{12}B^{12})a_t$$

According to this model, the INDEPENDENT paragraph for the 1960 intervention (I1960), will require a DFORDERS = 12 term, but the other two INDEPENDENT paragraphs (for summer and winter interventions) will not.

Input 2T.7

```
/ INPUT        VARIABLE IS 1.
               FORMAT IS STREAM.
               TITLE IS 'Los Angeles ozone data with
                         intervention'.

/ VARIABLE     NAME IS ozone.

/ TRANSFORM    month = KASE MOD 12. IF (month EQ 0) THEN
                                      month = 12.
               I1960 = 0. summer = 0.    winter = 0.
               IF (KASE GT 60) THEN I1960 = 1.
               IF (KASE GT 132 AND month GE 6 AND month LE 10)
                        THEN summer = 1.
               IF (KASE GT 132 AND summer EQ 0) THEN winter = 1.
/ END
(the data are the same as in Input 2T.1)
/ END

ARIMA          VARIABLE = ozone. DFORDER = 12.
               MAORDERS = '(1),(12)'.                          /

INDEP          VARIABLE = I1960.  TYPE = BINARY.  UPORDER = '(0)'.
               DFORDER = 12.                                    /

INDEP          VARIABLE = summer.TYPE = BINARY.
               UPORDER = '(0)'.                                 /
```

Input 2T.7 *(continued)*

```
INDEP          VARIABLE = winter.   TYPE = BINARY.
               UPORDER = '(0)'. /

ESTIMATION     RESIDUALS = rozone.                              /

ACF            VARIABLE IS rozone. MAXLAG IS 24.                /

PACF           VARIABLE IS rozone. MAXLAG IS 24.                /

END   /
```

Output 2T.7

—we omit portions of the output—

```
ARIMA      VARIABLE = OZONE. DFORDER = 12. MAORDERS = '(1),(12)'. /       [1]

THE COMPONENT HAS BEEN ADDED TO THE MODEL.

THE CURRENT MODEL HAS
OUTPUT VARIABLE = ozone
INPUT  VARIABLE = NOISE

 INDEP      VARIABLE = SUMMER.  TYPE = BINARY.  UPORDER = '(0)'.   /

THE COMPONENT HAS BEEN ADDED TO THE MODEL

THE CURRENT MODEL HAS
OUTPUT VARIABLE = OZONE
INPUT  VARIABLE = NOISE     I1960        [2]

INDEP      VARIABLE = WINTER.  TYPE = BINARY.  UPORDER = '(0)'.   /       [3]
```

—we omit a portion of the output—

```
 INDEP      VARIABLE = WINTER.  TYPE = BINARY.  UPORDER = '(0)'.   /      [4]
```

—we omit a portion of the output—

```
FIRST CASE NUMBER TO BE USED      =              1
LAST CASE NUMBER TO BE USED       =            216


ESTIMATION BY CONDITIONAL LEAST SQUARES METHOD

RELATIVE CHANGE IN RESIDUAL SUM OF SQUARES LESS THAN 0.1000E-02

SUMMARY OF THE MODEL                                                      [5]

OUTPUT VARIABLE -- OZONE
INPUT VARIABLES -- NOISE    I1960    SUMMER   WINTER

VARIABLE  VAR. TYPE  MEAN      TIME     DIFFERENCES
                                                12
OZONE      RANDOM             1- 216   (1-B   )
                                                12
I1960     NONRANDM            1- 216   (1-B   )

SUMMER     BINARY             1- 216

WINTER     BINARY             1- 216

PARAMETER VARIABLE   TYPE  FACTOR  ORDER ESTIMATE        ST. ERR.  T-RATIO   [6]
      1  OZONE        MA     1       1    -0.2997         0.0677    -4.43
      2  OZONE        MA     2      12     0.5922         0.0580    10.20
      3  I1960        UP     1       0    -1.262          0.2586    -4.88
      4  SUMMER       UP     1       0    -0.2616         0.0886    -2.95
      5  WINTER       UP     1       0    -0.8196E-01     0.0757    -1.08

RESIDUAL SUM OF SQUARES  =        145.878876
DEGREES OF FREEDOM       =               199                                 [7]
RESIDUAL MEAN SQUARE     =          0.733060


ESTIMATION BY BACKCASTING METHOD                                             [8]

RELATIVE CHANGE IN RESIDUAL SUM OF SQUARES LESS THAN 0.1000E-02


SUMMARY OF THE MODEL

OUTPUT VARIABLE -- OZONE
INPUT VARIABLES -- NOISE    I1960    SUMMER   WINTER

VARIABLE  VAR. TYPE  MEAN      TIME     DIFFERENCES
                                                12
OZONE      RANDOM             1- 216   (1-B   )
                                                12
I1960     NONRANDM            1- 216   (1-B   )

SUMMER     BINARY             1- 216

WINTER     BINARY             1- 216


PARAMETER VARIABLE   TYPE  FACTOR  ORDER ESTIMATE        ST. ERR.  T-RATIO
      1  OZONE        MA     1       1    -0.2580         0.0627    -4.11
      2  OZONE        MA     2      12     0.9202         0.0181    50.85
      3  I1960        UP     1       0    -1.359          0.1515    -8.97
      4  SUMMER       UP     1       0    -0.2123         0.0415    -5.12
      5  WINTER       UP     1       0    -0.8356E-01     0.0355    -2.35
```

```
RESIDUAL SUM OF SQUARES  =      103.280930
DEGREES OF FREEDOM       =             199
RESIDUAL MEAN SQUARE     =        0.519000
( BACKCASTS EXCLUDED )

 ACF         VARIABLE IS ROZONE.   MAXLAG IS 24.                   /

FIRST CASE NUMBER TO BE USED      =            13
LAST CASE NUMBER TO BE USED       =           216
NO. OF OBS. AFTER DIFFERENCING    =           204               [9]
MEAN OF THE (DIFFERENCED) SERIES  =       -0.0179
STANDARD ERROR OF THE MEAN        =        0.0499
T-VALUE OF MEAN (AGAINST ZERO)    =       -0.3576

AUTOCORRELATIONS

 1- 12      -.03  .11  .12  .09 -.06  .06 -.05 -.06  .04  .01  0.0  .05
 ST.E.       .07  .07  .07  .07  .07  .07  .07  .07  .07  .07  .07  .07

 13- 24     0.0   .01 -.06 -.01 -.08  0.0  .02 -.08 -.05  .04 -.08  .02
 ST.E.       .07  .07  .07  .07  .07  .07  .07  .07  .07  .07  .07  .08

PLOT OF AUTOCORRELATIONS

           -1.0 -0.8 -0.6 -0.4 -0.2  0.0  0.2  0.4  0.6  0.8  1.0
LAG    CORR. +----+----+----+----+----+----+----+----+----+----+
                                      I
  1  -0.030                       + XI  +
  2   0.106                       +  IXXX
  3   0.117                       +  IXXX
  4   0.090                       +   IXX +
  5  -0.063                       + XXI   +
  6   0.057                       +   IX  +
  7  -0.047                       +  XI   +
  8  -0.055                       +  XI   +
  9   0.039                       +   IX  +
 10   0.007                       +   I   +
 11  -0.005                       +   I   +
 12   0.048                       +   IX  +
 13   0.000                       +   I   +
 14   0.015                       +   I   +
 15  -0.059                       +  XI   +
 16  -0.009                       +   I   +
 17  -0.076                       + XXI   +
 18   0.004                       +   I   +
 19   0.016                       +   I   +
 20  -0.080                       + XXI   +
 21  -0.047                       +  XI   +
 22   0.045                       +   IX  +
 23  -0.084                       + XXI   +
 24   0.017                       +   I   +
```

[1] The "noise" component of the model is specified in the ARIMA paragraph. This is the same as in Example 2T.2.

[2] The effect of the 1960 intervention is specified in the first INDEPENDENT paragraph. The intervention variable (1960) is listed along with NOISE as one of the inputs of the model.

[3] The summer intervention variable is specified.

[4] The winter intervention variable is specified.

[5] The model is summarized in the ESTIMATION paragraph.

[6] Parameter estimates using CLS.

[7] The CLS residual mean square is larger than the backcasting residual mean square in **[8]** below. Therefore we will use the backcasting estimates in our final model.

[8] The backcasting estimates are somewhat different from the CLS estimates, particularly for the second MA parameter. The parameters are several standard errors away from zero, as required in the valid model. This model indicates that there was a 1.35 unit drop in ozone attributed to the 1960 intervention and a progressive (annual) 0.21 drop in summers starting with 1966. The progressive change in the winter is of the correct sign (negative), but is less substantial.

[9] Unlike Example 2T.2, none of the autocorrelations for the residuals (ROZONE) are significant, as required in a valid model. The residual mean is now less than 0.4 standard errors from zero.

The PACF for ROZONE is not shown as it is similar to the ACF.

Multiple-Input Transfer-Function Models

You can incorporate one or more time series in a model to predict the value of another series, by using a transfer function (akin to a linear regression function) that specifies the relationship between the predictor series and the output series. Transfer functions may be used to predict the present value of one series from the past or present value of another series, called a "leading indicator." Or one series may be used as a covariate to remove some of the noise in another series and thereby allow a more sensitive test of the effect of an intervention.

We use the classic illustration of the transfer function in Examples 2T.8 through 2T.13, where we think of a continuing chemical reaction in equilibrium with some input series (gas feed in our example) "driving" some output series (CO_2 production). Another common example occurs in economics where the price of a commodity (input) is used to forecast future inventories (output).

The output series is given as $Y_t = f(X_t) + N_t$, where X_t is the predictor series, $f(X_t)$ is the transfer function we want to model and test, and N_t may be any noise process (including a full ARIMA model).

This model is very similar to the ARIMA intervention model; in fact, intervention models are a special case of transfer function models.

The general form of a transfer function is

$$f(X)_t = \frac{(U_0 + U_1B + \dots + U_sB^s)}{(1 - S_1B^1 - S_2B^2 - \dots - S_rB^r)} X_t$$

where $U(B)$ is the U polynomial as in intervention analysis, $S(B)$ is the S polynomial as in intervention analysis, and b is the lead interval (or "dead time") between X and its predictive effect on Y.

The b parameter may be excluded by adding b to the subscripts for U_i and to the exponents for the backshift operator B in the numerator:

$$f(X)_t = \frac{(U_{0+b}B^{0+b} + U_{1+b}B^{1+b} + \dots + U_{s+b}B^{s+b})}{(1 - S_1B^1 - S_2B^2 - \dots - S_rB^r)} X_t$$

The first notation has the advantage of drawing attention to the dead time b, while the latter has the advantage of clearly expressing the coefficients in terms of the X_t to which they apply.

Transfer function analysis has three main stages:

1. Pre-analyze both series to facilitate the calculation of statistics to be used in identifying the transfer function.
2. Identify the transfer function and calculate the residuals N_t.
3. Fit an ARIMA model to N_t and assess the adequacy of the model as a whole.

Transfer function analysis is more complex than intervention analysis, and transfer functions are more complex than most intervention functions. As a result, several tools have been developed to aid in identifying and fitting transfer functions.

The key to identifying the transfer function is a set of "impulse weights," which express the effect of a unit change of the predictor series upon the value of the output series at various lags. To define impulse weights, imagine expressing $f(X_t)$ as a function of a single polynomial in B and of X_t:

$$f(X_t) = v(B)X_t = (v_0 + v_1B + v_2B^2 + \dots)X_t$$

where $v(B)$ is algebraically equivalent to the $U(B)/S(B)$ polynomial ratio. When $S_1, S_2, \dots, S_r$ are all zero, $v(B) = U(B)$. Otherwise, $v(B)$ has an infinite number of terms, although as the lags increase, the v_i decrease exponentially and become negligible. Under this alternate representation of $f(X_t)$, the v_i are the "impulse weights" and give direct information about the relationship between the predictor and output series.

Of course, the impulse weights are not known, nor are their equivalents, the U and S values. However, under certain circumstances they may be estimated from the cross-correlation function CCF. The CCF(k) is the relationship between the predictor series X at time t and the output series Y at time $t + k$. Thus when $k > 0$, the CCF(k) is the correlation between the value of X and future values of Y. When $k < 0$, the CCF(k) is the correlation between X and the past value of Y. Like the ACF(k) and PACF(k), the CCF(k) is based on fewer observations and its reliability decreases as k increases. Unlike the ACF(k) and PACF(k), the CCF(k) is not symmetric about 0; in general, CCF($-k$) $\neq$ CCF(k).

The CCF is a measure of the relationship between X and Y. However, it is also affected by autocorrelation in X and/or Y. If X or Y contains autocorrelation or is not stationary about its mean, the CCF will not reflect the true relationship between X and Y. Before we calculate the CCF, therefore, we must remove autocorrelation.

Two methods have been suggested for removing autocorrelation before calculating the CCF:

1. "Prewhiten" (filter) the series. Recommened by Box and Jenkins, (1976); see below for a complete description of the process.
2. Fit separate ARIMA models to X_t and Y_t, and then calculate the CCF based on the residuals.

The first method is more widely used, although the issue is not completely resolved. Vandaele (1983) suggests using prewhitening for leading indicator models, and residual analysis for more exploratory studies. 2T allows you to use either method, although the program is structured to facilitate prewhitening.

Prewhitening involves five conceptual steps. We list them here and give more explanation in Examples 2T.8 through 2T.9. First, difference each series until it is stationary about its mean:

$$w_t = (1 - B)^d X_t$$
$$z_t = (1 - B)^{d'} Y_t$$

The degree of differencing **need not** be the same for the X_t and Y_t series. Before prewhitening, use a separate DIFFERENCE paragraph to make each series stationary.

Second, fit an ARIMA model to the differenced predictor series w_t:

$$w_t = \frac{\Theta(B)}{\Phi(B)} a_t$$

Third, use the inverse of the model to filter w_t (leaving the residuals a_t):

$$a_t = \frac{\Phi(B)}{\Theta(B)} w_t$$

Fourth, use the same inverse to filter (or prewhiten) z_t, yielding the prewhitened b_t:

$$b_t = \frac{\Phi(B)}{\Theta(B)} z_t$$

Fifth, cross-correlate the residuals a_t and the prewhitened b_t. This CCF reflects the impulse weights and, therefore, may be used to identify the transfer function.

The second method of removing autocorrelation is to fit separate ARIMA models to the predictor and output series, and then filter each series alone to identify the transfer function.

In Examples 2T.8 through 2T.13 we demonstrate identification and analysis of transfer functions. The problem is to control the output of an industrial process (from Box and Jenkins, 1976). The predictor series is the feed rate of a mixture of gases (including methane and oxygen); the output series is the concentration of carbon dioxide (CO_2). The data are described in Appendix D.

We first build ARIMA models for the two series, use the predictor series model to filter the output series, and compute cross-correlations (Example 2T.8). We then specify the transfer function (Example 2T.9), model the output series (Example 2T.10), and perform diagnostic checking (Example 2T.11). Finally, we demonstrate how to forecast the predictor series (Example 2T.12), calculate "psiweights" to be used in determining variance, and forecast the output variable (Example 2T.13).

Example 2T.8 Identifying the transfer function

Following Jenkins (1979), we begin by building an ARIMA model for the output series (CO_2), ignoring the predictor series (FEED). This model provides a basis for evaluating how much precision we gain by later incorporating the predictor transfer function series, and gives us a good idea of the noise function N_t after we have identified the transfer function.

In Input 2T.8, we specify models for both series, estimate parameters for the feed model, check the adequacy of the FEED model, and use the FEED model to FILTER the CO_2 series. We also compute the CCF and the transfer function weights. For more information on FILTER and other commands used in Input 2T.8, see "2T Commands."

Input 2T.8

Input data for both series

Identify and define a good model for CO_2 without using FEED for prediction

Erase CO_2 model before identifying FEED model

```
/ INPUT      VARIABLES ARE 2.
             FORMAT IS STREAM.

/ VARIABLE   NAMES ARE FEED, CO2.

/ END
 -0.109    53.800
  0.000    53.600
  .....    ......
/ END

ACF          VAR IS CO2.    MAXLAG IS 25.             /

PACF         VAR IS CO2.    MAXLAG IS 25.             /

ARIMA        VAR IS CO2.    ARORDERS ARE '(1,2)'.
             CONSTANT.                                /

ESTIMATION   RESID IS RY.   PCOR.                     /

ERASE MODEL.                                          /
```

Input 2T.8 *(continued)*

Identify and define a model for FEED without CO_2

Estimate parameters and store residuals

Evaluate the FEED model

Use the FEED model to filter the CO_2 series

Compute cross-correlations and transfer function weights

```
ACF         VAR IS FEED.   MAXLAG IS 25.          /

PACF        VAR IS FEED.   MAXLAG IS 25.          /

ARIMA       VAR IS FEED.
            ARORDERS ARE '(1,2,3)'.               /

ESTIMATION RESID IS RX.    PCOR.                  /

ACF         VAR IS RX.     MAXLAG IS 25.
            LBQ.                                  /

PACF        VAR IS RX.     MAXLAG IS 25.          /

ACF         VAR IS RX.     DFORDER IS 1.
            MAXLAG IS 25.                         /

FILTER      VAR IS CO2.    RESID IS RY.           /

CCF         VAR ARE RX, RY.
            MAXLAG IS 10.                         /
```

Output 2T.8

```
—we omit a portion of the output—

PARAMETER VARIABLE    TYPE  FACTOR  ORDER ESTIMATE        ST. ERR.  T-RATIO
      1  CO2          AR      1       1      1.807         0.0301     60.13
      2  CO2          AR      1       2     -0.8582        0.0301    -28.48     [1]
      3  CO2          MEAN    1       0     53.53          0.4410    121.38

—we omit a portion of the output—

RESIDUAL SUM OF SQUARES  =        43.635132
DEGREES OF FREEDOM       =              291
RESIDUAL MEAN SQUARE     =         0.149949                                   [2]
( BACKCASTS EXCLUDED )

—we omit a portion of the output—

ESTIMATION BY BACKCASTING METHOD

—we omit a portion of the output—

PARAMETER VARIABLE    TYPE  FACTOR  ORDER ESTIMATE        ST. ERR.  T-RATIO
      1  FEED         AR      1       1      1.977         0.0549     35.98
      2  FEED         AR      1       2     -1.376         0.0995    -13.83     [3]
      3  FEED         AR      1       3      0.3437        0.0549      6.26

RESIDUAL SUM OF SQUARES  =        10.439016
DEGREES OF FREEDOM       =              290
RESIDUAL MEAN SQUARE     =         0.035997
( BACKCASTS EXCLUDED )

—we omit a portion of the output—

FILTER VAR IS CO2.
       RESID IS RY.           /

FIRST CASE NUMBER TO BE USED     =          1
LAST CASE NUMBER TO BE USED      =        296

RESIDUAL SUM OF SQUARES  =      2634.490967                                   [4]
DEGREES OF FREEDOM       =              290
RESIDUAL MEAN SQUARE     =         9.084452
( BACKCASTS EXCLUDED )

VARIABLE CO2      IS FILTERED, RESULTS ARE STORED IN VARIABLE RY

CCF      VAR ARE RX,RY.
         MAXLAG IS 10.        /

FIRST CASE NUMBER TO BE USED     =          4
LAST CASE NUMBER TO BE USED      =        296

EFFECTIVE NUMBER OF CASES =  293

CORRELATION        OF RX           AND RY         IS     0.00

CROSS CORRELATIONS OF RX        (I) AND RY       (I+K)

  1- 10      .05 -.02 -.28 -.33 -.46 -.27 -.17 -.02  .03 -.05                 [5]
  ST.E.      .06  .06  .06  .06  .06  .06  .06  .06  .06  .06

CROSS CORRELATIONS OF RY        (I) AND RX       (I+K)

  1- 10     -.03  .01 -.05 -.02  0.0 -.12 -.03 -.09  0.0  .02
  ST.E.      .06  .06  .06  .06  .06  .06  .06  .06  .06  .06
TRANSFER FUNCTION WEIGHTS
```

```
        SCCF(X(I),Y(I+K))          SCCF(Y(I),X(I+K))
LAG     *SY/SX      *SX/SY      *SY/SX      *SX/SY

  0   -0.00307    -0.00083    -0.00307    -0.00083
  1    0.10435     0.02826    -0.05749    -0.01557
  2   -0.04691    -0.01270     0.01902     0.00515
  3   -0.54198    -0.14676    -0.09374    -0.02538
  4   -0.63374    -0.17161    -0.02964    -0.00803
  5   -0.87468    -0.23685    -0.00462    -0.00125
  6   -0.51318    -0.13896    -0.23049    -0.06241
  7   -0.32185    -0.08715    -0.04900    -0.01327
  8   -0.04741    -0.01284    -0.17109    -0.04633
  9    0.06093     0.01650     0.00222     0.00060
 10   -0.10428    -0.02824     0.04224     0.01144
```

[6]

```
WHERE X(I) IS THE FIRST SERIES, Y(I) THE SECOND
SERIES, SX THE STANDARD ERROR OF X(I), AND SY
THE STANDARD ERROR OF Y(I)
```

```
PLOT OF CROSS CORRELATIONS

          -1.0 -0.8 -0.6 -0.4 -0.2  0.0  0.2  0.4  0.6  0.8  1.0
LAG   CORR. +----+----+----+----+----+----+----+----+----+----+
                                     I
-10   0.022                       +  IX +
 -9   0.001                       +  I  +
 -8  -0.089                       +XXI  +
 -7  -0.025                       + XI  +
 -6  -0.120                       XXXI  +
 -5  -0.002                       +  I  +
 -4  -0.015                       +  I  +
 -3  -0.049                       + XI  +
 -2   0.010                       +  I  +
 -1  -0.030                       + XI  +
  0  -0.002                       +  I  +
  1   0.054                       +  IX +
  2  -0.024                       + XI  +
  3  -0.282                   XXXX+XXI  +
  4  -0.330                  XXXXX+XXI  +
  5  -0.455               XXXXXXXX+XXI  +
  6  -0.267                   XXXX+XXI  +
  7  -0.167                      X+XXI  +
  8  -0.025                       + XI  +
  9   0.032                       +  IX +
 10  -0.054                       + XI  +
```

[7]

[1] The ACF (not shown) for the output series gives no evidence of nonstationarity but appears to decay exponentially. The PACF (not shown) has spikes at lags 1 and 2. We therefore identify the model as ARIMA(2, 0, 0). Estimating the model yields significant estimates for both AR parameters and for the constant.

[2] The residual mean square is 0.150. We will use this value to evaluate the usefulness of our transfer function. Analysis of residuals confirms that our model is satisfactory.

[3] The next step is to model the predictor series. The ACF (not shown) appears to decay. The PACF (not shown) has three significant spikes, positive for lags 1 and 3 and negative for lag 2. Thus, we tentatively identify an ARIMA(3, 0, 0). We estimate the model and store the residuals in RX. The values for all parameters are significant, and analysis of residuals demonstrates the model to be satisfactory.

[4] Using the model in memory (for FEED), we FILTER the output CO_2 series, and store the results in RY.

[5] The CCF between the prewhitened RX and the filtered RY.

[6] The estimated impulse weights v_i between FEED and CO_2.

[7] A plot of the CCF.

Now we are ready to use the CCF to identify the transfer function. Recall that the general transfer function is

$$f(X_t) = \frac{U(B)}{S(B)} X_{t-b}$$

The first component to be identified is b, the dead time between X_t and its effect on Y_t. The parameter b may be zero, in which case X has an immediate correlation with Y; or b may be greater than zero, in which case the relationship is a function translated b points into the future. Thus, if $b = 6$, the level of the series X at time t affects Y at time $t + 6$. These values of b are reflected in the structure of the CCF. If the CCF has significant spikes at lag 0 and subsequent low lags, it corresponds to $b = 0$. On the other hand, if there are no significant CCF spikes until lag 6 or 7, this suggests that $b = 6$. As a general rule, set b equal to the first significant lag of the CCF.

Once b is known, identifying the transfer function is relatively straightforward. If the CCF contains spikes, it indicates U_i parameters. Exponential decay in the CCF indicates S_j parameters, but in all cases the transfer function must contain at least one U_i parameter. A spike at lag $b + i$ indicates a U_i parameter. A decay beginning at lag $b + i + j$ indicates an S_j parameter. Figure 2T.3 gives some examples of CCF and the transfer functions they indicate.

Note that j is the lag at which the decay is first evident. Thus, there is always a higher spike one lag before the j lag.

Figure 2T.3
Forms of CCF and associated $f(X_t)$ after McCleary and Hay (1980, p. 233)

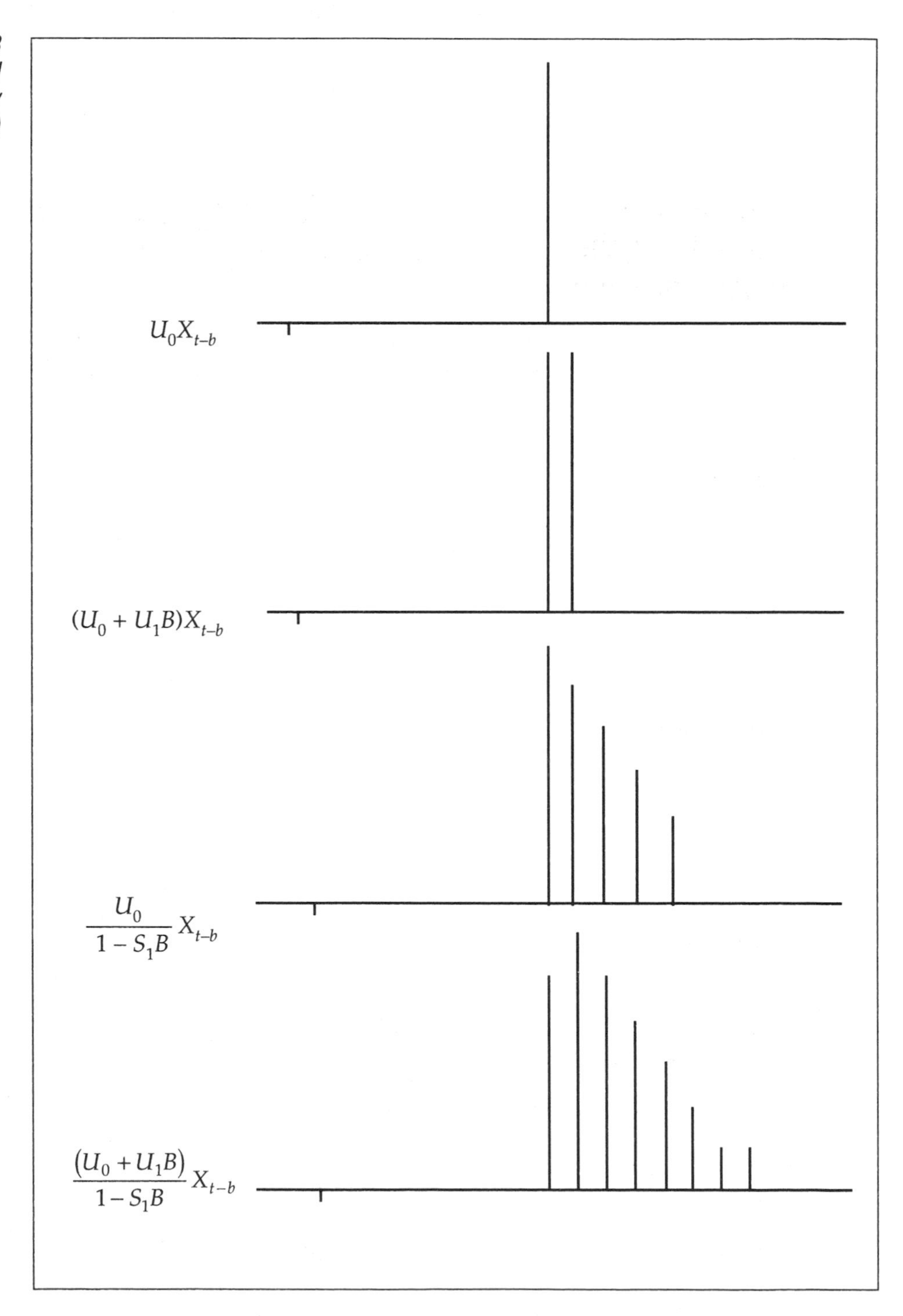

For the present case, the CCF(k) is not significant for $k < 3$. We therefore identify $b = 3$. The pattern of spikes at lags 3, 4, and 5 do not show any decay, but the CCF does appear to decay after lag 5. We tentatively identify the transfer function as

$$f(X_t) = \frac{(U_0 + U_1 B + U_2 B^2)}{(1 - S_1 B)} X_{t-3}$$

or

$$f(X_t) = \frac{(U_3 B_3 + U_4 B^4 + U_5 B^5)}{(1 - S_1 B)} X_t$$

The second formula expresses the coefficients of terms X_t, so we use it to derive the specifications of the INDEPENDENT paragraph in Input 2T.9, below.

Example 2T.9 Specifying the transfer function

Specifying a transfer function is very similar to specifying an intervention component. You use the INDEPENDENT paragraph for both. However, for transfer functions, you specify TYPE = RANDOM. (For more information on the INDEPENDENT paragraph, see 2T Commands.)

Input 2T.9 shows how we specify the transfer function for gas rate data. Because the CO_2 series is not differenced, the CONSTANT is the mean (see "Interpreting the constant" in Special Features).

Input 2T.9
(Add to the end of Input 2T.8)

Define transfer function (ARIMA, INDEPENDENT)

Estimate model parameters (ESTIMATION)

Assess model adequacy (ACF, PACF)

```
ERASE          MODEL.                                      /

ARIMA          VAR IS CO2.       CONSTANT.                 /

INDEPENDENT    VAR IS FEED. TYPE IS RANDOM.
                          SPORDERS ARE '(1)'.
                          UPORDERS ARE '(3,4,5)'.          /

ESTIMATION     RESID = RCO2A.   PCOR.                      /

ACF            VAR IS RCO2A     MAXLAG IS 25.              /

PACF           VAR IS RCO2A.    MAXLAG IS 25.              /
```

Output 2T.9

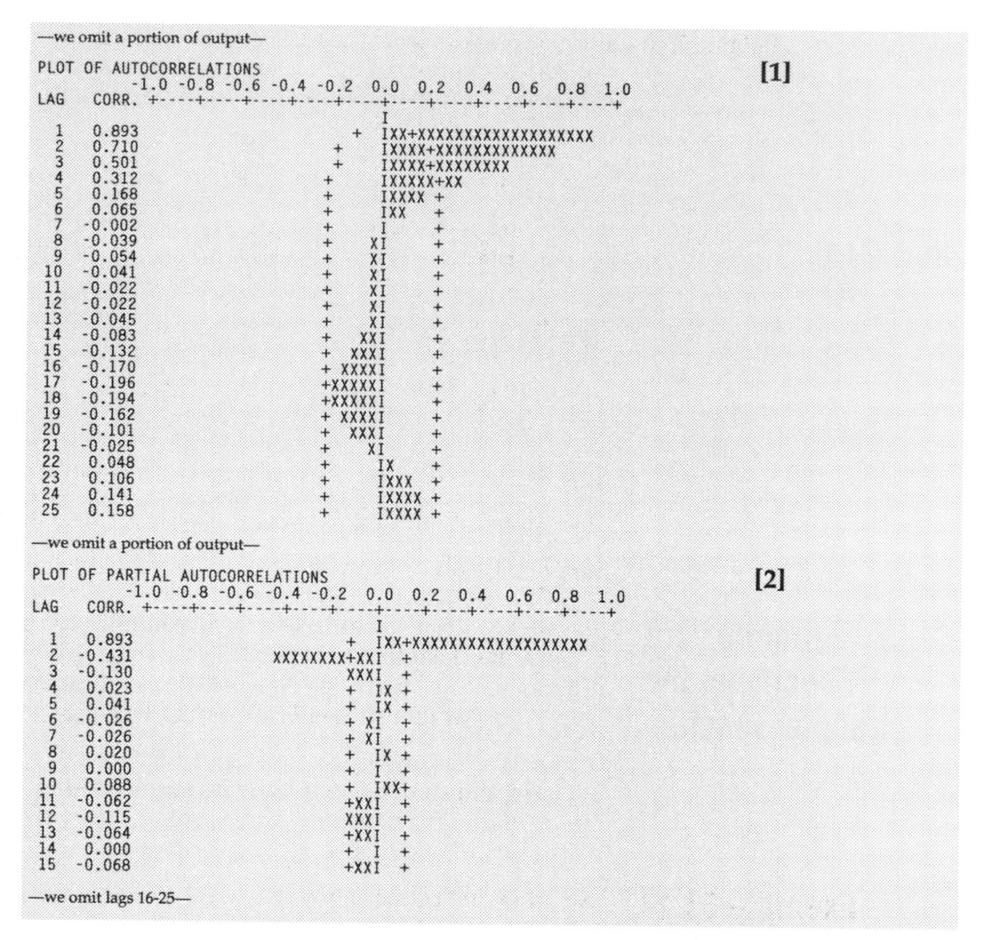

```
—we omit a portion of output—

PLOT OF AUTOCORRELATIONS                                                  [1]
          -1.0 -0.8 -0.6 -0.4 -0.2  0.0  0.2  0.4  0.6  0.8  1.0
LAG   CORR. +----+----+----+----+----+----+----+----+----+----+
                                     I
 1   0.893                       +   IXX+XXXXXXXXXXXXXXXXXXX
 2   0.710                        +    IXXXX+XXXXXXXXXXXXX
 3   0.501                        +    IXXXX+XXXXXXXX
 4   0.312                      +      IXXXXX+XX
 5   0.168                      +      IXXXX +
 6   0.065                      +      IXX    +
 7  -0.002                      +      I      +
 8  -0.039                      +     XI      +
 9  -0.054                      +     XI      +
10  -0.041                      +     XI      +
11  -0.022                      +     XI      +
12  -0.022                      +     XI      +
13  -0.045                      +     XI      +
14  -0.083                      +    XXI      +
15  -0.132                      +   XXXI      +
16  -0.170                       + XXXXI      +
17  -0.196                       +XXXXXI      +
18  -0.194                       +XXXXXI      +
19  -0.162                       + XXXXI      +
20  -0.101                      +   XXXI      +
21  -0.025                      +     XI      +
22   0.048                      +      IX     +
23   0.106                      +      IXXX   +
24   0.141                      +      IXXXX +
25   0.158                      +      IXXXX +

—we omit a portion of output—

PLOT OF PARTIAL AUTOCORRELATIONS                                          [2]
          -1.0 -0.8 -0.6 -0.4 -0.2  0.0  0.2  0.4  0.6  0.8  1.0
LAG   CORR. +----+----+----+----+----+----+----+----+----+----+
                                     I
 1   0.893                          +  IXX+XXXXXXXXXXXXXXXXXXX
 2  -0.431                XXXXXXXX+XXI  +
 3  -0.130                          XXXI  +
 4   0.023                          +  IX +
 5   0.041                          +  IX +
 6  -0.026                          + XI  +
 7  -0.026                          + XI  +
 8   0.020                          +  IX +
 9   0.000                          +  I  +
10   0.088                          +  IXX+
11  -0.062                          +XXI  +
12  -0.115                          XXXI  +
13  -0.064                          +XXI  +
14   0.000                          +  I  +
15  -0.068                          +XXI  +

—we omit lags 16-25—
```

[1] The ACF demonstrates a decay process.

[2] The PACF shows significant spikes at lags 1 and 2.

From the ACF and PACF of the residuals, we tentatively identify the transfer function N_t model as ARIMA (2, 0, 0). This corresponds to our original univariate model for the CO_2 series (see Example 2T.8).

Example 2T.10 Modeling the output series

Now that we have identified the N_t model, we are ready to model the output series CO_2.

Input 2T.10
(add to the end of Input 2T.9)

```
ERASE        MODEL.                                              /

ARIMA        VAR IS CO2. ARORDERS ARE '(1,2)'. CONSTANT.         /

INDEPENDENT  VAR IS FEED.     UPORDERS ARE '(3,4,5)'.
             TYPE IS RANDOM.  SPORDERS ARE '(1)'.                /

ESTIMATION   RESID = RCO2B.   PCOR.                              /
```

Output 2T.10

```
—we omit portions of output—
INDEPENDENT   VAR IS FEED.                                          [1]
              TYPE IS RANDOM.
              UPORDERS ARE '(3,4,5)'.
              SPORDERS ARE '(1)'.            /
—we omit portions of output—
PARAMETER VARIABLE   TYPE  FACTOR  ORDER ESTIMATE      ST. ERR.  T-RATIO
    1   CO2           AR      1      1      1.531       0.0476    32.18
    2   CO2           AR      1      2     -0.6322      0.0501   -12.62
    3   CO2          MEAN     1      0     53.37        0.1429   373.53
    4   FEED          UP      1      3     -0.5293      0.0749    -7.06   [2]
    5   FEED          UP      1      4     -0.3802      0.1030    -3.69
    6   FEED          UP      1      5     -0.5184      0.1075    -4.82
    7   FEED          SP      1      1      0.5492      0.0375    14.63

RESIDUAL SUM OF SQUARES   =        16.592497
DEGREES OF FREEDOM        =              282                            [3]
RESIDUAL MEAN SQUARE      =         0.058839
—we omit portions of output—
```

[1] We needed to specify the transfer function here because it was erased earlier.

[2] All parameter estimates are significant and lie within their bounds.

[3] The residual mean square of 0.059 is much less than the 0.150 obtained for the univariate ARIMA model (output 2T.8). Clearly, our addition of the predictor series enhanced our estimates of the CO_2 series. Based on this output, our model appears to be

$$(1 - 1.53B + 0.63B^2)\, Y_t = 53.4 + \frac{-0.53B^3 - 0.38B^4 - 0.52B^5}{1 - 0.55B} X_t + a_t$$

where the coefficients have been rounded.

Example 2T.11 Diagnostic Checking

An adequate transfer function model must meet two further requirements:

1. The residuals of the model must not differ from white noise. Thus, we perform the usual residual analysis on the model residuals.
2. The model residuals must not be correlated with the predictor series. This requirement is the same as in linear regression: the model must capture all possible predictor-output dependence. To evaluate the model with respect to this requirement, we obtain the CCF between the filtered predictor series and the model residuals. The result should not differ from white noise.

Input 2T.11
(add to the end of Input 2T.10)

```
ACF     VAR IS RCO2B.       MAXLAG IS 25.   LBQ.          /
PACF    VAR IS RCO2B.       MAXLAG IS 25.                 /
ACF     VAR IS RCO2B.       DFORDER IS 1. MAXLAG IS 25.   /
CCF     VAR ARE RX,RCO2B.   MAXLAG IS 10.                 /
```

Output 2T.11

```
—we omit portions of output—                                              [1]
            -1.0 -0.8 -0.6 -0.4 -0.2  0.0  0.2  0.4  0.6  0.8  1.0
LAG    CORR. +----+----+----+----+----+----+----+----+----+----+    [2]
                                      I
 1   -0.518             XXXXXXXXXX+XXI  +
 2    0.084                       +   IXX +
 3   -0.079                       + XXI   +
 4    0.012                       +   I   +
 5   -0.094                       + XXI   +
 6    0.136                       +   IXXX+
 7   -0.038                       +  XI   +
 8    0.053                       +   IX  +
 9   -0.128                       +XXXI   +
10    0.091                       +   IXX +
—we omit lags 11-25—
—we omit portions of output—
```

```
                -1.0 -0.8 -0.6 -0.4 -0.2  0.0  0.2  0.4  0.6  0.8  1.0
LAG    CORR.   +----+----+----+----+----+----+----+----+----+----+
                                          I
-10    0.050                           +  IX +
 -9    0.048                           +  IX +
 -8   -0.029                           + XI  +
 -7    0.015                           +  I  +
 -6   -0.094                           +XXI  +
 -5   -0.003                           +  I  +
 -4   -0.012                           +  I  +
 -3   -0.043                           + XI  +
 -2    0.036                           +  IX +
 -1   -0.044                           + XI  +
  0   -0.056                           + XI  +
  1    0.025                           +  IX +
  2   -0.014                           +  I  +
  3    0.003                           +  I  +
  4    0.004                           +  I  +
  5    0.009                           +  I  +
  6    0.006                           +  I  +
  7   -0.045                           + XI  +
  8    0.029                           +  IX +
  9    0.077                           +  IXX+
 10   -0.025                           + XI  +
```

[3]

[1] The ACF of the residuals (not shown) has no significant spikes. The $LBQ_{df=23} = 28.00$, which is not significant. The PACF has no significant spikes.

[2] The ACF of the first differences of the residuals shows the expected single spike at lag 1. The value –.52 is close to the theoretical value of –.50.

[3] The CCF has no significant spikes. Thus, we accept the transfer function model.

Example 2T.12 Forecasting the predictor series

Like an ARIMA model, a transfer function model is not an end in itself, but is generally intended for some use. A common use of transfer functions is to enhance the accuracy of forecasts by including exogenous information. To forecast the output series, we must first have values of the predictor series from which to predict. In this example, the predictor series (FEED) is deliberately manipulated and therefore known. However, in some cases you may need to forecast the predictor series. For example, suppose you had developed a model by which the month's unemployment rate was predicted by the previous month's inflation. To predict unemployment six months in the future, you would need to forecast the inflation rate (the predictor series) five months ahead. For purposes of illustration, therefore, we forecast the predictor series FEED.

In Input 2T.12 we set up the model for the FEED series, FORECAST it 25 time points ahead, and JOIN those time points to the data set so that they will be available for future use. The CENTERED command in the ARIMA paragraph directs 2T to subtract the mean from each observation in the series, so that the series will be centered at zero for analysis. During forecasting, 2T adds the mean back into the series. Thus, the forecasts shown in Output 2T.12 regress toward the mean (–0.0568, not shown).

Input 2T.12
(add to the end of Input 2T.11)

```
ERASE        MODEL.                                         /

ARIMA        VAR IS FEED. ARORDERS ARE '(1,2,3)'.
             CENTERED.                                      /

ESTIMATION   METHOD IS CLS.

FORECAST     CASES ARE 25.   JOIN.                          /
```

Output 2T.12

```
—we omit a portion of the output—
FORECAST        CASES ARE 25.  JOIN.          /

 FORECAST ON VARIABLE FEED      FROM TIME PERIOD   297

 PERIOD            FORECASTS     ST.  ERR.       ACTUAL            RESIDUAL
  297              -0.26486      0.18969
  298              -0.22880      0.41992
  299              -0.18105      0.63730
  300              -0.13724      0.80836
  301              -0.10394      0.92619
  302              -0.08199      0.99890
  303              -0.06937      1.03968
  304              -0.06317      1.06079
  305              -0.06075      1.07106
  306              -0.06016      1.07588
  307              -0.06019      1.07816
  308              -0.06024      1.07928
  309              -0.06009      1.07987
  310              -0.05974      1.08021
  311              -0.05926      1.08042
  312              -0.05876      1.08055
  313              -0.05829      1.08064
  314              -0.05791      1.08069
  315              -0.05761      1.08073
  316              -0.05739      1.08075
  317              -0.05723      1.08076
  318              -0.05713      1.08077
  319              -0.05705      1.08077
  320              -0.05700      1.08078
  321              -0.05696      1.08078

 STANDARD ERROR = 0.189690         BY CONDITIONAL METHOD
```

Example 2T.13 Calculating psiweights and forecasting the output series

If we are to know the variance of the output series forecasts, we must first know the variance of the predictor series. To obtain this information, we must solve the ARIMA model algebraically in terms of its noise component, the a_t. The coefficients of a model expressed in this manner are called "psiweights." Once an ARIMA model is in psiweight form, we can easily determine the variance of the forecasts for that series.

In Input 2T.13 we request calculation of 40 PSIWEIGHTS. 2T automatically stores the psiweights it calculates and uses them for future analyses. These psiweights are available until replaced by another PSIWEIGHT paragraph or removed by an ERASE PSIWEIGHT paragraph. The CHECK paragraph asks 2T to print out the psiweights.

Next we erase the previous ARIMA model (for FEED, the predictor series) to make way for the output series CO_2. Then we define the CO_2 ARIMA model and the transfer function, and request a FORECAST for 25 cases. The forecasts shown in Output 2T.13 embody the prediction from the FEED series.

Input 2T.13
(add to the end of Input 2T.12)

```
PSIWEIGHTS   MAXPSI = 40.                                        /

CHECK        PSIWEIGHTS.                                         /

ERASE        MODEL.                                              /

ARIMA        VAR IS CO2. ARORDERS ARE '(1,2)'. CONSTANT./

INDEPENDENT  VAR IS FEED.     UPORDERS ARE '(3,4,5)'.
             TYPE IS RANDOM. SPORDERS ARE '(1)'.                 /

ESTIMATION   METHOD IS CLS.                                      /

FORECAST     CASES ARE 25.                                       /

END          /
```

Output 2T.13

```
PSIWEIGHTS        MAXPSI = 40.            /

FIRST CASE NUMBER TO BE USED      =              1
LAST CASE NUMBER TO BE USED       =            321                    [1]
  33 PSI-WEIGHTS ARE STORED.

CHECK             PSIWEIGHTS.             /

VARIABLE =FEED               VARIANCE =    0.3322E-01      PSI-WEIGHTS ARE

 1.0000  1.9750  2.5273  2.6216  2.3833  1.9723  1.5202  1.1099  0.7800
 0.5367  0.3691  0.2589  0.1883  0.1428  0.1120  0.0897  0.0722  0.0577
 0.0456  0.0355  0.0273  0.0208  0.0157  0.0118  0.0089  0.0067  0.0051
 0.0039  0.0030  0.0023  0.0018  0.0014  0.0010
```

—we omit a portion of the output—

```
FORECAST ON VARIABLE CO2           FROM TIME PERIOD   297

PERIOD              FORECASTS       ST.  ERR.       ACTUAL          RESIDUAL        [2]
 297                56.53432        0.24257
 298                56.06361        0.44367
 299                55.63726        0.60791
 300                55.26081        0.73494
 301                54.87080        0.87299
 302                54.47784        1.12349
 303                54.11636        1.49970
 304                53.81697        1.91591
 305                53.59564        2.29099
 306                53.45303        2.58505
 307                53.37874        2.79241
 308                53.35663        2.92651
 309                53.36940        3.00721
 310                53.40168        3.05305
 311                53.44149        3.07804
 312                53.48066        3.09136
 313                53.51438        3.09843
 314                53.54050        3.10224
 315                53.55872        3.10436
 316                53.56983        3.10557
 317                53.57519        3.10628
 318                53.57630        3.10671
 319                53.57456        3.10697
 320                53.57118        3.10714
 321                53.56707        3.10723

STANDARD ERROR = 0.242567              BY CONDITIONAL METHOD
```

[1] The psiweight model can have an infinite number of parameters; indeed, this is guaranteed if the model had at least one AR parameter. Yet the bounds of stationarity determine that as k increases, the psiweight associated with a_{t-k} decreases exponentially. Thus, as k increases, the effect of a_t decreases to a point where we may ignore it with no real loss of precision.

2T stops storing psiweights when they fall below 0.001, even if the MAXPSI limit has not been reached. In this example we requested 40 psiweights; however, 2T stored only the 33 psiweights with values of 0.001 or greater. By default, 2T stores 200 psiweights (if all have values of 0.001 or greater).

[2] The standard errors of the forecasts increase as we forecast further into the future. Because this is always the case, you may want to check on forecast accuracy by beginning your forecasts with the last few real observations in the output series.

Special Features

Interpreting the constant

The meaning of the constant depends upon the ARIMA model under consideration; specifically, it depends upon:

- whether the series has been differenced
- whether the model includes autoregressive (AR) parameters

For an undifferenced series, the constant C corresponds to the intercept term (usually the mean of the series):

$$Y_t = C + a_t$$

However, if the series has been differenced, the intercept is removed, because it is added to Y_t and subtracted as part of Y_{t-1}. The constant in a differenced series is the average difference between adjacent observations, i.e., the trend of the series. For single differencing, for example, the trend corresponds to the slope of a straight line fit through the series. Testing the significance of the constant in a singly differenced series is equivalent to testing whether a linear trend (with nonzero slope) is present in the data.

Similarly, if there are no AR parameters, the model is

$$Y_t = C + \Theta(B)a_t$$

If the model contains AR terms, however, the model is

$$\Phi(B)Y_t = C + \Theta(B)a_t$$

or

$$Y_t = \frac{C}{\Phi(B)} + \frac{\Theta(B)}{\Phi(B)}a_t$$

To simplify discussion, we define

$$\mu = \frac{C}{\Phi(B)}$$

The constant estimated in an ARIMA model containing AR parameters is C, **not** μ. Consider the CO_2 series in Example 2T.8. From the output of the ACF paragraph (not shown), we find that the mean of the series (C) is 53.10. When we specify CONSTANT in the ARIMA paragraph, the mean is estimated as 53.53 (Output 2T.8, backcasting method), which is, for practical purposes, equivalent to C. Thus, specifying CONSTANT yields C. On the other hand, μ is 53.53/(1.807 + .8582) = 1045.5 and is not an indicator of the time series level.

If you bear in mind the relationship between C and μ, you can avoid much frustration in attempting to duplicate published analyses that report μ, often called Θ_0.

Specifying a model

Specifying a model in the ARIMA paragraph is straightforward, if you remember that the algebra of ARIMA models allows both multiplicative and non-multiplicative components to be included. The ARORDERS and MAORDERS sentences reflect the model's structure. Parameters within the same parentheses are added, while components in separate parentheses are multiplied. Multiplicative models are the norm in ARIMA modeling, especially for seasonal models (see "Building a seasonal ARIMA model").

For example, suppose we want to model a time series consisting of a store's monthly sales. The sentence ARORDERS = '(1, 12)' specifies

$$\begin{aligned}(1 - \Phi_1 B - \Phi_{12}B^{12})Y_t &= C + a_t \\ Y_t - \Phi_1 Y_t - \Phi_{12}Y_{t-12} &= C + a_t \\ Y_t &= C + \Phi_1 Y_{t-1} + \Phi_{12}Y_{t-12} + a_t\end{aligned}$$

Thus, this February's sales would be predicted by this January's sales, last February's sales, and the current random shock. In contrast, ARORDERS = '(1),(12)' specifies

$$\begin{aligned}(1 - \Phi_1 B)(1 - \Phi_{12}B^{12})Y_t &= C + a_t \\ (1 - \Phi_1 B - \Phi_{12}B^{12} + \Phi_1\Phi_{12}\, BB^{12})Y_t &= C + a_t \\ Y_t - \Phi_1 Y_{t-1} - \Phi_{12}Y_{t-12} + \Phi_1\Phi_{12}\, Y_{t-13} &= C + a_t \\ Y_t &= C + \Phi_1 Y_{t-1} + \Phi_{12}Y_{t-12} - \Phi_1\Phi_{12}Y_{t-13} + a_t\end{aligned}$$

The multiplicative model would include predictions from this January's and last February's sales and from the current random shock, with an additional prediction ($\Phi_1\Phi_{12}Y_{t-13}$) from last January's sales.

Nonmultiplicative components consist of a single polynomial with the intervening terms suppressed. For example, the term ARORDERS = '(1, 4)' is best conceptualized as an AR(4) model, with Φ_2 and Φ_3 both constrained to equal zero.

A similar situation occurs in differencing. A DFORDERS = 2 statement specifies seasonal differencing with $s = 2$ (i.e., $(1 - B^2)$), **not** second-order differencing $(1-B)^2$. Second-order differencing is specified by DFORDERS = 1, 1. The difference between these two models is clear:

DFORDERS = 2. $(1 - B)Y_t = Y_t - Y_{t-2}$

DFORDERS = 1, 1. $(1 - B)(1 - B)Y_t = (1 - 2B + B^2)Y_t = Y_t - 2Y_{t-1} + Y_{t-2}$

Before you can specify a transfer function, intervention, or ARIMA time series model in 2T, you must put the model in the following form:

Output series = (transfer or intervention) + noise

or

∇Yt	=	$(U\nabla X_t)/S$	+	$(\Theta/\Phi)a_t$
ARIMA paragraph		INDEPENDENT paragraph		ARIMA paragraph

where Y_t = the output series at time t; X_t = the predictor (input) series or intervention component ($X_t = I_t$) at time t; and a_t = random noise ("shock") at time t.

Note that when you specify ∇ (DFORDERS) in the ARIMA paragraph, you must also include the same ∇ in the INDEPENDENT paragraph; 2T does not automatically specify it (see Intervention Analysis).

For ARIMA models (without intervention or transfer function), you can reduce the form to

$$\nabla Y_t = (\Theta/\Phi)a_t$$

To specify nonzero means (for undifferenced Y_t) or trends (for differenced Y_t), use CONSTANT in the ARIMA paragraph. If there is more than one X_t, or I_t, add additional INDEPENDENT paragraphs corresponding to the additional terms. Tables 2T.1 and 2T.2 give a translation of various model components into BMDP instructions.

Table 2T.1: ARIMA components

Component Name	Component	Expanded Polynomial	BMDP Instructions
Difference	∇	$(1 - B^c)(1 - B^c)\ldots$	DFORDER = $c_1, c_2, \ldots$
Autoregressive	Φ	$(1 - \Phi_1 B^c - \Phi_2 B^c - \ldots)$	ARORDER = '($c_1, c_2, \ldots$)'.
Autoregressive (multiplicative)	Φ	$(1 - \Phi_{11} B^c - \Phi_{12} B^c - \ldots)(1 - \Phi_{21} B^c - \Phi_{22} B^c - \ldots)$	ARORDER= '($c_{11}, c_{12}, \ldots$), ($c_{21}, c_{22}, \ldots$)'.
Moving Average	Θ	$(1 - \Theta_1 B^c - \Theta_2 B^c - \ldots)$	MAORDER= '($c_1, c_2, \ldots$)'.
Moving Average (multiplicative)	Θ	$(1 - \Theta_{11} B^c - \Theta_{12} B^c - \ldots)(1 - \Theta_{21} B^c - \Theta_{22} B^c - \ldots)$	MAORDER = '($c_{11}, c_{12}, \ldots$), ($c_{21}, c_{22}, \ldots$)'.
Mean			CONSTANT. (if no differencing is requested)
Trend			CONSTANT. (if differencing is requested)

Table 2T.2: Independent (transfer or intervention) components

Component Name	Component	Expanded Polynomial	BMDP Instructions
Difference	∇	$(1 - B^c)(1 - B^c) \ldots$	DFORDER = $c_1, c_2, \ldots$.
U polynomial	U	$(U_1 B^c + U_2 B^c + \ldots)$	UPORDER = $'(c_1, c_2, \ldots)'$.
U polynomial (multiplicative)	U	$(U_{11} B^c + U_{12} B^c + \ldots)(U_{21} B^c + U_{22} B^c + \ldots)\ldots$	UPORDER = $'(c_{11}, c_{12}, \ldots)\ (c_{21}, c_{22}, \ldots)'$.
S polynomial	S	$(1 - S_1 B^c - S_2 B^c - \ldots)$	SPORDER = $'(c_1, c_2, \ldots)'$.
S polynomial (multiplicative)	S	$(1 - S_{11} B^c - S_{12} B^c - \ldots)(1 - S_{21} B^c - S_{22} B^c - \ldots)\ldots$	SPORDER = $'(c_{11}, c_{12}, \ldots)\ (c_{21}, c_{22}, \ldots)\ldots'$.

***Note:** In the expanded polynomials in Tables 2T.1 and 2T.2, c_{ij} should be read as the jth constant in the ith polynomial. For example, c_{12} is the 2nd constant in the 1st polynomial.*

Running 2T in interactive and batch modes

Because building an ARIMA model usually requires several exploratory iterations you will probably want to run 2T interactively. When 2T encounters the slash at the end of a program-specific paragraph, it executes that paragraph immediately. This lets you interactively examine the results of your instructions, change some aspect of the model or ask for a diagnostic, and continue until you arrive at a satisfactory model.

Paragraphs beginning with a slash, such as the INPUT and VARIABLE paragraphs, are not executed interactively, since they are used to initially read in the data. It is best to execute these noninteractive paragraphs from the BMDP editor. In this way, you can correct the input instructions until the data are read in properly. But once the data are read, you may switch to a fully interactive mode in order to perform the exploratory model building. When in interactive mode, be sure to stop typing after the ending slash and press <return> in order to see the result before starting the next paragraph.

The **B** command is used to switch or "toggle" between the BMDP editor's command mode, the editor's input mode, and interactive (no editor) mode. Each time you type B and press return you will advance to the next mode, in the order stated. This is why the B command is called a toggle. Typing B the first time makes the switch from the editor's command mode to the editor's insert mode after the program executes the current input. Typing B a second time makes the switch from the editor's input mode to interactive mode after the program executes. Typing B again switches back to the editor's command mode (after execution).

You might want to run 2T in batch if, for example, you are almost sure your model is the correct one and you want to save time, or if you want to modify an existing version of a complete 2T run. (The BMDP.LOG file will contain a record of your last interactive run). To run 2T in batch, type all the instructions in an input file and then run the program as you would any other BMDP program. Be sure to type a slash at the end of every paragraph that follows the /END paragraph after the data, just as you would do in interactive mode.

Initial values for estimation

The initial value used in the ESTIMATION paragraph for the parameters in the polynomials AR, MA, U, and S is 0.1 or the value from the previous ESTIMATION paragraph. Erasing a model resets the initial values to 0.1.

2T Commands

The PRINT, SAVE, TPLOT, BLOCK, ACF, PACF, CCF, DIFFERENCE, ARIMA, INDEP, CHECK, ERASE, FILTER, ESTIMATION, PSIWEIGHT, and FORECAST paragraphs can be repeated in any order before the END paragraph. *Note that the paragraphs are executed one at a time, that the slash must come at the* ***end*** *of each paragraph, and that each paragraph must begin on a new line.* The END paragraph signals the end of the current analysis.

PRINT... /

Requests a data listing for the variables and cases specified. Optional. Note that you can control the format of a data listing by including a PRINT paragraph (with the slash **before** PRINT) with a FIELDS command before the first END paragraph (i.e., in the common BMDP instruction block). This feature allows you to specify the number of character positions, decimal places, and whether a field is alphanumeric. See Chapter 7 for a full explanation.

VARiables = list. `VAR = LOAN, MLOAN.`

State names or numbers of variables, residuals, or filtered series to be printed. Default: no variables printed.

CASEs = # list. `CASES = 35 TO 45.`

List sequence numbers of cases to be printed. Default: all cases printed.

SAVE ... /

Saves data in a BMDP file. Optional. See descriptions of CODE, UNIT, NEW, and CONTENT in Chapter 8. Forecasts may be saved by JOINING them with the original series (see FORECAST paragraph) before using the SAVE paragraph.

KEEP = list. `KEEP = CPZ.`

Identify the names or numbers of variables to be saved.

TPLOT... /

Plots data for one or more series, either in a common frame or in separate side by-side frames. Optional.

VARiables = list. `VAR IS DWORK.`

State names or numbers of the series (variables) to be plotted. Required for production of a plot.

CASEs = # list. `CASES = 1 TO 65.`

Sequence numbers of cases to be plotted. Default: all cases plotted.

MARK = #. `MARK = 10.`

Defines the interval of tick marks on the time axis. Default: 5.

MINimum = # list. `MIN = 100, 50.`

Sets lower limits for the variables to be plotted. Default: the minimum of each series.

MAXimum = # list. `MAX = 300, 150.`

Sets upper limits for the variables to be plotted. Default: the maximum of each series.

COMMon. `COMMON.`

Plots all variables in a single frame. If you omit COMMON or specify NO COMMON, the variables are plotted in separate frames.

SIZE = #. `SIZE = 60.`

Number of characters on the horizontal axis of each frame (that is, the axis that represents the variable being plotted, Y_t). The width specified includes the space for the vertical axis (that is, the axis that represents time). Thus SIZE = 60 may provide only 55 characters of horizontal space in the body of the plot and 5 characters for the vertical axis and its scale. Default: 45 (unless you specify LINESIZE in the PRINT paragraph).

SYMBols = character list. `SYMBOLS = J, F, M, A.`

Specify a sequence of characters to represent the variables or to represent cases for a single variable. If more than one variable is plotted in a frame, the first symbol represents the first variable, the second symbol represents the second variable, etc. The maximum number of symbols that may be used is 20. Default: A, B, C, ...

If only one variable is plotted, or a separate frame is used for each variable, the first symbol represents the first case, the second symbol represents the second case, etc. Default: A.

BLOCK... /

Defines blocks of complete data for analysis. Optional. Should be specified before the analysis paragraph (e.g., ACF, PACF, CCF, FILTER, ESTIMATION).

FIRST.
LAST.
LARGEST.
RANGE = #1, #2.
(one only)

`RANGE = 1, 60.`

Specify which block of observations is to be used. This specification remains in effect until changed by a later BLOCK paragraph. RANGE specifies a subset of a larger block. If the range has any missing data, 2T uses the largest block of complete observations within the range. Default: LARGEST.

LIST = list. `LIST = LOAN, PERIOD.`

Specifies names or numbers of variables to be scanned for missing data. 2T prints the range of case numbers for each block of complete cases and for the block of incomplete cases following it.

INDicate = variable. `INDICATE = MLOAN.`

Assigns a new binary missing value indicator variable that can be used for further study. The value of the new variable is 0 for any case having a missing value for any variable in the specified LIST, and 1 if data are complete for all listed variables. If LIST is not specified, 2T scans all variables currently being used.

ACF... /

Computes, prints, and plots sample autocorrelations of a series. Optional.

VARiable = variable. `VAR = ozone.`

Identifies the name or number of the series (variable) for which autocorrelations are to be computed.

DFORders = # list. `DFORDERS = 1, 4.`

Specify orders of differences to be performed before computing the ACF. **Example**: Differences taken at lags 1 and 4; i.e., $(1 - B)(1 - B^4)Y_t$. See "Specifying a model" in Special Features.

MAXLag = #. `MAXLAG = 25.`

Sets the maximum lag of the autocorrelations to be computed. The appropriate number depends upon the length (n) of the series; Box and Jenkins (1976) recommend $n/4$ as an upper limit. Default: 36.

PLOT. `NO PLOT.`

Plots the sample autocorrelations. Default: PLOT.

PRINt. `NO PRINT.`

Prints the sample autocorrelations. Default: PRINT.

CI. `NO CI.`

Plots the confidence interval of the sample autocorrelations (two standard deviations on either side of the estimate). Default: CI.

LBQ. `NO LBQ.`

Computes the Ljung-Box Q statistic. Default: NO LBQ.

PACF... /

Computes, prints, and plots sample partial autocorrelations of a series. Commands are identical to those for the ACF paragraph, except that LBQ is not available for the PACF.

See the ACF paragraph.

CCF... /

Computes, prints, and plots sample cross-correlations and transfer function weights between two series.

VARiables = variable, variable. `VAR = RX, RY.`

Specify names or numbers of the two series for which cross-correlations are to be computed. Note that the two series must have the same number of observations.

MAXLag = #.
PLOT.
PRINt.
CI.

See the ACF paragraph.

DIFFerence... /

A differencing polynomial can be specified in the ACF, PACF, ARIMA, and INDEPENDENT paragraphs. Use this paragraph if you want to store the series or if your instructions include a CCF paragraph.

OLD = variable. `OLD = WORK.`

Identifies the name or number of the series to be differenced.

NEW = variable. `NEW = DWORK.`

Assigns a name (not number) to the variable where the differenced series is to be stored. This variable may be new.

DFORder = # list. `DFORDER = 1, 4.`

Specifies the order of differences to be performed. Default: 1. See "Specifying a model" in Special Features.

FILTer... /

Uses the model specified in the ARIMA and INDEPENDENT paragraphs to filter a time series.

VARiable = variable. `VAR = ozone.`

Identifies the name or number of the series to be filtered. Default: the name in the last ARIMA paragraph.

RESIduals = variable. `RESIDUALS = rozone.`

Assigns a name to the variable where the filtered series is to be stored. Once stored, the residuals may be SAVED. Stored residuals remain in memory and consume space until the analysis is FINISHED. If you do not need to retain a set of residuals, you can reduce memory consumption by using the same variable name for the next set.

ARIMa... /

Specifies the ARIMA component of a time series model. The model used in the examples below is

$$(1-B)(1-B^4)(1-0.4B)Y_t = (1-0.3B)(1-0.2B^4)a_t$$

For an explanation of ARIMA notation and for instructions on specifying a model, see Special Features.

VARiable = variable. `VARIABLE = ozone.`

Identifies the name or number of the series (variable, Y_t) to be analyzed.

DFORders = # list. `DFORDERS = 1, 4.`

Specify orders of differences in the difference polynomial.

ARORders = '(#,..., #),...,(#,...,#)'. `ARORDERS = '(1)'.`

Specify orders of parameters in the AR polynomial.

ARVAlues = # list. `ARVA = 0.4.`

Specify values (or initial values) of the parameters in the AR polynomial. Default: 0.1 or the value from the latest ESTIMATION.

MAORders = '(#,..., #),...,(#,...,#)'. `MAORDERS = '(1), (4 )'.`

Specify orders of parameters in the MA polynomial.

MAVAlues = # list. `MAVALUES = 0.3. 0.2.`

Specify values (or initial values) of the parameters in the MA polynomial. Default: 0.1 or the value from the latest ESTIMATION.

CONStant. `CONSTANT.`

When the model does not include differencing, CONSTANT adds a mean parameter to the model. When the model includes differencing to remove trend, CONSTANT adds the trend. When the model includes differencing to remove stochastic drift, the CONSTANT is zero and can therefore be omitted. Default: NO CONSTANT. See "Interpreting the constant" in Special Features.

CVAL = #. `CVAL = 2.5.`

Specifies an initial value for the constant. Default: zero.

CENTERED. `CENTERED.`

Subtracts the mean from each observation in the series, so the series will be centered at zero for analysis.

INDEpendent... /

Specifies an intervention or transfer function component. The INDEPENDENT paragraph must follow an ARIMA paragraph.

When the intervention X_t (transfer function)

$$\frac{0.5B^3 + 0.4B^4}{1 - 0.3B} X_t$$

is added to the model illustrated above for the ARIMA paragraph, the complete model becomes

$$(1 - B)(1 - B^4)(1 - 0.4B)Y_t = \frac{0.5B^3 + 0.4B^4}{1 - 0.3B}(1 - B)(1 - B^4)(1 - 0.4B)X_t$$

$$+ (1 - 0.3B)(1 - 0.2B^4)a_t$$

The program automatically applies the autoregressive polynomial previously specified in the ARIMA paragraph to the numerator of the transfer function component. However, to apply the difference polynomial to the transfer function, you must specify it in the INDEPENDENT paragraph.

Note that the alternative form found in many texts for the above model is

$$Y_t = \frac{(0.5B^3 + 0.4B^4)}{(1 - 0.3B)} X_t + \frac{(1 - 0.3B)(1 - 0.2B^4)}{(1 - B)(1 - B^4)(1 - 0.4B)} a_t$$

VARiable = variable. `VAR = loan.`

Required with INDEP paragraph. Identifies the name of the intervention variable.

DFORders = # list. `DFORDERS = 1, 4.`

Specify orders of differences in the difference polynomial.

SPORders = '(#,....,#)...,(#...,#)'. `SPORDERS = '(1)'.`

Specify parameter orders in the *S* polynomial. See Examples 2T.6 and 2T.8.

SPVAlues = # list. `SPVALUES = 0.3.`

Specify values (or initial values) of the S polynomial parameters. Default: 0.1 or the value from the latest ESTIMATION.

UPORders = '(#,..., #),...,(#,...,#)'. `UPORDERS = '(3, 4)'.`

Specify parameter orders in the *U* polynomial. See Examples 2T.6 and 2T.8.

UPVAlues = #list. `UPVALUES = 0.5, 0.4.`

Specify values (or initial values) of the *U* polynomial parameters. Default: 0.1 or the value from the latest ESTIMATION.

TYPE = (*one only*) `TYPE = RANDOM.`
BINARY.
NONRANDOM.
RANDOM.

Defines the type of independent variable: BINARY for a binary variable, NONRANDOM for other nonstochastic variables, and RANDOM for stochastic variables. Default: RANDOM.

REPLace. `REPLACE.`

Replaces the model for the component with the same variable name. Default: NO REPLACE.

CENTered. `CENTER.`

Removes the mean of the series. Default: NO CENTERED.

ESTImation... /

Estimates the parameters of a time series model specified in the ARIMA and INDEPENDENT paragraphs.

RESIduals = variable. `RESIDUALS = RWORK.`

Assigns a name to the variable where residuals for the output series are to be stored. Once stored, the residuals may be SAVED. Stored residuals remain in memory and consume space until the analysis is FINISHED. If you do not need to retain a set of residuals, you can reduce memory consumption by using the same variable name for the next set.

METHod = *(one only)* `METHOD = CLS.`
CLS.
BACKCASTING.

Specifies the estimation method. CLS uses the conditional least squares method only; BACKCASTING uses the backcasting method only. Default: both.

ONEStep = variable. `ONESTEP = OWORK.`

Assigns a name to the variable where one step-ahead forecasts are to be stored.

MAXIt = #. `MAXIT = 12.`

Defines the maximum number of iterations in the nonlinear estimation. Default: 20.

SCRSs = #. `SCRSS = 0.0001.`

Identifies a stopping criterion for the nonlinear estimation in terms of residual sum of squares (RSS). Default: 0.00005.

SCEStimates = #. `SCESTIMATES = 0.001.`

Identifies a stopping criterion for the nonlinear estimation in terms of parameter estimates. Default: 0.0001.

PCLS. `NO PCLS.`

Prints the results under conditional least squares estimation. If you specify NO PCLS, 2T displays the results of the backcasting method only. Default: PCLS.

PITEration. `PITE.`

Prints the parameter estimates at each iteration of nonlinear estimation. Default: NO PITE.

PCORrelation. `PCOR.`

Prints the correlation matrix of the parameter estimates. Default: NO PCORRELATION.

TOLerance = #. `TOL = 0.00001.`

Specifies tolerance limits for matrix inversion. Default: 0.001.

PSIWeight... /

Computes psiweights and variance of the time series specified in the ARIMA paragraph.

MAXPsi = #. `MAXPSI = 80.`

Defines the maximum number of psiweights to be computed. 2T stops storing psiweights when they fall below 0.001, even if the MAXPSI limit has not been reached. Default: 200.

FOREcast... /

Predicts future values of a time series.

CASEs = #. `CASES = 40.`

Identifies the number of forecasts to be made. Default: 36.

STARt = #. `START = 101.`

Assigns the starting point of the forecasts. Default: the last time period of the series plus one.

RMS = #. `RMS = 0.54.`

Specifies the residual mean square of the series to be forecast. Default: 2T computes the RMS by the conditional method.

JOIN. `JOIN.`

If *X* is the variable that contains the current series, stating JOIN adds ("joins") any forecast values to X. Default: NO JOIN.

JVAR = variable. `JVAR = VA.`

If *X* is the variable that contains the current series, specifying JVAR (with or without JOIN) stores joined forecast and original values in variable VA, leaving *X* unchanged. Thus the number of variables is increased in the data matrix. Default: joined values are not saved.

FVAR = variable. `FVAR = VB.`

If *X* is the variable that contains the current series, specifying FVAR (with or without JOIN) adds any forecast values to *X* and copies only the forecast values to VB. The original (non-forecast) values in *X* are not copied to VB. Stating both JVAR and FVAR works as above for VA and VB and leaves *X* unchanged. Default: forecast values are not saved.

CHECK... /

Prints the specified time series model or computed psiweights.

MODEl. `MODEL.`

Prints the specified time series model.

PSIWeight. `PSIWEIGHT.`

Prints the computed psiweights.

ERASE... /

Erases the time series model or the computed psiweights.

MODEl. `MODEL.`

Erases the most recently specified time series model. Required if you want to use more than one model.

PSIWeight. `PSIW.`

Erases the computed psiweights.

Order of Instructions

■ indicates required paragraph

```
■ / INPUT
  / VARIABLE
  / TRANSFORM
  / PRINT
  / SAVE
■ / END
  data
    PRINT ... /
    SAVE ... /
    TPLOT ... /
    BLOCK .... /
    ACF ... /
    PACF ... /                 Repeat for
    CCF ... /                  subproblems
    DIFFERENCE ... /
    FILTER ... /
    ARIMA ... /
    INDEPENDENT ... /
    ESTIMATION ... /
    PSIWEIGHT ... /
    FORECAST ... /
    CHECK ... /
    ERASE ... /
    END /
    FINISH /
```

Repeat for additional problems. See Multiple Problems, Chapter 10. (applies to the block from / INPUT through END /)

Summary Table for Commands Specific to 2T

	Paragraphs Commands	Defaults	Multiple Problems	See
	PRINT... /			
▲	VARiables = list.	none	–	2T.4
	CASEs = # list.	all cases	–	2T.4
	SAVE... /			
▲	KEEP = list.	none	–	Cmds
	TPLOT... /			
▲	VARiable = list.	none	–	2T.1
	CASEs = # list.	all cases	–	Cmds
	MARK = #.	5	–	Cmds
	MINimum = # list.	series minimum	–	Cmds
	MAXimum = # list.	series maximum	–	Cmds
		Plot in one common frame		
	COMMon.	NO COMMon.	–	Cmds
	SIZE = #.	45	–	2T.1
▲	SYMBols = character list.	A,B,C, ...	–	2T.1

Summary Table (continued)

	Paragraphs Commands	Defaults	Multiple Problems	See
	Plot in separate frames			
	COMMon.	NO COMMon.	–	Cmds
	SIZE = #.	depends on # of frames	–	Cmds
▲	SYMBols = character list.	A	–	Cmds
	BLOCK... /			
	(one only) FIRST, LAST, LARGEST, *or* RANGE = #1, #2.	LARGEST.	–	2T.4
▲	LIST = list.	all used variables	–	2T.4
▲	INDIcate = variable.	none	–	2T.4
	ACF... /			
▲	VARiable = variable.	none	–	2T.1
▲	DFORders = # list.	none	–	2T.1
▲	MAXLag = #.	36	–	2T.1
	PLOT.	PLOT.	–	Cmds
	PRINt.	PRINT.	–	Cmds
	CI.	CI.	–	Cmds
▲	LBQ.	NO LBQ.	–	2T.2
	PACF... /			
▲	VARiable = variable.	none	–	2T.1
	DFORders = # list.	none	–	2T.1
▲	MAXLag = #.	36	–	2T.1
	PLOT.	PLOT.	–	Cmds
	PRINt.	PRINT.	–	Cmds
	CI.	CI.	–	Cmds
	CCF... /			2T 8
▲	VARiable = variable, variable.	none	–	
▲	MAXLag = #.	36	–	2T.8
	PLOT.	PLOT.	–	Cmds
	PRINt.	PRINT.	–	Cmds
	CI.	CI.	–	Cmds
	DIFFerence... /			
▲	OLD = variable.	none	–	Cmds
▲	NEW = variable.	none	–	Cmds
▲	DFORder = # list.	1	–	Cmds
	FILTer... /			
	VARiable = variable.	VAR in ARIMA	–	2T.8
	RESIduals = variable.	none	–	2T.8
	ARIMa... /			
▲	VARiable = variable.	none	–	2T.2
▲	DFORders = # list.	none	–	2T.2
▲	ARORders = '(#,...,#), ...,(#,...,#)'.	none	–	2T.4
	ARVAlues = # list.	1	–	Cmds

Summary Table (continued)

	Paragraphs Commands	Defaults	Multiple Problems	See
	MAORders = '(#,...,#), ...,(#,...,#)'.	none	–	2T.2
	MAVAlues = # list.	.1	–	Cmds
▲	CONStant.	NO CONSTANT.	–	2T.8
	CVAL = #.	0	–	Cmds
	CENTERED.	NO CENTERED.	–	Cmds
	INDEpendent... /			
▲	VARiable = variable.	none	–	2T.6
▲	DFORders = # list.	none	–	2T.6
▲	SPORders = '(#,...,#), ...,(#,...,#)'.	none	–	2T.6
	SPVAlues = # list.	.1	–	Cmds
▲	UPORders = '(#,...,#), ...,(#,...,#)'.	none	–	2T.6
	UPVAlues = # list.	.1	–	Cmds
	TYPE = (*one only*) BINARY, NONRANDOM, RANDOM.	RANDOM	–	2T.6
	REPLace.	NO REPLACE.	–	Cmds
	CENTered.	NO CENTERED.	–	Cmds
	ESTImation... /			
▲	RESIduals = variable.	none	–	2T.2
▲	METHod = (*one only*) CLS, BACKCASTING.	both	–	2T.4
	ONEStep = variable.	none	–	Cmds
	MAXIt = #.	20	–	Cmds
	SCRSs = #.	.00005	–	Cmds
	SCEStimates = #.	.0001	–	Cmds
	PCLS.	PCLS.	–	Cmds
	PITEration.	NO PITE.	–	Cmds
▲	PCOR.	NO PCOR.	–	2T.2
	TOLerance = #.	.001	–	Cmds
	PSIWeight... /			
	MAXPsi = #.	200	–	2T.13
	FOREcast... /			
▲	CASEs = #.	36	–	2T.3
▲	STARt = #.	$N + 1$	–	2T.3
	RMS = #.	computed	–	2T.3
▲	JOIN.	NO JOIN.	–	2T.4
	JVAR = VA.	none	–	Cmds
	FVAR = VB.	none	–	Cmds
	CHECK.../			
▲	MODEl.	none	–	2T.2
	PSIWeight.	none	–	Cmds
	ERASE... /			
▲	MODEl.	none	–	2T.6
	PSIWeight.	none	–	Cmds

Key: ▲ Frequently used command or paragraph
– Default reassigned

2V

Analysis of Variance and Covariance with Repeated Measures

2V performs analysis of variance or covariance for a wide variety of fixed-effects and repeated-measures designs. You can analyze models that have grouping factors, within factors, or both. Grouping factors are also called between-groups or whole-plot factors. Within-subjects factors are also called trial, split-plot, repeated measures, or simply within factors. The grouping and the within factors must be crossed (not nested) in 2V. Group sizes may be unequal for combinations of grouping factors, but each subject must have a response for every combination of within factors. You can analyze both complete and incomplete fixed-effects factorial designs, including Latin square designs, incomplete block designs, and fractional factorial designs. When there are within factors, 2V will perform an orthogonal decomposition of the within factors in addition to the overall analysis of variance. The program can also perform contrasts over within factors.

Program 9D serves as a helpful adjunct to 2V; it can provide plots of group means and means of repeated measures, as well as repeated measures plots for individual subjects. For a fixed-effects two-way analysis of variance with detailed data screening, see program 7D. For mixed models with equal cell sizes, see program 8V; for unbalanced mixed models, see 3V. For a multivariate as well as a univariate approach to repeated measures analysis of variance, see program 4V. When some subjects in a repeated-measures design have data missing, see program 5V. 5V also allows alternative structures for the within-subjects covariance matrix. See Volume 2 for descriptions of 3V, 4V, 5V, and 8V.

In this chapter, we present examples of a variety of designs for the analysis of variance and covariance with repeated and nonrepeated factors. (For a general introduction to these subjects see Cox and Snell, 1981; Dunn and Clark, 1987; and Kirk, 1982.) One of the examples may be similar to your design. If so, you may want to skip the discussion of the other examples. We describe several examples in detail to explain the sums of squares and test statistics computed by 2V. If your example has any grouping factors, we recommend that you read

Example 2V.1, and if it has any within factors we recommend that you read the introduction to repeated measures preceding Example 2V.6 and 2V.7. The latter two examples both illustrate the same design with one grouping and one within factor; in 2V.7, results for an orthogonal decomposition are added.

Where to Find It

Examples

Special Features

Fixed Effects Factorial Designs (Grouping Variables Only)

2V performs an analysis of variance or covariance on fixed effects factorial designs with one or more grouping factors. The grouping factors must be crossed and not nested. You can obtain the correct sums of squares for nested factors by addition (see Dunn and Clark, 1987, for examples of this addition), but the desired *F* tests are not computed.

Analysis of variance is used to test null hypotheses about group means. When there are two or more grouping factors, 2V tests null hypotheses about equality of main effects for each factor, and about interactions between factors (that is, effects of combinations of factors not predictable by summing the main effects for each factor).

To introduce factorial designs, we borrow the two-way analysis of variance example from Chapter 7D (Example 7D.3). The effects of sex and education on average income are assessed using data collected in a community survey. The design has a 2 x 4 structure: 2 levels of sex and 4 levels of education (high school dropouts, high school graduates, college dropouts, and college graduates). Our hypotheses are the same as those tested in Example 7D.3: there are no differences in average income (1) between males and females and (2) between groups of respondents with varying levels of education. We also test a third hypothesis that the effect of educational level on income is the same for males as for females; i.e., there is no interaction between sex and education.

Using notation μ_{ij} to indicate the true underlying mean of the cell for sex i and educational level j, the hypotheses above also can be written as:

$$H_1: \mu_{11} + \mu_{12} + \mu_{13} + \mu_{14} = \mu_{21} + \mu_{22} + \mu_{23} + \mu_{24}$$
$$H_2: \mu_{11} + \mu_{21} = \mu_{12} + \mu_{22} = \mu_{13} + \mu_{23} = \mu_{14} + \mu_{24}$$
$$H_3: \mu_{11} - \mu_{21} = \mu_{12} - \mu_{22} \text{ , etc.}$$

To interpret an interaction between factors, it is helpful to look at a plot of the cell means (see program 9D for miniplots of cell means).

Figure 2V.1 shows three hypothetical outcomes of the income study. Average income is plotted on the vertical axis of each frame, while level of education is on the horizontal axis. Incomes for males and females are denoted by m and f, respectively. In plot (a) there is no interaction; the difference in average income between males and females is about the same for each of the four educational groups. In plot (b), there is a strong interaction. For the poorly educated groups there is little difference in average income between males and females. As education level increases, the differences between male and female means becomes larger. In plot (c), there also may be an interaction (the significance depends on the within-cell variability). For example, as we move from education level 1 to 2, the change in average income for the males appears much larger than that for the females, etc.

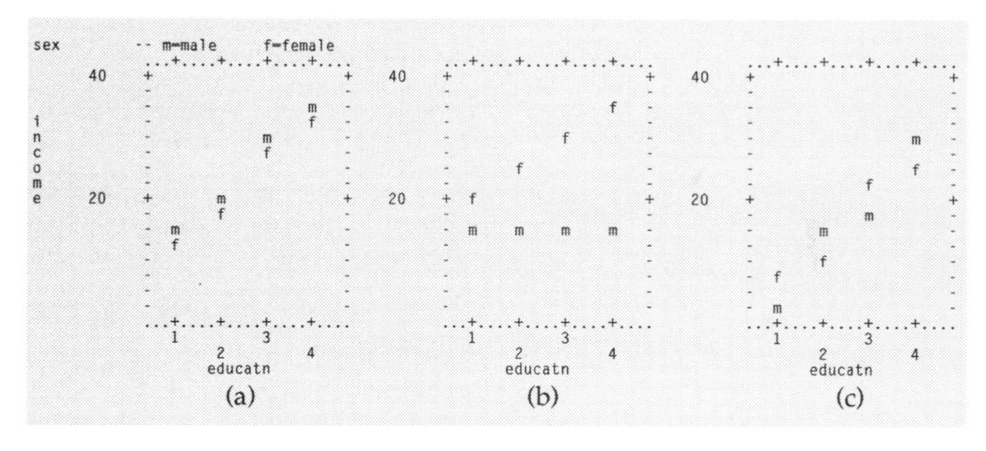

Figure 2V.1 Miniplots of cell means for three hypothetical situations: (a) no interaction between sex and education; (b) and (c) interactions present

The hypotheses tested by 2V are not dependent on the group sizes, so the number of subjects in each group need not be equal. See Herr (1986) for a discussion of the different approaches to hypothesis testing with unequal n. In 2V we use what Herr calls the standard parametric model for the analysis of unbalanced factorial designs. Computationally, the sums of squares used in testing these hypotheses are obtained by taking the difference in residual sums of squares for regression models fit with and without specific terms. (See Appendix B.20 for more details, and program 4V, Volume 2, for tests of other hypotheses about differences between means.)

The analysis of variance assumes that the responses/observations within each of the treatment groups are normally distributed with a common variance. You can use the histograms in program 7D to study the within-cell distributions and the Levene test to identify unequal variances. In general, the F test is reasonably robust to nonnormality, provided that the shapes of the distributions are similar. The F test is also robust to moderate violations of the assumption of homogeneity of variance. However, when the variances and the sample sizes are unequal, tests of significance may be affected markedly, and it may be necessary to transform the data or use robust procedures (see program 7D for selecting transformations and robust analysis of variance tests).

We include five examples in this section. The first is a simple two-way analysis of variance, with three levels of each grouping factor. The second is an analysis of variance with two covariates. The third is a Latin square design. The fourth is an incomplete block design, and the fifth is a fractional factorial design.

Example 2V.1 Analysis of variance with two grouping factors

For our first example, we use 2V to analyze a two-way factorial design described in Kirk (1982, pp. 353–359). A large metropolitan city agency is evaluating a human relations course for new police officers. The agency wants to find out whether the amount of human relations training given to the officers, and the type of neighborhood to which they are assigned, affect their attitudes toward members of minority groups. The dependent variable is a test score measuring the officers' attitude toward minority groups (higher scores indicate a positive attitude). The first grouping variable is the officers' assigned beat: upper-, middle-, or lower-class neighborhood. The second is the amount of human relations training they received: 5, 10, or 15 hours. If we name the grouping factors beat and training, we have the following 3 x 3 design:

Figure 2V.2
A 3 x 3 design

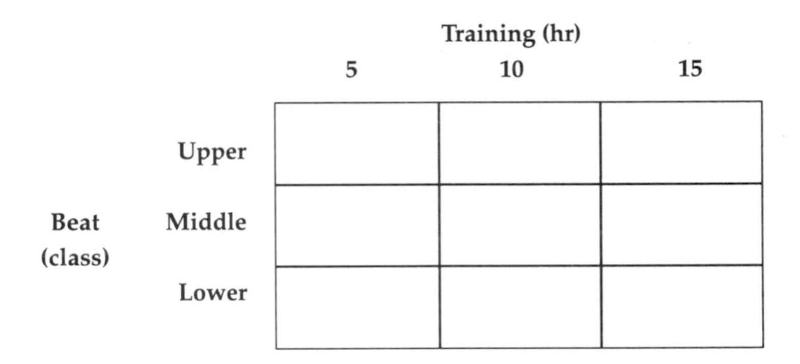

The formal model can be written as

$$y_{ijk} = \mu + \alpha_i + \gamma_j + (\alpha\gamma)_{ij} + \varepsilon_{ijk}$$

where α_i signifies the fixed effect of beat (inferences will be made only to these three types of beats), γ_j signifies the fixed effect of training time, and $(\alpha\gamma)_{ij}$ the possible interaction of beat and training time. ε_{ijk} is the random deviation of y_{ijk} from the population mean response for each combination of beat and training time and is assumed to be normally distributed.

We organize our data by writing one record for each officer with four entries: an ID number, codes to indicate type of neighborhood and length of training received, and attitude score (for a complete listing of the data, see Appendix D).

In Input 2V.1, the INPUT and VARIABLE paragraphs are common to all BMDP programs and are described in Chapters 3 and 4. The FILE command tells the program where to find the data and is used on systems like VAX and the IBM PC. For IBM mainframes see UNIT, Chapter 3. The GROUP paragraph (see Chapter 5) identifies the grouping factors and assigns names to the levels of the factors, in this case for BEAT and TRAINING.

A DESIGN paragraph is required in 2V to specify the dependent variables to be analyzed. Here the DEPENDENT variable is ATTITUDE toward minority groups.

Input 2V.1

```
/ INPUT      VAR = 4.
             FORMAT = FREE.
             FILE = 'police.dat'.

/ VARIABLE   NAMES = id, beat, training, attitude.

/ GROUP      VARIABLE = beat, training.
             CODES(beat) = 1, 2, 3.
             NAMES(beat) = upper, middle, lower.
             CODES(training) = 5, 10, 15.

/ DESIGN     DEPENDENT = attitude.

/ END
```

Output 2V.1

```
 CASE    1        2        3       4                                          [1]
  NO. ID       BEAT     TRAINING ATTITUDE
----- -------- -------- -------- --------
    1     1.00 UPPER        5.00    24.00
    2     2.00 UPPER        5.00    33.00
    3     3.00 UPPER        5.00    37.00
    4     4.00 UPPER        5.00    29.00
    5     5.00 UPPER        5.00    42.00
    6     6.00 UPPER       10.00    44.00
    7     7.00 UPPER       10.00    36.00
    8     8.00 UPPER       10.00    25.00
    9     9.00 UPPER       10.00    27.00
   10    10.00 UPPER       10.00    43.00

NUMBER OF CASES READ. . . . . . . . . . . . . .          45                   [2]

   VARIABLE        STATED VALUES FOR             GROUP CATEGORY     INTERVALS   [3]
  NO.   NAME    MINIMUM MAXIMUM MISSING    CODE  INDEX   NAME       .GT.   .LE.
  ---- --------  ------- ------- -------  ------ ----- --------   ------ ------

   2  BEAT                                 1.000    1  UPPER
                                           2.000    2  MIDDLE
                                           3.000    3  LOWER

   3  TRAINING                             5.000    1  *5
                                           10.00    2  *10
                                           15.00    3  *15

*** N O T E *** THE PROGRAM CREATED CATEGORY NAMES BEGINNING WITH ASTERISKS(*).
-------------------------------------------------------------------------------

GROUPING VARIABLE. . . BEAT
                               CATEGORY    FREQUENCY
                               --------    ---------
                               UPPER           15
                               MIDDLE          15
                               LOWER           15

GROUPING VARIABLE. . . TRAINING
                               CATEGORY    FREQUENCY
                               --------    ---------
                               *5              15
                               *10             15
                               *15             15

DESCRIPTIVE STATISTICS OF DATA                                                [4]
----------- ---------- -- ----

 VARIABLE    TOTAL          STANDARD  ST.ERR    COEFF SMALLEST LARGEST
 NO. NAME     FREQ.  MEAN     DEV.    OF MEAN   OF VAR  VALUE   VALUE    RANGE

   1 ID         45  23.000  13.134  1.9579   .57104  1.0000  45.000  44.000
   4 ATTITUDE   45  35.000  10.892  1.6237   .31120  10.000  64.000  54.000
```

—we omit portions of output here—

```
 DESIGN SPECIFICATIONS                                                        [5]
 ---------------------
          GROUP =   2   3
          DEPEND =  4

  GROUP STRUCTURE          [6]

  BEAT      TRAINING     COUNT
  upper     *5             5
  upper     *10            5
  upper     *15            5
  middle    *5             5
  middle    *10            5
  middle    *15            5
  lower     *5             5
  lower     *10            5
  lower     *15            5
```

```
 CELL MEANS           FOR  1-ST DEPENDENT VARIABLE     [7]
 -------------------
                                                                                                              MARGINAL
BEAT     -  upper      upper       upper       middle      middle      middle      lower       lower       lower
TRAINING-   *5         *10         *15         *5          *10         *15         *5          *10         *15

ATTITUDE    33.00000   35.00000    38.00000    30.00000    31.00000    36.00000    20.00000    40.00000    52.00000    35.00000

COUNT              5          5           5           5           5           5           5           5           5          45
STANDARD DEVIATIONS  FOR  1-ST DEPENDENT VARIABLE
-------------------
BEAT     -  upper      upper       upper       middle      middle      middle      lower       lower       lower
TRAINING-   *5         *10         *15         *5          *10         *15         *5          *10         *15

ATTITUDE     6.96419    8.80341     9.51315     6.96419     6.96419     9.51315     7.51665     6.20484     7.96869


A N A L Y S I S   O F   V A R I A N C E  FOR   THE   1-ST DEPENDENT VARIABLE                 [8]
---------------------------------------
THE TRIALS ARE REPRESENTED BY THE VARIABLES:
ATTITUDE

        SOURCE                         SUM OF        D.F.        MEAN          F        TAIL
                                       SQUARES                   SQUARE                 PROB.

        MEAN                         55125.00000       1      55125.00000   882.00    0.0000
        BEAT                           190.00000       2         95.00000     1.52    0.2324
        TRAINING                      1543.33333       2        771.66667    12.35    0.0001
        BT                            1236.66667       4        309.16667     4.95    0.0028
  1     ERROR                         2250.00000      36         62.50000
```

[1] Prints the first ten cases of data. Group names are printed if specified.

[2] Number of cases read. Only cases containing acceptable values for all variables specified in the GROUP paragraph are used. An acceptable value is one that is not missing or out of range. In addition, if CODES are specified for any GROUPING factor, a case is included only if the value of the GROUPING factor is equal to a specified CODE.

[3] The grouping information is printed as interpreted from the codes and names specified in the GROUP paragraph.

[4] This table gives all descriptive statistics divided by group. This table includes:

- Mean
- Standard Deviation
- Standard Error of the Mean
- Coefficient of Variation
- Smallest value
- Largest value
- Range

[5] Interpretation of the DESIGN and GROUP paragraphs for analysis. This panel verifies that there are two grouping variables (variables 2 and 3) and one dependent variable (variable 4).

[6] Frequency of observations in each cell. The panel shows that the number of subjects in each of the nine groups is five.

[7] The mean, frequency, and standard deviation for each cell of the dependent variable. Higher values are associated with more positive feelings toward minority groups. You can see that within each of the three beats, groups with more hours of human relations training have higher scores. Further, the highest attitude scores are found in groups in the lower-class beat with the two higher levels of human relations training. This suggests that there may be an interaction between the police beat and length of training.

[8] The analysis of variance table, with the sum of squares, degrees of freedom, mean square, *F* ratio, and associated *p*-value for each effect in the design. In this case, both the main effect of length of training and the interaction are highly significant (*p*-values much less than .01). When the interaction is significant, caution must be exercised in making statements about the significance of a main effect. In order to understand what is happening, it is wise to study the interaction by plotting cell means. You may find that you need to use the levels of one factor to stratify the analysis, or you can request tests of simple effects in program 4V.

In Figure 2V.3 below, we use 9D to plot cell means for the attitude score on the vertical axis, and length of training on the horizontal axis. Symbols indicate the type of police beat: upper class (U); middle class (M); and lower class (L). The average attitude scores for officers assigned to middle- and upper-class beats tends to change little as the length of training increases, whereas the scores of those assigned to the lower-class beat increases markedly with training. The significant interaction is a result of this difference.

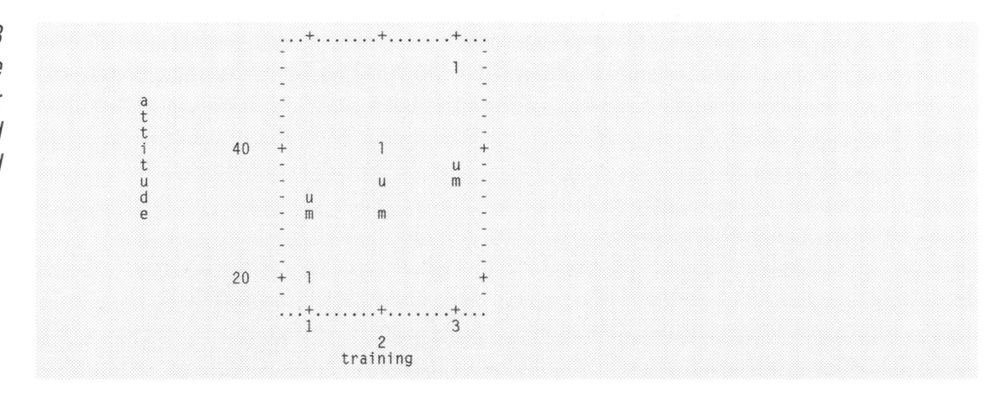

Figure 2V.3 Attitude scores by course length for three police beats: lower (l), middle (m), and upper (u) class neighborhood

Example 2V.2 Analysis of variance with two grouping factors and two covariates

When a background variable or pretest (independent variable or covariate) is strongly related to the outcome variable (dependent variable), an analysis of covariance may increase the precision of comparisons between treatments by reducing the within-group variability in the outcome variable due to the influence of the covariates, and by adjusting for the effects of any initial imbalances between the groups.

In addition to the usual assumptions required for an analysis of variance, an analysis of covariance rests on the additional assumptions that:

1. within each group, the dependent variable has a **linear** relationship with the covariates,
2. the slope of the regression line for each covariate is the same in each group (i.e., the lines are parallel).

These assumptions can be checked by plotting the dependent variable against every covariate for each group using program 6D, and by a direct test of the assumption of homogeneity or parallelism of regressions using program 1V.

In this example, we use 2V to analyze the Exercise data (see Appendix D). These data include pulse rates before and after exercise for 40 subjects. We want to find out whether post-exercise pulse rates differ depending on gender and smoking status after adjusting for age and pre-exercise pulse rate. If we name our grouping factors sex and cigarettes, we have a 2 x 2 design:

Figure 2V.4
A 2 x 2 design

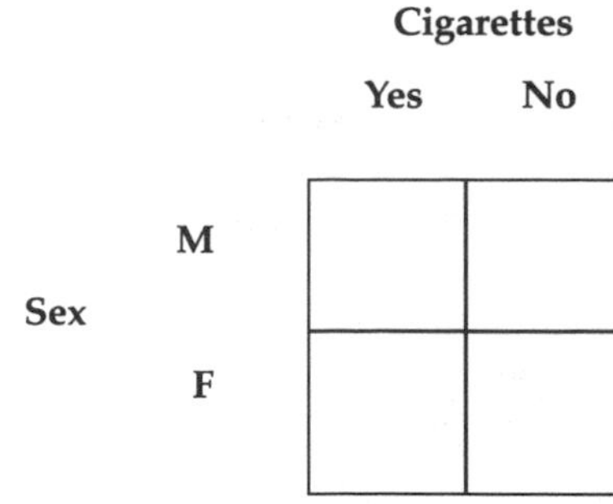

Input 2V.2 shows the BMDP instructions for this analysis. Each record in the file EXERCISE.DAT contains an ID number, codes for sex and smoking status, age, pre-exercise pulse rate, post-exercise pulse rate, and name. We also use a TRANSFORM paragraph (see Chapter 6) to correct an erroneous value in the data (a post-exercise pulse rate was recorded as 265 instead of 165). Our grouping variables are sex and smoking status, which are specified and defined in the GROUP paragraph (SEX and CIGARETS). The CODES 1 and 2 are used for both variables. We use the variable name cigarets instead of smoking because 2V uses the first letter of the name to label the output. Variables with the same first letter are assigned unique labels by the program.

In the DESIGN paragraph we identify PULSE_2 as the measurement analyzed (the DEPENDENT variable). The COVARIATE command identifies the covariates AGE and PULSE_1. Last, we include a PRINT paragraph to request the predicted values for each case, along with the residuals (the difference between each case's actual and predicted post-exercise pulse rates).

Input 2V.2

```
/ INPUT        VAR = 8.
               FORMAT= FREE.
               FILE = 'exercise.dat'.

/ VAR          NAMES = id, sex, cigarets, age, pulse_1,
                       pulse_2, name1, name2.
               LABEL = name1, name2.

/ TRANSFORM    IF (pulse_2 EQ 265) THEN pulse_2 = 165.

/ GROUP        VARIABLE = sex, cigarets.
               CODES(sex, cigarets) = 1, 2.
               NAMES(sex) = male, female.
               NAMES(cigarets) = yes, no.

/ DESIGN       DEPENDENT = pulse_2.
               COVARIATES = age, pulse_1.

/ PRINT        RESIDUAL.

/ END
```

Output 2V.2

— we omit portions of output from this analysis —

```
DESIGN SPECIFICATIONS                                          [1]
---------------------
         GROUP -    2   3
         DEPEND -   6
         COVAR -    4   5
```

— we omit portions of output —

```
CELL MEANS          FOR  1-ST COVARIATE                                              [2]
--------------------
                                                                   MARGINAL
sex     -  male         male         female       female
cigarets-  yes          no           yes          no

age          25.70000    26.83333    27.66667    26.75000    26.65000

COUNT          10           12           6           12           40

STANDARD DEVIATIONS  FOR  1-ST COVARIATE
--------------------
sex     -  male         male         female       female
cigarets-  yes          no           yes          no

age           6.63409     6.57590     8.93682     7.16208

CELL MEANS          FOR  2-ND COVARIATE
--------------------
                                                                   MARGINAL
sex     -  male         male         female       female
cigarets-  yes          no           yes          no

pulse_1      73.60000    70.83333    82.00000    80.00000    75.95000

COUNT          10           12           6           12           40

STANDARD DEVIATIONS  FOR  2-ND COVARIATE
--------------------
sex     -  male         male         female       female
cigarets-  yes          no           yes          no

pulse_1      12.10326     4.54939    10.73313     7.62770

CELL MEANS          FOR  1-ST DEPENDENT VARIABLE
--------------------
                                                                   MARGINAL
sex     -  male         male         female       female
cigarets-  yes          no           yes          no

pulse_2     132.60000   128.33333   155.16667   143.58333   138.00000

COUNT          10           12           6           12           40

STANDARD DEVIATIONS  FOR  1-ST DEPENDENT VARIABLE
--------------------
sex     -  male         male         female       female
cigarets-  yes          no           yes          no

pulse_2       7.54542    13.09638     6.70572     7.12816

A N A L Y S I S   O F   V A R I A N C E  FOR   THE   1-ST DEPENDENT VARIABLE         [3]
---------------------------------------

THE TRIALS ARE REPRESENTED BY THE VARIABLES:
pulse_2

   SOURCE                      SUM OF      D.F.      MEAN         F       TAIL
                               SQUARES               SQUARE               PROB.

   sex                        1876.65312     1     1876.65312    29.76   0.0000
   cigarets                    447.75015     1      447.75015     7.10   0.0117
   sc                          170.63701     1      170.63701     2.71   0.1092
   age                         429.82752     1      429.82752     6.82   0.0133
   pulse_1                     311.51315     1      311.51315     4.94   0.0330
   ALL COVARIATES             1038.93010     2      519.46505     8.24   0.0012
 1 ERROR                      2143.88657    34       63.05549

   REG. COEFF.                  ESTIMATE         STD. ERROR  T-VALUE  P-VALUE         [4]
   age                          -0.50542           0.19358    -2.61   0.4967
   pulse_1                       0.35074           0.15780     2.22   0.0082

 CASE  LABEL sex      cigarets       PREDICTD     RESIDUAL                             [5]

    1  ALLAN    male     yes         125.85263     0.14737
    2  MARY     female   yes         157.63861    -3.63861
    3  BILL     male     no          125.34692     2.65308
    4  LINDA    female   no          148.05803     6.94198
    5  MICHAEL  male     yes         132.30984    -4.30984
```

— we omit the predicted values and residuals for cases 6 through 35 —

```
   36  WILLIAM  male     yes         124.33636     7.66365
   37  KYLE     male     no          125.31639    -9.31639
   38  BEN      male     yes         140.72769    -2.72769
   39  GREG     male     no          129.58638    12.41362
   40  RICHARD  male     no          126.44054     5.55946

 ERROR       SUM OF          RECOMPUTED          RELATIVE
 TERM        SQUARES       FROM RESIDUALS         ERROR
   1        2143.88657       2143.88657          0.00000

ADJUSTED CELL MEANS  FOR  1-ST DEPENDENT VARIABLE                                      [6]
--------------------
                                                                 MARGINAL
sex     -  male         male         female       female
cigarets-  yes          no           yes          no
pulse_2    132.94409   130.22063   153.55852   142.21336   138.00000
COUNT          10           12           6           12          40

STANDARD ERRORS OF ADJUSTED CELL MEANS FOR  1-ST DEPENDENT VARIABLE                    [7]
---------------------------------------
sex     -  male         male         female       female
cigarets-  yes          no           yes          no

pulse_2       2.54497     2.43060     3.38518     2.37980
```

[1] Interpretation of the design, with its assignment of variables (by number) as GROUPING factors, DEPENDENT variable, and COVARIATES.

[2] Cell means and standard deviations for the covariates and dependent variable. An examination of the means of the second covariate shows that the females had higher resting pulse rates than the males. The cell means for the dependent variable, post-exercise pulse, appear higher for females, regardless of smoking status. Smokers, regardless of sex, have higher post-exercise pulse rates.

[3] An analysis of variance table is printed, identifying the sum of squares, degrees of freedom, mean square, *F* ratio, *p*-value for each effect in the design, including covariates, and the regression coefficients for the two covariates. (Since this is not an experiment in which individuals were randomly assigned to groups, the results must be interpreted with caution.) The model for this analysis of covariance can be written as

$$E(y_{ij}) = \mu + \alpha_i + \gamma_j + (\alpha\gamma)_{ij} + \beta_1 x_{1ij} + \beta_2 x_{2ij}$$

where α_i and γ_j represent the main effects sex and cigarette; $(\alpha\gamma)_{ij}$ the interaction of sex and cigarette; and β_1 and β_2 are the coefficients of the covariates age and pre-exercise pulse. For this problem we test six hypotheses: (1) there are no differences in adjusted postexercise pulse due to sex; (2) there are no differences due to smoking status; (3) the effect of smoking status on the adjusted pulse rate is the same for males as for females (the interaction $\alpha\gamma$ in the model is zero); (4) the slope β_1 for age is zero; (5) the slope β_2 for pre-exercise pulse is zero; (6) the slopes for the combined effect of age and pre-exercise pulse are zero.

The hypotheses tested are independent, but the sums of squares may not be orthogonal. For example, the sums of squares for each covariate do not add to the sums of squares for both covariates. The sum of squares used in the test of each hypothesis (each analysis of variance component, each covariate by itself, and both covariates together) can be obtained as the difference in the residual sums of squares of two models; one in which all covariates and effects are fitted and the other in which the effect or covariate(s) of interest is set to zero.

In this example, after adjusting for age and pre-exercise pulse, we reject the first two hypotheses. There are differences in post-exercise pulse due to sex ($p < .00005$) and smoking ($p = .0117$). The interaction (sc) between the adjusted effects is not significant ($p = .1092$). The covariates are significant individually and when combined. Note that the regression coefficient for post-exercise pulse on age is negative, indicating that the older subjects in this sample tended to have lower values. Note that when covariates are included in the model, the line in the ANOVA table labeled MEAN is omitted.

[4] The regression coefficients are estimates of the coefficients for the covariates when all effects and covariates in the model are fitted.

[5] Predicted values and residuals are printed for each case. The residuals are the difference between each observed value and the adjusted cell mean (for definition, see **[5]**). The predicted values and residuals may be saved in a BMDP File (see Chapter 8) for later use in plots or other analyses.

[6] Cell means for post-exercise pulse adjusted for the linear effects of the covariates age and pre-exercise pulse. The adjusted cell means provide a sample estimate of the PULSE_2 means measured at the overall mean of age and pre-exercise pulse for each cell. The adjusted mean for cell (i,j) is

$$\hat{\mu}_{ij} = \bar{y}_{ij} + \hat{\beta}_1(\bar{x}_1 - \bar{x}_{1ij}) + \hat{\beta}_2(\bar{x}_2 - \bar{x}_{2ij})$$

where $\bar{y}_{ij}$, $\bar{x}_{1ij}$, and $\bar{x}_{2ij}$ are means computed for cell (i,j); $\bar{x}_1$ and $\bar{x}_2$ are means computed for all cases; and $\hat{\beta}_1$ and $\hat{\beta}_2$ are the regression coefficient estimates. The regression coefficients are assumed to be constant across groups.

[7] The standard error of the adjusted cell means for the post exercise pulse variable. The standard error of the mean is standard deviation of the sampling distribution. Essentially, this is a measure of the variability of a measure over repeated samplings.

Example 2V.3 Latin square design

A Latin square design provides a way of estimating the main effects of three different factors, each with k levels, using only k^2 subjects. By comparison, the corresponding full factorial design requires k^3 subjects. Latin square designs are often used to provide balance for nuisance factors such as order of administration. Because interactions cannot be estimated adequately, the Latin square design rests on the assumption that interactions are negligible.

We analyze data from a study evaluating the durability of four types of rubber compounds used in tires (Kirk, 1982; here we use data only from the first Latin square and add names for the autos). The other two factors in the design are type of automobile (lemon, nogo, heap, junker) and wheel position (right-front, right-rear, left-front, left-rear). The dependent variable is the thickness of tread remaining after 10,000 miles of driving. A diagram of the 4 x 4 Latin square is shown in Figure 2V.5. See Example 9D.2 for a miniplot of these data.

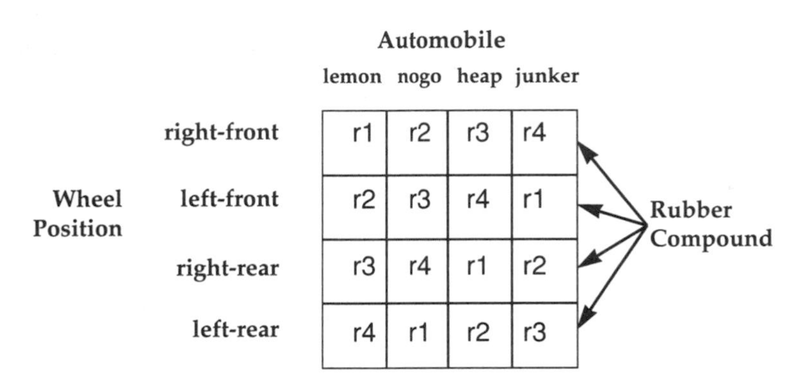

Figure 2V.5 Latin square design for rubber compound experiment

The 2V instructions for a Latin square resemble those used in Input 2V.1 for the full factorial analysis of variance. Each record in the file TIRE.DAT contains information for a tire: codes for the GROUPING variables, RUBBER, WHEEL, and AUTO, and the DEPENDENT variable TREAD. These are listed in the GROUP paragraph, as well as the NAMES assigned to the numeric CODES (1 to 4) that indicate brand of AUTO and WHEEL position.

For a Latin square design we must instruct 2V to test main effects only by using either INCLUDE or EXCLUDE in the DESIGN paragraph.

- Use INCLUDE to list the effects to estimate and test. In this example, the main effects are RUBBER, WHEEL, and AUTO. The numbers in the INCLUDE list correspond to the order of the grouping variables listed in the VARIABLE sentence of the GROUP paragraph. They do not correspond to the input order in the VARIABLE paragraph.

or

- Use EXCLUDE to list those effects that you do not want. In this example, we want to exclude four interactions: RUBBER x WHEEL, RUBBER x AUTO, WHEEL x AUTO, and TIRE x WHEEL x AUTO. We would write EXCLUDE = 12, 13, 23, 123. Notice that there are no spaces between the numbers designating the factors in each interaction.

The choice between INCLUDE and EXCLUDE is simply a matter of convenience. You can specify any main effects or interactions to be included or excluded depending on the analysis you want.

Input 2V.3

```
/ INPUT       VARIABLES = 4.
              FORMAT = FREE.
              FILE = 'tire.dat'.

/ VARIABLE    NAMES = rubber, wheel, auto, tread.

/ GROUP       VARIABLE = rubber, wheel, auto.
              CODES(rubber,wheel,auto) = 1, 2, 3, 4.
              NAMES(wheel) = r_front, l_front, r_rear,
                             l_rear.
              NAMES(auto) = lemon, nogo, heap, junker.

/ DESIGN      DEPENDENT = tread.
              INCLUDE = 1, 2, 3.
/ END
```

Output 2V.3

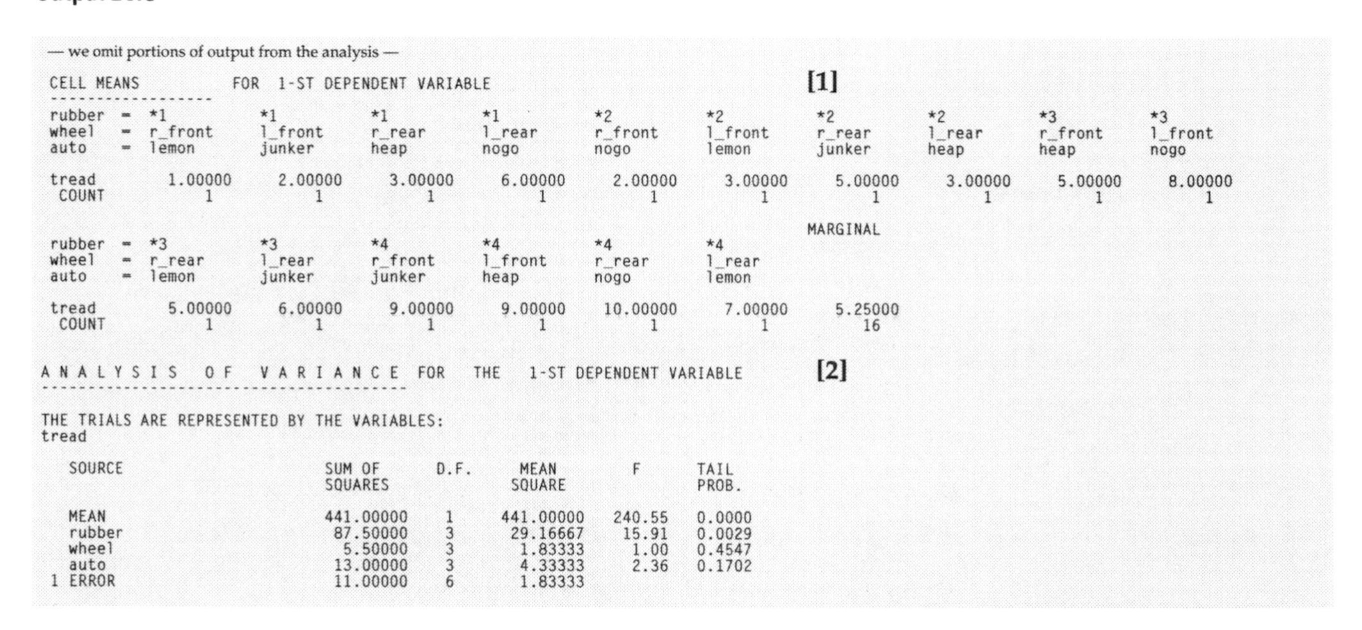

— we omit portions of output from the analysis —

```
CELL MEANS          FOR  1-ST DEPENDENT VARIABLE                                        [1]
------------------
rubber  -  *1         *1         *1         *1         *2         *2         *2         *2         *3         *3
wheel   -  r_front    l_front    r_rear     l_rear     r_front    l_front    r_rear     l_rear     r_front    l_front
auto    -  lemon      junker     heap       nogo       nogo       lemon      junker     heap       heap       nogo

tread        1.00000    2.00000    3.00000    6.00000    2.00000    3.00000    5.00000    3.00000    5.00000    8.00000
 COUNT             1          1          1          1          1          1          1          1          1          1

                                                                             MARGINAL
rubber  -  *3         *3         *4         *4         *4         *4
wheel   -  r_rear     l_rear     r_front    l_front    r_rear     l_rear
auto    -  lemon      junker     junker     heap       nogo       lemon

tread        5.00000    6.00000    9.00000    9.00000   10.00000    7.00000    5.25000
 COUNT             1          1          1          1          1          1         16

A N A L Y S I S   O F   V A R I A N C E  FOR   THE   1-ST DEPENDENT VARIABLE           [2]
----------------------------------------

THE TRIALS ARE REPRESENTED BY THE VARIABLES:
tread

   SOURCE                           SUM OF      D.F.     MEAN        F       TAIL
                                    SQUARES              SQUARE              PROB.

   MEAN                           441.00000     1     441.00000   240.55   0.0000
   rubber                          87.50000     3      29.16667    15.91   0.0029
   wheel                            5.50000     3       1.83333     1.00   0.4547
   auto                            13.00000     3       4.33333     2.36   0.1702
 1 ERROR                           11.00000     6       1.83333
```

[1] 2V prints the means for the 16 cells along with the group structure of the Latin square. Since we have only one observation per cell, these means are the original data. Each ordering of levels of the three grouping variables by row and column is unique. Note that since we did not name the levels of RUBBER, 2V uses the codes as category names, each preceded by an asterisk.

[2] The analysis of variance table. We see that the treatment effect, RUBBER, is significant ($p = .0029$), while the two nuisance variables are not significant. Since there is only one case per cell in this example, we have no within-cell error term; the mean square of the residuals is used as an error term to test the treatment and nuisance effects.

Example 2V.4 Incomplete block design

A full factorial design may require many cases to evaluate all treatment combinations. However, if the experimenter is mainly interested in estimating and testing main effects plus a few lower-order interactions, and can safely assume that other effects are negligible, an incomplete block design or a fractional factorial (see Example 2V.5) can provide the desired information with a fraction of the sample size. In these *confounded factorial designs*, the experimenter confounds specific interactions with specific main effects or other interactions. We first illustrate how to specify the DESIGN paragraph for an incomplete block design.

John (1971, p. 135) discusses a partially confounded incomplete block design used for an agricultural experiment to test the effects of three chemicals on potato crop yield. The three factors, each with two levels (yes and no), are sulphate of ammonia, sulphate of potash, and nitrogen. The dependent variable is the yield in pounds for each plot of potatoes. The experiment uses eight blocks (plots of ground). Each block contains four subplots, each receiving one combination of the levels of the three treatment factors. In Data Set 2V.1, we record the data as five variables: the first variable is the block number, the next three are the levels (0 or 1) of the three treatment factors, and the last is the dependent variable (yield). Thus, four of the 2^3 treatment combinations are assigned to each block in such a way that across the eight blocks each treatment combination is evaluated three times. (See Example 9D.3 for a plot of these data.)

Data Set 2V.1
Crop yield experiment

```
1 0 0 0 101   3 0 0 0 106   5 0 0 0  87   7 0 0 0 131
1 1 0 1 373   3 1 1 0 306   5 1 0 1 324   7 1 0 0 103
1 0 1 1 398   3 0 0 1 324   5 0 1 0 279   7 0 1 1 445
1 1 1 0 291   3 1 1 1 449   5 1 1 1 471   7 1 1 1 437
2 0 0 1 312   4 0 1 0 272   6 0 0 1 323   8 0 0 1 324
2 1 0 0 106   4 1 0 0  89   6 1 0 0 128   8 1 0 1 361
2 0 1 0 265   4 0 1 1 407   6 0 1 1 423   8 0 1 0 302
2 1 1 1 450   4 1 0 1 338   6 1 1 0 334   8 1 1 0 272
```

To specify the grouping criteria we add the line in the GROUP paragraph

```
/ GROUP    VARIABLE = BLOCK, AMMONIA, POTASH, NITROGEN.
```

To specify the DESIGN, we submit the dependent variable:

```
/ DESIGN   DEPENDENT = YIELD.
```

As in the Latin square design of Example 2V.3, the effects that we want included in the model must be specified. Otherwise the model has more parameters than can be estimated. For the above example we might specify

```
INCLUDE = 1, 2, 3, 4, 23, 24, 34, 234.
```

Using this specification, the model contains the main effect for blocking, and all main effects and interactions of the three treatment factors. The model does not contain any interactions of the block effect with treatment effects. (Note that the order of the factors must correspond to the order of names in the GROUP statement.) Alternatively, we could obtain the same analysis by specifying the effects we want to exclude from the analysis:

```
EXCLUDE = 12, 13, 14, 123, 124, 1234.
```

Input 2V.4 shows the complete setup using INCLUDE. We do not show output for this example.

Input 2V.4

```
/ INPUT       VARIABLES = 5.
              FORMAT = FREE.

/ VARIABLE    NAMES = block, ammonia, potash, nitrogen,
                      yield.

/ GROUP       VARIABLE = block, ammonia, potash, nitrogen.
              CODES(ammonia, potash, nitrogen) = 0, 1.
              NAMES(ammonia, potash, nitrogen) = yes, no.
```

Input 2V.4
(continued)

```
/ DESIGN     DEPENDENT = yield.
             INCLUDE = 1, 2, 3, 4, 23, 24, 34, 234.
/ END
 data
/ END
```

Example 2V.5 Fractional factorial design

In fractional factorial designs, you use a *fraction* of the number of treatment combinations for a full factorial design. For example, a 4 x 4 x 4 x 4 factorial design has 256 treatment combinations. If you use a one-fourth fractional replication, you limit the size of the experiment to 64 treatment combinations. The size of the experiment is reduced by confounding the main effects and interactions of interest with those effects that can be assumed to be negligible.

John (1971, p. 154) gives a numerical example of a fractional factorial design where the dependent variable is a score representing the octane requirement of a car, and the four independent variables are qualitative factors affecting that requirement. (Although John does not specify names for the factors, we assign names to help visualize the design.) Suppose five variables are recorded for each car, including codes for

- Compression—high vs. normal engine compression
- Number of Valves/Cylinder—two vs. four
- Engine Size—small vs. large
- Overall Weight—1900 vs. 3000 lb

and the dependent variable OCTANE requirement. The data are displayed in Data Set 2V.2; Figure 2V.6 shows a diagram of the design.

Data Set 2V.2
Octane experiment

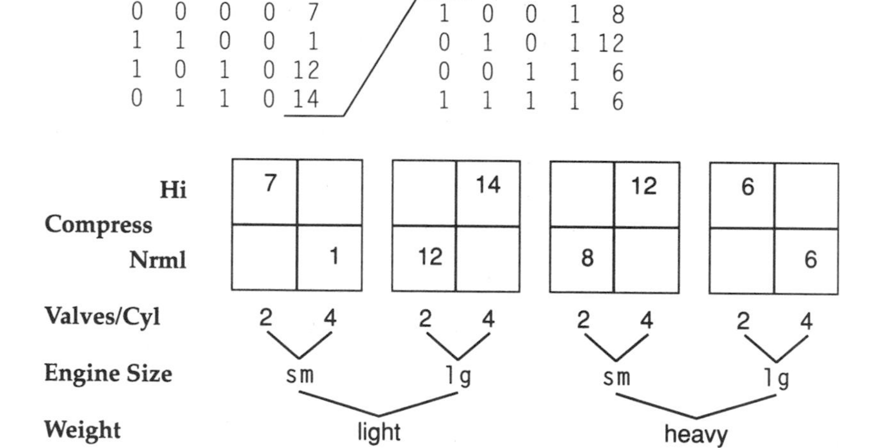

```
0  0  0  0  7      1  0  0  1  8
1  1  0  0  1      0  1  0  1 12
1  0  1  0 12      0  0  1  1  6
0  1  1  0 14      1  1  1  1  6
```


Figure 2V.6
Fractional factorial design for octane experiment

Input 2V.5 shows the design specification for the fractional factorial. We name the four grouping factors COMPRESS, VALVES, SIZE, and WT, and the dependent variable OCTANE. In addition, we must specify the effects to be tested: all main effects (they are confounded with third-order interactions) and three two-factor interactions (each is confounded with a second two-factor interaction). We describe the model to 2V in the DESIGN paragraph as:

```
INCLUDE ARE 1, 2, 3, 4, 12, 13, 23.
```

Input 2V.5

```
/ INPUT      VARIABLES = 5.
             FORMAT = FREE.
             FILE = 'octane.dat'.

/ VARIABLE   NAMES = compress, valves, size, wt, octane.

/ GROUP      VARIABLE = compress, valves, size, wt.
             CODES(compress, valves, size, wt) = 0, 1.
             NAMES(compress) = high, normal.
             NAMES(valves) = '2', '4'.
             NAMES(size) = small, large.
             NAMES(wt) = light, heavy.

/ DESIGN     DEPEND = octane.
             INCLUDE = 1, 2, 3, 4, 12, 13, 23.
/ END
```

Because we estimate seven effects and there are only eight cases, no degrees of freedom are left to calculate an error term. When we run Input 2V.5, no F ratios and p-values are printed in the analysis of variance table. In such cases, the experimenter would need an independent estimate of the error to calculate the F ratios and corresponding p-values or would need to include fewer than seven terms in the model. We do not show output.

Repeated Measures Designs

In repeated measures models, the same variable is measured several times for each subject (or case). For example, suppose we record each subject's body weight every week for six weeks. If there were only two weekly measurements for each subject, our analysis would be a paired-comparison t test (see program 3D). However, if we want to study three or more measures, we need to use a repeated measures analysis of variance.

Several types of studies and experiments can lead to a repeated measures design. These include *time course or growth curve studies;* an example is the amount of food left in an animal's stomach at intervals following ingestion of a radiolabeled meal. Another type of repeated measures study is a *comparison of treatments made in sequence.* In those studies, each experimental subject receives several different stimuli, treatments, or doses and, thus, several responses are recorded for each subject. A third type of repeated measures study is a *true split-plot study,* named for the agricultural experiments where treatments are assigned to plots in a field. Other examples are assignment of treatments to the four teats of a cow's udder, division of serum or tissue samples into aliquots, and assignment of treatments to different members of a litter.

Explanation of a repeated measures analysis. In order to best understand the 2V output, it is helpful to understand the quantities that 2V derives from your data. (Readers familiar with repeated measures analyses should skip to Example 2V.6.) Suppose there are only two measures, y_1 and y_2 for each subject. You could analyze these data using a paired-comparison t test in 3D, or you could do the calculations yourself. Begin by computing the difference $y_1 - y_2$ for each subject, as in Table 2V.1. Then, calculate the average of the differences, $\bar{d}$, and the standard deviation of the differences, s_d, and put the results into the formula for the test statistic. In a similar way, 2V derives values from your data and uses them in analysis of variance computations to test within-subject changes as well as differences between groups of subjects. The values derived from your data are called *orthogonal polynomial components.*

Table 2V.1
Computing a paired-comparison t test

Subject	y_1	y_2	Difference
1	10	8	2
2	13	14	–1
.	.	.	.
n	11	7	4

Suppose we expand the data in Table 2V.1 so that there are two groups of subjects (e.g., smokers and nonsmokers) or one grouping factor with repeated measures y_1 and y_2. For each subject, 2V computes the sum

$$p_0 = \frac{(y_1 + y_2)}{\sqrt{2}}$$

and the difference

$$p_1 = \frac{(y_1 - y_2)}{\sqrt{2}}$$

(the denominator in the above equation is chosen to make the sum of squares of the coefficients equal to one). 2V uses p_0 and p_1 to compute two analysis of variance tables. In the first table, 2V uses the values of p_0 to test the group effect (this is equivalent to a two-sample t test comparing the smoker and nonsmoker means of p_0). A test that the grand mean of p_0 is zero (labeled MEAN in the 2V output) is printed at the top of this table. For the second table, the values of p_1 are used to test whether there is no change across values of the within factor (i.e., that the mean of p_1 is zero). Next, a test of the *interaction* between the grouping factor and the within factor is reported (i.e., the equality of means for p_1 between the two groups).

Suppose there is one grouping factor and a within-subject factor with 3 levels (see Table 2V.2). 2V derives three components (p_0, p_1, and p_2):

$$p_0 = \frac{(y_1 + y_2 + y_3)}{\sqrt{3}}$$

$$p_1 = \frac{(y_1 + 0y_2 - y_3)}{\sqrt{2}}$$

$$p_2 = \frac{(y_1 - 2y_2 + y_3)}{\sqrt{6}}$$

The changes for each subject are now decomposed into two orthogonal components: a *linear* component, p_1, and a *quadratic* component, p_2. In Table 2V.2, we show calculations for p_0, p_1, and p_2 (we omit the denominator). 2V uses these derived values to compute two analysis of variance tables:

1. p_0 is used to test between-group differences. When there are more grouping factors, p_0 is used as the dependent variable in the usual factorial design computations.

2. p_1 and p_2 are used to test differences or changes across the within-factor and also to test the interaction of the within factor with the grouping factor.

Table 2V.2
Orthogonal polynomial breakdown for three repeated measures

Smoker	Subject	Data			Orthogonal Polynomials		
		y_1	y_2	y_3	p_0	p_1	p_2
yes	1	9	8	7	24	2	0
	2	7	5	3	15	4	0
	3	8	6	3	17	5	–1
no	4	3	2	1	6	2	0
	5	4	2	1	8	3	+1
	6	3	2	1	6	2	0
	7	5	4	2	11	3	–1

An analysis of variance is performed for both p_1 and p_2, but the individual analyses are not printed unless ORTHOGONAL is requested in the DESIGN paragraph (significant results for p_1 indicate a linear increase, or decrease, across the three measures; significance for p_2 indicates changes that follow a quadratic curve). 2V always reports pooled results for p_1 and p_2 where the sums of squares for each effect are added. Similarly, the pooled error sum of squares is the sum of the error sum of squares for p_1 and p_2.

When each subject has four measures, the orthogonal decomposition for the within-subject tests is:

$$p_1 = \frac{(3y_1 + y_2 - y_3 - 3y_4)}{\sqrt{20}} \qquad \text{(linear)}$$

$$p_2 = \frac{(y_1 - y_2 - y_3 + y_4)}{\sqrt{4}} \qquad \text{(quadratic)}$$

$$p_3 = \frac{(y_1 - 3y_2 + 3y_3 - y_4)}{\sqrt{20}} \qquad \text{(cubic)}$$

The number of polynomials is always one less than the number of repeated measures. (Note that the coefficients for the linear component are the same as those used in Example 7D.5 to test a linear increase in average income across four ordered levels of education.) Again, in 2V, the sum of the four measures for each subject, p_0, is used in the between groups part of the design (see **[4]** in Output 2V.6), and p_1, p_2, and p_3 are pooled for the within-subject tests (see **[5]** in Output 2V.6 and **[5]** in Output 2V.7).

Assumptions for Repeated Measures. In the usual fixed effects analysis of variance model, assumptions are made that the observations are independent and normally distributed, and that the variance is the same in all cells. In the repeated measures model, measures made on the same subject are usually correlated. The independence assumption is replaced by sphericity assumptions.

There are two ways to check the assumptions necessary for a valid test of the within factor main effects and interactions. The first applies to the data themselves, and the second, to the values from the orthogonal polynomial breakdown. The assumptions are satisfied if the data satisfy the conditions of *compound symmetry*. That is, the measures have the same variance and the correlation between the measurements for any two levels of the within factor is equal to the correlation between any other two levels. Compound symmetry is a sufficient but not necessary condition. A somewhat less restrictive condition is to evaluate the *symmetry* of the *orthogonal polynomials*. They should be independent and have equal variance. This is a sufficient and necessary condition. For checking equality of variance of the orthogonal polynomials, 2V prints the sums of squares associated with each polynomial. The correlation of the components is also printed (see **[2]** in Output 2V.6). A sphericity test found in Anderson (1958, p. 259) is printed below the display. The sphericity test has low power for small sample sizes. For large sample sizes, the test is likely to show significance although the effect on the analysis of variance may be negligible. This test can be very sensitive to outliers. If there is a failure to meet the symmetry assumption you may

- use the *F* tests for the pooled orthogonal polynomials but reduce the degrees of freedom. There are two methods for reducing the degrees of freedom (see discussion of the Greenhouse-Geisser and Huynh-Feldt methods at end of Appendix B.20).
- use the single degree-of-freedom *F* tests associated with each individual orthogonal polynomial because each test is valid whether or not the sphericity condition is satisfied. The disadvantage is that the error degrees of freedom for each test are reduced. (And of course, if the levels of your within factor are **not** ordered, these tests may not be of interest).
- use program 5V, because it allows other structures on the covariance matrix of the repeated measures.

Note also that when the repeated measure has only two levels, the symmetry assumption is not required.

In the following examples, we illustrate only designs with one grouping factor. If your repeated measures design has more than one grouping factor, refer to Examples 2V.1 to 2V.5 to set up the group part of the design and BMDP Technical Report #83 for further information.

Example 2V.6 One grouping and one within factor

In this example, we analyze growth data for the jaws of 11 girls and 16 boys from Potthoff and Roy (1964). For each subject, the distance from the center of the pituitary to the pterygomaxillary fissure is recorded at the ages of 8, 10, 12, and 14 years. We are interested in characterizing the growth curve and determining whether it differs between the sexes. The grouping factor is sex, with two levels, and the repeated or within factor is age, with four levels. The data for this study are displayed in Data Set 2V.3. For this design, there are three hypotheses:

H_1: there are no differences between male and female children in jaw size.

H_2: there are no differences in jaw size for children measured at ages 8, 10, 12, and 14.

H_3: growth in jaw size between the specified ages does not differ for male vs. female children.

(For tests of other hypotheses for these data, see program 5V in Volume 2.)

Data Set 2V.3
Jaw growth study

Sex	Age 8	10	12	14
1	26.0	25.0	29.0	31.0
1	21.5	22.5	23.0	26.5
1	23.0	22.5	24.0	27.5
1	25.5	27.5	26.5	27.0
1	20.0	23.5	22.5	26.0
1	24.5	25.5	27.0	28.5
1	22.0	22.0	24.5	26.5
1	24.0	21.5	24.5	25.5
1	23.0	20.5	31.0	26.0
1	27.5	28.0	31.0	31.5
1	23.0	23.0	23.5	25.0
1	21.5	23.5	24.0	28.0
1	17.0	24.5	26.0	29.5
1	22.5	25.5	25.5	26.0
1	23.0	24.5	26.0	30.0
1	22.0	21.5	23.5	25.0
2	21.0	20.0	21.5	23.0
2	21.0	21.5	24.0	25.5
2	20.5	24.0	24.5	26.0
2	23.5	24.5	25.0	26.5
2	21.5	23.0	22.5	23.5
2	20.0	21.0	21.0	22.5
2	21.5	22.5	23.0	25.0
2	23.0	23.0	23.5	24.0
2	20.0	21.0	22.0	21.5
2	16.5	19.0	19.0	19.5
2	24.5	25.0	28.0	28.0

In the next example (2V.7), we show how to use orthogonal polynomials to test hypotheses about the shape of the growth curve.

In Input 2V.6, the INPUT, VARIABLE, and GROUP paragraphs are the same as in other BMDP programs (see Chapters 3, 4, and 5). Each input record contains data for one child with his/her four jaw measurements and a code for sex, so we read five variables. The names of the four repeated measures are listed with the DEPEND command. The LEVEL command, used **only** for repeated measures models, specifies the number of repeated measures (see Example 2V.8 if you have more than one within factor).

Input 2V.6

```
/ INPUT       VARIABLES = 5.
              FORMAT IS FREE.
              FILE IS 'potthoff.dat'.

/ VARIABLE    NAMES = sex, age_8, age_10, age_12, age_14.

/ GROUP       VARIABLE = sex.
              CODES(sex) = 1, 2.
              NAMES(sex) = male, female.

/ DESIGN      DEPENDENT = age_8 to age_14.
              LEVEL = 4.
              NAME = age.
/ END
```

Output 2V.6

—we omit portions of output—

```
DESIGN SPECIFICATIONS                                                  [1]
---------------------
          GROUP  =   1
          DEPEND =   2   3   4   5
          LEVEL  =   4


GROUP STRUCTURE

 sex           COUNT
 male            16
 female          11

SUMS OF SQUARES AND CORRELATION MATRIX OF THE                          [2]
ORTHOGONAL COMPONENTS POOLED FOR ERROR  2 IN ANOVA TABLE BELOW

          59.16733          1.000
          26.04119          0.286    1.000
          62.91932          0.074    0.032    1.000

SPHERICITY TEST APPLIED TO ORTHOGONAL COMPONENTS - TAIL PROBABILITY       0.2001

CELL MEANS             FOR  1-ST DEPENDENT VARIABLE                    [3]
-------------------
                                             MARGINAL
     sex        =  male          female
           age

age_8          1   22.87500     21.18182     22.18519
age_10         2   23.81250     22.22727     23.16667
age_12         3   25.71875     23.09091     24.64815
age_14         4   27.46875     24.09091     26.09259

     MARGINAL    24.96875     22.64773     24.02315

     COUNT          16           11           27

STANDARD DEVIATIONS   FOR  1-ST DEPENDENT VARIABLE
-------------------
     sex        =  male          female
           age

age_8          1    2.45289      2.12453
age_10         2    2.13600      1.90215
age_12         3    2.65185      2.36451
age_14         4    2.08542      2.43740
```

```
A N A L Y S I S   O F   V A R I A N C E  FOR   THE   1-ST DEPENDENT VARIABLE
-----------------------------------------------

THE TRIALS ARE REPRESENTED BY THE VARIABLES:
age_8    age_10    age_12    age_14

        SOURCE                     SUM OF        D.F.       MEAN          F      TAIL     GREENHOUSE    HUYNH      [6]
                                   SQUARES                  SQUARE               PROB.     GEISSER      FELDT
                                                                                            PROB.       PROB.
        MEAN                    59118.50189        1    59118.50189  3910.84   0.0000
        sex          [4]          140.46486        1      140.46486     9.29   0.0054
  1     ERROR                     377.91477       25       15.11659

        age          [5]          209.43697        3       69.81232    35.35   0.0000     0.0000       0.0000
        as                         13.99253        3        4.66418     2.36   0.0781     0.0878       0.0781
  2     ERROR                     148.12784       75        1.97504

ERROR
TERM              EPSILON FACTORS FOR DEGREES OF FREEDOM ADJUSTMENT

                     GREENHOUSE-GEISSER     HUYNH-FELDT
  2                       0.8672              1.0000
```

[1] The design statement specifies the grouping variable, four levels of a repeated factor, and four dependent variables.

[2] The test for violation of the sphericity condition is not significant ($p > .2$) for these data. Note that the correlations of the orthogonal components are small (.03 to .29). See Example 2V.7 for further explanation of this portion of the output.

[3] The means and standard deviations at each of the four age levels for each of the two sexes are shown. We see that jaw size increases at each age level, and that the means are greater for males than for females.

[4] Analysis of variance table for p_0. The grouping factor, SEX, is significant ($p = 0.0054$), with males having a larger average jaw measurement across ages than females.

[5] Analysis of variance table for the pooled orthogonal polynomials p_1, p_2, p_3. (See Example 2V.7 for the breakdown of the individual orthogonal components.) The repeated or within factor, AGE, is highly significant ($p < 0.00005$), indicating that the jaw growth differs significantly across AGE. The test for the interaction of SEX with AGE is reported on the line labeled 'as'. The results are marginal ($p = .0781$). Highly significant results would indicate that the jaw growth pattern for boys and girls differs. See additional comments in Output 2V.7.

If there had been a second grouping factor, for example, TREATMENT (drugs A, B, and a placebo), tests of other interactions would be displayed; TREATMENT with AGE (labeled at), and AGE by SEX by TREATMENT (ast).

[6] The Greenhouse-Geisser and Huynh-Feldt adjusted p-values provide conservative tests of the repeated measures factor. Since the sphericity assumption was not rejected for these data, the p-values differ little from the unadjusted tail probability. See the commentary on Output 2V.7 for more details.

Example 2V.7 Miniplots and breaking the within factor into orthogonal components

When the levels of the within factor are ordered—increasing ages, drug doses, etc.—you may want to test hypotheses that assess whether there is evidence of a linear trend, whether the slopes of the groups differ, whether there is any evidence of nonlinearity, and if so, whether it is quadratic or higher order. If you state ORTHOGONAL in the DESIGN paragraph, 2V breaks the overall sum of squares for the within factors into orthogonal polynomials and computes an appropriate error term for each component. The advantage of this approach is that the F tests for each component are valid whether or not the sphericity condition is satisfied. The disadvantage is that error degrees of freedom for each test are reduced. (Note that the individual components are appropriate when the levels of the within factor are equally spaced. Example 2V.8 shows how to specify other spacings for the levels with the POINT statement.)

In Input 2V.7 we add ORTHOGONAL to the DESIGN paragraph and request a MINIPLOT of the cell means in the PLOT paragraph. (9D has other miniplots.)

Input 2V.7
(we change the DESIGN paragraph of Input 2V.6 and add a PLOT paragraph)

```
/ DESIGN   DEPENDENT = age_8 to age_14.
           LEVEL = 4.
           NAME = age.
           ORTHOGONAL.

/ PLOT     MINI.
```

Output 2V.7

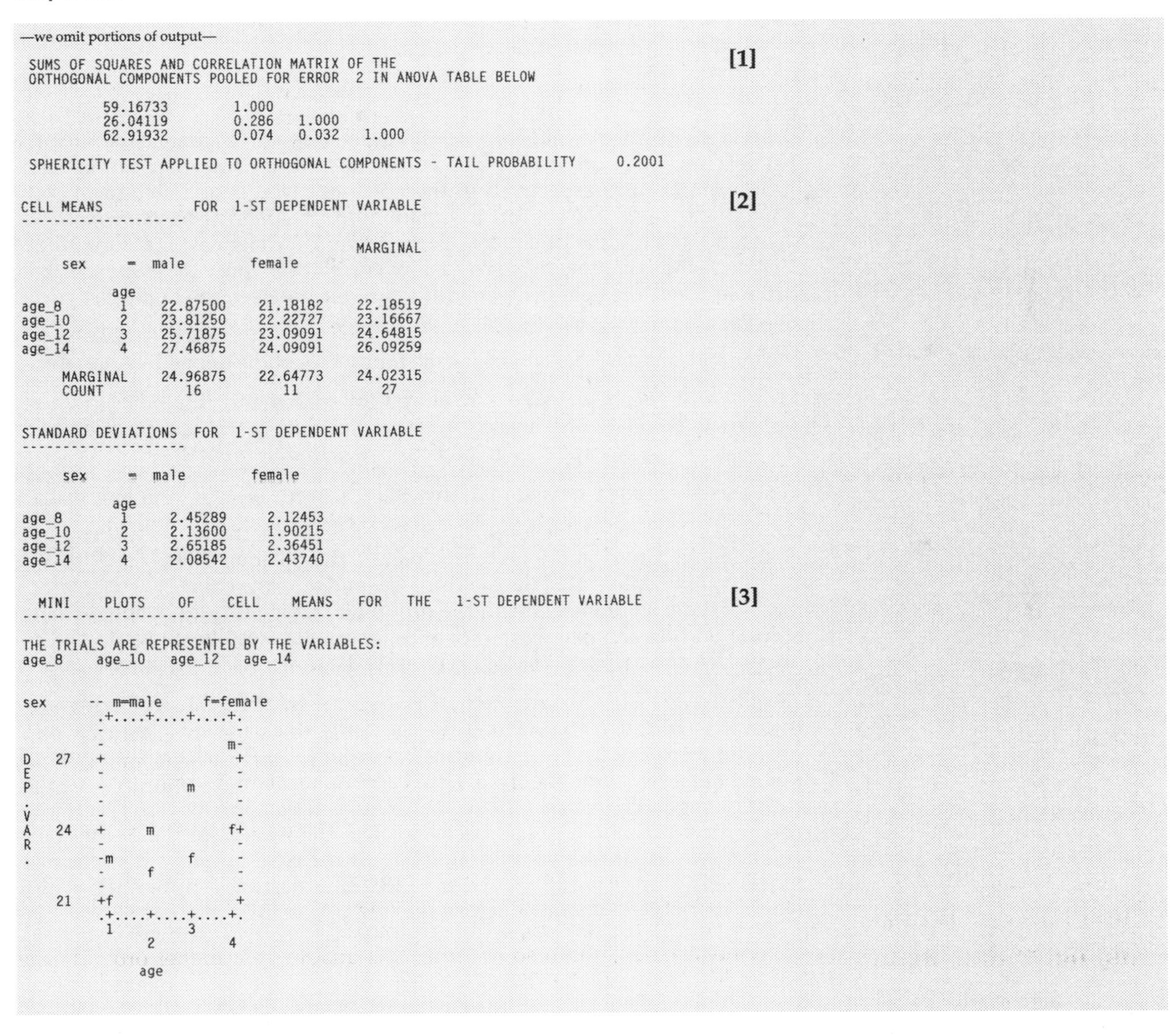

```
—we omit portions of output—
                                                                            [1]
 SUMS OF SQUARES AND CORRELATION MATRIX OF THE
 ORTHOGONAL COMPONENTS POOLED FOR ERROR  2 IN ANOVA TABLE BELOW

          59.16733          1.000
          26.04119          0.286    1.000
          62.91932          0.074    0.032    1.000

 SPHERICITY TEST APPLIED TO ORTHOGONAL COMPONENTS - TAIL PROBABILITY     0.2001

CELL MEANS          FOR  1-ST DEPENDENT VARIABLE                            [2]
--------------------
                                                MARGINAL
     sex      =  male         female

             age
age_8         1     22.87500     21.18182     22.18519
age_10        2     23.81250     22.22727     23.16667
age_12        3     25.71875     23.09091     24.64815
age_14        4     27.46875     24.09091     26.09259

     MARGINAL       24.96875     22.64773     24.02315
     COUNT             16           11           27

STANDARD DEVIATIONS  FOR  1-ST DEPENDENT VARIABLE
--------------------

     sex      =  male         female

             age
age_8         1      2.45289      2.12453
age_10        2      2.13600      1.90215
age_12        3      2.65185      2.36451
age_14        4      2.08542      2.43740

  MINI   PLOTS   OF   CELL   MEANS   FOR   THE   1-ST DEPENDENT VARIABLE    [3]
----------------------------------------

THE TRIALS ARE REPRESENTED BY THE VARIABLES:
age_8    age_10   age_12   age_14

sex      -- m=male     f=female
         .+....+....+....+.
         -                -
         -               m-
D   27   +                +
E        -                -
P        -          m     -
.        -                -
V        -                -
A   24   +     m         f+
R        -                -
.        -m         f     -
         -     f          -
         -                -
    21   +f               +
         .+....+....+....+.
          1         3
               2         4

               age
```

```
A N A L Y S I S   O F   V A R I A N C E  FOR   THE   1-ST DEPENDENT VARIABLE
-------------------------------------------

THE TRIALS ARE REPRESENTED BY THE VARIABLES:
age_8    age_10   age_12   age_14

          SOURCE                       SUM OF       D.F.          MEAN          F        TAIL     GREENHOUSE    HUYNH
                                       SQUARES                    SQUARE                 PROB.     GEISSER      FELDT
                                                                                                    PROB.       PROB.
          MEAN              [4]      59118.50189      1      59118.50189   3910.84    0.0000
          sex                          140.46486      1        140.46486      9.29    0.0054
 1        ERROR                        377.91477     25         15.11659

          a(1)              [5]        208.26600      1        208.26600     88.00    0.0000
          a(1)s                         12.11415      1         12.11415      5.12    0.0326
          ERROR                         59.16733     25          2.36669

          a(2)                           0.95881      1          0.95881      0.92    0.3465
          a(2)s                          1.19955      1          1.19955      1.15    0.2935
          ERROR                         26.04119     25          1.04165

          a(3)                           0.21216      1          0.21216      0.08    0.7739
          a(3)s                          0.67883      1          0.67883      0.27    0.6081
          ERROR                         62.91932     25          2.51677

          age                          209.43697      3         69.81232     35.35    0.0000     0.0000       0.0000
          as                            13.99253      3          4.66418      2.36    0.0781     0.0878       0.0781
 2        ERROR                        148.12784     75          1.97504

ERROR
TERM                EPSILON FACTORS FOR DEGREES OF FREEDOM ADJUSTMENT

                       GREENHOUSE-GEISSER     HUYNH-FELDT
 2                           0.8672              1.0000
```

[1] Error sums of squares and correlation matrix of the orthogonal components. The left column lists the error sums of squares for the linear, quadratic, and cubic components, respectively. If these values diverge too much from each other, then we would suspect a violation of the sphericity assumption. To the right of the sums of squares is the lower off-diagonal of a correlation matrix of the components; zero or low correlations provide evidence that the data meets the sphericity condition (in our data set, the highest correlation is 0.286). The sphericity condition is not rejected ($p = .2001$).

[2] Means and standard deviations for the four levels of the dependent variable.

[3] Miniplot of mean jaw size across age for males and females. Jaw size appears to be increasing linearly with age for both males and females, and the slope of the line appears greater for males than for females.

[4] Analysis of variance table for p_0. This is the same as **[4]** in Example 2V.6.

[5] Analysis of variance table for p_1, p_2, and p_3. Results for the individual components are displayed in the first three panels: the notation a(1) is used for the linear component p_1; a(2), the quadratic p_2; and a(3), the cubic p_3. The interactions of these components with the grouping factor, SEX, are denoted in the second line of each panel as a(1)s, a(2)s, and a(3)s. Results for p_1, p_2, and p_3 are pooled and displayed last. This panel is the same as **[5]** in Example 2V.6.

Results for the linear component of jaw size with age, a(1), are highly significant ($F = 88.0$, $p < .00005$). Thus, not only does growth change across age, but there is a highly significant linear increase. Note that the sums of squares for this component (208.266) accounts for more than 99% of the total change across the within factor AGE (208.266/209.43697). If the assumption of symmetry had been violated, the results for the linear component would still be meaningful. The linear AGE by SEX interaction is marginally significant ($F = 5.12$, $p = .0326$), indicating a difference in the slopes for the boys and girls across age (see plot in **[3]**).

Example 2V.8 One grouping and two within factors

In this example, we analyze an experiment designed to test the effectiveness of a new aspirin substitute for arthritis pain. Three patients with hip pain and three with shoulder pain are studied. (Note that the number of subjects per group need not be equal.) Each subject reports to the clinic on Monday for three weeks, when a different dose of the medication is administered (1, 2, or 4 mg, chosen so that log dose is equally spaced). Patients are tested for range of motion at times 2, 4, 6, and 10 hours after receiving the drug. Hypothetical range of motion scores are shown in Data Set 2V.4.

Data Set 2V.4
Arthritis study

Dose		1.0				2.0				4.0			
Time		2	4	6	10	2	4	6	10	2	4	6	10
	1	27	32	39	28	38	44	53	43	53	55	60	49
Hip	1	29	31	36	21	31	34	41	35	42	47	48	43
	1	37	44	47	33	53	55	58	44	64	64	69	62
	2	23	31	33	19	33	41	48	43	47	50	48	53
Shoulder	2	17	28	31	20	26	30	37	32	43	42	45	47
	2	27	37	40	27	38	44	49	33	58	56	60	61

In Input 2V.8, the INPUT and GROUP paragraphs are the same as in other BMDP programs. In the VARIABLE paragraph, we name the grouping variable, JOINT and the 12 range of motion measurements recorded for each patient. We choose names to denote the DOSE-TIME combination; e.g., d1_t2 denotes the response to a 1 mg dose at 2 hours after administration.

In the GROUP paragraph, we specify JOINT as the grouping factor and specify the names and codes for JOINT, while the DESIGN paragraph specifies the 12 DEPENDENT variables for analysis. The LEVELS of DOSE and TIME are entered as 3 and 4, respectively, because the levels of DOSE change most slowly. (Note that the first letters of the within factor names DOSE and TIME must be unique, or 2V will substitute the labels R, S, T, ... , etc. in the analysis of variance table.)

In the DESIGN paragraph we request a number of specific types of comparisons. We request an ORTHOGONAL polynomial breakdown to assess whether there are trends in the range of motion scores related to dose and whether the effects tend to wear off during the 10 hours of monitoring. Dose levels are equally spaced on a log scale, but the times are not, so we use a POINT statement to specify the actual levels for TIME. Next, we request group contrasts with the GCONT command and we specify contrast coefficients. Finally, we request a BONFERRONI pairwise trial comparisons test, to test the differences between specific trial levels means.

In the PLOT paragraph we request MINIPLOTS of average scores for each group at each time and dose. The means for each response level are plotted along the *y*-axis. The default plots generate one position on the *x*-axis for each level of the first repeated factor specified in the LEVEL command (DOSE). However, we want a separate plot for each dose level, with TIME on the *x*-axis, so we request TIME. (If we specify TRIAL = TIME, DOSE, we would get an additional set of plots with DOSE on the *x*-axis.) The miniplots use the first letter of each group name (SHOULDER and HIP) as plot symbols.

Input 2V.8

```
/ INPUT       VARIABLES = 13.
              FORMAT IS FREE.
              FILE IS 'pain.dat'.

/ VARIABLE    NAMES = joint, d1_t2, d1_t4, d1_t6, d1_t10,
                      d2_t2, d2_t4, d2_t6, d2_t10, d4_t2,
                      d4_t4, d4_t6, d4_t10.

/ GROUP       VARIABLE = joint.
              CODES(joint) = 1, 2.
              NAMES(joint) = hip, shoulder.

/ DESIGN      DEPENDENT ARE 2 TO 13.
              NAMES ARE DOSE, TIME.
              LEVELS ARE 3, 4.
              ORTHOGONAL.
              POINT(2) = 2, 4, 6, 10.
```

Input 2V.8
(continued)

```
                   GCONT(joint) = JOINTLIN.
                   JOINTLIN = 1, -1.
                   BON.

 / PLOT            MINI.
                   TRIAL = time.
 / END
```

Output 2V.8

```
DESIGN SPECIFICATIONS
---------------------
          GROUP =    1                                                          [1]
          DEPEND =   2   3   4   5   6   7   8   9  10  11  12  13
          LEVEL =    3   4
```

— we omit portions of output from this analysis —

```
SUMS OF SQUARES AND CORRELATION MATRIX OF THE                                  [2]
ORTHOGONAL COMPONENTS POOLED FOR ERROR  2 IN ANOVA TABLE BELOW

        118.83333          1.000
        109.72222          0.136   1.000

SPHERICITY TEST APPLIED TO ORTHOGONAL COMPONENTS - TAIL PROBABILITY     0.9702

SUMS OF SQUARES AND CORRELATION MATRIX OF THE
ORTHOGONAL COMPONENTS POOLED FOR ERROR  3 IN ANOVA TABLE BELOW

          55.03175         1.000
          11.38240         0.253   1.000
           7.69697        -0.037   0.628   1.000

SPHERICITY TEST APPLIED TO ORTHOGONAL COMPONENTS - TAIL PROBABILITY     0.4617

SUMS OF SQUARES AND CORRELATION MATRIX OF THE
ORTHOGONAL COMPONENTS POOLED FOR ERROR  4 IN ANOVA TABLE BELOW

          30.71905         1.000
           4.65368         0.306   1.000
          16.46061         0.588  -0.566   1.000
          94.03968         0.162  -0.738   0.845   1.000
          10.44012         0.258   0.116   0.210   0.394   1.000
           1.57576        -0.510   0.418  -0.866  -0.891  -0.657   1.000

SPHERICITY TEST APPLIED TO ORTHOGONAL COMPONENTS - TAIL PROBABILITY     0.0000

CELL MEANS           FOR  1-ST DEPENDENT VARIABLE                              [3]
------------------
                                                   MARGINAL
          JOINT    =    hip           SHOULDER

           DOSE  TIME

D1_T2       1     1    31.00000     22.33333     26.66667
D1_T4       1     2    35.66667     32.00000     33.83333
D1_T6       1     3    40.66667     34.66667     37.66667
D1_T10      1     4    27.33333     22.00000     24.66667
D2_T2       2     1    40.66667     32.33333     36.50000
D2_T4       2     2    44.33333     38.33333     41.33333
D2_T6       2     3    50.66667     44.66667     47.66667
D2_T10      2     4    40.66667     36.00000     38.33333
D4_T2       3     1    53.00000     49.33333     51.16667
D4_T4       3     2    55.33333     49.33333     52.33333
D4_T6       3     3    59.00000     51.00000     55.00000
D4_T10      3     4    51.33333     53.66667     52.50000

          MARGINAL     44.13889     38.80556     41.47222
          COUNT            3            3            6

STANDARD DEVIATIONS  FOR  1-ST DEPENDENT VARIABLE
--------------------

          JOINT    =  hip           SHOULDER

           DOSE  TIME

D1_T2       1     1     5.29150      5.03322
D1_T4       1     2     7.23418      4.58258
D1_T6       1     3     5.68624      4.72582
D1_T10      1     4     6.02771      4.35890
D2_T2       2     1    11.23981      6.02771
D2_T4       2     2    10.50397      7.37111
D2_T6       2     3     8.73689      6.65833
D2_T10      2     4     4.93288      6.08276
D4_T2       3     1    11.00000      7.76745
D4_T4       3     2     8.50490      7.02377
D4_T6       3     3    10.53565      7.93725
D4_T10      3     4     9.71253      7.02377
```

```
MINI   PLOTS   OF   CELL   MEANS   FOR   THE   1-ST DEPENDENT VARIABLE
-----------------------------------------

THE TRIALS ARE REPRESENTED BY THE VARIABLES:                              [4]
D1_T2    D1_T4    D1_T6    D1_T10   D2_T2    D2_T4    D2_T6    D2_T10
D4_T2    D4_T4    D4_T6    D4_T10

JOINT   -- h=hip      S=SHOULDER
         .+....+....+....+.+....+....+....+.+....+....+....+.
         -                -                -                -
         -                -                -          h     -
D        -                -                -     h          -
E  52.5  +                +                +h         S    *+
P        -                -          h     -S   S          -
.        -                -     h    S     -                -
V        -          h     -h              h-                -
A        -                -     S          -                -
R  35.0  +    h    S     +               S+                +
.        -h   S          -S               -                -
         -               h-                -                -
         -                -                -                -
         -S              S-                -                -
         .+....+....+....+.+....+....+....+.+....+....+....+.
          1         3      1         3      1         3
               2         4      2         4      2         4

              TIME             TIME             TIME

DOSE    :         1                2                3

A N A L Y S I S   O F   V A R I A N C E  FOR   THE   1-ST DEPENDENT VARIABLE
-----------------------------------------

THE TRIALS ARE REPRESENTED BY THE VARIABLES:
D1_T2    D1_T4    D1_T6    D1_T10   D2_T2    D2_T4    D2_T6    D2_T10
D4_T2    D4_T4    D4_T6    D4_T10
```

	SOURCE		SUM OF SQUARES	D.F.	MEAN SQUARE	F	TAIL PROB.	GREENHOUSE GEISSER PROB.	HUYNH FELDT PROB.
	MEAN	[5]	123836.05556	1	123836.05556	217.95	0.0001		
	joint		512.00000	1	512.00000	0.90	0.3962		
1	ERROR		2272.77778	4	568.19444				
	d(1)	[6]	5830.02083	1	5830.02083	196.24	0.0002		
	d(1)j		13.02083	1	13.02083	0.44	0.5441		
	ERROR		118.83333	4	29.70833				
	d(2)		9.50694	1	9.50694	0.35	0.5877		
	d(2)j		7.56250	1	7.56250	0.28	0.6273		
	ERROR		109.72222	4	27.43056				
	dose		5839.52778	2	2919.76389	102.20	0.0000	0.0000	0.0000
	dj		20.58333	2	10.29167	0.36	0.7083	0.7046	0.7083
2	ERROR		228.55556	8	28.56944				
	t(1)	[7]	0.12857	1	0.12857	0.01	0.9276		
	t(1)j		37.64444	1	37.64444	2.74	0.1734		
	ERROR		55.03175	4	13.75794				
	t(2)		843.37698	1	843.37698	296.38	0.0001		
	t(2)j		5.10859	1	5.10859	1.80	0.2513		
	ERROR		11.38240	4	2.84560				
	t(3)		44.55000	1	44.55000	23.15	0.0086		
	t(3)j		10.91364	1	10.91364	5.67	0.0759		
	ERROR		7.69697	4	1.92424				
	time		888.05556	3	296.01852	47.93	0.0000	0.0002	0.0000
	tj		53.66667	3	17.88889	2.90	0.0790	0.1297	0.0790
3	ERROR		74.11111	12	6.17593				
	dt(1,1)	[8]	35.72917	1	35.72917	4.65	0.0972		
	dt(1,1)j		8.14821	1	8.14821	1.06	0.3612		
	ERROR		30.71905	4	7.67976				
	dt(1,2)		187.09091	1	187.09091	160.81	0.0002		
	dt(1,2)j		47.41126	1	47.41126	40.75	0.0031		
	ERROR		4.65368	4	1.16342				
	dt(1,3)		0.07576	1	0.07576	0.02	0.8986		
	dt(1,3)j		1.33636	1	1.33636	0.32	0.5993		
	ERROR		16.46061	4	4.11515				
	dt(2,1)		12.90179	1	12.90179	0.55	0.4999		
	dt(2,1)j		0.55734	1	0.55734	0.02	0.8851		
	ERROR		94.03968	4	23.50992				
	dt(2,2)		12.38167	1	12.38167	4.74	0.0950		
	dt(2,2)j		7.65440	1	7.65440	2.93	0.1620		
	ERROR		10.44012	4	2.61003				
	dt(2,3)		8.18182	1	8.18182	20.77	0.0104		
	dt(2,3)j		1.30909	1	1.30909	3.32	0.1424		
	ERROR		1.57576	4	0.39394				
	dt		256.36111	6	42.72685	6.49	0.0004	0.0272	0.0031
	dtj		66.41667	6	11.06944	1.68	0.1686	0.2507	0.2062
4	ERROR		157.88889	24	6.57870				

EPSILON FACTORS FOR DEGREES OF FREEDOM ADJUSTMENT

ERROR TERM	GREENHOUSE-GEISSER	HUYNH-FELDT
2	0.9804	1.0000
3	0.5437	1.0000
4	0.2940	0.6398

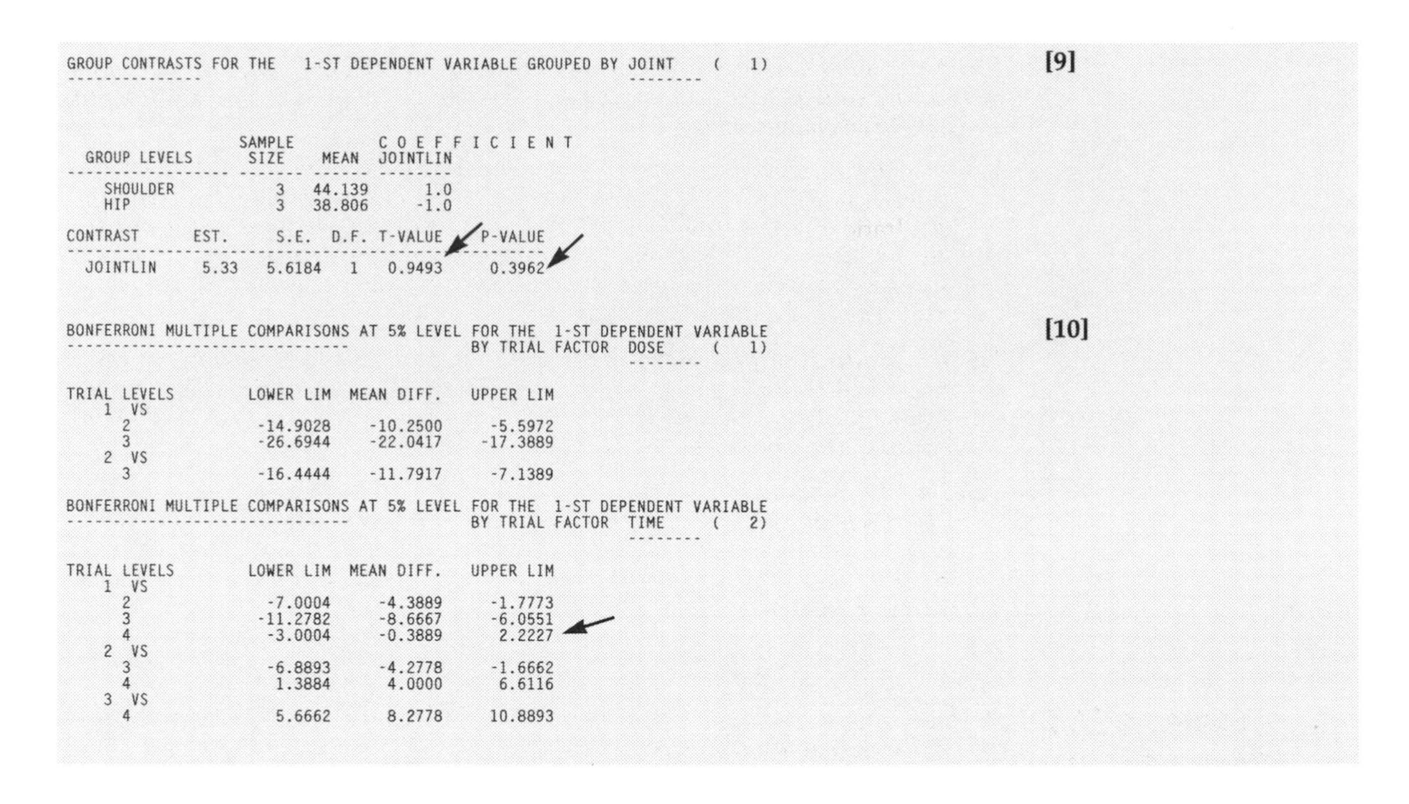

```
GROUP CONTRASTS FOR THE   1-ST DEPENDENT VARIABLE GROUPED BY JOINT    (   1)          [9]
---------------                                                  --------

                        SAMPLE             C O E F F I C I E N T
  GROUP LEVELS           SIZE     MEAN    JOINTLIN
 -----------------     -------   ------   --------
    SHOULDER               3    44.139       1.0
    HIP                    3    38.806      -1.0

CONTRAST       EST.      S.E.   D.F.  T-VALUE    P-VALUE
---------------------------------------------------------
  JOINTLIN     5.33    5.6184    1    0.9493     0.3962


BONFERRONI MULTIPLE COMPARISONS AT 5% LEVEL FOR THE  1-ST DEPENDENT VARIABLE          [10]
------------------------------                BY TRIAL FACTOR  DOSE      (   1)
                                                             --------

TRIAL LEVELS         LOWER LIM  MEAN DIFF.    UPPER LIM
   1  VS
      2              -14.9028    -10.2500      -5.5972
      3              -26.6944    -22.0417     -17.3889
   2  VS
      3              -16.4444    -11.7917      -7.1389

BONFERRONI MULTIPLE COMPARISONS AT 5% LEVEL FOR THE  1-ST DEPENDENT VARIABLE
------------------------------                BY TRIAL FACTOR  TIME      (   2)
                                                             --------

TRIAL LEVELS         LOWER LIM  MEAN DIFF.    UPPER LIM
   1  VS
      2               -7.0004     -4.3889      -1.7773
      3              -11.2782     -8.6667      -6.0551
      4               -3.0004     -0.3889       2.2227
   2  VS
      3               -6.8893     -4.2778      -1.6662
      4                1.3884      4.0000       6.6116
   3  VS
      4                5.6662      8.2778      10.8893
```

[1] The design specifications describe the grouping variable, three levels of DOSE, four levels of TIME, and 12 dependent measures.

[2] The error sums of squares of the orthogonal components for DOSE, TIME, and the DOSE x TIME interaction are shown, along with the lower off-diagonal of the correlation matrix for those error sums of squares. We see that for DOSE (error 2) and TIME (error 3), the sphericity assumption is not rejected, even though there are marked differences in the sums of squares for each orthogonal component (sample sizes are small). The results of the sphericity test for DOSE x TIME (orthogonal components pooled for "Error 4") are highly significant; thus, we would need to adjust the *p*-value when looking for an overall effect of DOSE x TIME or DOSE x TIME x JOINT.

[3] The means and standard deviations at each combination of the one grouping and two repeated factors.

[4] Miniplots of DOSE, TIME, and JOINT. The letters h and s represent categories of the grouping variable, JOINT; asterisks show overlap between h and s. The means of the dependent variable, range of motion, are plotted against time for each dose level. We note that average range of motion increases for each dose level. At the 1 and 2 mg dose levels, range of motion increases steadily up to 6 hours, and then drops sharply at 10 hours. At the 4 mg dose level (the third plot, labeled DOSE = 3), range of motion follows the same pattern for the shoulder group, but remains fairly constant across time for the hip group.

[5] Between groups analysis of variance table for p_0. There is no overall difference in average range of motion scores for patients with problem hip joints or patients with shoulder pain ($p = .3962$).

[6] Analysis of variance table for the linear and quadratic components (p_1 and p_2) for DOSE averaged across TIME. There is a significant linear increase in average motion scores ($p = .0002$). We do not find the hip and shoulder patients to have different slopes ($p = .5441$). Pooled results for the linear and quadratic components appear at the bottom of this panel. The ERROR labeled as 2 is ERROR 2 described in **[2]** above.

[7] Analysis of variance table for the linear, quadratic, and cubic components (p_1, p_2, and p_3) for TIME averaged across DOSE. There is a significant quadratic change across TIME ($F = 296.38$, $p = .0001$).

[8] Analysis of variance table for interactions between dose and time effects. There is a highly significant interaction of the linear effect of dose with the quadratic effect of time—dt(1,2). This indicates that the time patterns vary for the three doses (see miniplot in **[4]**), suggesting that we should report time effects separately for each dose. We perform a contrast on two levels of the DOSE variable in Example 2V.9.

[9] The group contrasts for the between group factor (JOINT). This test indicates that there is no significant difference between the two different groups (SHOULDER and HIP), with a *t*-value of 0.993 and a $p = 0.3962$. In this case, the group contrast is the same as the *F* test in **[5]**.

[10] The 95% Bonferroni confidence intervals for pairwise mean comparisons of the trial factor for the two grouping variables (DOSE, TIME). In these confidence intervals, all numbers show significant differences between the trial level means, with the exception of the difference for the TIME factor between level one and level four.

Example 2V.9 Contrast over a within factor

In this example, we test a hypothesis about differences between levels of a within factor. Example 2V.8 suggested that time effects may vary for each level of dose in the arthritis pain data. We need to test differences between dose levels. So, in this example we perform a contrast on two levels of DOSE. The CONTRAST command in the **DESIGN** paragraph allows us to perform within groups contrasts.

Our research question regarding this data is as follows:

- Does the range of motion for patients change significantly between the first and second drug doses?

Our null hypothesis is that range of motion does not change significantly after subjects take the second dose of medication. We supply contrast coefficients for the effect D1_D2, which measures the difference between first and second dose values.

In Input 2V.9, the contrast name is specified in the DESIGN paragraph as CONTRAST = D1_D2 (difference between DOSE1 and DOSE2) with contrast coefficients entered as a list. The order of the coefficients must match the level order. We specify our contrast as

```
d1_d2 = 1, -1, 0.
```

You can specify additional contrasts for different hypotheses about the data. For example, if another hypothesis was that range of motion does not change significantly between second and third dose levels, we could add the following contrast and contrast coefficients:

```
CONTRAST(dose) = d2_d3.
d2_d3 = 0, 1 -1.
```

Input 2V.9
(replace the DESIGN paragraph of Input 2V.8 with the following)

```
/ DESIGN   DEPENDENT = d1_t2 TO d4_t10.
           NAMES = dose, time.
           LEVEL = 3,4.
           CONTRAST(dose) = d1_d2.
           d1_d2 = 1, -1, 0.
```

Output 2V.9

—we omit output found in Output 2V.8—

```
  1-ST DEPENDENT VARIABLE ANALYSIS

GROUPING VARIABLE(S)    1 JOINT

TRIAL  LEVELS           DEPENDENT           CONTRAST                          [1]
DOSE      TIME          VARIABLE          D1_D2

   1        1            2 D1_T2            1.0
   1        2            3 D1_T4            1.0
   1        3            4 D1_T6            1.0
   1        4            5 D1_T10           1.0
   2        1            6 D2_T2           -1.0
   2        2            7 D2_T4           -1.0
   2        3            8 D2_T6           -1.0
   2        4            9 D2_T10          -1.0
   3        1           10 D4_T2            0.0
   3        2           11 D4_T4            0.0
   3        3           12 D4_T6            0.0
   3        4           13 D4_T10           0.0

CONTRAST(DOSE      ) = D1_D2

MEANS
-----

JOINT      = HIP           SHOULDER      MARGINAL                             [2]
time
  1          -9.66667      -10.00000      -9.83333
  2          -8.66667       -6.33333      -7.50000
  3         -10.00000      -10.00000     -10.00000
  4         -13.33333      -14.00000     -13.66667
MARGINAL
            -10.41667      -10.08333     -10.25000

STANDARD DEVIATIONS
-------------------

JOINT      = HIP           SHOULDER
time
  1           7.09460       1.00000
  2           4.93288       4.04145
  3           4.58258       4.58258
  4           2.08167       9.16515

ANALYSIS OF VARIANCE FOR CONTRAST(dose       ) = d1_d2
--------------------

SOURCE                     SUM OF     DEGREES OF        MEAN          F       TAIL    [3]
                           SQUARES      FREEDOM         SQUARE                 PROB.

D1_D2                   1260.75000        1          1260.75000     40.21    0.0032
DJ                         0.33333        1             0.33333      0.01    0.9228
ERROR                    125.41667        4            31.35417

DT                        58.41667        3            19.47222      2.41    0.1176
DTJ                        4.16667        3             1.38889      0.17    0.9133
ERROR                     96.91667       12             8.07639
```

[1] 2V prints a summary panel of the contrasts. We see the coefficient associated with each level of DOSE and TIME. Coefficients correspond with those stated in DESIGN paragraph instructions.

[2] The means and standard deviations at each combination of a TIME level and grouping factor (JOINT).

[3] The ANOVA table for our contrast is printed. In this table, 2V reports the sum of squares, degrees of freedom, mean square, *F* value, and *p* value for the effect (D1_D2) and interactions among the effect, the grouping factor, and the other within factor (TIME). Under the SOURCE column, D represents the effect D1_D2, not the within factor DOSE. Also, J represents the grouping factor JOINT and T represents the within factor TIME.

Tail probabilities indicate that we may reject our null hypothesis. The *p* value for the contrast between the first and second doses (.0032) is significant at the 0.01 level. The values for the interaction between the contrast and the other factors are not significant. Therefore, we can safely conclude that range of motion in tested patients does change significantly between the first and second doses of medication.

Example 2V.10 One grouping and one within factor with a covariate constant across trials

In this example, we evaluate a 12-week course to improve a student's College Entrance Exam score. Our study design includes two groups of students: one completing the course and another not taking any exam preparation course. Both groups of students take the exam once before the course begins and a second time one month after the course ends. To control for added variability of the exam scores due to differing IQs among the students, we use IQ as a covariate. That is, we can study differences in exam scores after adjusting for IQ. The scores are the combined Verbal and Mathematical scores, and are shown in Data Set 2V.5.

Data Set 2V.5
College entrance exam study

Course	IQ	Exam Score1	Exam Score2
yes	104	1000	1130
	132	1300	1490
	140	1260	1450
	119	1000	1080
no	112	940	920
	120	1160	1210
	132	1290	1330
	122	1060	1030

In Input 2V.10, the GROUPING variable, COURSE, consists of two groups (yes = took course, no = did not take course), and the repeated factor EXAM has two levels (score1 and score2). IQ, the COVARIATE, remains constant over trials. Since we use the same covariate for both scores, we must list it twice with the COVARIATE command.

Input 2V.10

```
/ INPUT      VARIABLES = 4.
             FORMAT IS FREE.
             FILE IS 'exam.dat'.

/ VARIABLE   NAMES = course, iq, score1, score2.

/ GROUP      VARIABLE = course.
             CODES(course) = 1, 2.
             NAMES(course) = yes, no.

/ DESIGN     DEPENDENT ARE score1, score2.
             LEVEL = 2.
             NAME IS exam.
             COVARIATE IS iq, iq.

/ END
```

Output 2V.10

—we omit a portion of this output—

```
CELL MEANS          FOR  1-ST COVARIATE
-------------------

                                             MARGINAL
     COURSE  =  YES          NO

           EXAM
IQ            1  123.75000   121.50000   122.62500
IQ            2  123.75000   121.50000   122.62500

     MARGINAL    123.75000   121.50000   122.62500
     COUNT               4           4           8
```

```
STANDARD DEVIATIONS  FOR  1-ST COVARIATE
-------------------

     COURSE  =  YES          NO

           EXAM
IQ           1    15.75595     8.22598
IQ           2    15.75595     8.22598

CELL MEANS           FOR  1-ST DEPENDENT VARIABLE
-------------------

                                           MARGINAL
     COURSE  =  YES          NO

           EXAM
SCORE1       1  1140.00000  1112.50000  1126.25000
SCORE2       2  1287.50000  1122.50000  1205.00000

     MARGINAL  1213.75000  1117.50000  1165.62500
     COUNT              4           4           8

STANDARD DEVIATIONS  FOR  1-ST DEPENDENT VARIABLE
-------------------

     COURSE  =  YES          NO

           EXAM
SCORE1       1   162.48077   148.63266
SCORE2       2   212.34798   182.82505

A N A L Y S I S   O F   V A R I A N C E  FOR   THE   1-ST DEPENDENT VARIABLE
--------------------------------------

THE TRIALS ARE REPRESENTED BY THE VARIABLES:
SCORE1   SCORE2

   SOURCE                       SUM OF       D.F.     MEAN           F        TAIL
                                SQUARES                SQUARE                  PROB.

   COURSE                     19180.74742     1     19180.74742     0.88    0.3907
   1-ST COVAR                265598.20958     1    265598.20958    12.22    0.0174
 1 ERROR                     108689.29042     5     21737.85808

   EXAM                       24806.25000     1     24806.25000    22.09    0.0033
   EC                         18906.25000     1     18906.25000    16.84    0.0063
 2 ERROR                       6737.50000     6      1122.91667

   REG. COEFF.                  ESTIMATE          STD. ERROR   T-VALUE   P-VALUE
   1-ST COVAR                   11.83725            3.38646       3.50    0.0174

ADJUSTED CELL MEANS  FOR  1-ST DEPENDENT VARIABLE
-------------------

                                           MARGINAL
     COURSE  =  YES          NO

           EXAM
SCORE1       1  1126.68310  1125.81690  1126.25000
SCORE2       2  1274.18310  1135.81690  1205.00000

     MARGINAL  1200.43310  1130.81690  1165.62500
     COUNT              4           4           8

STANDARD ERRORS OF ADJUSTED CELL MEANS FOR  1-ST DEPENDENT VARIABLE
--------------------------------------
     COURSE  =  YES          NO

           EXAM
SCORE1       1    73.81720    73.81720
SCORE2       2    73.81720    73.81720
```

The results for Input 2V.10 are displayed in Output 2V.10. In the between groups ANOVA table, we do not find a significant difference in the total exam scores by course adjusted by the covariate IQ (p = .3907). IQ is predictive of average score across test-retest (p = .0174). The within subject analysis of variance table shows that the change in EXAM score is highly significant (F = 22.09, p = .0033), with students' scores improving on the second exam. The interaction between EXAM and COURSE (ec) is also significant (F = 16.84, p = .0063); those students that took the 12-week preparation class improved more on the average than those students who did not. The estimated between subjects regression coefficient (11.837) is obtained by assuming that the true coefficient is the same in both groups of students. Note that when the model includes covariates, the test for MEAN is not reported.

Caution: *In Example 2V.2, we stated that in addition to the requirements of an analysis of variance, an analysis of covariance requires that*

1. within each group, the dependent variable have a linear relationship with the covariates, and

2. the slope of the regression line for each covariate be the same in each group. In this example, 11.837 is the estimate of this common slope.

These assumptions also apply for a repeated measures ANCOVA.

Example 2V.11 One grouping and two within factors with a covariate changing across trials

In this example we use a covariate that changes across dose trials. We use the range of motion data from Examples 2V.8 and 2V.9. Here we include a range of motion measure taken just prior to each drug administration as a covariate. The data are shown in Data Set 2V.6. The columns labeled with X are the covariates.

Data Set 2V.6 Arthritis study with covariate changing across trials

Dose		1.0					2.0					4.0				
Time		X	2	4	6	10	X	2	4	6	10	X	2	4	6	10
Patient																
	1	21	27	32	39	28	27	38	44	53	43	33	53	55	60	49
Hip	1	17	29	31	36	21	20	31	34	41	35	28	42	47	48	43
	1	29	37	44	47	33	32	53	55	58	44	39	64	64	69	62
	2	10	23	31	33	19	17	33	41	48	43	29	47	50	48	53
Shoulder	2	12	17	28	31	20	16	26	30	37	32	26	43	42	45	47
	2	21	27	37	40	27	23	38	44	49	33	30	58	56	60	61

Input 2V.10 is similar to Input 2V.8; however, we change our INPUT FILE to 'NEWPAIN.DAT', modify the INPUT VARIABLES and VARIABLE NAMES statements to include the three covariates (BASE1, BASE2, and BASE4), and add a COVARIATES sentence. We use each of the three covariates four times (to match their respective levels of time).

```
COVARIATES = base1, base1, base1, base1, base2, base2,
             base2, base2, base4, base4, base4, base4.
```

Thus we have 12 covariates for the 12 dependent variables. We omit the ORTHOGONAL polynomial breakdown to save space.

Input 2V.11

```
/ INPUT    VARIABLES = 16.
           FORMAT IS FREE.
           FILE IS 'newpain.dat'.

/ VAR      NAMES = joint, base1, d1_t2, d1_t4, d1_t6, d1_t10,
                          base2, d2_t2, d2_t4, d2_t6, d2_t10,
                          base4, d4_t2, d4_t4, d4_t6, d4_t10.

/ GROUP    VARIABLE = joint.
           CODES(joint) = 1, 2.
           NAMES(joint) = hip, shoulder.

/ DESIGN   DEPENDENT ARE 3 TO 6, 8 TO 11, 13 TO 16.
           NAMES ARE dose, time.
           LEVELS ARE 3, 4.
           COVARIATES ARE base1, base1, base1, base1,
                          base2, base2, base2, base2,
                          base4, base4, base4, base4.
/ END
```

Output 2V.11

```
— we omit portions of output —
DESIGN SPECIFICATIONS                              [1]
---------------------
            GROUP  =   1
            DEPEND =   3   4   5   6   8   9  10  11  13  14  15  16
            COVAR  =   2   2   2   2   7   7   7   7  12  12  12  12
            LEVEL  =   3   4
— we omit portions of output —
A N A L Y S I S   O F   V A R I A N C E  FOR   THE   1-ST DEPENDENT VARIABLE
---------------------------------------------

THE TRIALS ARE REPRESENTED BY THE VARIABLES:
D1_T2     D1_T4     D1_T6     D1_T10    D2_T2     D2_T4     D2_T6     D2_T10
D4_T2     D4_T4     D4_T6     D4_T10

        SOURCE                     SUM OF     D.F.       MEAN       F      TAIL    GREENHOUSE    HUYNH      REGRESSION
                                   SQUARES                SQUARE            PROB.     GEISSER     FELDT    COEFFICIENTS
                                                                                      PROB.       PROB.
        joint          [2]       168.73052     1      168.73052     3.27   0.1681
        1-ST COVAR              2118.15825     1     2118.15825    41.10   0.0077                                1.3618
  1     ERROR                    154.61953     3       51.53984

        dose           [3]       514.09530     2      257.04765     8.00   0.0156      0.0163      0.0156
        dj                        23.96334     2       11.98167     0.37   0.7017      0.6982      0.7017
        1-ST COVAR                 3.55480     1        3.55480     0.11   0.7492                               -0.1768
  2     ERROR                    225.00076     7       32.14297

        time                     888.05556     3      296.01852    47.93   0.0000      0.0002      0.0000
        tj                        53.66667     3       17.88889     2.90   0.0790      0.1297      0.0790
  3     ERROR                     74.11111    12        6.17593

        dt                       256.36111     6       42.72685     6.49   0.0004      0.0272      0.0031
        dtj                       66.41667     6       11.06944     1.68   0.1686      0.2507      0.2062
  4     ERROR                    157.88889    24        6.57870

 ERROR               EPSILON FACTORS FOR DEGREES OF FREEDOM ADJUSTMENT
 TERM
                        GREENHOUSE-GEISSER    HUYNH-FELDT
  2                           0.9804             1.0000
  3                           0.5437             1.0000
  4                           0.2940             0.6398

 POOLED REGRESSION COEFFICIENTS                                       [4]
   1-ST COVARIATE            1.14988

ADJUSTED CELL MEANS  FOR  1-ST DEPENDENT VARIABLE
-------------------

                                                 MARGINAL
          JOINT    =  HIP          SHOULDER

           DOSE  TIME

D1_T2        1     1    32.78870     33.32106     33.05488
D1_T4        1     2    37.45537     42.98772     40.22154
D1_T6        1     3    42.45537     45.65439     44.05488
D1_T10       1     4    29.12203     32.98772     31.05488
D2_T2        2     1    37.85585     38.33825     38.09705
D2_T4        2     2    41.52252     44.33825     42.93039
D2_T6        2     3    47.85585     50.67158     49.26372
D2_T10       2     4    37.85585     42.00492     39.93039
D4_T2        3     1    42.14004     44.22276     43.18140
D4_T4        3     2    44.47338     44.22276     44.34807
D4_T6        3     3    48.14004     45.88943     47.01474
D4_T10       3     4    40.47338     48.55610     44.51474

          MARGINAL      40.17820     42.76625     41.47222
          COUNT            3            3            6

*** N O T E *** THE ADJUSTED STANDARD ERRORS WILL NOT BE PRINTED BECAUSE COVARIATES ARE NOT CONSTANT ACROSS TRIALS.
```

[1] The design specification. The covariates are paired with their respective dependent variables.

[2] The between-subjects analysis of variance table. We find no differences between hip and shoulder JOINT after adjusting for the covariate. The estimated between-subjects regression coefficient is 1.36177, and the corresponding F statistic is 41.10 ($p = .0077$). There is a significant linear relationship, at each dose, between the covariate and movement score averaged across time.

[3] The within-subjects sum of squares DOSE, TIME, and DOSE/TIME interaction (dt). In Output 2V.10, there were no within-subject covariate adjustments, since the covariate was constant over the trial factor. In this example, the covariate is constant over TIME but changes over DOSE. Thus, a within-subject regression coefficient of –0.17676 and corresponding F statistic (0.11) are reported for DOSE but not for TIME or the DOSE/TIME interaction. If the covariate had changed over both DOSE and TIME, then a separate TIME and DOSE/TIME regression coefficient and F statistic would be reported.

[4] The estimate of the regression coefficient is made assuming that all between and within coefficients are equal.

Special Features

Miniplots of cell means

You can plot means of the dependent variable against levels of a trial factor by stating MINI. Miniplots use the VARIABLE and DEPEND specifications from the GROUP and DESIGN paragraphs (or FORM command) respectively. The layout of the plots is determined by the order in which the design is stated; however, you can alter the layout of the plots by using TRIAL.

The PLOT structure is defined as follows:

y-axis – the means of the dependent variable (specified by the DEPENDENT command in the DESIGN paragraph).

x-axis – number of repeats of the first within factor designated by the LEVELS command. You may change the default by using the TRIAL command to specify which repeated factor defines the *x*-axis. If you list more than one repeated factor with TRIAL, a separate set of plots is printed for each.

symbols – the first letters of the GROUP NAMES of the first grouping variable.

frames – separate frames are formed for the each combination of the remaining grouping variables and repeated factors.

Program 9D can also plot means for both factorial and repeated measures designs; however, only one repeated factor is allowed (you can specify more than one grouping factor). See Example 9D.4 for information on how to organize plots with one frame per subject.

Box-Cox diagnostic plots

You can state MEAN to print Box-Cox diagnostic plots for determining the appropriate transformation for stabilizing cell variances. In the Box-Cox plot, the logarithms of robust estimates of the standard deviations (*y*-axis) are plotted against the logarithms of the cell means. Please see Chapter 7D, Example 7D.6, for details on using the Box-Cox diagnostic.

Split-plot designs

2V performs a split-plot analysis when the two grouping factors are main plot or whole plot treatments and the within factor is the subplot treatment (see Figure 2V.7). Often, one of the main plot effects is a blocking factor. The repeated measures model contains an interaction of the blocking effect with the subplot treatment. If this interaction is assumed to be zero and the design is balanced (equal observations in each cell), you can pool the sum of squares of this interaction with the error. Snedecor and Cochran (1967, p. 370) consider a split-plot design with randomized blocks.

The model is

$$Y_{ijk} = \mu + M_i + B_j + \varepsilon_{ij} + T_k + (MT)_{ik} + \delta_{ijk}$$

where M is a main plot treatment, B a blocking effect, T a subplot treatment, ε the main plot error, and δ the subplot error. If the main plot treatment M is recorded as the first variable, the blocking effect B as the second, and subplots as the levels (for example, 4) of the within factor, the GROUP and DESIGN paragraph is

```
/ GROUP    VARIABLE = 1, 2.
/ DESIGN   DEPENDENT = 3 TO 6.
           LEVEL = 4.
           EXCLUDE = 12.
```

Figure 2V.7
Split-plot design

Main	Block	Treatment t_1	t_2	t_3	t_4
1	1	y_{111}	y_{112}	y_{113}	y_{114}
1	1	.	.	.	.
1	1	.	.	.	.
1	2	.	.	.	.
1	2	.	.	.	.
1	2	.		.	.
2	1	.	.	.	.
2	1	.	.	.	.
2	1	.	.	.	.
2	2	.	.	.	.
2	2	.	.	.	.
2	2	y_{221}	y_{222}	y_{223}	y_{224}

Since this split-plot model does not include $(MB)_{ij}$ and $(BT)_{jk}$, we can pool the sum of squares for *BT* with that for δ (the ERROR in the last panel). The interactions *MB* and *MBT* are not estimated because *MB* is EXCLUDED. We do not show output.

You can also specify grouping factors **only** and choose the proper error term. To do this, the data records need to be *unpacked* or reorganized so that each measure is on a separate record with codes to indicate (1) the factor MAIN; (2) the factor BLOCK; and a new code added to indicate (3) the TREATMENT level. For the above model you would use

```
/ GROUP     VARIABLE = 1, 2, 3.
/ DESIGN    DEPENDENT = 4.
            INCLUDE = 1, 2, 3, 12, 13.
```

where in the GROUP paragraph variable 1 is the main plot treatment, 2 the blocking factor, 3 the subplot treatment, and in the DESIGN paragraph 4 is the dependent variable. Here, we INCLUDE three main effects and two interactions. The interactions are between (1) main plot treatments and blocks, and (2) main plot and subplot treatments. However, you must select the proper error term for testing main plot treatment and blocking effects. 2V's tests are for a fixed effects model. They work for subplot treatment effects and the interaction of main and subplot treatment effects, but not for main plot or blocking effects.

Covariates

In 2V, covariates may be used in designs with (1) between subject factors only (see Example 2V.2), and (2) repeated measures where the covariate is constant across trials (see Example 2V.10) **or** changes across trials (see Example 2V.11). Use the COVARIATE command to list the covariates. When there are no repeated measures, the covariates may be listed in any order. In repeated measures designs the COVARIATES list must correspond to the variables in the DEPENDENT list. In Example 2V.10, the covariate IQ **does not change across trials** (SCORE1 and SCORE2), so we specify

```
DEPENDENT = score1, score2.
COVARIATES = IQ, IQ.
```

in the DESIGN paragraph. As a more complex example, suppose we have two repeated factors, DOSE and TIME, and two covariates, AGE and RBC, constant across trials. In Figure 2V.8, we display a sample data record. (There are two doses and measurements taken at 3 times for each.) The respective DEPENDENT and COVARIATES lists are:

```
DEP =    d1_t1,   d1_t2,   d1_t3,   d2_t1,   d2_t2,   d2_t3.
COV = age,RBC, age,RBC, age,RBC, age,RBC, age,RBC, age,RBC.
```

Thus, in the COVARIATES list, we repeat AGE and RBC for each measure in the

DEPENDENT list. The number of entries in the COVARIATES list equals the number of entries in the DEPENDENT list (when there is one covariate), or a multiple of the number in the DEPENDENT list (when there are two or more covariates).

Figure 2V.8
Sample data record for design with covariates constant across trials

		Dose1			Dose2		
		Time1	Time2	Time3	Time1	Time2	Time3
age	RBC	d1_t1	d1_t2	d1_t3	d2_t1	d2_t2	d2_t3
35	438	235	256	546	454	412	788

As an example of **covariates that change across trials,** suppose we again collect measurements at three times for each of two doses. We measure RBC separately for each dose, and, in addition, collect calcium values at the same time. (We now have two values for each covariate: RBC1 and RBC2, and CALC1 and CALC2.) We would then use the following DEPENDENT and COVARIATES lists in our input:

```
DEP = d1_t1,       d1_t2,       d1_t3,
      d2_t1,       d2_t2,       d2_t3.
COV = RBC1,CALC1,RBC1,CALC1,RBC1,CALC1,
      RBC2,CALC2,RBC2,CALC2,RBC2,CALC2.
```

Predicted values and residuals

Predicted values and residuals can be printed or saved in a BMDP File to be further analyzed or plotted by other BMDP programs (e.g., 6D). When there are no within factors, there is only one predicted value for each case (the estimate of its expected value). The residual is the difference between the observed and predicted values. When there are within factors, two sets of predicted values can be estimated in terms of

- the orthogonal polynomials used in the computation of the sums of squares; a predicted value and residual is computed for each orthogonal polynomial and for products of orthogonal polynomials corresponding to interactions.
- the means and deviations from the means.

For within factors, you can choose between the two sets of parameters by stating, in the DESIGN paragraph, either

`RESIDUALS ARE ORTH.` or `RESIDUALS ARE MEAN.`

RESIDUALS ARE MEAN is the default. The predicted values and residuals are requested in the PRINT paragraph. (The default is MEAN.) You can SAVE the residuals with the data in a BMDP File (see Chapter 8). To save the data without the residuals specify

```
NO RESIDUAL.
```

in the SAVE paragraph.

When the option MEAN is used, the residuals and predicted values are computed both for individual cells and for all marginals of the within factors. Therefore, the number of pairs of predicted values and residuals is the product $(1 + L_1)(1 + L_2)\ldots$, where L_1 is the number of levels of the first within factor, etc. When the orthogonal polynomial option is selected, the number of pairs of predicted values and residuals is the product $L_1L_2\ldots$. See BMDP Technical Report #83 for details on order, names, and interpretation of residuals.

Case weights

The user can supply case weights when the variances of the residuals are known to differ. Different choices are appropriate depending on what is known about the variances. For example, when the cases that are input are actually averages of groups of cases, the case weights are usually the numbers of observations used in defining the averages. Case weights affect the between-groups comparisons but not the within-groups comparisons, as only one weight value is specified per case. If the weight for a case is equal to zero, the case is not used in the computations of any statistics except predicted values and residuals.

To specify case weights, you need to add a sentence to the VARIABLE paragraph defining the name or subscript of the variable containing the case weights; e.g.,

```
WEIGHT = wtvar.
```

The total number of degrees of freedom is equal to the number of cases with positive case weight.

Using the FORM command to specify the design

You may use either of two ways to identify your experimental design. In Examples 2V.1 through 2V.11 we used one or more of the following commands in the GROUP paragraph:

VARIABLE – to specify the grouping factors.

CODES – to specify codes for grouping factors (cutpoints if continuous).

NAMES – to identify the names of the grouping factors.

In Examples 2V.1 through 2V.11, the following commands were used in the DESIGN paragraph:

DEPENDENT – to identify the measures analyzed in factorial and repeated measures designs.

LEVEL – to list the number of levels for each within subjects factor.

COVARIATES – to list covariates.

A second way to specify the design is to use the FORM command in the DESIGN paragraph. The FORM command uses a compact notation to tell how to use each variable on the input record. List one symbol to identify the role of each variable in the order it appears in the data: use G for a grouping variable; Y for a dependent variable; X for a covariate; and D to indicate a variable that is not used. You must account for all variables including those generated by transformations—but only up to the last variable actually used in an analysis. In other words, if you have 25 variables on a record, and you are only using the first ten, you do not specify 15D at the end of the FORM statement.

Grouping Factors Only. In Example 2V.1, each data record for a two-way ANOVA design included an ID, the two grouping variables BEAT and TRAINING, followed by the dependent variable ATTITUDE. The FORM command is

```
FORM = 'D,2G,Y'.
```

Covariates. In Example 2V.2, we performed a two-way analysis of variance with two grouping factors and two covariates. There were 8 variables on each data record: ID (not used), SEX and CIGARETS (grouping variables), AGE and PULSE_1 (covariates), PULSE_2 (dependent variable), and NAMES (two alphanumeric variables that are not used). The FORM command for that example is

```
FORM = 'D,2G,2X,Y'.
```

Note how we omit 2D for the names, because they occur after the last variable used.

Repeated Factors. When you use a FORM command to specify designs with repeated factors, you need to use parentheses to indicate the levels of the repeated factor(s). Example 2V.6 is an analysis of variance with one grouping factor (SEX), and one repeated factor (AGE) with four repeated levels (8, 10, 12,

14). The order of the variables on each record is SEX, AGE_8, AGE_10, AGE_12, and AGE_14. The FORM command is

```
FORM = 'G,4(Y)'.
```

With more than one repeated factor, we nest parentheses. In Example 2V.8, the arthritis medication experiment, we had one grouping (JOINT) and two repeated factors (DOSE and TIME). We show a sample data record with annotations in Figure 2V.9. A data record in this study consists of 13 entries. The first entry is a code for JOINT (the grouping factor). Next are the response measures, recorded here as three sets of four numbers. When there is more than one repeated factor, we nest parentheses so that the levels of the fastest moving factor are nested closest to the dependent variable indicator (Y). Since DOSE is the slowest-moving repeated factor and TIME the fastest-moving one, we write the FORM command as

```
FORM = 'G,3(4(Y))'.
```

Figure 2V.9
Sample data record for design with one grouping and two repeated factors

Dose	1				2				4			
Time	2	4	6	10	2	4	6	10	2	4	6	10
1	23	28	37	46	35	40	52	57	47	45	53	60

In Example 2V.11, we added a covariate to Example 2V.8; the covariate changed across each level of DOSE. Therefore, the FORM command would appear as

```
FORM = 'G,3(X,4(Y))'.
```

See the *BMDP User's Digest* for additional examples of how to use the FORM command.

Repeated measures for multiple dependent variables

You can request a repeated-measures analysis of each of several dependent variables. For example, if we had three dependent variables U, V, and W recorded in the following form:

```
G  U1  V1  W1  U2  V2  W2  U3  V3  W3  U4  V4  W4
```

The design could be specified as

```
/ GROUP    VARIABLE = 1.

/ DESIGN   DEPEND = 2 TO 13.
           LEVEL = 4.
```

The design could also be specified as

```
/ DESIGN   FORM = 'G,4(3Y)'.
```

The expression 4(3Y) means that each of three variables has been measured four times and that the first measurements of all three variables are recorded in the data matrix before any of the second measurements.

Confidence intervals for trial cell means

It is now possible to produce a printout of the 95% confidence intervals for trial cell means within each group as well as 95% confidence intervals of population adjusted cell means within each group. Use the CONFIDENCE command alone in the PRINT paragraph to create horizontal confidence limits. If you prefer a vertical format, use the CONFIDENCE command with the VERTICAL command in the PRINT paragraph.

2V Commands

/ VARiable

WEIGht = variable. `WEIGHT = WTVAR.`

Specify the name or number of the variable containing the case weights. Default: no case weights.

New Feature

MMV = #. `MMV = 10.`

Specify the maximum number of missing values or values out of range that will be allowed before a case is excluded from the analysis. Default: use all acceptable values.

/ GROUP

New syntax

VARiable = list. `VARIABLE = SEX, CIGARET.`

List names or numbers of the GROUPING variables (between subject factors). If you prefer, you can still specify a grouping variable with the GROUPING command in the DESIGN paragraph as described in the 1990 *BMDP Manual*. You must specify CODES or CUTPOINTS in the GROUP paragraph if the variable takes on more than 10 distinct values. Groups must have unique first letters in order for the names to appear in the analysis of variance table; if not, the groups are labeled G, H,..., etc. in the table. **Example:** The variables SEX and CIGARET contain values for separating the cases into cells (groups). See Volume 1, Chapter 5, for details on the GROUP paragraph.

/ DESIGN

Required. Specify design by either DEPEND, LEVEL, and COVA, or FORM. If FORM is used, the rest are ignored. See INCLUDE and EXCLUDE to specify the appropriate error term for testing special models.

FORM = 'symbol list'. `FORM = '2(4(Y)), D, G'.`

Specify the FORM of the design using symbols G (grouping variable), Y (dependent variable), X (covariate), and D (variable not used). See Special Features.

DEPendent = list. `DEPEND = SCORE_1, SCORE_2.`

Indicate names or numbers of the dependent variables, i.e., the variables that are the measurements or repeated measures values being tested. **Example:** SCORE_1 and SCORE_2 are the dependent variables. For repeated measures designs, use with LEVEL. If LEVEL is omitted, the 2 variables are analyzed separately.

LEVel = # list. `LEVEL = 2, 4.`

Specify the number of levels of the trial factors for repeated measures variables. The first number is that for the most slowly changing factor, the second for the next most slowly changing, etc.

NAME = list. `NAME = DAY, AMPM.`

List names of the trial factors (up to four characters each) for labeling the output. The order of these names must match the numbers in the LEVEL command; e.g., the first NAME is for the first trial factor specified in the LEVEL command, or the leftmost trial factor in the FORM command, etc. Names of trial factors and grouping factors must have unique first characters in order to be used as output labels. (Only the first character is used in the ANOVA table.) Use when more than one trial factor is specified. Default: R,S,T,... are used.

COVariates = list. `COV = IQ, AGE.`

State names or numbers of the covariates. In an analysis of variance without repeated factors, the covariates can be listed in any order. In a repeated measures analysis of variance, the COVARIATES list must correspond to the DEPENDENT list to indicate the specific covariates assigned to each dependent variable. Each covariate is listed several times if it does not change across trials. Note that you can use either COVARIATE = AGE, AGE, AGE, for example, or the equivalent expression COVARIATE = 3*AGE. See Special Features for examples.

INCLude = # list. `INCLUDE = 1, 2, 3, 12, 13.`
EXCLude = # list. `EXCLUDE = 23, 123.`

Specify terms to be included or excluded from the model for the grouping factors. Use with designs that do not fit all possible interactions of the grouping factors, e.g., Latin squares, incomplete blocks, and fractional factorials. Specify the main effects and interactions to be included in the model **or** the interactions to be excluded. The grouping factors are identified by the numbers 1, 2, 3 (the order as defined in DESIGN GROUP or FORM—not the number of the grouping variable). Interactions are specified as 12, 13, 23, 123, etc. **Example:** The model will include 3 main effects (1, 2, and 3) and 2 interactions (between factors 1 and 2 and factors 1 and 3). The alternative specification is that the 2nd order interaction, 23, and the 3rd order interaction, 123, are assumed to be zero and are part of the error.

ORTHogonal. `ORTH.`

Prints a separate analysis of variance table for each orthogonal polynomial. The orthogonal decomposition into linear, quadratic, etc. components is of special interest when the levels of the trial factor are ordered.

POINt(*j*) = # list. `POINT(2) = 1, 3, 4.`

Indicate spacing for the levels of the trial factor *j*, where *j*=1 is the first factor in the LEVEL or FORM command, *j*=2 the second, etc. The numbers are the values assigned to the levels of the trial factor to obtain unequal spacing in the orthogonal decomposition. **Example:** Measurements for the 2nd trial factor were taken at 1, 3, and 4 o'clock. If POINT is not specified, the analysis assumes 1, 2, and 3 o'clock (equal spacing).

CONTrast(factor name) = list.

```
CONT(DOSE) = D1, D2.
D1 = 1, -1, 0, 0.
D2 = 1, 0, -1, 0.
```

Specify names for within group contrasts. Specify coefficients for each CONTRAST listed as shown below. Default: no within group contrasts.

New Feature

GCONTrast (factor name) = list.

```
GCONT(JOINT) = JOINTLIN.
JOINTLIN = 1, -1.
```

Specify names for between group contrasts. Specify coefficients for each CONTRAST listed as shown below. Default: no between group contrasts.

SYMmetry. `NO SYM.`

Unless you state NO SYM, 2V prints a test for sphericity of the orthogonal polynomial decomposition, using the error sum of squares of trial factor.

MEAN. `NO MEAN.`

Unless you specify NO MEAN, 2V prints the mean and standard deviation.

RESIdual = *(one only).* `RESID = ORTH.`
MEAN, ORTH.

Predicted values and residuals can be estimated two ways for repeated measures models: (1) using each orthogonal polynomial and products of orthogonal polynomials corresponding to interactions, and (2) using means and deviations from means. The specified residuals and predicted values can be PRINTED and SAVED. See Special Features and BMDP Technical Report #83 for more information. Default: MEAN.

PRINt. `PRINT.`

Prints the residuals and predicted values (see RESIDUAL).

TOLerance = #. `TOL = .001.`

Specify the tolerance limit, which is used to avoid pivoting when there will be a loss of accuracy. This rarely needs to be changed. Default: .01.

New Feature

BONferroni. `BON.`

Prints the 95% Bonferroni confidence intervals for pairwise mean comparisons between trial factor levels. Default: NO BON.

/ PLOT

MEAN. `MEAN.`

Prints a Box-Cox diagnostic plot to determine the appropriate transformation for stabilizing cell variances. A logarithmic scale is used for plots of the cell standard deviation versus the cell mean. See program 7D for more information.

Miniplot options

Use to plot cell means for repeated measures designs in up to 8 plots per page (all with the same horizontal and vertical scale). Use MINI alone or specify GROUP and/or TRIAL (to obtain plots you must specify one or more of these commands). See program 9D for more information on plotting means with a repeated factor, and for plotting means for designs with no repeated measures.

MINIplot. `MINI.`

Plots the dependent variable means against levels of a trial factor. Miniplots use the VARIABLE and DEPEND specifications from the GROUP and DESIGN paragraphs (or FORM command) respectively. The layout of the plots is determined by the order in which the design is stated. However, you can alter the layout of the plots by using the TRIAL command. The PLOT structure is defined as follows:

y-axis – the means of the dependent variable (specified by the DEPENDENT command in the DESIGN paragraph).

x-axis – number of repeats of the first within factor designated by the LEVELS command in the DESIGN paragraph. You may change the default by using the TRIAL command to specify which repeated factor defines the *x*-axis. If you list more than one repeated factor with TRIAL, a separate set of plots is printed for each.

symbols – the first letters of the GROUP NAMES of the **first** GROUPING variable.

frames – separate frames are formed for each combination of the remaining grouping variables and repeated factors.

TRIAL = repeated measures factor(s) for *x*-axis. `TRIAL = TIME.`

Use to alter the default plot layout described for MINI. List name(s) or number(s) of repeated measures factors to use as labels for the *x*-axis. A separate miniplot layout is formed for each trial factor listed with the levels of that trial factor appearing on the *x*-axis.

GROUP = list. `GROUP = A, B.`

Use to alter the plot symbols described for MINI. List names or numbers of DESIGN GROUPING variables used as plot symbols. A separate miniplot layout is formed for each variable listed (all cell means from the DESIGN paragraph specification appear in each layout). The plot symbols identify the groups for this variable—the remaining grouping variables are used to form plot frames.

SIZE = #1, #2. `SIZE = 10, 12.`

Specify the size of each miniplot frame as the number of characters for the horizontal axis (#1) and the number of lines for the vertical axis (#2). State the width first. The default plot is 15 characters wide, 12 lines high. The minimum plot frame allowed is 9 characters wide, 10 lines high.

/ **PRINT**

MARGinal. `MARGINAL.`

Print unweighted marginal means for each term in the ANOVA table (cell frequencies are not used as weights in the computations).

RESIdual. `RESIDUAL.`

Prints the predicted values and residuals after the data for each case. See the RESIDUAL command in the DESIGN paragraph to select which residuals to print.

New Feature

CASE = #. `CASE = 50.`

Specify the number of cases for which data are printed. Values flagged by MISS, MAX, and MIN in the VARIABLE paragraph specifications are replaced by MISSING, GT MAX, and LT MIN. Default: 10.

New Feature

MINimum. `MIN.`

Prints data for cases in which at least one variable has a value less than its lower limit as specified in the VARIABLE paragraph. Default: such cases are not printed unless you specify DATA.

New Feature

MAXimum. `MAX.`

Prints data for cases in which at least one variable has a value greater than its upper limit as specified in the VARIABLE paragraph. Default: such cases are not printed unless you specify DATA.

New Feature

MISSing. `MISS.`

Prints data for cases that have at least one variable with a missing value. Missing values in the input may be identified in several ways: (1) an asterisk (the default missing value character), (2) a missing value code identified using MISS in the VARIABLE paragraph, (3) a character identified as MCHAR in the INPUT paragraph, (4) blanks (applies only to input read with FORTRAN fixed format). Default: such cases are not printed unless you specify DATA.

New Feature

METHod = VARiable *or* **CASE.** `METH = CASE.`

Use to determine how the printout is organized. VARIABLE is the default value.

VARIABLE – Prints for each case only as many variables as fill the print line (up to ten variables) and begins the next line with the same variables for the next case. Thus the data for each variable are in an easy-to-scan column. When the program finishes printing the first subset of variables for all cases, it continues with another subset.

CASE – Prints the values of all variables for one case (possibly filling several print lines) before printing data for the next case.

New Feature

CONFidence. `CONF.`

State CONF to obtain a printout of confidence intervals of trial cell means within each group as well as the confidence intervals for population adjusted cell means within each group. Unless you specify VERT, the confidence intervals will be printed in a horizontal format. In order to print the confidence intervals, the MEAN option in the PRINT paragraph must be in effect. Default: no confidence intervals.

New Feature

VERTical. `VERT.`

State VERT to obtain a vertical format printout of confidence intervals for trial cell means within each group. This command can only be used with both the MEAN and CONF commands in the PRINT paragraph. Default: horizontal confidence intervals.

MEAN. NO MEAN.

Unless you specify NO MEAN, 2V prints the mean, standard deviation, and frequency of each used variable.

New Feature

EXTReme. NO EXTR.

Unless you state NO EXTR, 2V prints the minimum and maximum values for each variable used.

New Feature

SKewness. SK.

Prints skewness and kurtosis for each used variable. No skewness or kurtosis printed.

New Feature

EZSCores. EZSC.

Prints z-scores for the minimum and maximum values of each used variable. Default: no z-scores.

New Feature

ECASe. ECAS.

Prints the case numbers containing the minimum and maximum values of each variable. Default: No extreme cases numbers printed.

/ SAVE

Required only if a BMDP File is created. See description of CODE, UNIT, and NEW in Chapter 8. An additional command for 2V is RESIDUAL.

RESIdual. NO RESI.

When a BMDP File is made, the residuals and predicted values are saved after the data for each case. When there are no within factors, there is only one predicted value for each case (the estimate of its expected value). When there are within factors, two sets of predicted values can be estimated in terms of (1) the orthogonal polynomials used in the computation of the sums of squares, and (2) the means and deviations from the means. To save data without the residuals, state NO RESIDUAL. When DESIGN RESIDUAL = MEAN is specified, the residuals and predicted values are computed for each cell and for all marginals of the trial factors. When DESIGN RESIDUAL = ORTH is selected, the number of pairs of predicted values and residuals is the product $L_1L_2\ldots$, where each L_i is the number of levels of trial factor i.

Order of Instructions

■ indicates required paragraph

■ / INPUT
/ VARIABLE
/ GROUP
/ TRANSFORM
/ SAVE
■ / DESIGN
/ PLOT
/ PRINT
■ / END
data

Repeat for additional problems.
See Multiple Problems, Chapter 10.

Summary Table for Commands Specific to 2V

	Paragraphs Commands	Defaults	Multiple Problems	See
	/ **VARiable**			
	WEIGht = variable.	no case weights	●	SpecF
	MMV = #.	use all acceptable values	●	Cmds
	/ **GROUP**			
▲	VARiable = variables.	no grouping	–	Cmds
■	/ **DESIGN**			
	FORM = 'symbol list'.	none	–	SpecF
	DEPendent = list.	none	–	2V.1
	LEVel = # list.	none	–	2V.6
	COVariates = list.	none	–	2V.2, 2V.10–11
	INCLude = # list.	none	–	2V.3–2V.5
	EXCLude = # list.	none	–	2V.3, 2V.4
	NAME = list.	R, S, T,..., etc.	–	2V.10
	ORTHogonal.	none	●	2V.7
	POINt(*j*) = # list.	Assumes equal spacing.	–	2V.8
	CONTrast(factor) = list.	no within groups contrasts	–	2V.9
	GCONT(factor) = list.	no between groups contrasts	–	2V.8
	contrast name = # list.	(required with CONTRAST, GCONTRAST)	–	2V.9
	SYMmetry *or* NO SYM.	SYM.	●	2V.6–2V.8
	MEAN *or* NO MEAN.	MEAN.	●	2V.1–2V.7
	RESIdual = (*one only*) MEAN, ORTH.	MEAN.	●	SpecF
	PRINT.	NO PRINT.	●	Cmds
	TOLerance = #.	.01	●	Cmds
	BONferroni.	NO BON.	●	2V.8
▲	/ **PLOT**			
	MEAN. (Box-Cox plots)	NO MEAN.	●	SpecF
	MINIplot.	NO MINI.	●	2V.7, 2V.8
	TRIAL = list.	If PLOT MINI or PLOT GROUPS specified, labels on *x*-axis are those used in first trial factor.	–	2V.8
	GROUP = list.	If MINI and/or TRIAL specified, plot symbols taken from 1st GROUPING variable in DESIGN.	–	Cmds
	SIZE = #1, #2.	15, 12	●	Cmds
	/ **PRINT**			
	MARGinal.	NO MARGINAL.	●	Cmds
	RESIdual.	NO RESIDUAL.	●	2V.2
	CASE = #.	10	●	Cmds
	MISSing.	NO MISS.	●	Cmds
	MINimum.	NO MIN.	●	Cmds
	MAXimum.	NO MAX.	●	Cmds
	METHod = VARiables *or* CASE.	VARIABLES	–	SpecF
	CONFidence.	NO CONFidence intervals	●	Cmds
	VERTical.	Horizontal confidence intervals	●	Cmds
	MEAN *or* NO MEAN.	MEAN.	●	Cmds
	EXTReme *or* No EXTR.	EXTReme. values printed		
	SKewness.	NO SKewness.	●	Cmds
	EZSCores.	NO EZSC. (skewness and kurtosis)	●	Cmds
	ECASe.	NO ECASe. (Case #'s of Min and Max printed)	●	Cmds

Summary Table (continued)

Paragraphs Commands	Defaults	Multiple Problems	See
/ **SAVE**			
RESIdual.	RESIdual.	●	Cmds

Key:
- ■ Required paragraph or command
- ▲ Frequently used paragraph or command
- ● Value retained for multiple problems
- – Default reassigned

Appendix A
Problem Size

Contents

A.1 Increasing the capacity of a BMDP program

If your problem does not exceed the capacity of the program (and most problems do not), this appendix will not be needed. In this section we describe how to increase the capacity of the program. In the following section (A.2) we give ***approximate*** formulas from which you can estimate the capacity needed. The amount of space that you have for arrays is reported in BMDP output.

When the number of locations needed for an analysis exceeds the number printed in the output line "THE AMOUNT OF SPACE REQUESTED IS # INTEGER WORDS", the capacity of the program must be increased.

On **VAX/VMS systems,** increase the program capacity by stating

```
LEN = n
```

on the command line after the instructions calling a BMDP program and specifying input and output files. *n* is the number of words that will be used for dynamic storage.

On **IBM OS** and similar systems, additional array space is obtained by increasing the REGION parameter on the EXEC system card.

```
// EXEC BIMED,PROG=BMDPxx,REGION=nnnnk
```

where *nnnnk* is thousands of bytes (4 byte/word), and M (the number of locations as determined from the formula in Appendix A.2) is in words.

The total REGION size should also accommodate the program, which (on the average) takes 190K of memory.

When using **BMDP PC**, the number of words of dynamic storage (M) is fixed at 16,000 and cannot be increased. However, problem size in BMDP/386 Dynamic can be increased by adding additional RAM memory until a maximum of 16 megabytes of memory is installed on the machine.

A.2 Program limitations

The amount of data analyzed by BMDP varies from program to program. It usually depends on the number of variables analyzed and sometimes is restricted by the number of cases and groups. In our discussion, we provide general guidelines regarding the number of variables, cases, and groups. We add more specific limits for the number of storage locations (M) in computer memory (computer words). We base our discussion on $M \leq 16{,}000$ words, the upper limit for BMDP PC. To check how many words of storage are available on your system, look for the line

```
NUMBER OF INTEGER WORDS OF MEMORY FOR STORAGE  . . . . .   19998
```

in the interpretation of the BMDP instructions at the beginning of your output. On some systems you can increase this upper limit (see Appendix A.1).

Messages about insufficient space. If a program runs out of space you will get a message like the following:

```
UNABLE TO ALLOCATE SUFFICIENT SPACE FOR ARRAY PYXR.
THE AMOUNT OF SPACE REQUESTED IS 1766 INTEGER WORDS.
AVAILABLE STORAGE SPACE MUST BE INCREASED TO AT LEAST 20913.
INCREASE SPACE USING THE LEN=### OPTION ON THE COMMAND LINE.
```

The amount of suggested increased space may not be enough, because the space report might have been printed early in the computations. A later routine may require more space.

General comments. All programs allot memory locations for **basic information** such as variable names, transformations, control language, etc. Consider that the basic information requires roughly $8V_t + 2000$ memory locations (see Table A.1 for notation used). Additional space is required for **grouping information. Plots** reuse locations used in the analyses, and are formed one at a time as there is room available.

Note that our discussion focuses on memory locations, not disk space. Many programs do not require that data be held in memory. The limit on cases for those programs depends on the amount of storage available on disk on your system.

Table A.1
Notation used in problem size discussion

V_i	– number of variables read as input
V_t	– number of variables after transformations (V_i plus any variables added by transformation)
V_{use}	– number of variables used in analysis (V_t if no USE list is present)
C_{use}	– number of cases used in analysis (e.g., number of complete cases with USE set to one by case selection)
G_n	– number of groups (number of CODES specified plus the number of intervals defined by CUTPOINTS, plus 10 times the number of grouping variables for which CODES or CUTPOINTS are not specified)
GV_n	– number of grouping or category variables
P	– space required for one plot: $(L_n + 11)(C_n + 15)/4$, where L_n is number of lines, C_n number of columns
B	– space required for basic information; roughly $V_t + 2000$

What to do when available space is exceeded. If your system does not allow you to request more space, you must do less in one computer run. This may require some preliminary exploratory runs to see if you can eliminate some variables, cases, number of categories per grouping variable, plots, etc. Check the guidelines below for factors affecting the space requirements of a specific analysis. You might consider the following:

Variables. For programs where M is a function of V_{use}, begin by using the USE list in the VARIABLE paragraph to decrease the number of variables. By running your analysis for several different subsets of variables, you may find that several variables are not useful and can be deleted from your final analysis. Or you may try to find a meaningful summary measure (e.g., a total or a slope) for a subset of variables and use that in later analyses. To obtain the summary measure, you could use program 4M to compute factor scores from highly correlated subsets, or you could compute summary measures using the multiple argument functions in the TRANSFORM paragraph. In a time series or repeated measures problem you might try a preliminary analysis using every other measure.

The worst case scenario is that you might have to read fewer variables. With fixed format it is easy to skip data fields or read only the first part of a record. With FREE format you can use RECLEN in the INPUT paragraph to limit the number of columns scanned on each record. Otherwise you might have to use a word processor or editor to delete variables from the file. If you can read your file into program 1D, you can use the KEEP command in the SAVE paragraph to write out a file with fewer variables.

Cases. If there is no special ordering among the cases in your file, you might read only the first, say, 80% of the cases using the CASE command in the INPUT paragraph. Or you might try a random subsample (or even 2 or 3 random subsamples; see Chapter 6).

Groups. Be sure to specify CODES or CUTPOINTS in the GROUP paragraph, because many programs assume 10 levels for each grouping variable unless otherwise specified.

Plots. Use the SIZE command to make plots smaller, or request fewer plots. Also, restrict the number of groups.

Space Allocation for Volume 1 Programs

	Limits		
	Cases	Variables	Approximate formula and comments
1D	none	V_t, V_{use}	$M = B + 5V_t + 4GV_n + G_n(6V_{use})$ Sorting uses available space.
2D	none	V_t, V_{use} where	$M = B + 9V_t + 3V_{use} + 2ND$ ND = total number of distinct values for each variable
3D	none	V_t, V_{use} where	$M = B + 7V_t + 7V_{use} + 29G_n + TV_n(11G_n + 2TV_n + 4)$ TV_n = number of matched pairs or of one-group or two-group variables
5D	none	V_t where	$M = B + 6V_t + 4G_n + G_nP_n + 24G_t + I_tT_n + 4F_n + P$ G_t = total number of groups specified by GROUP command F_n = number of frequency tables P_n = number of plot variables I_t = total number of intervals for all histograms T_n = number of histogram types Space is allocated for one PLOT paragraph at a time.

6D	none	V_t	$M = B + 6V_t + 2V_tG_n + 4G_n +$
			$\Sigma_{all\ plots}(3XV_n + 3YV_n + 14NP) + P$
		where	XV_n = number of x variables
			YV_n = number of y variables
			NP = total number of plots in all PLOT paragraphs
			If symbols are requested, XV_nYV_n is added for each plot.
			Space is allocated for all plots in each problem.
7D	C_{use}	V_t	$M = B + 7V_t + C_{use}(V_{use} + 14) + 4G_n + G_nGV_n$
			Add $P + 3G_n$ if a plot is specified.
			If contrasts are requested, additional space is required.
9D	none	V_t, V_{use}	$M = B + 2V_t + 2V_{use} + 15GV_n + 2GV_n(G_n + 40) + 250G_n$
4F	none	V_t	$M = B + 24V_t + 5C_n + 15M_g + 20T_t + 7GV_n + 34TP_n$
			If structural zeros are used, add $T_tC_n^2$
			If tables are saved, add $5G_n$.
			If loglinear modeling is performed, add
			$IT_n + 40FP_n + M_n[9 + GV_n(ST + 4)(1 + L)]$
		where	C_n = number of cells input
			M_g = maximum group size
			T_t = total number of all tables created
			TP_n = number of TABLE paragraphs
			IT_n = maximum number of iterations
			FP_n = number of FIT paragraphs
			M_n = number of terms in the model
			ST = maximum step size
			L = 1 if marginals requested, 0 otherwise
4M	none	V_t, V_{use}	$M = B + 22V_{use} + 8V_t + 2MF^2$ if loadings or factor scores only are input.
			$M = B + 23V_{use} + 14V_t + V^2_{use}(MF + 9) + V_{use}(2MF + 8)$ if data are input.
		where	MF = maximum number of factors
			Add 2050 if more than one factor loading.
			Add P if a plot is requested and there is more than one factor.
7M	C_{use}	V_t, V_{use}	$M = B + 3V_t + 27V_{use} + V^2_{use} + 4C_{use} +$
			$20G_n + 6G_n^2 + G_n(5V_{use} + C_{use})$
			Add $2C_{use}$ if plot or canonical scores are requested.

2R	none	V_t V_{use}	$M = B + 12V_t + 34V_{use} + V^2_{use} + 260$ If plots are requested, add $P + 4VP_n + 2PR_n + 2XY_n + V^2_{use}/2$
		where	VP_n = number of variables to plot PR_n = number of PREP plots XY_n = number of $x-y$ plots If simulations or information about EXTREME values are requested, additional space is required.
9R	C_{use}	V_t, V_{use}	$M = B + 8V_t + 6V_{use} + C_{use}(V_{use} + 9) + CL_n(V_t + 3V_{use} + 6) + P$
		where	CL_n = number of clusters
AR	Cuse	V_t, V_{use}	$M = B + 12V_t + V_{use} + 2C_{use}(V_{use} + 200) + 300$ Add P if plots are specified.
3S	C_{use}	V_t, V_{use}	$M = B + 13V_t + 5G_n + C_{use}(V_t + 5) + V_{use}[(3S_1 + 3S_2 + 3W + K)(V_{use} + 1)/2]$
		where	$S_1 = 1$ if sign test requested, 0 otherwise $S_2 = 1$ if Spearman test requested, 0 otherwise $W = 1$ if Wilcoxon test requested, 0 otherwise $K = 1$ if Kendall test requested, 0 otherwise
2T	C_{use}	V_t, V_{use}	$M = B + 4V_t + 10V_{use} + C_{use}(V_{use} + 9) + P_n(B_n + 3P_n + 12) + PW_n + 200$
		where	B_n = number of cases in a block P_n = number of parameters estimated PW_n = sum of the number of psiweights in a problem
2V	none	V_t	$M = B + 21V_t + 33GV_n + 500$ If contrasts are used, add $CT_n^2 + 6CT_n$
		where	CT_n = total number of contrasts If trial factors are used, the total is increased by a complex function of the trial factor levels and the number of independent and dependent variables.

Appendix B
Formulas and Computational Methods

Contents

Appendix B.1 General formulas and formulas for weighting options

Descriptive statistics

Most programs compute the mean, standard deviation, and frequency for each variable. In addition, several BMDP programs compute other univariate statistics. We review the definitions of these statistics below.

Let $x_1, x_2, \ldots, x_N$ be the observed (acceptable) values for a variable in the cases used in an analysis. Then N is the *sample size*, frequency, or count for the variable. The *mean*, $\bar{x}$ is defined as

$$\bar{x} = \sum x_j / N$$

(Other estimates of location, such as the median and more robust estimates, are available in 2D and 7D.)

The *standard deviation*, s, is

$$s = [\sum (x_j - \bar{x})^2 / (N-1)]^{1/2}$$

The *variance* is s^2 and the *standard error of the mean* is $s/(N^{1/2})$. The *coefficient of variation* is the ratio of the standard deviation to the mean, $s/\bar{x}$. If a variable has a very small coefficient of variation, loss of computational accuracy can result due to the limited accuracy with which a number can be represented internally in the computer.

Most analyses require that the distribution of the data be normal, or at least symmetric. A measure of symmetry is *skewness*, and a measure of long-tailedness is *kurtosis*. The BMDP programs compute skewness, g_1, as

$$g_1 = \sum (x_j - \bar{x})^3 / (Ns^3)$$

and kurtosis, g_2, as

$$g_2 = \sum (x_j - \bar{x})^4 / (Ns^4) - 3$$

If the data are from a normal distribution, the standard error of g_1 is $(6/N)^{1/2}$ and of g_2 is $(24/N)^{1/2}$. A significant nonzero value of skewness is an indication of asymmetry—a positive value indicates a long right tail, a negative value a long left tail. A value of g_2 significantly greater than zero indicates a distribution that is longer-tailed than the normal. We recommend that you also examine histograms when using these statistics, for they are sensitive to a few extreme values.

The smallest observed (acceptable) value, x_{min}, and the largest, x_{max}, are printed by several programs. The *range* is $(x_{max} - x_{min})$. The smallest and largest *standard scores* (z-scores, z_{min} and z_{max} respectively) are also printed by some programs. We define z_{min} and z_{max} as

$$z_{min} = (x_{min} - \bar{x})/s \quad \text{and} \quad z_{max} = (x_{max} - \bar{x})/s$$

Covariances and correlations

The *covariance* between two variables, x and y, is

$$\text{cov}(x, y) = \sum (x_j - \bar{x})(y_j - \bar{y}) / (N-1)$$

The *correlation*, r, between two variables, is

$$r = \frac{\text{cov}(x, y)}{s_x s_y} = \frac{\sum (x_j - \bar{x})(y_j - \bar{y})}{\sqrt{\sum (x_j - \bar{x})^2 \sum (y_j - \bar{y})^2}}$$

This is also called the product-moment correlation coefficient.

The correlation can also be computed after adjusting for the linear effect of one or more variables. For example, we may want the correlation between x and y

adjusted for z and w (sometimes referred to as correlation at a fixed value of z and w). This is called the *partial correlation* coefficient between x and y given z and w. It is equivalent to fitting separate regression equations in z and w to x and y, and computing the correlation between the residuals from the two regression lines.

A *multiple correlation* coefficient (R) is the maximum correlation that can be attained between one variable and a linear combination of other variables. This is the correlation between the first variable and the predicted value from the multiple regression of that variable on the other variables.

R^2 is the proportion of variance of the first variable explained by the multiple regression relating it to the other variables.

Formulas for case frequencies, case weights, and regression weights

Three parameters that affect computation of statistics are used in BMDP programs: case frequencies, case weights, and regression weights. See Chapter 4 for a description of how these features work and which programs use them. The same analysis may have more than one of these to adjust for case frequency, sampling design, and random error structure. If a case has a zero or negative frequency or case weight, it is excluded from the calculations. Cases with zero or negative regression weight are excluded from the regression calculations, but they will be included in the descriptive statistics if their case frequency and case weights are positive. The effect of weighting on computations is described below.

Let x_i be the ith case's value for variable x, f_i be the ith case frequency, c_i the ith case weight, and w_i the ith regression weight. Then the descriptive mean and standard deviation reported by most BMDP programs is given by

$$\bar{x} = \frac{\sum f_i c_i x_i}{\sum f_i c_i} \qquad s = \sqrt{\frac{\sum f_i c_i (x_i - \bar{x})^2}{(\sum f_i) - 1}}$$

If a variable is redesignated as a case weight instead of a frequency weight, the above formulas show that estimates of the mean remain unaffected, but estimates of the variance will change.

In regression programs, the descriptive mean and standard deviation also use the formulas above. The sample correlation between x and y is given by

$$r_{xy} = \frac{\sum f_i c_i w_i (x_i - \tilde{x})(y_i - \tilde{y})}{\sqrt{\sum f_i c_i w_i (x_i - \tilde{x})^2 \sum f_i c_i w_i (y_i - \tilde{y})^2}}$$

$$\text{where } \tilde{x} = \frac{\sum f_i c_i w_i x_i}{\sum f_i c_i w_i} \text{ and } \tilde{y} = \frac{\sum f_i c_i w_i y_i}{\sum f_i c_i w_i}$$

If there are p parameters, the mean squared error is given by

$$\text{MSE} = \frac{\sum f_i c_i w_i (y - \hat{y})^2}{(\sum f_i) - p}$$

where $\hat{y}$ is the predicted outcome. Note however that

$$f_i \equiv 1 \quad \text{and} \quad c_i \equiv 1$$

in 3R and AR, since these two programs do not allow case frequencies and case weights. Similarly

$$c_i \equiv 1 \quad \text{and} \quad w_i \equiv 1$$

in program LR.

Using matrix notation, the estimated regression coefficient vector $\hat{\beta}$ in linear regression programs is given by

$$\hat{\beta} = (\mathbf{X'WX})^{-1}\mathbf{X'WY}$$

where **X** is the matrix of independent variables with their means (of the form given by $\tilde{x}$) subtracted, **Y** is the dependent variable vector with mean (as given by $\tilde{y}$) subtracted, and **W** is a diagonal matrix with $f_i c_i w_i$ as the ith diagonal element.

Weights in 4V. Weighting in 4V is by cell (in an ANOVA design) rather than by case, although these cell weights are sometimes chosen to reflect sampling.

Weights in 2V. The WEIGHT option in 2V is equivalent to the CWEIGHT option in other programs.

Weights in 1V and 8V. There are no weighting options in programs 1V and 8V.

Weights in 3R and AR. The REGRESSION WEIGHT option has remained the same in programs 3R and AR. Obtaining estimates via iterative reweighting will work as before.

Computational accuracy

The computer represents each number by a binary sequence of limited accuracy. As a result there can be a loss of accuracy in certain computations, such as matrix inversion. Loss of accuracy is especially pronounced if a variable has a small coefficient of variation $(s/\bar{x})$ or if a variable has a very high multiple correlation with other variables. All programs represent data values in single precision. Some programs do computations in double precision; these programs are the ones whose computations are most likely to be affected by a loss of accuracy if the computations are done in single precision.

Appendix B.2 Method of provisional means

Many BMDP programs compute the mean, variance or sum of squares, skewness, kurtosis, and covariance. All but the mean require that deviation about the mean be used in the computations. If the data are kept in computer memory it is possible to first compute the mean and then compute each deviation. However, this restricts the number of cases that can be analyzed.

A provisional means algorithm is used to avoid keeping the data in memory or reading the data twice. This algorithm produces more accurate results than a formula such as $\sum x_j^2 - (\sum x_j)^2/N$ for the sum of squares.

Means and variances

In the provisional means algorithm, the mean $\bar{x}$ and sum of squares of deviations (S^2) for each variable are computed recursively as

$$W_0 = 0$$
$$\bar{x}_0 = 0$$
$$s_0^2 = 0$$
$$W_j = W_{j-1} + w_j$$
$$\bar{x}_j = \bar{x}_{j-1} + (x_j - \bar{x}_{j-1}) \cdot w_j / W_j$$
$$S_j^2 = S_{j-1}^2 + (x_j - \bar{x}_{j-1})\, w_j\, (x_j - \bar{x}_j)$$

where

x_j is the jth observation

w_j is the case frequency, case weight, or regression weight for case j, or the product of any combination of the three (if no case frequency, case weight, or regression weight is specified, w_j is equivalent to 1)

W_j is the sum of the weights for the first j cases

$\bar{x}_j$ is the mean of the first j cases

S_j^2 is the sum of squares of deviations for the first j cases

Only cases with acceptable data and a nonzero weight are used.

The mean is $\bar{x}_N$ and the variance is computed from s_N^2 by dividing by the proper degrees of freedom.

Skewness and kurtosis

Skewness (g_1) and kurtosis (g_2) are computed in a similar manner when no case weights are specified, or all case weights are either zero or one; Spicer (1972) describes a similar algorithm.

Covariances and correlations

All BMDP programs (except 3D, 8D, and AM) that compute the covariance between two variables use only cases for which both variables have acceptable values. Most programs are even more restrictive and use only cases that contain acceptable values for all the variables in the covariance matrix. We now describe how covariances are computed in all programs except 8D and AM, which use a slightly modified algorithm.

The formula for the covariance between variable x and y is as described above.

The covariance is computed recursively as

$$u_j = f_j c_j$$

$$U_j = U_{j-1} + u_j$$

$$\bar{x}_j = \bar{x}_{j-1} + (x_j - \bar{x}_{j-1})\, u_j / U_j$$

$$\bar{y}_j = \bar{y}_{j-1} + (y_j - \bar{y}_{j-1})\, u_j / U_j$$

and

$$S^2_{xy_j} = S^2_{xy_{j-1}} + (x_j - \bar{x}_{j-1})\, u_j (y_j - \bar{y}_j)$$

where

U_j is the sum of the case frequencies times case weights of the first j cases

$\bar{x}_j$ is the mean of the first j cases for variable x

$S^2_{xy_j}$ is the sum of squares of cross-products for the first j cases

Then the covariance between variables x and y is

$$S^2_{xy_N} / \sum f_j - 1$$

the variance of variable x is

$$S^2_{xx_N} / \sum f_j - 1$$

and the correlation between variables x and y is

$$S^2_{xy_N} / (S^2_{xx_N} S^2_{yy_N})^{1/2}$$

Each program description specifies the cases that are used in the computations.

Note: Weights are described in Chapter 4 and Appendix B.1.

Appendix B.3 Random numbers

You can generate both uniform and normal random numbers, either by BMDP instructions in the TRANSFORM paragraph or by FORTRAN statements in the BIMEDT subroutine. Below we describe both random number generators and how to use them in FORTRAN statements. In Chapter 6 we describe how to use them in BMDP instructions.

Uniform random number generator

The FORTRAN code used in the uniform random number generator is from "Algorithm AS 183: An Efficient and Portable Pseudo-random Number Generator" (Wichman and Hill, 1982). The algorithm uses three simple multiplicative congruential generators. Each uses a prime number for its modulus and a primitive root for its multiplier. The three results are added, and the fractional part is taken. This generator has been thoroughly tested. For further discussion and test results, see Wichman, B. A. and Hill, I. D. (1982) A pseudo-random number generator. *NPL Report* DITC 6/82.

The integer I that you specify in the BMDP command Y = RNDU(I) is the initial value of I. Each new random number is normalized to lie between zero and one. The BMDP RNDU(I) function can be accessed as a FORTRAN function RANDOM(I). You must specify an initial value for I in a data statement; e.g.,

```
DATA I/52345/
```

and then use RANDOM(I) in your FORTRAN statements. That is,

```
Y = RANDOM(I)
```

stores the random number in Y. Y can be a subscripted variable, such as X(3), or it can be used in an expression to compute the value for a variable, such as

```
X(2) = .7 * RANDOM(I) - .35
```

Normal random numbers

Pairs of normal random numbers are generated from pairs of uniform random numbers by the Box and Muller (1958) method. That is, if r_1 and r_2 are two uniform random numbers, then

$$z_1 = (-2 \ln r_1)^{1/2} \cos (2\pi r_2) \quad \text{and}$$
$$z_2 = (-2 \ln r_1)^{1/2} \sin (2\pi r_2)$$

are a pair of statistically independent random numbers from a normal distribution with mean zero and variance one. The uniform random number generator described above is used to generate the r_1, r_2 values.

In BMDP instructions, RNDG(I) is the function which generates normal random values, where I in RNDG(I) has the same role as in RNDU(I).

The normal random number generator can be a part of your FORTRAN transformation statements. Its calling sequence is

```
CALL RANDG(N,Z,XBAR,SIGMA,I)
```

where

N	is the number of normal random numbers that you want
Z	is the vector containing the random numbers
XBAR	is the mean of the normal population (i.e., 0.0 if random numbers from the standard normal distribution are wanted)
SIGMA	is the standard deviation of the normal population (i.e., 1.0 for the standard normal distribution)

I is a large odd integer used to start the generator

In BMDP instructions the I in RNDG(I) has the same role as in RNDU(I).

Appendix B.4 Program 1D

Computational method

1D computes the means and standard deviations in single precision by the provisional means algorithm described in Appendix B.2.

Appendix B.5 Program 2D

Computational method

2D reads the data for each variable and constructs the frequency table of distinct values. All statistics are computed from the frequency table (not from the original data). Therefore, when the data are rounded or truncated, the statistics are affected. The formulas for the statistics use the frequency of each distinct value. First the mean is computed, and then 2D calculates the variance, skewness, and kurtosis using differences from the mean.

Robust estimators

2D includes robust estimators. The psi-function for the biweight estimator is:

$$\psi(u) = u \cdot w(u)$$

where the weight function $w(u)$ is

$$w(u) = \begin{cases} (1-u^2)^2 & \text{for } |u| \le 1 \\ 0 & \text{elsewhere} \end{cases}$$

$$u = \left(\frac{x_i - \text{estimate}}{k \cdot \text{MAD}} \right)$$

with $k = 6.0$ and the scale estimate MAD is the median of the absolute deviations from the median.

The psi-function for the Hampel estimate is

$$\psi(u; a, b, c) = (\text{sign}\, u) \begin{cases} |u| & 0 \le |u| < a \\ a & a \le |u| < b \\ \dfrac{c - |u|}{c - b} a & b \le |u| < c \\ 0 & c \le |u| \end{cases}$$

with $(a,b,c) = 1.7, 3.4, 8.5$. ($u = (x_i - \text{estimate})/\text{MAD}$).

Shapiro and Wilk's W statistic (Royston, 1982)

Let $m = (m_1, \ldots, m_n)$ denote the vector of expected values of standard normal order statistics, and let $V = (v_{ij})$ be the corresponding n x n covariance matrix; that is

$$E(x_i) = m_i \; (i = 1, \ldots, n) \quad \text{and} \quad \text{cov}(x_i, x_j) = v_{ij} \; (i,j, = 1, \ldots, n)$$

where $x_1 < x_2 < \ldots < x_n$ is an ordered random sample from a standard normal distribution $N(0,1)$. Suppose $y' = (y_1, \ldots, y_n)$ is a random sample on which the W test of normality is to be carried out, ordered $y_{(1)} < y_{(2)} < \ldots < y_{(n)}$. Then

$$W = \left[\sum_1^n a_i y_{(i)} \right]^2 \sum_1^n (y_i - \bar{y})^2$$

where

$$a' = (a_1, \ldots, a_n) = m'V^{-1}[(m'V^{-1})(V^{-1}m)]^{-1/2}$$

The coefficients $\{a_i\}$ are the normalized "best linear unbiased" coefficients tabulated for $n \le 20$ by Sarhan and Greenberg (1956). Shapiro and Wilk (1965) offer a satisfactory approximation for a which improves with increasing sample size N, which we adopt. By definition, a has the property $a'a = 1$. Let $a^* = m'V^{-1}$; approximations $\hat{a}^*$ for a^* are

$$\hat{a}_i^* = \begin{cases} 2m_i & i = 2, 3, \ldots, n-1 \\ \left(\dfrac{\hat{a}_i^2}{1 - 2\hat{a}^2} \sum_2^{n-1} \hat{a}_i^{*2} \right)^{1/2} & i = 1, i = n \end{cases} \quad (1)$$

where

$$\hat{a}_1^2 = \hat{a}_n^2 = \begin{cases} g(n-1) & n \le 20 \\ g(n) & n > 20 \end{cases}$$

and

$$g(n) = \frac{\Gamma\left[\frac{1}{2}(n+1)\right]}{\sqrt{2\Gamma\left[\frac{1}{2}n + 1\right]}}$$

Using Stirling's formula, $g(n)$ may be approximated and simplified:

$$g(n) = \left[\frac{6n+7}{6n+13}\right]\left[\frac{\exp(1)}{n+2}\left[\frac{n+1}{n+2}\right]^{n-2}\right]^{1/2} \quad (2)$$

For $n = 6$, the exact value $g(n)$ is 0.39166 compared with an approximate value given by (2) of 0.39164. We have used the approximations (1) and (2) throughout the range $7 \le n \le 2000$, but exact values of the $|a_i|$ for $n < 7$.

Appendix B.6 Program 3D

Computational method

Means, standard deviations, and covariances are computed in single precision using the method of provisional means (Appendix B.2).

Test statistics

***t* (pooled variances)**. When the population variances in the two groups are assumed to be equal, the estimate of the variance can be obtained by pooling (averaging) the two estimates of the variance. This is the standard Student's t test. This two-sample t statistic is

$$t = \frac{(\bar{x}_1 - \bar{x}_2)}{s_p\left(\dfrac{1}{n_1} + \dfrac{1}{n_2}\right)^{1/2}}$$

where

$$s_p^2 = \frac{(n_1 - 1)s_1^2 + (n_2 - 1)s_2^2}{n_1 + n_2 - 2}$$

The degrees of freedom are $n_1 + n_2 - 2$. When n_1 and n_2, the sums of the case frequencies, differ and the two population variances are not equal, the p value obtained using the pooled variance t test may be either incorrectly large or incorrectly small. (See Yuen and Dixon, 1973, and Yuen, 1974.)

t (separate variances). This is a two-sample t statistic in which the variance of each group is estimated separately. It is appropriate whenever the population variances are not assumed to be equal (see explanation of F below). The t statistic is

$$t = \frac{(\bar{x}_1 - \bar{x}_2)}{\left(\frac{s_1^2}{n_1} + \frac{s_2^2}{n_2}\right)^{1/2}}$$

where the subscripts 1 and 2 refer to two groups. The degrees of freedom are approximated (Welch, 1947) by

$$f = \left[\frac{c^2}{n_1 - 1} + \frac{(1-c)^2}{n_2 - 1}\right]^{-1}$$

where

$$c = \frac{\text{var}(\bar{x}_1)}{\text{var}(\bar{x}_1 + \bar{x}_2)} = \frac{s_1^2/n_1}{s_1^2/n_1 + s_2^2/n_2}$$

Trimmed t test. 3D trims one observation from the top and one from the bottom of each group. The t test (separate and pooled) is calculated using a Winsorized standard deviation to adjust for deletion of these cases (see Dixon and Massey, 1983, and Dixon and Tukey, 1968).

Levene F (for variability). 3D computes Levene's test of equality of variance between the two groups. The Levene F test (Brown and Forsythe, 1974b) is obtained by performing an analysis of variance on the absolute deviations of each case from its cell mean. We do not use the Bartlett test to test the equality of variances because it is sensitive to departures from normality (Armitage, 1971, p. 143). The Levene test should be used with caution when sample sizes are small (see Church and Wike, 1976). All p-values quoted correspond to two-sided tests of significance (i.e., to test inequality in either direction).

Nonparametric tests. For an explanation of the nonparametric statistics, see program 3S. The same procedures are used for tied ranks in 3D and 3S.

Formulas used for multivariate tests

Let $\bar{\mathbf{X}}$ be a $p \times 1$ column vector that contains the means of p variables. Let Σ be the $p \times p$ true covariance matrix of the p variables.

The **Mahalanobis distance for a single sample**, with no data missing, is

$$D^2 = \bar{\mathbf{X}}'\Sigma^{-1}\bar{\mathbf{X}} = \sum_j \sum_k \bar{X}_j \bar{X}_k \sigma_{jk} \qquad (1)$$

where both sums are from 1 to p and σ_{jk} is the (j,k)th element of Σ^{-1}. If $\bar{X}_j$ and $\bar{X}_k$ are means of X_j and X_k computed using available cases, and have null hypothesized underlying mean 0, then

$$E(\bar{X}_j^2) = \sigma_{jj}/m_j \qquad E(\bar{X}_k^2) = \sigma_{kk}/m_k$$

$$E(\bar{X}_j \bar{X}_k) = \sigma_{jk} m_{jk}/m_j m_k \qquad (2)$$

where $m_j(m_k)$ is the number of cases for which X_j (X_k) is observed, and m_{jk} is the number of cases with both X_j and X_k observed.

Hotelling's T^2. When no data are missing, $m_j = m_k = m_{jk} = n$ for all j and k. In this case Hotelling's T^2 is given by

$$T^2 = nD^2$$

Under multivariate normality and the null hypothesized underlying mean of zero, T^2 has an asymptotic χ^2 distribution with p degrees of freedom. When no data are missing, the quantity

$$F = (n - p)T^2/p(n - 1)$$

has an F distribution under the null hypothesis with p and $n - p$ degrees of freedom.

When data are missing, the statistic D^2 in (1) is modified. Define $\Gamma = (\lambda_{jk})$ such that

$$\lambda_{jk} = n \cdot \text{cov}(\bar{X}_j, \bar{X}_k) \tag{3}$$

From (2), $\lambda_{jj} = \sigma_{jj}(n_1/m_j)$ and $\lambda_{jk} = \sigma_{jk}(n_1 m_{jk}/m_j m_k)$. Then the modified Mahalanobis distance and Hotelling's T^2 are

$$D^2_{\text{inc}} = \bar{\mathbf{X}}'\Gamma^{-1}\bar{\mathbf{X}} \qquad T^2_{\text{inc}} = nD^2_{\text{inc}}$$

where the mean $\bar{\mathbf{X}}$ is based on available cases. Then T^2_{inc} has a chi-squared distribution on $v \le p$ degrees of freedom, if the underlying means are zero. Thus for unknown Σ we replace T^2 by $\hat{T}^2_{\text{inc}}$, obtained from T^2_{inc} by replacing values of Σ by estimates $\hat{\Sigma}$. This statistic has the correct chi-squared distribution asymptotically.

The Mahalanobis distance for two groups. For comparing two groups with means $\bar{\mathbf{X}}_j$ $(j = 1, 2)$ and true pooled covariance matrix $\mathbf{C}$, the Mahalanobis distance in the absence of missing values is

$$D^2 = (\bar{\mathbf{X}}_1 - \bar{\mathbf{X}}_2)'\mathbf{C}^{-1}(\bar{\mathbf{X}}_1 - \bar{\mathbf{X}}_2) \tag{4}$$

The Hotelling T^2 statistic is

$$T^2 = \frac{D^2}{(1/n_1 + 1/n_2)} \tag{5}$$

where n_j is the sample size in group j and the F statistic under the null hypothesis is

$$F = \frac{(n_1 + n_2 - p - 1)}{(n_1 + n_2 - 2)\,p} T^2$$

with p and $n_1 + n_2 - p - 1$ degrees of freedom. In large samples note that

$$T^2 \cong \chi^2_v \tag{6}$$

the chi-squared distribution with v degrees of freedom, if the underlying population means are equal.

When data are missing, the analogous statistic to $\hat{T}^2_{\text{inc}}$ is easily developed for comparing groups of size n_1 and n_2. Redefine $\Gamma = (\lambda_{jk})$ such that

$$\text{cov}(\bar{X}_j^{(1)} - \bar{X}_j^{(2)}, \bar{X}_k^{(1)} - \bar{X}_k^{(2)}) = \lambda_{jk}(1/n_1 + 1/n_2)$$

where $\bar{X}_j^{(k)}$ is the mean of variable j in group k $(j = 1, \ldots, v;\ k = 1, 2)$. Then

$$\lambda_{jj} = \sigma_{jj}(1/m_j^{(1)} + 1/m_j^{(2)})\,(1/n_1 + 1/n_2)^{-1}$$

$$\lambda_{jk} = \sigma_{jk}\left(\frac{m_{jk}^{(1)}}{m_j^{(1)} m_k^{(1)}} + \frac{m_{jk}^{(2)}}{m_j^{(2)} m_k^{(2)}}\right)\left(\frac{1}{n_1} + \frac{1}{n_2}\right)^{-1}$$

where the sample sizes m_j, m_k, and m_{jk} are superscripted by group. The modified Mahalanobis distance and Hotelling T^2 are

$$D^2_{\text{inc}} = (\bar{\mathbf{X}}^{(1)} - \bar{\mathbf{X}}^{(2)})'\Gamma^{-1}(\bar{\mathbf{X}}^{(1)} - \bar{\mathbf{X}}^{(2)})$$

$$T^2_{\text{inc}} = (1/n_1 + 1/n_2)^{-1} D^2_{\text{inc}}$$

with elements of Σ in Γ replaced by estimates $\tilde{\Sigma}$.

Formulas used for correlations

The method of computing correlations depends on whether COMPLETE cases or all acceptable data are used. Normally, for each correlation 3D uses all cases for which both variables being correlated are present. When you specify COMPLETE, 3D uses only those cases for which all variables specified in the VARIABLE USE list are present. When COMPLETE cases are used, the Pearson correlation between two variables (i and k) is

$$r_{ik} = \frac{\sum_j f_j c_j (x_{ij} - \bar{x}_i)(x_{kj} - \bar{x}_k)}{\left[\sum_j f_j c_j (x_{ij} - \bar{x}_i)^2 \sum_j f_j c_j (\bar{x}_{kj} - x_k)^2\right]^{1/2}}$$

where the summations are over all complete cases in the group.

When all available data are used, the covariance of variables i and k is computed using complete pairs of observations; i.e.,

$$\text{cov}_{ik} = \sum_j f_j c_j (x_{ij} - \bar{x}_{i(k)})(x_{kj} - \bar{x}_{k(i)}) / (N_{ik} - 1)$$

where $\bar{x}_{i(k)}$ is the mean of variable i for cases that also contain variable k. The summation is over the cases in the group for which data for both variables are present (N_{ik} cases). The correlation is computed as

$$r_{ik} = \text{cov}_{ik} / (s_i s_k)$$

where s_i and s_k are the standard deviations for each variable (that is, they are based on all acceptable data for the variable in the group). It is therefore possible that the correlation matrix based on all acceptable data will not be positive definite (for a definition of positive definite, see Scheffé, 1959, p. 398), and that a computed correlation will be greater than 1 or less than –1. This method is also used in the COVPAIR option in program 8D.

QQ plots

Let $x_1, \ldots, x_m$ and $y_1, \ldots, y_n$ be two sets of observations from the random variables x and y respectively, and let $x_{(1)}, \ldots, x_{(m)}$ and $y_{(1)}, \ldots, y_{(n)}$ be the sorted data values of x and y. If $n = m$ the QQ plot of x against y simply plots $x_{(i)}$ against $y_{(i)}$. Now suppose that $n > m$. In this case we use the entire set of sorted values from the smaller data set and interpolate a corresponding set of quantiles from the larger data set as follows.

Against the point $x_{(i)}$ we want to plot the $i/(m + 1)$ quantile of y. To obtain the point we compute y such that

$$\frac{\nu}{n+1} = \frac{i}{m+1}$$

or

$$\nu = i\frac{n+1}{m+1}$$

If ν is an integer, we simply plot $y_{(\nu)}$ against $x_{(i)}$, and if $\nu = j + \theta$ where j is its integer part and θ is its fractional part, we plot $x_{(i)}$ against $(1 - \theta)y_{(j)} + y_{(j+1)}$.

Appendix B.7 Program 5D

Computational method

Computations of the means and standard deviations are in single precision by the method of provisional means (Appendix B.2).

Probability plots

For the normal probability plot, the data values are ordered before plotting: the vertical axis corresponds to the expected normal value based on the rank of the observation. Let $x_{(1)}, x_{(2)}, \ldots$ represent the data values after ordering from smallest to largest. The subscript (j) is the rank order of the observation. If N is the total frequency, the vertical plotting position corresponds to the expected normal value for the relative rank (j out of N) of the observation. The expected normal value is computed as

$$\Phi^{-1}[(3j-1)/(3N+1)]$$

where Φ^{-1} is the inverse Gaussian cumulative distribution function. For the detrended normal probability plot, the vertical scale represents the differences between the expected normal values and the standardized values of the observations. That is, each observation is transformed into a standardized value by subtracting the mean and dividing by the standard deviation; i.e., $z_{(j)} = (x_{(j)} - \bar{x})/s$. We then compute

$$\Phi^{-1}[(3j-1)/(3N+1)] - z_{(j)}$$

For the half-normal plot, the expected value of the jth observation (after ordering) is estimated by

$$\Phi^{-1}[(3N+3j-1)/(6N+1)]$$

Appendix B.8 Program 6D

Computational method

6D computes the means, standard deviations, and covariances in single precision by the method of provisional means (Appendix B.2).

When MINIMUMS and MAXIMUMS are not specified in the PLOT paragraph for any variable to be plotted, the data are first read to find the observed minimums and maximums and then reread to construct the scatterplots. Thus, stating the minimums and maximums improves the program's efficiency.

Calculation of p*-values associated with* r

The procedure for calculating p-values associated with r in 6D is to calculate Z', divide by the standard deviation, and look up in a table.

$$Z' = \frac{1}{2}\ln\left[\frac{1+r}{1-r}\right]$$

$$\text{standard deviation} = \frac{1}{\sqrt{N-3}}$$

Appendix B.9 Program 7D

Computational method

Means, standard deviations, and covariances are computed in single precision using the method of provisional means (Appendix B.2).

The subroutine for the two-way analysis of variance computations transforms the sample size matrix, the cell mean matrix, and the sum of values of the dependent variable to the matrix of cross-products between all design variables and the dependent variable. This matrix is then swept to give the usual analysis of variance tests as in BMDP 2V. The computations here are substantially less than would be used if the matrix of cross-products were computed directly from the design matrix. Note also that the interaction dummy variables are swept only once (i.e., they are not swept and reverse swept), thereby achieving even greater efficiency. This routine requires data for all cells.

Explanation of test statistics

One-way designs

The one-way analysis of variance (ANOVA) is computed as

$$F = \mathrm{MS}_{\mathrm{between}} / \mathrm{MS}_{\mathrm{error}}$$

where

$$\mathrm{MS}_{\mathrm{between}} = \frac{\sum_{i=1}^{g} n_i (\bar{x}_i - \bar{x})^2}{g-1}$$

and

$$\mathrm{MS}_{\mathrm{error}} = s_p^2 = \frac{\sum_{i=1}^{g} \sum_{k=1}^{n_i} (x_{ki} - \bar{x}_i)^2}{N-g}$$

with

g = number of groups (levels)
i = group number
k = number of the case within the group
n_i = number of cases in group i
N = total number of cases

The Welch statistic is defined as

$$W = \frac{\sum_i w_i [(\bar{x}_i - \tilde{x})^2 / (g-1)]}{1 + \frac{2(g-2)}{g^2-1} \sum_i \left[\left(1 - \frac{w_i}{u}\right)^2 (n_i - 1) \right]}$$

where $w_i = n_i / s_i^2$ $\quad u = \sum_i w_i$ $\quad$ and $\quad \tilde{x} = \sum w_i \bar{x}_i / u$

When all population means are equal (even if the variances are unequal), W is approximately distributed as an F statistic with $g - 1$ and f degrees of freedom, where f is implicitly defined as

$$1/f = \left(\frac{3}{g^2 - 1}\right) \sum_i [(1 - w_i/u)^2 / (n_i - 1)]$$

The Brown-Forsythe statistic (1974a, b) is

$$F^* = \sum_i n_i (\bar{x}_i - \bar{x})^2 / \sum_i (1 - n_i/N) s_i^2$$

Critical values are obtained from the F distribution with $g - 1$ and f degrees of freedom where f is implicitly defined by the Satterthwaite approximation

$$\frac{1}{f} = \sum_i c_i^2 / (n_i - 1)$$

and

$$c_i = \frac{(1 - n_i/N) s_i^2}{\sum_i (1 - n_i/N) s_i^2}$$

When there are only two groups, both F^* and W reduce to the separate variance t test computed by program 3D. W and F^* weight the sums of squares in the numerator differently.

The Levene statistic for testing equal variability among cells. Using the absolute values of the deviations from the group means as data, the Levene F statistic is computed as the one-way analysis of variance F defined above (Levene, 1960).

Robust standard deviations for a group are found by

$$\text{Robust S.D.} = \frac{\sum_{k=1}^{n} |x_k - \bar{x}|}{n} \sqrt{\frac{n\pi}{2(n-1)}}$$

where N = number of cases in group.

Two-way designs

Two-way analysis of variance (ANOVA) is computed with

$$F_{\text{rows}} = \text{MS}_{\text{rows}} / \text{MS}_{\text{error}}$$

where

$$\text{MS}_{\text{rows}} = \text{SS}_{\text{rows}} / (g-1)$$

and

$$\text{SS}_{\text{rows}} = \text{sum of squares due rows (see below)}$$

$$F_{\text{columns}} = \text{MS}_{\text{columns}} / \text{MS}_{\text{error}}$$

where

$$\text{MS}_{\text{columns}} = \text{SS}_{\text{columns}} / (h-1)$$

$$F_{\text{interaction}} = \text{MS}_{\text{interaction}} / \text{MS}_{\text{error}}$$

where

$$\text{MS}_{\text{interaction}} = \text{SS}_{\text{interaction}} / [(g-1)(h-1)]$$

and

$$\text{MS}_{\text{error}} = \frac{\sum_{i=1}^{g} \sum_{j=1}^{h} \sum_{k=1}^{n_{ij}} (x_{ijk} - \bar{x}_{ij})^2}{N - gh}$$

letting "rows" be the first grouping factor and "columns" the second, then

g = number of row groups
h = number of column groups
i = row group number
j = column group number
k = within group case number
n_{ij} = number of cases in cell formed by intersection of row i and column j
N = total number of cases

The sums of squares (SS) in each formula can be obtained as the difference in fitting two regression models. In one model the effects corresponding to the rows, columns, and interactions are fitted. In the second model the effects corresponding to row or column or interaction (whichever is being tested) are set to zero.

The Brown-Forsythe statistic. Brown and Forsythe (1974a, b) show that the formulas for their one-way analysis of variance test (see above) can be obtained by combining L orthonormal contrasts across the cell means. That is, each contrast partitions a single degree of freedom from the between-groups sum of squares. They then extend this idea to the two-way design. The tests for rows, columns, and the interaction are each stated as a set of contrasts on the cell means. When cell sizes are unequal, a set of contrasts is generated that spans the parameter space of the hypothesis to be tested. A Gram-Schmidt orthonormalization routine is used to orthonormalize each set of contrasts.

The Levene statistic uses the absolute deviation of each observation from its group mean as data. The cell mean for cell i is

$$\bar{z}_i = \sum_{j=1}^{n_i} \frac{|x_{ij} - \bar{x}_i|}{n_i}$$

with standard deviation defined as the square root of the sum of squares about z_i divided by $(n_i - 1)$.

Multiple comparisons and contrasts

The Bonferroni method determines significance of the pairwise mean difference when the following confidence interval (L) does not contain zero:

$$L = t\,(\text{tabled}) \sqrt{s_p^2 \left(\frac{1}{n_i} + \frac{1}{n_j} \right)}$$

where s_p^2 is the error MS from the ANOVA table, and the t table is entered with the within d.f. and the $1 - \alpha/2m$ percentile (m is the number of possible pairs). This adjusts the α level to insure that the probability of finding one or more comparisons significant when the overall null hypothesis is true is less than α.

The Tukey studentized range method uses

$$L = q\,(\text{tabled})\, s_p / \sqrt{n}$$

where, for a specified α, q is the percentile point of the studentized range distribution with $k(n - 1)$ for parameters. (k is the total number of means available for comparison, and n is the sample size.) For the unequal sample size situation, 7D uses the Tukey-Cramér adjustment (harmonic mean).

The Scheffé procedure uses

$$L = \sqrt{s_p^2 (k-1)\, F\,(\text{tabled}) \left(\frac{1}{n_i} + \frac{1}{n_j} \right)}$$

Dunnett's procedure is determined by

$$\bar{x}_i - \bar{x}_0 \pm |d|\, s_p \sqrt{2/n}$$

where for a specified α, d is the percentile point of a distribution of a nameless statistic that has limited tables (Miller, 1981), and $\bar{x}_0$ is the mean of the control group. Sample sizes can be unequal if the control group has the largest sample size.

The Student-Newman-Keuls procedure involves an r-mean significance level equal to α for each group of r ordered means; the largest difference, which involves means that are $r = p - 1$ steps apart, is tested first at the level of significance. Then, means $p - 2$ steps apart are tested at the α level, and so on. The equation is

$$(W_r) = q_{\alpha;\, r,\, v} \sqrt{\frac{s_p}{n}}$$

where r is the number of steps separating ordered means, v is the degrees of freedom associated with s_p, and n is the sample size. For unequal sample sizes, the Tukey-Cramér adjustment is used.

Duncan's multiple range test uses the same ordered means and sequential testing procedure as the Student-Newman-Keuls procedure, but adjusts α at each subset size with $1 - (1 - \alpha)^{r-1}$. The equation is the same as the Student-Newman-Keuls. Miller (1981) prefers the Student-Newman-Keuls procedure, and argues that the Duncan test violates the spirit of simultaneous inference.

Contrasts among group means. The associated F statistic is:

$$F_{\text{contrast}} = \frac{\left(\sum_{i=1}^{g} c_i \bar{x}_i\right)^2}{s_p^2 \sum_{i=1}^{g} \frac{c_i^2}{n_i}}$$

where

c_i = contrast coefficient for group i

s_p^2 = error mean square

Robust procedures based on trimming and Winsorizing

The robust procedures (Fung and Lee, 1985; Lee and Fung, 1983; Yuen, 1974; Yuen and Dixon, 1973) allow you to specify p percent to trim from each end of the within-cell distributions. Before presenting formulas that allow fractional trimming/Winsorizing, we define a trimmed mean and a Winsorized mean when exactly 2 cases have been trimmed/Winsorized from each end of a sample of 9 observations.

Let $x_1, x_2, \ldots, x_q$ be an ordered sample of size 9. The two-level trimmed mean and the two-level Winsorized mean are, respectively,

$$\bar{x}_{t2} = \sum_{j=3}^{7} x_j / 5$$

$$\bar{x}_{w2} = \frac{\sum_{j=3}^{7} x_j + 2(x_3 + x_7)}{9}$$

For p percent ($0 < p < .25$), the **p-trimmed mean** and the **p-Winsorized mean** for the ordered sample $x_1, x_2, \ldots, x_n$ are

$$\bar{x}_{tp} = \frac{\sum_{j=k+1}^{n-k} x_j + (1 - \varepsilon)(x_k + x_{n-k+1})}{n - 2pn}$$

$$\bar{x}_{wp} = \frac{\sum_{j=k+1}^{n-k} x_j + k[(1 - \varepsilon)(x_k + x_{n-k+1}) + \varepsilon(x_{k+1} + x_{n-k})]}{n}$$

where

$k = [pn] + 1$ with $[pn]$ denoting the largest integer less than or equal to pn

$\varepsilon = pn - [pn]$

The *p*-Winsorized sum of squared deviations is then

$$\mathrm{SSD}_{wp} = \sum_{j=k+1}^{n-k} (x_j - \bar{x}_{wp})^2 + k\{ [(1-\varepsilon)x_k + \varepsilon x_{k+1} - \bar{x}_{wp}]^2 + [(1-\varepsilon)x_{n-k+1} + \varepsilon x_{n-k} - \bar{x}_{wp}]^2 \}$$

The Winsorized standard deviations in the TRIM panel (Example 7D.7) are

$$s_w = \sqrt{\mathrm{SSD}_{wp} / (n - 2m - 1)}$$

where m is the smallest integer greater than or equal to pn.

Note that for levels 0, 1, 2, 3, $\varepsilon = 0$.

The one-way analysis of variance with trimming is computed as

$$F_t(p) = \frac{\sum_{i=1}^{g} h_i (\bar{x}_{tpi} - \bar{x}_{tp})^2 / (g-1)}{\sum_{i=1}^{g} \mathrm{SSD}_{wpi} / (H - g)}$$

where

g = the number of groups

h_i = $n_i(1 - 2p)$

N = $\sum_i n_i$

H = $\sum_i h_i = N(1 - 2p)$

$\bar{x}_{tpi}$ = the p-trimmed mean of the ith group

$\bar{x}_{tp}$ = the p-Winsorized sum of squared deviations of the ith group

The trimmed modification of the Welch statistic is

$$w_t(p) = \frac{\sum_{i=1}^{g} w_i (\bar{x}_{tpi} - \tilde{x}_{tp})^2 / (g-1)}{1 + \frac{2(g-2)}{(g^2-1)} \sum_{i=1}^{g} \frac{(1 - w_i/w)^2}{h_i - 1}}$$

where

w_i = h_i / s^2_{wpi}

s^2_{wpi} = $\mathrm{SSD}_{wpi} / (h_i - 1)$

w = $\sum_i w_i$

$\tilde{x}_{tp}$ = $\sum_i w_i \bar{x}_{tpi} / w$

The denominator degrees of freedom f_t are

$$1/f_t = \left(\frac{3}{g^2 - 1}\right) \sum_{i=1}^{g} \frac{(1 - w_i/w)^2}{h_i - 1}$$

The trimmed modification of the Brown-Forsythe statistic is

$$F_t^*(p) = \frac{\sum_{t=1}^{g} h_i (\bar{x}_{tpi} - \bar{x}_{tp})^2}{\sum_{i=1}^{g} (1 - h_i/H)\, s^2_{wpi}}$$

with denominator degrees of freedom defined as

$$1/f^* = \sum v_i^2/(H_i - 1)$$

where

$$v_i = \frac{(1 - h_i/H)\, s_{wpi}^2}{\sum_i (1 - h_i/H)\, s_{wpi}^2}$$

Appendix B.10 Program 9D

Computational method

9D computes means and standard deviations in single precision by a provisional means algorithm (Appendix B.2). It copies the data to a temporary file and determines how many variables (V = 1, 2, 3, or 4) to plot and analyze at one time. It then rereads the data from the temporary file as many times as necessary to plot and analyze all the variables.

Appendix B.11 Program 4F Tests and measures in the two-way frequency table

Notation

Let (a_{ij}; $i = 1, \ldots, R$; $j = 1, \ldots, C$) be frequency counts in the R x C two-way contingency table. Let

$$r_i = \sum_j a_{ij} \qquad c_j = \sum_i a_{ij} \qquad \text{and} \qquad N = \sum_i \sum_j a_{ij}$$

represent the row totals, column totals, and table total respectively. Also let

$$A_{ij} = \sum_{k>i} \sum_{l>j} a_{kl} + \sum_{k<i} \sum_{l<j} a_{kl}$$

$$D_{ij} = \sum_{k>i} \sum_{l<j} a_{kl} + \sum_{k<i} \sum_{l>j} a_{kl}$$

$$P = \sum_i \sum_j a_{ij} A_{ij}$$

$$Q = \sum_i \sum_j a_{ij} D_{ij}$$

Therefore, A_{ij} is the sum of the frequencies in cells in which both indices are greater, or both indices are less than (i, j), and D_{ij} is the sum of frequencies in cells in which one index is greater and the other is less than (i, j). Hence P is twice the total number of agreements in the order of the subscripts between all pairs of observations, and Q is twice the total number of disagreements.

Asymptotic standard errors (ASEs)

Two ASEs are computed for each measure. One (labelled ASE1 in the output) is appropriate to use in setting confidence limits on the parameters. It is obtained by the method of Goodman and Kruskal (1972) based on multinomial sampling conditional on the total table frequency N (except for the Goodman and Kruskal τ and λ^* and the tetrachoric r_t). This ASE is denoted s_1 since it is the ASE derived by assuming the alternate hypothesis H_1 is true; i.e., the measure is not zero.

The second ASE, denoted s_0, is a modification by Brown and Benedetti (1977a) that is more appropriate to test the null hypothesis H_0 that the measure is zero. In the output, T-VALUE is the ratio of the statistic to this ASE (s_0).

The formulas to compute s_1 (ASE1) for Yule's Q and Y and ln(cross-product ratio) appear in Kendall and Stuart (1967); and for γ, Somer's D, λ, λ^*, τ appear in Goodman and Kruskal (1963, 1972). The formula for s_1 in the tetrachoric r_t is from Pearson (1913). The remaining ASEs, including all those under the null hypothesis, are derived according to Brown and Benedetti (1977a).

Guidelines to the use of these ASEs both to test if the measure is zero and to set confidence limits for the measure are given by Brown and Benedetti (1977a, 1976).

All but three ASEs are derived based on multinomial sampling conditional on the total table frequency. Two exceptions are the Goodman and Kruskal τ and λ^*, whose ASEs are derived based on multinomial sampling conditional on the row or column totals. The third is the tetrachoric correlation.

The formulas for s_1 and s_0 follow the definition of the measure. Many measures can be written as a ratio v/w where v is the numerator and w the denominator. The notation v and w is used in the formulas for s_1 and s_0.

Statistics

Chi-square (χ^2)

This is the classical chi-square test of independence between two categorical variables:

$$\chi^2 = \sum_i \sum_j \frac{(a_{ij} - e_{ij})^2}{e_{ij}} \qquad \text{df} = (R-1)(C-1)$$

where

$e_{ij} = r_i c_j / N$. The significance level of χ^2 is also given.

Likelihood ratio chi-square (G^2)

This is a test of independence between rows and columns based upon a likelihood ratio approach. In general the results are similar to that of the usual χ^2, but this one is computationally more expensive.

$$G^2 = 2 \sum_i \sum_j a_{ij} \ln (a_{ij}/e_{ij}) \qquad \text{df} = (R-1)(C-1)$$

where

$e_{ij} = r_i c_j / N$. The significance level of G^2 is also given.

Yates' corrected chi-square (χ_y^2)

When the contingency table is 2 x 2, a chi-square that includes a correction for continuity is also computed:

$$\chi_y^2 = \begin{cases} \dfrac{N\left(|a_{11}a_{22} - a_{12}a_{21}| - \dfrac{N}{2}\right)^2}{r_1 r_2 c_1 c_2} & \text{if } |a_{11}a_{22} - a_{12}a_{21}| > \dfrac{N}{2} \\ 0 & \text{otherwise} \end{cases}$$

The significance level of χ_y^2 is also given.

Fisher's exact probability

For the 2 x 2 table, the exact probability of any configuration under the assumption of independence between rows and columns can easily be calculated.

$$\Pr(a_{11}, a_{12}, a_{21}, a_{22}) = \frac{r_1! \, r_2! \, c_1! \, c_2!}{N! \, a_{11}! \, a_{12}! \, a_{21}! \, a_{22}!}$$

conditioned on the row and column totals.

The **direction** of the 1-tail probability is chosen as follows:

If $a_{11}a_{22} \le a_{12}a_{21}$ choose the minimum of a_{11} and a_{22}; otherwise, choose the minimum of a_{12} and a_{21}. For example, let cell (1,1) be the cell so chosen. Then the probability that a value of cell (1,1) is equal to or less than a_{11} is

$$P_1 = \sum_{x=0}^{a_{11}} \Pr(a_{11} - x, a_{12} + x, a_{21} + x, a_{22} - x)$$

Assuming that a_{11} was chosen, the other tail probability is computed as follows:

$$P_2 = \sum \Pr(a_{11} + y, a_{12} - y, a_{21} - y, a_{22} + y)$$

where the sum is over y such that

$$\Pr(a_{11}, a_{12}, a_{21}, a_{22}) \ge \Pr(a_{11} + y, a_{12} - y, a_{21} - y, a_{22} + y)$$

and $y > 0$; i.e., over all terms in the 2nd tail whose probability does not exceed that of the observed outcome $(a_{11}, a_{12}, a_{21}, a_{22})$

$$\text{PROB}(1 - \text{TAIL}) = p_1$$

$$\text{PROB}(2 - \text{TAIL}) = p_1 + p_2$$

If a one-tail test is desired, you must check to see whether the tail chosen for PROB(1 – TAIL) corresponds to the tail appropriate for your test. If it does not, then the p-value for your test is ≥ 0.5, which is a nonsignificant result.

Phi (ϕ)

The value of χ^2 depends not only on the relative frequency in each cell but also on the total sample size N. A measure that is independent of N is

$$\phi = (\chi^2/N)^{1/2}$$

For a 2 x 2 table, ϕ may be positive or negative according to the formula

$$\phi = (a_{11}a_{22} - a_{12}a_{21})/(r_1 r_2 c_1 c_2)^{1/2}$$

The significance level of the test that $E(\phi) = 0$ may be obtained from the corresponding significance level for χ^2.

Maximum value of phi (ϕ_{max})

In a 2 x 2 table, the maximum possible absolute value of ϕ given the row and column totals is less than or equal to 1. This value is calculated.

If $a_{11}a_{22} < a_{12}a_{21}$, then $\phi_{max} = -[(r_1c_1)/(r_2c_2)]^{1/2}$ or $-[(r_2c_2)/(r_1c_1)]^{1/2}$

whichever is less than 1 in absolute value

otherwise $\phi_{max} = [(r_1c_2)/(r_2c_1)]^{1/2}$ or $[(r_2c_1)/(r_1c_2)]^{1/2}$

whichever is less than 1.

Note that if $\phi < 0$, then ϕ_{max} is the lower bound; i.e., $\phi_{max} < 0$.

Contingency coefficient (C)

Another transformation of χ^2 which does not depend on the sample size N is

$$C = [\chi^2/(N + \chi^2)]^{1/2} = [\phi^2/(1 + \phi^2)]^{1/2}$$

The significance level of the test for $E(C) = 0$ may be obtained from the significance level of χ^2.

Maximum value of the contingency coefficient (C_{max})

C is always less than or equal to one. For the 2 x 2 table this maximum value is calculated

$$C_{max} = [\phi^2_{max}/(1 + \phi^2_{max})]^{1/2}$$

Cramér's V

Another transformation of χ^2 (which reduces to ϕ for the 2 x 2 table) is

$$V = \left[\frac{\chi^2}{N(m-1)}\right]^{1/2} \qquad \text{where } m = \text{minimum } (R, C)$$

(See Bishop, Fienberg, and Holland, 1975, p. 386.)

Yule's Q and Y

For a 2 x 2 table, Yule defined the measure of association Q

$$Q = \frac{a_{11}a_{22} - a_{12}a_{21}}{a_{11}a_{22} + a_{12}a_{21}}$$

The standard errors of Q are

$$s_1^2 = \frac{1}{4}(a - Q^2)^2\left(\frac{1}{a_{11}} + \frac{1}{a_{12}} + \frac{1}{a_{21}} + \frac{1}{a_{22}}\right)$$

$$s_0^2 = \frac{1}{4}\left[\frac{N^3}{(a_{11}+a_{12})\,(a_{11}+a_{21})\,(a_{12}+a_{22})\,(a_{21}+a_{22})}\right]$$

Yule also defined a measure of colligation Y

$$Y = \frac{\sqrt{a_{11}a_{22}} - \sqrt{a_{12}a_{21}}}{\sqrt{a_{11}a_{22}} + \sqrt{a_{12}a_{21}}}$$

The standard errors of Y are

$$s_1^2 = \frac{1}{16}(1 - Y^2)^2\left(\frac{1}{a_{11}} + \frac{1}{a_{12}} + \frac{1}{a_{21}} + \frac{1}{a_{22}}\right)$$

$$s_0^2 = \frac{1}{16}\left[\frac{N^3}{(a_{11}+a_{12})\,(a_{11}+a_{21})\,(a_{12}+a_{22})\,(a_{21}+a_{22})}\right]$$

Both Q and Y are –1 when a_{11} or a_{22} is equal to zero, 1 when a_{12} or a_{21} is equal to zero, and zero when $a_{11}a_{22} = a_{12}a_{21}$ (i.e., the cross-product ratio is one).

Cross-product ratio (α)

For a 2 x 2 table, the cross-product (or odds) ratio is defined as

$$\alpha = (a_{11}a_{22})/(a_{12}a_{21})$$

This is 1 when χ^2 is 0, and 0 or ∞ when one of the cells has zero frequency. Since its distribution is extremely asymmetric under the hypothesis of independence between rows and columns, but that of $\ln(\alpha)$ is symmetric, the LN(CROSS-PRODUCT RATIO) is also given as well as its ASE.

The standard errors of $\ln(\alpha)$ are

$$s_1^2 = \frac{1}{a_{11}} + \frac{1}{a_{12}} + \frac{1}{a_{21}} + \frac{1}{a_{22}}$$

$$s_0^2 = \frac{N^3}{(a_{11}+a_{12})\,(a_{11}+a_{21})\,(a_{12}+a_{22})\,(a_{21}+a_{22})}$$

Relative risk and the Mantel-Haenszel statistics

Suppose we have k 2 x 2 tables. Let $a_{11i}, a_{12i}, a_{21i}, a_{22i}$ represent the four frequencies in the ith table, and let N_i be the total frequency in the ith table.

Then the Mantel-Haenszel statistic (Mantel and Haenszel, 1959) is defined as

$$R_{MH} = \frac{\sum_i (a_{11i}a_{22i}/N_i)}{\sum_i (a_{12i}a_{21i}/N_i)}$$

A test of the null hypothesis that the expected value of R_{MH} is 1 is provided by

$$\chi^2 = \frac{[\sum (a_{11i}a_{22i} - a_{12i}a_{21i})/N_i - 0.5]^2}{\sum r_{1i}r_{2i}c_{1i}c_{2i}/(N_i^2(N_i - 1))}$$

where r_{1i}, r_{2i} are the row totals and c_{1i}, c_{2i} are the column totals. This statistic has approximately a chi-square distribution with one degree of freedom.

The weighted combination of ln(risk) is defined as

$$\ln R_W = \sum (w_i \ln \alpha_i) / \sum w_i$$

where α_i is the cross-product for the ith table and

$$1/w_i = 1/a_{11i} + 1/a_{12i} + 1/a_{21i} + 1/a_{22i}$$

When one of the cell frequencies in a table is zero, all cell frequencies in that table are incremented by 0.5 in the calculation of R_W and $1/w_i$.

The variance of ln R_W is $1/\sum w_i$. Therefore approximate confidence limits for R_W are given as

$$\exp[\ln(R_W) - 2(1/\sum w_i)^{1/2}] \quad \text{and} \quad \exp[\ln(R_W) + 2(1/\sum w_i)^{1/2}]$$

The chi-square test for homogeneity is

$$\chi^2 = \sum w_i(\ln \alpha_i)^2 - (\sum w_i)(\ln R_W)^2$$

This is distributed approximately as a chi-square with $(k - 1)$ degrees of freedom, where k is the number of tables.

Tetrachoric correlation (r_t)

The tetrachoric correlation is the correlation of a bivariate normal distribution that exactly duplicates the four cell probabilities in a 2 x 2 table. It is found by solving the following equations implicitly for r_t.

a	b
c	d

$$\Phi(z_1) = (a + c)/N$$

$$\Phi(z_2) = (a + b)/N$$

$$\int_{-\infty}^{z_2}\int_{-\infty}^{z_1} \phi(z_1, z_2, r_t)\, dz_1 dz_2 = \frac{a}{N}$$

where $\Phi(z)$ is the Gaussian cdf with zero mean and unit variance and $\phi(z_1, z_2, r_t)$ is the bivariate normal with mean zero, variance 1, and correlation r_t.

The bivariate normal integral is approximated by an infinite series in r_t (Everitt, 1910) and r_t is found implicitly by iteration. When the infinite series does not converge within 100 terms ($|r_t| > 0.9$), Gaussian quadrature is used to evaluate the integral.

If a single cell is zero, the tetrachoric correlation can be defined as unity (or –1.0). We prefer to modify the cell to be 1/2 and then solve for r_t. The tetrachoric correlation should then be considered as any value from the computed value up to unity. Our justification is that the pattern of frequencies is duplicated by the choice of r_t; yet since frequencies are ordinal, the underlying probability is estimated with uncertainty equal to 1/(2N) in each cell. Therefore, setting the zero cell to 1/2 yields a lower bound for the estimate of the correlation (Brown and Benedetti, 1977b).

The formula for s_1^2 is from Pearson (1913).

$$s_1^2 = \frac{1}{N^3}\left(\frac{1}{\phi(z_1,z_2,r_t)}\right)^2\left[\frac{(a+d)(b+c)}{4} + (a+c)(b+d)\Phi_2^2 + (a+b)(c+d)\Phi_1^2 + 2(ad-bc)\Phi_1\Phi_2 - (ab-cd)\Phi_2 - (ac-bd)\Phi_1\right]$$

$$s_0^2 = \left(\frac{1}{\phi(z_1,z_2,0)}\right)^2\frac{(a+b)(a+c)(b+d)(c+d)}{N^5}$$

where

$\Phi_1 = \Phi[(z_1 - r_t z_2)/(1 - r_t^2)^{1/2}] - 0.5$ and $\Phi_2 = \Phi[(z_2 - r_t z_1)/(1 - r_t^2)^{1/2}] - 0.5$

and

$$\phi(z_1,z_2,r_t) = \frac{1}{2\pi(1-r_t^2)^{1/2}}\exp\left[-\frac{z_1^2 - 2r_t z_1 z_2 + z_2^2}{2(1-r_t^2)}\right]$$

Gamma (γ)

Gamma is a measure of the monotone relationship between two ordered variables (Goodman and Kruskal, 1954). It is estimated by

$$\gamma = (P - Q)/(P + Q)$$

The standard errors of γ are

$$s_1^2 = 16\left[\sum\sum a_{ij}(QA_{ij} - PD_{ij})^2\right]/(P+Q)^4$$

$$s_0^2 = 4\left[\sum\sum a_{ij}(A_{ij} - D_{ij})^2 - (P-Q)^2/N\right]/(P+Q)^2$$

Kendall's Tau-B (τ_b)

Kendall's τ_b is a correlation measure of the monotone relationship between two ordered variables that compares the number of agreements in the ordering of the indices between pairs of observations with the number of disagreements. Siegel (1956) gives the following estimate when many ties are expected.

$$\tau_b = \frac{(P-Q)}{\left[\left[N(N-1) - \sum_i r_i(r_i - 1)\right]\left[N(N-1) - \sum_j c_j(c_j - 1)\right]\right]^{1/2}} \equiv \frac{v}{w}$$

The standard errors of τ_b are

$$s_1^2 = \left[\sum\sum a_{ij}[2w(A_{ij} - D_{ij}) + \tau_b(r_i(N^2 - \sum c_j^2) + c_j(N^2 - \sum r_i^2))]^2 - N\tau_b^2(2N^2 - \sum r_i^2 - \sum c_j^2)^2\right]/w^4$$

$$s_0^2 = 4\left[\sum\sum a_{ij}(A_{ij} - D_{ij})^2 - \frac{(P-Q)^2}{N}\right]/w^2$$

where $w'_{ij} = -\frac{N}{W}[r_i(N^2 - \sum c_j^2) + c_j(N^2 - \sum r_i^2)]$

Stuart's Tau-C (τ_c)

A similar measure to τ_b which uses a different denominator is τ_c. It is estimated by

$$\tau_c = (P - Q)/N^2 \cdot [m/(m-1)]$$

where $m = \min(R, C)$.

The two standard errors of τ_c are similar

$$s_1^2 = s_0^2 = \frac{4}{N^4}\left[\frac{m}{m-1}\right]^2 \sum\sum a_{ij}(A_{ij} - D_{ij})^2$$

Product moment correlation (r)

The ordinary correlation coefficient is computed using the cell entries as frequencies and cell indices as the observed values.

$$r = \frac{\sum_i \sum_j a_{ij}(i - \bar{i})(j - \bar{j})}{\left[\sum_i r_i (i - \bar{i})^2 \sum_j c_j (j - \bar{j})^2\right]^{1/2}} \equiv \frac{v}{w}$$

where $\bar{i} = \sum_i ir_i/N$ and $\bar{j} = \sum_j jc_j/N$

The standard errors of r are

$$s_1^2 = \frac{1}{w^4}\sum\sum a_{ij}\left[w(i - \bar{i})(j - \bar{j}) - \frac{v}{2w}\left[(i - \bar{i})^2 \sum c_j (j - \bar{j})^2 + (j - \bar{j})^2 \sum r_i (i - \bar{i})^2\right]\right]^2$$

$$s_0^2 = \frac{1}{w^2}\left[\sum\sum a_{ij}(i - \bar{i})^2 (j - \bar{j})^2 - \frac{v^2}{N}\right]$$

Spearman rank correlation coefficient (r_s)

The Spearman rank correlation coefficient, r_s, is computed using the cell entries as frequencies and the cell indices to order the cells. The formula for ties (Siegel, 1956) has been rewritten in our notation.

$$r_s = \frac{\sum_i \sum_j a_{ij}\left(\sum_{k<i} r_k + \frac{r_i}{2} - \frac{N}{2}\right)\left(\sum_{l<j} c_l + \frac{c_j}{2} - \frac{N}{2}\right)}{\frac{1}{12}\left[N^3 - N - \sum_i (r_i^3 - r_i)\right]\left[N^3 - N - \sum_j (c_j^3 - c_j)\right]^{1/2}}$$

The standard errors of r_s are

$$s_1^2 = \frac{1}{N^2 w^4}\sum\sum a_{ij}\left(wv'_{ij} - vw'_{ij} - \frac{w}{N}\sum\sum a_{ij}v'_{ij} + \frac{v}{N}\sum\sum a_{ij}w'_{ij}\right)^2$$

$$s_0^2 = \frac{1}{N^2 w^2}\sum\sum a_{ij}\left(v'_{ij} - \frac{1}{N}\sum\sum a_{ij}v'_{ij}\right)^2$$

where $R(i) = 2\sum_{k<i} r_k + r_i - N$ and $C(j) = 2\sum_{l<j} c_l + c_j - N$

$$v = \frac{1}{4}\sum\sum a_{ij}R(i)C(j) \text{ and } w = \frac{1}{12}\left[(N^3 - \sum r_i^3)(N^3 - \sum c_j^3)\right]^{1/2}$$

$$v'_{ij} = N\left[\frac{1}{4}R(i)C(j) + \frac{1}{4}\sum_{j'} a_{ij'}C(j') + \frac{1}{4}\sum_{i'} a_{i'j}R(i') + \frac{1}{2}\sum_{j'}\sum_{i'>i} a_{i'j'}C(j') + \frac{1}{2}\sum_{i'}\sum_{j'>j} a_{i'j'}R(i')\right]$$

$$w'_{ij} = -\frac{N}{96w}\left[(N^3 - \sum r_i^3)c_j^2 + (N^3 - \sum c_j^3)r_i^2\right]$$

Somers' D (d_{asym})

Somers' D is an asymmetric version of τ_b. It is appropriate when one ordered index is to be predicted from knowledge of the second ordered index.

$$d_{j|i} = (P-Q)/(N^2 - \sum_i r_i^2) \text{ and } d_{i|j} = (P-Q)/(N^2 - \sum_j c_j^2)$$

The standard errors of $d_{j|i}$ are

$$s_1^2 = \frac{4}{(N^2 - \sum r_i^2)^4} \sum\sum a_{ij} [(N^2 - \sum r_i^2)(A_{ij} - D_{ij}) - (N - r_i)(P - Q)]^2$$

$$s_0^2 = \frac{4}{(N^2 - \sum r_i^2)^2} \sum\sum a_{ij} (A_{ij} - D_{ij})^2$$

The ASEs of $d_{i|j}$ are obtained by permuting the indices.

Tau-asymmetric (τ_{asym})

Goodman and Kruskal (1954) propose another measure of association in addition to τ. This measure, τ_{asym}, compared random proportional prediction of an index with conditional proportional prediction based upon knowledge of the second index.

$$\tau_{j|i} = \frac{N\sum_i\sum_j (a_{ij}^2/r_i) - \sum_j c_j^2}{N^2 - \sum_j c_j^2} \equiv \frac{v}{w} \qquad \tau_{i|j} = \frac{N\sum_i\sum_j (a_{ij}^2/c_j) - \sum_i r_i^2}{N^2 - \sum_i r_i^2}$$

The ASEs of $\tau_{j|i}$ (or $\tau_{i|j}$) are obtained by assuming multinomial sampling in each row (or column).

$$s_1^2 = \frac{4}{w^4} \Bigg[w^2 N^2 \sum\sum \frac{a_{ij}^3}{r_i^2} + (v-w)^2 \sum_j c_j^3 + 2w(v-w) N \sum\sum \frac{a_{ij}^2 c_j}{r_i}$$
$$- w^2 N^2 \sum_i \frac{1}{r_i^3} (\sum_j a_{ij}^2)^2 - (v-w)^2 \sum_i \frac{1}{r_i} (\sum_j a_{ij} c_j)^2$$
$$+ 2w(v-w) N \sum_i \frac{1}{r_i^2} (\sum_j a_{ij}^2)(\sum_j a_{ij} c_j) \Bigg]$$

The ASEs of $\tau_{i|j}$ are obtained by permuting the indices.

Lambda-asymmetric (λ_{asym})

Lambda is a measure of predictive association that measures the degree of success with which one index can be used to predict the second index (Goodman and Kruskal, 1954). No ordering of indices is assumed.

$$\lambda_{j|i} = \frac{\sum_i \max_j a_{ij} - \max_j c_j}{N - \max_j c_j} \qquad \lambda_{i|j} = \frac{\sum_j \max_i a_{ij} - \max_i r_i}{N - \max_i r_i}$$

The standard error of $\lambda_{j|i}$ is

$$s_1^2 = \frac{1}{(N - c_l)^2} \left[\sum\sum a_{ij} (\delta_{ij}^{in} - \delta_j^l + \lambda \delta_j^l)^2 - N\lambda_{j|i}^2 \right]$$

where l is the index of column such that c_l is the maximum

$\delta_{ij}^{in} = 1$ if $j = n$ and zero otherwise

$\delta_j^l = 1$ if $l = j$ and zero otherwise

The ASEs for $\lambda_{i|j}$ are obtained by permuting the indices.

Lambda-symmetric (λ)

This is a symmetric version of the previous statistic.

$$\lambda = \frac{\sum_i \max_j a_{ij} + \sum_j \max_i a_{ij} - \max_j c_j - \max_i r_i}{2N - \max_j c_j - \max_i r_i}$$

The standard error of λ is

$$s_1^2 = \frac{1}{(2N - c_l - r_k)^2} \{ \sum \sum a_{ij} [(\delta_{ij}^{in} + \delta_{ij}^{mj} - \delta_j^l - \delta_i^k) + \lambda (\delta_j^l + \delta_i^k)]^2 - 4N\lambda^2 \}$$

where, in addition to the notation for λ_{asym},

k is the index of rows such that r_k is the maximum

$\delta_{ij}^{mj} = 1$ if $i = m$ and zero otherwise

$\delta_i^k = 1$ if $k = i$ and zero otherwise

Lambda-star-asymmetric (λ^*_{asym})

A modification of lambda-asymmetric that is not affected by differing row (or column) marginals is λ^* where

$$\lambda^*_{j|i} = \frac{\sum_i (\max_j a_{ij}) / r_i - \max_j \sum_i (a_{ij}/r_i)}{R - \max_j \sum_i (a_{ij}/r_i)} \equiv \frac{v}{w}$$

$$\lambda^*_{i|j} = \frac{\sum_j (\max_i a_{ij}) / c_j - \max_i \sum_j (a_{ij}/c_j)}{C - \max_i \sum_j (a_{ij}/c_j)}$$

The ASE for $\lambda^*_{j|i}$ (or $\lambda^*_{i|j}$) are derived based on multinomial sampling in each row (or column).

$$s_1^2 = \frac{1}{w^2} \sum \sum \frac{a_{ij}}{r_i^2} [(\delta_{ij}^{in} - \delta_j^l) + \lambda^*_{j|i} \delta_j^l]^2 - \sum_i \frac{1}{r_i^3} [(a_{in} - a_{il}) + \lambda^*_{j|i} a_{il}]^2$$

where

a_{in} is the maximum frequency in row i

l is the column j for which $\sum_j a_{ij}/r_i$ is a maximum

The ASEs of $\lambda^*_{i|j}$ are obtained by permuting the indices.

Uncertainty coefficient-asymmetric

This is an asymmetric measure of association between two unordered variables based on an information theory approach that reflects the relative reduction of uncertainty about one factor when the second factor is known.

$$U_{j|i} = \frac{U_j + U_i - U_{ij}}{U_j} \equiv \frac{v}{w} \qquad U_{i|j} = \frac{U_j + U_i - U_{ij}}{U_i}$$

where

$$U_{ij} = -\sum_i \sum_j \frac{a_{ij}}{N} \log \frac{a_{ij}}{N} \qquad U_j = -\sum_j \frac{c_j}{N} \log \frac{c_j}{N} \qquad U_i = -\sum_i \frac{r_i}{N} \log \frac{r_i}{N}$$

The standard error of $U_{j|i}$ is

$$s_1^2 = \frac{1}{N^2 w^4} \sum \sum a_{ij} \left[(U_i - U_{ij}) \log \frac{c_j}{N} + U_j \left(\log \frac{a_{ij}}{N} - \log \frac{r_i}{N} \right) \right]^2$$

The ASEs of $U_{i|j}$ are obtained by permuting the indices.

Uncertainty coefficient-symmetric

A symmetric measure related to the above asymmetric statistic is

$$U = \frac{U_j + U_i - U_{ij}}{(U_j + U_i)/2}$$

This measure has been normalized so that U may range between zero and one. U is frequently defined without the 1/2 in the denominator (see Brown, 1975). The ASE of U is

$$s_1^2 = \frac{4}{N^2 (U_j + U_i)^4} \sum\sum a_{ij} \left[U_{ij} \left(\log \frac{r_i}{N} + \log \frac{c_j}{N} \right) - (U_j + U_i) \log \frac{a_{ij}}{N} \right]^2$$

McNemar's test of symmetry (χ^2_{MC})

This is a test of symmetry which is calculated for square tables ($R = C$) only.

$$\chi^2_{MC} = \sum_{i<j}\sum \frac{(a_{ij} - a_{ji})^2}{a_{ij} + a_{ji}}$$

The degrees of freedom are ordinarily $R(R - 1)/2$; however, in the presence of certain patterns of missing data, it equals the number of terms in the summation below, in which the sum of the off-diagonal elements in a term is nonzero.

$$\text{df} = \sum\sum \frac{(n_{ij} - n_{ji})^2}{n_{ij} + n_{ji}}$$

Test for marginal homogeneity

In the case where the hypothesis of symmetry is rejected, the weaker hypothesis of marginal homogeneity may be of interest. This hypothesis postulates that the row marginal probabilities are equal to the corresponding column marginal probabilities. This may be formulated as follows:

$$H_0: p_{i.} = p_{.i} \text{ (for } i = 1, 2, \ldots, r)$$

For the general $r \times r$ table the test statistic is given by:

$$\chi^2 = \mathbf{d}'\mathbf{V}^{-1}\mathbf{d}$$

where **d** is a column vector of *any* $(r - 1)$ of the differences $d_1, d_2, \ldots, d_r$, and $d_i = n_{i.} - n_{.i}$ and **V** is an $(r - 1) \times (r - 1)$ matrix of the corresponding variances and covariances of the d_i's and has elements:

$$v_{ii} = n_{i.} + n_{.i} - 2n_{ii}$$

$$v_{ij} = -(n_{ij} + n_{ji})$$

If H_0 is true then χ^2 has a chi-square distribution with $(r - 1)$ df (Everitt, 1977).

Kappa

A test of reliability computed for square, two-way tables.

$$\kappa = \frac{p_o - p_e}{1 - p_e}$$

where

$p_o = \sum_i a_{ii}/N$ proportion observed agreement

$p_e = \sum_i r_i c_i/N^2$ proportion expected agreement

The standard errors of κ are

$$s_1 = \frac{\sqrt{A+B-C}}{(1-p_e)\sqrt{N}}$$

where

$A = \sum_i p_{ii}\,[1-(p_{i.}+p_{.i})(1-\hat{\kappa})]^2$

$B = (1-\hat{\kappa})^2 \sum\sum p_{ij}(p_{.i}+p_{j.})^2$

$C = [\hat{\kappa} - p_e(1-\hat{\kappa})]^2$

and

$$s_0 = \frac{1}{(1-p_e)\sqrt{N}}\sqrt{p_e + p_e^2 - \sum_i p_{i.}p_{.i}(p_{i.}+p_{.i})}$$

where

$p_{i.} = \sum_j a_{ij}/N$ proportion in row i

$p_{.j} = \sum_i a_{ij}/N$ proportion in column j

Test of linear regression

Let the frequency table be 2 x k (2 rows and k columns) and a set of k coefficients (z_j) assigned, one to each column. The z_j's are either CODE values or, when CUTPOINTS are assigned, the sequence 1, 2, ... The regression of a_{ij}/c_j on z_j is performed, and the following χ^2 test with 1 df is computed (Cochran, 1954).

$$\chi^2 = \left(\frac{\left[\sum c_j \left(\frac{a_{1j}}{c_j} - \frac{r_1}{N}\right)(z_j - \tilde{z})\right]^2}{\sum c_j (z_j - \tilde{z})^2} \cdot \frac{N^2}{r_1 r_2} \right)$$

where

$\tilde{z} = \sum c_j z_j / N$

Appendix B.12 Program 4F The log-linear model

Each log-linear model is fitted by the iterative proportional fitting algorithm starting from an initial fit matrix. The initial fit matrix is (1) entirely ones if EMPTY cells or INITIAL values are not specified for the table; (2) a table of zeros for the EMPTY cells and ones for the remaining cells; or (3) the set of INITIAL values specified by the user (ones for the cells that have no initial value specified).

The notation of 4F is used to describe a log-linear model. That is

$$\ln F_{ijkl} = \theta + \lambda_i^A + \lambda_j^B + \lambda_k^C + \lambda_l^D + \lambda_{ij}^{AB} + \lambda_{ik}^{AC} + \lambda_{il}^{AD} + \lambda_{jk}^{BC} + \lambda_{jl}^{BD} + \lambda_{kl}^{CD} + \lambda_{ijk}^{ABC} + \lambda_{ijl}^{ABD} + \lambda_{ikl}^{ACD} + \lambda_{jkl}^{BCD} + \lambda_{ijkl}^{ABCD}$$

where f_{ijkl} and F_{ijkl} are the observed and expected values of cell (i, j, k, l) in an I x J x K x L contingency table. Also let

$$y_{ijkl} = \ln F_{ijkl}$$

for all cells.

Computation of degrees of freedom

When fitting the models of full order (the SIMULTANEOUS option) or before fitting any other log-linear model, 4F computes the number of estimable parameters (EP) for each possible effect Z (Z can represent a main effect or interaction). The formula for degrees of freedom (df) is

df = total number of cells – number of cells with expected values equal to zero – $\sum$ EP

where the summation is over all effects in the model being fitted.

When there are no zeros in the marginal subtable corresponding to an effect Z, the number of estimable parameters for Z is

$$\mathrm{EP} = (I-1)^{\delta^{ZA}} (J-1)^{\delta^{ZB}} (K-1)^{\delta^{ZC}} (L-1)^{\delta^{ZD}}$$

where

$\delta^{ZA} = 1$ if the effect Z contains index A

$\delta^{ZA} = 0$ otherwise

with similar definitions for δ^{ZB}, δ^{ZC}, and δ^{ZD}. The formula is the same as that used for effects in a balanced ANOVA.

When there are zeros in the marginal subtable corresponding to Z, the number of estimable parameters (EP′) is

EP′ = EP – number of zeros in the marginal subtable corresponding to Z

+ number of zeros in all subtables formed by collapsing the marginal subtable over each index in Z (in turn)

– number of zeros in all subtables formed by collapsing the marginal subtable over each pair of indices in Z (in turn)

+ ... – ... (until there are no more zeros)

Note: The above formula for degrees of freedom is valid when the table is not separable with respect to the model analyzed; that is, the table cannot be separated into two or more parts, each of which can be analyzed separately. If estimates of the parameters are requested and 4F obtains the estimates by Method 1 below, the degrees of freedom printed in the output for the model can be checked as follows: if the number of pivots is printed, the number of degrees of freedom = number of cells – number of cells with expected value equal to zero – number of pivots. If this answer differs from that printed in the output, the table is separable; the answer obtained using the number of pivots should be used in place of the degrees of freedom in the output.

Appendix B.13 Program 4F Parameter estimates and standard errors for log-linear models

Computation of parameter estimates and their standard errors

The parameter estimates and their standard errors are obtained by one of two methods. The first method is used unless (1) the model is direct and there are no expected values equal to zero (and the variance-covariance matrix of the parameter estimates was not requested) or (2) there is insufficient space to form the variance-covariance matrix of the parameter estimates. When either of these two exceptions holds, the second method (the "delta" method) will be used.

Method 1: Weighted least-squares

Form a design matrix **X** having p columns and n rows where p is the number of parameters to be estimated in the log-linear model (including the constant term) and n is the number of cells in the table that are not defined to be structural zeros. Each column in **X** consists of 1's, 0's and -1's; if there are no structural zeros, the sum of each column is zero except for the vector corresponding to the constant term.

Then the log-linear model can be written as

$$\mathbf{Y} = \mathbf{X}'\lambda$$

where λ is a column vector of parameters, and **Y** is a column vector containing $\ln F_{ijkl}$.

Let **W** be a diagonal matrix containing $1/F_{ijkl}$ on the diagonal; i.e., **W** is the asymptotic variance-covariance matrix of **Y**. The parameter estimates are obtained as

$$\hat{\lambda} = (\mathbf{X'WX})^{-1}\mathbf{X'WY}$$

and the variance-covariance matrix of the parameter estimates is $(\mathbf{X'WX})^{-1}$. The standard errors of $\hat{\lambda}$ are the square roots of the diagonal elements of $(\mathbf{X'WX})^{-1}$.

The computations are performed by setting up the matrix

$$\begin{bmatrix} \mathbf{X'WX} & \mathbf{X'WY} \\ \mathbf{Y'WX} & \mathbf{Y'WY} \end{bmatrix}$$

and sweeping the matrix on $\mathbf{X'WX}$ (see Appendix B.16).

When not all parameters are estimable (one or more design vectors are aliased with other design vectors), 4F prints the number of pivots actually performed as well as a message that not all pivots were performed. This number can be used to validate the number of degrees of freedom (see above). 4F will either print a zero for each parameter that is not estimated or will omit printing the parameter entirely. When printing the variance-covariance matrix, nonestimable parameters are omitted entirely.

Method 2: The "delta" method

The parameter estimates $\hat{\lambda}$ for each effect Z are obtained by applying the usual ANOVA formulas to $y_{ijkl} = \ln F_{ijkl}$. For example, let Z represent the interaction ABC, then

$$\hat{\lambda}_{ijk}^{ABC} = y_{ijk.} - y_{ij..} - y_{i.k.} - y_{.jk.} + y_{i...} + y_{.j..} + y_{..k.} - y_{....}$$

where a dot represents the mean over the omitted index.

4F computes the parameter estimates by using the following formula that is equivalent to the one above:

$$\hat{\lambda}_{ijk}^{ABC} = (1/IJKL)(IJKy_{ijk+} - IJy_{ij++} - IKy_{i+k+} - JKy_{+jk+} + Iy_{i+++} + Jy_{+j++} + Ky_{++k+} - y_{++++})$$

where + signifies the total over the omitted index.

The asymptotic variance of Z (say ABC) in a saturated model can be written as

$$AV(\hat{\lambda}_{i'j'k'l'}^{Z}) = (1/IJKL)^2 \sum_{ijkl} \frac{\left[(I-1)^{\delta_{ii'}^{ZA}} (J-1)^{\delta_{jj'}^{ZB}} (K-1)^{\delta_{kk'}^{ZC}} (L-1)^{\delta_{ll'}^{ZD}}\right]^2}{F_{ijkl}}$$

where $\delta_{ii'}^{ZA}$ is 1 if A is an index in Z and $i = i'$; zero otherwise. Similar definitions are used for the other δs.

4F uses the following equivalent formula in order to compute the asymptotic variance of $\hat{\lambda}$ in a saturated model:

$$\begin{aligned} AV(\hat{\lambda}_{i'j'k'l'}^{Z}) = (1/IJKL)^2 [&I(I-2)J(J-2)K(K-2)w_{ijk+} + I(I-2)J(J-2)w_{ij++} \\ &+ I(I-2)K(K-2)w_{i+k+} + J(J-2)K(K-2)w_{+jk+} \\ &+ I(I-2)w_{i+++} + J(J-2)w_{+j++} + K(K-2)w_{++k+} + w_{++++}] \end{aligned}$$

where $w_{ijkl} = 1/F_{ijkl}$ and + represents the sum over an index.

When the model is direct but not saturated, Lee (1977) describes how the asymptotic variance can be obtained as the sum and difference of asymptotic variances of the same term in a series of saturated models. 4F implements Lee's method whenever the model is direct.

When the model is not direct and there is not enough space to form the variance-covariance matrix, modifications of the "delta" method are used although it is known that the answers are not exact. If the model is indirect, 4F finds a direct model that includes the indirect model as a subset and computes the asymptotic variances using the direct model; this provides an upper bound for the asymptotic variances. When there are zero expected values, 4F first estimates the expected values for cells with structural zeros from the model fitted to the data in all other cells. Then the parameters and standard errors are estimated using the "completed" data; the resulting estimates of the variances are underestimates of the asymptotic variances. When there are zero expected values due to the model (and not as a result of declaring structural zeros), both the parameter estimates and their standard errors may be seriously affected.

Appendix B.14 Program 4M

Computational method

Single precision is used for reading data and performing transformations. All other computations are performed in double precision. Means, standard deviations, and covariances are computed by the method of provisional means. When case weights are used, the statistics are adjusted so that the average nonzero weight is one.

The description of many of the formulas is given in the program description. The inverse of the correlation or covariance matrix is computed by stepwise pivoting. Eigenvalues and eigenvectors are computed using routines from EISPACK (Garbow et al., 1977 and Smith et al., 1974). The method of shading the correlation matrix is similar to that of Ling (1973).

Maximum likelihood factor analysis is discussed by Lawley and Maxwell (1971). The algorithm used here was developed by R. I. Jennrich and P. F. Sampson at the Health Sciences Computing Facility, UCLA.

Let **R** be the sample correlation matrix. The normal theory maximum likelihood estimates are computed for the factor loading matrix Λ and the unique standard deviation matrix Ψ (square roots of one minus the communality for each variable) in the factor analytic decomposition

$$\Sigma = \Lambda\Lambda' + \Psi^2$$

where Σ is the population correlation matrix. Here Ψ is diagonal and the loadings Λ are canonical (Rao, 1955).

The program uses a Newton-Raphson iteration

$$\Delta\Psi = -(f''_{ij})^{-1}\,(f'_i)$$

to minimize a function f of Ψ which is a negative slope affine transformation of the conditional maximum of the likelihood for the sample given Ψ (cf. Jennrich and Robinson, 1969; Clarke, 1970; Joreskog and van Thillo, 1971).

In terms of the eigenvalues

$$\gamma_1 \leq \gamma_2 \leq \ldots \leq \gamma_p$$

and vectors $\mathbf{w}_1, \ldots, \mathbf{w}_p$ of $\Psi\mathbf{R}^{-1}\Psi$, the function

$$f = \sum_{m=k+1}^{p} (\log\gamma_m + \gamma_m^{-1} - 1)$$

where k is the number of factors and p the number of variables.

The required derivatives $f'_i = \delta f / \delta\Psi_i$ and $f''_{ij} = \delta^2 f / \delta\Psi_i \delta\Psi_j$ are given by

$$f'_i = \sum_{m=k+1}^{p} (1 - \gamma_m^{-1})\, w_{im} v_{im}$$

and

$$f''_{ij} = \sum_{m=k+1}^{p} \nu_{im}\nu_{jm} \sum_{n=1}^{k} \frac{\gamma_m + \gamma_n - 2}{\gamma_m - \gamma_n} w_{in}w_{jn} + \frac{1}{2}\delta_{ij} \sum_{m=k+1}^{p} (3\gamma_m^{-1} - 1)\,\nu_{im}^2$$

where

$$\nu_m = 2\gamma_m^{-1}\mathbf{R}^{-1}\Psi w_m$$

If, as frequently occurs during the initial steps, (f''_{ij}) is not positive definite, it is replaced by a matrix of approximate second order derivatives

$$f''_{ij} \approx \left(\sum_{m=k+1}^{p} \nu_{im}\nu_{jm}\right)\left(\sum_{m=k+1}^{p} w_{im}w_{jm}\right)$$

If the value of f is not decreased by the step $\Delta\Psi$, the step is halved (i.e., $\Delta\Psi$ is replaced by $^1/_2\,\Delta\Psi$) and halved again until the value of f decreases or ten halvings fail to produce a decrease. Stepping continues until all values of $\Delta\Psi_i$ are less than the convergence criterion or the number of steps equals its prescribed maximum.

Using the converged value of Ψ, the eigenvalue problem

$$\Psi^2 u_m = \gamma_m \mathbf{R} u_m\,,\ m = 1, \ldots, p$$

is solved and the estimated columns of Λ are computed using

$$\lambda_m = \rho_m \mathbf{R} u_m\,,\ m = 1, \ldots, k$$

where $\rho_m = (1 - \gamma_m)^{1/2}$ is the estimated canonical correlation. The advantage of this parameterization is that it converges smoothly even in Heywood cases—when some of the diagonal components of Ψ are zero.

Carmines' theta

Carmines' theta is computed as follows:

$$\theta = (N/(N-1))(1 - 1/\lambda_1)$$

where N is the number of items in the total factor analysis and λ_1 is the largest (i.e., the first) eigenvalue.

Rotating the factors

4M provides eight commands for factor rotation. The default VARIMAX orthogonal rotation is used in Example 4M.1 and the DQUART direct quartimin in Example 4M.3. For the rotation commands ORTHOG (orthogonal) and DOBLI (direct oblimin), you may specify the value of the rotation parameter GAMMA (Γ). Note that most users use the default value.

Rotations are performed to minimize the *simplicity criterion G* (Harman, 1967, Chapters 14 and 15).

$$G = \sum_{i \neq j}\left[\sum_{k=1}^{p} a_{ki}^2 a_{kj}^2 - \frac{\Gamma}{p}\left(\sum_{k=1}^{p} a_{ki}^2\right)\left(\sum_{k=1}^{p} a_{kj}^2\right)\right]$$

where

- Γ = GAMMA, a parameter discussed below
- i, j = $1, \ldots, m$
- m = number of factors
- p = number of variables
- (a_{ij}) = the matrix of factor loadings for orthogonal and direct oblimin rotation and is the factor structure for indirect oblimin

Although GAMMA = 0 (the preassigned value) is almost always desirable, you can change it. For *orthogonal* rotation, low values of gamma (e.g., GAMMA = 0) emphasize "simplifying" the rows of a loadings matrix, and large values (e.g., GAMMA = 1) emphasize simplifying the columns (Harman, 1967, p. 304). For *direct oblimin*, increasing gamma causes factors to be *more* highly correlated (more oblique), but positive values for gamma may result in convergence problems (Jennrich, 1978).

For other methods of rotation, the value of GAMMA is preset. For varimax rotation (the default), GAMMA is set to one; for equamax rotation, it is set to the number of factors divided by two ($m/2$). When the LJIFFY method is requested, the ORTHOB option is automatically assigned. When principal components is used, rotation may not be necessary. To omit rotation, specify METHOD = NONE. More details concerning methods of rotation are found in Jennrich and Sampson (1966), Harman (1967), Kaiser (1970), and Jennrich (1978).

The rotation terminates when the maximum number of iterations is reached (MAXIT) or when the relative change in the value of G (defined above) between two successive iterations is less than CONSTANT. Kaiser's normalization (Harman, 1967, p. 306) is performed unless NO NORMAL is specified in the ROTATE paragraph.

Appendix B.15 Program 7M

Computational method

The data are read in single precision. All computations are performed in double precision. The computational procedure is described below. See also Jennrich (1977b).

We use the following notation:

p = number of variables available

q = number of variables entered at a given step

t = total number of groups

g = number of groups used to define the discriminant functions

n_i = number of cases in group i

n = total number of cases in the g defining groups

x_{ijr} = value of variable r in case j of group i

h = number of hypotheses (contrasts)

h_{ki} = coefficient for group i in hypothesis k

p_i = prior probability for group i

Assume for simplicity that the first g of the t groups are used to define the classification functions.

Step 1

The data are read and the method of provisional means (Appendix B.2) is used to compute the group means:

$$\bar{x}_{ir} = \sum_{j=1}^{n_i} x_{ijr}/n_i \qquad \begin{array}{l} i = 1,\ldots,t \\ r = 1,\ldots,p \end{array}$$

group standard deviations:

$$s_{ir} = \left(\sum_{j=1}^{n_i} (x_{ijr} - \bar{x}_{ir})^2 / (n_i - 1) \right)^{1/2} \qquad \begin{array}{l} i = 1,\ldots,t \\ r = 1,\ldots,p \end{array}$$

and pooled within groups sums of cross-product deviations:

$$w_{rs} = \sum_{i=1}^{g} \sum_{j=1}^{n_i} (x_{ijr} - \bar{x}_{ir})(x_{ijs} - \bar{x}_{is}) \qquad \begin{array}{l} r = 1,\ldots,p \\ s = 1,\ldots,p \end{array}$$

The latter are used to compute the within group correlations:

$$r_{ij} = w_{ij} / (w_{ii} w_{jj})^{1/2} \qquad \begin{array}{l} i = 1,\ldots, p \\ j = 1,\ldots, p \end{array}$$

Step 2

Let $\mathbf{H} = (h_{ki})$ be the h x g matrix of hypothesis contrasts. If no contrasts are specified, h is set to $g - 1$ and

$$h_{ki} = \begin{cases} 1 & i \le k \\ -k & i = k+1 \\ 0 & \text{otherwise} \end{cases}$$

These contrasts test the equality of all g group means.

The stepwise procedure is defined in terms of the matrices

$$\mathbf{W} = (w_{rs})$$

and

$$\mathbf{M} = \mathbf{W} + \bar{\mathbf{X}}'\mathbf{H}'(\mathbf{H}\mathbf{N}^{-1}\mathbf{H}')^{-1}\mathbf{H}\bar{\mathbf{X}}$$

where $\bar{\mathbf{X}} = (\bar{x}_{ir})$ is a g x p matrix, and $\mathbf{N}$ is the diagonal matrix $[n_1, \ldots, n_g]$ of group sizes. The entry and removal of variables is defined in terms of the results of sweeping on the diagonal elements of **W** and **M** (the computations themselves proceed somewhat more efficiently).

Assuming for simplicity that the first q variables have already been swept, write

$$\mathbf{W} = \begin{bmatrix} \mathbf{W}_{11} & \mathbf{W}_{12} \\ \mathbf{W}_{21} & \mathbf{W}_{22} \end{bmatrix} \qquad \mathbf{M} = \begin{bmatrix} \mathbf{M}_{11} & \mathbf{M}_{12} \\ \mathbf{M}_{21} & \mathbf{M}_{22} \end{bmatrix}$$

where $\mathbf{W}_{11}$ and $\mathbf{M}_{11}$ are q x q. At each step let

$$\mathbf{A} = \begin{bmatrix} -\mathbf{W}_{11}^{-1} & \mathbf{W}_{11}^{-1}\mathbf{W}_{12} \\ \mathbf{W}_{21}\mathbf{W}_{11}^{-1} & \mathbf{W}_{22}(-\mathbf{W}_{21})\mathbf{W}_{11}^{-1}\mathbf{W}_{12} \end{bmatrix}$$

$$\mathbf{B} = \begin{bmatrix} -\mathbf{M}_{11}^{-1} & \mathbf{M}_{11}^{-1}\mathbf{M}_{12} \\ \mathbf{M}_{21}\mathbf{M}_{11}^{-1} & \mathbf{M}_{22}(-\mathbf{M}_{21})\mathbf{M}_{11}^{-1}\mathbf{M}_{12} \end{bmatrix}$$

B is not actually computed, since only the diagonal elements are needed. These diagonal elements are computed from the matrix

$$\begin{bmatrix} \mathbf{A} & \mathbf{T} \\ \mathbf{T}' & \mathbf{C} \end{bmatrix}$$

which is defined at step zero to be

$$\begin{bmatrix} \mathbf{W} & \bar{\mathbf{X}} \\ \bar{\mathbf{X}}' & 0 \end{bmatrix}$$

and is updated at each step by sweeping or reverse sweeping the diagonal elements of **A**. The diagonal elements of **B** are computed using the fact that

$$\mathbf{B} = \mathbf{Q}'\mathbf{Q} + \mathbf{A}$$

where

$$\mathbf{Q} = (\mathbf{H}(\mathbf{N}^{-1} - \mathbf{C})\mathbf{H}')^{-1/2}\mathbf{H}'\mathbf{T}'$$

The following statistics are computed at each step

(1) F values for testing differences between each pair of groups:

$$F_{ij} = \left(\frac{(n-g-q+1)\,n_i n_j}{q\,(n_i+n_j)} D^2_{ij}\right) / (n-g) \qquad i,j = 1,\ldots,g$$

where

$$D^2_{ij} = (n-g)\,(\bar{\mathbf{X}}_i - \bar{\mathbf{X}}_j)'\mathbf{W}_{11}^{-1}(\bar{\mathbf{X}}_i - \bar{\mathbf{X}}_j)$$

is the (squared) Mahalanobis distance between groups i and j and where $\bar{\mathbf{X}}_i$ is the vector of means for group i for the q variables that have been entered. D^2_{ij} is computed as

$$(n-g)(2c_{ij} - c_{ii} - c_{jj})$$

(2) F *values for each variable*

If variable r has been entered,

$$F_r = \frac{a_{rr} - b_{rr}}{b_{rr}} \cdot \frac{n-g-q+1}{h}$$

with h and $n-g-q+1$ degrees of freedom.

If variable r has not been entered,

$$F_r = \frac{b_{rr} - a_{rr}}{a_{rr}} \cdot \frac{n-g-q}{h}$$

with h and $n-g-q$ degrees of freedom.

(3) Wilks' Λ-statistic for the hypothesis defined by **H**

$$\Lambda = \det(\mathbf{W}_{11})/\det(\mathbf{M}_{11})$$

with $(q, h, n-g)$ degrees of freedom. Λ is computed by initially setting it equal to one and updating it at each step by multiplying its previous value by a_{rr}/b_{rr}, where r is the index of the variable entered or removed at the step.

(4) The F *approximation to Λ (Rao, 1973, p. 556)*

$$F = \frac{1-\Lambda^{1/s}}{\Lambda^{1/s}} \cdot \frac{m^* s + 1 - hq/2}{hq}$$

where

$$m^* = n - g - \frac{(q-h+1)}{2}$$

$$s = \begin{cases} \left(\dfrac{h^2q^2-4}{h^2+q^2-5}\right)^{1/2} & h^2+q^2 \neq 5 \\ 1 & h^2+q^2 = 5 \end{cases}$$

The numbers of degrees of freedom for F are hq and $(m^*s + 1 - hq/2)$. The approximation is exact if either h or q is 1 or 2.

(5) Tolerance values

$$t_r = a_{rr}/w_{rr}\,,\ r = q+1, \ldots, p$$

Step 3

To move from one step to the next, a variable is removed or added according to the first of the following rules which applies.

Rule 1 If METHOD = 1 is specified in the DISCRIMINANT paragraph and one or more entered variables are available and have F values less than the F-to-remove threshold, the one with the smallest F value is removed.

Rule 2 If METHOD = 2 is specified and one or more entered variables are available, the one with the smallest F value is removed if by its removal Wilks' Λ will be smaller than it was when the same number of variables were previously entered.

Rule 3 If one or more nonentered variables are available, with tolerance above the tolerance threshold, and are either at a forced level or have F values above the F-to-enter threshold, the one with the highest F value is entered.

Variables in forced levels are available only if their level is equal to the working level, if they are not entered, and if they have tolerance above the tolerance threshold.

Variables in nonforced levels are considered available only if their level is less than equal to the working level and, for nonentered variables, if their tolerance is above the tolerance threshold. (See Appendix B.16 for a discussion of tolerance.)

The working level begins at 1 and moves to the next level when none of the rules apply. If the specified maximum level has already been reached, the specified maximum number of steps has been reached, or the specified maximum number of variables has been entered; the stepping terminates.

Step 4

When the stepping is complete, or when the number of variables entered is equal to one of the numbers indicated in the PRINT paragraph, the following are computed and printed except for (2) and (3) which are printed only when the stepping is complete:

(1) Group classification function coefficients, which are defined as

$$\hat{\beta}_i = (n-g)\,\mathbf{W}_{11}^{-1}\bar{\mathbf{X}}_i \qquad (\text{a}\,q \times 1\,\text{vector}) \qquad i = 1,\ldots,g$$

and the corresponding constants

$$\hat{\alpha}_i = \ln\,p_i - (n-g)\,\bar{\mathbf{X}}'_i\mathbf{W}_{11}^{-1}\bar{\mathbf{X}}_i/2 \qquad i = 1,\ldots,g$$

where

p_i is the prior probability for group i

Note that $\mathbf{W}_{11}^{-1}\bar{\mathbf{X}}_i$ is computed as a submatrix of $\mathbf{T}$ and $\bar{\mathbf{X}}'_i\mathbf{W}_{11}^{-1}\bar{\mathbf{X}}_i = -c_{ii}$. (For additional discussion, see Classification Functions.)

(2) The squared Mahalanobis distance of case j *in group* i *from the mean of group* k

$$D_{ijk}^2 = (n-g)\sum_{r=1}^{q}\sum_{s=1}^{q}(x_{ijr}-\bar{x}_{kr})\,a_{rs}\,(x_{ijs}-\bar{x}_{ks}) \qquad \begin{array}{l} i = 1,\ldots,t \\ j = 1,\ldots,n_i \\ k = 1,\ldots,g \end{array}$$

or if requested in the PRINT paragraph, the cross validation or jackknife distance ${}_*D_{ijk}^2$, obtained by withholding this case from all computations except the evaluation of the distance function. The computing formulas are

$D_{ijk}^2 = (n-g)e$

$$_*D^2_{ijk} = (n-1-g)\{e + a^2 / b - \delta_{ik} / [n_i(n_i - 1)]\}$$

where

$e \quad = f + 2(c_i - c_k) - c_{ii} + c_{kk}$

$a \quad = \delta_{ik} / n_k + c_i - c_k - c_{ik} + c_{ii}$

$b \quad = 1 - 1/n_i - f$

$\delta_{ik} \quad$ = the Kronecker delta

$c_k \quad = \mathbf{X}'_{ij}\mathbf{W}_{11}^{-1}\bar{\mathbf{X}}_k$ (where $\mathbf{W}_{11}^{-1}\bar{\mathbf{X}}_k$ has been computed as a submatrix of $\mathbf{T}$)

$c_{ij} \quad = -\bar{\mathbf{X}}'_i\mathbf{W}_{11}^{-1}\bar{\mathbf{X}}_j$ (already computed in $\mathbf{C}$)

$$f = \sum_{r=1}^{q}\sum_{s=1}^{q} (x_{ijr} - \bar{x}_{ir})\, a_{rs}\, (x_{ijs} - \bar{x}_{is})$$

$$= (\mathbf{X}_{ij} - \bar{\mathbf{X}}_i)'\mathbf{W}_{11}^{-1}(\mathbf{X}_{ij} - \bar{\mathbf{X}}_i)$$

$\mathbf{X}_{ij} \quad = (x_{ij1}, \ldots, x_{ijq})'$

(3) The posterior probability that case j *from group* i *came from group* k

$$P_{ijk} = p_k \exp(-D^2_{ijk}/2) \Big/ \left(\sum_{r=1}^{g} p_r \exp(-D^2_{ijr}/2) \right) \qquad \begin{array}{l} i = 1,\ldots,t \\ j = 1,\ldots,n_i \\ k = 1,\ldots,g \end{array}$$

If JACKKNIFE is requested, the D's are replaced by $_*D$'s in the above formula. A classification matrix that gives the number of cases n_{ik} in group i whose posterior probability p_{ijk} was largest for group k is printed.

Step 5

The eigenvalue problem

$$\mathbf{H}\bar{\mathbf{X}}\mathbf{W}_{11}^{-1}\bar{\mathbf{X}}'\mathbf{H}'u_i = \mathbf{H}(-\mathbf{C})\mathbf{H}'u_i = \lambda_i \mathbf{H}\mathbf{N}^{-1}\mathbf{H}'u_i$$

where $\bar{\mathbf{X}}$ is a g x q matrix of variable means for all groups, is solved for eigenvalues $\lambda_1 \geq \ldots \geq \lambda_h$ and eigenvectors $u_1, \ldots, u_h$ normalized so that

$$u'_i\mathbf{H}\mathbf{N}^{-1}\mathbf{H}'u_i = 1$$

The coefficients $\hat{\gamma}_i$ of the ith canonical discriminant function defined by $\mathbf{H}$ are given by

$$\hat{\gamma}_i = \mathbf{W}_{11}^{-1/2}\bar{\mathbf{X}}'\mathbf{H}'u_i\sqrt{\frac{n-g}{\lambda_i}} \qquad i = 1,\ldots,h$$

If $h < q$ and if required, $q - h$ additional canonical functions will be computed corresponding to zero eigenvalues. For each $i = h + 1, \ldots, q$ in turn, a random g-vector $\mathbf{V}_i$ is generated and orthogonalized by Gramm-Schmidt to $\hat{\gamma}_1,\ldots,\hat{\gamma}_{i-1}$.

Then

$$\hat{\gamma}_i = (\mathbf{W}_{11}^{-1/2}\mathbf{V}_i) / (\mathbf{V}'_i\mathbf{W}_{11}^{-1}\mathbf{V}_i)^{1/2} \qquad i = h+1,\ldots,q$$

The cumulative proportion of explained dispersion is computed as

$$v_i = \sum_{j=1}^{i}\lambda_j \Big/ \sum_{j=1}^{h}\lambda_j \qquad i = 1,\ldots,h$$

The canonical variables (values of the canonical functions)

$$f_{ijr} = \sum_{s=1}^{q} \hat{\gamma}_{rs}(x_{ijs} - \bar{x}_s) \qquad \begin{array}{l} i = 1,\ldots,t \\ j = 1,\ldots,n_i \end{array}$$

for $r = 1, 2$ are computed and plotted. All requested values of f_{ijr} are written in a BMDP File.

The canonical correlations are computed from the eigenvalues as

$$\rho_i = \sqrt{\frac{\lambda_i}{1+\lambda_i}}$$

Eigenvalues and eigenvectors are computed using routines from the EISPACK system (Garbow, et al., 1977, and Smith et al., 1974).

Classification functions

The classification functions from 7M can be used to classify new cases into the groups used in the analysis. This can be done by defining a new group and rerunning the analysis specifying that only the original groups be used to define the classification functions.

Another method is to use the classification functions directly on new data (e.g., through transformations in any BMDP program, or your own computer program) as follows. For each case, compute the classification score for each group from the classification function coefficients (multiply the data by the coefficients and add the constant term). The case is classified into the group for which the classification score is highest.

You can also convert the classification scores to posterior probabilities: let g be the number of groups and s_{ij} be the classification score for the ith case for the jth group, then the posterior probability that case i belongs to group j is

$$P_{ij} = \frac{\exp(s_{ij})}{\sum_{k=1}^{g} \exp(s_{ik})}$$

Standardized canonical variables

7M reports the standardized coefficients for canonical variables; that is, what the coefficients would be if all variables had compatible scale (within group variance). Each coefficient γ_{ij} (the coefficient of variable x_j to obtain the canonical variable v_i) is multiplied by $[w_{jj}/(n-g)]^{1/2}$, the pooled within group variance of variable x_j.

Appendix B.16 Program 2R

Computational method

All computations are performed in single precision. Means, standard deviations, and covariances are computed by the method of provisional means (Appendix B.2). The independent variables are entered into the regression equation or removed from it in a stepwise manner according to the specified (or preassigned) options that you select.

A multiple linear regression equation can be written as

$$E(y_j - \bar{y}) = \beta_1(x_{ij} - \bar{x}_1) + \ldots + \beta_p(x_{pj} - \bar{x}_p)$$

In matrix notation this can be written as

$$E(\mathbf{Y}') = \mathbf{X}'\beta'$$

where

$\mathbf{Y} = (y_j - \bar{y})$ is a row vector of length N

$\mathbf{X} = (x_{ij} - \bar{x}_i)$ is a $p \times N$ matrix

$\beta = (\beta_i)$ is a row vector of length p

Then the estimate $\mathbf{b}$ of β is

$$\mathbf{b}' = (\mathbf{X}\,\mathbf{X}')^{-1}\mathbf{X}\,\mathbf{Y}'$$

In this section we explain how $\mathbf{b}$ is obtained numerically by "sweeping" the matrix $\mathbf{X}\,\mathbf{X}'$ of cross products of deviations. Sweeping is a method of inverting the matrix of cross products of deviations so that at each step (as each variable is entered into the regression equation), the coefficients $\mathbf{b}$, the partial correlations, the multiple correlation and tolerance, and the residual sum of squares are computed as part of the matrix inversion (see also Jennrich, 1977a).

Sweeping and estimates of the coefficients

In BMDP programs, regression equations are usually computed by sweeping a matrix $\mathbf{C}$ of sums of cross products of deviations:

$$\mathbf{C} = \sum_{1}^{N} (\mathbf{x}_j - \bar{\mathbf{x}})'(\mathbf{x}_j - \bar{\mathbf{x}})$$

where $\mathbf{x}_j$ denotes the data for case j, $\bar{\mathbf{x}}$ denotes the vector of sample means, and N is the sample size. If weights are used, $\mathbf{C}$ is the weighted sum of cross-products. If a ZERO intercept is used, $\mathbf{C}$ is the sum of products $\Sigma \mathbf{x}_j \mathbf{x}'_j$.

A basic reference for sweeping is Dempster (1969, p. 62). In BMDP, sweeping is used both to obtain regression equations and to do matrix inversion. Sweeping is most easily described in the context of stepwise regression (2R).

Suppose we have p independent variables so that $\mathbf{C}$ is a $p + 1 \times p + 1$ matrix with dependent variable as variable $p + 1$. At step one of stepwise regression, an independent variable is chosen, say variable k. The matrix $\mathbf{C}$ is swept on variable k to produce a matrix $\mathbf{C}^*$, where

$C^*_{ij} = C_{ij} - C_{ik}C_{kj}/C_{kk}$ for $i \neq k, j \neq k$

$C^*_{kj} = C_{kj}/C_{kk}$

$C^*_{ik} = C_{ik}/C_{kk}$

$C^*_{kk} = -1/C_{kk}$

At this point, $C^*_{p+1,k}$ is the *regression coefficient* for predicting the dependent variable $p + 1$ from variable k. (The intercept is computed separately.) Also, $C^*_{p+1,p+1}$ is the *residual sum of squares* from regressing the dependent variable $(p + 1)$ on variable k. The *partial correlation* of a pair of variables i and j for $i \neq k$ and $j \neq k$ is

$$r_{ij.k} = \frac{C^*_{ij}}{\sqrt{C^*_{ii}C^*_{jj}}}$$

The estimate of the variance of the regression coefficient is proportional to $C^*_{p+1,p+1}C^*_{kk}$. In the computer, C^*_{ij} is stored in the same place that C_{ij} was stored, so $\mathbf{C}$ is overwritten as $\mathbf{C}^*$ is computed.

At step two, we choose another variable, say l, and sweep as we did for variable k. At this point $C_{p+1,p+1}$ contains the residual sum of squares for regressing the dependent variable $p + 1$ on variables k and l; $C_{p+1,k}$ and $C_{p+1,l}$ are the regression coefficients; the partial correlation of variables i and j (i and $j \neq k$ or l) is

$$\frac{C_{ij}}{\sqrt{C_{ii}C_{jj}}}$$

The estimated covariance matrix of the regression coefficients (assuming an intercept) is

$$\frac{-C_{p+1,p+1}}{n-3} \qquad \begin{pmatrix} C_{kk} & C_{kl} \\ C_{lk} & C_{ll} \end{pmatrix}$$

where –3 is replaced by –2 if there is no intercept. The standard errors of the regression coefficient are obtained by taking square roots of the diagonal elements of the covariance matrix of the regression coefficients.

At an arbitrary step, suppose that sweeping has been performed for variables 1 to k. Let

$$\mathbf{C} = \begin{pmatrix} -\mathbf{A}_{kk} & \mathbf{B'}_{kq} \\ \mathbf{B}_{qk} & \mathbf{D}_{qq} \end{pmatrix}$$

where $q = p - k + 1$. **A** is the inverse of the cross product of deviations for the variables that have been swept (entered into the equation). **B** contains the *regression coefficients* for predicting variables $k + 1$ to $p + 1$ from variables 1 to k. The estimate of the *covariance matrix of the regression coefficients* for predicting variable i ($k < i \leq p + 1$) is

$$\frac{C_{ii}}{N-k-1}\mathbf{A}$$

where the term –1 is dropped if there is no intercept. **D** is the sum of cross-products of the residuals of the variables that have not been swept, regressed on the variables that have been swept; the diagonal of **D** contains the residual sum of squares for each nonswept variable. The *partial correlations* for variables $k + 1$ to $p + 1$ adjusting for the linear effects of variables 1 to k are obtained by converting **D** to a correlation matrix. Thus the variables for which sweeping has been performed can be considered as independent variables, and the remaining variables are dependent variables. Sweeping a variable changes a variable from being dependent to independent. Sweeping is also reversible.

If a reverse sweep is performed, a variable changes its status from independent to dependent.

The inverse of a matrix is obtained by sweeping all variables. More precisely, the inverse is equal to the negative of the result of sweeping all variables.

Tolerance

Sometimes the matrix **C** is singular or nearly singular. In this case **C** cannot be inverted or cannot be inverted with satisfactory numerical accuracy. A deeper problem, say in the context of regression, is that the regression coefficients have poor statistical properties whenever the covariance matrix of the independent variables is nearly singular, since the estimated variance of the regression coefficient for a particular independent variable is inversely proportional to one minus the squared multiple correlation of that independent variable with all other independent variables. One minus this squared multiple correlation is called the tolerance for that variable. The reciprocal of the tolerance is called the variance inflation factor. In practice, if one independent variable has a high squared multiple correlation with the other independent variables, it is extremely unlikely that the independent variable in question contributes significantly to the prediction equation. For these reasons, BMDP programs always perform regression analysis (and matrix inversion) in a stepwise manner in such a way that at any step no variable is added to the list of independent variables if either

- its squared multiple correlation with already included independent variables exceeds one minus the tolerance limit
- including the variable would cause the squared multiple correlation for an already included variable to exceed one minus the tolerance limit (see Frane, 1977).

Variables that do not pass the tolerance test are considered redundant. When the solution of an ill-conditioned problem is necessary, a program like 9R (with METHOD = NONE) that uses double precision should be used.

Appendix B.17 Program AR

Computational method

The computations are performed in double precision.

AR finds the minimum of the function

$$Q(\mathbf{p}) = \Sigma w(y - f(\mathbf{x}, \mathbf{p}))^2$$

(where the summation is over all cases) subject to constraints each of which can be written as

$$\mathbf{b}'_l\mathbf{p} \geq c_l$$

Such constraints include MINIMUMS and MAXIMUMS on the parameters as well as the additional CONSTRAINTS and corresponding LIMITS, all of which are specified in the PARAMETER paragraph.

For each case

$\mathbf{x}$	is the vector of values of the independent variable
y	is the value of the dependent variable
w	is the case weight
$f(\mathbf{x}, \mathbf{p})$	is the value of the function evaluated using the values of the parameters $\mathbf{p} = (p_1, p_2, \ldots, p_m)$

If prior knowledge of the parameters is used, the function minimized is

$$Q(P) + \sum r_i (P_i - \tilde{P}_i)^2$$

where the summation is over the number of parameters, r_i represents the ridge values, and P_i represents the prior estimates of the parameters.

Overview of the procedure

If there are no boundary constraints, each iteration consists of approximating $f(\mathbf{x},\mathbf{p})$ by a linear function of $\mathbf{p}$ and solving the resulting linear regression problem to get a new estimate of $\hat{\mathbf{p}}$. The linear function $l(\mathbf{x}, \mathbf{p})$ equal to $f(\mathbf{x},\mathbf{p})$ at $\mathbf{p}^{(1)}, \ldots, \mathbf{p}^{(m^*+1)}$ (estimates of $\hat{\mathbf{p}}$ computed in previous $m^* + 1$ iterations, where m^* is the number of parameters to be estimated) is used to approximate $f(\mathbf{x},\mathbf{p})$. The $\mathbf{p}$ which minimizes

$$Q^*(\mathbf{p}) = \Sigma w(y - l(\mathbf{x}, \mathbf{p}))^2$$

is found and used to replace the oldest member of the set $\mathbf{p}^{(1)}, \ldots, \mathbf{p}^{(m^*+1)}$, and the revised set of parameter vectors is passed to the next iteration.

If there are boundary constraints on the parameters, the procedure described above is modified so that in each iteration an attempt is made to find a new estimate that satisfies all of the boundary constraints and decreases the value of Q^*. Unlike the unconstrained case, if $f(\mathbf{x},\mathbf{p})$ happens to be a linear function of $\mathbf{p}$, the constrained minimum may not be located in one iteration. Within each iteration the boundary constraints are divided into two sets—the active constraints (those for which $\mathbf{b}'_l\,\mathbf{p}^{(m+1)} = c_l$) and the inactive constraints ($\mathbf{b}'_l\,\mathbf{p}^{(m+1)} \geq c_l$). The direction of the new estimate is constrained to be along boundaries formed by some of the active constraints if not doing so would lead to the violation of some of the active constraints. Then if necessary, the step length is shortened so that all of the inactive constraints are satisfied by the new estimate. LaGrange multipliers are used to compute the constrained minimum of Q^*.

Generating the starting values

The $m + 1$ starting parameter vectors needed by the algorithm are generated from the INITIAL and DELTA parameters in the PARAMETER paragraph as follows:

$$p_i^{(k)} = \text{INIT}(i) + \delta_{ik}\text{DELTA}(i)$$

where δ_{ik} is the Kronecker delta.

If any parameters are designated as FIXED in the PARAMETER paragraph, only $m^* + 1$ starting values are computed, where m^* is the number of parameters to be estimated. If any of the $\mathbf{p}^{(k)}$ violate a boundary constraint, the magnitude of DELTA(k) is decreased so that $\mathbf{p}^{(k)}$ lies halfway between $\mathbf{p}^{(m^*+1)}$ and the nearest boundary.

The regression function is evaluated for each starting vector. Then the starting vectors are renumbered so that

$$Q(\mathbf{p}^{(k)}) \geq Q(\mathbf{p}^{(k+1)}) \text{ for } k = 1, \ldots, m^*$$

Description of one iteration when there are no boundary constraints

Let

$\mathbf{y}$ = n component vector of values of the dependent variable

$\mathbf{f}(\mathbf{p})$ = n component vector valued function of values of the regression function for each case

$\Delta\mathbf{P}$ = $m^* \times m^*$ matrix with columns $\mathbf{p}^{(k)} - \mathbf{p}^{(m^*+1)}$

$\Delta\mathbf{F}$ = $n \times m^*$ matrix whose columns are $\mathbf{f}(\mathbf{p}^{(k)}) - \mathbf{f}(\mathbf{p}^{(m^*+1)})$

$\mathbf{W}$ = $n \times n$ diagonal matrix of case weights

Step 1

Form the $(m^* + 1) \times (m^* + 1)$ matrix

$$\mathbf{A} = \begin{pmatrix} \Delta\mathbf{F}'\mathbf{W}\Delta\mathbf{F} & \Delta\mathbf{F}'\mathbf{W}\,[\mathbf{y} - \mathbf{f}(\mathbf{p}^{(m^*+1)})] \\ [\mathbf{y} - \mathbf{f}(\mathbf{p}^{(m^*+1)})]'\mathbf{W}\Delta\mathbf{F} & 0 \end{pmatrix}$$

Step 2

A stepwise regression modification of the Gauss-Jordan algorithm for matrix inversion is used to pivot on the first m^* diagonal elements of **A**. At each step the index of the pivoting element is the r which maximizes $a^2_{m^*+1,r}/a_{rr}$ among all of the unpivoted a_{rr} such that the TOLERANCE limit is not violated.

Step 3

Let α denote the vector containing the first m^* elements of the $(m^* + 1)$th column of **A**. If any of the first m^* diagonal elements of **A** were not used as pivots, the corresponding elements of α are set equal to zero. δ is computed from $\delta = \Delta\mathbf{P}\alpha$. Then $\mathbf{p}^{(m^*+2)} = \mathbf{p}^{(m^*+1)} + \delta$, and $\mathbf{f}(\mathbf{p}^{(m^*+2)})$ are computed.

Step 4

An attempt is made to find a point that decreases $Q(\mathbf{p})$. If $Q(\mathbf{p}^{(m^*+2)}) > Q(\mathbf{p}^{(m^*+1)})$, Q is evaluated at a sequence of points (up to step HALVING) in which the step length is successively halved and the direction alternates between δ and $-\delta$.

The set of parameter vectors to be passed to the next iteration is determined as follows:

Let r be the first subscript such that $|\alpha_r| > .00005$. Then for $k = r, \ldots, m^* + 1$ let

$$\mathbf{p}_{\text{new}}^{(k)} = \mathbf{p}_{\text{old}}^{(k+1)}$$

and if $r > 1$ for $k = 1, \ldots, r - 1$

$$\mathbf{p}_{\text{new}}^{(k)} = (\mathbf{p}_{\text{old}}^{(k)} + \mathbf{p}_{\text{old}}^{(m^*+2)})/2$$

This completes one iteration. Iterations are repeated until the convergence criterion is satisfied, or the maximum number of iterations is reached.

Boundary constraints

For each constraint a "tolerance" is computed as

$$t_l = \sum_{i=1}^{m} |b_{li}\text{DELTA}(i)| \cdot \text{TOLERANCE}$$

A constraint will be considered satisfied as an equality (i.e., active) if

$$|\mathbf{b}'_l\mathbf{p} - c_l| \le t_l$$

An inequality constraint will be considered satisfied by $\mathbf{p}$ if

$$\mathbf{b}'_l\mathbf{p} - t_l \ge c_l$$

At each iteration the constraints are divided into active and inactive constraints. Let s denote the number of active constraints at the current iteration.

For each active constraint, $\mathbf{A}$ (in Step 1 above) is augmented with a row and column whose first m^* elements are the components of the vector $\mathbf{b}'_l\Delta\mathbf{P}$ and remaining elements are zero. Pivoting as described in Step 2 is performed on this augmented matrix. Values of $\mathbf{b}'_l\delta$ are found among the last s elements of the $(m^* + 1)$th column of $\mathbf{A}$. To constrain δ so that $\mathbf{p}^{(m^*+1)} + \delta$ will satisfy the active inequality constraints, pivoting is performed as follows:

a. For $r = m^* + 2, \ldots, m^* + s + 1$ pivot on $a_{r,r}$ if

- r has not been used as a pivot index and
- $a_{m^*+1,r} < -t_{(r-m^*-1)}$

Repeat until no r satisfies the criteria used for pivoting.

b. If $a_{m^*+1,m^*+1}/D < \text{TOL}$, pivot on $a_{r,r}$ and return to part a, where r denotes the first integer such that

- r was used as a pivot index in part a
- $a_{m^*+1,r}/a_{rr} > t_{(r-m^*-1)}$

Compute $\delta^* = \Delta\mathbf{P}\alpha$, and $\delta = d\delta^*$ where $0 < d \le 1$ and d is the largest number such that $\mathbf{p}^{(m^*+2)} = \mathbf{p}^{(m^*+1)} + \delta$ satisfies all of the inactive constraints. Then compute $\mathbf{f}(\mathbf{p}^{(m^*+2)})$

Step 4 is performed as described above except:

- If δ points in a direction interior to any active boundary, the direction is not reversed.
- The step length in the reverse direction is shortened if not doing so would violate an inactive constraint.

Asymptotic standard deviations and correlations

The asymptotic covariance matrix of **p** is estimated by

$$\hat{\Sigma} = s^2 \Delta\mathbf{P}\,(\Delta\mathbf{F}'\mathbf{W}\Delta\mathbf{F})^{-1}\Delta\mathbf{P}'$$

where $\Delta\mathbf{F}$ and $\Delta\mathbf{P}$ are the values in the last iteration

$$s^2 = Q(\hat{\mathbf{p}})/(n' - m^+)$$

where

n' = number of cases with nonzero weight

m^+ = number of independent (pivoted) parameter estimates

$\hat{\mathbf{p}}$ = the vector that gives the smallest values of Q if this is one of the last m^* + 1 estimates or the final value

or if MEANSQ is specified

$$s^2 = \text{MEANSQ}$$

The standard deviation of the predicted value $\hat{f}(\mathbf{x},\mathbf{p})$ is estimated by

$$s(\mathbf{h}(\mathbf{x})'(\Delta\mathbf{F}'\mathbf{W}\Delta\mathbf{F})^{-1}\mathbf{h}(\mathbf{x}))^{1/2}$$

where $h_k(\mathbf{x}) = f(\mathbf{x}, \mathbf{p}^{(k)}) - f(\mathbf{x}, \mathbf{p}^{(m^*+1)})$

Functions of parameters

When functions of parameters $g_i(\mathbf{p})$ ($i = 1, 2, \ldots, q$) are specified, the m^* x 1 vector **p** of parameters is replaced by the $(m^* + q)$ x 1 vector $\bar{\mathbf{p}}$ given by

$$\bar{\mathbf{p}} = \begin{bmatrix} \mathbf{p} \\ \mathbf{g}(\mathbf{p}) \end{bmatrix}$$

where $\mathbf{g}(\mathbf{p}) = (g_1(\mathbf{p}), g_2(\mathbf{p}), \ldots, g_q(\mathbf{p}))'$.

The matrix $\Delta\mathbf{P}$ is replaced by a $(m^* + q)$ x $(m^* + q)$ matrix $\Delta\bar{\mathbf{P}}$ with kth column given by $\bar{\mathbf{p}}^{(k)} - \bar{\mathbf{p}}^{(m^*+q+1)}$, $k = 1, 2, \ldots, m^* + q$.

$$\Delta\bar{\mathbf{P}} = \begin{pmatrix} \Delta\mathbf{P}_1 & \Delta\mathbf{P}_2 \\ \Delta\mathbf{G}_1 & \Delta\mathbf{G}_2 \end{pmatrix}$$

Similarly, $\Delta\mathbf{F}$ is replaced by an n x $(m^* + q)$ matrix $\Delta\bar{\mathbf{F}}$ whose kth column is given by $\mathbf{f}(\mathbf{p}^{(k)}) - \mathbf{f}(\mathbf{p}^{(m^*+q+1)})$ and can be partitioned into

$$\Delta\bar{\mathbf{F}} = (\Delta\mathbf{F}_1 \quad \Delta\mathbf{F}_2)$$

The matrix **A** is formed as before using $\Delta\bar{\mathbf{F}}$ in place of $\Delta\mathbf{F}$ and $\mathbf{f}(\mathbf{p}^{(m^*+q+1)})$ in place of $\mathbf{f}(\mathbf{p}^{(m^*+1)})$. As before, pivoting is done on the first m^* diagonal elements of **A** such that tolerance and boundary constraints are respected as described above. Let **B** be the matrix that results from this pivoting. (Note that for diagonal elements not used as pivots, AR sets the corresponding elements of the last row of **B** to zero). Let $\mathbf{B}_{11}$ be the $(m^* + q)$ x $(m^* + q)$ submatrix of **B** with the last row and column removed. The asymptotic covariance matrix of p is given by

$$s^2 \Delta\bar{\mathbf{P}}\mathbf{B}_{11}\Delta\bar{\mathbf{P}}'$$

This symmetric matrix can be partitioned as

$$s^2 \begin{pmatrix} \Delta\mathbf{P}_1(\Delta\mathbf{F}'_1\mathbf{W}\Delta\mathbf{F}_1)^{-1}\Delta\mathbf{P}'_1 & \Delta\mathbf{P}_1(\Delta\mathbf{F}'_1\mathbf{W}\mathbf{F}_1)^{-1}\Delta\mathbf{G}'_1 \\ & \Delta\mathbf{G}_1(\Delta\mathbf{F}'_1\mathbf{W}\Delta\mathbf{F}_1)^{-1}\Delta\mathbf{G}'_1 \end{pmatrix}$$

As $\Delta\bar{\mathbf{P}}$ approaches zero, this estimates the full covariance matrix for $\bar{\mathbf{p}}$.

Priors and ridging

When prior values P_i of the parameters and corresponding ridge values r_i are provided in the PARAM paragraph, the matrix $\mathbf{\Delta PR\Delta P}$ is added to the $\mathbf{\Delta F'W\Delta F}$ component in the **A** matrix, where **R** is a diagonal matrix with elements r_i. In addition, $r_i(P_i - P_I)$ is added to the ith element of α (i.e., the ith row of the last column of **A**) before pivoting. See Special Features for hints on determining the r_i.

The algorithm used in AR

AR uses a pseudo-Gauss-Newton iterative algorithm (see Ralston and Jennrich, 1978). Instead of using (as does 3R) the partial derivatives of the function with respect to the parameters evaluated at the current parameter estimates, AR uses the partial derivatives of the equation of the plane spanned by the previous m + 1 estimates of the m parameters. In other words, 3R's Gauss-Newton algorithm uses the slopes of a tangent plane, while AR uses the slopes of a secant plane. The algorithm decides that a solution has been reached if either the maximum number of iterations allowed is reached (default 50), or both of the following conditions are met:

a) $$\max_{1 \le i \le m} \frac{\left| p_i^{(k+1)} - p_i^{(k)} \right|}{\max\left(1, p_i^{(k+1)}\right)} < \varepsilon_1$$

b) $$\max \frac{\mathrm{RSS}^{(k)} - \mathrm{RSS}^{(k+1)}}{\max\left(1, \mathrm{RSS}^{(k)}\right)} < \varepsilon_2$$

where $p_i^{(k)}$ is the value of the ith parameter at the kth iteration and $\mathrm{RSS}^{(k)}$ is the residual sum of squares at the kth iteration. The default values for ε_1 and ε_2 are 10^{-4} and 10^{-5} respectively.

For a given iteration, if the sum of squared residuals as first computed exceeds that of the previous iteration, then the increment size is halved, and the sum of squared residuals is recomputed and tested against the previous iteration's result. (In addition to halving, AR may also perform step reversal; see above.) This HALVING is repeated until the sum of squared residuals is less than that of the previous iteration, or until the maximum number of halvings is reached. At each iteration AR prints the number of increment halvings performed. (The first m + 1 iteration numbers are labeled zero, where m is the number of parameters.)

TOLERANCE for a parameter is defined as one minus the squared multiple correlation of a given independent variable with all other independent variables in the function, where each "independent variable" in this context is the partial derivative of the regression function with respect to a particular parameter. AR uses secant approximations to the partial derivatives to compute tolerance. The program stops if the tolerance value becomes too low, indicating near-singularity in the covariance matrix or independent variables. By default, the minimum tolerance value is set to 10^{-8}.

Appendix B.18 Program 3S

Computational method

All computations are performed in single precision. Means and standard deviations are computed by the method of provisional means (Appendix B.2).

Explanation of test statistics

Kruskal-Wallis one-way analysis of variance (Kruskal and Wallis, 1952). Let N observations be classified by a GROUPING variable into k groups; the jth group contains n_j observations. All the values for each variable are ranked from 1 to N;

tied values are assigned the average rank of the tied values. Let R_j be the sum of the ranks for group j. The Kruskal-Wallis statistic is

$$H = \frac{2}{N(N+1)} \sum_{j=1}^{k} \frac{R_j^2}{n_j} - 3(N+1)$$

If values are tied, H is modified to

$$H' = H / \left[1 - \frac{\sum^* (t_s^3 - t_s)}{N^3 - N} \right]$$

where t_s is the number of observations tied with a single value, and $\sum^*$ is the sum over all distinct values for which a tie exists. The level of significance of H is obtained from the chi-square distribution with $k - 1$ degrees of freedom. H has approximately a chi-square distribution when the minimum group size n_j is greater than five. For smaller sample sizes you can find exact probabilities in Siegel (1956, Table O, pp. 282-283).

For multiple comparisons, z is computed as described below for the Mann-Whitney test, except that (1) the subscripts 1 and 2 are replaced by i and j, corresponding to the groups being compared; (2) the estimate for σ is based on all of the subgroups, not just groups i and j (see Hollander and Wolfe, 1973); and (3) z is compared with the tabled $z_{\alpha'/2}$, where $\alpha' = 2\alpha/(k(k-1))$.

The Mann-Whitney rank-sum test. The nonparametric version of the two sample t test for independent samples. When there are only two groups, the Mann-Whitney (Wilcoxon) statistic U is

$$U = R_1 - \frac{n_1(n_1+1)}{2}$$

where R_1 is the sum of ranks for the first group and n_1 is the number of observations in that group. The ranks of tied observations are averaged and assigned to each tied observation. Note that when there are only two groups, the Mann-Whitney statistic U is reported when Kruskal-Wallis is requested. The level of significance of U is the same as that of H, since the Mann-Whitney (Wilcoxon) statistic is a transformation of the Kruskal-Wallis statistic when $k = 2$. It is known that

$$E(R_1) = (^1/_2)\, n_1(N+1)$$

and

$$\mathrm{VAR}(R_1) = \frac{n_1 n_2 (N+1)}{12} - \frac{n_1 n_2 \sum (t_s^3 - t_s)}{12N(N-1)}$$

where $N = n_1 + n_2$ and t_s is defined as in the Kruskal-Wallis test. For n_1 and n_2 both greater than 10, the statistic

$$z = \frac{R_1 - E(R_1)}{\sqrt{\mathrm{VAR}(R_1)}}$$

is approximately normally distributed, with mean 0 and variance 1. For small sample sizes, you may want to consult a table of significance values.

The sign test. For each case, 3S calculates the difference between the values of the variables in each pair and records the number of differences that are positive (N_+) and negative (N_-). Let N_{min} be the lesser of N_+ and N_-. Let the total number of nonzero differences be $N_T = N_+ + N_-$. When N_T is less than or equal to 100, the two-tail level of significance under the null hypothesis that the + and – signs of the differences are equally probable (each sign has probability 1/2) is

$$P = (1/2)^{N_T - 1} \sum_{j=0}^{N_{min}} \frac{N_T!}{j!\,(N_T - j)!}$$

For large samples, the normal approximation is used to find the two-tail level of significance.

The Wilcoxon signed-rank test. Let d_j be the difference between the two matched variables for pair j. The absolute values of all nonzero d_j's are ranked; the average rank is assigned when two or more $|d_j|$'s are tied. The sum of the ranks associated with positive differences R_+ and the sum of the ranks associated with negative differences R_- are calculated. The statistic R_{min} is the lesser of R_+ and R_-. 3S computes the z statistic using a normal approximation for the distribution of R_{min}.

$$z = \left[R_{min} - \frac{N_T(N_T+1)}{4}\right] \Big/ \left[\frac{N_T(N_T+1)\,(2N_T+1)}{24}\right]^{1/2}$$

where N_T is the total number of nonzero differences between the values in each pair. Exact probabilities are calculated if there are ≤ 8 nonzero differences or where the smaller sum of ranks is ≤ 2.5.

For small sample sizes ($N \leq 20$), you can find exact levels of significance in Dixon and Massey (1983, Table A-19). The smaller sum of like-signed ranks in the 3S output corresponds to the signed rank statistic T_α in the table. The program output displays the level of significance based on the z statistic.

Friedman's two-way analysis of variance. Let R_i be the sum of the ranks for the ith variable in a randomized block design. The Friedman statistic is

$$\chi_r^2 = \left[\frac{2}{Nk(k+1)} \sum_{i=1}^{k} R_i^2\right] - 3N(k+1)$$

where k is the number of variables and N is the number of cases. The level of significance of χ_r^2 is obtained from the chi-square distribution with $k - 1$ degrees of freedom. The levels of significance are appropriate except for very small sample sizes. For more exact probabilities for the Friedman test with small sample sizes, see Siegel (1956, Table N, pp. 280-281). For multiple comparisons, declare groups i and j to differ significantly at overall α if

$$|R_i - R_j| \geq z_{\alpha'/2}\sqrt{Nk(k+1)/6}$$

where

$$\alpha' = \frac{2\alpha}{k(k-1)}$$

Kendall's coefficient of concordance. Using the same notation as above, Kendall's coefficient of concordance is

$$W = \chi_r^2/[N(k-1)]$$

The level of significance of W is the same as that of the Friedman χ_r^2.

Kendall rank correlation. Let $r_1, r_2, \ldots, r_N$ represent the ranks of the values of one variable and $s_1, s_2, \ldots, s_N$ the corresponding ranks of the second variable. The Kendall rank correlation τ_b (tau b) is defined as

$$\tau_b = \frac{P - Q}{N(N-1)}$$

where

P = twice the number of pairs of rankings such that both $r_j > r_k$ **and** $s_j > s_k$

(agreements in rank order), and

Q = twice the number of pairs of rankings such that both $r_j > r_k$ **and** $s_j < s_k$ (disagreements in rank order).

When rankings are tied, the formula for τ_b is

$$\tau_b = \frac{P - Q}{\sqrt{[N(n-1) - T_1]\,[N(N-1) - T_2]}}$$

where $T_j = \Sigma^* (t_{ij}^3 - t_{ij})$ and t_{ij} is the number of observations tied with a single value for variable j, and the sum Σ^* is over all distinct sets of ties. T_1 is the total for the first variable and T_2 for the second variable. When $N > 10$, τ_b is approximately normally distributed with mean $\mu_\tau = 0$ under the null hypothesis, so

$$z = \frac{\tau_b}{\sqrt{\frac{2(2N+5)}{9N(N-1)}}}$$

is approximately normally distributed with zero mean and unit variance (Siegel, 1956). You may consult a standard statistical table for significance level.

Spearman rank correlation. The Spearman rank correlation coefficient is defined as

$$r_s = 1 - 6D(N^3 - N) \text{ where } D = \Sigma(r_j - s_j)^2$$

with r_j and s_j defined as for the Kendall τ_b. When rankings are tied, r_s is modified to

$$r_s = \frac{A + B - D}{2\sqrt{AB}} \quad \text{where} \quad A = \frac{(N^3 - N - T_1)}{12} \quad \text{and} \quad B = \frac{(N^3 - N - T_2)}{12}$$

with T_1 and T_2 defined as for the Kendall rank correlation. When $N \geq 10$, the Spearman rank correlation coefficient may be tested by the t test with $N - 2$ degrees of freedom, where

$$t = r_s \sqrt{\frac{N-2}{1 - r_s^2}}$$

Many statistics texts provide r_s significance level tables for $N < 10$.

Appendix B.19 Program 2T

Computational method

All computations are done in single precision.

Sample autocorrelation function (SACF)

Let $Y(t)$, $t = 1, 2, \ldots, n$ be the time series to be studied. The SACF of the series at lag k, r_k is

$$r_k = c_k / c_0$$

where

$$c_k = \frac{1}{n} \sum_{t=1}^{n-k} (Y(t) - \bar{Y})\,(Y(t+k) - \bar{Y}) \qquad k = 0,1,2,\ldots$$

$$\bar{Y} = \frac{1}{n} \sum_{t=1}^{n} Y(t)$$

The standard error of the SACF r_k is computed by

$$\text{STE}(r_k) = \frac{\left(1 + 2\sum_{l=0}^{k-1} r_l^2\right)^{1/2}}{\sqrt{n}} \qquad k = 1,2,\ldots$$

and the portmanteau lack-of-fit test statistic Q at lag k, following Ljung and Box (1978), is computed by

$$Q(k) = n(n+2)\sum_{l=1}^{k} r_l/(n-1)$$

$Q(k)$ is approximately distributed in χ^2_{k-p} where p is the number of ARIMA parameters in the model.

Sample partial autocorrelation function (SPACF)

The SPACF of a series at lag k, $\hat{\phi}_{kk}$ can be computed by the following recursive technique, where

$$\hat{\phi}_{kk} = \begin{cases} r_1 & k = 1 \\ \dfrac{r_k - \sum_{j=1}^{k-1} \hat{\phi}_{k-1,j} r_{k-j}}{1 - \sum_{j=1}^{k-1} \hat{\phi}_{k-1,j} r_j} & k = 2,3,\ldots \end{cases}$$

and $\qquad \hat{\phi}_{kj} = \hat{\phi}_{k-1,j} - \hat{\phi}_{kk}\hat{\phi}_{k-1,k-j} \qquad j = 1,2,\ldots,k-1$

The standard error of $\hat{\phi}_{kk}$ is approximately equal to $1/(n^{1/2})$.

Sample cross correlation function (SCCF)

Let $X(t)$, $Y(t)$, $t = 1, 2, \ldots, n$ be the two time series to be analyzed. The SCCF between $\{X(t)\}$ and $\{Y(t)\}$ at lag k is

$$r_{xy}(k) = C_{xy}(k) / \sqrt{C_{xx}(0)\, C_{yy}(0)}$$

where

$$C_{xy}(k) = \begin{cases} \dfrac{1}{n}\sum_{t=1}^{n-k} (X(t) - \bar{X})(Y(t+k) - \bar{Y}) & k = 0,1,2,\ldots \\ \dfrac{1}{n}\sum_{t=1}^{n+k} (Y(t) - \bar{Y})(X(t-k) - \bar{X}) & k = 0,-1,-2,\ldots \end{cases}$$

and $\qquad \bar{X} = \dfrac{1}{n}\sum_{t=1}^{n} X(t) \qquad \bar{Y} = \dfrac{1}{n}\sum_{t=1}^{n} Y(t)$

The standard error of $r_{xy}(k)$ is approximately equal to $1/(n-k)^{1/2}$ if $X(t)$ and $Y(t)$ are independent and one is a white noise series.

Model estimation

2T uses a Gauss-Newton method to perform nonlinear estimation of a time series model. This section describes the computation of the residuals ($a(t)$), which is the key computation in the model estimation.

Consider the following model

$$D(B)\,Y(t) = \sum_{k=1}^{m} \frac{U_k(B)\,D_k(B)}{S_k(B)} X_k(t) + \varepsilon(t) \tag{1}$$

$$\varepsilon(t) = \frac{c}{\eta(B)} + \frac{\theta(B)}{\phi(B)} a(t) \tag{2}$$

where

$\eta(B) = 1$ if $D(B) = 1$ and $\eta(B) = \phi(B)$ if $D(B) \neq 1$.

In (1) and (2), $\{Y(t)\}$ is some appropriate transformation of the observed time series $(Z(t)\}$, and $D(B)Y(t)$ is a stationary series; $U_k(B)D_k(B)/\delta_k(B)$ characterizes the effect of an exogenous variable $X_k(t)$ on $D(B)Y(t)$. $\{X_k(t)\}$ can be a stochastic or nonstochastic variable. $\{X_k(t)\}$ must be independent of $\{\varepsilon(t)\}$, and $D_k(B)X_k(t)$ must be stationary if it is a stochastic variable. $U_k(B)$, $S_k(B)$, $D_k(B)$, and $\eta(B)$ are products of a polynomial of B which is explained later; $\varepsilon(t)$ denotes stochastic variation not attributable to a known exogenous variable; $\{\varepsilon(t)\}$ follows an ARMA process. $\theta(B)$, $\phi(B)$, and $D(B)$ are also products of a polynomial of B that characterize the ARMA process, and c is a constant parameter.

The difference polynomial has an algebraic form as

$$\prod_{j=1}^{J} (1 - B^{s_j}) \qquad \text{with } s_j \geq 1$$

where s_j is called "order of difference."

The MA, AR and S polynomials can be expressed as

$$\prod_{j=1}^{J} (1 - \zeta_{j1} B^{d_{j1}} - \ldots - \zeta_{jr_j} B^{d_{jr_j}}) \qquad \text{with } d_{jk} \geq 1$$

where ζ represents θ, ϕ, or S.

The U polynomial can be written as

$$\prod_{j=1}^{J} (u_{j0} + u_{j1} B^{d_{j1}} + \ldots + u_{jr_j} B^{d_{jr_j}}) \qquad \text{with } d_{jk} \geq 1$$

The d_{jk} is the power of B which will also be referred to as the order of the jk-th parameter.

The model in (1) can also be written as

$$y(t) = \sum_{k=1}^{m} \frac{U_k(B)}{S_k(B)} x_k(t) + \varepsilon_t \tag{3}$$

with $y(t) = D(B)Y(t)$ and $x_k(t) = D_k(B)X_k(t)$.

To compute the residual $a(t)$'s, we need to obtain $\varepsilon(t)$'s first. The $\varepsilon(t)$'s can be computed as follows:

Let

$$\alpha_k(t) = \frac{U_k(B)}{S_k(B)} x_k(t) \tag{4}$$

which implies

$$S_k(B)\,\alpha_k(t) = U_k(B)\,x_k(t) \tag{5}$$

The $\alpha_k(t)$'s can be computed recursively by (5). The computation of the above difference equation is started from a value of t for which all previous $x_k(t)$'s are known. (Box and Jenkins (1976) recommend a starting value of t for which all previous $x_k(t)$'s and $y(t)$'s are known). Thus $\alpha_k(t)$ in (5) is calculated from $t = v + 1$ onwards, where v is the sum of the highest orders of the polynomials $S_k(B)$ and $D_k(B)$. For unknown $\alpha_k(t)$'s (with $t < v + 1$), they are set to 0 if the input series is centered, and to $\bar{x}U_k(B)/S_k(B)$ if it is not centered ($\bar{x}$ is the sample mean of $x_k(t)$ series). Note that these starting values for $\alpha_k(t)$'s may not be appropriate if the series $x_k(t)$ is not a random variable with constant mean. When the input series $x_k(t)$ is a binary variable (i.e., a variable consisting of zeros and ones), the unknown $\alpha_k(t)$'s are always set to zero.

After $\alpha_k(t)$'s, $k = 1, \ldots, m$ are computed, we can obtain $\varepsilon(t)$:

$$\varepsilon(t) = y_t - \sum_{k=1}^{m} \alpha_k(t) \qquad (6)$$

Once $\varepsilon(t)$'s are obtained, we can compute $a(t)$'s, by (2). The model in (2) can also be expressed as

$$\phi(B)\tilde{\varepsilon}(t) = \theta(B)a(t) \qquad (7)$$

where $\tilde{\varepsilon}(t) = \varepsilon(t) - c/\eta(B)$.

To compute $a(t)$'s from (7), we may use either a conditional or backcasting method.

For the conditional method we set starting values for $a(t)$'s to zero and then proceed directly with the forward recursions to obtain $a(t)$'s. For the backcasting method, we first compute $\psi(B)$, with $\psi(B)$ satisfying $\phi(B)\psi(B) = \theta(B)$. From the results we decide the number of backcasts. After the starting values are backcast, we can proceed with forward recursions to obtain $a(t)$'s. The detailed computations and illustrated examples can be found in Box and Jenkins (1976). Note that all polynomials of B are expanded in each related computation.

Forecasting

For the time series model in (1) and (2), the minimum mean square error (MMSE) forecast of $Y(T + l)$ at time T, denoted by $\hat{Y}(T,l)$, can be obtained as follows:

1. Compute $\alpha_k(t) = [U_k(B)/S_k(B)]x_k(t)$, $k = 1, 2, \ldots, m$, $t = 1, 2, \ldots, T, T+1, \ldots, T+m$

where m is the number of forecasts to be made at time T. The computations at this stage are the same as those described in the previous section. Note that computing $\alpha_k(t)$ with $t > T$ may require $x_k(t)$'s, $t > T$. Therefore, the $x_k(t)$'s must be appropriately provided before this stage of computations. After $\alpha_k(t)$, $k = 1, \ldots, m$, $t = 1, 2, \ldots, T$, are computed, the noise series $\varepsilon(t)$ can be obtained as below:

$$\varepsilon(t) = y(t) - \sum_{k=1}^{m} \alpha_k(t) \qquad t = 1,2,\ldots,T$$

Using the conditional method for computing residuals, we can then obtain $a(t)$, $t = 1, 2, \ldots, T$.

2. Using (2), the MMSE forecasts of $\varepsilon(T + l)$, also denoted by $\hat{\varepsilon}(T,l)$, can be obtained.

3. The MMSE forecast of $y(T + l)$, denoted by $\hat{y}(T,l)$, can be expressed as

$$\sum_{k=1}^{m} \alpha_k(T+l) + \hat{\varepsilon}(T,l)$$

Using the relation $y(t) = D(B)Y(t)$, the MMSE forecast of $Y(T + l)$ can be readily computed.

In the above three stages of computations, the first stage is the same as described in the previous section, and the third stage is straightforward. Below, we explain some details for the second stage of the computations.

The ARIMA model in (2) may be expressed as

$$\phi(B)\tilde{\varepsilon}(t) = \theta(B)a(t) \qquad \tilde{\varepsilon}(t) = \varepsilon(t) - c/\eta(B)$$

or

$$(1 - \phi_1 B - \ldots - \phi_p B^p)\tilde{\varepsilon}(t) = (1 - \theta_1 B - \ldots - \theta_q B^q)a(t)$$

where

$$1 - \phi_1 B - \ldots - \phi_p B^p = \phi(B)$$

and

$$1 - \theta_1 B - \ldots - \theta_q B^q = \theta(B)$$

The MMSE forecast of $\hat{\varepsilon}(T+l)$, denoted by $\hat{\tilde{\varepsilon}}(T,l)$, can be expressed as

$\hat{\tilde{\varepsilon}}(T,l) = \phi_1 E(\tilde{\varepsilon}(T+l-1)) + \ldots + \phi_p E(\tilde{\varepsilon}(T+l-p)) - \theta_1 E(a(T+l-1)) + \ldots + \theta_q E(a(T+l-q))$

where

$$E\tilde{\varepsilon}(t) = \begin{cases} \tilde{\varepsilon}(t) & t \le T \\ \hat{\tilde{\varepsilon}}(T, t-T) & t > T \end{cases}$$

and

$$Ea(t) = \begin{cases} a(t) & t \le T \\ 0 & t > T \end{cases}$$

Thus the MMSE forecast of $\varepsilon(T+l)$ is $\hat{\varepsilon}(T,l) = c/\eta(B) + \hat{\tilde{\varepsilon}}(T,l)$.

When the $X_k(t)$'s are all nonstochastic variables, the variance of $\hat{Y}(T,l)$ is

$$\mathrm{Var}(\hat{Y}(T,l)) = \sigma_a^2\{1 + \psi_1^2 + \psi_2^2 + \ldots + \psi_{l-1}^2\}$$

where $\psi_1, \psi_2, \ldots$ are coefficients satisfying the following equation:

$$\phi(B)D(B)(1 + \psi_1 B + \psi_2 B^2 + \ldots) = \theta(B)$$

When $X_k(t)$ is a stochastic variable and follows the ARIMA model

$$D_x(B)X_k(t) = \frac{c_x}{\eta_x(B)} + \frac{\theta_x(B)}{\phi_x(B)}e_t \qquad e_t \sim \text{iid } N(0,\sigma_e^2)$$

the variance of $\hat{Y}(T,l)$ will be increased by an amount of $\sigma_e^2\{v_0 + v_1^2 + v_2^2 + \ldots + v_{l-1}^2\}$ if we set $X_k(T+l) = \hat{X}_k(T,l)$ and if $X_k(t)$ is independent of other input variables. The coefficients $v_0, v_1, \ldots$ satisfy the following equation:

$$S_k(B)D(B)\phi_x(B)D_x(B)(v_0 + v_1 B + v_2 B^2 + \ldots) = U_k(B)D_k(B)\theta_x(B)$$

or

$$S_k(B)D(B)(v_0 + v_1 B + v_2 B^2 + \ldots) = U_k(B)D_k(B)(1 + \psi_{x1}B + \psi_{x2}B^2 + \ldots)$$

where $\psi_{x1}, \psi_{x2}, \ldots$ satisfy $D_x(B)\phi_x(B)(1 + \psi_{x1}B + \psi_{x2}B^2 + \ldots) = \theta_x(B)$

Nonlinear estimation

2T uses the Gauss-Newton method to perform nonlinear estimation. This part of the program is partially based on a nonlinear estimation subroutine, GAUS-

HAUS, developed at the Academic Computing Center of the University of Wisconsin, Madison. The algorithm of the nonlinear estimation is described below.

Consider the following nonlinear model:

$$Y(t) = f(\xi(t), \beta) + a(t)\ , t = 1, 2, \ldots, n \tag{8}$$

$$a(t) \sim \text{iid } N(0, \sigma_a^2)$$

where $\beta = [\beta_1, \beta_2, \ldots, \beta_p]'$ is a vector of parameters and $\xi_t = [\xi_{t1}, \xi_{t2}, \ldots, \xi_{tk}]$ is a set of independent variables corresponding to the observation. Using the vector notations

$$\mathbf{Y} = [Y_1, Y_2, \ldots, Y_n]'$$

and

$$\mathbf{f}(\beta) = [f(\xi_1,\beta), (\xi_2,\beta), \ldots, f(\xi_n,\beta)]'$$

the main computation problem is to minimize

$$S(\beta) = (\mathbf{Y} - \mathbf{f}(\beta))'(\mathbf{Y} - \mathbf{f}(\beta)) \tag{9}$$

as a function of β. If β_0 is an initial guess, the first order Taylor series expansion about β_0 is

$$\mathbf{f}(\beta) \doteq \mathbf{f}(\beta_0) + \mathbf{X}\delta \tag{10}$$

$$\text{where } \mathbf{X} = \{X_{tj}\}_{nxp} \qquad \text{with } X_{tj} = \left.\frac{\partial f(\xi_t,\beta)}{\partial \beta_j}\right|_{\beta = \beta_0} \qquad \begin{array}{l} t = 1,2,\ldots,n \\ j = 1,2,\ldots,p \end{array}$$

and $\delta = \beta - \beta_0$

The sum of squares in (9) is then approximately equal to

$$\dot{S}(\beta) \doteq (\gamma - \mathbf{x}\delta)'(\gamma - \mathbf{x}\delta) \tag{11}$$

where $\gamma = \mathbf{Y} - \mathbf{f}(\beta_0)$. Hence $\dot{S}(\beta)$ is minimized when

$$\delta = (\mathbf{X}'\mathbf{X})^{-1}\mathbf{X}'\mathbf{Y}\text{, if } (\mathbf{X}'\mathbf{X})^{-1} \text{ exists} \tag{12}$$

Then, by definition of δ, the new guess is $\beta_1 = \beta_0 + \delta$, and the next iteration can be started by expanding about β_1. The method is originally due to Gauss (1821). Frequently the approximation (10) is not sufficiently accurate, making $\dot{S}(\beta)$ a poor approximation for $S(\beta)$. In fact, it may happen that $S(\beta_1) > S(\beta_0)$, which is contrary to the objective. Thus we need some way of systematically controlling the size of the region over which the linear approximation of $f(\xi_t, \beta)$ is allowed to hold, and hence controlling the size of the correction vector δ. This is done by halving the vector δ and recomputing $S(\beta_1)$ until $S(\beta_1)$ becomes $\leq S(\beta_0)$.

Appendix B.20 Program 2V

Computational method

All computations are performed in double precision. Means, standard deviations and covariances are computed by the method of provisional means (Appendix B.2).

We consider the case of several group factors, $g_1, \ldots, g_a$, and several trial factors $t_1, \ldots, t_b$. Let y_{st} be the response of subject s to trial combination $t = (t_1, \ldots, t_b)$. Similarly let x_{stv} be the vth covariate value for subject s on trial combination t; and for ease of exposition, let

$$x_{s,t,p+1} = y_{st}$$

It is assumed that the response and the covariates have been measured for each subject on each trial combination, although the covariates may be constant across some or all trials.

Step 1

For each subject s and variable v, the values x_{stv} across trials are transformed into orthogonal polynomial components $\tilde{x}_{sdv}$. Here, $d = (d_1, \ldots, d_b)$ denotes a degree combination. Specifically

$$\tilde{x}_{sdv} = \sum_{t_1} \cdots \sum_{t_b} \pi_{d_1 t_1} \cdots \pi_{d_b t_b} x_{stv}$$

where $\pi_{d_j t_j}$ = the value of the orthogonal polynomial of degree d_j at t_j.

Each trial factor t_j is assumed to range over equally spaced points unless otherwise specified by a POINT command.

If trial contrasts are specified, then each matrix of orthogonal polynomials is augmented by column(s) of the contrast coefficients for the corresponding index J.

The resulting X_{sdv} then contains portions that can be identified with each contrast and the contrast's interactions with the other trial factors. Steps 2 through 8 are performed reporting only the non-contrast part.

Step 2

For each degree combination d, an analysis of covariance is performed on the data $\tilde{x}_{sdv}$, producing a sum of cross products $P_{rdv_1v_2}$ for each pair of variables x_{v_1} and x_{v_2} and each analysis of variance component r defined by the group factors $g_1 \ldots g_a$. This step and the next are described in greater detail below.

Step 3

The sums of cross products $P_{rdv_1v_2}$ are used to produce a standard analysis of covariance for each degree d.

Step 4

The cross products $P_{rdv_1v_2}$ are pooled over all degree combinations $d = (d_1, \ldots, d_b)$ that belong to the same analysis of variance component c defined by the trial factors $t_1, \ldots, t_b$. For this purpose, c may be represented by a list of zeros and ones so that

$$c = (0, 1, 0, 1)$$

represents a $t_2 \times t_4$ interaction. With this convention

$$P_{rcv_1v_2} = \sum_{d \in c} P_{rdv_1v_2}$$

where $d \in c$ whenever $d = (d_1, \ldots, d_b)$ and $c = (c_1, \ldots, c_b)$ have exactly the same zeros.

Step 5

For each analysis of variance component c defined by the trial factors, the pooled cross products $P_{rcv_1v_2}$ are used to produce a standard analysis of covariance.

Step 6

Let e denote the error component defined by the group effects. The pooled cross products $P_{ecv_1v_2}$ are further pooled across the trial components c. This pooling is weighted by the reciprocals of the error mean square MS_{ec} reported in the previous step. To be more specific, the pooled cross products are given by

$$P_{ev_1v_2} = \sum_{c} MS_{ec}^{-1} P_{ecv_1v_2}$$

Step 7

Let $\tilde{P}_{ev_1v_2}$ denote the result of performing pivots on the first p diagonal elements of $P_{ev_1v_2}$ (Jennrich and Sampson, 1968). Then

$$\hat{\beta}_v = \tilde{P}_{e,v,p+1}$$

are the pooled regression coefficient estimates that are used to compute the adjusted group means

$$\bar{y}^*_{gt} = \bar{y}_{gt} + \sum_{v=1}^{p} \hat{\beta}_v (\bar{x}_v - \bar{x}_{gtv})$$

for each group combination $g = (g_1, \ldots, g_a)$ and trial combination $t = (t_1, \ldots, t_b)$. Here $\bar{x}_{gtv}$ and $\bar{y}_{gt}$ denote the average of x_{stv} and y_{st} over all subjects s in group combination g; and $\bar{x}_v$ denotes the average of x_{stv} over all subjects s and trial combinations t.

Step 8

Using the regression coefficient estimates

$$\hat{\beta}_{cv} = \tilde{P}_{e,c,v,p+1}$$

obtained after pivoting in Step 5 and the analysis of variance component residuals x^*_{sdv} obtained by subtracting the group effects from the results $\tilde{x}_{sdv}$ of Step 1, the transformed residuals

$$\tilde{e}_{sd} = x^*_{s,d,p+1} - \sum_{v=1}^{p} x^*_{sdv} \hat{\beta}_{cv}$$

are computed for each subject s and degree combination d. Thus for each value of d, c is chosen so $d \in c$. The transformed fits are

$$\tilde{\nu}_{sd} = \tilde{y}_{sd} - \tilde{e}_{sd}$$

To obtain the untransformed fits and residuals, extend the definition of $\pi_{d_jt_j}$ so that it equals π_{d_j1} whenever $t_j = 0$ and let

$$\pi_{dt} = \pi_{d_1t_1}\pi_{d_2t_2}\ldots\pi_{d_bt_b}$$

Then the untransformed fits and residuals are given by

$$\hat{\nu}_{st} = \sum_{d \in t} \pi_{dt} \tilde{\nu}_{sd}$$

$$\hat{e}_{st} = \sum_{d \in t} \pi_{dt} \tilde{e}_{sd}$$

where $d \in t$ means that d has exactly the same zeros as t (as in Step 4).

Step 9

Means and standard deviations subscripted by the other trial factors and the grouping factors are reported for each trial contrast specified. Finally, the sums of squares for the contrast and its interactions with the other trial and group factors (which were computed in Steps 2, 3, and 4) are used to report an analysis of variance table.

Covariance analysis

As indicated in the previous section, the core computations are covariance analyses; many of them are executed, and all of them are similar in form. Although the covariance analyses are applied to transformed data, there is no need to make that distinction here, nor do the trial indices need to appear, because the analyses are applied only across groups. Accordingly, let

$$g_1, \ldots, g_a, x_1, \ldots, x_p, x_{p+1}$$

denote the group factors and variables. As above, $y = x_{p+1}$ is the dependent variable. The fitted model form is

$$y = \mu_{r_1} + \ldots + \mu_{r_q} + \beta_1 x_1 + \ldots + \beta_p x_p + e$$

where $r_1, \ldots, r_q$ are all possible subsets of the factors $g_1, \ldots, g_a$ unless otherwise specified by an INCLUDE or EXCLUDE command. A typical r_j has the form $r_j = (g_{i_1}, \ldots, g_{i_k})$. For each factor g_i in r_j, the component

$$\mu_{r_j} = \mu_{g_{i_1} \ldots g_{i_k}}$$

is assumed to sum to zero over all values of g_i corresponding to fixed values of all the other g_i in r_j.

Sums of cross products (SCP)

This corresponds to Step 2 above. For each $v_1, v_2 = 1, \ldots, p + 1$ and each $r = r_1, \ldots, r_q$, the residual cross products $P_{rv_1v_2}$ for the residuals in x_{v_1} and x_{v_2} after fitting all components $\mu_{r_1} \ldots \mu_{r_q}$ except μ_r are obtained together with the residual cross products $P_{ev_1v_2}$ after fitting all components $\mu_{r_1} \ldots \mu_{r_q}$. This is done by linear regression after generating the appropriate dummy variables to represent each component $\mu_{r_1} \ldots \mu_{r_q}$.

To simplify the analysis, it is assumed that the observed levels of the grouping factors are sufficient to uniquely define the components $\mu_{r_1} \ldots \mu_{r_q}$. The computation proceeds by computing all cross products for all variables and dummy variables across all subjects. The residual cross products $P_{rv_1v_2}$ are obtained by means of pivots on the appropriate diagonal elements of this large cross product matrix.

Report

This corresponds to Steps 3, 5, and 7 above. For each $r = r_1, \ldots, r_q$, view $P_{rv_1v_2}$ as a $(p + 1) \times (p + 1)$ matrix and let $\tilde{P}_{rv_1v_2}$ be the result of pivoting on the first p diagonal elements of $P_{rv_1v_2}$. If it is not possible to do this without encountering a diagonal zero (which can happen when the original covariates in the General Procedure are constant across trials) as many pivots as possible are performed. The pivoted cross products $\tilde{P}_{rv_1v_2}$ are used to report an analysis of covariance.

The error sums of squares

$$SS_e = \tilde{P}_{e,p+1,p+1}$$

have degrees of freedom

$$df_e = n - m - p'$$

where

- n = total number of subjects times the number of orthogonal polynomials (degree combinations d) that are pooled for the covariance analysis
- m = total number of dummy variables generated above
- p' = number of covariates for which it was possible to perform the pivots in the previous paragraph

The sum of squares due to the analysis of variance component $\mu_r = r_1, \ldots, r_q$ is

$$SS_r = \tilde{P}_{r,p+1,p+1} - SS_e$$

and has degrees of freedom df_r equal to the number of dummy variables required to represent μ_r. For each covariate x_v which was not eliminated because it corresponds to a diagonal zero, the sum of squares is

$$SS_{x_v} = (\tilde{P}_{e,v,p+1})^2 / |P_{evv}|$$

and has one degree of freedom $df_{x_v} = 1$. Let $r = r_1, \ldots, r_q, e$, or any noneliminated covariate $x_1, \ldots, x_p$. The mean square for the analysis of covariance component r is

$$MS_r = SS_r / df_r$$

and the F statistic for the corresponding hypothesis is

$$F_r = MS_r / MS_e$$

Greenhouse-Geisser and Huynh-Feldt adjustments

For repeated measures designs, 2V provides a test of the sphericity assumption (Anderson, 1958); however, this test is sensitive to nonnormality in the data (especially outliers), lacks power for small sample sizes, and may reject for minor departures from sphericity when N is large. Box (1954) developed a measure ε, of the degree to which the covariance matrix, Σ, departs from the sphericity assumption. For a response covariance matrix with entries σ_{ij}

$$\varepsilon = \frac{k^2(\bar{\sigma}_{jj} - \bar{\sigma}_{..})^2}{(k-1)\left(\sum\sum \sigma_{j'j}^2 - 2k\sum \bar{\sigma}_{j.}^2 + k^2\bar{\sigma}_{..}^2\right)}$$

where

$\bar{\sigma}_{jj}$ = mean of entries on main diagonal of Σ
$\bar{\sigma}_{..}$ = mean of all entries in Σ
$\bar{\sigma}_{j.}$ = mean of entries in row j of Σ

The quantity ε equals 1.0 if Σ satisfies the sphericity condition, for a design with n subjects and one trial factor with k levels. Box (1954) showed that under the null hypothesis, the F statistic has approximately an F distribution with $(k-1)\varepsilon$ and $(n-1)(k-1)\varepsilon$ degrees of freedom, where ε is defined above.

Greenhouse and Geisser (1959) proposed estimation of ε by substitution of the sample covariance matrix for the population matrix in the above equation and comparing the computed F statistic to a critical value of F with $\hat{\varepsilon}(k-1)$ and $\hat{\varepsilon}(k-1)(n-1)$ degrees of freedom.

Huynh and Feldt (1976) proposed a transformation of this estimate of ε to reduce its conservative bias. (Either estimate of ε may be poor for small n.) 2V provides output for both the Greenhouse-Geisser and Huynh-Feldt ε correction factors and p values. See BMDP Technical Report #83 for more details.

Repeated Measures Analysis and Multiple Comparisons

The problem of assessing the decrement in statistical significance attributable to multiple comparisons arises in repeated measures designs as well as in the analysis of other experimental designs.

The repeated measures design is characterized by 2 types of factors, usually labeled as grouping factors and trial factors. Factors of the grouping type are crossed with factors of the trial type. This characterizing aspect of repeated measures consists of characteristic covariance structure. We will illustrate the features and the resolutions of the multiple comparisons aspect of statistical significance analysis in terms of the special case of a signed trial factor and grouping factors of the form: a single group factor together with subjects nested within groups.

Let $y_{ijk} = \mu_{ik} + \varepsilon_{ijk}$ be the response to the kth level of the trial factor for the jth subject in the ith group, where μ_{ik} are fixed parameters and ε_{ijk} are "random" components:

$$E(\varepsilon_{ijk}) = 0$$

$$\operatorname{cov}(\varepsilon_{ijk}, \varepsilon_{i'j'k'}) = \begin{cases} \sigma^2_{kk'} & i' = i,\ j' = j \\ 0 & \text{otherwise} \end{cases}$$

Subjects and groups are considered statistically independent, but a covariance structure is allowed for response to the various levels of the trial factor. (In the case for which the term "repeated measures" applies, the trial factor consists of evaluations at a fixed set of times). We distinguish the special subcase in which all elements have the same variance and correlated variables have the same value of the correlation coefficient; i.e.,

$$\sigma^2_{kk'} = \begin{cases} \sigma^2 & k = k' \\ \rho\sigma^2 & k \neq k' \end{cases}$$

This assumption is called the sphericity assumption.

We assume further that there are n_g groups, n_s subjects within each group, and n_t levels of the trial factor. There are accordingly $N = n_g n_s n_t$ observations. Let Σ be the within-subject covariance matrix and $\sigma_{..}$ be the sum of all elements of the covariance matrix. The analysis of variance table for the analysis is given as

Source	**d.f.**	**Expected Mean Square**
Among groups	$n_g - 1$	$\sigma^2_A + n_s n_t \dfrac{\sum(\bar{\mu}_{i.} - \bar{\mu}_{..})^2}{n_g - 1}$
Among subjects within groups (error A)	$n_g(n_s - 1)$	σ^2_A
Among levels of trial factor	$n_t - 1$	$\sigma^2_B + n_g n_s \dfrac{\sum(\bar{\mu}_{.k} - \bar{\mu}_{..})^2}{n_t - 1}$
Trial by group	$(n_g - 1)(n_t - 1)$	$\sigma^2_B + n_s \dfrac{\sum(\mu_{ik} - \bar{\mu}_{i.} - \bar{\mu}_{.k} + \bar{\mu}_{..})}{(n_g - 1)(n_t - 1)}$
Trial by subject within groups	$n_g(n_s - 1)(n_t - 1)$	σ^2_B
Total	$n_g n_s n_t - 1$	

where $\sigma^2_A = \dfrac{\sigma_{..}}{n_t}$ and $\sigma^2_B = \left[\operatorname{Tr}(\Sigma) - \dfrac{\sigma_{..}}{n_t}\right] \dfrac{1}{(n_t - 1)}$.

If sphericity holds, $\sigma^2_B = (1 - \rho)\sigma^2 \qquad \sigma^2_A = [1 + (n_t - 1)\rho]\sigma^2$.

The estimate for a contrast among grouping factors is

$$\sum c_i \bar{y}_{i..}$$

with standard error

$$\text{s.e.} = \sqrt{\frac{s_A^2 \sum c_i^2}{n_s n_t}}$$

If the sphericity assumption holds, the Bonferroni confidence intervals are, for a maximum of m comparisons,

$$\sum c_k \bar{y}_{..k} \pm t_{1-\frac{\alpha}{2m}} s_B \sqrt{\frac{\sum c_k^2}{n_g n_s}} \quad , \text{ df } = n_g (n_s - 1)(n_t - 1)$$

Confidence intervals of cell means

$$\mu_{ik} = \bar{y}_{i.k} \pm t_0 s / \sqrt{n_s}$$

where

$\bar{y}_{i.k}$ is the sample mean of the (i, k) cell over all j

t_0 is the $\alpha/2$ upper quantile of Student's t-distribution with $n_s - 1$ degrees of freedom

s is the sample standard deviation of the (i, k) cell

The estimated standard error of the adjusted cell means

Covariate is assumed to be constant across trials. Let $\hat{\beta} = (\hat{\beta}_1, \hat{\beta}_2, \ldots, \hat{\beta}_p)$ be the within-cell estimated regression coefficients, and $\bar{x}_{gv}$ = average covariate for group g. Then

$$\text{s.e.} = \left(\frac{s_A^2}{n_s} + (\bar{\mathbf{x}}_{gv} - \bar{\mathbf{x}}_v)^T (\hat{\text{cov}}(\hat{\beta}))(\bar{\mathbf{x}}_{gv} - \bar{\mathbf{x}}_v) \right)^{1/2}$$

where $\hat{\text{cov}}(\hat{\beta})$ is the estimated covariance matrix of the regression coefficients.

Confidence intervals of population adjusted cell means

$$\mu_{ik} = \hat{\mu}_{ik} \pm t_0 \cdot \text{s.e.}$$

where

$\hat{\mu}_{ik}$ is the adjusted cell mean of the (i, k) cell

s.e. is the estimated standard error of the adjusted cell mean of the (i, k) cell

Appendix C
Using FORTRAN Transformations

User-supplied FORTRAN subroutines can be added to BMDP to provide special transformations, to include a user-specified function to be evaluated, or to add features to the program. (This feature is unavailable on the IBM PC.)

The BMDP command procedure allows you to write specified user subroutines before running a BMDP program. These may be added to a BMDP program from FORTRAN source code or linked from an object module.

The following are the user-suppliable FORTRAN subroutines:

- TRANSF, a data transformation subroutine, may be supplied to any of the BMDP programs.
- P3RFUN and PARFUN, which define the nonlinear functions to which you want to fit the data, may be used with 3R and AR, respectively.
- FPARM, which defines the functions of parameters, may be supplied to AR.
- DIFEQ, which defines the differential equations to be solved, may be supplied to AR. See Technical Report #85 on the use of AR to solve differential equation systems.

All of the above are used infrequently, since they may be replaced by the TRANS, FUN, FPARM, and DIFEQ paragraphs. Using the paragraph form will take somewhat more time, especially with a large amount of data or very complex functions.

The following three subroutines, infrequently used, do not have paragraph equivalents:

P7DFUN – allows you to augment the analysis performed by 7D
P4FFIT – allows you to augment the analysis performed by 4F
P4FTAB – allows you to augment the analysis performed by 4F

FORTRAN rules

FORTRAN transformations follow the rules of your locally available FORTRAN compiler, not the rules for BMDP instructions. For example, there are no periods at the end of statements. Variable names are not recognized in the FORTRAN subroutine and cannot be used in place of X(#). You must check for missing values; they are not automatically excluded from the transformation computations.

If new variables are added by the transformation, ADD = #. must be stated in the VARIABLE paragraph. BMDP functions are not available in FORTRAN, although some functions are available in both BMDP and FORTRAN. In some cases, the names differ (e.g., the BMDP functions LOG and LN correspond to the FORTRAN functions ALOG10 and ALOG). FORTRAN transformations can be used like their BMDP counterparts for case selection and deletion. The commands OMIT and DELETE cannot be used in the FORTRAN subroutine. To randomly select cases or generate data you can call FORTRAN functions that generate uniform or normal random numbers (Appendix B.3).

Any legitimate FORTRAN statements, such as READ, WRITE, DATA, DIMENSION, FUNCTION, and SUBROUTINE, can be used. The FORTRAN subroutine in which the transformations are included starts with

```
SUBROUTINE TRANSF(X, KASE, NPROB, USE, NVAR, XMIS)
DIMENSION X(NVAR)
```

and ends with

```
RETURN
END
```

Those four statements are supplied by BMDP if you add FORTRAN statements via the TRANSF parameter.

The following values should not be changed:

KASE – sequence number of case being processed.

NPROB – sequence number of the problem (INPUT paragraph) being processed.

NVAR – the greater of the number of variables read or the number of variables after transformation (i.e., the number of variables read plus the number of variables ADDED by transformation).

XMIS – missing value code used internally by the BMDP program. Values less than the lower limit or greater than the upper limit specified for a variable are recoded to values –2*XMIS and 2*XMIS, respectively. The values are recoded BEFORE transformations are processed unless AFTER is specified in the VARIABLE paragraph.

You can change the values of X and USE.

X – vector containing the data for the case being processed. The values for all variables read in the data are present. X(1) is the value of the first variable, X(2) the second, etc. The values of X can be modified to contain the transformed values. Variables created (added to input variables) by transformation are initialized to XMIS for each case, unless RETAIN is specified in the VARIABLE paragraph, asking to retain these values from the previous settings.

USE – flag for case selection. Its interpretation is identical to that of USE in the TRANSFORM paragraph. If USE is positive (1), the case is used in the analysis; if zero (0), the case is not used but is included in a saved file if one is created; if negative (-1), the case is not used or copied into a saved file. USE = -100 indicates that data reading should terminate as if an end-of-file had occurred.

System instructions

When you specify transformations in FORTRAN statements, the system instructions must be altered. The subroutine containing your FORTRAN statements must be integrated with the previously compiled BMDP program to form a modified program that will transform the data and perform the analysis. (Note: this option is not available on IBM PC and some other systems.

VAX/VMS. Command line features for adding your subroutine are TRANSF, FUN, and LINK.

TRANSF – Identifies the file containing user-supplied FORTRAN code for transformation. The SUBROUTINE, DIMENSION, RETURN, and END statements are automatically added to the FORTRAN statements. Hence, they should not be in the file containing the subroutine.

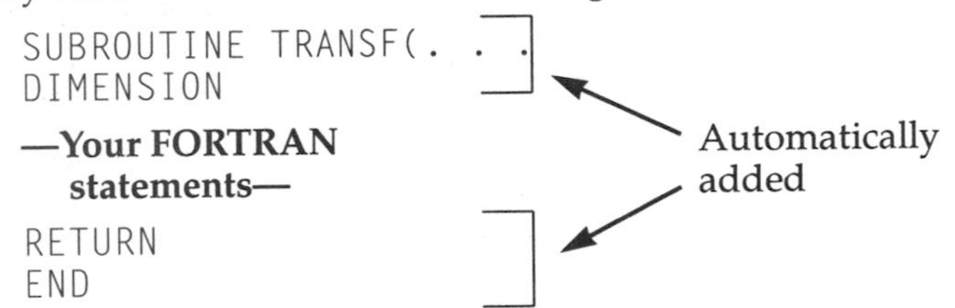

If you do not specify a file type, the suffix FOR is used.

```
BMDP 6D IN=monkey.inp TRANSF=mysub <Return>
```

P3RFUN, PARFUN – Use in programs 3R and AR to identify the file containing user-supplied FORTRAN code for the function. As shown for TRANSF, above, the appropriate SUBROUTINE, DIMENSION, RETURN, and END statements are automatically added to the FORTRAN statements. Hence, they should not be in the file MYSUB. If you do not specify a filetype, the suffix .FOR is used.

```
BMDP 3R IN=monkey.inp FUN=mysub.inp <Return>
```

LINK – Use to replace one or more BMDP subroutines with your own. LINK points to the file containing the complete object code for the subroutine(s). Note that the SUBROUTINE, DIMENSION, RETURN, and END statements are not provided with LINK, since the file is object code. If you do not specify a filetype, the suffix .OBJ is used. LINK may be used if compile time is large, instead of both TRANSF and FUN.

```
BMDP 6D IN=monkey.inp LINK=mysub.obj <Return>
```

IBM OS interactive run. To invoke program 1D, for example, in interactive mode you would use the following instructions:

BMDP 1D TRANSF(filename)

To use program AR, you would use

BMDP AR FUN(filename)

Note that SUBROUTINE DIMENSION, RETURN, and END statements are automatically added to the text in file, filename.

IBM OS Batch run. To compile a subroutine and execute the BMDP program on IBM OS systems, you must specify three system instructions (JCL) before your FORTRAN statements. (Note the T in BIMEDT. 3R is the BMDP program being run.) The SUBROUTINE and DIMENSION instructions at the beginning and the RETURN and END instructions at the end must be omitted. They are ordinarily supplied by the BIMEDT procedure for IBM OS and similar systems.

```
// user identification instruction (e.g., job card)
// EXEC BIMEDT,PROG=BMDP3R
// TRANSF DD *
(FORTRAN statements)
// GO.SYSIN DD *
(BMDP instructions and data)
//
```

The FORTRAN statements are followed by a system instruction:

//GO.SYSIN DD *. This is followed by BMDP instructions and the data, if included with the instructions. These are terminated by a final system instruction: //. If data are read from a separate file, or if a BMDP File is created, the system instructions describing the data location are placed immediately before //GO.SYSIN DD*.

Variations exist for the system instructions, so check with your system manager for those required.

Other systems: For IBM PC MS/DOS, this feature is not available; for UNIX systems, see the UNIX User Guide and/or installation guide; on other systems, ask your BMDP consultant.

Special subroutines

A powerful feature of BMDP is its open-ended analytical capability, using special user-supplied subroutines. This section describes the VMS form of these subroutines.

Note that the following subroutines begin with a subroutine declaration and end with the two statements, RETURN and END, if you are using the LINK option:

TRANSF (available with all programs). This is the subroutine used for specifying FORTRAN transformations. (See previous pages.)

P3RFUN (3R). See the FORTRAN appendix for Volume 2 of this manual.

PARFUN (AR). Used to specify the nonlinear regression function and its derivatives.

```
      SUBROUTINE PARFUN(F, P. X. N, KASE, NVAR. NPAR, IPASS,
     1                  XLOSS. ITER)
      DIMENSION P(NPAR), X(NVAR)
CB-SNG
        DOUBLE PRECISION F, P, X, XLOSS
CE-SNG
```

where F – function value (output)
P – current value of parameters (input)
X – current case (input)
N – code number for function (input)
KASE – current case number (input)
NVAR – number of variables (input)
NPAR – number of parameters (input)
IPASS – index of pass (input)
XLOSS – loss value (output) not used here
ITER – iteration number

DIFEQ (AR). Used to specify differential equations when they are used to implicitly define a regression equation.

FPARM (AR). Used to specify functions of parameters.

```
      SUBROUTINE FPARM (p, G)
      DOUBLE PRECISION P, G
      DIMENSION P(1), G(1)
```

—the functions in terms of the parameters P(*j*); i.e., G(*i*) =*f*(P(1), . . .)

```
      RETURN
      END
```

where G(*i*) is the value of the function
P(*j*) is the jth parameter

P7DFUN (7D). Can be used to augment the analysis performed by 7D, since it is called once for each set of histograms.

```
      SUBROUTINE P7DFUN (INDV, DVNAME, NGRP,  GMEAN, SIZE,
     1                   SS,     U,     V,     W,     X,
     2                   GNAME, VNAME, INDEX, GMEAN2, SIZ2,
     3                   SS2,  GNAM1, GNAM2, VNAM1, VNAM2,
     4                   INDX1, INDX2, NR, NC)
      DIMENSION GMEAN(NGRP), SIZE(NGRP), SS(NGRP), U(NGRP),
     1          V(NGRP), W(NGRP), X(NGRP), GMEAN2(NR, NC),
     2          SIZ2(NR, NC), SS2(NR. NC), GNAME(NGRP),
     3          GNAM1(NR, NC), GNAM2(NR, NC)
CB+ONE
C     INTEGER DVNAME, GNAME, VNAME. GNAM1, GNAM2, VNAM1, VNAM2
CE+ONE
CB-ONE
      DOUBLE PRECISION DVNAME, GNAME, VNAME, GNAM1, GNAM2,
     1                 VNAM1, VNAM2
CE-ONE
CB-SNG
      DOUBLE PRECISION GMEAN, SS, U, V, W, X, GMEAN2,SS2
CE-SNG
```

where
INDV – index of the dependent variable
DVNAME – name of the dependent variable
NGRP – number of groups (1-way ANOVA)
GMEAN – group means (1-way ANOVA)
SIZE – group sample sizes (1-way ANOVA)
SS – group sums of squares (1-way ANOVA)
U, V, W, X – scratch vectors
GNAME – group names (1-way ANOVA)
VNAME – name of grouping variable (1-way ANOVA)
INDEX – index of grouping variable (1-way ANOVA)
GMEAN2 – group means (2-way ANOVA, no empty cells)
SIZ2 – group sample sizes (2-way ANOVA, no empty cells)
SS2 – group sums of squares (2-way ANOVA, no empty cells)
GNAM1 – group names of first grouping variable (2-way ANOVA, no empty cells)
GNAM2 – group names of second grouping variable (2-way ANOVA, no empty cells)
VNAM1 – name of first grouping variable (2-way ANOVA, no empty cells)
VNAM2 – name of second grouping variable (2-way ANOVA, no empty cells)
INDX1 – index of first grouping variable (2-way ANOVA, no empty cells)
INDX2 – index of second grouping variable (2-way ANOVA, no empty cells)
NR – number of levels of second grouping variable (2-way ANOVA, no empty cells)
NC – number of levels of first grouping variable (2-way ANOVA, no empty cells)

P4FFIT (4F). Allows you to directly access the observed and expected (fitted) values of the crosstables as they are analyzed. This allows you to calculate new statistics or to write the observed and expected values to a separate file. It is called once for each FIT paragraph as long as MYSUB is stated in the FIT paragraph. The VAX/VMS form is

```
      SUBROUTINE P4FFIT(TABLE, FIT, FITB, NCELL, LEVELA, NODIMA,
     1                 LZERO, CONFIG, NCONF, IMODEL)
      DIMENSION TABLE(NCELL),FIT(NCELL),FITB(NCELL).
     1          LEVELA(NODIMA)
      INTEGER CONFIG(NODIMA,NCONF)
```

—Your FORTRAN statements—

```
      RETURN
      END
```

where TABLE – table of observed values, stored as a vector

FIT	–	table of fitted (expected) values
FITB	–	fit matrix if there are structural zeros
NCELL	–	number of cells in a table
LEVELA	–	vector containing the number of levels for each dimension in the table
NODIMA	–	number of dimensions (factors) in the table
LZERO	–	logical, whether structural zeros in table
CONFIG	–	"configuration," not ordinarily used except by BMDP staff
NCONF	–	number of columns in CONFIG
IMODEL	–	model number

P4FTAB (4F). Can be used to augment the analysis performed by 4F. It is called once for each table as long as MYSUB is stated in the PRINT paragraph. The VAX form is

```
      SUBROUTINE P4FTAB(TABLE, FITB, NCELL,LEVELA, NODIA,
     1                 LZERO. NTABLE, LEVCD)
      DIMENSION TABLE(NCELL),FITB(NCELL),LEVELA(NODIMA)
      LOGICAL ZERO
```

—Your FORTRAN statements—

```
      RETURN
      END
```

where

TABLE	–	table of observed values
FITB	–	fit matrix if there are structural zeros
NCELL	–	number of cells in a table
LEVELA	–	vector containing the number of levels for each dimension in the table
NODIMA	–	number of dimensions (factors) in the table
LZERO	–	logical, whether structural zeros in table
LEVCD	–	level of conditioning variable

Appendix D
Data

This Appendix describes the longer data sets used in the examples in this volume. Many of the short data sets are listed in the programs where they are used. The data set names are presented in alphabetical order; the chapters in which the data are used in examples are in parentheses. The data sets listed below are available on disk or tape. Contact BMDP for details.

Contents

D.1 Air pollution data

The air pollution data, AIRPOL.DAT, use information from 60 U.S. metropolitan areas (McDonald and Schwing, 1973; Henderson and Velleman, 1981). For each record, the data include the following:

Variables

1,2. name – city name split into two variables, four characters each
3. rain – mean annual precipitation in inches
4. educatn – median school years completed for those over 25 in 1960 SMSA

5. pop_den – population/mile^2 in urbanized area in 1960
6. nonwhite – percentage of urban area population that is nonwhite
7. nox – relative pollution potential of oxides of nitrogen, NO_x
8. so2 – relative pollution potential of sulfur dioxide, SO_2
9. mortalty – total age-adjusted mortality rate, expressed as deaths per 100,000

1 2	3	4	5	6	7	8	9
akronOH	36	11.4	3243	8.8	15	59	921.9
albanyNY	35	11.0	4281	3.5	10	39	997.9
allenPA	44	9.8	4260	0.8	6	33	962.4
—we omit intervening 54 cases—							
worctrMA	45	11.1	3678	1.0	3	8	895.7
yorkPA	42	9.0	9699	4.8	8	49	911.8
youngsOH	38	10.7	3451	11.7	13	39	954.4

D.2 Exercise data

In these data, EXERCISE.DAT, 40 persons were asked to take their pulse, run one mile, and take their pulse a second time. We use two versions of the data: one with subject names and one without. We show the data with names here; the data without names are shown in Chapter 2 without case labels. This dataset is also referred to without the label variables as PULSE2.DAT in examples shown in Program 3D.

Variables

1. id – identiflcation number
2. sex – male(1), female(2)
3. smoke – yes(1), no(2)
4. age – subject's age in years
5. pulse_1 – pre exercise pulse rate
6. pulse_2 – post-exercise pulse rate
7. name_1 – first four characters of subject name (not in PULSE2.DAT)
8. name_2 – next four characters of subject name (not in PULSE2.DAT)

1	2	3	4	5	6	7 8
1	1	1	31	62	126	ALLAN
2	2	1	20	78	154	MARY
3	1	2	28	64	128	BILL
4	2	2	29	96	155	LINDA
5	1	1	21	66	128	MICHAEL
6	2	1	27	96	265	CATHY
7	1	2	21	68	120	HARVEY
8	2	2	42	72	138	JENEE
9	2	1	22	88	160	JEAN
10	1	1	28	90	144	FREDDY
11	2	2	21	82	140	PAT
12	1	2	22	74	134	MARK
13	2	1	43	66	148	SUSAN
14	2	2	19	68	142	DENISE
15	1	1	23	92	134	JOHN
16	1	2	41	68	112	DAVID
17	1	2	24	76	158	ROBERT
18	2	2	21	86	146	ALISON
19	2	1	21	88	156	JILL
20	1	1	20	66	132	JACKSON
21	1	1	38	70	122	ARTHUR

22	1	2	20	80	136	SAMUEL
23	2	1	33	76	148	AMY
24	2	2	25	78	148	ANNIE
25	2	2	37	76	136	JANE
26	2	2	22	80	158	BETH
27	1	2	32	68	116	CHRIS
28	1	2	22	70	120	FRANCIS
29	1	1	22	68	126	ERNIE
30	1	1	19	70	144	BERTRAM
31	2	2	21	86	144	NANCY
32	1	2	26	72	126	BRUCE
33	2	2	32	84	136	MARGE
34	2	2	24	72	142	BARBARA
35	2	2	28	80	138	JENNY
36	1	1	34	62	132	WILLIAM
37	1	2	35	74	116	KYLE
38	1	1	21	90	138	BEN
39	1	2	21	66	142	GREG
40	1	2	30	70	132	RICHARD

D.3 Fatness data

These data, FATNESS.DAT, describe a group of 32 girls born in Berkeley, California, between January 1928 and June 1929.

Variables

1. id – identification number
2. sex – identifies sex (coded 0 and 1)
2. wt_2 – weight in kg at age 2
3. ht_2 – height in cm at age 2
4. wt_9 – weight in kg at age 9
5. ht_9 – height in cm at age 9
6. leg_9 – leg circumference at age 9
7. strong_9 – strength measure at age 9
8. wt_18 – weight in kg at age 18
9. ht_18 – height in cm at age 18
10. leg_18 – leg circumference at age 18
11. strong_18 – strength measure at age 18
12. fatness – a measure of fatness on a 7-point scale, determined from a photograph at age 18: slender(1), fat(7)

1	2	3	4	5	6	7	8	9	10	11	12
331	12.6	83.8	33.0	136.5	29.0	57.0	71.2	169.6	38.8	107.0	6.0
334	12.0	86.2	34.2	137.0	27.3	44.0	58.2	166.8	34.3	130.0	5.0
335	12.0	85.1	28.1	129.0	27.4	48.0	56.0	157.1	37.8	101.0	5.0
351	12.7	88.6	27.5	139.4	25.7	68.0	64.5	181.1	34.2	149.0	4.0
352	11.3	83.0	23.9	125.6	24.5	22.0	53.0	158.4	32.4	112.0	5.0
353	11.8	88.9	32.2	137.1	28.2	59.0	52.4	165.6	33.8	136.0	4.0
354	15.4	89.7	29.4	133.6	26.6	58.0	56.8	166.7	32.7	118.0	4.5
355	10.9	81.3	22.0	121.4	24.4	44.0	49.2	156.5	33.5	110.0	4.0
356	13.2	88.7	28.8	133.6	26.5	58.0	55.6	168.1	34.1	104.0	4.5
357	14.3	88.4	38.8	134.1	31.1	57.0	77.8	165.3	39.8	138.0	6.5
358	11.1	85.1	36.0	139.4	28.2	64.0	69.6	163.7	38.6	108.0	5.5
359	13.6	91.4	31.3	138.1	27.6	64.0	56.2	173.7	34.2	134.0	3.5
361	13.5	86.1	33.3	138.4	29.4	73.0	64.9	169.2	36.7	141.0	5.0
362	16.3	94.0	36.2	139.5	28.0	52.0	59.3	170.1	32.8	122.0	4.5
364	10.2	82.2	23.4	129.8	22.6	60.0	49.8	164.2	30.0	128.0	4.0
365	12.6	88.2	33.8	144.8	28.3	107.0	62.6	176.0	35.8	168.0	5.0
366	12.9	87.5	34.5	138.9	30.5	62.0	66.6	170.9	38.8	126.0	5.0
367	13.3	88.6	34.4	140.3	31.2	88.0	65.3	169.2	39.0	142.0	5.0
368	13.4	86.9	38.2	143.8	29.8	78.0	65.9	172.0	35.7	132.0	5.5
369	12.7	86.4	31.7	133.6	27.5	52.0	59.0	163.0	32.7	116.0	5.5

370	12.2	80.9	26.6	123.5	27.2	40.0	47.4	154.5	32.2	112.0	4.0
371	15.4	90.0	34.2	139.9	29.1	71.0	60.4	172.5	35.7	137.0	4.0
372	12.7	94.0	27.7	136.1	26.7	30.0	56.3	175.6	34.0	114.0	3.0
373	13.2	89.7	28.5	135.8	25.5	76.0	61.7	167.2	35.5	122.0	4.5
374	12.4	86.4	30.5	131.9	28.6	59.0	52.4	164.0	34.8	121.0	5.0
376	13.4	86.4	39.0	130.9	29.3	38.0	58.4	161.6	33.0	107.0	6.5
377	10.6	81.8	25.0	126.3	25.0	50.0	52.8	153.6	33.4	140.0	5.0
380	12.7	91.4	29.8	135.5	27.0	57.0	67.4	173.5	34.5	123.0	5.0
382	11.8	88.6	27.0	134.0	26.5	54.0	56.3	166.2	36.2	135.0	4.5
383	13.3	86.4	41.4	138.2	32.5	44.0	82.8	162.8	42.5	125.0	7.0
384	13.2	94.0	41.6	142.0	31.0	56.0	68.1	168.6	38.4	142.0	5.5
385	15.9	89.2	42.4	140.8	32.6	74.0	63.1	169.2	37.9	142.0	5.5

D.4 Fidell data

The Fidell data, FIDELL.DAT, are from a survey of 465 women by L. S. Fidell and J. E. Prather (described in Tabachnick and Fidell, 1983, Appendix B). The data consist of psychological and demographic measures, of which we use 6 (the names are changed slightly). An asterisk is inserted in the positions where data are missing.

Variables

1. esteem – self-esteem measure coded from 1 to 22: low(1), high(22)
2. hap_stat – happiness with marital status coded from 1 to 48: low(1), high(48)
3. womenrol – attitude toward role of women on scale from 1 to 38: conservative(1-16), moderate(17-23), liberal(24 and up)
4. educatn – number of years of education
5. workstat – work status: paid work(0), homemaker(1 and 2)
6. marital – single(1), married(2)

1	2	3	4	5	6
14	23	14	12	1	2
13	38	18	12	0	2
9	39	12	12	0	2
—we omit intervening 459 cases—					
19	47	15	14	2	2
13	39	21	12	2	2
14	40	11	12	2	2

D.5 Fisher iris data

The Fisher (1936) iris data, FISHER.DAT, contain measurements of sepal length and width and petal length and width on samples of 50 irises from each of 3 species. The data for the four measurements are recorded to one-tenth of a centimeter.

Variables

1. sepal_l – sepal length
2. sepal_w – sepal width
3. petal_l – petal length
4. petal_w – petal width
5. iris – setosa(1), versicolor(2), virginica(3)

1	2	3	4	5
50	33	14	02	1
64	28	56	22	3
65	28	46	15	2
—we omit intervening 144 cases—				
67	30	50	17	2
63	33	60	25	3
53	37	15	02	1

D.6 Gas furnace data

These data from Box and Jenkins (1976) are used in 2T to illustrate control of the output of an industrial process. The first variable is the feed rate of a mixture of gases (including methane and oxygen); the second is the concentration of carbon dioxide (CO_2) output.

1	2
-0.109	53.8
0.000	53.6
0.178	53.5
—we omit 290 cases—	
0.017	57.8
-0.182	57.3
-0.262	57.0

D.7 Patient data

These data, PATIENT.DAT, are a survey of AIDS patients' reactions to their physicians conducted by Van Servellen based on a scale developed by Cope et al. (1986). The 14 items in the survey questionnaire measure patient attitudes about physician personality, demeanor, competence, and prescribed treatment. For a more complete explanation of the survey questions, see Chapter 4M. Note that these data must be read in 12F1 format, as there are no spaces between variables.

```
2232241122122
23232232233212
22334422422442
```

—we omit the intervening 62 cases—

```
1121111122111
2122211111221
1112111122111
```

D.8 Police data

These data, POLICE.DAT, are taken from Kirk (1982), and contain ratings of attitude toward minorities for officers working three different types of beat. Subgroups of officers on each beat are assigned to one of three human relations training classes, each with a different time duration.

Variables

1. id – identification number
2. beat – neighborhood: lower(1), middle(2), upper-class(3)
3. train – hours spent in human relations training
4. attitude – attitude toward minorities, from low(0) to high(75)

1	2	3	4
1	1	5	24
2	1	5	33
3	1	5	37
4	1	5	29
5	1	5	42
6	1	10	44
7	1	10	36
8	1	10	25
9	1	10	27
10	1	10	43
11	1	15	38
12	1	15	29
13	1	15	28

14	1	15	47
15	1	15	48
16	2	5	30
17	2	5	21
18	2	5	39
19	2	5	26
20	2	5	34
21	2	10	35
22	2	10	40
23	2	10	27
24	2	10	31
25	2	10	22
26	2	15	26
27	2	15	27
28	2	15	36
29	2	15	46
30	2	15	45
31	3	5	21
32	3	5	18
33	3	5	10
34	3	5	31
35	3	5	20
36	3	10	41
37	3	10	39
38	3	10	50
39	3	10	36
40	3	10	34
41	3	15	42
42	3	15	52
43	3	15	53
44	3	15	49
45	3	15	64

D.9 Survey data

The survey data, SURVEY.DAT, are from a study of depression in Los Angeles, using a subset of 294 respondents out of the original 1000 (Afifi and Clark, 1990). The data include mental and physical health variables and demographic variables (sex, age, education, and income).

Variables

1. id – identification number
2. sex – male(1), female(2)
3. age – years at last birthday
4. marital – never married(1), married(2), divorced(3), separated(4), widowed(5)
5. educatn – less than high school(1), some high school(2), finished high school(3), some college(4), flnished bachelor's degree(5), finished master's degree(6), flnished doctorate(7)
6. employ – employment: full time(1), part time(2), unemployed(3), retired(4), houseperson(5), in school(6), other(7)
7. income – $ thousands/yr
8. religion – Protestant(1), Catholic(2), Jewish(3), none(4), other(5)

Responses for the following 20 depression items indicate that the subject felt this way less than 1 day last week(0), 1–2 days(1), 3–4 days(2), or 5–7 days(3). (Codes are reversed for variables 16 through 19.)

9. blue	19. enjoy
10. depress	20. bothered
11. lonely	21. no_eat
12. cry	22. effort
13. sad	23. badsleep
14. fearful	24. getgoing
15. failure	25. mind
16. asgood	26. talkless
17. hopeful	27. unfriend
18. happy	28. dislike

29. total – sum of the above depression scores: lowest score possible(0), highest(60)
30. case – normal(0), depressed(1)
31. drink – regular drinker: yes(1), no(2)
32. healthy – general health: excellent(1), good(2), fair(3), poor(4)
33. doctor – has regular physician: yes(1), no(2)
34. meds – doctor has given prescription: yes(1), no(2)
35. bed days – spent entire day(s) in bed in last 2 months: no(0), yes(1)
36. illness – acute illness in last 2 months: no(0), yes(1)
37. chronic – chronic illness in last year: no(0), yes (1)

1	2	3	4	5	6	7	8	9	10	11	12	13	14	15	16	17	18	19
1	2	68	5	2	4	4	1	0	0	0	0	0	0	0	0	0	0	0
2	1	58	3	4	1	15	1	0	0	1	0	0	0	0	0	0	0	0
3	2	45	2	3	1	28	1	0	0	0	0	1	0	0	0	0	0	0
—we omit intervening 288 cases—																		
292	1	64	2	4	1	55	3	0	0	0	0	0	1	1	3	0	1	1
293	1	43	3	6	1	28	1	0	0	0	0	0	0	0	0	0	0	0
294	2	58	1	3	4	9	1	1	0	0	0	0	0	0	0	1	1	1

1	20	21	22	23	24	25	26	27	28	29	30	31	32	33	34	35	36	37
1	0	0	0	0	0	0	0	0	0	0	0	2	2	1	1	0	0	1
2	1	0	0	1	0	1	0	0	0	4	0	1	1	1	1	0	0	1
3	0	0	1	1	1	0	0	0	0	4	0	1	2	1	1	0	0	0
—we omit intervening 288 cases—																		
292	1	0	0	1	0	0	0	0	0	9	0	1	2	1	2	0	0	1
293	0	0	0	0	0	0	2	0	0	2	0	2	1	1	2	0	1	1
294	2	0	2	1	0	0	0	0	1	10	0	2	3	1	1	0	0	1

D.10 Tire data

The tire data, TIRE.DAT, are from an industrial study in Kirk (1982) that evaluates the durability of several types of rubber compounds used in automobile tires. Each record includes information for one tire.

Variables

1. rubber – rubber compound, coded 1 to 4.
2. wheel – wheel position: r_front(1), l_front(2), r_rear(3), l_rear(4)
3. auto – brand of automobile: lemon(1), nogo(2), heap(3), junker(4)
4. tread – remaining after 10,000 miles

1	2	3	4
1	1	1	1
2	1	2	2
3	1	3	5
4	1	4	9
2	2	1	3
3	2	2	8
4	2	3	9
1	2	4	2
3	3	1	5
4	3	2	10
1	3	3	3
2	3	4	5
4	4	1	7
1	4	2	6
2	4	3	3
3	4	4	6

D.11 Ulcer data

The ulcer data, ULCER.DAT, are from a study of 157 ulcer patients taking part in a two-phase trial (Ippoliti et al., 1983). We inserted an asterisk in positions where data are missing. The codes 9 or 99 are also used to flag missing values.

Variables

1. treatmnt – one of two active drugs, coded 1 or 2
2. age
3. sex
4. fam_hist – number of family members with ulcer disease
5. cigarets – number of packs of cigarettes smoked per day
6. nitepain – presence of night pain at study entry: no(0), yes(1)
7. outcome – final outcome at end of two-month trial
8. sqrttime – square root of months since first ulcer diagnosis
9. b_acid – square root of basal acid output
10. peakacid – square root of peak acid output
11. trouble – previous ulcer complications: yes(0), no(1)
12. log_time – logarithm of time since first ulcer diagnosis
13. painfreq – a pain score at study entry

1	2	3	4	5	6	7	8	9	10	11	12	13
2	58	1	0	99	0	1	16.248	*	*	1	2.422	0
1	60	1	1	0	1	3	20.785	*	*	1	2.635	2
1	30	1	0	99	1	1	1.000	*	*	0	0.000	14
—we omit intervening 151 cases—												
1	24	1	0	1	1	1	9.165	2.646	6.083	0	1.924	5
1	50	1	0	1	0	2	13.856	2.646	6.481	1	2.283	1
1	71	1	0	0	0	6	7.416	1.000	3.000	1	1.740	4

D.12 Werner blood chemistry data

The Werner et al. (1970) blood chemistry data, WERNER.DAT, include information for 188 age-matched pairs of women. One member of each pair does not use birth control pills and the other does. Albumin, calcium, and uric acid are recorded to the nearest tenth of a unit. Other measurements are to the nearest unit. Note that for the following data, we inserted an asterisk (*) in the positions where data are missing.

Variables

1. id – identification number
2. age – in years
3. height – in inches
4. weight – in pounds
5. brthpill – birth control pill usage: no(1), yes(2)
6. cholstrl – cholesterol level
7. albumin – albumin level
8. calcium – calcium level
9. uricacid – uric acid level

1	2	3	4	5	6	7	8	9
2381	22	67	144	1	200	43	98	54
1946	22	64	160	2	600	35	*	72
1610	25	62	128	1	243	41	104	33
—we omit intervening 182 cases—								
575	53	65	140	2	220	40	107	46
2271	54	66	158	1	305	42	103	48
39	54	60	170	2	220	35	88	63

Appendix E
Special Topics

Contents

E.1 Quick and dirty Monte Carlo

Sometimes it is desirable to get a quick idea of the sensitivity of a statistic to departures from the usual underlying assumptions. The following setup can be used in 3D to assess the sensitivity of the ordinary two-sample *t* test to the assumption of equal variances. 3D reports both the usual Student's *t* and a *t* test based on separate variances.

```
/ INPUT     VARIABLES ARE 0.
            CASES ARE 100.

/ TRANSF    group = KASE GT 75.
            var_1 = RNDG(17948).
            var_2 = RNDG(4981).
            IF (group EQ 1) THEN
               (var_1 = var_1 * 2.
                var_2 = var_2 * 2.).

/ GROUP     VARIABLE = group.

/ END
```

In this example, data are generated randomly for two statistically independent variables (var_1 and var_2) and two groups ($N_1 \approx 75$, $N_2 \approx 25$) with the variances in the second group four times those of the first group. The first variable is the grouping variable.

More complex Monte Carlo experiments can be made by using the BIMEDT procedure. In the following example, 100 statistically independent variables are generated for two groups. The p-values for the t tests obtained for these 100 variables should constitute a random sample from the uniform distribution. However, since the variances are unequal, the p-values will not have a uniform distribution.

```
/ INPUT      VARIABLES = 0. CASES = 100.

/ TRANS      GRP = 1.
             SIGMA = 1.
             IF (KASE GT 75)
                 THEN (GRP = 2. SIGMA = 2.).
             FOR J = 1 TO 100.
                 % X|J = SIGMA * RNDG(12349).  %

/ GROUP      VARIABLE = grp.

/ END
```

These methods are not intended to be used for extensive Monte Carlo experiments, since special purpose programs would be more efficient.

E.2 Computing predictions

The equations of a model may be evaluated on a set of cases other than those used to determine the parameters of the model. This is accomplished by assigning zero weight to cases that will not be allowed to affect the computations of the regression coefficients.

Suppose for the first 100 cases the dependent variable is recorded, and for the next 50 you wish to predict the dependent variable. Your input could be

```
/ INPUT      VARIABLES = 5.
             FORMAT = FREE.
             CASES = 150.

/ VARIABLE   WEIGHT = WT.

/ TRANSFORM  WT = KASE LE 100.

/ REGRESS    ...

/ END
.
.
(data matrix)
.
```

The above example sets the weight variable (WT) equal to 1 for the first 100 cases and to 0 for the next 50 cases. If knowledge of the outcome variable is coded in the data by some other scheme, appropriate coding can be devised to accommodate the scheme. For example, if the dependent variable is zero only when it is not known, one can use the transformation

```
WT = 1.0.
IF(Y EQ 0)THEN WT = 0.
```

to set the weight to zero or one.

E.3 Analysis of binary data in 2V

Introduction

The outcome variable in many studies is binary or dichotomous. Yes/ No,– True/False, Accept/Reject, Purchase/Do Not Purchase, for example, all denote only two possible results. The proportion of positive responses is called the rate of response and can vary from 0 to 1.

We outline a portion of the weighted least squares (WLS) approach of Grizzle, Starmer, and Koch (1969) for binary data. A discussion of this approach can be found in Cox (1970). We briefly mention the maximum likelihood logit method, which is also known as the log-linear model.

The following example is taken from Forthofer and Lehnen (1981). The outcome is the result of a pulmonary function test (PFT), classified as either normal or abnormal in 479 people. This part of the study considers two possible influences on the test—smoking status and lead level in ambient air. The data are given in Figure E.l.

Weighted least squares on proportions

Since we wish to analyze the binary response, and the unequal variances of the groups preclude the usual analysis of variance, we can think of each group as being characterized by its abnormality rate. There are eight rates, since there are 4 x 2 = 8 combinations of smoking and lead exposure. Each of these rates is thought of as one case. Our hypotheses involve the equality of rates among smoking levels, and between high and low lead levels.

An analysis of variance applied to the rates should use case weights, because their variances differ greatly. The variance of a rate, p, is given in many textbooks as $p(1 - p)/n$, where p is the proportion based on n cases. The weight for each of the eight cases should be the reciprocal of the variance.

Figure E.1
Data for smoking status and lead level study

Smoking	Outcome	Lead High	Lead Low	Total
NEVER	NORMAL	33	160	193
	ABNORMAL	3	4	7
FORMER	NORMAL	12	49	61
	ABNORMAL	2	6	8
LIGHT	NORMAL	21	75	96
	ABNORMAL	2	6	8
HEAVY	NORMAL	16	84	100
	ABNORMAL	3	3	6

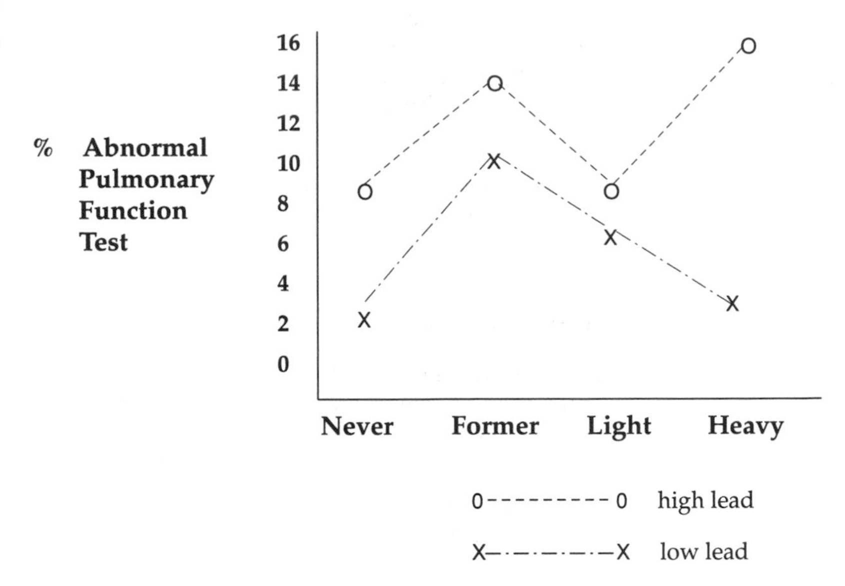

Figure E.2
Pulmonary function test by smoking and lead exposure

The BMDP instructions for 2V are given in Input E.1, the results in Output E.1. Note that we have not reproduced the mean squares, the *F* value, or the probability, since these are not relevant to this method of analysis. We also ignore the line for the mean. In a sense, we are "tricking" the program by ignoring some of the usual output.

Input E.1

```
 / INPUT      FORMAT = STREAM.
              VARIABLES = 4.

 / VARIABLE   NAMES = lead, smoking, normal, abnormal.
              WEIGHT = wt.

 / GROUP      CODES(lead) = 1,2.
              NAMES(lead) = high, low.
              CODES(smoking) = 1 TO 4.
              NAMES(smoking) = never, former, light, heavy.

 / TRANSFORM N = normal + abnormal. P = abnormal/N.
              wt = N/(P*Q).          Q = 1 - P.

 / DESIGN     GROUPING = lead, smoking.
              DEPENDENT = P.
 / END
1 1   33    3       1   2    12    2
2 1   160   4       2   2    49    6
1 3   21    2       1   4    16    3
2 3   75    6       2   4    84    3
 / END
```

Output E.1

ANALYSIS OF VARIANCE TABLE

SOURCE	SUM OF SQUARES	DEGREES OF FREEDOM
lead	2.05676	1
smoking	4.55968	3
ERROR	0.83550	3

The sum of squares column in Output E.1 gives the weighted least squares chi-square values. The chi-square labeled ERROR is the goodness-of-fit test statistic. Since the observed chi-square of .83 is less than the critical value of 7.81 for the 5% level (for three degrees of freedom, according to a chi-square table), we do not have evidence of a lack of fit. It can be said that our lead and smoking model appears to adequately describe the data. Since the chi-square for lead is less than 3.84, it is not statistically significant. The smoking effect is not significant since it is also less than the critical value.

Weighted least squares on logits

Some statisticians feel that it has been convincingly argued that a log-linear, or logit model is better suited to binomial data. A clear discussion of this is given in Fleiss (1981). The changes necessary to use 2V for this are minimal, as you can see from Input E.2.

The transformation paragraph is used to calculate each logit as the natural log of p divided by q. The weight is the reciprocal of the variance, which is itself the sum of two reciprocals.

Input E.2

```
 / INPUT       FORMAT = STREAM.
               VARIABLES = 4.

 / VARIABLE    NAMES = lead, smoking, abnormal, normal.
               WEIGHT = WT.

 / GROUP       CODES(lead) = 1, 2.
               NAMES(lead) = high, low.
               CODES(smoking) = 1 TO 4.
               NAMES(smoking) = never, former, light, heavy.

 / TRANSFORM   N = normal + abnormal.
               P = abnormal/N. Q = 1-P.
               LOGIT = LN(P/Q).
               WT = 1/(1/normal + 1/abnormal).

 / DESIGN      GROUPING = lead, smoking.
               DEPENDENT = LOGIT.

 / END
1  1   33  3      1  2  12  2
2  1  160  4      2  2  49  6
1  3   21  2      1  4  16  3
2  3   75  6      2  4  84  3
 /END
```

Output E.2

SOURCE	SUM OF SQUARES	DEGREES OF FREEDOM
lead	4.36593	1
smoking	5.70725	3
ERROR	2.19830	3

The results of the weighted least squares analysis of logits also give chi-square values in the sum of squares column, as seen in Output E.2.

In this example, the difference between measuring distances in linear or log-linear units is substantial, since the LEAD effect now would be called significant at the 5% level. Collapsing the data across smoking history, 12 percent of those with high lead exposure had an abnormal PFT, while the tests of 5 percent of those with low lead exposure were abnormal.

The difference in testing significance between rates and logits is the attention focused on changes near the extremes. If the rates all fall between 30 and 70%, there is little difference. However, the logit scale considers a shift from 10 to 5% much more difficult (and therefore larger) than the change from 40 to 35%.

Many of the other desirable output features of 2V are also useful here—especially printing predicted and residual values for each of the eight groups.

Cautions

Weighted least squares is an asymptotic (large-sample) method that allows the chi-square values to be tested with standard tables if the sample size is large enough. But how large is large enough? If all of the proportions and group sizes meet the criteria $np \geq 5$ and $n(1 - p) \geq 5$, then all is well. The less stringent rule suggested by Cochran (1954) can probably be used with safety.

Maximum likelihood

The maximum likelihood method of estimation is an alternative to weighted least squares. See Program LE, Maximum Likelihood Estimation, in Volume 2. The maximum likelihood method is also available in BMDP in either LR for logistic regression or 4F contingency tables. These two programs use iterative reweighting to do maximum likelihood estimation. They both give the same answer, but it is not necessarily the answer obtained by using the weighted least squares approach described above. This is hardly surprising, since the weighted least squares approach is not maximum likelihood.

The advantage of using LR rather than 4F is that it is direct; i.e., it allows for both categorical predictors (smoking and lead) as well as continuous ones. Thus, we also have the analogs to analysis of covariance and multiple regression for data with binary outcome variables. LR functions in a stepwise fashion, so it can assist in model building.

In 4F, the chi-squares and their p-values for the maximum likelihood solution are given in the table of partial association. They are identified as interactions with the outcome variable.

References

Afifi, A. A., and S. P. Azen. 1979. *Statistical Analysis: A Computer Oriented Approach.* 2d ed. New York: Academic Press.

Afifi, A. A., and V. Clark. 1990. *Computer-aided Multivariate Analysis.* Belmont, California: Lifetime Learning Publications.

Anderson, T. W. 1958. *An Introduction to Multivariate Statistical Analysis.* New York: Wiley.

Andrews, D. F., P. J. Bickel, F. R. Hampel, P. J. Huber, W. H. Rogers, and J. W. Tukey. 1972. *Robust Estimates of Location: Survey and Advances.* Princeton: Princeton University Press.

Armitage, P. 1971. *Statistical Methods in Medical Research.* Oxford: Blackwell Scientific Publications.

Atkinson, A. C. 1985. *Plots, Transformations, and Regression.* Oxford: Oxford University Press.

Atkinson, A. C. 1982. Regression diagnostics, transformations and constructed variables. *J. R. Statist. Soc. B.* 44:1–36.

Belsley, D. A., E. Kuh, and R. E. Welsch. 1980. *Regression Diagnostics: Identifying Influential Data and Sources of Collinearity.* New York: Wiley.

Benedetti, J. K., and M. B. Brown. 1976. Alternate methods of building log-linear models. *Proceedings of the 9th International Biometrics Conference.* 2:209–227.

Benedetti, J. K., and M. B. Brown. 1978. Strategies for the selection of log-linear models. *Biometrics* 34:680–686.

Bentler, P. M. 1989. *EQS Structural Equations Program Manual.* Los Angeles: BMDP Statistical Software, Inc.

Bishop, Y., S. Fienberg, and P. Holland. 1975. *Discrete Multivariate Analysis: Theory and Practice.* Cambridge, Mass.: MIT Press.

Bliss, C. I. 1967. *Statistics in Biology.* Vol. 1. New York: McGraw-Hill.

BMDP Data Manager Manual. 1990. Los Angeles: BMDP Statistical Software, Inc.

BMDP PC-90 User's Guide. 1990. Los Angeles: BMDP Statistical Software, Inc.

BMDP 386 Dynamic User's Guide. 1992. Los Angeles: BMDP Statistical Software, Inc.

Bollen, K. A. 1989. *Structural Equations with Latent Variables.* New York: John Wiley.

Bourque, L. B., & Clark, V. A. 1992. *Processing Data: The Survey Example,* NU 85. Newbury Park: Sage.

Bowker, A. H., and G. J. Lieberman. 1972. *Engineering Statistics.* Englewood Cliffs, New Jersey: Prentice-Hall.

Box, G.E.P. 1954. Some theorems on quadratic forms applied in the study of analysis of variance problems. II. Effects of inequality of variance and of correlation between errors in the two-way classification. *Ann. Math. Stat.* 25:484–498.

Box, G.E.P., and D. R. Cox. 1964. Analysis of transformations. J. *Royal Statist. Soc.*, Series B. 26:211–252.

Box, G.E.P., and G. M. Jenkins. 1976. *Time Series Analysis: Forecasting and Control.* Rev. ed. San Francisco: Holden Day.

Box, G.E.P., and M. A. Muller. 1958. A note on the generating of random normal deviates. *Ann. Math. Statist.* 29:610–613.

Box, G.E.P., and D. A. Pierce. 1970. Distribution of residual autocorrelations in autoregressive-integrated moving average time series models. J. *Amer. Statist. Assoc.* 64: 1509.

Box, G.E.P., and G. C. Tiao. 1975. Intervention analysis with application to economic and environmental problems. *J. Amer. Statist. Assoc.* 70:70–79

Brown, M. B. 1975. The asymptotic standard errors of some estimates of uncertainty in the two-way contingency table. *Psychometrika* 40:291–296.

Brown, M. B. 1976. Screening effects in multidimensional contingency tables. *Appl. Statist.* 25:37–46.

Brown, M. B., M. Doron, and A. Laron. 1974. Approximate confldence limits for the concentration of insulin in radioimmunoassays. *Diabetologia* 10:23–25.

Brown, M. B., and A. B. Forsythe. 1974a. The small sample behavior of some statistics which test the equality of several means. *Technometrics* 16:129–132.

Brown, M. B., and A. B. Forsythe. 1974b. Robust tests for the equality of variances. J. *Amer. Statist. Assoc.* 69:364–367.

Brown, M. B., and C. Fuchs. 1983. On maximum likelihood estimation in sparse contingency tables. *Computational Statistics and Data Analysis* 1:3–15.

Brown, M. B., and J. K. Benedetti. 1976. Asymptotic standard errors and their sampling behavior for measures of association and correlation in the two-way contingency table. I. Testing the null hypothesis; II. Power and confidence limits. *BMDP Technical Report #23.* Los Angeles: BMDP Statistical Software, Inc.

Brown, M. B., and J. K. Benedetti. 1977a. Sampling behavior of tests for correlation in two-way contingency tables. *J. Amer. Statist. Assoc.* 72:309–315.

Brown, M. B., and J. K. Benedetti. 1977b. On the mean and variance of the tetrachoric correlation coefficient. *Psychometrika* 42:347–355.

Bryson, M., and M. Johnson. 1981. The incidence of monotone likelihood in the Cox model. *Technometrics* 23:381–383.

Campbell, D. T., and J. C. Stanley. 1966. *Experimental and Quasi-experimental Designs for Reseach.* Chicago: Rand McNally.

Carmines, E. G., and R. A. Zeller. 1979. Reliability and validity assessment. In *Quantitative Applications in the Social Sciences*, ed. J. L. Sullivan. Beverly Hills: Sage Publications.

Chambers J. M., W. S. Cleveland, B. Kleiner, and P. A. Tukey. 1983. *Graphical Methods for Data Analysis.* Belmont, California: Wadsworth International Group.

Chatfield, C. 1984. *The Analysis of Time Series.* New York: Chapman and Hall.

Chatterjee, S., and A. S. Hadi. 1986. Influential observations, high leverage points, and outliers in linear regression. *Statistical Science* 1:3.

Church, J. D., and E. L. Wike. 1976. The robustness of homogeneity of variance tests for asymmetric distributions: a Monte Carlo study. *Bulletin of the Psychometric Society* 7:417–420.

Clarke, M.R.B. 1970. A rapidly converging method for maximum likelihood factor analysis. *Brit. J. Math. Statist. Psych.* 23:43–52.

Cleary, J. P., and H. Levenbach. 1982. *The Professional Forecaster.* Belmont, California: Lifetime Learning Publications.

Cochran, W. G. 1954. Some methods for strengthening the common χ^2 tests. *Biometrics* 10:417–441.

Colquhoun, D. 1971. Lectures on Biostatistics: *An Introduction to Statistics with Applications in Biology and Medicine.* London: Clarendon Press.

Conover, W. J. 1980. *Practical Nonparametric Statistics.* 2d ed. New York: Wiley.

Cook, R. D., and S. Weisberg. 1982. *Residuals and Influence in Regression.* New York: Chapman and Hall.

Cook, R. D., and S. Weisberg. 1990. Confidence curves in nonlinear regression. *J. Amer. Stat. Assoc.* 85:544-551.

Cope, D. W., L. S. Linn, B. Leake, and P. A. Barrett. 1986. Modification of resident's behavior by preceptor feedback of patient satisfaction. *J. General Internal Med.* 1:394–398.

Cox, D. R. 1970. *The Analysis of Binary Data.* London: Chapman and Hall.

Cox, D. R., and E. J. Snell. 1981. *Applied Statistics: Principles and Examples.* London: Chapman and Hall.

Cramér, H. 1946. *Mathematical Models of Statistics.* Princeton: Princeton University Press.

Daniel, C., and F. S. Wood. 1971. *Fitting Equations to Data.* New York: Wiley.

Daniel C. 1959. Use of half-normal plots in interpreting factorial two-level experiments. *Technometrics* 1:311–341.

Dempster, A. P. 1969. *Elements of Continuous Multivariate Analysis.* San Francisco: Addison-Wesley.

Dixon, W. J., and F. J. Massey, Jr. 1983. *Introduction to Statistical Analysis.* 4th ed. New York: McGraw-Hill.

Dixon, W. J., and J. W. Tukey. 1968. Approximate behavior of the distribution of Winsorized t (Trimming/Winsorization 2). *Technometrics* 10:83–98.

Draper, N. R., and H. Smith. 1981. *Applied Regression Analysis.* 2d ed. New York: Wiley.

Draper, N. R., and J. A. John. 1981. Influential observations and outliers in regression. *Technometrics* 23:1.

Dunn, O. J. 1964. Multiple comparisons using rank sums. *Technometrics* 6:241–252.

Dunn, O. J., and V. A. Clark. 1987. *Applied Statistics: Analysis of Variance and Regression.* 2d ed. New York: Wiley.

Elashoff, J. D. 1986. Analysis of repeated measures designs. *BMDP Technical Report #83.* Los Angeles: BMDP Statistical Software, Inc.

Elashoff, J. D., T. J. Reedy, and J. H. Meyer. 1982. Analysis of gastric emptying data. *Gastroenterology* 83:1306–1312.

Everitt, B. S. 1977. *The Analysis of Contingency Tables.* London: Chapman and Hall.

Everitt, P. F. 1910. Table of the tetrachoric functions for fourfold correlation tables. *Biometrika* 7:437–451.

Fienberg, S. E. 1977. T*he Analysis of Cross-Classified Categorical Data.* Cambridge, Mass.: MIT Press.

Finney, D. J. 1978. *Statistical Method in Biological Assay.* 3d ed. New York: Macmillan.

Fisher, R. A. 1936. The use of multiple measurements in taxonomic problems. *Annals of Eugenics* 7:179–188.

Fleiss, J. L. 1981. S*tatistical Methods for Rates and Proportions.* 2d ed. New York: Wiley.

Forsythe, A. B., L. Engelman, R. Jennrich, and P.R.A. May. 1973. A stopping rule for variable selection in multiple regression. *J. Amer. Statist. Assoc.* 68:75–77.

Forthofer, R. N., and R. G. Lehnen. 1981. *Public Program Analysis: A New Categorical Data Approach.* Belmont, Ca.: Lifetime Learning Publications.

Frane, J. W. 1977. A note on checking tolerance in matrix inversion and regression. *Technometrics* 19:513–514.

Frane, J. W. and M. A. Hill. 1987. Annotated computer output for factor analysis: A supplement to the writeup for computer program BMDP4M. *BMDP Technical Report #8.* Los Angeles: BMDP Statistical Software, Inc.

Friedman, M. 1937. The use of ranks to avoid the assumption of normality implicit in the analysis of variance. *J. Amer. Statist. Assoc.* 32:688–689.

Fung, K. Y., and H. Lee. 1985. Behavior of trimmed *F. Sankhya* B47:186–201.

Garbow, B. S., J. M. Boyle, J. J. Dongarra, C. B. Moler, eds. 1977. *Matrix Eigensystem Routines: EISPACK Guide Extension.* New York: Springer-Verlag.

Glass, G. V., V. L. Willson, and J. M. Gottman. 1975. *Design and Analysis of Time-Series Experiments.* Boulder, Col.: Colorado Associated University Press.

Goodman, L. A. 1968. The analysis of cross-classified data: independence, quasi-independence, and interaction in contingency tables with or without missing cells. *J. Amer. Statist. Assoc.* 63:1091–1131.

Goodman, L. A., and W. H. Kruskal. 1954. Measures of association for cross-classiflcation. *J. Amer. Statist. Assoc.* 49:732–764.

Goodman, L. A., and W. H. Kruskal. 1963. Measures of association for cross-classification. III. Approximate sampling theory. *J. Amer. Statist. Assoc.* 58:310–364.

Goodman, L. A., and W. H. Kruskal. 1972. Measures of association for cross-classification. IV. Simplification of asymptotic variances. *J. Amer. Statist. Assoc.* 67:415–421.

Greenhouse, S. W., and S. Geisser. 1959. On methods in the analysis of profile data. *Psychometrika* 24:95–112.

Greenland, S. 1983. Tests for interaction in epidemiologic studies: A review and a study of power. *Stat. Med.* 2:243–51.

Grizzle, J. E., C. F. Starmer, and G. G. Koch. 1969. Analysis of categorical data by linear models. *Biometrics* 25:489-504.

Haberman, S. J. 1972. Log-linear fit for contingency tables, algorithm AS 51. *Appl. Statist.* 21:218–224.

Haberman, S. J. 1973. The analysis of residuals in cross-classified tables. *Biometrics* 29:205–229.

Harman, H. H. 1967. *Modern Factor Analysis.* 2d ed. Chicago: University of Chicago Press.

Hastings, C., Jr. 1955. *Approximations for digital computers.* Princeton: Princeton University Press.

Hawkins, D. M. 1974. The detection of errors in multivariate data using principal components. *J. Amer. Statist. Assoc.* 69:340–344.

Henderson, H. V. and P. F. Velleman. 1981. Building multiple regression models interactively. *Biometrics* 391–411.

Herr, D. G. 1986. On the history of ANOVA in unbalanced, factorial designs: the first 30 years. *The American Statistician* 40:265–270.

Hettmansperger, T. 1984. *Statistical Inference Based on Ranks.* New York:Wiley.

Hibbs, D. A., Jr. 1974. Problems of statistical estimation and causal inference in time-series regression models. In *Sociological Methodology,* 1973–1974, ed. L. Costner. San Francisco: Jossey-Bass.

Hollander, M., and D. A. Wolfe. 1973. *Nonparametric Statistical Methods.* New York: Wiley.

Huynh, H., and L. S. Feldt. 1970. Conditions under which mean square ratios in repeated measurements designs have exact F-distributions. *J. Amer. Statist. Assoc.* 65:1582–1589.

Huynh, H., and L. S. Feldt. 1976. Estimation of the Box correction for degrees of freedom from sample data in randomized block and split-plot designs. *J. Ed. Stat.* 1:69–82.

Ippoliti, A., J. Elashoff, J. Valenzuela, R. Cano, H. Frankl, M. Samloff, and R. Karetz. 1983. Recurrent ulcer after successful treatment with cimetidine or antacid. *Gastroenterology* 85(4):875–880.

Jacquez, J. A. 1985. *Compartmental Analysis in Biology and Medicine.* 2d ed. Ann Arbor: University of Michigan Press.

Jenkins, G. M. 1979. *Practical Experiences with Modelling and Forecasting Time Series.* St. Helier, Jersey, Channel Islands: Gwilym Jenkins & Partners (Overseas) Ltd.

Jennrich, R. I., and P. B. Bright. 1976. Fitting systems of linear differential equations using computer-generated exact derivatives. *Technometrics* 18:385–392.

Jennrich, R. I., and R. H. Moore. 1975. Maximum likelihood estimation by means of nonlinear least squares. *BMDP Technical Report #9.* Los Angeles: BMDP Statistical Software, Inc.

Jennrich, R. I., and S. M. Robinson. 1969. A Newton-Raphson algorithm for maximum likelihood factor analysis. *Psychometrika* 34:111–123.

Jennrich, R. I., and P. F. Sampson. 1966. Rotation for simple loadings. *Psychometrika.* 31:313–323.

Jennrich, R. I., and P. F. Sampson. 1968. Application of stepwise regression to nonlinear least squares estimation. *Technometrics* 10:63–67.

Jennrich, R. I., 1977a. Stepwise regression. In *Statistical Methods for Digital Computers,* K. Enslein, A. Ralston, and H. S. Wilf, eds. New York: Wiley.

Jennrich, R. 1.1977b. Stepwise discriminant analysis. In *Statistical Methods for Digital Computers.* K. Enslein, A. Ralston, and H. S. Wilf, eds. New York: Wiley.

Jennrich, R. I. 1978. Admissible values of gamma in direct oblimin rotation. *Psychometrika* 44:173–177.

John, P.W.M. 1971. *Statistical Design and Analysis of Experiments.* New York: MacMillan.

Joreskog, K. G., and M. van Thillo. 1971. New rapid algorithms for factor analysis by unweighted least squares, generalized least squares and maximum likelihood. *Research Memorandum* 71-S. Princeton: Educational Testing Service.

Kaiser, H. F. 1970. A second-generation Little Jiffy. *Psychometrika* 35:401–415.

Kendall, M. G., and A. Stuart. 1967. *The Advanced Theory of Statistics.* Vol. 2. 2d ed. New York: Hafner.

Kirk, R. E. 1982. *Experimental Design.* 2d ed. Monterey, California: Brooks/Cole Publishing Co.

Kruskal, W. H., and W. A. Wallis. 1952. Use of ranks in one-criterion variance analysis. 1. *Amer. Statist. Assoc.* 47:583–621.

Lachenbruch, P., and R. M. Mickey. 1968. Estimation of error rates in discriminant analysis. *Technometrics* 10:1–11.

Landaw, E., P. Sampson, and J. Toporek. 1982. Advanced nonlinear regression in BMDP: comparison to ad hoc dummy data procedures. In *Proceedings of the Statistical Computing Section, American Statistical Association* 228–233.

Larsen, W. A., and S. J. McCleary. 1972. The use of partial residual plots in regression analysis. *Technometrics* 14:781–790.

Lawless. 1982. *Statistical Models and Methods for Lifetime Data.* J. Wiley and Sons.

Lawley, D. N., and A. E. Maxwell. 1971. *Factor Analysis as a Statistical Method.* 2d ed. New York: American Elsevier.

Lee, S. K. 1977. On the asymptotic variances of u-terms in log-linear models of multidimensional contingency tables. *J. Amer. Statist. Assoc.* 72:412–419.

Lee, H., and K. Y. Fung. 1983. Robust procedures for multi-sample location problems with unequal group variances. *J. Statist. and Computer Simulation* 18:125–143.

Lehmann, E. L. 1975. *Nonparametrics: Statistical Methods Based on Ranks.* Oakland: Holden-Day.

Levenbach, H., and J. P. Cleary. 1981. *The Beginning Forecaster.* Belmont, California: Lifetime Learning Publications.

Levene, H. 1960. Robust tests for equality of variance. In *Contributions to Probability and Statistics,* ed. I. Olkin. Palo Alto: Stanford University Press.

Ling, R. F. 1973. A computer-generated aid for cluster analysis. *Communications of ACM.* 16:355–361.

Ljung, G. M., and G.E.P. Box. 1978. On a measure of lack of fit in time series models. *Biometrika* 65:297.

Mantel, N., and W. Haenszel. 1959. Statistical aspects of the analysis of data from retrospective studies of disease. *J. Natl. Cancer Inst.* 22:719–748.

McCleary, R., and R. A. Hay. 1980. *Applied Time Series Analysis for the Social Sciences.* Beverly Hills: Sage Publications.

McDonald, G. C., and R. C. Schwing. 1973. Instabilities of regression estimates relating air pollution to mortality. *Technometrics* 15:463–481.

McKennell, A. C. (1977). Attitude Scale Construction, pp. 183–220 in C. A. O'Muircheataugh and Payne (eds) *Exploring Data Structures, Vol. 1: The Analysis of Survey Data.* New York: John Wiley.

Mickey, M. R., O. J. Dunn, and V. Clark. 1967. Note on the use of stepwise regression in detecting outliers. *Comp. Biomed. Res.* 1:105–111.

Miller, R. G., Jr. 1981. *Simultaneous Statistical Inference.* New York: McGraw-Hill.

Milliken, G. A., and D. E. Johnson. 1984. *Analysis of Messy Data*. Belmont, California: Lifetime Learning Publications.

Morrison, A. S., M. M. Black, C. R. Lowe, B. McMahon, and S. Y. Yuasa. 1973. Some international differences in histology and survival in breast cancer. *Intl. I. Cancer.* 11:261–267.

Morrison, D.F. 1967. *Multivariate Statistical Methods*. New York: McGraw-Hill.

Oja, H. 1983. New tests for normality. *Biometrika* 70:297–299.

Pankratz, A. 1983. *Forecasting with Univariate Box-Jenkins Models: Concepts and Cases*. New York: Wiley.

Pearson, K. 1913. On the probable error of a coefficient of correlation as found from a fourfold table. *Biometrika* 9:22–27.

Plackett, R. L. 1974. *The Analysis of Categorical Data*. London: Griffin.

Potthoff, R. F., and S. N. Roy. 1964. A generalized multivariate analysis of variance model useful especially for growth curve problems. *Biometrika* 51:313–326.

Pratt, J. W. 1964. Robustness of some procedures for the two-sample location problem. *J. Amer. Statist. Assoc.* 59:665–680.

Prentice and Marek. 1979. A Qualitative discrepancy between censored data rank tests. *Biometrics* 35, 861–67.

Ralston, M. L., and R. I. Jennrich. 1978. DUD, a derivative free algorithm for nonlinear least squares. *Technometrics* 20:7–14.

Ralston, M. L., R. I. Jennrich, P. F. Sampson, F. K. Uno, E. Landaw, J. D Elashoff, J. Gornbein, B. Leake, and L. Engelman. 1987. Fitting pharmacokinetic models with program AR: new instructions for PCs and mainframes. *BMDP Technical Report #85*. Los Angeles: BMDP Statistical Software, Inc.

Rao, C. 1955. Estimation and tests of significance in factor analysis. *Psychometrika* 20:93–111.

Rao, C. R. 1973. *Linear Statistical Inference and its Application*. 2d ed. New York: Wiley.

Royston, J. P. 1982. An extension of Shapiro and Wilk's W test for normality to large samples. *Applied Statistics* 31:115–24.

Ryan, Joiner, and Ryan. 1976. *Minitab Student Handbook*.

Sarhan, A. E., and B. G. Greenberg. 1962. *Contributions to ORDER statistics*. New York: Wiley.

Scheffé, H. 1959. *The Analysis of Variance*. New York: Wiley.

Servellen, V. 1988. Patient reactions to their physicians. Personal communication.

Shapiro, S. S., and M. B. Wilk. 1965. An analysis of variance test for normality (complete samples). *Biometrika* 52:591–611.

Siegel, S. S.,1956. *Nonparametric Statistics for the Behavioral Sciences*. New York: McGraw-Hill.

Smith, B. T., J. M. Boyle, J. J. Dongarra, B. S. Garbow, Y. Ikebe, V. C. Klema, C. B. Moler, eds. 1974. *Matrix Eigensystem Routines: EISPAC Guide*. New York: Springer Verlag.

Snedecor, G. W., and W. G. Cochran. 1967. *Statistical Methods*, 6th ed. Ames, Iowa: Iowa State University Press.

Spicer, C. C. 1972. Algorithm AS 52. Calculation of power sums of deviations about the mean. *Appl. Statist.* 21:226–227.

Stevens, W. L. 1951. Asymptotic regression. *Biometrics* 7:247.

Sugiyama, Y., T. Yamada, and N. Kaplowitz. 1982. Identification of hepatic Z-protein in a marine elasmobranch, Platyrhinoides triseriata. *Biochem. J.* 203:377-381. (Data used supplied in personal communication from last author.)

Tabachnick, B. G., and L. S. Fidell. 1983. *Using Multiuariate Statistics*. New York: Harper and Row.

Theil, H. 1971. *Principles of Economics*. New York: Wiley.

Tukey, J. W. 1977. *Exploratory Data Analysis*. Reading, Mass.: Addison Wesley.

Vandaele, W. 1983. *Applied Time Series and Box-Jenkins Models*. San Diego: Academic Press.

Weisberg, S. 1985. *Applied Linear Regression*. 2d ed. New York: Wiley.

Welch, B. L. 1947. The generalization of Student's problem when several different population variances are involved. *Biometrika* 34:28–35.

Welsch, R., and E. Kuh. 1977. *Linear Regression Diagnostics.* Sloan School of Management Working Paper. Cambridge, Mass.:MIT. 923–977.

Werner, M., R. Tolls, J. Hultin, and J. Mellecker. 1970. Sex and age dependence of serum calcium, inorganic phosphorus, total protein, and albumin in a large ambulatory population. In *Fifth Internntional Congress on Automation, Advances in Automated Analysis* 2:59–65.

Wichmann, B.A., and I. D. Hill. 1982. Algorithm AS 183: An efficient and portable pseudo-random number generator. *Applied Statistics* 31:188–190.

Yuen, K., and W. Dixon. 1973. The approximate behavior and performance of the two-sample trimmed *t*. *Biometrika* 61:369–374.

Yuen, K. 1974. The two-sample trimmed *t* for unequal population variances. *Biometrika* 61 :165–170.

Index

A

B

C

D

E

F

G

H

I

J

K

L

M

O

P

Index

Q

R

S

T

U

V

W

X

Y

Z